WITHDRAWN

Rayner Joel
BSc(Eng) London, CEng, FIMechE, FIMarE

Basic Engineering Thermodynamics

Fifth Edition

An imprint of **Pearson Education**

Harlow, England • London • New York • Reading, Massachusetts • San Francisco • Toronto • Don Mills
Ontario • Sydney • Tokyo • Singapore • Hong Kong • Seoul • Taipei • Cape Town • Madrid • Mexico City
Amsterdam • Munich • Paris • Milan

Pearson Education Limited
Edinburgh Gate
Harlow
Essex CM20 2JE
England

and Associated Companies throughout the world

Visit us on the World Wide Web at:
http://www.pearsoneduc.com

First published as *Heat Engines* 1960
Second edition 1966
Third edition 1971
Fourth edition 1987
Fifth edition 1996
10 9 8 7 6
06 05 04 03

British Library Cataloguing in Publication Data
A catalogue entry for this title is available from the British Library

ISBN 0-582-25629-1

Set by 32 in Times Roman 9 1/4/12pt
Printed in China
GCC/06

And it shall come to pass afterward, that I will pour out my spirit upon all flesh; and your sons and your daughters shall prophesy, your old men shall dream dreams, your young men shall see visions.

JOEL, Chapter 2, Verse 28

Contents

Preface

With the new fifth edition, there has again been an opportunity to sift through the text and to rearrange, add to and also remove material. An update has also been carried out where appropriate. The new arrangement roughly separates into two sections. The earlier chapters are concerned with basic engineering thermodynamic theory whereas the later chapters introduce practical applications.

Concepts of engineering thermodynamics warrant serious thought and study. Modern industrial society relies almost entirely upon their application to supply its basic energy requirements. Without the various power-generating plants and engines, industrial society would collapse. Failure of electrical generating plant would mean the failure of the electronics industry and all its consequences.

There is heavy reliance on the use of fossil fuels which have no immediate replacement. There are sources of renewable energy, such as wind and solar power, but their effect is small compared with nuclear fuels and the fossil fuels, coal, gas and oil. The most serious problem is the rate at which energy is consumed. Much atmospheric damage is caused by exhaust emissions from burning fossil fuels. Better and moderated use of fundamental world resources must be addressed for the future.

This book will fit most curriculums that include the study of basic engineering thermodynamics. It is a serious subject, both necessary and rewarding, for perhaps it can help to preserve and enhance the quality of life on earth.

Leigh-on-Sea 1996 R.J.

Acknowledgements

The author thanks the following manufacturing organisations for their illustrations and information:

- Babcock International plc for Figs. 10.2, 10.3, 10.4, 10.5
- Ford Motor Company for Figs. 16.12, 16.13, 16.14, 16.15, 16.16, 16.17, 16.18, 16.20, 16.21
- GEC Alsthom Power Generation for Fig. 13.8
- MAN B&W Diesel A/S for Figs. 16.9, 16.22
- E. Reader plc for Fig. 11.1
- Rolls-Royce plc for Fig. 16.32
- Siemens AG for Figs. 16.27, 16.28, 16.29, 16.30, 16.31

Data concerning new ecologically friendly refrigerants from:

- Du Pont de Nemour International SA, also associated with British Oxygen Company (BOC)
- ICI Chemical and Polymers Limited
- Rhône-Poulenc Chemicals

Data concerning fuels has been taken from:

- *Modern Petroleum Technology*, edited by G.D. Hobson, Applied Science Publishers on behalf of the Institute of Petroleum, Great Britain.
- *Technical Data on Fuel*, by J.W. Rose and J.R. Cooper, British National Committee World Energy Conference, distributed by Scottish Academic Press.
- *Technology of Gasoline*, edited by E.G. Hancock, The Society of Chemical Industry and Blackwell Scientific.

List of symbols

See also BS 5775: Specification for quantities, units and symbols.

A area
BDC bottom dead centre
BS British Standard
BMEP brake mean effective pressure
b.p. brake power
C temperature (Celsius)
C velocity
CHP combined heat and power
CI compression ignition
CN cetane number
c specific heat capacity (sometimes massic heat capacity)
c_p specific heat capacity at constant pressure
c_v specific heat capacity at constant volume
$\tilde{C}_p$ *(or* C_p*)* molar heat capacity at constant pressure
C_v *(or* C_v*)* molar heat capacity at constant volume
d diameter
E total energy
F force
FI fuel injection
f.p. friction power
g gravitational acceleration
H enthalpy
h specific enthalpy (sometimes massic enthalpy)
IC internal combustion
IMEP indicated mean effective pressure
i.p. indicated power
K Kelvin temperature (Celsius absolute)
k thermal conductivity; steam isentropic index
L (l) length; stroke
M (m) mass

$\tilde{M}$	molar mass
$\dot{m}$	mass flow rate
N	rotational speed
n	polytropic index; number of
ON	octane number
p (P)	power; pressure
p_m	mean effective pressure
p_b	brake mean effective pressure
p_i	indicated mean effective pressure
Q	heat
$\dot{Q}$	rate of heat transfer
R	characteristic gas constant
$\tilde{R}$ (or R_0)	molar gas constant
r	radius; expansion ratio
r_p	pressure ratio
r_v	volume ratio
S	entropy
s	specific entropy
SFC	specific fuel consumption
T	absolute temperature
TDC	top dead centre
t	temperature
U	internal energy; surface or overall heat transfer coefficient
u	specific internal energy (sometimes massic internal energy)
V	volume
$\dot{V}$	volume flow rate
v	specific volume
W	work; load; weight
$\dot{W}$	work transfer rate
x	dryness fraction
Z	height above datum

α (alpha)	angle; absorptivity
β (beta)	angle
γ (gamma)	ratio of specific heats, c_p/c_v for gases
ε (epsilon)	emissivity; effectiveness
η (eta)	efficiency
θ (theta)	temperature difference
ρ (rho)	density
σ (sigma)	Stefan–Boltzmann constant
ϕ (phi)	relative humidity; angle
ψ (psi)	percentage saturation
ω (omega)	specific humidity

Chapter 1

General introduction

1.1 Thermodynamics – its meaning and relevance

The word *thermodynamics* derives from two Greek ideas: *thermē* meaning hot or heat, and *dynamikos* originally meaning power or powerful and now the study of matter in motion.

Thus, thermodynamics is the study of heat related to matter in motion. Much of the study of engineering (or applied) thermodynamics is concerned with work producing or utilising machines such as engines, turbines and compressors together with the working substances used in such machines. Their development has given us the ability to create our modern industrial society.

It is important to understand that without thermodynamic engines – petrol engines, gas turbines, steam turbines, etc. – modern industrial society could not survive. But this raises many ethical questions, questions which need to be addressed. For example, there is the exploitation of natural finite global resources, e.g. oil and coal; there is the massive atmospheric pollution which occurs; there is always considerable political debate and conflict concerning energy resources and reserves.

It will be seen, therefore, that as humankind insists, at present, on the continuance of industrial societies, a study of engineering thermodynamics remains essential.

There are certain topics which are common and fundamental to many sections of the study of engineering thermodynamics. It is the purpose of this general introduction to investigate these common fundamentals. In the chapters which follow, it will be seen how they apply and are related to other sections of the study.

1.2 Working substance

All thermodynamic systems require some working substance in order that the various operations required of each system can be carried out. The working substances are, in general, fluids which are capable of deformation in that they can readily be expanded and compressed. The working substance also takes part in energy transfer. For example, it can receive or reject heat energy or it can be the means by which work is done. Common examples of working substances used in thermodynamic systems are air and steam.

1.3 Pure substance

A pure substance is a single substance or mixture of substances which has the same consistent composition throughout. In other words, it is a homogeneous substance and its molecular structure does not vary. For example, steam and water, or a mixture of steam and water can be considered as pure substances. Each has the same molecular or chemical structure through its mass. Air in its gaseous and liquid form is a pure substance. However, during the liquefaction process of air, which is a mixture of gases, mainly oxygen and nitrogen, the oxygen and nitrogen liquefy at different temperatures. Until all the air becomes liquid, the relative concentrations of oxygen and nitrogen (and other gases) in the liquid that has formed will therefore be different from those of the original air. The relative concentrations will also differ between the condensing vapour and the original air. Thus the air in these circumstances ceases to be a pure substance.

The importance of the concept of a pure substance in this work is that the condition, or state, of a pure substance can be completely defined by any two independent properties of the substance. For example, if the pressure and volume of a fixed mass of oxygen is known, then its temperature and such other properties as will be discussed later are also completely known.

1.4 Macroscopic and microscopic analysis

If the properties of a particular mass of a substance, such as its pressure, volume and temperature, are analysed, then the analysis is said to be **macroscopic**. This is the method of analysis usually used by the engineer and is the type of analysis used therefore in the study of heat engines and engineering thermodynamics. If, however, an analysis is made in which the behaviour of the individual atoms and molecules of a substance are under investigation, then the analysis is said to be **microscopic**. Some studies in nuclear physics would be of a microscopic nature, such as the atomic structure of a fissionable material like uranium.

1.5 Properties and state

In the macroscopic analysis of a substance any characteristic of the substance which can be observed or measured is called a **property** of the substance. Examples of properties are pressure, volume and temperature. This type of property which is dependent upon the physical and chemical structure of the substance is called an **internal** or **thermostatic** property. Other types of thermostatic properties will be discussed later.

If a value can be assigned to a property then it is said to be a **point function** because its value can be plotted on a graph. Properties which are independent of mass, such as temperature and pressure, are said to be **intensive** properties. Properties which are dependent upon mass, such as volume and energy in its various forms, are called **extensive** properties.

If a property can be varied at will, quite independently of other properties, then the property is said to be an **independent** property. The temperature and pressure of a gas, for example, can be varied quite independently of each other and thus, in this case, temperature and pressure are independent properties.

It will be found, however, when discussing the formation of a vapour, that the temperature at which a liquid boils depends upon the pressure at which the formation of the vapour is occurring. Here, if the pressure is fixed then the temperature becomes dependent upon the pressure. Hence the pressure is an independent property but the temperature is a **dependent** property.

A knowledge of the various thermostatic properties of a substance defines the **state** of the substance. If a property, or properties, are changed, then the state is changed.

Properties are independent of any process which any particular substance may.have passed through from one state to another, being dependent only upon the end states. In fact, a property can be identified if it is observed to be a function of state only.

Since, at a particular state, a substance will have certain properties which are functions of that state, then there will be certain relationships which exist between them. These property relationships will be investigated in the text.

A property which includes a function of time, used to define a rate at which some interaction can occur, such as the transfer of mass, momentum or energy, is referred to as a **transport property**. Examples of transport properties are thermal conductivity and viscosity.

1.6 Specific quantity

In the discussion of properties it was suggested that those properties which were associated with the mass of a substance are called extensive properties. For convenience, at times, it is useful to discuss the properties of unit mass of a substance. To indicate that this is the case, the word *specific* is used to prefix the property.

Thus, the specific volume of a substance, at some particular state, is the volume occupied by unit mass of the substance at that particular state. Other specific quantities will be discussed in the text.

1.7 Temperature

Temperature describes the degree of hotness or coldness of a body.

The subject of temperature investigation is called **thermometry**.

Many attempts have been made in the past to lay down a scale of temperature. The work has culminated in the generally accepted use of two temperature scales, **Fahrenheit** and **Celsius**. The Fahrenheit scale is named after its German inventor, Daniel Gabriel Fahrenheit (1686–1736) of Danzig (now Gdańsk, Poland). The Celsius scale (often referred to as the centigrade scale) is named after Anders Celsius (1701–1744), a Swedish astronomer born at Uppsala.

The Celsius scale is the temperature scale which is most commonly used worldwide. The Fahrenheit scale is generally becoming progressively phased out. The customary temperature scale adopted for use with the SI system of units is the Celsius scale. For customary use, the lower fixed point is the temperature of the melting of pure ice, commonly referred to as the freezing point. This point is designated 0 °C. The upper fixed point is the temperature at which pure water boils and this is designated 100 °C. In the past, this customary temperature scale has been referred to as the centigrade scale. The use of the word *centigrade* is now discouraged, the accepted reference now being that of the Celsius scale. Of interest, the freezing and boiling points of pure water are designated 32 °F and 212 °F, respectively, on the Fahrenheit scale.

It will be shown later, when dealing with the properties of solids, liquids and vapours, that the temperature at which a liquid freezes or boils depends upon the pressure exerted at the surface of the liquid. This temperature increases as the pressure increases in the case of boiling and slightly decreases with increase of pressure in the case of freezing. To standardise the freezing and boiling temperature on a thermometric scale, one must therefore standardise the pressure at which the freezing or boiling occurs. This pressure is taken as 760 mm of mercury which is called the **standard atmospheric pressure** or the **standard atmosphere**, being a mean representative pressure of the atmosphere.

Figure 1.1 shows the way the customary Celsius scale is divided up. The lower fixed point is 0 °C and the upper fixed point is 100 °C; there are 100 Celsius degrees between them. These 100 Celcius degrees are together called the **fundamental interval**.

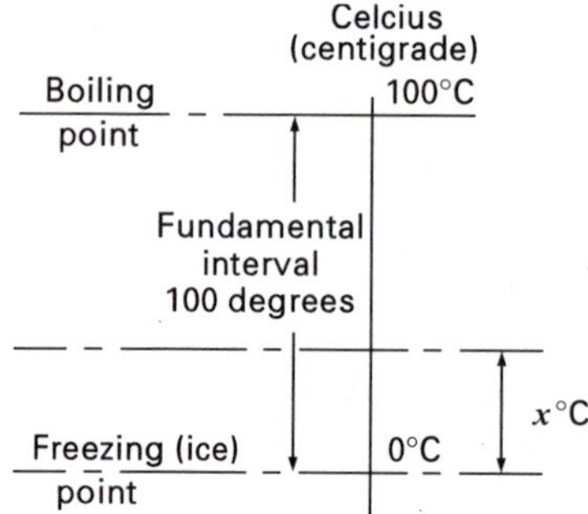

Fig. 1.1 Celsius scale of temperature

In the above discussion it will be noted that the choice of the fixed points was of an arbitrary nature. The freezing and boiling points of water were chosen for convenience. Other points on the **International Temperature Scale** are then chosen and referred to the originally conceived scale. The original choice of fixed points was arbitrary, so the Celsius scale is sometimes called the **normal**, the **empirical**, the **customary** or the **practical** temperature scale.

Since the Celsius scale is only a part of the more extensive **thermodynamic**, or **absolute** temperature scale, it is sometimes called a **truncated thermodynamic** scale. Subsequent work will show that there is the possibility of an absolute zero of temperature which will then suggest an **absolute temperature scale**.

An absolute zero of temperature would be the lowest temperature possible and therefore this would be a more reasonable temperature to adopt as the zero for a temperature scale. The absolute thermodynamic temperature scale is called the **Kelvin** scale. It was devised by Lord Kelvin, a British scientist, in about 1851.

The Kelvin unit of temperature is called the **kelvin** and is given the symbol K. A temperature, T, on the Kelvin scale is written T K, not T °K. The kelvin has the same magnitude as the Celsius degree for all practical purposes.

The absolute zero of temperature appears impossible to reach in practice. However, its identity is defined by giving to the triple point of water a value of 273.16 kelvin (273.16 K). The triple point is defined in Chapter 4. With the absolute zero so defined, the zero of the Celsius thermodynamic scale is defined as 0 °C = 273.15 K.

Thus

$$t = T - 273.15 \qquad [1]$$

where t = temperature on the Celsius thermodynamic scale = t °C
T = temperature on the Kelvin thermodynamic scale = T K

From equation [1]

$$T = t + 273.15 \text{ (can use 273 for most calculations)} \qquad [2]$$

By choosing the zero of the Celsius thermodynamic scale as 0 °C = 273.15 K, this approximates very closely to the customary Celsius scale and thus 0 °C on the customary Celsius scale is very nearly equal to 0 °C on the Celsius thermodynamic scale. Also, 100 °C on the customary Celsius scale is very nearly equal to 100 °C on the Celsius thermodynamic scale.

1.8 Pressure

Pressure is defined as force per unit area. Thus, if a force F is applied to an area A, and if this force is uniformly distributed over the area, then the pressure P exerted is given by the equation

$$P = \frac{F}{A} \qquad [1]$$

If F = force in newtons (N) and A = area in square metres (m^2) then the unit of pressure becomes the newton/metre^2 (N/m^2), which is the basic unit of pressure in the SI system of units. This unit of pressure is sometimes called the **pascal** (Pa).

Common multiples of this basic unit of pressure will be the kilonewton/metre^2 (1 $kN/m^2 = 10^3$ N/m^2) and the meganewton/metre^2 (1 $MN/m^2 = 10^6$ N/m^2). The **bar** may also be commonly used (1 bar = 10^5 N/m^2) as may also the **hectobar** (1 hbar = 10^2 bar = 10^7 N/m^2).

The bar is sometimes seen as being useful in that it is very nearly equal to one standard atmosphere.

$$1 \text{ standard atmosphere} = 1.013\,25 \text{ bar} = 0.101\,325 \text{ MN/m}^2 \qquad [2]$$

If a force is applied to a solid then it will be transmitted through the solid in the direction of application of the force.

Figure 1.2(a) shows a solid being pressed against a fixed wall by means of a force F. If the contact area is A then the pressure set up at the contact surface = F/A and it is normal to the contact surface.

On the other hand, Fig. 1.2(b) shows a piston of area A enclosing a fluid in a cylinder. If a force F is now applied to the piston then a pressure $P = F/A$ will be set up in the fluid. Unlike the solid, however, where this pressure would be transmitted in the line of action of the applied force, in the fluid, this pressure $P = F/A$ is set up in all directions in the cylinder. Any vessel then, in which there is a fluid under pressure, must be capable of withstanding the pressure in all directions. The fact that the pressure distribution in a fluid does occur in all directions is easily demonstrated by blowing up a balloon which swells up in all directions.

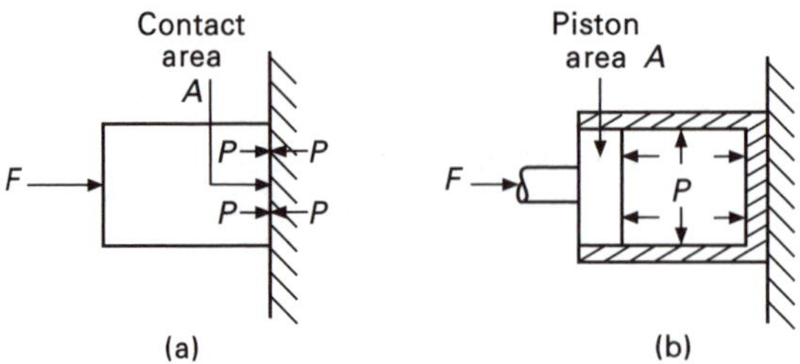

Fig. 1.2 Concept of pressure: (a) of a solid (b) using a piston

1.9 Volume

Volume is a property associated with cubic measure. The unit of volume is the cubic metre (m^3) together with its multiples and submultiples. Sometimes the litre (l) may be used. 1 litre = 1 cubic decimetre ($1\ dm^3 = (10^{-1}\ m)^3$).

If the volume of a substance increases then the substance is said to have been expanded. If the volume of a substance decreases then the substance is said to have been compressed. Specific volume is given the symbol v. The volume of any mass, other than unity, is given the symbol V.

1.10 Phase

When a substance is of the same nature throughout its mass, it is said to be in a **phase**. Matter can exist in three phases; solid, liquid and vapour (or gas). If the matter exists in only one of these forms then it is in a **single phase**. If two phases exist together then the substance is in the form of a **two-phase mixture**. Examples of this are when a solid is being melted into a liquid or when a liquid is being transformed into a vapour. In a single phase the substance is said to be **homogeneous**. If it is two-phase it is said to be **heterogeneous**. A heterogeneous mixture of three phases can exist. This is covered by the section on the triple point during the discussion on steam (see section 4.4).

1.11 Two-property rule

Now that the concept of properties, state and phase have been indicated, it is possible to write down the two-property rule.

If two independent properties of a pure substance are defined, then all other properties, or the state of the substance, are also defined. If the state of the substance is known then the phase or mixture of phases of the substance are also known.

The idea of the two-property rule was suggested in section 1.3.

1.12 Process

When the state of a substance is changed by means of an operation or operations having been carried out on the substance, then the substance is said to have undergone a **process**. Typical processes are the expansion and compression of a gas or the conversion of water into steam. A process can be analysed by an investigation into the changes which occur in the properties of a substance, and the energy transfers which may have taken place.

1.13 Cycle

If processes are carried out on a substance such that, at the end, the substance is returned to its original state, then the substance is said to have been taken through a **cycle**. This is commonly required in many engines. A sequence of events takes place which must be repeated and repeated. In this way the engine continues to operate. Each repeated sequence of events is called a cycle.

1.14 The constant temperature process

This is a process carried out such that the temperature remains constant throughout the process. It is often referred to as an **isothermal** process. Particular cases of the constant temperature process will be dealt with in the text.

1.15 The constant pressure process

This is a process carried out such that the pressure remains constant throughout the process. It is often referred to as an **isobaric** or **isopiestic** process. Particular cases of the constant pressure process will be dealt with in the text.

1.16 The constant volume process

This is a process carried out such that the volume remains constant throughout the process. It is often referred to as an **isometric** or **isochoric** process. Particular cases of the constant volume process will be dealt with in the text.

1.17 Energy

Energy is defined as that capacity a body or substance possesses which can result in the performance of work. Here, work is defined, as in mechanics, as the result of moving a force through a distance. The presence of energy can only be observed by its effects and these can appear in many different forms. An example where some of the forms in which energy can appear is in the motor car.

The petrol put into the petrol tank must contain a potential chemical form of energy because, by burning it in the engine and through various mechanisms, it propels the motor car along the road. Thus work, by definition, is being done because a force is being moved through a distance.

As a result of burning the petrol in the engine, the general temperatures of the working substances in the engine, and the engine itself, will be increased and this increase in temperature must initially have been responsible for propelling the motor car.

Due to the increase in temperature of the working substances then, since the motor car is moved and work is done, the working substance at the increased temperature must have contained a form of energy resultant from this increased temperature. This energy content resultant from the consideration of the temperature of a substance is called **internal energy** (see section 1.23).

Some of this internal energy in the working substances of the engine will transfer to the cooling system of the engine because the cooling water becomes hot. A transfer of energy in this way, because of temperature differences, is called **heat transfer** (see section 1.24).

The motor car engine will probably have an electric generator, or alternator, which is rotated by the engine and is used to charge the battery. The battery, by its construction and chemical nature, stores energy which can appear at the battery terminals as electricity. The electricity from the battery can be used to rotate the engine starter which, in turn, rotates and starts the engine. By rotating the engine to start it, the electric motor must be doing work; thus electricity must have the capacity for doing work, hence it is a form of energy.

To stop the motor car the brakes are applied. After the motor car has stopped, the brake drums are hot, so the internal energy of the brake drum materials must have been increased. This internal energy increase resulted from the stopping of the motor car, so there must have been a type of energy which the motor car possessed while it was in motion. This energy of motion is called **kinetic energy**.

It will be seen that energy can appear in many forms, and through the action of various devices, it can be converted from one form into another.

All the possible forms of energy have not been discussed here.

More will be said about energy, and its various forms, later in the text.

1.18 Work

If a system exists in which a force at the boundary of the system is moved through a distance, then **work** is done by or on the system (see Chapter 2). As soon as the force ceases to be moved, it ceases to do any work. Work is therefore a **transient quantity**; it describes a process by which a force is moved through a distance. Work, being a transient quantity, is therefore not a property.

Work is given the symbol W. If it is required to indicate a rate at which work is being done then a dot is placed over the symbol W. Thus

$$\dot{W} = \text{work done/unit time}$$

1.19 Work and the pressure–volume diagram

Consider Fig. 1.3. In the lower half of the diagram is shown a cylinder in which a fluid at pressure P is trapped using a piston of area A. The fluid here is the system.

From this

$$\text{Force on piston} = \text{pressure} \times \text{area} = PA \qquad [1]$$

Let this force PA be just sufficient to overcome some external load.

Now let the piston move back a distance L along the cylinder while at the same time the pressure of the fluid remains constant. The force on the piston will have remained constant.

$$\begin{aligned}\text{Work done} &= \text{force} \times \text{distance}\\ &= PA \times L \qquad [2]\end{aligned}$$

This equation could be rearranged to read

$$\text{Work done} = P \times AL$$

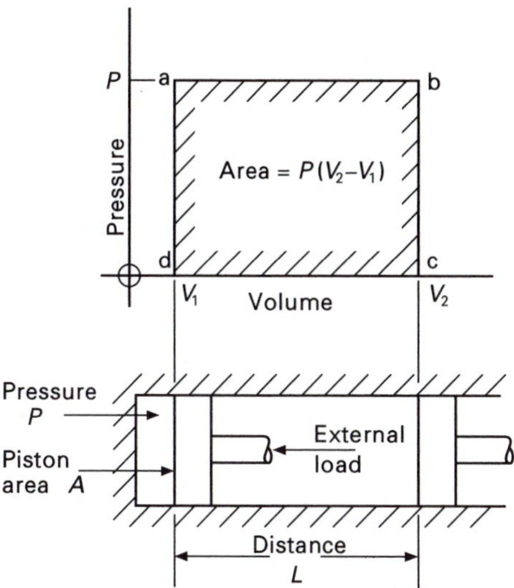

Fig. 1.3 Work and the *P–V* diagram

But

AL = volume swept out by the piston, called the **swept** or **stroke** volume
$= (V_2 - V_1)$

$$\therefore \quad \text{Work done} = P(V_2 - V_1) \qquad [3]$$

Above the diagram of the piston and cylinder is shown a graph of the operation plotted with the axes of pressure and volume. Such a graph is called a *P–V* diagram, sometimes said to be illustrated on a *P–V* plane. The graph appears as horizontal straight line *ab* whose height is at pressure *P* and whose length is from original volume V_1 to final volume V_2.

Now consider the area *abcd* under this graph.

$$\text{Area} = P(V_2 - V_1) \qquad [4]$$

But this is the same as the right-hand side of equation [3].

Hence it follows that the area under a *P–V* diagram gives the work done. This can be shown to be true by an analysis of the units.

If the pressure is in newtons (N/m^2) and the volume is in cubic metres (m^3) then, by equation [3], the work done is given by the product of pressure and change in volume.

$$\therefore \quad \text{Units of work done} = \frac{\text{N}}{\text{m}^2} \times \text{m}^3 = \text{N m} \qquad [5]$$

The unit, N m (newton-metre) is the unit of work, the joule.

1 N m = 1 J

Now the graph shown in Fig. 1.3 illustrates the particular case of constant pressure expansion.

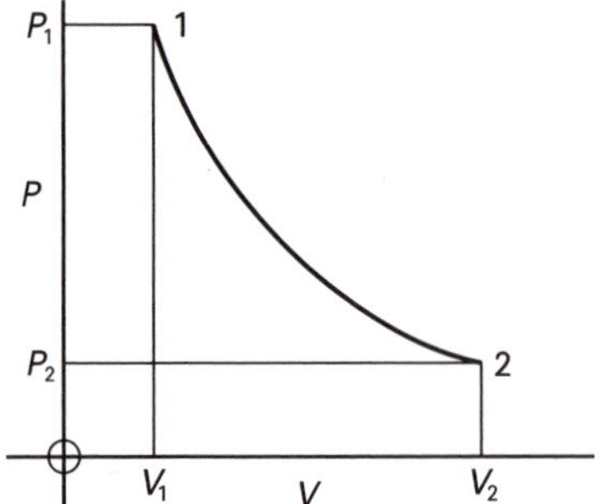

Fig. 1.4 Usual *P–V* diagram

Consider Fig. 1.4. Here is shown a *P–V* diagram of the type usually obtained when an expansion takes place in a thermal engine. It is now a curve with original pressure and volume P_1 and V_1, respectively. The final pressure and volume are P_2 and V_2, respectively. Both the pressure and the volume have changed in this case. What of the area under this graph and will it still give the work done?

Figure 1.5 shows the same graph but this time it has been divided up into small rectangles. The area of each small rectangle represents work done as has been shown. The sum of all the areas of these small rectangles would therefore approximate very closely to the area under the graph and hence the work done. The greater the number of rectangles then the more nearly equal are the sum and the actual area, hence the actual work done. If the number of rectangles were made infinitely great, the sum would, in fact, equal the actual area, which would then give the actual work done. Now this is exactly what happens when the area is solved by the use of the integral calculus.

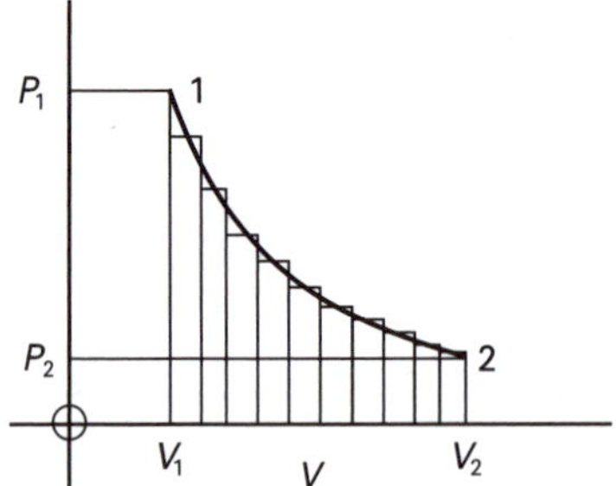

Fig. 1.5 Area division of *P–V* diagram

Consider Fig. 1.6. This is the same *P–V* diagram. Consider some point in the expansion, X say, where the pressure is P and the volume is V. Let there be an elemental expansion δV from this volume V.

Then

$$\text{Work done during the elemental expansion} = P\,\delta V \qquad [6]$$

The total work done will be obtained by summing all the elemental strips of width δV from volume V_1 to volume V_2.

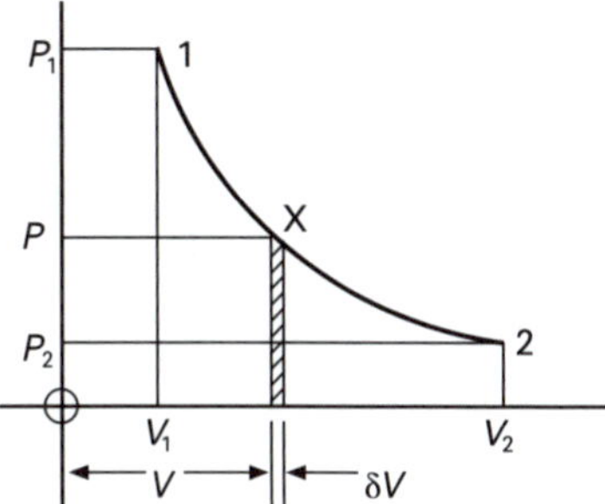

Fig. 1.6 Area calculation on *P–V* diagram

$$\therefore \quad \text{Total work done} = \sum_{V=V_1}^{V=V_2} P\,\delta V \qquad [7]$$

Now, if an infinite number of strips are taken, in which case δV becomes infinitely small (this is written as $\delta V \rightarrow 0$), then

$$\text{Work done} = \int_{V_1}^{V_2} P\,\mathrm{d}V \qquad [8]$$

Nothing more will be said about this equation at present. It will be met later when it will be solved for particular cases. Its limitations will be discussed in Chapter 6 on thermodynamic reversibility.

It is important to note that the discussion has concentrated on expansion. In engines, however, many cases of compression are encountered. The compression is really the reverse of an expansion. What has been said about expansion, therefore, applies equally well to compression. The compression curve plotted on a *P–V* diagram has the same general shape as an expansion curve except that it is in a reverse direction. This means that the volume decreases while the pressure increases. The area under the curve, given by equation [6], gives the work done.

It should be noted the area for an expansion is positive, indicating that work is obtained from an expansion.

The area for a compression is negative, indicating that work must be done on the working fluid in order to compress it.

Example 1.1 *A fluid in a cylinder is at a pressure of 700 kN/m². It is expanded at constant pressure from a volume of 0.28 m³ to a volume of 1.68 m³. Determine the work done.*

SOLUTION

$$\begin{aligned}\text{Work done} &= W = P(V_2 - V_1)\\ &= 700 \times 10^3 \times (1.68 - 0.28)\\ &= 7 \times 10^5 \times 1.4\\ &= (9.8 \times 10^5)\ \text{N m}\\ &= (9.8 \times 10^5)\ \text{J}\\ &= 980\ \text{kJ}\\ &= \mathbf{0.98\ MJ}\end{aligned}$$

1.20 The polytropic process $PV^n = C$, a constant

Changes of state of working substances in thermodynamic systems are often brought about by the expansion or compression of the working substance.

Suppose that an experiment is conducted on a mass of working substance such that an expansion takes place changing the state from state 1 to state 2. Let the pressure change from P_1 to P_2 and the volume from V_1 to V_2. Assume that arrangements are made to record the pressure and volume as the experiment proceeds. From the results obtained, if values of pressure and volume are plotted on a P–V graph, they produce a smooth curve as shown in Fig. 1.7.

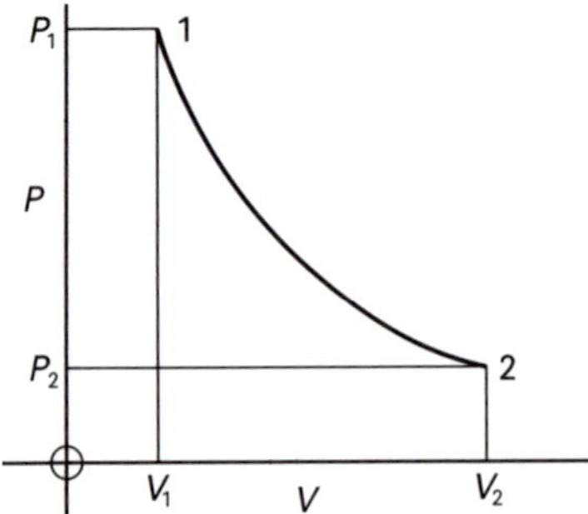

Fig. 1.7 General P–V curve

Simply by inspection of the curve, it is not directly possible to tell whether there is a law connecting pressure and volume for the expansion carried out. However, suppose now that $\log P$ is plotted against $\log V$. The graph obtained is as shown in Fig. 1.8. This is much better for the graph appears as a straight line and is of the form

$$\log P = -n \log V + \log C \qquad [1]$$

where $-n$ = slope of the line
$\log C$ = intercept on the $\log P$ axis

Now equation [1] can be rewritten

$$\log P + n \log V = \log C$$

or taking antilogs

$$PV^n = C\text{, a constant} \qquad [2]$$

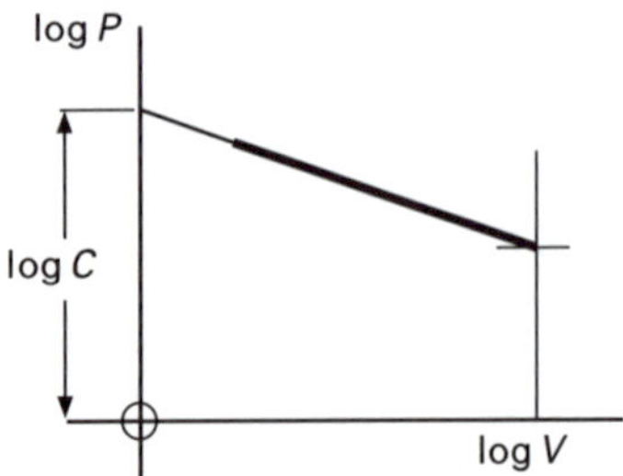

Fig. 1.8 Plot of $\log P$ against $\log V$

Further experiments on different substances taking different quantities of substance and also including the case of compression as well as expansion will yield a similar result. Equation [2] may therefore be considered as the law for the general case of expansion or compression of a substance. This general case of expansion or compression of a substance according to the law $PV^n = C$ is called a **polytropic expansion** or **compression** or a **polytrope**. It should be noted the value of the constant C will change with each change of condition, so also will the value of n, which is called the **index** of the expansion or compression, or the **polytropic exponent**.

Since all conditions of state during the expansion or compression lie on the curve $PV^n = C$, it follows that

$$P_1V_1^n = P_2V_2^n = P_3V_3^n = P_4V_4^n = \ldots, \text{ etc.} \quad [3]$$

where 1, 2, 3, 4, etc., represent different conditions of state taken during the expansion or compression.

To give an idea of the value of the index n, it will generally lie within the range 1 to 1.7. For most cases, however, it will probably lie more closely within the range 1.2 to 1.5.

Further, note that if $n = 0$ the equation becomes

$$PV^0 = C, \text{ a constant} \quad [4]$$

and since $V^0 = 1$ then equation [4] becomes

$$P = C, \text{ a constant} \quad [5]$$

which indicates a constant pressure process.

Also, the equation $PV^n = C$ can be rearranged to read

$$P^{1/n}V = C, \text{ a constant} \quad [6]$$

This is obtained by taking the nth root of both sides. Now if $n = \infty$ then $P^{1/n} = P^{1/\infty} = P^0 = 1$, in which case, equation [6] becomes

$$V = C, \text{ a constant} \quad [7]$$

which indicates a constant volume process.

Example 1.2 *0.112 m^3 of gas has a pressure of 138 kN/m^2. It is compressed to 690 kN/m^2 according to the law* $PV^{1.4} = C$. *Determine the new volume of the gas.*

SOLUTION

Since the gas is compressed according to the law $PV^{1.4} = C$, then

$$P_1V_1^{1.4} = P_2V_2^{1.4}$$

$$\therefore \quad \frac{P_1}{P_2} = \left(\frac{V_2}{V_1}\right)^{1.4} \quad \text{or} \quad \frac{V_2}{V_1} = \left(\frac{P_1}{P_2}\right)^{1/1.4}$$

from which

$$V_2 = V_1\left(\frac{P_1}{P_2}\right)^{1/1.4} = V_1\sqrt[1.4]{\frac{P_1}{P_2}}$$

$$= 0.112 \times \sqrt[1.4]{\frac{138}{690}} = 0.112\sqrt[1.4]{\tfrac{1}{5}}$$

$$= \frac{0.112}{\sqrt[1.4]{5}} = \frac{0.112}{3.157}$$

$$= \mathbf{0.035\,5\ m^3}$$

1.21 Work and the polytropic process

If a substance is to be compressed from a lower pressure to a higher pressure then work will be required in order to carry out the compression. When the substance is at the new high pressure, it has the potential to expand and, in expanding, do some work. It is important to determine the magnitude of this quantity of work. It has already been shown that work done is given by the area under a P–V diagram of an expansion or compression; now the problem is to determine the area under a curve of the form $PV^n = C$. Consider Fig. 1.9; it shows a P–V graph of an expansion according to the law $PV^n = C$ from state P_1, V_1 to new state P_2, V_2.

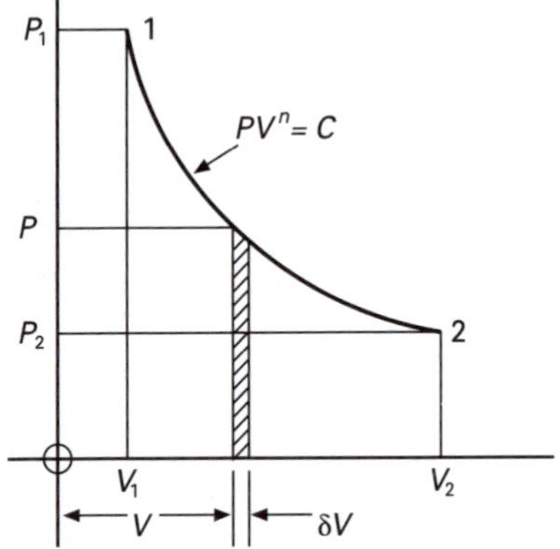

Fig. 1.9 Graph for $PV^n = C$

Consider a point on the curve at which the pressure is P and the volume is V. Let the gas expand from this point by very small volume δV according to the law $PV^n = C$. The work done during this very small expansion is very nearly equal to $P\delta V$. In the limit, as $\delta V \to 0$, the area, and hence the work done $= P\,\mathrm{d}V$. For the whole expansion from 1 and 2

$$\text{Work done} = \int_{V_1}^{V_2} P\,\mathrm{d}V \qquad [1]$$

Now $PV^n = C$, or

$$P = CV^{-n} \qquad [2]$$

Substituting equation [2] in equation [1]

$$\text{Work done} = C\int_{V_1}^{V_2} V^{-n}\,dV \qquad [3]$$

Integrating

$$\text{Work done} = \frac{C}{-n+1}\left[V^{-n+1}\right]_{V_1}^{V_2}$$

$$= \frac{C}{-n+1}\left[V_2^{-n+1} - V_1^{-n+1}\right]$$

$$= \frac{C}{-n+1}\left[V_2^{-n}V_2 - V_1^{-n}V_1\right]$$

$$= \frac{P_2V_2 - P_1V_1}{-n+1} \quad \text{(from equation [2])}$$

Multiplying top and bottom by −1

$$\text{Work done} = \frac{P_1V_1 - P_2V_2}{n-1} \qquad [4]$$

Equation [4] will apply equally well to an expansion or a compression.

By reading P_1, V_1 as the original conditions and P_2, V_2 as the final conditions it will be found that for an expansion, the work done is positive, meaning that work is done by the substance.

For a compression, however, again reading P_1, V_1 as the original conditions and P_2, V_2 as the final conditions, it will be found that the work done is negative, meaning that the work must be done on the substance.

Example 1.3 *0.014 m^3 gas at a pressure of 2070 kN/m^2 expands to a pressure of 207 kN/m^2 according to the law* $PV^{1.35} = C$. *Determine the work done by the gas during the expansion.*

SOLUTION

The work done during a polytropic expansion is given by

$$\text{Work done} = \frac{P_1V_1 - P_2V_2}{n-1}$$

In this problem V_2 is, as yet, unknown and must therefore be calculated.

$$\text{Now } P_1V_1^n = P_2V_2^n$$

$$\therefore \quad V_2 = V_1\left(\frac{P_1}{P_2}\right)^{1/n}$$

or

$$V_2 = 0.014 \times \left(\frac{2070}{207}\right)^{1/1.35} = 0.014 \times \sqrt[1.35]{10} = 0.014 \times 5.05$$

$$= \mathbf{0.077\ m^3}$$

$$\therefore \quad \text{Work done} = \frac{(2070 \times 10^3 \times 0.014) - (207 \times 10^3 \times 0.077)}{1.35 - 1}$$

$$= \frac{10^3}{0.35}(29 - 15.95) = \frac{10^3}{0.35} \times 13.05$$

$$= (37.3 \times 10^3)\ \text{N m}$$
$$= (37.3 \times 10^3)\ \text{J} \quad (1\ \text{N m} = 1\ \text{J})$$
$$= \mathbf{37.3\ kJ}$$

1.22 Work and the hyperbolic process

The hyperbolic process is a particular case of the polytropic process, $PV^n = C$, for $n = 1$.

Thus, the law for the hyperbolic process is

$$PV = C \qquad [1]$$

This law, if plotted on a P–V diagram will appear as a rectangular hyperbola, hence its name.

For a hyperbolic change from state 1 to state 2, from equation [1]

$$P_1V_1 = P_2V_2 \qquad [2]$$

An expression for the work done during a polytropic process has already been determined. This has been shown to be

$$\text{Work done} = \frac{P_1V_1 - P_2V_2}{n - 1} \qquad [3]$$

In the case of a hyperbolic process $P_1V_1 = P_2V_2$; substituting this in equation [3] together with the fact that $n = 1$ gives

$$\text{Work done} = \frac{P_1V_1 - P_2V_2}{n - 1} = \frac{0}{0}$$

This is indeterminate.

Now if the law $PV = C$ is plotted on a P–V graph there is a definite area beneath the curve. So it appears that a start from first principles is necessary to determine this area.

Consider Fig. 1.10. By similar analysis to that given for the polytropic expansion

$$\text{Work done} = \int_{V_1}^{V_2} P\,\mathrm{d}V \qquad [4]$$

In this case, however $PV = C$, hence $P = C/V$ and substituting this in equation [4] gives

$$\text{Work done} = C\int_{V_1}^{V_2} \frac{\mathrm{d}V}{V} = C\Big[\ln V\Big]_{V_1}^{V_2}$$

$$= C[\ln V_2 - \ln V_1]$$

$$= C\ln\frac{V_2}{V_1} = PV\ln\frac{V_2}{V_1} \qquad [5]$$

since $PV = C$. Here ln represents $\log_e$.

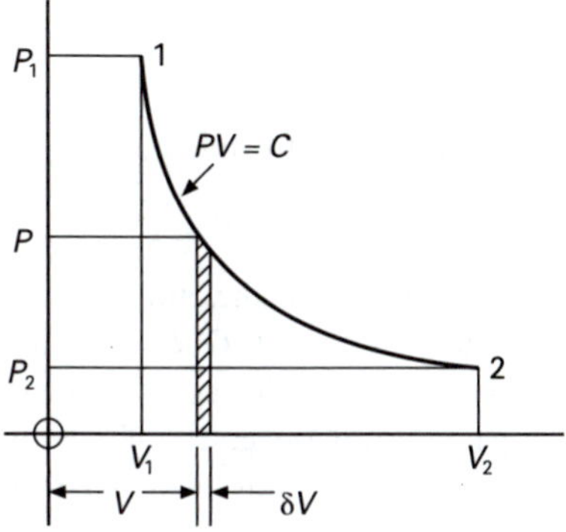

Fig. 1.10 Graph for $PV = C$

Now V_2/V_1 is the expression ratio, often designated by the letter r, so

$$\text{Work done} = PV \ln r \qquad [6]$$

Example 1.4 *A gas is compressed hyperbolically from a pressure and volume of 100 kN/m² and 0.056 m³, respectively, to a volume of 0.007 m³. Determine the final pressure and the work done on the gas.*

SOLUTION

Since the gas is compressed hyperbolically

$$P_1V_1 = P_2V_2 \text{ or } P_2 = P_1\frac{V_1}{V_2} = 100 \times \frac{0.056}{0.007} = 100 \times 8$$

$$= \mathbf{800\ kN/m^2}$$

$$\text{Work done} = PV \ln r = PV \ln \frac{V_2}{V_1}$$

$$= 100 \times 10^3 \times 0.056 \ln \frac{0.007}{0.056}$$

$$= -100 \times 10^3 \times 0.056 \ln \frac{0.056}{0.007}$$

$$= -5.6 \times 10^3 \ln 8$$

$$= -5.6 \times 10^3 \times 2.079$$

$$= -(11.64 \times 10^3)\ \text{N m} \quad (10^3\ \text{N m} = 1\ \text{kJ})$$

$$= \mathbf{-11.64\ kJ}$$

The work done on the gas is 11.64 kJ.

1.23 Internal energy

If a hot body is placed in contact with a cold body then the temperature of the hot body begins to fall while the temperature of the cold body begins to rise. To account for this it is said that the hot body gives up heat, hence its temperature falls, and the cold body receives this heat, hence its temperature rises.

Observing this fact, some early investigators around in the eighteenth century considered that heat must have properties similar to those of a fluid. One such fluid was called caloric. Thus, if a body was heated, caloric was said to have passed from the source of heat supply into the body, hence it became hot. Conversely, if a body cooled, it lost some of its caloric. Since the weight of the body was unaffected by being heated or cooled, it was considered that caloric was a weightless fluid and was said to fill the minute spaces or pores of the body.

Another such fluid, frigoric, was used to explain the phenomenon of cold. It was said to be composed of minute darts of frost. If one's hand is placed on a piece of ice, for example, not only does it feel cold but it can feel somewhat painful. It was suggested that the reason for this was that the dart-like particles of frigoric were being transferred from the ice into the hand and thus the hand became cold accompanied by a sense of pain.

Yet another fluid which accounted for the heating effect produced by a fire was called phlogiston. To produce its heating effect the fire was said to have given up phlogiston to the bodies being heated. As a matter of interest, when it was found that a gas, which is now known as oxygen, was directly associated with the phenomenon of combustion, it was called dephlogisticated air. The reason for this gas, being the only one associated with the production of fire, was therefore the only gas without phlogiston; as it received phlogiston, fire resulted and heat was produced.

Nearly all the known phenomena of heating and cooling could be explained by the introduction of such fluids as caloric, frigoric and phlogiston.

These theories were eventually found to be false. Notable among those who showed them to be wrong was Count Rumford of Munich. Count Rumford was really an American citizen, born near Boston in 1753. His real name was Benjamin Thompson and he had to leave America for his part in the rebellion of the British Colonies. He settled in Bavaria where he became the superintendent of an arsenal in Munich. He was rewarded with the title Count Rumford for his services in this respect.

During his work in the arsenal, he noticed that when boring a cannon, the material of the cannon became extremely hot; this did not seem to tie up very well with the caloric theory which was widely accepted at the time. He accordingly conducted an experiment in an attempt to settle this matter. Instead of using a sharp boring tool he used a blunt one which was rubbed against a mass of about 51 kg of gunmetal. After some 960 revolutions the temperature of the gunmetal had risen by about 39 °C. A minute quantity of gunmetal had been rubbed off during the experiment.

Here then was an experiment in which there had been no hot source to supply heat, by yielding up some of its caloric in heating the gunmetal, yet the gunmetal had, in fact, become extremely hot. It might have been suggested that the caloric, originally in the metallic dust that was rubbed off, had been left behind in the main bulk of gunmetal and was therefore responsible for the temperature rise. The amount of metallic dust was so small and the quantity of heat developed so large that Count

Rumford concluded this was impossible. It also appeared that as long as the rubbing, or friction, of the boring tool continued, heat would continue to be produced, hence the supply of heat generated in this way was inexhaustible. Since the heat was generated as a result of motion, Rumford therefore suggested that heat was in some way the result of the motion of the particles which make up a body.

There was nothing really new about this idea. It had been the view taken by some philosophers from very early times.

After its short accepted life, during the eighteenth century, the supposed fluid nature of heat was dropped. It then became the accepted theory that heat was a manifestation of the degree of agitation of the very minute particles (atoms and molecules, to be discussed later) which make up a body. Part of the old philosophy remains, however, for it is common practice to refer to such things as 'the flow of heat' and 'quantity of heat' which still suggest a fluid nature of heat.

It is important, however, to think a little more about the theory that heat is a result of the degree of agitation of the particles which make up a body. If a particle is in motion, it will possess a kinetic energy, which is a function of the velocity at which the particle is moving. It appears, in general, the greater the kinetic energy that can be imparted to the particles which make up a body, the higher the temperature of that body will become.

It has now become clear that the store of energy which results from the random motion of the atoms and molecules of a body would be far better referred to as **internal energy**, leaving the term **heat** to be used to describe that energy transfer process which results from a temperature difference.

At any one particular state, the atoms and molecules will have a particular overall degree of random motion and, in a pure substance, this degree of random motion will be the same each time the substance returns to that state. The degree of random motion must therefore be a property. Internal energy is a function of the degree of random motion, so it must be a property.

Count Rumford's experiment showed that, in his particular case, an increase in internal energy content resulted in an increase in temperature. This is always the case in a single-phase system. Count Rumford's single-phase system was the gunmetal cannon, and note here that the internal energy increase was the result of a blunt boring tool being rubbed against the cannon. The energy transfer in this case was really a work transfer, work having been done against friction.

It must be noted, however, that an internal energy increase does not always result in an increase in temperature. It will be shown during the discussion on two-phase systems that when the phase is being changed from one to another, such as water into steam, the temperature will remain constant. Here the internal energy increases at constant temperature; the increase in internal energy is necessary to carry out the degree of separation of the molecules to change the water into steam. The same general situation arises during the change of a solid into a liquid.

It has been stated that the internal energy of a substance results from the motion of its atoms or molecules. In a fluid, the atoms and molecules have rather greater motions than solids; in fact, they move about freely (rather more freely in the case of gases). This means that the atoms and molecules will be constantly impinging upon the walls of any container. Now the impact of a particle on a wall means that a force will be imparted t that wall. The constant bombardment of the walls of the containing vessel results i total average force on each wall. When this average force is reduced to that which occurs on unit area of the wall, it is called the **pressure** on the wall.

Again, in the above discussion it has been noted that the internal energy content is the result of the motion of the atoms and molecules which make up a body. Further, it was noted that an increase in internal energy content generally results in an increase in temperature. As the temperature of a body falls, the motion of the atoms and molecules therefore reduces, and the internal energy content also reduces. So it is reasonable to assume that a condition exists in which the atoms and molecules of a body are completely at rest, in which case the internal energy content would then be zero and the temperature would have reached its absolute zero. The idea of an absolute zero of temperature was mentioned in section 1.7.

In the case of internal energy, specific internal energy is designated u, the internal energy of any mass, other than unity, is designated U.

1.24 Heat

The discussion on internal energy suggested how bodies were once believed to contain heat. This is now not considered as being the case; the internal store of energy is now called internal energy, which is a property.

However, it was further suggested that, during an energy transfer process which results from the temperature difference between one body and another, the energy so transferred is called **heat**. The heat, having been transferred, will then disperse into other forms of energy, such as internal energy or work, the disposal being a function of the system employed.

Note that heat is a transient quantity; it describes the energy transfer process through a system boundary resulting from temperature different. If there is no temperature difference, there is no heat transfer.

And since the term *heat* is used to describe a transfer process, heat energy ceases to exist when the process finishes. Thus heat is not a property.

Heat energy is given the symbol Q. To indicate a rate of heat transfer, a dot is placed over the symbol, thus

$$\dot{Q} = \text{heat transfer/unit time}$$

1.25 Specific heat capacity

For unit mass of a particular substance at a temperature t, let there be a change of temperature δt brought about by a transfer of heat δQ.

The specific heat capacity, c, of the substance at temperature t is defined by the ratio $\delta Q/\delta t$. Thus

$$c = \frac{\delta Q}{\delta t} \qquad [1]$$

In the limit, as $\delta t \to 0$, then

$$c = \frac{dQ}{dt} \qquad [2]$$

Specific heat capacity is generally found to vary with temperature. For example, the specific heat capacity of water falls slightly from a temperature of 0 °C to a minimum of about 35 °C and then begins to rise again.

Specific heat capacity can also vary with pressure and volume. This is particularly true of compressible fluids such as gases (see Chapter 5).

It is common practice to use an average value of specific heat capacity within a given temperature range.

This average value is then used as being constant within the temperature range, so equation [2] can be rewritten

$$c = \frac{Q}{\Delta t} \qquad [3]$$

where Q = heat transfer/unit mass, J/kg
Δt = change in temperature, K

Note that from equation [3] the basic unit for specific heat capacity is **joules/kilogram kelvin** or **J/kg K**; multiples such as kilojoules/kilogram kelvin (kJ/kg K) may also be used.

A particular application of specific heat capacity arises from the use of water as a measuring device in calorimetry. Temperature measurements during a calorimetric experiment are made while the pressure of the water remains constant.

A process in which the pressure remains constant is said to be **isobaric**. The specific heat capacity in this case is therefore said to be the **isobaric specific heat capacity** and is written c_p.

The table gives a few examples of average specific heat capacities of some solids and liquids. The specific heat capacity of gases is dealt with separately in Chapter 5.

Table of average specific heat capacities

Solid	Specific heat capacity (J/kg K)	Solid	Specific heat capacity (J/kg K)	Liquid	Specific heat capacity (J/kg K)
Aluminium	915	Lead	130	Benzene	1 700
Brass	375	Nickel	460	Ether	2 300
Cast iron	500	Steel	450	Ethanol	2 500
Copper	390	Tin	230	Paraffin	2 130
Crown glass	670	Zinc	390	Mercury	140

Example 1.5 *5 kg of steel, specific heat capacity 450 J/kg K, is heated from 15 °C to 100 °C. Determine the heat transfer.*

SOLUTION

From equation [1]

$$\begin{aligned}\text{Heat required} &= mc(t_2 - t_1)\\ &= 5 \times 450 \times (100 - 15)\\ &= 5 \times 450 \times 85\\ &= 191\,250 \text{ J}\\ &= \mathbf{191.25\ kJ}\end{aligned}$$

Example 1.6 *A copper vessel of mass 2 kg contains 6 kg of water. If the initial temperature of the vessel plus water is 20 °C and the final temperature is 90 °C, how much heat is transferred to accomplish this change, assuming there is no heat loss? Take the specific heat capacity of water to be 4.19 kJ/kg K.*

SOLUTION

From the table of average specific heat capacities

$$\text{Specific heat capacity of copper} = 390\ \text{J/kg K}$$

$$\begin{aligned}\text{Heat required by copper vessel} &= 2 \times 390 \times (90 - 20)\\ &= 2 \times 390 \times 70\\ &= 54\,600\ \text{J}\\ &= \mathbf{54.6\ kJ}\end{aligned}$$

$$\begin{aligned}\text{Heat required by water} &= 6 \times 4.19 \times (90 - 20)\\ &= 6 \times 4.19 \times 70\\ &= \mathbf{1759.8\ kJ}\end{aligned}$$

$$\begin{aligned}\text{Heat to vessel} + \text{water} &= 54.64 + 1759.8\\ &= \mathbf{1814.4\ kJ}\end{aligned}$$

Example 1.7 *An iron casting of mass 10 kg has an original temperature of 200 °C. If the casting loses heat to the value 715.5 kJ, determine the final temperature.*

SOLUTION

From the table of average specific heat capacities

$$\text{Specific heat capacity of cast iron} = 500\ \text{J/kg K}$$

$$\text{Heat transferred from casting} = mc(t_2 - t_1) \qquad [1]$$

$$\begin{aligned}\text{Heat transferred} &= -715.5\ \text{kJ}\\ &= -715\,500\ \text{J}\end{aligned}$$

Note the negative sign, indicating a heat loss.
From equation [1]

$$-715\,500 = 10 \times 500 \times (t_2 - 200)$$

$$\therefore \quad t_2 = 200 - \frac{715\,500}{10 \times 500} = 200 - 143.1$$

$$= \mathbf{56.9\ °C}$$

Example 1.8 *A liquid of mass 4 kg has its temperature increased from 15 °C to 100 °C. Heat transfer into the liquid to the value 714 kJ is required to accomplish the increase in temperature. Determine the specific heat capacity of the liquid.*

SOLUTION

Heat transfer required $= Q = mc(t_2 - t_1)$

$$\therefore \quad c = \frac{Q}{m(t_2 - t_1)} \quad \text{and} \quad Q = 714 \text{ kJ} = 714\,000 \text{ J}$$

$$= \frac{714\,000}{4(100 - 15)} = \frac{714\,000}{4 \times 85}$$

$$= \frac{714\,000}{340}$$

$= 2100$ J/kg K
$=$ **2.1 kJ/kg K**

1.26 Calorimetry

Calorimetry is the determination of standard thermal quantities such as specific heat capacity and the calorific value of fuels (see section 1.30 and Chapter 8).

Further reference will be made in the text as it applies to particular cases.

1.27 The adiabatic process

If a process is carried out in a system such that there is no heat transferred into or out of the system (i.e. $Q = 0$) then the process is said to be **adiabatic**. Such a process is not really possible in practice, but it can be closely approached.

If a system is sufficiently thermally insulated, heat transfer can be considered as negligible and the process or processes within the system can be considered as being adiabatic. Alternatively, if a process is carried out with sufficient rapidity, there will be little time for heat transfer. Thus if a process is rapid enough, it can be considered as being effectively adiabatic.

The implications of any particular process being considered as adiabatic will be dealt with in the text.

1.28 Relationship between heat and work

Figure 1.11 shows two containers each containing a mass of water m and each having a thermometer inserted such that temperature measurement can be made. In each case, the mass of water is the system. And for this discussion, any other fluid of mass m could also be considered as the system.

In (a) it is arranged that an external heater can transfer heat energy Q through the system boundary into the water. In (b) it is arranged that a paddle-wheel is immersed in water such that external paddle or stirring work W is done when the wheel is rotated. In each case it is assumed that there is no energy loss from the system.

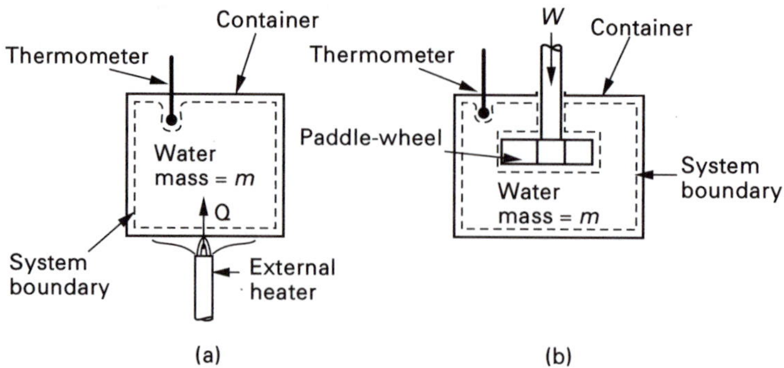

Fig. 1.11 Temperature rise by (a) supplying heat (b) doing work

Consider the arrangement in (a). It is common experience to heat water in some containing vessel by means of some external heating device. Let the initial temperature as recorded on the thermometer be t_1, and after heating, in which heat energy Q is transferred into the water, let the final temperature be t_2.

Consider, now, the arrangement in (b). The container once again contains a mass of water m but in this case a paddle-wheel is introduced into the water. It is common experience that friction makes things warm. The simple experience of rubbing one's hands together in a brisk manner will show this. In the case under consideration it is possible to rotate the paddle-wheel against the frictional resistance of the water. Assume that the initial temperature of the water is t_1 and, after doing an amount of work W on the paddle-wheel, the final temperature is t_2.

Now a similar effect has been produced in both cases (a) and (b) in that a mass of water m starting at a temperature t_1 has experienced a rise in temperature $(t_2 - t_1)$.

Case (a) used a heat transfer to produce an effect; case (b) used a work transfer to produce the same effect.

The conclusion must be that there is a relationship between heat and work.

If the unit of energy is the same for both work and heat, since the same effect was produced in each case, the relationship is of the form

$$W = Q \qquad [1]$$

The unit of energy in the SI system of units is the joule (J).

$$1\ \mathrm{J} = 1\ \mathrm{N\,m}$$

which has the units $\mathrm{kg}\ \dfrac{\mathrm{m}}{\mathrm{s}^2} \times \mathrm{m} = \mathrm{kg}\ \dfrac{\mathrm{m}^2}{\mathrm{s}^2}$

From equation [1], since $W = Q$, then the unit of energy for both work and heat is the **joule**, named after James Prescott Joule (1819–1889), an English physicist.

In older systems of units, heat energy was defined using water as a reference substance. This is now abandoned and in its place is the energy unit, the joule. In some calorimetric devices, however, water is used as a means of measurement.

1.29 Enthalpy

It has been shown that internal energy, pressure and volume are properties. During subsequent discussion a particular combination of these properties will often appear. The combination is in the form $u + Pv$ and, because this combination has a particular significance in some processes, it is given a name. The name is **enthalpy** and is given the symbol h. Thus, $h = u + Pv$. Note that, since pressure, volume and temperature are properties, their combination is also a property, so enthalpy is a property. Specific enthalpy is designated h. The enthalpy of any mass other than unity is designated H.

1.30 The principle of the thermodynamic engine

The thermodynamic engine is a device in which energy is supplied in the form of heat and some of this energy is transformed into work. It would be ideal if all the energy supplied was transformed into work. Unfortunately, no such complete transformation process exists or, as will be shown in Chapter 6, can possibly exist.

The usual process in the engine can be followed by reference to Fig. 1.12.

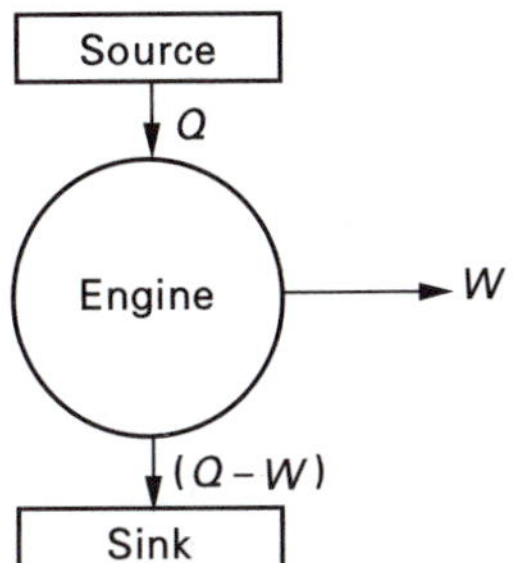

Fig. 1.12 Thermodynamic engine

With all engines there must be a source of supply of heat and, with any quantity of heat Q supplied from the source to the engine, an amount W will successfully be converted into work. This will leave a quantity of heat $(Q - W)$ to be rejected by the engine into the sink.

The ratio

$$\frac{W}{Q} = \frac{\text{Work done}}{\text{Heat received}} \qquad [1]$$

is called the **thermal efficiency** and determines what fraction of the heat input has actually been successfully converted into work output. It will be evident that the object in all engines is, or should be, to make the thermal efficiency as near to unity as possible.

All engines use a working substance as the means of carrying out the conversion of heat energy into work. The heat energy is usually obtained by burning a fuel or by thermonuclear reaction. This heat energy is suitably transferred into the working substance and, consequently, the pressure and temperature of the substance are usually raised above that of the surroundings. In this condition the substance is capable of doing work. For example, it could be enclosed by using a piston in a cylinder; if the piston were free to move, it would be pushed down the cylinder and work would be done as the substance expanded. The substance would lose some of its energy in doing this work. When the substance has performed as much work as is practically possible, it could be removed from the cylinder and rejected to the sink. By returning the piston to its original position and then introducing some more high-energy-containing substance, the process could be repeated. This is what happens in any piston engine. The intake and rejection processes of the working substance are intermittent in this case.

In the majority of turbine engines, however, the working substance passes through in a continuous flow.

There are two possibilities with regard to the introduction of the energy into the working substance which, in most cases, is either a vapour or a gas.

The first possibility is to transfer heat into the substance outside the engine and then to pass the high-energy-containing substance over into the engine. This is the usual process carried out when using steam as the working substance which is formed outside the engine in a boiler and is then passed to the engine. This is a case of a vapour being used as the working substance.

The second possibility is to introduce the energy directly into the working substance in the engine. This is the usual process carried out in petrol, oil and gas engines in which the fuel is introduced directly into, and burnt in, the engine cylinders. When this is the case, the engines are called **internal combustion engines**, IC engines. Each method naturally has its own complexity. More will be said in later chapters.

Now a further note about thermal efficiency. It has already been stated that the process in the engine is that of receiving heat, converting some of it into work and then rejecting the remainder. So it appears that, neglecting losses, the difference between the heat received and the heat rejected is equal to the work done, or

$$\text{Heat received} - \text{Heat rejected} = \text{Work done} \qquad [2]$$

Now

$$\text{Thermal } \eta = \frac{\text{Work done}}{\text{Heat received}} \quad \text{(see equation [1])} \qquad [3]$$

(η, Greek letter eta, is the symbol usually used for efficiency.)

Using equation [2] in [3]

$$\text{Thermal } \eta = \frac{\text{Heat received} - \text{Heat rejected}}{\text{Heat received}} \qquad [4]$$

$$= 1 - \frac{\text{Heat rejected}}{\text{Heat received}} \qquad [5]$$

From equation [3]

$$\text{Work done} = \text{Heat received} \times \text{Thermal } \eta \qquad [6]$$

Also

$$\text{Heat received} = \frac{\text{Work done}}{\text{Thermal } \eta} \qquad [7]$$

Now from equation [4]

$$\text{Thermal } \eta = \frac{\text{Heat received} - \text{Heat rejected}}{\text{Heat received}}$$

$$\therefore \quad \text{Heat received} \times \text{Thermal } \eta = \text{Heat received} - \text{Heat rejected}$$

From which

$$\begin{aligned} \text{Heat rejected} &= \text{Heat received} - (\text{Heat received} \times \text{Thermal } \eta) \\ &= (1 - \text{Thermal } \eta) \times \text{Heat received} \qquad [8] \end{aligned}$$

Strictly, equations [1] to [8] will only apply to a system in which heat and work only transfer across the system boundary, a heat engine in fact (see section 1.31). For other systems, such as an IC engine, the thermal efficiency may be defined as

$$\text{Thermal } \eta = \frac{\text{Work done}}{\text{Energy received}} \qquad [9]$$

It is useful to note here that the amount of energy liberated by a fuel when burnt is defined by its **calorific value** (see Chapter 8).

The calorific value of a fuel is defined as the amount of energy liberated by burning unit mass or volume of the fuel. Thus, if by burning 1 kg of petrol, 43 MJ/kg are liberated then the calorific value of the petrol is 43 MJ/kg. Special calorimeters have been developed for the determination of the calorific value of fuels. They are described in Chapter 8.

Example 1.9 *A petrol engine uses 20.4 kg of petrol per hour of calorific value 43 MJ/kg. The thermal efficiency of the engine is 20 per cent. Determine the power output of the engine and the energy rejected/min.*

SOLUTION

$$20.4 \text{ kg petrol/h} = \frac{20.4}{3600} \text{ kg/s}$$

$$\text{Energy liberated petrol} = \left(\frac{20.4}{3600} \times 43 \times 10^6\right) \text{ J/s}$$

Of this, only 20 per cent is successfully transformed into power output.

$$\therefore \quad \text{Power output} = \left(\frac{20.4}{3600} \times 43 \times 10^6 \times 0.2\right) \text{ J/s}$$

$$= \left(\frac{20.4 \times 43 \times 0.2}{3.6} \times 10^3\right) \text{ W}$$

$$= \left(\frac{20.4 \times 43 \times 0.2}{3.6}\right) \text{ kW}$$

$$= \mathbf{48.7\ kW}$$

Energy rejected = (1 − thermal η) × Energy received (see equation [8])

$$= (1 - 0.2)\left(\frac{20.4}{60} \times 43 \times 10^6\right) \text{ J/min}$$

$$= \left(0.8 \times \frac{20.4}{60} \times 43 \times 10^6\right) \text{ J/min}$$

$$= \left(0.8 \times \frac{20.4}{60} \times 43\right)$$

$$= \mathbf{11.7\ MJ/min}$$

It may be thought that a thermal efficiency of 20 per cent seems extremely low. It is unfortunately the case that engine thermal efficiencies are very low; more on this later when dealing with engines and plant.

Example 1.10 *A steam plant uses 3.045 tonne of coal per hour. The steam is fed to a turbine whose output is 4.1 MW. The calorific value of the coal is 28 MJ/kg. Determine the thermal efficiency of the plant.*

SOLUTION

$$3.045 \text{ tonne/h} = 3.045 \times 10^6 \text{ megagram/h}$$

$$= 3045 \text{ kg/h} = \frac{3045}{3600} = 0.846 \text{ kg/s}$$

Energy liberated by coal = $(0.846 \times 28 \times 10^6)$ J/s
Power output from turbine = (4.1×10^6) W
= (4.1×10^6) J/s

$$\text{Thermal } \eta = \frac{\text{Power output}}{\text{Energy liberated by coal}}$$

$$= \frac{4.1 \times 10^6}{0.846 \times 28 \times 10^6}$$

$$= \mathbf{0.173} \text{ or } \mathbf{17.3\%}$$

1.31 The heat engine

Heat is defined as that transfer of energy which results from a difference in temperature, so a **heat engine** must be an engine in which a transfer of heat occurs. If heat is introduced into a system and, as the result of a cyclic process, some work appears from that system, together with some heat rejection from the system, then this is a heat engine. This is illustrated in Fig. 1.12. In practice, such an engine is the closed-circuit steam turbine plant of a power station.

On the other hand, the open-circuit internal combustion engine, such as a petrol engine, is strictly not a heat engine; fuel and air are admitted across the system boundary, combustion is internal, as the name implies, and combustion products and heat are rejected, with some work crossing the system boundary.

However, nearly all thermodynamic engines are colloquially referred to as heat engines.

1.32 Mechanical power

Power is defined as the rate of doing work, or

$$\text{Power} = \frac{\text{Work done}}{\text{Time taken}} \qquad [1]$$

If the unit of work is the joule (J) and the time taken is in seconds (s), equation [1] gives the unit of power as

$$\frac{\text{J}}{\text{s}} \text{ or joule/second.}$$

The rate of doing work of 1 joule/second is called the **watt** (W), thus

$$1 \text{ W} = 1 \text{ J/s} \qquad [2]$$

The watt is named after James Watt (1736–1819) of steam engine fame.

In section 1.33 it will be shown that the unit of electrical power is also the watt, thus giving a very convenient comparison between mechanical and electrical power.

Example 1.11 *At a speed of 50 km/h the resistance to motion of a car is 900 N. Neglecting losses, determine the power output of the engine of the car at this speed.*

SOLUTION

$$\text{Speed} = 50 \text{ km/h} = \frac{50 \times 10^3}{3600} \text{ m/s}$$

$$\text{Power} = \text{Work done/s} = \text{Resistance to motion (N)} \times \text{speed (m/s)}$$

$$= \left(\frac{900 \times 50 \times 10^3}{3600}\right) \text{N m/s}$$

$$= (12.5 \times 10^3) \text{ N m/s} = (12.5 \times 10^3) \text{ J/s}$$

$$= (12.5 \times 10^3) \text{ W}$$

$$= \mathbf{12.5\ kW}$$

The power output of the engine is 12.5 kW.

1.33 Electrical power

The use of electricity is now so widespread that it is essential to have a knowledge of electrical power. The fact that electrical energy can be converted into mechanical energy can be readily observed in the electric motor. Again, electrical energy can be converted into thermal energy using the common electric heater.

Since electrical energy can readily be converted into work, electrical energy input to an electrical circuit is sometimes referred to as electric work transfer.

The effort which drives electricity through an electric circuit is called the **potential difference**, symbol V. This effort is usually supplied by a generator or a battery. The unit of potential difference is called the **volt** (V). An instrument called a voltmeter is made to measure potential difference. To measure the potential difference of a generator or battery a voltmeter is connected across the terminals of the generator or battery.

The quantity of electricity being driven round a circuit is called the **current**, symbol I. The unit of current is the **ampere** or **amp** (A). An instrument called the ammeter is made to measure electric current. In order to measure the current, the ammeter is connected in the circuit such that the current must flow through it.

Figure 1.13 illustrates the connections of the voltmeter and ammeter into an electric circuit. Note that the voltmeter is connected across the circuit, thus measuring the potential difference. If any electrical device is connected across a circuit in this manner it is said to be connected in parallel.

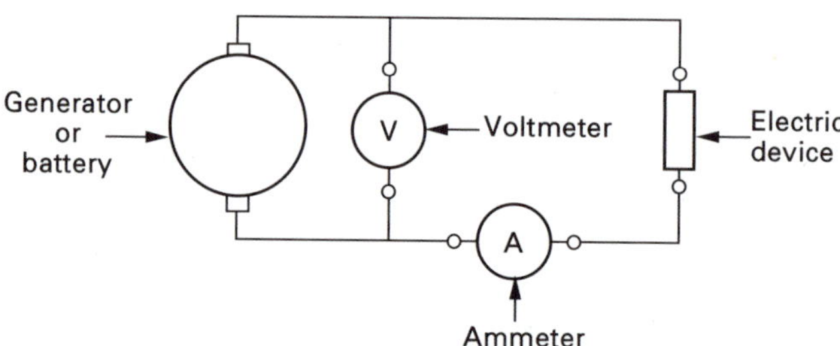

Fig. 1.13 Simple electric circuit

The ammeter, on the other hand, is connected actually in the circuit such that the current must flow through it and hence the ammeter will measure the current. If any device is connected actually in an electric circuit, such as the ammeter, it is said to be connected in series.

In some generators, and in all batteries, the current delivered to any circuit is always in the same direction. The connections to either the generator or battery are made by means of terminals. Current flowing in one direction is said to be **direct current**, abbreviated d.c. One of the terminals of either the generator or battery is said to be positive, marked +, and the other is said to be negative, marked −. Direct current is always considered as flowing from the positive to the negative terminal. The generator of direct current electricity is referred to as a d.c. generator.

Other generators generate electricity in which the current is continuously changing its direction. Such current is called **alternating current**, abbreviated a.c. In this case, neither terminal can be designated as positive or negative; both are continuously changing in polarity. The type of meters to measure potential difference and current are different in design in this case but they measure potential difference in volts and current in amps, as before. The generator to develop alternating current electricity is referred to as an **alternator**. Most electric power developed in power stations is a.c., and in the United Kingdom the standard number of current direction changes is 50 per second. Each change from positive to negative and back is called a cycle. Thus, in the above case, the current is said to have 50 cycles per second, which is called the current **frequency**. Now a frequency of 1 cycle per second is called **1 hertz** (Hz). Hence a frequency of 50 cycles per second = 50 c/s = 50 Hz.

The unit of power in an electric circuit is called the **watt**, and this is the rate of working in an electrical circuit whose potential difference is 1 volt with a current flow of 1 amp.

Thus for any circuit

$$VI = W \qquad [1]$$

where V = potential difference in volts
I = current in amps
W = power in watts

The unit of electric power, the watt, has the same magnitude as that of the unit of mechanical power (1 W = 1 J/s). This is arranged by the choice of the units of potential difference and current.

Example 1.12 *An engine drives an electric generator and 8 per cent of its power is lost in the transmission to the generator. The generator has an efficiency of 95 per cent and its electrical output is at 230 volts. It delivers a current of 60 amps. Determine the power output of the engine.*

SOLUTION

$$\begin{aligned}\text{Power delivered by generator} &= VI \text{ watts}\\ &= (230 \times 60)\\ &= 13\,800 \text{ W}\\ &= \mathbf{13.8\ kW}\end{aligned}$$

But the generator is only 95 per cent efficient.

$$\therefore \quad \text{Power input from engine} = \frac{13.8}{0.95} = \mathbf{14.53\ kW}$$

Also, 8 per cent of the engine power is lost in transmission. Hence, 14.55 kW represents 92 per cent of the available engine output.

$$\therefore \quad \text{Power output from engine} = \frac{14.53}{0.92} = \mathbf{15.79\ kW}$$

Example 1.13 *A 4-kilowatt heater operates at 230 volts. Determine the current taken in amps.*

SOLUTION

$VI = W$ where W is in watts

$$\therefore \quad I = \frac{W}{V} = \frac{4 \times 1000}{230} = \mathbf{17.4\ amps}$$

Example 1.14 *A power station output is 500 megawatts. During a test it is found that this represents 28 per cent of the energy put into the plant by burning coal in the boilers. The coal used liberates 29.5 MJ/kg. Determine the mass of coal burnt by the power station in 1 hour.*

SOLUTION

$500 \text{ MW} = (500 \times 10^6) \text{ W}$

This represents 28 per cent of the energy available from the coal.

$$\therefore \quad \text{Energy from coal} = \frac{(500 \times 10^6)}{0.28} \text{ W}$$

$$= \frac{(500 \times 10^6)}{0.28} \text{ J/s}$$

$$= \frac{(500 \times 10^6 \times 3600)}{0.28} \text{ J/h}$$

The coal liberates 29.5 MJ/kg $= (29.5 \times 10^6)$ J/kg

$$\therefore \quad \text{Mass of coal used/h} = \frac{500 \times 10^6 \times 3600}{0.28 \times 29.5 \times 10^6}$$

$$= \frac{5 \times 10^2 \times 3.6 \times 10^3}{0.28 \times 29.5}$$

$$= (2.18 \times 10^5) \text{ kg}$$
$$= 218\,000 \text{ kg}$$
$$= \mathbf{218\ t\ (tonne)}$$

(1 t = 1 tonne = 1 megagram = 10^3 kg)

Questions

1. The temperature of 4.5 kg of water is raised from 15 °C to 100 °C at constant atmospheric pressure. Determine the heat transfer required. Take the specific heat capacity of water to be 4.18 kJ/kg K.

   [1598.9 kJ]

2. A gas turbine plant delivers an output of 150 MW. The gas consumption is 55 000 m^3/h. The calorific value of the gas used is 38.3 MJ/m^3. Determine the thermal efficiency of the plant.

   [25.6%]

3. A car has a mass of 1600 kg. It has an engine which develops 35 kW when travelling at a speed of 70 km/h. Neglecting losses, determine the resistance to motion in N/kg.

   [1.25 N/kg]

4. A power station has an output of 800 MW and the thermal efficiency is 28 per cent. Determine the coal consumed in tonne per hour if the calorific value of the coal is 31 MJ/kg.

   [331.8 tonne/h]

5. A diesel engine uses 54.5 kg of fuel oil per hour of calorific value 45 MJ/kg. The thermal efficiency of the engine is 25 per cent. Determine the power output of the engine in kilowatts.

   [170.3 kW]

6. An engine rejects 1260 MJ/h when running at a thermal efficiency of 22 per cent. The calorific value of the fuel used is 42 MJ/kg. Determine the power output of the engine in kilowatts and the mass of fuel used per hour.

   [98.7 kW; 38.46 kg/h]

7. 14.5 litres of gas at a pressure of 1720 kN/m^2 is contained in a cylinder. It is expanded at constant pressure until its volume becomes 130.5 litres. Determine the work done by the gas.

   [199.5 kJ]

8. A quantity of steam, the original pressure and volume being 140 kN/m^2 and 150 litres respectively, is compressed to a volume of 30 litres, the law of compression being $PV^{1.2} = C$. Determine the final pressure and the work done.

   [966 kN/m^2; -39.9 kJ]

9. A quantity of gas has an initial pressure of 2.72 MN/m^2 and a volume of 5.6 dm^3 (l). It is expanded according to the law $PV^{1.35} = C$ down to a pressure of 340 kN/m^2. Determine the final volume and the work done.

   [26.15 dm^3; 18.2 kJ]

Chapter 2

Systems

2.1 General introduction

All physical things in nature have some form of boundary whose shape in general identifies it as a particular object. Inside its boundary there are various features which have particular characteristics and functions. This internal arrangement is called a **system**. Outside the boundary of the object are the surroundings, and the reaction between the system and surroundings in general controls the behaviour pattern of the object. A human being and a tree are systems. Heat engines and allied arrangements, which are the concern here, are other systems. It is not necessary that at any one time a complete object need be under investigation. Only part may be under study and this part may then be considered as the system. In other words, a system can be defined as a particular region which is under study. It is identified by its boundary around which are the surroundings.

The boundary need not be fixed. For example, a mass of gas (the system) may expand, so the boundary in this case will modify and interactions will occur with the surroundings at the boundary. If the mass of a system remains constant, the system is a **closed system**. If, on the other hand, the mass of a system changes, or is continuously changing, the system is an **open system**. For example, an air compressor is an open system since air is continuously streaming into and out of the machine, in other words, air mass is crossing its boundary. This is called a **two-flow open system**.

Another example is air leaving a compressed air tank. This would be a **one-flow open system** since air is only leaving the tank and none is entering. In any system, energies such as work and heat could be arranged to cross the boundary. Closed and open systems are illustrated in Fig. 2.1.

2.2 Control volume

If the volume of a system under study remains constant then this volume is called the **control** volume. The control volume is bounded by the control surface.

A control volume and its surface are illustrated in Fig. 2.2. It is shown as a fixed volume enclosing a steam turbine and condenser. Various masses and energies can be investigated as they cross the control surface into, or out of, the control volume. The control volume is similar in concept to the open system. In the case of the control volume both the volume and position are fixed, whereas with an open system the volume could change both in size and position. The air compressor given as an example of an open system in section 2.1 could also be considered as a control volume.

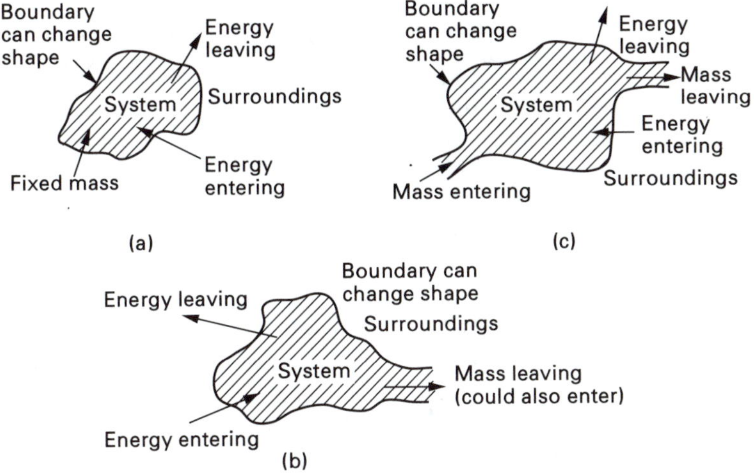

Fig. 2.1 Systems: (a) closed; (b) open, one-flow; (c) open, two-flow

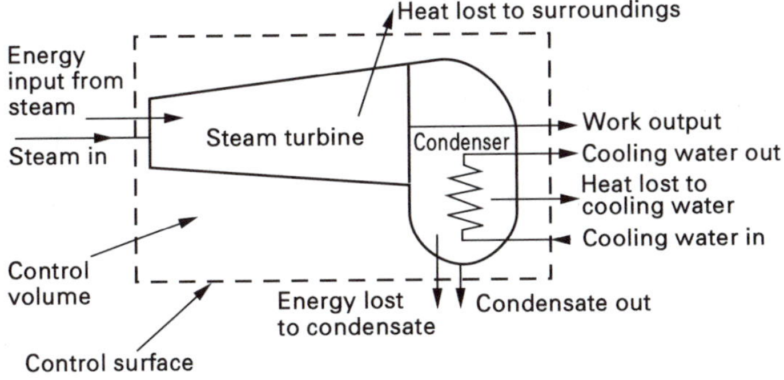

Fig. 2.2 Control volume

2.3 *The conservation of energy*

The concept of energy was discussed in section 1.17. From this discussion it appears that, by designing suitable devices, one form of energy can be transformed into another. In a power station, the potential chemical energy in the fuel produces a high-temperature furnace. Heat energy is transferred from the furnace into the steam being formed, which is passed into a turbine where some of it is converted into work. The work is put into an alternator where some is converted into electrical energy. The electricity generated is then passed out of the station to the public, who use it in various devices to produce heat, light and power. Not all the energy put into the furnaces of the power station ultimately appears as electrical energy. There are many losses through the plant, as indeed there are in any power plant.

However, it is found that, in any energy transformation system, if all the energy forms are totalled, including any losses which may have occurred, the sum is always equal to the energy input.

Written as an equation, this becomes

$$\begin{matrix}\text{Initial energy of}\\ \text{the system}\end{matrix} + \begin{matrix}\text{Energy entering}\\ \text{the system}\end{matrix} = \begin{matrix}\text{Final energy of}\\ \text{the system}\end{matrix} + \begin{matrix}\text{Energy leaving}\\ \text{the system}\end{matrix}$$

Naturally, all the energies must be expressed in the same units.

The fact that the total energy in any one energy system remains constant is called the principle of the **conservation of energy**. This states that energy can neither be created nor destroyed; it can only be changed in form.

As a further point, from work carried out in the field of nuclear physics, it appears there is some relationship between energy and matter. This has been made manifest by the fact that, during a nuclear reaction, some of the energy released can only be accounted for by reference to the loss of nuclear matter which has occurred during the reaction. Thus it appears that matter and energy are related in some way. From this, the conservation of energy should strictly be modified to the conservation of energy and matter. But any matter–energy transformation that occurs outside the nuclear field is extremely small, if it occurs at all. Thus in the absence of nuclear reaction, all energy transformation is discussed using the principle of the conservation of energy.

2.4 *Energy forms in thermodynamic systems*

Various energy forms can exist in thermodynamic systems. In some systems they may all be present. In other systems only some may be present. The various forms of energy appearing in thermodynamic systems are listed below. The basic unit of energy, in all forms, is the joule (J). Multiples such as the kilojoule (kJ) or the megajoule (MJ) are often used.

2.4.1 *Gravitational potential energy*

If the fluid is at some height Z above a given datum level, as a result of its mass, it possesses gravitational potential energy with respect to that datum. Thus, for unit mass of fluid, in the close vicinity of the earth

$$\begin{aligned}\text{Potential energy} &= gZ\\ &\approx 9.81Z\end{aligned}$$

2.4.2 *Kinetic energy*

If the fluid is in motion then it possesses kinetic energy. If the fluid is flowing with velocity C then, for unit mass of fluid

$$\text{Kinetic energy} = \frac{C^2}{2}$$

2.4.3 *Internal energy*

All fluids store energy. The store of energy within any fluid can be increased or decreased as the result of various processes carried out on or by the fluid. The energy stored within a fluid which results from the internal motion of its atoms and molecules is called its **internal energy** and was discussed in section 1.23. It is usually designated by the letter U. The internal energy of unit mass of fluid is called the **specific internal energy** and is designated by u.

2.4.4 Flow or displacement energy

Any volume of fluid entering or leaving a system must displace an equal volume ahead of itself in order to enter or leave the system, as the case may be. The displacing mass must do work on the mass being displaced, since the movement of any mass can only be achieved at the expense of work.

Figure 2.3 illustrates a part of a system which can be considered as being the entry or exit of the system. Consider unit mass of fluid entering or leaving the system. Let the fluid be at uniform pressure P and enter or leave the system a distance l over a uniform area A.

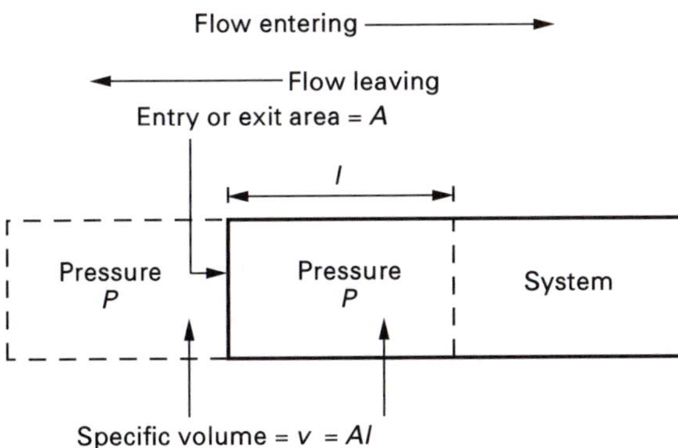

Fig. 2.3 System entry, system exit

Let the specific volume of the fluid be v, equal to the volume of unit mass.

$$\text{Work done} = \text{force} \times \text{distance} \quad [1]$$

$$\begin{aligned} \text{Force} &= \text{pressure} \times \text{area} \\ &= P \times A \\ &= PA \quad [2] \end{aligned}$$

$$\begin{aligned} \therefore \quad \text{Work done} &= PA \times l \\ &= P \times Al \\ &= Pv \quad (\text{since } Al = v) \quad [3] \end{aligned}$$

This is called **flow work** or **displacement work**.

- At entry it is energy received by the system.
- At exit it is energy lost by the system.

2.4.5 Heat received or rejected

In any system a fluid can have a direct reception or rejection of heat energy transferred through the system boundary. Designated by Q, it must be taken in its algebraic sense.

- If heat is received then Q is positive.
- If heat is rejected then Q is negative.
- If heat is neither received nor rejected then $Q = 0$.

2.4.6 External work done

In any system, a fluid can do external work or have external work done on it by transfer across the system boundary. Designated by W, it must be taken in its algebraic sense.

- If external work is done by the fluid then W is positive.
- If external work is done on the fluid then W is negative.
- If no external work is done on or by the fluid then $W = 0$.

2.5 The closed system

The idea of the closed system was discussed in section 2.1. Consider the possible forms of energy which can be associated with such a system. The substance enclosed by the system will possess internal energy U, resulting from the motion of its atoms or molecules.

Furthermore, if the substance within the system is some type of fluid or gas, it may experience some degree of turbulence. Any part of the substance, considered separately, could therefore be experiencing its own changes in gravitational potential energy (PE) when referred to some datum level and also changes in kinetic energy (KE).

There could also be some other random energies such as electrical, magnetic and surface tension, but these forms of random energy will be neglected here as being insignificantly small.

In a closed system, the sum of all the energies possessed by the contained substance is called the **total energy** (E).

$$E = U + \sum \text{PE} + \sum \text{KE} \qquad [1]$$

where $\sum$PE and $\sum$KE are the summation of the separate local gravitational potential and kinetic energies.

Consider now a process carried out on a closed system. Let

E_1 = initial total energy of the contained substance
E_2 = final total energy of the contained substance
Q = heat transferred to or from the substance in the system
W = work transferred to or from the substance in the system

Then by the principle of conservation of energy (see section 2.3)

$$\begin{matrix}\text{Initial energy of}\\ \text{the system}\end{matrix} + \begin{matrix}\text{Energy entering}\\ \text{the system}\end{matrix} = \begin{matrix}\text{Final energy of}\\ \text{the system}\end{matrix} + \begin{matrix}\text{Energy leaving}\\ \text{the system}\end{matrix}$$

or in this case

$$E_1 + Q = E_2 + W \qquad [2]$$

from which

$$Q = (E_2 - E_1) + W \qquad [3]$$

also

$$Q - W = E_2 - E_1 \qquad [4]$$

Note that in equation [2] it has been assumed that heat has been transferred into the system, so Q is positive. Had the heat transfer been out of the system, Q would have been negative.

Similarly, the work transfer has been assumed to be from the system, so W is positive. Had work been transferred into the system, W would have been negative.

Note also, from equation [4], that Q and W are the energy forms which are responsible for the change of E in a closed system. Equation [4] is as statement of the first law of thermodynamics (see Chapter 3).

Example 2.1 *In a process carried out on a closed system, the heat transferred into the system was 2500 kJ and the work transferred from the system was 1400 kJ. Determine the change in total energy, and state whether it is an increase or a decrease.*

SOLUTION

From equation [4]

$$\begin{aligned} E_2 - E_1 &= Q - W \\ &= 2500 - 1400 \\ &= \mathbf{1100\ J} \end{aligned}$$

This is positive, so there is an increase in total energy.

Example 2.2 *In a process carried out on a closed system, the work transferred into the system was 4200 kJ and the increase in the total energy of the system was 3500 kJ. Determine the heat transferred and state the direction of transfer.*

SOLUTION

From equation [3]

$$\begin{aligned} Q &= (E_2 - E_1) + W \\ &= 3500 + (-4200) \quad \text{(}-4200\text{ because work is transferred into the system)} \\ &= 3500 - 4200 \\ &= \mathbf{-700\ kJ} \end{aligned}$$

This is negative, so heat is transferred from the system.

2.6 The non-flow energy equation

Section 2.5 discussed energies associated with the closed system.

The sum of all the energies possessed by the contained substance was called the total energy (E), where

$$E = U + \sum \text{PE} + \sum \text{KE}$$

If the contained substance is considered to be at rest, there is no turbulence; in this case the random potential and kinetic energies will be zero.

Thus, for a substance at rest, the contained energy will be only the internal energy U.

In this case, equation [2] of section 2.5 becomes

$$U_1 + Q = U_2 + W \qquad [1]$$

from which

$$Q = (U_2 - U_1) + W = \Delta U + W \qquad [2]$$

Also

$$Q - W = U_2 - U_1 = \Delta U \qquad [3]$$

where

$$\Delta U = U_2 - U_1$$

Because the system is closed, there is no flow of substance into or out of the system.

Therefore the process in a closed system is called a **non-flow process**, and equation [3] is referred to as the **non-flow energy equation** (NFEE). A typical non-flow process is the expansion or compression of a substance in a cylinder.

Equations [1], [2] and [3] are further statements of the first law of thermodynamics (see Chapter 3).

Example 2.3 *During the working stroke of an engine the heat transferred out of the system was 150 kJ/kg of working substance. The internal energy also decreased by 400 kJ/kg of working substance. Determine the work done and state whether it is work done on or by the engine.*

SOLUTION

From equation [3], the non-flow energy equation

$$Q = \Delta u + W \quad (\Delta u \text{ because energies/kg are given})$$

From this

$$\begin{aligned} W &= Q - \Delta u \\ &= -150 - (-400) \quad (-400 \text{ because there is a decrease in internal energy}) \\ &= -150 + 400 \\ &= \mathbf{250\ kJ/kg} \end{aligned}$$

This is positive, so is work done by the engine per kilogram of working substance.

2.7 The open system

Section 2.1 discussed the concept of the open system. This section examines the two-flow open system, in which an equal mass of fluid per unit time is both entering and leaving the system, called continuity of mass flow.

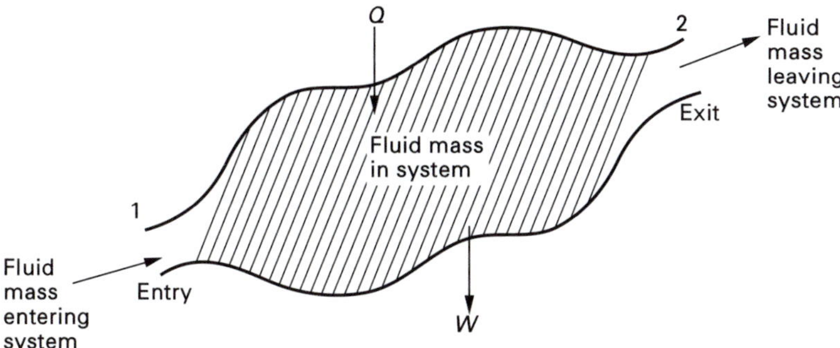

Fig. 2.4 Open system

Consider the two-flow open system illustrated in Fig. 2.4. The forms of energy which will be associated with the moving fluid mass entering the system will be

$$\text{Internal energy} = U_1$$
$$\text{Displacement or flow energy} = P_1 V_1 \quad \text{(see section 2.4)}$$
$$\text{Kinetic energy} = KE_1$$
$$\text{Gravitational potential energy} = PE_1$$

The forms of energy which will be associated with the fluid mass leaving the system will be

$$\text{Internal energy} = U_2$$
$$\text{Displacement or flow energy} = P_2 V_2$$
$$\text{Kinetic energy} = KE_2$$
$$\text{Gravitational potential energy} = PE_2$$

As the fluid mass enters the system, let the total energy of the fluid mass actually in the system be E_{S1}.

After passing through the system, as the fluid mass leaves the system, let the total energy of the fluid mass remaining in the system be E_{S2}.

In its passage through the system let the fluid mass transfer heat be Q and the transfer work be W.

For the two-flow open system

$$\text{Energy of the fluid mass entering the system} = U_1 + P_1 V_1 + KE_1 + PE_1$$

$$\text{Energy of the fluid mass leaving the system} = U_2 + P_2 V_2 + KE_2 + PE_2$$

By the principle of conservation of energy

$$\begin{matrix}\text{Initial energy of} \\ \text{the system}\end{matrix} + \begin{matrix}\text{Energy entering} \\ \text{the system}\end{matrix} = \begin{matrix}\text{Final energy of} \\ \text{the system}\end{matrix} + \begin{matrix}\text{Energy leaving} \\ \text{the system}\end{matrix}$$

or

$$E_{S1} + U_1 + P_1V_1 + KE_1 + PE_1 + Q = E_{S2} + U_2 + P_2V_2 + KE_2 + PE_2 + W \quad [1]$$

From section 1.29

$$U + PV = \text{H} = \text{Enthalpy} \quad [2]$$

So equation [1] can be written

$$E_{S1} + H_1 + KE_1 + PE_1 + Q = E_{S2} + H_2 + KE_2 + PE_2 + W \quad [3]$$

From equation [3]

$$Q - W = (E_{S2} - E_{S1}) + (H_2 - H_1) + (KE_2 - KE_1) + (PE_2 - PE_1) \quad [4]$$

2.8 The steady-flow energy equation

In a steady-flow system it is considered that the mass flow rate of fluid or substance throughout the system is constant. It is further considered that the total energy of the fluid mass in the system remains constant.

If this is the case then

$$E_{S1} = E_{S2} \quad [1]$$

hence

$$E_{S2} - E_{S1} = 0 \quad [2]$$

Using this, equation [4] in section 2.7 becomes

$$Q - W = (H_2 - H_1) + (KE_2 - KE_1) + (PE_2 - PE_1) \quad [3]$$

This is known as the **steady-flow energy equation** (SFEE).

To put more detail into the steady-flow energy equation, consider equation [1] of section 2.7 and assume that $E_{S1} = E_{S2}$ as before, then

$$U_1 + P_1V_1 + KE_1 + PE_1 + Q = U_2 + P_2V_2 + KE_2 + PE_2 + W \quad [4]$$

This equation is for any mass flow rate. It is often convenient, however, to consider the flow of unit mass through a system, in which case specific quantities are used and equation [4] becomes

$$u_1 + P_1v_1 + KE_1 + PE_1 + Q = u_2 + P_2v_2 + KE_2 + PE_2 + W \quad [5]$$

Figure 2.5 illustrates a steady-flow open system into which a fluid flows with pressure P_1, specific volume v_1, specific internal energy u_1 and velocity C_1. The entry is at height Z_1 above a datum level.

In its passage through the system, specific heat energy Q and specific work W are transferred into or out of the system. The fluid leaves the system with pressure P_2, specific volume v_2, specific internal energy u_2 and velocity C_2. The exit is at height Z_2 above the datum level.

Using equation [5] the steady-flow energy equation for the system becomes

$$u_1 + P_1v_1 + \frac{C_1^2}{2} + gZ_1 + Q = u_2 + P_2v_2 + \frac{C_2^2}{2} + gZ_2 + W \quad \text{(see section 2.4)} \quad [6]$$

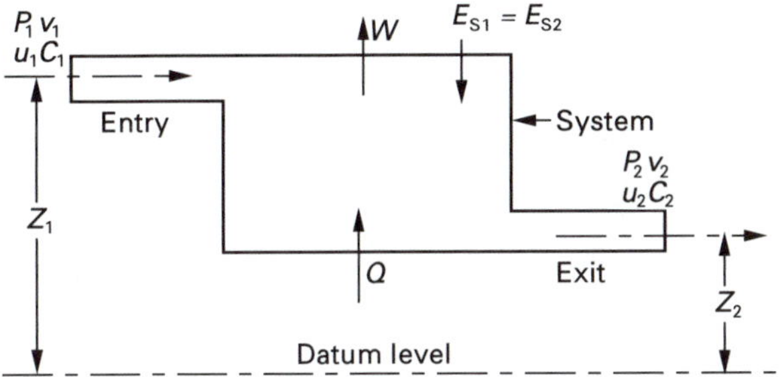

Fig. 2.5 Steady-flow open system

In thermodynamic systems, any changes in gravitational potential energy are mostly small compared with other energy forms. The gZ terms are therefore neglected.

Equation [6] then becomes

$$u_1 + P_1 v_1 + \frac{C_1^2}{2} + Q = u_2 + P_2 v_2 + \frac{C_2^2}{2} + W \qquad [7]$$

Also, since $u_1 + P_1 v_1 = h =$ specific enthalpy (see section 1.29), equation becomes [7]

$$h_1 + \frac{C_1^2}{2} + Q = h_2 + \frac{C_2^2}{2} + W \qquad [8]$$

Note that the various forms of the steady-flow energy equation, such as equations [6], [7] and [8], are further statements of the first law of thermodynamics (see Chapter 3).

2.9 Continuity of mass flow

During the discussion on the steady-flow open system in section 2.8, reference was made to the continuity of mass flow.

For a fluid substance flowing through a steady-flow open system, the mass flow rate through any section in the system must be constant.

At any section in the system, let

$\dot{m}$ = mass flow, kg/s
v = specific volume, m^3/kg
A = cross-sectional area, m^2
C = velocity, m/s

Consider the volume of fluid substance passing the section per second $= \dot{V}$ m^3/s. Now

$$\dot{V} = \dot{m} v \qquad [1]$$

also

$$\dot{V} = AC \qquad [2]$$

From this

$$\dot{V} = \dot{m}v = AC \qquad [3]$$

so

$$\dot{m} = \frac{AC}{v} \qquad [4]$$

This mass flow rate must be constant at all sections of the system for steady-flow. (Note that to indicate a rate of flow a dot is placed over the appropriate symbol. Thus, $\dot{m}$ indicates mass flow rate and $\dot{V}$ indicates volume flow rate.)

Example 2.4 *In a steady-flow open system a fluid substance flows at the rate of 4 kg/s. It enters the system at a pressure of 600 kN/m², a velocity of 220 m/s, internal energy 2200 kJ/kg and specific volume 0.42 m³/kg. It leaves the system at a pressure of 150 kN/in², a velocity of 145 m/s, internal energy 1650 kJ/kg and specific volume 1.5 m³/kg. During its passage through the system, the substance has a loss by heat transfer of 40 kJ/kg to the surroundings.*

Determine the power of the system, stating whether it is from or to the system. Neglect any change of gravitational potential energy.

SOLUTION

The steady-flow energy equation for the system is

$$u_1 + P_1v_1 + \frac{C_1{}^2}{2} + Q = u_2 + P_2v_2 + \frac{C_2{}^2}{2} + W$$

From this

$$W = (u_1 - u_2) + (P_1v_1 - P_2v_2) + \left(\frac{C_1{}^2 - C_2{}^2}{2}\right) + Q$$

Working in kilojoules (kJ)

$$\text{Specific work} = W = (2200 - 1650) + (600 \times 0.42 - 150 \times 1.5) + \left(\frac{220^2 - 145^2}{2 \times 10^3}\right) - 40$$

(Q is −40 kJ/kg because the heat transfer is a loss from the system.)

$$\therefore \quad W = 550 + (252 - 225) + \frac{(48\,400 - 21\,025)}{2 \times 10^3} - 40$$

$$= 550 + 27 + \frac{27\,375}{2 \times 10^3} - 40$$

$$= 550 + 27 + 13.69 - 40$$

$$= \mathbf{550.69\ kJ/kg}$$

This is positive, so power is output from the system.

For a fluid substance flow rate of 4 kg/s

$$\text{Power output from the system} = 550.69 \times 4$$

$$= 2202.75 \text{ kJ/s}$$

$$= \mathbf{2202.75\ kW} \quad (1 \text{ kJ/s} = 1 \text{ kW})$$

Example 2.5 *Lead is extruded slowly through a horizontal die. The pressure difference across the die is 154.45 MN/m². Assuming there is no cooling through the die, determine the temperature rise of the lead. Assume the lead is incompressible and to have a density of 11 360 kg/m³ and a specific heat capacity of 130 J/kg K.*

SOLUTION

Consider the steady-flow energy equation applied to the given extrusion process.

Since the die is horizontal then there will be no change in gravitational potential energy from inlet to outlet. Hence the potential energy terms can be neglected.

The velocity is also low, in which case any change in kinetic energy will be small enough to be sensibly neglected.

There is no cooling and no external work is done, so they too can be neglected.

From this the energy equation now becomes

$$P_1V_1 + U_1 = P_2V_2 + U_2$$

so

$$U_2 - U_1 = P_1V_1 - P_2V_2 \qquad [1]$$

Capital letters are used to indicate that a mass other than unity is being considered.

If a substance experiences a temperature rise, the energy associated with this temperature rise is given by

$$\text{mass} \times \text{specific heat capacity} \times \text{temperature rise} = mct \qquad [2]$$

The lead in this case is assumed to be incompressible, so no energy transfer will be associated with work of expansion or compression.

Hence, the energy transfer associated with temperature rise in this case will appear as a change of internal energy only

$$\therefore \quad mct = U_2 - U_1 \qquad [3]$$

Also, since it is assumed that the lead is incompressible

$$V_1 = V_2 = V, \text{ say} \qquad [4]$$

Substituting equations [3] and [4] in equation [1]

$$mct = V(P_1 - P_2)$$

from which

$$t = \frac{V(P_1 - P_2)}{mc} \qquad [5]$$

and considering 1 m³ of lead

$$t = \frac{1 \times (154.45 \times 10^6)}{11\ 360 \times 130}$$

$$= 104.6 \text{ K}$$
$$= \mathbf{104.6\ °C}$$

The temperature rise of the lead is 104.6 °C.

Example 2.6 *Air passes through a gas turbine system at the rate of 4.5 kg/s. It enters the turbine system with a velocity of 90 m/s and a specific volume of 0.85 m^3/kg. It leaves the turbine system with a specific volume of 1.45 m^3/kg. The exit area of the turbine system is 0.038 m^2. In its passage through the turbine system, the specific enthalpy of the air is reduced by 200 kJ/kg and there is a heat transfer loss of 40 kJ/kg. Determine*

(a) the inlet area of the turbine in m^2

(b) the exit velocity of the air in m/s

(c) the power developed by the turbine system in kilowatts

(a)

At inlet

$$\dot{m} = \frac{A_1 C_1}{v_1} \quad \text{(see section 2.9 equation [4])}$$

$$\therefore \quad A_1 = \frac{\dot{m} v_1}{C_1}$$

$$= \frac{4.45 \times 0.85}{90}$$

$$= \mathbf{0.042\ m^2}$$

The inlet area is 0.042 m^2.

(b)

At exit

$$\dot{m} = \frac{A_2 C_2}{v_2}$$

$$\therefore \quad C_2 = \frac{\dot{m} v_2}{A_2}$$

$$= \frac{4.5 \times 1.45}{0.038}$$

$$= \mathbf{171.71\ m/s}$$

The exit velocity is 171.71 m/s.

(c)

The steady-flow energy equation for this system is

$$h_1 + \frac{C_1{}^2}{2} + Q = h_2 + \frac{C_2{}^2}{2} + W \quad \text{(see section 2.8, equation [8])}$$

from which

$$W = (h_1 - h_2) + \frac{(C_1{}^2 - C_2{}^2)}{2 \times 10^3} \quad \text{(energy in kJ/kg)}$$

$$= 200 + \frac{(90^2 - 171.71^2)}{2 \times 10^3} - 40 \quad \text{(loss by heat transfer)}$$

$$= 200 + \frac{(8100 - 29\,484.3)}{2 \times 10^3} - 40$$

$$= 200 - \frac{21\,384.3}{2 \times 10^3} - 40$$

$$= 200 - 10.7 - 40$$
$$= \mathbf{149.3\ kJ/kg}$$

$\therefore$ Power developed $= 149.3 \times 4.5$
$= 671.85$ kJ/s
$=$ **671.85 kW** (1 kJ/s = 1 kW)

Questions

1. In a non-flow process there is a heat transfer loss of 1055 kJ and an internal energy increase of 210 kJ. Determine the work transfer and state whether the process is an expansion or a compression.

   [−1265 kJ, a compression]

2. In a non-flow process carried out on 5.4 kg of fluid substance, there is a specific internal energy decrease of 50 kJ/kg and a work transfer from the substance of 85 kJ/kg. Determine the heat transfer.

   [189 kJ, a gain]

3. Air enters a gas turbine system with a velocity of 105 m/s and has a specific volume of 0.8 m^3/kg. The inlet area of the gas turbine system is 0.05 m^2. At exit the air has a velocity of 135 m/s and has a specific volume of 1.5 m^3/kg. In its passage through the turbine system, the specific enthalpy of the air is reduced by 145 kJ/kg and the air also has a heat transfer loss of 27 kJ/kg. Determine
   (a) the mass flow rate of the air through the turbine system in kg/s
   (b) the exit area of the turbine system in m^2
   (c) the power developed by the turbine system in kW

   [(a) 6.56 kg/s; (b) 0.073 m^2; (c) 750.46 kW]

4. Air is compressed by a rotary compressor in a steady-flow process at a rate of 1.5 kg/s. At entry, the air has a specific volume of 0.9 m^3/kg and has a velocity of 80 m/s. At exit, the air has a specific volume of 0.4 m^3/kg and has a velocity of 45 m/s. In its passage through the compressor, the specific enthalpy of the air is increased by 110 kJ/kg and it experiences a heat transfer loss of 20 kJ/kg. Determine
   (a) the inlet and exit areas of the compressor in m^2
   (b) the power required to drive the compressor in kW

   [(a) 0.0169 m^2, 0.013 3 m^2; (b) −191.72 kW]

Chapter 3

The laws of thermodynamics

3.1 General introduction

The laws of thermodynamics are really statements of thermodynamic behaviour. They are natural laws which are based on observable phenomena. They are designated as laws because they have never been shown to be contradicted.

3.2 The zeroth law (law number zero)

This law is concerned with thermal equilibrium. It states

> If two bodies are separately in thermal equilibrium with a third body then they must be in thermal equilibrium with each other.

Thus, in Fig. 3.1, if bodies B and C are in thermal equilibrium with body A then bodies B and C must be in thermal equilibrium with each other.

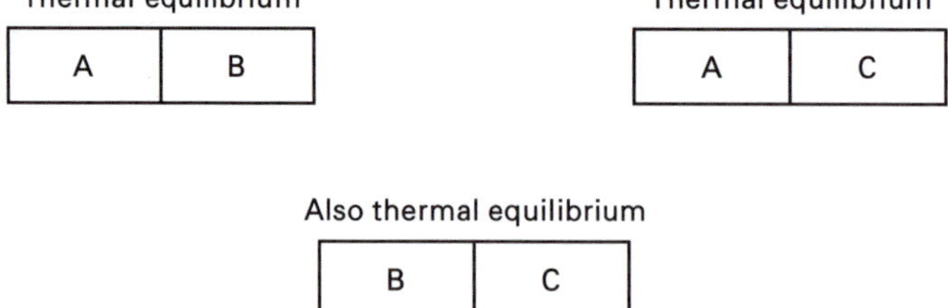

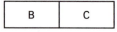

Fig. 3.1 Thermal equilibrium

Thermal equilibrium means there is no change of state and hence the zeroth law implies that the bodies A, B and C will all be at the same temperature and, furthermore, that all bodies, if in thermal equilibrium, will be at the same temperature.

As an example, this situation arises when taking a temperature using a mercury-in-glass thermometer. When the thermometer is steady, it is assumed that the mercury, the glass container and the body whose temperature is being measured, are all at the same temperature, so they are in thermal equilibrium.

The concept of this law was generally developed after the enunciation of the first, second and third laws of thermodynamics (see sections 3.3, 3.4 and 3.5). However, the law is concerned with thermal equilibrium, a type of base, as it were. The other laws have concern for possible work and energy transfers, plus the further possibility of temperature difference, so they are involved with change. It is thus considered that the law involved with thermal equilibrium should logically precede the other laws, so it is called the zeroth law (law number zero).

3.3 The first law of thermodynamics

Section 1.28 established a relationship between heat and work of the form

$$W = Q \qquad [1]$$

where W = work transfer
Q = heat transfer

The fact that there is a relationship between heat and work, as in equation [1], is a statement of the **first law of thermodynamics**.

Equation [1] is not meant to imply that if a certain amount of work is done on a system, it is all converted into heat or, conversely, if a certain amount of heat is supplied to a system, it is all converted into work.

Equation [1] simply means that, if some work is converted into heat or some heat is converted into work, the relationship between the heat and the work so converted will be of the form $W = Q$.

It is possible to convert work completely into heat by friction, for example. The reverse process of converting heat completely into work is impossible, as will be discussed in Chapter 6 (see also section 3.4).

A further extension of equation [1] appears in work with thermodynamic cycles. The cycle was discussed in section 1.13. To complete a cycle, a working substance is taken through a sequence of events and is returned to its original state. If the working substance is returned to its original state, its final properties are identical with its original properties before the cycle.

If work is transferred during the cycle then, since there is no final change in the properties of the working substance, the energy to provide the work must have been transferred as heat and must exactly equal the work. During some processes in a cycle, work will be done by the substance; during other processes, work is done on the substance. Similarly, during some processes, heat is transferred out of the substance; during others, it is transferred into the working substance.

Thus, for a cycle, since there is no net property change

$$\text{Net heat transfer} - \text{Net work transfer} = 0 \qquad [2]$$

or

$$\text{Net heat transfer} = \text{Net work transfer} \qquad [3]$$

or

$$\sum Q = \sum W \qquad [4]$$

or

$$\oint Q = \oint W \qquad [5]$$

The symbol $\oint$ means the summation round the cycle.

Equations [2], [3], [4] and [5] are further statements of the first law of thermodynamics. The first law implies that, in a cycle, there must be heat transfer for there to be work transfer.

For example, an engine which could provide work transfer without heat transfer would violate the first law because it would create energy. This is contrary to the principle of conservation of energy (see section 2.3). No violation of the first law has been shown.

An engine which could provide work transfer without heat transfer would run forever; in other words, it would have perpetual motion! Such an engine would have what is sometimes called **perpetual motion of the first kind**. Another impossible perpetual motion engine is discussed in section 3.4 and Chapter 6.

Consider now a closed system which does not execute a cycle. If a process is carried out on a substance in a closed system such that there is both heat and work transfer, it is not necessarily the case that the algebraic sum of these energy transfers is zero. If this is the case, then

$$\sum Q \neq \sum W \qquad [6]$$

Now, the principle of conservation of energy states that

$$\text{Energy in} = \text{Energy out} \quad \text{(see section 2.3)} \qquad [7]$$

Thus, if the heat and work transfers are not equal, any energy difference must have been added to the substance or have been lost from the substance.

This again introduces the concept of internal energy, energy residing within the substance (see sections 1.23, 2.4, 2.6). With the inclusion of internal energy, equation [1] becomes

$$Q = \Delta U + W \qquad [8]$$

where Q = heat transfer
ΔU = change of internal energy
W = work transfer

Equation [8] is the non-flow energy equation (see section 2.6); it is another statement of the first law of thermodynamics.

Section 2.8 developed the steady-flow energy equation for an open system (see equations [6], [7] and [8] of section 2.8).

The steady-flow energy equation was shown to be of the form

$$u_1 + P_1 v_1 + \frac{C_1^2}{2} + gZ_1 + Q = u_2 + P_2 v_2 + \frac{C_2^2}{2} + gZ_2 + W \qquad [9]$$

where u = specific internal energy
$P_1 v_1$ = specific flow work
C = velocity (gives specific kinetic energy)
Z = height above given datum level (gives specific gravitational potential energy)
Q = heat transfer
W = work transfer

The steady-flow energy equation is a further statement of the first law of thermodynamics.

3.4 The second law of thermodynamics

The second law of thermodynamics is a directional law in that it states that heat transfer will occur of its own accord down a temperature gradient as a natural phenomenon.

Heat transfer can be made to transfer up a gradient but not without the aid of external energy (see Chapter 18).

Natural heat transfer down a temperature gradient degrades energy to a less valuable level. A limit occurs when temperatures become equal, thus there is thermal equilibrium (see section 3.2).

From the first law of thermodynamics, for a cycle, and hence for engines because engines must work in cycles in order to continue in operation

$$\text{Net work transfer} = \text{Net heat transfer} \qquad [1]$$

or

$$\sum W = \sum Q \quad \text{(see section 3.3)} \qquad [2]$$

But experience always shows that

$$\text{Net work transfer} < \text{Net heat transfer} \qquad [3]$$

or

$$\sum W < \sum Q \qquad [4]$$

Since the work transfer is less than the heat transfer

$$\sum Q - \sum W > 0 \qquad [5]$$

and has some positive value.

This means that some heat transfer must be rejected and is lost. Therefore there must always be some inefficiency.

Note that, from the second law of thermodynamics, unless there is a temperature difference, there is no heat transfer. Equations [1], [2], [3], [4] and [5] require that there is heat transfer in order that there shall be work transfer. Thus it is implied that there can be no work transfer unless there is a temperature difference.

The concept of the second law of thermodynamics were put together in the past in various forms, notably as follows.

Sadi Carnot (1796–1832)

Whenever a temperature difference exists, motive power can be produced.

Strictly, Carnot's proposition was not given as a statement of the second law of thermodynamics.

Carnot's concept of heat was in error in that, at that time, heat was thought to have the properties of a fluid which flowed into or out of a body. It was suggested that as the result of this fluid flow, a body became hotter if it flowed in and colder if it flowed out. Heat was not considered as an energy transfer process occurring naturally as the result of temperature difference, now the accepted definition (see sections 1.23 and 1.24). However, the suggestion that temperature difference is the prerequisite of the ability to produce motive power is correct.

The statement by Sadi Carnot is a positive statement in that it declares when it is possible to produce motive power. The statements which follow are negative because they declare impossibilities.

Rudolf Clausius (1822–1888)

> It is impossible for a self-acting machine, unaided by any external agency, to convey heat from a body at a low temperature to one at a higher temperature.

The implication of the Clausius statement is that, unless external energy is made available, heat transfer up a gradient of temperature is impossible.

The fact that heat transfer can be made to occur up a temperature gradient is made manifest in the refrigerator. However, the refrigerator is not self-acting. It requires external energy in order that it can operate (see Chapter 17).

Lord Kelvin (1824–1907)

> We cannot transfer heat into work merely by cooling a body already below the temperature of the coldest surrounding objects.

Lord Kelvin (William Thomson) implies that when a body reaches the temperature of the coldest surrounding objects no further heat transfer is possible, hence no further work transfer is possible.

Max Planck (1858–1947)

> It is impossible to construct a system which will operate in a cycle, extract heat from a reservoir, and do an equivalent amount of work on the surroundings.

According to Planck, the complete conversion of heat transfer into work transfer is an impossibility. The inference is that there must always be some heat transfer rejection, which is a loss from the system.

Kelvin–Planck

> It is impossible for a heat engine to produce net work in a complete cycle if it exchanges heat only with bodies at a single fixed temperature.

The Kelvin–Planck combination implies that it is not possible to produce work transfer if a heat engine system is connected only to a single heat energy source or reservoir which is at a single fixed temperature. Note that if such a heat engine system were possible it would have perpetual motion! This arrangement is said to have **perpetual motion of the second kind**. No such arrangement exists (see section 3.3 for perpetual motion of the first kind).

In summary, the implications of the second law of thermodynamics are as follows:

- Heat transfer will only occur, and will always naturally occur, when a temperature difference exists, and always naturally down the temperature gradient.
- If, due to temperature difference, there is heat transfer availability, then work transfer is always possible. However, there is always some heat transfer loss.

- Temperature can be elevated but not without the expenditure of external energy. Elevation of temperature cannot occur unaided.
- There is no possibility of work transfer if only a single heat energy source or reservoir at a fixed temperature is available.
- No contradiction of the second law of thermodynamics has been demonstrated.
- If work transfer is supplied to a system, it can all be transformed in heat energy.

Examples of work being transformed into heat energy are seen in the cases of friction and the generation of electricity. But heat energy transfer cannot all be transformed into work transfer. There will always be some loss. Thus work transfer appears to have a higher transfer value than heat transfer. It is important to attenuate this last statement because it is usual that work transfer is only made available by the expenditure of heat transfer.

From the second law of thermodynamics it follows that, in order to run all the engines and devices in use at the present time and to maintain and develop modern industrial society, a supply of suitable fuels is absolutely essential. It is by burning and consuming these fuels that the various working substances (e.g. air and steam) have their temperatures raised above the temperature of their surroundings, thus enabling them to release energy by heat transfer in a natural manner according to the second law of thermodynamics.

By virtue of the second law of thermodynamics it is essential that all fuels should be used as efficiently as possible in order that fuel stocks may be preserved for as long as possible. It must always be remembered that, once energy has been degraded by heat transfer down a temperature gradient, further energy is only made available at the expense of further fuel.

3.5 *The third law of thermodynamics*

This law is concerned with the level of availability of energy. Section 1.23 discussed the concept of internal energy. This section suggests that the internal energy of a substance results from the random motion of its atoms and molecules. Furthermore, this motion is also associated with temperature, and from this develops the idea of an absolute zero of temperature when all random motion ceases.

Chapter 7 develops the concept of **entropy** and shows how it is associated with temperature and with the availability of thermal energy.

For a substance, if the random translational, rotational and vibrational types of motion of its constituent atoms and molecules are reduced to zero, the substance is considered to become perfectly crystalline and the energies associated with these forms of motion will be reduced to zero. Thus, the energy within the substance is reduced to the **ground state**. This neglects the energy within the basic atomic structure of the substance, associated with electrons, neutrons and other particles.

These considerations led to the development of the third law of thermodynamics:

> At the absolute zero of temperature the entropy of a perfect crystal of a substance is zero.

Chapter 4

Steam and two-phase systems

4.1 General introduction

Before investigating the formation and properties of steam it will be useful to discuss the various forms which matter can take and the relationship between them. Matter can take the forms, solid, liquid, vapour or gas, and many substances can exist in any one of these forms.

Consider, for example, a metal. In its natural state, metal is solid. If it is heated, at some temperature the metal will melt and become a liquid. Further transfer of heat, to the now liquid metal, will ultimately transform the liquid into a vapour and finally a gas. If the temperature is now reduced, the gaseous metal will pass back through all the stages it passed through until it finally becomes a solid once again.

Each change from one form to another is called a change of **phase**, and each change of phase is accomplished by the addition or extraction of heat. The temperature at which the changes of phase take place will vary according to the substance being used.

A further point to be considered is that a change of phase is accompanied by a change of volume.

Generally, the change of volume which accompanies a change from solid to liquid is not very great. On the other hand, the change of volume during the change from liquid to vapour or gas can be very large. The ability of a fluid readily to expand or contract is the requisite feature for successful operation of a thermodynamic engine. Thus both vapours and gases can be used in thermodynamic engines, but the technique for vapours is somewhat different from the technique for gases. The distinction between a vapour and a gas will be made late in this chapter, and Chapter 5 is devoted to the properties of gases. For now the focus is on vapours.

A vapour results from a change of phase of a liquid due to a transfer of heat. The bulk of the liquid is generally very much smaller than the bulk of the vapour formed. It follows, therefore, that any liquid which can be easily obtained and handled can be used as the generator of the vaporous working substance for use in an engine.

Water is such a liquid. It is in abundant supply, can be easily handled and readily turned into its vaporous phase called steam. Consequently, the change of phase from water into steam deserves closer investigation. But before doing this, it is useful to note that any other liquid undergoing a change of phase into a vapour will follow the same general features as the water–steam transformation. The difference lies in the pressures, temperatures and energy quantities at which the various phenomena occur.

A system in which a liquid is being transformed into a vapour is a **two-phase system**. The mixture of liquid and vapour is a **two-phase mixture**.

4.2 The formation of steam

In the following discussion it will be assumed that the water and ultimately the steam are in some suitable container which can accommodate any changes of state. The container is called a **boiler**, and the way in which it accommodates the changes will be discussed in Chapter 10. Steam is almost invariably formed at constant pressure, so that it is a good place to begin.

If a mass of water is heated then, like all other substances, its temperature increases. There is also a small increase in volume. For a time, these are the only changes which take place. After a while, small bubbles appear, clinging to the side of the containing vessel. They are soon released then float to the surface and disappear. These bubbles are the dissolved gases being driven off; as well as being able to dissolve some solids, water can also dissolve some of the atmospheric gases. Further heating produces further temperature rise but, apart from this, there is no other apparent external change. Soon, however, signs of internal activity appear.

Small steam bubbles are formed on and near the heating surface; they rise a little through the water and collapse. Their density is lower than the surrounding water, so they rise through it. The surrounding water is cooler so it extracts some energy from the steam bubble, which immediately collapses. This collapse of steam bubbles is the reason for the singing of a kettle. The temperature continues to rise with the transfer of heat and the bubble activity increases correspondingly. Finally, the water mass is at such a temperature that the steam bubbles are able to rise completely through the water, escaping from its surface. The water mass is now in an extremely turbulent state; it is **boiling**. A rather more technical term used for boiling is to say that the water is in a state of **ebullition**.

But what of the temperature now? As soon as boiling commenced the temperature ceased to rise, remaining at what is commonly called the **boiling point**. This is important because, while boiling continues, the temperature will remain constant, independent of the quantity of heat transferred to the water. In fact, so long as there is water present, it is apparently impossible to increase the temperature beyond the boiling point. A name is given to this boiling point; it is called **saturation temperature**.

And what of the nature of the steam being produced? The boiling water is now in great turbulence as a result of the steam bubbles formed, forcing their way up through the water to break through the surface. The turbulence can be increased or decreased by increasing or decreasing heat energy supply. As the steam breaks away from the water surface it will carry with it small droplets of water. The larger droplets will tend to gravitate back to the water surface, but the smaller droplets will continue on their way with the steam. Steam with these small droplets of water in suspension is called **wet steam**. Steam formed from a water mass will always be wet to a greater or lesser extent; wetness generally depends on the turbulence occurring in the water. It is impossible to obtain dry steam while water is present.

Continuing the heating process will produce more and more wet steam until eventually the whole of the water mass disappears. The temperature, by the way, has remained constant at saturation temperature. The water droplets in suspension make the wet steam visible. Steam itself is a transparent vapour, but the inclusion of water droplets in suspension gives it the white cloudy appearance. What really is being seen is the cumulative effect of the water droplets reflecting light.

Further transfer of heat to the wet steam will convert the suspended water droplets into steam and finally a state will be reached when all the water has been turned into steam. The steam is then called **dry saturated steam**. It has now lost its visible characteristic, it has become completely transparent; this condition marks the end of the constant temperature intermediate phase.

Still further transfer of heat to the now dry saturated steam produces a temperature rise and the steam now becomes **superheated steam**. This is the last phase in the transformation of water into steam. It thus appears there are three distinct stages in the production of steam from water.

4.2.1 Stage 1

Stage 1 is the warming phase in which the temperature of the water increases up to saturation temperature. The energy required to produce this temperature rise is called the **liquid enthalpy**.

4.2.2 Stage 2

Stage 2 takes place at constant temperature; it is when the water is transformed into steam. Stage 2 begins with all water at saturation temperature and ends with all dry saturated steam at saturation temperature. Between these two extremes, the steam formed will always be wet steam. The energy required to produce the total change from all water into all steam is called the **enthalpy of evaporation**. It is sometimes colloquially known as **latent heat** because no temperature rise is produced in this stage.

4.2.3 Stage 3

Stage 3 begins when all dry saturated steam has been formed at saturation temperature. Further transfer of heat produces superheated steam which is accompanied by a rise in temperature. The amount of energy added in the superheat phase is called the **superheat enthalpy**.

Note that temperature increase (or it could be decrease) only takes place during the transfer of heat when a substance is in a single phase. In this case there has to be all water or all dry steam before the temperature changes.

The phenomenon of temperature change happens with all substances but it only occurs in a single phase, be it solid, liquid or vapour. If a two-phase mixture exists (solid–liquid or liquid–vapour), the temperature remains constant until a complete change from one phase to another has been completed.

A word here about the three phases of matter – solid, liquid and vapour (or gas) – as they are related to atomic or molecular activity. In the solid phase the atoms or molecules of a substance oscillate about a mean position. As the temperature increases, the degree of oscillation increases until the atoms or molecules are able to overcome interatomic or intermolecular attractions. When this occurs, the substance becomes fluid and the atoms or molecules are now able to move freely but not independent of the main mass. The substance is now a liquid. The temperature remains constant while the change from solid to liquid occurs; energy increase is necessary to accelerate the atoms or molecules to the velocity required to produce freedom of movement. When the substance is all liquid, energy increase produces an increase in temperature once again and the atoms or molecules move about faster and faster (see section 1.23 on internal energy). If the atoms or molecules of a liquid move about faster and faster as the temperature increases, the liquid must become turbulent. Eventually, some atoms or molecules must reach escape velocity; they are able to escape from the liquid mass at its free surface because their velocity is sufficient to overcome all internal interatomic or intermolecular attractions. This is the phenomenon of boiling; once again the temperature remains constant while all atoms or molecules absorb enough energy to attain escape velocity. Eventually, all atoms or molecules attain this velocity and the whole mass is now a vapour. From here on, further energy increases will increase the atomic or molecular velocities, causing the temperature to rise.

4.3 Saturation temperature and pressure

The previous discussion centred round the fact that steam is usually produced at constant pressure. This pressure can be higher or lower, as the case may be. Boiling will occur at saturation temperature, and it is found that saturation temperature depends upon the pressure exerted at the surface of the water. In other words, it depends upon the pressure at which the steam is being formed. Chapter 1 discussed the fixed points of the conventional thermometer; it explained that the upper fixed point is the boiling point of water, which is taken as 100 °C. But this is only the case when atmospheric pressure is 760 mm Hg. If atmospheric pressure is increased, the boiling point (or saturation temperature) increases. Conversely, if the pressure decreases then so does the saturation temperature.

Figure 4.1 shows the type of curve obtained when saturation temperature is plotted against absolute pressure. It is called the liquid–vapour equilibrium line.

It will be noted that the rate of increase of saturation temperature is not as great at the higher pressures as at the lower pressures.

4.4 The triple point

Figure 4.1 represents changes in saturation temperature with pressure for steam. Figure 4.2 enlarges this plot for the region of low temperatures and low pressures. The line dividing the liquid and vapour phases is the liquid–vapour equilibrium line; the line dividing the solid and liquid phases is the solid–liquid equilibrium line. The two lines join at point 3, the **triple point**.

The solid–liquid equilibrium is shown as horizontal, indicating there is little change in the solid–liquid (melting) point as a result of change in pressure. Actually, increase in pressure very slightly depresses the freezing point of water up to about 200 MN/m^2. The liquid–vapour equilibrium line shows the increase in saturation temperature with pressure increase as already discussed.

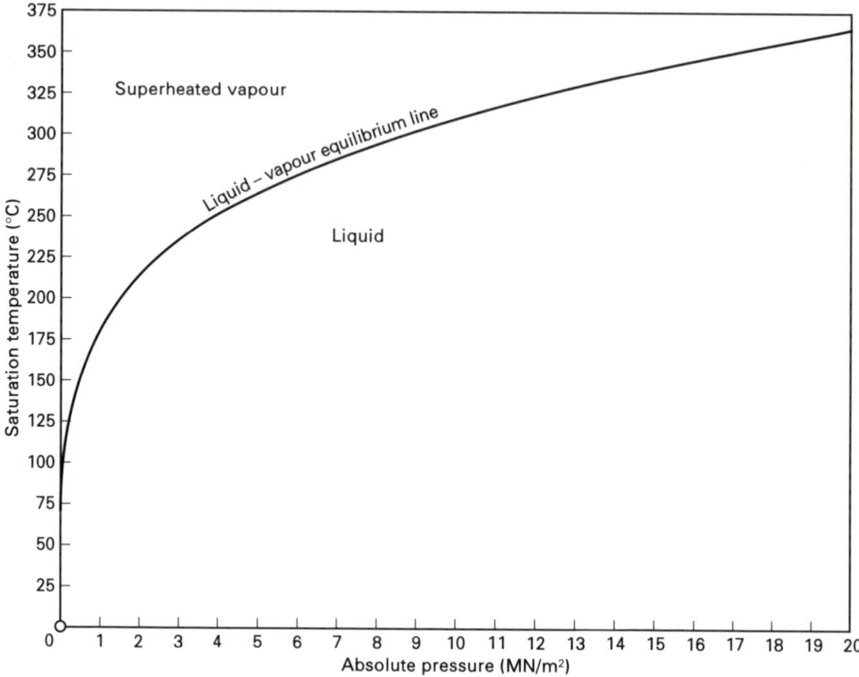

Fig. 4.1 Liquid–vapour equilibrium line

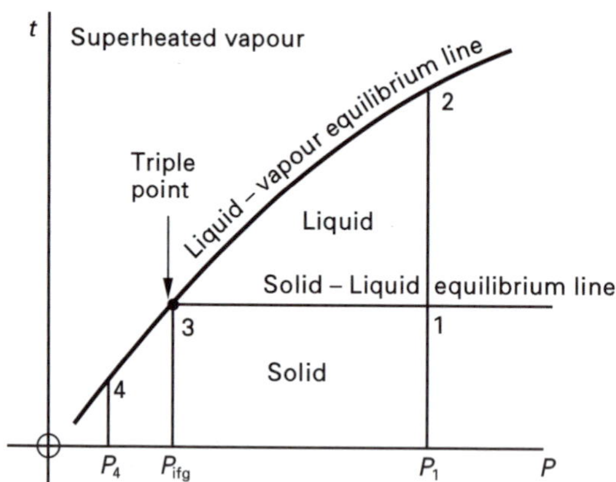

Fig. 4.2 Triple point

Referring to Fig. 4.2, at a pressure P_1 and at temperatures below point 1, the substance is solid (ice). At point 1 a change from solid into liquid takes place (melting point) and the solid is in equilibrium with the liquid. At temperatures above point 1 and below point 2 the substance is all liquid (water). At point 2 a change from liquid into vapour takes place (boiling point) and the liquid is in equilibrium with the vapour. Above point 2 the substance is all vapour (superheated steam, see section 4.2).

At the triple point pressure P_{ifg}, and at temperatures below point 3, the substance is all solid. At point 3, the triple point, solid, liquid and vapour can exist in equilibrium, hence the name. Above point 3 the substance is all vapour.

At pressure $P_4 (< P_{ifg})$, and at temperatures below point 4, the substance is all solid. At point 4 a direct change from solid into vapour takes place and the solid is in equilibrium with the vapour. This direct change from solid into vapour is called **sublimation**. At temperatures above point 4 the substance is all vapour.

4.5 Enthalpy and the formation of steam at constant pressure

Consider unit mass of a substance and let heat energy Q be transferred at constant pressure P, thus changing the state of the substance from specific internal energy u_1 and specific volume v_1 to specific internal energy u_2 and specific volume v_2.

By the non-flow energy equation

$$Q = \Delta u + W \qquad [1]$$

For this case

$$Q = (u_2 - u_1) + P(v_2 - v_1) \qquad [2]$$
$$= (u_2 + Pv_2) - (u_1 + Pv_1) \qquad [3]$$

or

$$Q = h_2 - h_1 = \text{change of specific enthalpy} \qquad [4]$$

since

$$h = u + Pv = \text{specific enthalpy} \qquad [5]$$

In section 4.2 it was mentioned that steam is almost invariably formed at constant pressure, so it follows that heat energy transferred during the formation of steam (or some other vapour), at constant pressure, appears as a change of enthalpy in the steam.

4.6 Enthalpy tables

Enthalpy values during the formation of steam and other vapours at constant pressure are commonly set out in tabular form. In the case of steam, preparation of tables began in about the middle of the nineteenth century, but most of the work has been carried out since 1900. Notable early names in this connection are Callendar, and Keenan and Keyes. There are now many others. Values shown in tables are the result of experimental investigation.

Most tables are made out for the formation of unit mass (1 kg) of steam or vapour. Enthalpy tables may be very extensive, but abridged tables in common use give an appropriate selection of values.

4.7 Reference state for tables

In 1956 the Fifth International Conference on Properties of Steam recommended that the liquid at the triple point (water) should be made the datum state for steam tables.

At this datum, it was proposed that specific internal energy and specific entropy (see Chapter 7) should be considered as zero. The triple point for water is at a pressure $P_{ifg} = 0.611\,2$ kN/m^2 and a temperature $t_{ifg} = 0.01$ °C (273.16 K).

It should be noted that enthalpy values as given in tables are not absolute values. They are simply change values from the reference state. Above the reference state values of enthalpy will be positive. Below the reference state values of enthalpy, if given, will be negative. A similar situation arises with temperature measurement. Above the zero of the conventional temperature scale, recordings will be positive but below this zero, recordings will be negative.

Similar sets of tables are prepared for vapours other than steam. Each vapour has its own reference state.

4.8 Liquid enthalpy

The first phase in the production of steam is the warming of water to saturation temperature, t_f. The energy added to the water in this phase is called **liquid enthalpy**. For unit mass of steam, specific liquid enthalpy is written h_f. The accurate value of h_f at any given saturation temperature corresponding to a particular pressure is given in steam tables.

The units of h_f are usually given as kJ/kg. In the absence of tables, an approximate value for h_f is given by

$$h_f = 4.186\,8 t_f \text{ kJ/kg where } t_f \text{ is in °C.}$$

This equation is only approximately correct at lower pressures and temperatures. For example, at 1 standard atmosphere = 0.101 35 MN/m^2, $t_f = 100$ °C

$$\therefore \quad h_f \approx 4.186\,8 \times 100 = \mathbf{418.68\ kJ/kg}$$

From the tables, the accurate value is

$$h_f = \mathbf{417.5\ kJ/kg}$$

The calculated value is not very far out in this case.

As another example, at 1 MN/m^2, $t_f = 179.9$ °C

$$\therefore \quad h_f \approx 4.186\,8 \times 179.9 = \mathbf{753.2\ kJ/kg}$$

From tables, the accurate value is

$$h_f = \mathbf{762.2\ kJ/kg}$$

There is some error in this case and the error becomes even greater at higher pressures and temperatures. It is always preferable to refer to tables for accurate values.

4.9 Enthalpy of evaporation

The specific enthalpy of evaporation is written h_{fg}. As with specific liquid enthalpy, it can be looked up in tables. The evaporation of a liquid into vapour takes place at constant saturation temperature, as already stated. Thus h_{fg} is added at constant saturation temperature t_f.

Starting with unit mass of liquid at temperature t_f, the addition of h_{fg} will transform the liquid into dry saturated vapour, also at temperature t_f. The units of h_{fg} are commonly kJ/kg. Enthalpy of evaporation used to be called latent heat; this term has now largely been dropped.

4.10 *Enthalpy of dry saturated vapour*

The specific enthalpy of dry saturated vapour is written h_g. In order that a vapour shall become dry saturated, firstly the liquid enthalpy must be introduced and to this must be added the enthalpy of evaporation. Thus

$$h_g = h_f + h_{fg}$$

h_g is commonly given in kJ/kg.

Example 4.1 *Determine the specific liquid enthalpy, specific enthalpy of evaporation and specific enthalpy of dry saturated steam at 0.5 MN/m².*

SOLUTION
Look up the values in steam tables.

Pressure (MN/m^2)	Saturation temperature t_f (°C)	Specific enthalpy (kJ/kg)		
		h_f	h_{fg}	h_g
0.50	151.8	640.1	2 107.4	2 747.5

Thus

Specific liquid enthalpy = **640.1 kJ/kg**
Specific enthalpy of evaporation = **2107.4 kJ/kg**
Specific enthalpy of dry saturated steam = **2747.5 kJ/kg**

Note that

$$h_g = h_f + h_{fg}$$
$$= 640.1 + 2107.4 = \mathbf{2747.5\ kJ/kg}$$

Note also that saturation temperature = **151.8 °C**.

Example 4.2 *Determine the saturation temperature, specific liquid enthalpy, specific enthalpy of evaporation and specific enthalpy of dry saturated steam at a pressure of 2.04 MN/m².*

SOLUTION
If 2.04 MN/m^2 is looked up in abridged steam tables it will be found that a line is not given for this pressure. There is a line for 2.0 MN/m^2 and another for 2.1 MN/m^2. When this occurs, interpolated values for 2.04 MN/m^2 must be extracted. It is accurate enough to assume that values between those actually given behave in a linear manner. Proportional values are thus extracted between the given values. For the case given, the two lines, 2.1 MN/m^2 and 2.0 MN/m^2 are written down, top pressure first.

Pressure (MN/m^2)	Saturation temperature t_f (°C)	Specific enthalpy (kJ/kg)		
		h_f	h_{fg}	h_g
2.1	214.9	920.0	1 878.2	2 798.2
2.0	212.4	908.6	1 888.6	2 797.2
Difference +0.1	+2.5	+11.4	−10.4	+1.0
Require +0.04	$\frac{0.04}{0.1} \times 2.5 = 1.0$	$\frac{0.04}{0.1} \times 11.4 = +4.56$	$\frac{0.04}{0.1} \times -10.4 = -4.16$	$\frac{0.04}{0.1} \times 1.0 = +0.4$
Adding 2.04	213.4	913.16	1 884.44	2 797.6

Hence, at 2.04 MN/m^2

Saturation temperature = **213.4 °C**

Specific liquid enthalpy = **913.16 kJ/kg**

Specific enthalpy of evaporation = **1884.44 kJ/kg**

Specific enthalpy of dry saturated steam = **2797.6 kJ/kg**

4.11 Enthalpy of superheated vapour

From dry saturated condition, a vapour receives superheat and its temperature rises above saturation temperature, t_f. It has now entered the superheat phase.

The difference between the superheat vapour temperature, t, and the saturation temperature, t_f, is called the **degree of superheat**. Thus

$$\text{Degree of superheat} = (t - t_f)\ \text{K} \qquad [1]$$

The enthalpy added during the superheat phase is the superheat enthalpy. The total enthalpy of superheated vapour will be the sum of the enthalpy of dry saturated vapour and the superheat enthalpy, or

$$h = h_g + \text{superheat enthalpy} \qquad [2]$$

An approximation of the value of the superheat enthalpy can be found as follows. Let

c_p = specific heat capacity of superheated vapour at constant pressure

then

$$\text{Specific superheat enthalpy} = c_p(t - t_f) \qquad [3]$$

Hence, from equations [2] and [3]

$$h = h_g + c_p(t - t_f) \qquad [4]$$

An average value of c_p for superheated steam is 2.093 4 kJ/kg K.

Accurate values of h are given in tables; equation [4] is used as an approximation only.

Example 4.3 *Determine the specific enthalpy of steam at 2 MN/m^2 and with a temperature of 250 °C.*

SOLUTION

At 2 MN/m^2, from tables, $t_f = \mathbf{212.4\ °C}$

The steam must therefore be superheated because its temperature is above t_f.

Degree of superheat = 250 − 212.4 = **37.6 K**

The specific enthalpy can be looked up in steam tables under the heading **superheated states**. From tables

Specific enthalpy of steam at 2 MN/m^2 with a temperature of 250 °C = **2902 kJ/kg**

Alternatively

$$\begin{aligned} h &= h_f + c_p(t - t_f) \\ &= 2797.2 + (2.093\,4 \times 37.5) \\ &= 2797.2 + 78.7 \\ &= \mathbf{2875.9\ kJ/kg} \end{aligned}$$

Note that this gives an approximation only.

Example 4.4 *Determine the specific enthalpy of steam at a pressure of 2.5 MN/m^2 and with a temperature of 320 °C.*

SOLUTION

Looking up steam tables shows that at 2.5 MN/m^2 the saturation temperature is 223.9 °C. The steam is therefore superheated.

Degree of superheat = 320 − 223.9 = **96.1 K**

The specific enthalpy could be estimated using

$$\begin{aligned} h &= h_g + c_p(t - t_f) \\ &= 2800.9 + (2.093\,4 \times 96.1) \\ &= 2800.9 + 201.18 \\ &= \mathbf{3002.08\ kJ/kg} \end{aligned}$$

However, a more accurate value can be interpolated from tables giving specific enthalpy against temperature. Looking up these tables will show that neither the pressure of 2.5 MN/m^2 nor the temperature of 320 °C are given. Interpolation for both pressure and temperature is therefore required. A note is made of values of specific enthalpy on either side of the pressure and temperature.

	Pressure (MN/m^2)	2	4
		Specific enthalpy (kJ/kg)	
	Temperature (°C)		
	325	3 083	3 031
	300	3 025	2 962
Difference	+25	+58	+69
Require	+20	$\frac{20}{25} \times 58 = +46.4$	$\frac{20}{25} \times 69 = +55.2$
Adding	320	3 071.4	3 017.2

This gives values of specific enthalpy at the temperature of 320 °C.

An interpolation is now required for the pressure.

	Pressure (MN/m^2)	Specific enthalpy (kJ/kg)
	4	3 017.2
	2	3 071.4
Difference	+2	−54.2
Require	+0.5	$\frac{0.5}{2} \times -54.2 = -13.6$
Adding	2.5	3 057.8

Thus, the specific enthalpy of steam at a pressure of 2.5 MN/m^2 and with a temperature of 320 °C = **3057.8 kJ/kg**.

Note that the estimated value of 3002.08 kJ/kg was not very accurate.

4.12 Wet vapour and dryness fraction

The vapour produced at saturation temperature in the transformation stage will contain liquid droplets in suspension for as long as there is liquid present. Vapour so produced is called **wet vapour**. It should be noted that vapour, as such, is dry. If the vapour is made wet by liquid droplets in suspension, it is important to know the degree of wetness.

Any mass of wet vapour will consist of some dry saturated vapour and some liquid droplets in suspension.

The ratio

$$\frac{\text{Mass of dry saturated vapour}}{\text{Mass of wet vapour containing the dry saturated vapour}}$$

is called the **dryness fraction**, symbol x.

Consider 1 kg of wet vapour of dryness fraction x. The 1 kg will be made up of x kg of dry saturated vapour at saturation temperature t_f together with $(1 - x)$ kg of liquid droplets in suspension, also at saturation temperature t_f. Evidently, then, only x kg have received the enthalpy of evaporation.

$$\therefore \quad \text{Specific enthalpy of evaporation of wet vapour} = xh_{fg} \qquad [1]$$

and

$$\text{Specific enthalpy of wet vapour} = h = h_f + xh_{fg} \qquad [2]$$

Example 4.5 *Determine the specific enthalpy of wet steam at a pressure of 70 kN/m² and having a dryness fraction of 0.85.*

SOLUTION

$$\begin{aligned} h &= h_f + xh_{fg} \\ &= 376.8 + (0.85 \times 2283.3) \quad \text{(values of } h_f \text{ and } h_{fg} \text{ from tables)} \\ &= 376.8 + 1945 \\ &= \mathbf{2321.8\ kJ/kg} \end{aligned}$$

4.13 Temperature–enthalpy diagram for a vapour

Figure 4.3 illustrates the type of curve obtained if temperature is plotted against enthalpy for the constant pressure formation of steam. Consider some moderately low-pressure formation of steam. At 273.16 K (0.01 °C) the enthalpy of water is considered as being zero. Starting at this point on the temperature axis and plotting the liquid phase will produce a curve such as AB. At B the specific liquid enthalpy h_f has been introduced for the pressure under consideration, and saturation temperature t_f has been reached. From here on, specific enthalpy of evaporation h_{fg} is introduced from B to C at constant temperature t_f.

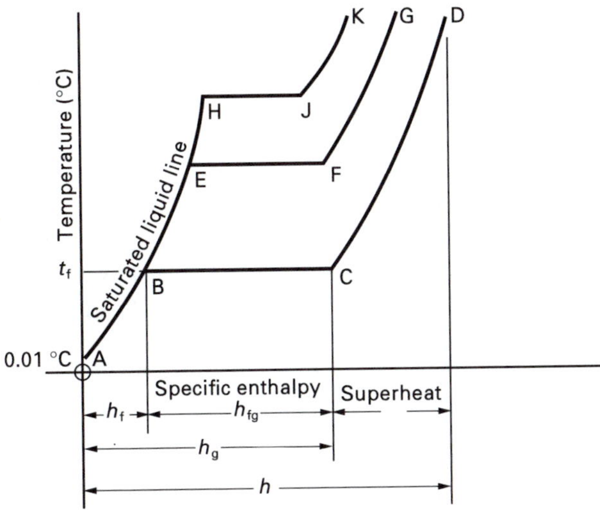

Fig. 4.3 *T–s* diagram for a vapour

The superheat phase then follows from C in which superheat is added and the temperature rises, producing a curve such as CD. Note that the three stages of formation are very clearly shown on this graph. If the formation pressure is increased, it produces curves like AEFG and AHJK. The line ABEH is called the **saturated liquid line**.

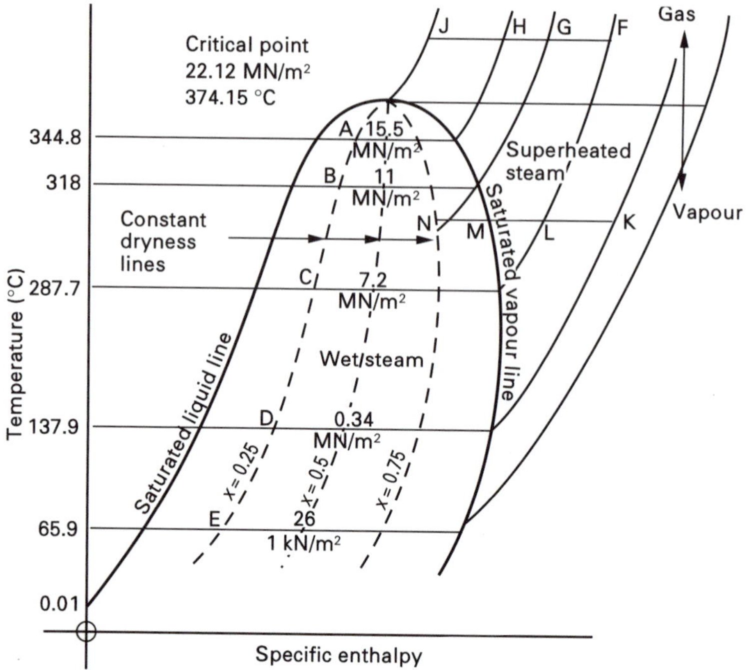

Fig. 4.4 Complete *T–h* diagram for a vapour

Dry saturated steam points C, F and J appear to lie on a smooth curve. If a wide range of pressures are considered and the results plotted at suitable pressure intervals, the complete temperature–enthalpy diagram is obtained (Fig. 4.4).

In this diagram it will be noted that the dry saturated steam points have been joined by a smooth curve, called the **saturated vapour line**. This line and the saturated liquid line enclose an area in which the steam is wet. To the right of the saturated vapour line the steam is superheated. Inside the wet steam area can be drawn lines of constant dryness (shown dotted). These lines are obtained by joining such points as A, B, C, D and E, where the dryness is the same.

It is very important to realise that the liquid line and the saturated vapour line may be continued upward until eventually they meet. The point at which the two lines join is called the **critical point**. This point is found to occur at a pressure of 22.12 MN/m^2 and a temperature of 374.15 °C. As the pressure increases toward the **critical pressure**, the required enthalpy at evaporation is reduced until it finally becomes zero at the critical point.

This implies that, at the critical point, the water changes directly into dry saturated steam. The critical point is really the division between behaviour as a vapour and behaviour as a gas. When a gas is compressed it does not liquefy in the normal course of events. This can be explained with reference to the temperature–enthalpy diagram and the critical point. The diagram shows that liquid, in this case water, can only exist at temperatures below the critical point.

Consider the case of a compression above the critical point and, for simplicity, let the compression be at constant temperature. Such a compression would appear as FGHJ, for example, moving from right to left. Inspection shows that, no matter how much compression takes place, no liquefaction occurs or can occur. On the other hand, if compression at constant temperature takes place below the critical point, it will appear as KLMN, for example. This shows that compression passes out of the superheat region into the wet region, thus liquefaction is taking place.

As long as the temperature is above the critical temperature, no change of phase occurs during compression or expansion; the behaviour is similar to that of a gas. In fact, at temperatures higher than the critical temperature, the vapour becomes a gas.

It should again be noted here that, although the above discussion has concentrated on the production of a temperature–enthalpy diagram for steam, similar diagrams can be developed for other vapours. The temperature–enthalpy diagram and others which will follow are called **phase diagrams**.

4.14 Volume of steam

As with all other substances, the volume of water and of steam increases as the temperature increases. In the lower pressure ranges the volume of water is very small compared with the volume of steam it produces. The main change of volume occurs in the evaporation stage. But as the pressure and temperature approach the critical point, the change in volume decreases until, at the critical point, there is no change of volume from the water phase to the dry steam phase. However, the vast majority of applications for steam use a pressure below the critical pressure, so the three stages of formation have to be considered.

4.15 Volume of water

Steam tables quote the specific volume of water. At saturation temperature, for a given pressure, the specific volume of water is tabulated as v_f m^3/kg. Alternatively, specific volume may be tabulated at a particular temperature against various pressures.

Example 4.6 *Determine the specific volume of water at saturation temperature for a pressure of 4.0 MN/m^2.*

SOLUTION

Look up the values in steam tables.

Pressure (MN/m^2)	Saturation temperature (°C)	Specific volume v_f(m^3/kg)
4.0	250.3	0.001 252

Example 4.7 *Determine the specific volume of water at a temperature of 175 °C and a pressure of 4.0 MN/m².*

SOLUTION
Look up the values in steam tables.

Pressure (MN/m^2)	2	4	6
	Specific volume (m^3/kg)		
Temperature (°C)			
150	0.001 089	0.001 088	0.001 087
175	0.001 120	0.001 119	0.001 116
200	0.001 156	0.001 153	0.001 152

So the specific volume of water at a temperature of 175 °C and a pressure of 4.0 MN/m^2 is 0.001 119 m^3/kg.

4.16 Volume of dry saturated steam

The specific volume of dry saturated steam is tabulated against its corresponding saturation temperature and pressure in steam tables and is designated as v_g. Volume is given in m^3/kg.

4.17 Volume of wet steam

Consider 1 kg of wet steam at dryness fraction x. This steam will be made up of x kg of dry saturated steam and $(1 - x)$ kg of water in suspension.

$$\therefore \text{ Volume of wet steam} = v = xv_g + (1 - x)v_f$$

At the lower pressures the volume v_f is very small compared with the volume v_g, and $(1 - x)$ is generally small compared with x.

Hence the term $(1 - x)v_f$ can be sensibly neglected, so

$$\text{Volume of wet steam} = v = xv_g$$

Note that from this

$$x = \frac{v}{v_g}$$

This equation can sometimes be used to determine the dryness fraction.

Example 4.8 *Determine the specific volume of wet steam of dryness fraction 0.9 at a pressure of 1.25 MN/m².*

SOLUTION
From tables, at 1.25 MN/m^2, $v_g = 0.156\,9$ m^3/kg

$$\therefore \quad v = xv_g = 0.9 \times 0.156\,9 = \mathbf{0.141\,2\ m^3/kg}$$

4.18 *Volume of superheated steam*

Steam tables quote the specific volume of superheated steam, either with pressure against actual steam temperature or with pressure against degree of superheat. Volume is given in m^3/kg.

Example 4.9 *Determine the specific volume of steam at a pressure of 2 MN/m^2 and with a temperature of 325 °C.*

SOLUTION

Steam tables give pressures and temperatures near to the required state.

Pressure (MN/m^2)	1	2	4
	Specific volume (m^3/kg)		
Temperature (°C)			
300	0.257 7	0.125 5	0.058 8
325	0.270 3	0.132 1	0.062 8
350	0.282 5	0.138 5	0.066 4

This shows that the specific volume of steam at a pressure of 2 MN/m^2 and with a temperature of 325 °C = **0.132 1 m^3/kg**.

Saturation temperature at 2 MN/m^2 = 212.4 °C. The actual steam temperature is 325 °C, so the steam is superheated. The degree of superheat is 325 − 212.4 = 112.6 K. Some superheat steam tables give the degree of superheat instead of the actual steam temperature.

4.19 *Density of steam*

If the specific volume of any quality steam is v, then

$$\text{Density} = \rho = \frac{1}{v}\ \text{kg/m}^3 \qquad [1]$$

Thus the density of superheated steam in Example 4.9 is given by

$$\text{Density} = \frac{1}{0.132\,1} = \mathbf{7.57\ kg/m^3}$$

If the density, ρ, is known, then

$$\text{Specific volume} = v = \frac{1}{\rho}\ \text{m}^3/\text{kg} \qquad [2]$$

Example 4.10 *Steam 0.95 dry at a pressure of 0.7 MN/m^2 is supplied to a heater through a pipe of 25 mm internal diameter; the velocity in the pipe is 12 m/s. Water enters the heater at 19 °C, the steam is blow into it and the mixture of water and condensate leaves the heater at 90 °C. Calculate*

(a) the mass of steam entering the heater in kg/h

(b) the mass of water entering the heater in kg/h

Extract from tables

Pressure (MN/m^2)	Sat. temp. t_f (°C)	Spec. enthalpy (kJ/kg)			Spec. vol. v_g (m^3/kg)
		h_f	h_{fg}	h_g	
0.7	165	697.1	2 064.9	2 762.0	0.273

(a)
Specific volume of steam 0.95 dry and at pressure 0.7 MN/m^2 is

$$xv_g = 0.95 \times 0.273 = \mathbf{0.259\ m^3/kg}$$

$$\text{Volume of steam passing/s} = \left[\frac{\pi}{4} \times (25 \times 10^{-3})^2 \times 12\right] \text{m}^3$$

$$\text{Volume of steam passing/h} = \left[\frac{\pi}{4} \times (25 \times 10^{-3})^2 \times 12 \times 3600\right] \text{m}^3$$

$$\therefore \quad \text{Mass of steam entering/h} = \frac{\pi \times 25^2 \times 10^{-6} \times 12 \times 3600}{4 \times 0.259}$$

$$= \mathbf{81.9\ kg}$$

(b)
Specific enthalpy of steam entering heater is

$$\begin{aligned} h_f + xh_{fg} &= 697.1 + (0.95 \times 2064.9) \\ &= 697.1 + 1961.7 \\ &= \mathbf{2658.8\ kJ/kg} \end{aligned}$$

Enthalpy gained by water = Enthalpy lost by steam
Let $\dot{m}$ = mass of water per hour

From steam tables

at 90 °C, hf = **376.8 kJ/kg**
at 19 °C, hf = **79.8 kJ/kg**

Hence

$$\dot{m}(376.8 - 79.8) = 81.9\ (2658.8 - 376.8)$$

$$\dot{m} = \frac{81.9 \times (2658.8 - 376.8)}{(376.8 - 79.8}$$

$$= \frac{81.9 \times 2282}{297}$$

$$= \mathbf{629.28\ kg/h}$$

4.20 The pressure–volume diagram for a vapour

Figure 4.5 shows another phase diagram for a vapour; this time the axes are pressure and specific volume. It plots a whole series of isotherms (lines of constant temperature). At a temperature less than the critical temperature ($T < T_c$) an isotherm will appear as ABCD. Note that points such as B will generate the saturated liquid line, whereas points such as C will generate the saturated vapour line. These two lines will again join at the top at the critical point F. The area enclosed by the two lines will be the wet vapour area (liquid–vapour mixture area). The critical temperature isotherm ($T = T_c$) is shown dotted as line EFG. Between the critical temperature isotherm and the saturated vapour line the vapour is superheated. At temperatures above critical ($T > T_c$), the isotherms gradually lose their discontinuity and eventually become smooth curves such as HJ. The behaviour then becomes that of a gas.

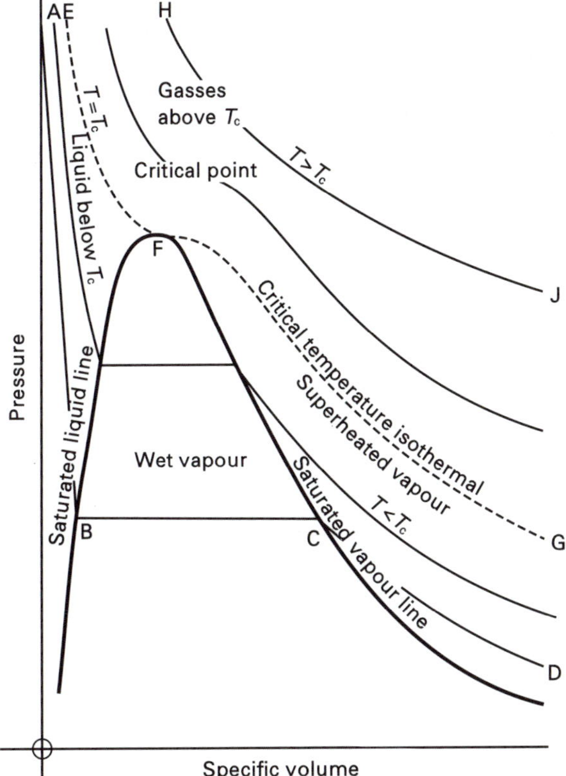

Fig. 4.5 *P–V* phase diagram

4.21 The internal energy of vapours

Heat energy transferred during the formation of vapour constant pressure appears as a change of enthalpy in the vapour. And it has been shown that tables of properties, such as enthalpy and specific volume, are prepared for a variety of vapours, including steam. Now

$$h = u + Pv \quad \text{(see section 4.5)} \qquad [1]$$
$$\therefore \quad u = h - Pv \qquad [2]$$

where u = specific internal energy, J/kg
h = specific enthalpy, J/kg
P = absolute pressure, N/m^2
v = specific volume, m^3/kg

Thus, by using tables of properties and by suitable substitution into equation [2], the specific internal energy of a vapour at a particular state can be determined. Note that some tables actually tabulate specific internal energy.

Example 4.11 *1.5 kg of steam originally at a pressure of 1 MN/m^2 and temperature 225 °C is expanded until the pressure becomes 0.28 MN/m^2. The dryness fraction of the steam is then 0.9. Determine the change of internal energy which occurs.*

SOLUTION
At 1 MN/m^2 and 225 °C, from tables

$$h_1 = \mathbf{2886\ kJ/kg}$$
$$v_1 = \mathbf{0.2198\ m^3/kg}$$

$$\therefore \quad u_1 = h_1 - P_1v_1 = 2886 - 1 \times \frac{10^6}{10^3} \times 0.2198$$
$$= 2886 - 219.8$$
$$= \mathbf{2566.2\ kJ}$$

At 0.28 MN/m^2 and dryness fraction 0.9

$$h_2 = h_{f2} + xh_{fg2}$$
$$= 551.4 + (0.9 \times 2170.1)$$
$$= 551.4 + 1953.09$$
$$= \mathbf{2504.49\ kJ/kg}$$
$$v_2 = xv_{g2} = 0.9 \times 0.646$$
$$= \mathbf{0.581\ 4\ m^3/kg}$$

$$\therefore \quad u_2 = h_2 - P_2v_2 = 2504.49 - \left(0.28 \times \frac{10^6}{10^3} \times 0.581\ 4\right)$$
$$= 2504.49 - 162.8$$
$$= \mathbf{2341.69\ kJ/kg}$$

Hence

$$u_2 - u_1 = 2341.69 - 2566.2$$
$$= \mathbf{-324.51\ kJ/kg}$$

This is a loss.
For 1.5 kg

$$\text{Loss of internal energy} = -324.51 \times 1.5$$
$$= \mathbf{-486.77\ kJ}$$

4.22 *Throttling*

If a gas or steam (or any other vapour) is passed through a fine orifice, as shown in Fig. 4.6, it is said to have been throttled. The gas or steam will pass from the high-pressure side to the low-pressure side, so throttling is a flow condition when applied to the steady-flow energy equation developed in section 2.8.

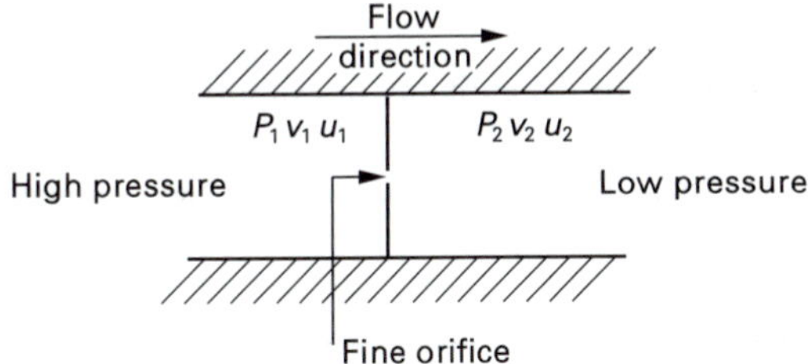

Fig. 4.6 Throttling

The steady-flow energy equation was shown to be

$$gZ_1 + u_1 + P_1v_1 + \frac{C_1^2}{2} + Q = gZ_2 + u_2 + P_2v_2 + \frac{C_2^2}{2} + W \quad [1]$$

Now in throttling there will be no change in potential energy, so the terms gZ can be neglected. Also there will be little or no change in kinetic energy, so the terms $C^2/2$ can be neglected. And, theoretically, there will be no heat transfer to or from the surroundings (the system is adiabatic, $Q = 0$); no external work is done. The terms Q and W can therefore be neglected.

Hence, the energy equation for the throttling process becomes

$$u_1 + P_1v_1 = u_2 + P_2v_2 \quad [2]$$

or

$$h_1 = h_2 \quad [3]$$

That is

Specific enthalpy before throttling = Specific enthalpy after throttling [4]

Also, more generally, for any mass of gas or steam

Enthalpy before throttling = Enthalpy after throttling [5]

4.23 *The effect of throttling on a vapour*

The pressure on the downstream side of a throttle orifice is lower than on the upstream side, so the liquid enthalpy after throttling can be less than before throttling. But the total enthalpy is the same before and after throttling.

Let h_1 = specific enthalpy before throttling

h_2 = specific enthalpy after throttling

h_{f1} = liquid enthalpy before throttling

h_{f2} = liquid enthalpy after throttling

Now $h_1 = h_2$, from section 4.23, so

$$\text{if } h_{f2} < h_{f1} \text{ then } h_1 - h_{f1} < h_2 - h_{f2}$$

The quantity $h - h_f$ is the enthalpy available to the enthalpy of evaporation. Hence the enthalpy of evaporation after throttle > the enthalpy of evaporation before throttle. From this, if $h_2 - h_{f2} > h_{fg2}$ then the vapour after throttle becomes superheated.

Example 4.12 *Steam at 1.4 MN/m² and of dryness fraction 0.7 is throttled to 0.11 MN/m². Determine the dryness fraction of the steam after throttling.*

SOLUTION

For a throttle

$$h_1 = h_2$$

and in this case

$$h_{f1} + x_1 h_{fg1} = h_{f2} + x_2 h_{fg2}$$

$$830.1 + (0.7 \times 1957.7) = 428.8 + (x_2 \times 2250.8)$$

$$\therefore \quad x_2 = \frac{[830.1 + (0.7 \times 1957.7] - 428.8}{2250.8}$$

$$= \frac{(830.1 + 1370.4) - 428.8}{2250.8}$$

$$= \frac{1771.7}{2250.8}$$

$$= \mathbf{0.787}$$

Note that the steam becomes drier in this case.

Throttling does not always dry the steam; this is illustrated in Fig. 4.7. Dry saturated steam at about 4.0 MN/m² is throttled and immediately moves into the wet region until, at about 3.0 MN/m², it reaches its maximum wetness. This is because at about 3.0 MN/m² the enthalpy of dry saturated steam is at a maximum. After 3.0 MN/m², continuing the throttle, the steam becomes drier and, at about 2.4 MN/m², it again becomes dry saturated. Throttling to pressures lower than 2.4 MN/m² produces superheated steam in this case.

This ability to produce superheated steam, given appropriate initial pressure and dryness, is exploited in the throttling calorimeter.

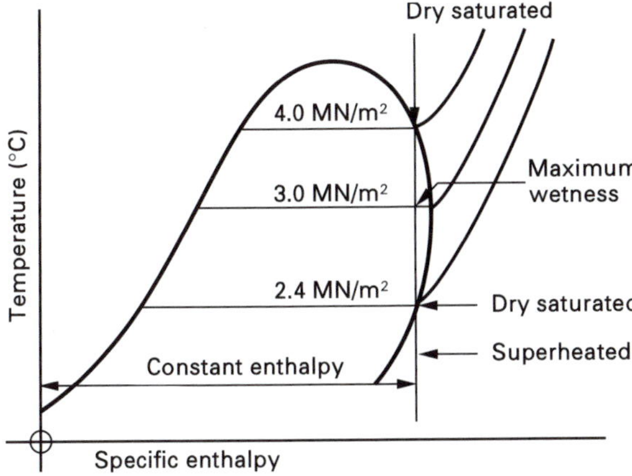

Fig. 4.7 Throttling and steam quality

Example 4.13 *Steam at a pressure of 2 MN/m², a temperature of 300 °C and flowing at a rate of 2 kg/s is throttled to a pressure of 800 kN/m². It is then mixed with steam at a pressure of 800 N/m² and 0.9 dry which flows at a rate of 5 kg/s. Determine*

(a) the condition of the resulting steam mixture

(b) the internal diameter of the pipe to convey the steam at 2 MN/m² if the velocity of the steam is limited to 15 m/s

(a)

Specific enthalpy of steam at a pressure of 2 MN/m^2 and a temperature of 300 °C = **3025 kJ/kg**.

At a pressure of 2 MN/m^2 the saturation temperature, t_f = **212.4 °C**. So at 300 °C the steam is superheated.

For the throttling process

$$\text{Specific enthalpy before throttling} = \text{Specific enthalpy after throttling} = \mathbf{3025\ kJ/kg}$$

$$\therefore \quad \text{Total enthalpy of 2 kg of this steam} = 2 \times 3025 = \mathbf{6050\ kJ}$$

Specific enthalpy of steam at a pressure of 800 kN/m^2 and with a dryness fraction of 0.9 is

$$720.9 + (0.9 \times 2046.5) = 720.9 + 1841.85 = \mathbf{2562.75\ kJ/kg}$$

$$\therefore \quad \text{Total enthalpy of 5 kg of this steam} = 5 \times 2562.75 = \mathbf{12\,813.75\ kJ}$$

$$\text{Total mass of mixture} = 2 + 5 = \mathbf{7\ kg/s}$$

$$\text{Total enthalpy of steam mixture} = 6050 + 12\,813.75 = \mathbf{18\,863.75\ kJ/s}$$

$$\therefore \quad \text{Specific enthalpy of steam mixture} = 18\,863.75/7 = \mathbf{2694.82\ kJ/kg}$$

At a pressure of 800 MN/m^2, h_g = **2767.5 kJ/kg** so the steam mixture is wet.
Now, for wet steam

$$h = h_f + xh_{fg} \quad \text{(see section 4.12)}$$

from this

$$x = \frac{h - h_f}{h_{fg}}$$

$$= \frac{2694.82 - 720.9}{2046.5}$$

$$= \frac{1973.92}{2046.5}$$

$$= \mathbf{0.964}$$

The condition of the resulting mixture is 0.964 dry.

(b)
Now

$$\dot{m} = \frac{AC}{v} \quad \text{(see section 2.9)}$$

where A = area of cross-section, m^2
C = velocity, m/s
$\dot{m}$ = mass flow, kg/s

$$\therefore \quad A = \frac{\dot{m}v}{C} = \frac{2 \times 0.125\,5}{15}$$

$$= \mathbf{0.016\,7\ m^2}$$

$$\therefore \quad \frac{\pi d^2}{4} = 0.016\,7 \times 10^6 \text{ where } d = \text{diameter in mm}$$

From this

$$d^2 = \frac{0.016\,7 \times 10^6 \times 4}{\pi} = 0.021\,26 \times 10^6\ \text{mm}^2$$

$$d = \sqrt{(0.021\,26 \times 10^6)} = \mathbf{145.8\ mm}$$

4.24 The determination of dryness fraction

4.24.1 The separating calorimeter

The separating calorimeter (Fig. 4.8) is really a mechanical device, not a thermodynamic device. It has already been indicated that wet steam consists of dry saturated steam and suspended water droplets. The density of water is higher than the density of dry saturated steam. If a mass of wet steam is rotated, the water in suspension will move outwards, by centrifugal action, and will separate from the dry steam. This is the principle of the separating calorimeter.

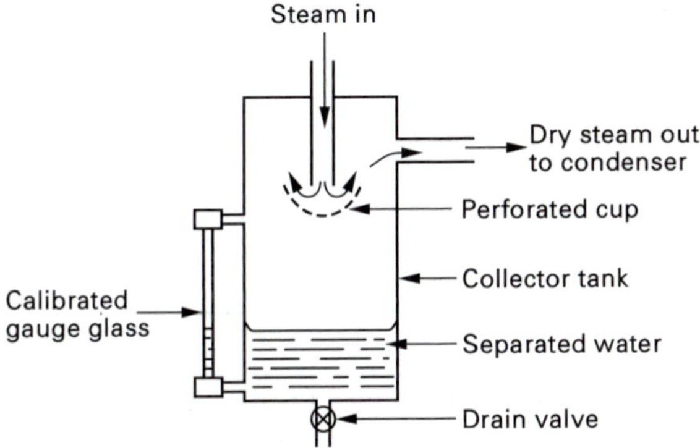

Fig. 4.8 Separating calorimeter

Figure 4.8 shows a collector tank that is fed with the steam under test. The entry steam pipe feeds into the top of a perforated cup suspended in the collector tank. A calibrated gauge glass fits into the side of the collector tank. A dry steam exit is provided from the side of the tank at the top and a drain valve is provided at the bottom.

In operation, the steam passes into the calorimeter and is rapidly forced to change its direction when it hits the perforated cup. This introduces vortex motion into the steam and the water separates out by centrifugal action. Some drains through the perforated cup, some falls as large droplets and some precipitates on the walls of the tank and will drain down. All will collect at the bottom of the tank, where the level will be recorded by the gauge glass. The dry steam will pass out of the apparatus into a small condenser for collection as condensate. The perforated cup shown in Fig. 4.8 is just one of many devices used to create a vortex in the steam.

Several precautions should be taken when using the separating calorimeter. It must be adequately warmed up before starting any measurement, otherwise condensation will occur on the interior of the apparatus, which will introduce an error into the results. As far as is possible, care must be taken to ensure that the steam does not come into contact with the water which has already been separated, otherwise a certain amount of condensation will occur and this too will affect the results.

The calorimeter should have adequate thermal insulation to prevent condensation due to heat loss. Theoretically the system should be adiabatic ($Q = 0$).

It is found in practice that not all the water is separated out; some passes out with the assumed dry steam. Consequently, this apparatus can only give a close approximation of the dryness fraction of the steam.

From the results obtained, let

M = mass of dry steam condensed
m = mass of suspended water separated in calorimeter in same time

then

$$\text{Dryness fraction} = x = \frac{\text{Mass of dry steam}}{\text{Mass of wet steam containing dry steam}}$$

$$= \frac{M}{M + m}$$

Example 4.14 *Estimate the dryness fraction of the steam entering a separating calorimeter if the separated water collected is 0.2 kg and the mass of condensate in the same time is 1.8 kg.*

SOLUTION

$$x = \frac{M}{M+m} = \frac{1.8}{1.8+0.2} = \frac{1.8}{2} = \mathbf{0.9}$$

It can be noted here that a device called a **separator** is often fitted in series into a wet steam main to help improve the steam quality (i.e. to make the steam dryer). It works in a similar way to the separating calorimeter.

One such separator takes the full steam flow and induces a rapid U-turn change of direction to the wet steam. This rapid change of direction precipitates some of the suspended water from the steam. The precipitated water can then be removed through a steam trap after it has drained to the bottom of the separator.

4.24.2 The throttling calorimeter

Figure 4.9 shows the throttling calorimeter. Steam is drawn from the main through a sampling tube placed across the steam main. The tube is perforated by many small holes and its end is sealed. Steam is forced through the small holes to obtain a representative sample across the main. The sampling tube could be placed in any direction across the main. But the steam to be analysed is wet, so the suspended water tends to gravitate to the bottom of a horizontal steam main. A vertical sampling tube will therefore pick off steam from the driest at the top to the wettest at the bottom. Hence the average sample, which then proceeds to the throttling calorimeter, will be more truly representative than for any other orientation of the sampling tube. If the steam main is vertical, the dryness will be near enough constant across the main, so the sampling tube can be placed in any direction.

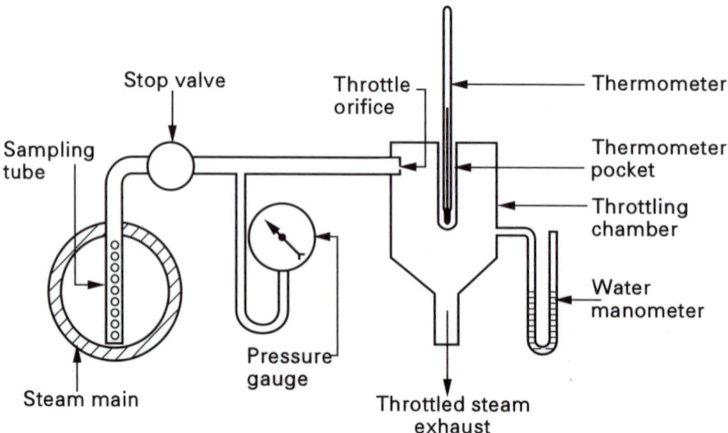

Fig. 4.9 Throttling calorimeter

From the sampling tube the steam under test proceeds, via a stop valve and a pressure gauge, to the throttle orifice of the calorimeter. The stop valve is provided in order that the calorimeter can be isolated from the main when not in use. After throttling, the steam passes into the throttle chamber of the calorimeter where its pressure can be determined using a water manometer and its temperature determined using a thermometer. The steam then passes away to exhaust, either to atmosphere or to a small condenser, where it can be collected as condensate and its mass determined after condensation. The mass of the condensate need not be determined when using the calorimeter alone. It will be noted that the manometer is a water manometer because the pressure after throttling is usually similar to atmospheric pressure. Mercury would be much too dense for this application; it would not be sensitive enough to record the small pressure change that occurs.

To operate the throttling calorimeter the stop valve is fully opened, ensuring that the steam does not experience a partial throttle as it passes through the valve. Steam is then allowed to pass through the apparatus for a while in order that pressure and temperature conditions become steady. As a general check on whether the steam is being throttled to superheat condition, remember that the pressure after throttle will not be greatly different from atmospheric pressure. Saturation temperature in this case will be round about 100 °C. If the temperature after throttle is somewhat above 100 °C, it can be taken that the steam is being superheated.

After conditions have become steady, the gauge pressure before throttling is read from the pressure gauge. The temperature and gauge pressure after throttle are recorded from the thermometer and manometer, respectively. The barometric pressure is also recorded. Let

$$\text{Gauge pressure before throttle} = P \text{ kN/m}^2$$
$$\text{Barometric height} = h \text{ mm Hg}$$

then

$$\text{Absolute pressure before throttle} = P_1 = (P + 0.133\,4h) \text{ kN/m}^2$$

Let h_{f1} = specific liquid enthalpy before throttle, kJ/kg
h_{fg1} = specific enthalpy of evaporation before throttle, kJ/kg
x_1 = unknown dryness fraction before throttle

then

$$\text{Specific enthalpy before throttle} = (h_{f1} + x_1 h_{fg1}) \text{ kJ/kg}$$

For the condition after throttle, let

$$\text{Manometer height} = h_m \text{ mm } H_2O$$

then

$$\text{Absolute pressure after throttle} = P_2 = 0.133\,4\left(h \pm \frac{h_m}{13.6}\right) \text{ kN/m}^2$$

Let the Celsius temperature after throttle be t_2. Then, from steam tables, at pressure P_2 and temperature t_2

$$\text{Specific enthalpy after throttle} = h_2 \text{ kJ/kg}$$

For a throttle

Specific enthalpy before throttle = Specific enthalpy after throttle

$$\therefore \quad h_{f1} + x_1 h_{fg1} = h_2$$

Hence

$$x_1 = \frac{h_2 - h_{f1}}{h_{fg1}}$$

It is important to realise that, after throttling, the steam must be superheated. If it is not, then an unknown dryness fraction will exist after throttling as well as the dryness fraction to be determined before throttling. So unless the steam become superheated after throttling, the calorimeter is of no value. This means that the throttling calorimeter cannot be used to determine the dryness fraction of moderately wet steam. The limit of its use is when the steam after throttle is just dry saturated, so that

$$x_1 = \frac{h_{g2} - h_{f1}}{h_{fg1}}$$

where h_{g2} = specific enthalpy of dry saturated steam at pressure P_2 after the throttle, kJ/kg

Dryness fractions below this value will produce wet steam after throttle, so no calculation can be made. Strictly, for successful operation, there must be some degree of superheat after throttling, because when the steam is dry saturated it is at saturation temperature. Any quality of wet steam is also at saturation temperature, so if the temperature after throttle were found to be saturation temperature, it would be impossible to tell whether the steam had been dry saturated or wet. However, the theoretical dryness limit is as determined by the above equation.

Example 4.15 *The dryness fraction of steam at a pressure of 2.2 MN/m^2 is measured using a throttling calorimeter. After throttling, the pressure in the calorimeter is 0.13 MN/m^2 and the temperature is 112 °C. Determine the dryness fraction of the steam at 2.2 MN/m^2.*

Extract from steam tables

Saturated steam

Pressure (MN/m^2)	Sat. temp. t_f (°C)	Spec. enthalpy (kJ/kg)		
		h_f	h_{fg}	h_g
2.2	217.2	931	1 870	2 801

Superheated steam

Pressure (MN/m^2)	Sat. temp. (°C)	Spec. enthalpy (kJ/kg)	
		h_g	at 150 °C
0.1	99.6	2 675	2 777
0.5	111.4	2 693	2 773

SOLUTION

For a throttling process

Specific enthalpy before throttling = Specific enthalpy after throttling

After throttling, the steam is at a pressure of 0.13 MN/m^2 and a temperature of 112 °C. To determine the specific enthalpy of this steam, linear interpolation is required from the superheat steam tables.

	Pressure (MN/m^2)	Sat. temp. t_f (°C)	Spec. enthalpy (kJ/kg)	
			h_g	at 150 °C
	0.15	111.4	2 693	2 773
	0.1	99.6	2 675	2 777
Difference	+0.05	+11.8	+18	−4
Require	+0.03	$+11.8 \times \frac{0.03}{0.05} = +7.08$	$+18 \times \frac{0.03}{0.05} = +10.8$	$-4 \times \frac{0.03}{0.05} = -2.4$
Adding	0.13	106.68	2 685.8	2 774.6

The last line of the table uses linear interpolation to determine the specific enthalpy of the steam at 0.13 MN/m^2 and 112 °C. The steam is superheated because the saturation temperature at the pressure of 0.13 MN/m^2 has been determined as being at 106.68 °C.

Thus, specific enthalpy at the pressure of 0.13 MN/m^2 and with a temperature of 112 °C is

$$h_2 = 2685.2 + (2774.6 - 2685.8)\frac{(112 - 106.68)}{(150 - 106.68)}$$

$$= 2685.2 + 88.8 \times \frac{5.32}{43.32}$$

$$= 2685.2 + 10.9$$

$$= \mathbf{2696.1\ kJ/kg}$$

Now, for wet steam, before the throttle

$$h_1 = h_{f1} + x_1 h_{fg1} \text{ where } x_1 = \text{required dryness fraction}$$

$$= 931 + (x_1 \times 1870)$$

For the throttling process

$$h_1 = h_2$$

$$\therefore \quad 931 + (x_1 \times 1870) = 2696.1$$

$$x_1 = \frac{2696.1 - 431}{1870}$$

$$= \frac{1765.1}{1870}$$

$$= \mathbf{0.944}$$

Example 4.16 *What is the minimum dryness fraction which can theoretically be determined using a throttling calorimeter if the steam to be tested is at a pressure of 1.8 MN/m*^{2}*? The pressure after throttling is 0.11 MN/m*^{2}*.*

SOLUTION

The theoretical dryness fraction occurs when the steam is just dry saturated after throttling (see section 4.25). At 0.11 MN/m^2 and just dry saturated the specific enthalpy = **2680 kJ/kg**.

At 1.8 MN/m^2

$$\text{Specific enthalpy} = 885 + (x \times 1912)\ \text{kJ/kg}$$

where x = minimum dryness fraction

Now for the throttling process

Specific enthalpy before throttling = Specific enthalpy after throttling

Hence

$$885 + (x \times 1912) = 2680$$

from which

$$x = \frac{2680 - 885}{1912}$$

$$= \mathbf{0.939}$$

4.25 *Various non-flow processes with steam*

Steam involved in a non-flow process can be expanded or compressed. It can also be associated with heat and work transfer during the process.

It follows, therefore, that the non-flow energy equation

$$Q = \Delta u + W$$

will apply to steam.

In the following cases, let

P_1 = original pressure — P_2 = final pressure

v_1 = original specific volume — v_2 = final specific volume

u_1 = original specific internal energy — u_2 = final specific internal energy

h_1 = original specific enthalpy — h_2 = final specific enthalpy

4.26 *The constant volume process (isochoric)*

The volume remains constant in this case, so

$$v_1 = v_2 = v \qquad [1]$$

And there is no change of volume, so there can be no external work done. Hence, $W = 0$ and the non-flow energy equation becomes

$$Q = \Delta u \qquad [2]$$

From this

$$\begin{aligned} Q &= u_2 - u_1 \\ &= (h_2 - P_2 v_2) - (h_1 - P_1 v_1) \\ &= (h_2 - h_1) - (P_2 v_2 - P_1 v_1) \\ &= (h_2 - h_1) - v(P_2 - P_1) \quad \text{(from equation [1])} \end{aligned} \qquad [3]$$

Q in this case is the transfer of heat per unit mass.

Example 4.17 *A closed vessel of 0.8 m³ capacity contains dry saturated steam at 360 kN/m². The vessel is cooled until the pressure is reduced to 200 kN/m². Calculate.*

(a) the mass of steam in the vessel
(b) the final dryness of the steam
(c) the amount of heat transferred during the cooling process

(a)
At 360 kN/m², $v_g = \mathbf{0.510\ m^3/kg}$

$$\therefore \quad \text{Mass of steam in vessel} = \frac{0.8}{0.51} = \mathbf{1.569\ kg}$$

(b)
At 200 kN/m², $v_g = \mathbf{0.885\ m^3/kg}$
The volume remains constant, so

Specific volume after cooling = Specific volume before cooling

$$\therefore \quad x \times 0.885 = 0.510$$

$$x = \frac{0.510}{0.885} = \mathbf{0.576}$$

The final dryness fraction of the steam is 0.576.

(c)
For a non-flow process, $Q = \Delta u + W$, and for a constant volume change $W = 0$

$$\therefore \quad Q = \Delta u$$

Hence, the heat transferred is equal to the change of internal energy. Now $u = h - Pv$, so at 360 kN/m², dry saturated

$$\begin{aligned} u_1 &= 2732.9 - (360 \times 0.510) \\ &= 2732.9 - 183.6 \\ &= \mathbf{2549.3\ kJ/kg} \end{aligned}$$

At 200 kN/m², dryness 0.576

$$\begin{aligned} h_2 &= 504.7 + (0.576 \times 2201.6) \\ &= 504.7 + 1268.1 \\ &= \mathbf{1772.8\ kJ/kg} \\ \therefore \quad u_2 &= 1772.8 - (200 \times 0.510) \\ &= 1772.8 - 102 \\ &= \mathbf{1670.8\ kJ/kg} \end{aligned}$$

$\therefore$ Change in specific internal energy $= u_2 - u_1$

$= 1670.8 - 2549.3$

$= \mathbf{-878.5\ kJ/kg}$ (a loss)

But there are 1.569 kg of steam in the vessel, so the amount of heat transferred during the cooling process is

$-878.5 \times 1.569 = \mathbf{-1378.4\ kJ}$ (a loss)

4.27 The constant pressure process (isobaric)

This process follows the normal processes already discussed in the formation of steam at constant pressure. The heat received or rejected is therefore equal to the change in enthalpy.

The non-flow energy equation therefore becomes

$$Q = \Delta u + W$$

or

$$h_2 - h_1 = (u_2 - u_1) + P(v_2 - v_1)$$

These are energy conditions per unit mass.

Example 4.18 *Steam at 4 MN/m² and dryness fraction 0.95 receives heat at constant pressure until its temperature becomes 350 °C. Determine the heat received by the steam per kilogram.*

SOLUTION

At 4 MN/m^2 and 0.95 dry

$$h_1 = 1087.4 + (0.95 \times 1712.9)$$
$$= 1087.4 + 1627.3$$
$$= \mathbf{2714.7\ kJ/kg}$$

At 4 MN/m^2 and temperature 350 °C

$$h_2 = \mathbf{3095\ kJ/kg}$$

The steam in this case is superheated because saturation temperature at 4 MN/m^2 = **250.3 °C**

$\therefore$ Heat received $= h_2 - h_1$

$= 3095 - 2714.7$

$= \mathbf{380.3\ kJ/kg}$

4.28 The hyperbolic process $PV = C$

In the hyperbolic process (Fig. 4.10) steam is assumed to be expanded or compressed according to the law $PV = C$. This is the law of the rectangular hyperbola, hence **hyperbolic process**. It was introduced in section 1.22 and will be encountered in Chapter 11 on steam engines.

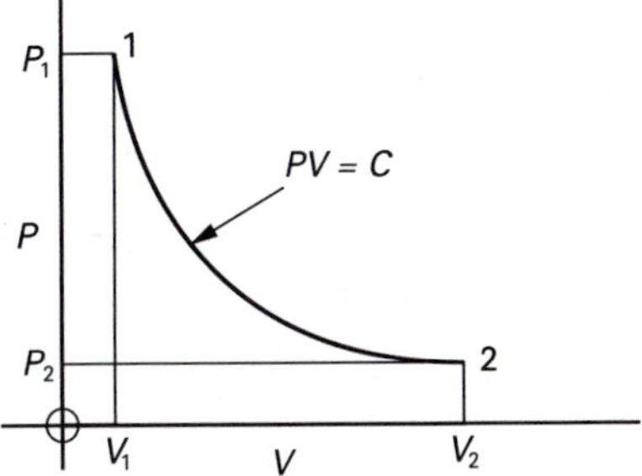

Fig. 4.10 Hyberbolic process, $PV = C$

Since $PV = C$, then

$$P_1V_1 = P_2V_2 \quad [1]$$

Also

$$\text{Work done} = W = PV\ln\frac{v_2}{v_1} \text{ for any mass} \quad [2]$$

$$= Pv\ln\frac{v_2}{v_1} \text{ for unit mass} \quad [3]$$

The non-flow energy equation becomes

$$Q = \Delta u + W$$

or

$$Q = (u_2 - u_1) + Pv\ln\frac{v_2}{v_1}$$

$$= (h_2 - P_2v_2) - (h_1 - P_1v_1) + Pv\ln\frac{v_2}{v_1}$$

$$= (h_2 - h_1) + Pv\ln\frac{v_2}{v_1} \quad [4]$$

because $P_1v_1 = P_2v_2$ from equation [1].

These are energy conditions for unit mass. But Fig. 4.10 has been illustrated using any mass because the volumes are written as V_1 and V_2.

The actual mass can be determined by the ratio V_1/v_1 or V_2/v_2 because v_1 and v_2 are specific volumes. Note that $v_2/v_1 = V_2/V_1$. Note also that Fig. 4.10 shows an expansion; for a compression, 1 and 2 are interchanged.

Example 4.19 *A quantity of dry saturated steam occupies 0.395 1 m^3 at 1.5 MN/m^2. Determine the condition of the steam*

(a) after isothermal compression to half its initial volume

(b) after hyperbolic ($PV = C$) compression to half its initial volume

In case (a) determine the heat rejected during the compression.

Extract from steam tables

Pressure (MN/m^2)	Sat. temp. t_f (°C)	Spec. enthalpy (kJ/kg)			Spec. vol. v_g (m^3/kg)
		h_f	h_{fg}	h_g	
1.5	198.3	844.7	1 945.2	2 789.9	0.131 7
3.0	233.8	1 008.4	1 793.9	2 802.3	0.066 6

(a)

v_g at 1.5 MN/m^2 = **0.131 7 m^3/kg**

$$\therefore \quad \text{Quantity of steam present} = \frac{0.3951\ 1}{0.131\ 7} = \mathbf{3\ kg}$$

The steam is operating in the evaporation region because the temperature remains constant.

$$\therefore \quad \frac{v_g}{2} = \frac{0.1317}{2} = \mathbf{0.065\ 9\ m^3/kg}$$

The final specific volume is 0.065 9 m^3/kg, hence final condition is at 1.5 MN/m^2 with a dryness fraction of

$$\frac{0.0659}{0.1317} = \mathbf{0.5}$$

$$\begin{aligned} \text{Specific enthalpy} &= 844.7 + (0.5 \times 1945{,}2) \\ &= 844.7 + 972.6 \\ &= \mathbf{1817.3\ kJ/kg} \end{aligned}$$

For 3 kg

$$\text{Enthalpy} = 3 \times 1817.3 = \mathbf{5451.9\ kJ}$$

The loss of heat during this process will be the loss of enthalpy of evaporation, changing from dry saturated steam to wet steam of dryness fraction 0.5 at constant temperature.

$$\therefore \quad \text{Heat loss} = 0.5h_{fg} = 0.5 \times 1945.2 = \mathbf{972.6\ kJ/kg}$$

For 3 kg

$$\text{Heat loss} = 972.6 \times 3 = \mathbf{2917.8\ kJ}$$

(b)

If the compression is according to the law $PV = C$, then $P_1V_1 = P_2V_2$

$$\therefore \quad P_2 = P_1\frac{V_1}{V_2} = 1.5 \times 2$$

$$= \mathbf{3.0\ MN/m^2}$$

Specific volume after compression = **0.065 9 m^3/kg**

At 3.0 MN/m^2

$$v_g = \mathbf{0.066\ 6\ m^3/kg}$$

So the dryness fraction after compression is

$$\frac{0.065\ 9}{0.066\ 6} = \mathbf{0.989}$$

Specific enthalpy $= 1008.4 + (0.989 \times 1793.9)$
$= 1008.4 + 1774.2$
$= \mathbf{2782.6\ kJ/kg}$

For 3 kg

Enthalpy $= 3 \times 2782.6$
$= \mathbf{8347.8\ kJ}$

4.29 The polytropic process $PV^n = C$

In the polytropic process (Fig. 4.11) the steam is assumed to be expanded or compressed according to the law $PV^n = C$. It was introduced in sections 1.20 and 1.21.

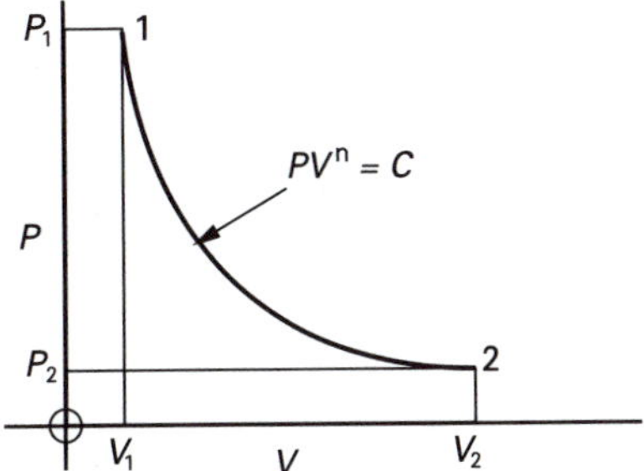

Fig. 4.11 Polytropic process, $PV^n = C$

Since $PV^n = C$, then

$$P_1V_1^n = P_2V_2^n \qquad [1]$$

Also

$$\text{Work done} = W = \frac{P_1V_1 - P_2V_2}{n-1} \quad \text{for any mass} \qquad [2]$$

$$= \frac{P_1v_1 - P_2v_2}{n-1} \quad \text{for unit mass} \qquad [3]$$

The non-flow energy equation becomes

$$Q = \Delta u + W$$

or

$$Q = (u_2 - u_1) + \left(\frac{P_1v_1 - P_2v_2}{n-1}\right)$$

$$= (h_2 - P_2v_2) - (h_1 - P_1v_1) + \left(\frac{P_1v_1 - P_2v_2}{n-1}\right)$$

$$= (h_2 - h_1) - (P_2v_2 - P_1v_1) + \left(\frac{P_1v_1 - P_2v_2}{n - 1}\right)$$

$$= (h_2 - h_1) - (P_1v_1 - P_1v_1) + \left(\frac{P_1v_1 - P_2v_2}{n - 1}\right)$$

$$= (h_2 - h_1) + \left(1 + \frac{1}{n - 1}\right)(P_1v_1 - P_2v_2)$$

$$= (h_2 - h_1) + \left(\frac{n - 1 + 1}{n - 1}\right)(P_1v_1 - P_2v_2)$$

$$Q = (h_2 - h_1) + \frac{n}{n - 1}(P_1v_1 - P_2v_2) \qquad [4]$$

These are energy conditions for unit mass. The actual mass can be determined by volume ratios V_1/v_1 or V_2/v_2 because v_1 and v_2 are specific volumes. Note that Fig. 4.11 shows an expansion; for a compression, 1 and 2 are interchanged.

Example 4.20 *A quantity of steam at a pressure of 2.1 MN/m² and 0.9 dry occupies a volume of 0.427 m³. It is expanded according to the law* $PV^{1.25}$ = *constant to a pressure of 0.7 MN/m². Determine*

(a) the mass of steam present
(b) the work transfer
(c) the change of internal energy
(d) the heat exchange between the steam and surroundings, stating the direction of transfer

Extract from steam tables

Pressure (MN/m²)	Sat. temp. t_f (°C)	Spec. enthalpy (kJ/kg)			Spec. vol. v_g (m³/kg)
		h_f	h_{fg}	h_g	
0.7	165	697.1	2 064.9	2 762.0	0.273
2.1	214.9	920.0	1 878.2	2 798.2	0.094 9

(a)
Specific volume of steam at 2.1 MN/m² and 0.9 dry is

$$v_1 = x_1v_{g1} = 0.9 \times 0.094\,9 = \mathbf{0.085\,4\ m^3/kg}$$

$$\therefore \quad \text{Mass of steam present} = \frac{0.427}{0.085\,4} = \mathbf{5\ kg}$$

(b)
For the expansion, $P_1v_1^{1.25} = P_2v_2^{1.25}$

$$\therefore \quad v_2 = v_1\left(\frac{P_1}{P_2}\right)^{1/1.25} = 0.085\,4 \times \left(\frac{2.1}{0.7}\right)^{1/1.25}$$

$$= 0.0854 \times 3^{1/1.25}$$
$$= 0.0854 \times 2.41$$
$$= \mathbf{0.205\,8\ m^3/kg}$$

The steam is wet after expansion; its dryness fraction is

$$x_2 = \frac{v_2}{v_{g2}} = \frac{0.205\,8}{0.273} = \mathbf{0.754}$$

$$\text{Work transfer} = \frac{P_1 v_1 - P_2 v_2}{n-1}$$

$$= \frac{10^3 \times (2.1 \times 0.085\,4 - 0.7 \times 0.205\,8)}{1.25 - 1}$$

$$= \frac{10^3 \times (0.179\,3 - 0.144\,1)}{0.25}$$

$$= 10^3 \times \frac{0.035\,2}{0.25} = 10^3 \times 0.140\,8$$

$$= \mathbf{140.8\ kJ/kg}$$

$$\therefore \quad \text{Work transfer for 5 kg} = 140.8 \times 5$$

$$= \mathbf{704\ kJ}$$

(c)

$$u_1 = h_1 - P_1 v_1$$

$$h_1 = h_{f1} + x_1 h_{fg1} = 920.0 + (0.9 \times 1878.2)$$

$$= 920.0 + 1690.4$$

$$= \mathbf{2610.4\ kJ/kg}$$

$$\therefore \quad u_1 = 2610.4 - \frac{10^6}{10^3} \times 2.1 \times 0.085\,4$$

$$= 2610.4 - 179.3$$

$$= \mathbf{2431.1\ kJ/kg}$$

$$u_2 = h_2 - P_2 v_2$$

$$h_2 = h_{f2} + x_2 h_{fg2} = 697.1 + (0.754 \times 2064.9)$$

$$= 697.1 + 1556.9$$

$$= \mathbf{2254.0\ kJ/kg}$$

$$\therefore \quad u_2 = 2254.0 - \frac{10^6}{10^3} \times 0.7 \times 0.205\,8$$

$$= 2254.0 - 144.1$$

$$= \mathbf{2109.9\ kJ/kg}$$

$$\therefore \quad \text{Change in internal energy} = u_2 - u_1$$

$$= 2109.9 - 2431.1$$

$$= \mathbf{-321.2\ kJ/kg} \quad \text{(a loss)}$$

For 5 kg of steam

$$\text{Loss of internal energy} = -321.2 \times 5 = \mathbf{-1606\ kJ}$$

(d)

$$Q = \Delta U + W$$

$$= -1606 + 704$$

$$= \mathbf{-902\ kJ} \quad \text{(a loss to the surroundings)}$$

4.30 The adiabatic process

In section 1.27 the adiabatic process was defined as a process carried out such that there is no heat transfer during the process, i.e. $Q = 0$.

It is possible to conceive of steam being expanded or compressed adiabatically. Such a process with steam will be a particular case of the law $PV^n = C$. The value of n will be the value which will satisfy the condition $Q = 0$.

Hence, for an adiabatic expansion or compression with steam from state 1 to state 2

$$P_1V_1^n = P_2V_2^n \qquad [1]$$

Also

$$\text{Work done} = \frac{P_1V_1 - P_2V_2}{n-1} \quad \text{for any mass} \qquad [2]$$

$$= \frac{P_1v_1 - P_2v_2}{n-1} \quad \text{for unit mass} \qquad [3]$$

The non-flow energy equation becomes

$$Q = \Delta u + W$$

and for an adiabatic process $Q = 0$

$$\therefore \quad 0 = \Delta u + W$$

or

$$W = -\Delta u \qquad [4]$$

This means that:

- Work transfer during an adiabatic expansion is done at the expense of the internal energy of the substance, in this case, steam.
- Work transfer during an adiabatic compression increases the store of internal energy of the substance.

From equation [4]

$$W = -\Delta u$$

or

$$\left(\frac{P_1v_1 - P_2v_2}{n-1}\right) = -(u_2 - u_1)$$

$$\left(\frac{P_1v_1 - P_2v_2}{n-1}\right) = (u_1 - u_2)$$

or

$$\left(\frac{P_1v_1 - P_2v_2}{n-1}\right) = (h_1 - P_1v_1) - (h_2 - P_2v_2) \qquad [5]$$

The approximate values of n for the adiabatic compression or expansion of steam are

$n = 1.13$ for wet steam
$n = 1.3$ for superheated steam

Energy conditions given above are for unit mass.

Example 4.21

(a) Determine the volume occupied by 1 kg of steam at a pressure of 0.85 MN/m^2 and having a dryness fraction of 0.97.

(b) This volume is expanded adiabatically to a pressure of 0.17 MN/m^2; the law of expansion is $PV^{1.13}$ = constant. Determine

(i) the final dryness fraction of the steam

(ii) the change of internal energy of the steam during the expansion

Extract from steam tables

Pressure (MN/m^2)	Spec. vol. v_g (m^3/kg)
0.17	1.031
0.85	0.226 8

(a)

$v_1 = x_1 v_{g1} = 0.97 \times 0.226\,8 = \mathbf{0.22\ m^2/kg}$

(b)

(i) $P_1 v_1^{1.13} = P_2 v_2^{1.13}$

$$\therefore \quad v_2 = v_1 \left(\frac{P_1}{P_2}\right)^{1/1.13}$$

$$= 0.22 \times \left(\frac{0.85}{0.17}\right)^{1/1.13}$$

$$= 0.22 \times 5^{1/1.13}$$
$$= 0.22 \times 4.15$$
$$= \mathbf{0.913\ m^3/kg}$$

At 0.17 MN/m^2, $v_{g2} = \mathbf{1.031\ m^3/kg}$

The steam is wet; its final dryness fraction is

$$x_2 = \frac{v_2}{v_{g2}} = \frac{0.913}{1.031} = \mathbf{0.886}$$

(ii) For an adiabatic expansion, $\Delta u = -W$

$$\therefore \quad u_2 - u_1 = \frac{-(P_1 v_1 - P_2 v_2)}{n - 1}$$

$$= -\frac{10^6}{10^3}\frac{(0.85 \times 0.22 - 0.17 \times 0.913)}{1.13 - 1}$$

$$= -10^3 \times \frac{(0.187 - 0.155)}{0.13} = 10^3 \times \frac{0.032}{0.13}$$

$$= -10^3 \times 0.246$$
$$= \mathbf{-246\ kJ/kg} \quad \text{(a loss of internal energy)}$$

4.31 The isothermal process

Section 4.2.2 described the evaporation stage during the formation of a vapour from a liquid. Throughout the evaporation stage the temperature remains constant at saturation temperature, t_f. Since the isothermal process is defined as a process carried out at constant temperature, the evaporation of liquid to vapour is an isothermal process. It is also carried out at constant pressure, so the energy involved is the enthalpy of evaporation. See sections 4.9 and 4.12 plus Example 4.19.

Questions

1. Determine the enthalpy, volume and density of 1.0 kg of steam at a pressure of 5 MN/m^2 and with a dryness fraction of 0.94.

   [2695.82 kJ/kg; 0.037 m^3/kg; 27.03 kg/m^3]

2. Determine the enthalpy, volume and density of 4.5 kg of steam at a pressure of 2 MN/m^2 and with a temperature of 300 °C.

   [13 612.5 kJ; 0.562 5 m^3; 7.97 kg/m^3]

3. Determine the specific enthalpy and the specific volume of steam at a pressure of 18 bar and with a temperature of 320 °C.

   [3075.16 kJ/kg; 0.149 m^3/kg]

4. Steam at a pressure of 28 kN/m^2 is passed into a condenser and leaves as condensate at a temperature of 59 °C. Cooling water circulates through the condenser at the rate of 500 kg/min. It enters at 15 °C and leaves at 30 °C. If the steam flow rate is 14 kg/min, determine the dryness fraction of the steam as it enters the condenser.

   [0.942]

5. Steam at a pressure of 1.25 MN/m^2 and with a dryness fraction of 0.96 flows through a steam main of 150 mm internal diameter with a velocity of 26 m/s. The steam is throttled to a pressure of 0.12 MN/m^2. After throttling, 5 kg of the steam is blown into a tank containing 98 kg of water, the original temperature of which is 16 °C. Take the specific heat capacity of superheated steam at constant pressure as 2.09 kJ/kg K and, neglecting heat losses, determine
   (a) the mass flow of steam through the steam main in kg/s
   (b) the temperature of the steam after throttling
   (c) the temperature of the water in the tank after receiving the 5 kg of blown steam

   Extract from steam tables

Pressure (MN/m^2)	Sat. temp. t_f (°C)	Spec. enthalpy (kJ/kg)			Spec. vol v_g (m^3/kg)
		h_f	h_{fg}	h_g	
0.12	104.8	439.4	2 244.1	2 683.4	1.428
1.25	189.8	806.7	1 977.4	2 784.1	0.156 9

   [(a) 3.04 kg/s; (b) 115.1 °C; (c) 48.54 °C]

6. A quantity of steam at a pressure of 3 MN/m^2 has a dryness fraction of 0.72. The steam occupies a volume of 0.4 m^3. Heat is transferred into the steam while the pressure remains constant at 3 MN/m^2 until the steam becomes dry saturated. The steam is then cooled at constant volume until the pressure becomes 1.8 MN/m^2. Determine
   (a) the heat transferred during the constant pressure process
   (b) the percentage of the heat transfer which appears as work transfer
   (c) the heat transferred during the constant volume process

Extract from steam tables

Pressure (MN/m^2)	Sat. temp. t_f (°C)	Spec. enthalpy (kJ/kg)			Spec. vol v_g (m^3/kg)
		h_f	h_{fg}	h_g	
1.8	207.1	884.6	1 910.3	2 794.8	0.110 3
3.0	233.8	1 008.4	1 793.9	2 802.3	0.066 6

[(a) 4189.1 kJ; (b) 11.14%; (c) −5704.6 kJ]

7. Water enters a heater at a temperature of 18 °C. Steam at a pressure of 1.2 bar and 0.95 dry is mixed with, and condensed in, the water. The mixture of water and condensate leaves the heater as hot water. The mass of steam is 8 per cent of the combined hot water leaving the heater. Neglecting heat transfer loss, determine the temperature of the hot water output. Take the specific heat capacity of water at constant pressure as 4.19 kJ/kg K.

Extract from steam tables

Pressure (bar)	Sat. temp. t_f (°C)	Spec. enthalpy (kJ/kg)		
		h_f	h_{fg}	h_g
0.12	104.8	439.4	2 244.1	2 683.4

[65.7 °C]

8. A quantity of steam at a pressure of 2 MN/m^2 has a volume of 0.75 m^3 and an enthalpy content of 21.5 MJ. Determine (a) the dryness fraction of the steam and (b) the mass of the steam. If some of this steam is throttled to a pressure of 1.2 MN/m^2, determine (c) the quality of the steam after throttling.

Extract from steam tables

Pressure (MN/m^2)	Sat. temp. t_f (°C)	Spec. enthalpy (kJ/kg)			Spec. vol v_g (m^3/kg)
		h_f	h_{fg}	h_g	
1.2	188.0	798.4	1 984.3	2 787.7	0.163 2
2.0	212.4	908.6	1 888.6	2 797.2	0.099 5

[(a) 0.943; (b) 8 kg; (c) 0.953]

9. Steam at a pressure of 2 MN/m^2 and with a temperature of 250 °C is expanded to a pressure of 0.4 MN/m^2; the law of expansion is $PV^{1.2} = C$. After expansion the steam is cooled at constant volume to a pressure of 0.2 MN/m^2. Determine
   (a) the condition of the steam after expansion according to the law $PV^{1.2} = C$
   (b) the heat transferred per kilogram of steam during the expansion
   (c) the condition of the steam after the constant volume cooling
   (d) the heat transferred during the constant volume cooling

[(a) 0.921; (b) −91 kJ/kg; (c) 0.481; (d) −921 kJ/kg]

10. Superheated steam at a pressure of 4 MN/m^2 and with a temperature of 375 °C enters a desuperheater at the rate of 5 tonne/h. Water at a temperature of 75 °C is sprayed into the superheated steam as it passes through the desuperheater to produce a combined mixture of output steam at a pressure of 4 MN/m^2 and with a dryness of 0.96. Determine
   (a) the mass of the output steam in tonne/h
   (b) the internal diameter of the steam pipe required to carry the output steam if the steam velocity is 20 m/s

[(a) 5.878 tonne/h; (b) 70.5 mm]

11. 2.5 kg of steam at a pressure of 100 kN/m^2 and with a dryness fraction of 0.96 is compressed hyperbolically to a pressure of 8.0 bar. Determine
    (a) the final condition of the steam
    (b) the heat transferred during the compression

[(a) 0.845; (b) −1182.5 kJ]

12. A throttling calorimeter is used to determine the quality of steam which is at a pressure of 2.2 MN/m^2. The pressure and temperature after throttling are 0.12 MN/m^2 and 109.6 °C, respectively. Determine
    (a) the dryness fraction of the steam at 2.2 MN/m^2
    (b) the least dryness fraction which can be theoretically determined under the given pressure conditions

    Take the specific capacity of superheated steam as 2.1 kJ/kg K.

Extract from steam tables

Pressure (MN/m^2)	Sat. temp. t_f (°C)	Spec. enthalpy (kJ/kg)		
		h_f	h_{fg}	h_g
0.12	104.8	439.4	2 244.1	2 683.4
2.2	217.2	931.0	1 868.1	2 799.1

[(a) 0.943; (b) 0.938]

13. Steam at a pressure of 1.5 bar and with a dryness fraction of 0.9 is compressed adiabatically to a pressure of 7 bar. The law of compression is $PV^{1.13} = C$. Determine
    (a) the final condition of the steam
    (b) the final density of the steam
    (c) the change of specific internal energy of the steam

Extract from steam tables

Pressure (bar)	Spec. vol. v_g (m^3/kg)
1.5	1.159
7	0.272 8

[(a) 0.98; (b) 3.75 kg/m^3; (c) +234.6 kJ/kg]

14. A refrigerant is at a pressure of 0.745 MN/m^2 and has a temperature of 45 °C. It is cooled at constant pressure until it becomes liquid at saturation temperature. It is then throttled down to a pressure of 0.219 MN/m^2. Determine
    (a) the heat transfer during the constant pressure cooling process per kilogram of refrigerant
    (b) the quality of the refrigerant after throttling

Extract from steam tables

Pressure (MN/m^2)	Sat. temp. t_f (°C)	Spec. enthalpy (kJ/kg)		
		h_f	h_{fg}	superheated by 20 K
0.219	−10	26.9	183.2	195.7
0.745	30	64.6	199.6	214.3

[(a) −146.03; (b) 0.241]

15. Steam at a pressure of 3 MN/m^2 and with a temperature of 250 °C is mixed with wet steam at a pressure of 3 MN/m^2 and with a dryness fraction of 0.92 in the ratio of 1:2.5 by mass. Determine
    (a) the quality of the steam mixture
    (b) the density of the steam mixture

    [(a) 0.952; (b) 15.77 kg/m^3]

16. Steam flows through a turbine at a rate of 3 kg/s. The steam enters the turbine at a pressure of 4 MN/m^2 and with a temperature of 350 °C. The pressure at exhaust from the turbine is 60 kN/m^2 and the dryness fraction is 0.92. Neglecting any change in kinetic energy of the steam and any heat transfer loss, determine
    (a) the theoretical power output from the turbine
    (b) the steam exit area at the turbine exhaust if the steam velocity at exit is 32 m/s

    [(a) 1875 kW; (b) 0.235 m^2]

Chapter 5

Gases and single-phase systems

5.1 General introduction

This is an investigation into single-phase systems. The single-phase being considered is that phase above the critical point when a substance is called a gas.

The wide use made of gases in the field of engineering makes it necessary to investigate their behaviour when they are heated, cooled, expanded or compressed. The beginning of any investigation, such as the behaviour of gases, is usually made by conducting experiments; from the results obtained laws are determined which govern their behaviour. The first two laws in this chapter were established by experiment.

5.2 Boyle's law

With any mass of gas it is possible to vary the pressure, volume and temperature. In this experiment it is arranged that the temperature of a fixed mass of gas remains constant while corresponding changes in pressure and volume are observed. An apparatus suitable for conducting such an experiment is illustrated in Fig. 5.1.

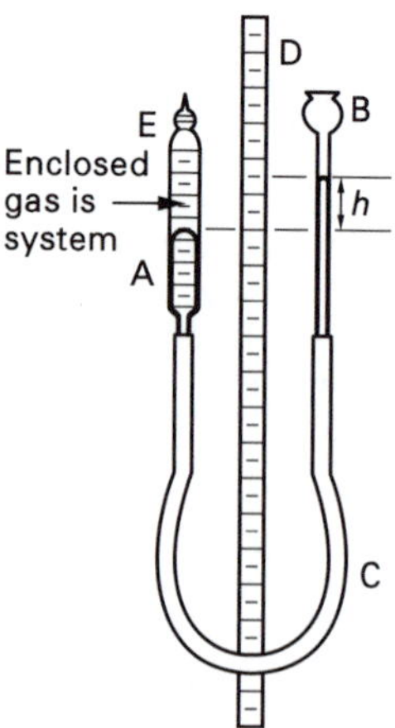

Fig. 5.1 Boyle's law apparatus

An inverted glass pipette A is connected to a glass thistle funnel B by means of a long rubber, or plastic, tube C. Both the pipette and the thistle funnel are mounted vertically such that they can be moved up or down on either side of a vertical scale D. With the tap E open, the apparatus is filled with a suitable quantity of mercury. It is possible to adjust the height of the mercury columns, and hence the volume of gas in the pipette – the system – by moving the thistle funnel up or down. If the tap is then closed, a fixed mass of gas is trapped in the apparatus and modification to the height of the thistle funnel will bring about pressure changes in the gas which will be accompanied by corresponding changes in volume of the gas.

The pipette is calibrated to read the volume of gas contained in it; the vertical scale serves to establish the difference in height h of the two mercury columns. The absolute pressure of the gas will be given by the sum of the height h and the barometer reading. In order to satisfy the condition that the temperature should remain constant, a period of time is allowed to elapse after every change of condition before any new readings are taken. After a suitable number of results are obtained, the corresponding values of absolute pressure and volume are plotted on a graph; the curve is shown in Fig. 5.2.

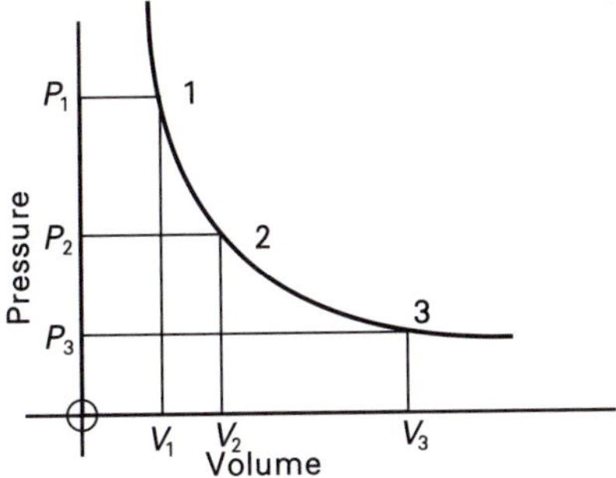

Fig. 5.2 Boyle's law graph

Taking any point on the curve, 1 say, the product of its corresponding pressure and volume P and V will equal some number, C say. Investigation of other points, such as 2 and 3, shows that, within the limits of experimental error, the products of their corresponding pressures and volumes also equal this same number, or

$$P_1V_1 = P_2V_2 = P_3V_3 = C, \text{ a constant} \qquad [1]$$

Further experiments at different fixed temperatures, with different fixed masses and with different gases yield the same result, although the constant C will be different with each quantity of gas, each fixed temperature and each type of gas.

From the results of this experiment, a general statement may be made:

> During a change of state of any gas in which the mass and the temperature remain constant, the volume varies inversely as the pressure.

Expressed mathematically

$$PV = C, \text{ a constant} \qquad [2]$$

This is known as Boyle's law, named after its discoverer, Robert Boyle (1627–1691), an English scientist. As a point of interest, a Frenchman, Edme Mariotte, made the same discovery at about the same time while working quite independently of Boyle.

The graph of the law $PV = C$ is a rectangular hyperbola. Note also that if $PV = C$, then $P = C/V$. If P is plotted against $1/V$ the result will be a straight line passing through the origin and of slope C as shown in Fig. 5.3. This method could be used as a check of the above results; by plotting P against $1/V$ and obtaining a straight line passing through the origin, $PV = C$ would be proved.

The temperature is constant during a process carried out according to Boyle's law, so the process is isothermal (see section 1.14).

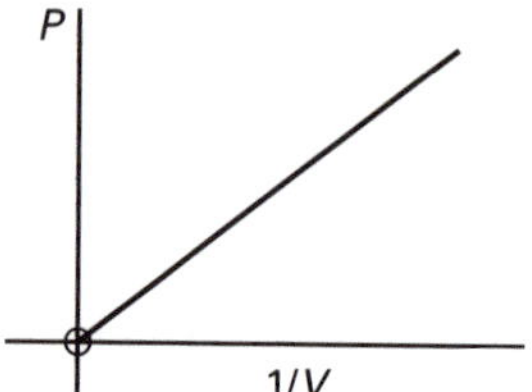

Fig. 5.3 Plot of P against $1/V$

Example 5.1 *During an experiment on Boyle's law, the original volume of air trapped in the apparatus, with the two mercury levels of the same, as 20 000 mm*^{3}*. The apparatus was then modified such that the volume of air became 17 000 mm*^{3}*, while the temperature remained constant. If the barometer reading was 765 mm Hg, what was the new pressure exerted on the air in mm Hg? Also, what was the difference in the two mercury column levels?*

SOLUTION

Since both levels of mercury are the same at the beginning, then

P_1 = atmospheric pressure = **765 mm Hg**

Now Boyle's law states that $PV = C$, a constant, from this

$$P_1V_1 = P_2V_2$$

$$\therefore \quad P_2 = P_1\frac{V_1}{V_2} = 765 \times \frac{20\,000}{17\,000}$$

$$= \mathbf{900\ mm\ Hg}$$

Notice that the pressure has been left in mm Hg during this part of the solution. This can be done because there is a pressure term on both sides of the equation. As long as both terms have the same units, the equality will hold. The final pressure $P_2 = 900$ mm Hg and the atmospheric pressure = 765 mm Hg, so

Difference in height of the two mercury columns = 900 − 765
= **135 mm**

This will be the height h which was mentioned in the work on Boyle's law apparatus.

Example 5.2 *A gas whose original pressure and volume were 300 kN/m² and 0.14 m³ is expanded until its new pressure is 60 kN/m² while its temperature remains constant. What is its new volume?*

SOLUTION
The temperature remains constant, so this is an expansion according to Boyle's law.

$$\therefore \quad P_1V_1 = P_2V_2 \quad \text{or} \quad V_2 = V_1\frac{P_1}{P_2}$$

$$\therefore \quad V_2 = 0.14 \times \frac{300}{60}$$

$$= \mathbf{0.7\ m^3}$$

5.3 *Charles' law and also absolute temperature*

Consider now an experiment in which the pressure of a fixed mass of gas is kept constant while the volume and temperature are varied. A simple piece of apparatus on which to conduct such an experiment is illustrated in Fig. 5.4.

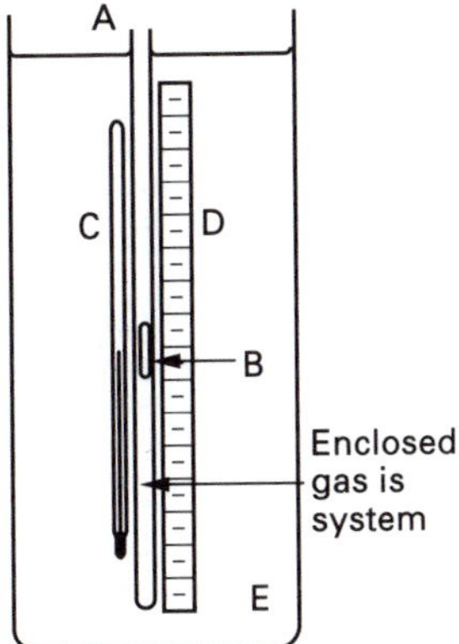

Fig. 5.4 Charles' law apparatus

A long glass tube A with one end sealed has a pellet of mercury B introduced; the pellet acts as a piston enclosing a fixed mass of gas, the system, in the end of the tube. Thermometer C and a volume scale D are attached along this tube and the assembly is immersed in a water bath E. The temperature of the water bath is then varied; this will be accompanied by changes in volume of the gas which will be registered by the mercury pellet moving along the glass tube. The pressure of the gas in the tube will remain constant because the open end of the tube is always presented to the same external pressure conditions. The corresponding gas volumes and temperatures observed during the conduct of the experiment are recorded and plotted on a graph. The graph obtained is a straight line, as illustrated in Fig. 5.5, showing a linear relationship between volume and temperature of a fixed mass of gas when the pressure remains constant. This takes the form:

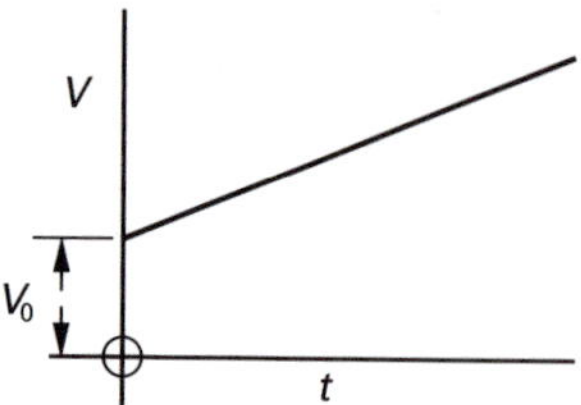

Fig. 5.5 Plot of V against t

$$V = Ct + V_0 \qquad [1]$$

where V = volume
t = temperature
C = slope
V_0 = intercept on V axis

Further experiments at different pressures, with different masses and with different gases give similar results. An interesting point, however, is that if all the straight lines obtained are extended back to cut the temperature axis, they all cut this axis at the same point. This is illustrated in Fig. 5.6.

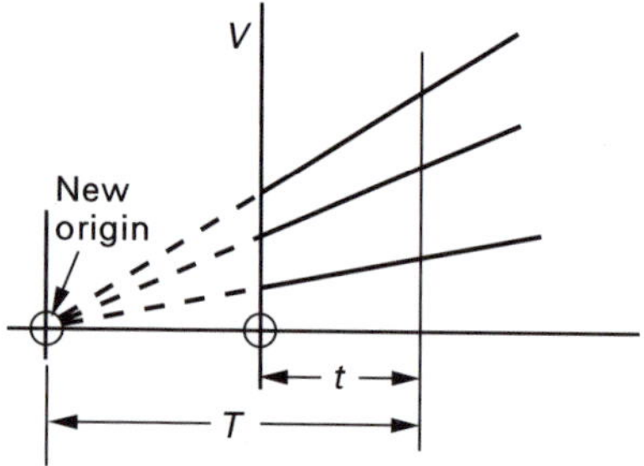

Fig. 5.6 Extended Charles' law plots

If this is the case then perhaps it would be better to use this point as a new origin and the law of the graph would then become

$$V = CT \qquad [2]$$

where T is the temperature recorded from the new origin. Equation [2] is of a better form than equation [1] because the constant V_0 is now absent. But reading temperature from this new origin has introduced a new temperature scale. Now the graph cuts the temperature axis at approximately -273 °C. The value of T in equation [2] is therefore determined by adding 273 °C to the value of t as recorded from the thermometer, or

$$T = (t + 273) = T\,\mathrm{K} \qquad [3]$$

Temperature recorded in this manner is called **absolute temperature** and the new zero is called the **absolute zero of temperature**. The reason for this can be seen by referring to Fig. 5.6. It will be noted that, at the new zero, all volumes have reduced to zero. No further reduction seems possible because there is nothing left to reduce in temperature! The fallacy of this argument is that extension of the straight lines to determine the new zero assumes that the gas remains as a gas in the low-temperature region. This is not true in practice because all gases, on being cooled, will eventually liquefy and then finally solidify, thus losing their properties as a gas. Experiments on the problem of an absolute zero of temperature have shown, however, that approximately −273 °C appears to be the lowest temperature possible, and is extremely difficult to reach.

From equation [2] it follows that

$$\frac{V}{T} = C, \text{ a constant} \qquad [4]$$

In words this may be stated as follows:

> During the change of state of any gas in which the mass and pressure remain constant, the volume varies in proportion with the absolute temperature.

This is known as Charles' law.

Of historic interest, the law dealt with above is attributed to a Frenchman, Jacques A. Charles (1746–1823). It is also interesting to note that another Frenchman, Joseph-Louis Gay-Lussac (1778–1850), made the same discovery at about the same time.

The concept of the **absolute scale of temperature** has already been discussed in section 1.7.

Example 5.3 *During an experiment on Charles' law, the volume of gas trapped in the apparatus was 10 000 mm^3 when the temperature was 18 °C. The temperature of the gas was then raised to 85 °C. Determine the new volume of gas trapped in the apparatus if the pressure exerted on the gas remained constant.*

SOLUTION

Now according to Charles' law

$$\frac{V}{T} = C, \text{ a constant}$$

From this

$$\frac{V_1}{T_1} = \frac{V_2}{T_2} \qquad [1]$$

In order to use this equation the temperatures T_1 and T_2 must be absolute temperatures.

$$\therefore \quad T_1 = 18 + 273 = \mathbf{291\ K}$$

and

$$T_2 = 85 + 273 = \mathbf{358\ K}$$

From equation [1]

$$V_2 = V_1 \frac{T_2}{V_1} \qquad [2]$$

or

$$V_2 = 10\,000 \times \frac{358}{291}$$

$$= \mathbf{12\,302\ mm^3}$$

Example 5.4 *A quantity of gas whose original volume and temperature are 0.2 m³ and 303 °C, respectively, is cooled at constant pressure until its volume becomes 0.1 m³. What will be the final temperature of the gas?*

SOLUTION

Again, this is a change according to Charles' law.

$$\therefore \quad \frac{V_1}{T_1} = \frac{V_2}{T_2}$$

$$T_2 = T_1 \frac{V_2}{V_1} \qquad [1]$$

The temperature is in degrees Celsius this time.

$$\therefore \quad T_1 = 303 + 273 = 576\ \text{K}$$

and from equation [1]

$$T_2 = 576 \times \frac{0.1}{0.2}$$

$$= 288\ \text{K}$$

$$\therefore \quad t_2 = 288 - 273 = \mathbf{15\ °C}$$

5.4 *The characteristic equation of a perfect gas*

In the previous two experiments it was arranged that, in each case, one of the three conditions of state, pressure, volume and temperature, remained constant while a law connecting variations in the other two was established. An investigation must now be made into the more general change of state of a gas which neither pressure, volume nor temperature remains constant. Consider a gas whose original state is pressure P_1, volume V_1 and temperature T_1, and let this gas pass through a change of state such that its final state is P_2, V_2 and T_2. Inspection of Fig. 5.7 will show that there are an infinite number of paths which connect states 1 and 2 when the process is shown on a P–V diagram.

The concern at the moment, however, is not in how the state changed from 1 to 2, but in the fact that since states 1 and 2 can exist for the same mass of gas, is there any law connecting them? This being the case, a choice of path from 1 to 2 is quite arbitrary, and it is therefore reasonable to assume a path about which something is already known. Boyle's and Charles' laws supply the answer. Figure 5.8 shows that it is quite possible to move from 1 to 2 by first carrying out a Boyle's law change down to some intermediate state A, say, then carry out a Charles' law change to the final condition.

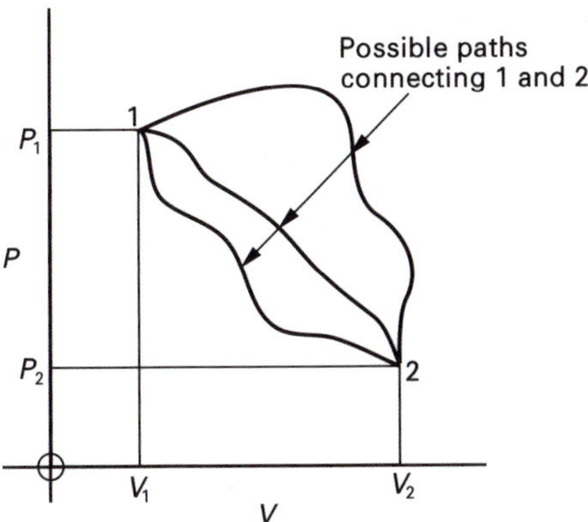

Fig. 5.7 Three of the infinite number of paths from 1 to 2

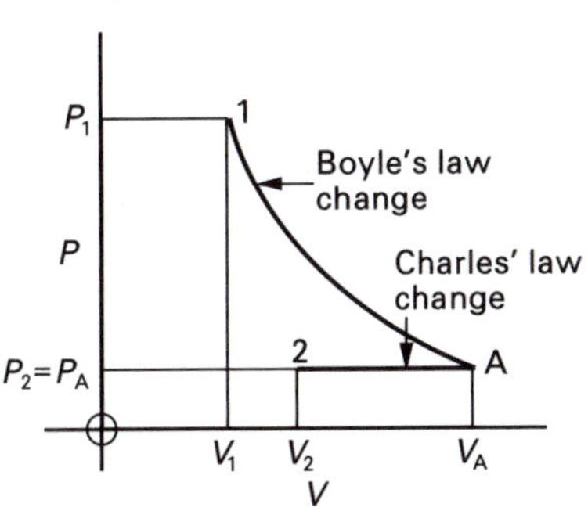

Fig. 5.8 Getting from 1 to 2 in two known stages

Consider the Boyle's law change from 1 to A. In this case the temperature remains constant at T_1. Also

$$P_1 V_1 = P_A V_A \qquad [1]$$

All the pressure change must take place during this process because there will be no change in pressure during the Charles' law process which follows. In this case $P_A = P_2$.

$\therefore$ Equation [1] becomes $P_1 V_1 = P_2 V_2$

or

$$V_A = \frac{P_1 V_1}{P_2} \qquad [2]$$

Consider now the Charles' law change from A to 2. In this case the pressure remains constant at P_2. Also

$$\frac{V_A}{T_A} = \frac{V_2}{T_2} \qquad [3]$$

During the Boyle's law change from 1 to A the temperature remained constant.

$$\therefore \quad T_A = T_1$$

from which equation [3] becomes

$$\frac{V_A}{T_1} = \frac{V_2}{T_2} \qquad [4]$$

But

$$V_A = \frac{P_1 V_1}{P_2}$$

from equation [2] and substituting this in equation [4]

$$\frac{P_1 V_1}{P_2 T_1} = \frac{V_2}{T_2}$$

from which

$$\frac{P_1 V_1}{T_1} = \frac{P_2 V_2}{T_2} \qquad [5]$$

Now any change of state from state 1 would produce a similar result, and hence equation [5] could be extended to read

$$\frac{P_1 V_1}{T_1} = \frac{P_2 V_2}{T_2} = \frac{P_3 V_3}{T_3} = \frac{P_4 V_4}{T_4} = \ldots, \text{etc.} \qquad [6]$$

where 3 and 4 represent other new conditions of state of the same mass of gas.

From equation [6] it follows that for any fixed mass of gas, changes of state are connected by the equation

$$\frac{PV}{T} = \text{a constant} \qquad [7]$$

Sooner or later it will be necessary to know the actual mass of gas used during any particular process.

Let v = volume of 1 kg of gas, the **specific volume** (see section 1.9).

Then from equation [7]

$$\frac{Pv}{T} = \text{a constant} \qquad [8]$$

When 1 kg of gas is considered, this constant is written R and is called the **characteristic gas constant**, sometimes, the **specific gas constant**. So for 1 kg of gas

$$\frac{Pv}{T} = R \qquad [9]$$

Now consider the case when there are m kg of gas. Multiply both sides of equation [9] by m, then

$$\frac{P(mv)}{T} = mR$$

But mv is the total volume of the gas being used, V, so for m kg of gas it follows that

$$\frac{PV}{T} = mR$$

or

$$PV = mRT \qquad [10]$$

This is known as the **characteristic equation of a perfect gas.**

The units of R can be obtained from equation [9]. If pressure is in N/m^2, specific volume is in m^3/kg and temperature in K, then

$$\frac{Pv}{T} = R = \frac{N}{m^2} \times \frac{m^3}{kg} \times \frac{1}{K} = \frac{N\,m}{kg\,K} = \frac{J}{kg\,K}$$

For air, the value of R is usually of the order 0.287 kJ/kg K.

Actually the value of R is numerically equal to the work done when 1 kg of gas is heated at constant pressure through 1 degree rise of temperature. This can be shown as follows. Consider 1 kg of gas at original state P_1, V_1, T_1 and let it be heated at constant pressure through 1 degree. The new state will then be P_1, V_2, $(T_1 + 1)$.

Now from the characteristic equation

$$P_1V_1 = RT_1 \qquad [11]$$

and

$$P_1V_2 = R(T_1 + 1) \qquad [12]$$

Subtracting equation [11] from equation [12] gives

$$P_1(V_2 - V_1) = R \qquad [13]$$

This is equal to the area under the graph of the process plotted on a P–V diagram and this has been shown to be equal to the work done (see section 1.19).

Section 5.8 contains a further reference to the characteristic gas constant R and its relationship with the specific heat capacity of a gas.

A further point to note about this work on the characteristic equation is that it is now possible to predict the behaviour of a gas if the volume remains constant. If this is the case then $V_1 = V_2$, and from equation [5], since $P_1V_1/T_1 = P_2V_2/T_2$, it follows that

$$\frac{P_1}{T_1} = \frac{P_2}{T_2} \qquad [14]$$

It may be wondered why use the term *perfect gas*? Very accurate experiment shows that actual gases do not obey the above gas laws exactly. The deviation is very small, however, and can be sensibly neglected in all general calculations. A perfect gas may be defined as a gas which obeys the gas laws exactly. A further extension of the use of the characteristic equation of a perfect gas is made in section 8.24.

Example 5.5 *A gas whose original pressure, volume and temperature were 140 kN/m², 0.1 m³ and 25 °C, respectively, is compressed such that its new pressure is 700 kN/m² and its new temperature is 60 °C. Determine the new volume of the gas.*

SOLUTION

By the characteristic equation

$$\frac{P_1V_1}{T_1} = \frac{P_2V_2}{T_2} \quad \text{and} \quad T_1 = 25 + 273 = \mathbf{298\ K}$$

$$\text{also} \quad T_2 = 60 + 273 = \mathbf{333\ K}$$

$$\therefore \quad V_2 = \frac{P_1}{P_2}\frac{T_2}{T_1}V_1 = \frac{140}{700} \times \frac{333}{298} \times 0.1 = \mathbf{0.022\ 3\ m^3}$$

Example 5.6 *A quantity of gas has a pressure of 350 kN/m² when its volume is 0.03 m³ and its temperature is 35 °C. If the value of* R = *0.29 kJ/kg K, determine the mass of gas present. If the pressure of this gas is now increased to 1.05 MN/m² while the volume remains constant, what will be the new temperature of the gas?*

SOLUTION

By the characteristic equation

$$PV = mRT$$

$$\therefore \quad m = \frac{PV}{RT} = \frac{350 \times 10^3 \times 0.3}{0.29 \times 10^3 \times 308} = \mathbf{0.118\ kg}$$

For the second part of the problem

$$\frac{P_1V_1}{T_1} = \frac{P_2V_2}{T_2} \quad \text{and in this case, } V_1 = V_2$$

$$\therefore \quad \frac{P_1}{T_1} = \frac{P_2}{T_2} \quad \text{or} \quad T_2 = T_1\frac{P_2}{P_1} = 308 \times \frac{1.05 \times 10^6}{0.35 \times 10^6}$$

$$= 308 \times 3 \quad (T_1 = 35 + 273 = 308\ \text{K})$$

$$= \mathbf{924\ K}$$

$$\therefore \quad t_2 = 924 - 273 = \mathbf{651\ °C}$$

5.5 *The internal energy of a gas and Joule's law*

The internal energy term appears in the steady-flow energy equation and the non-flow energy equation so an investigation is required into the way it applies to a gas.

Joule carried out an experiment on this subject from which he concluded:

> The internal energy of a gas is a function of temperature only and is independent of changes in pressure and volume.

This is known as **Joule's law**.

A sketch of Joule's apparatus is shown in Fig. 5.9. Two copper vessels A and B were connected together as shown and were isolated from each other by means of a gas tap C. Vessel A was filled with compressed air to a pressure of about 21 atmospheres (about 2.1 MN/m^2) and vessel B was exhausted to a condition of vacuum. This assembly was immersed in a water bath D and temperature recordings were made by means of a thermometer E. After leaving the apparatus for some time, in order to let the temperature conditions become steady, the gas tap C was opened and some air from vessel A expanded into vessel B while the pressure dropped, eventually to stabilise at some new common pressure. The volume was then equal to the total volume of vessels A and B. During this process no change in temperature was observed.

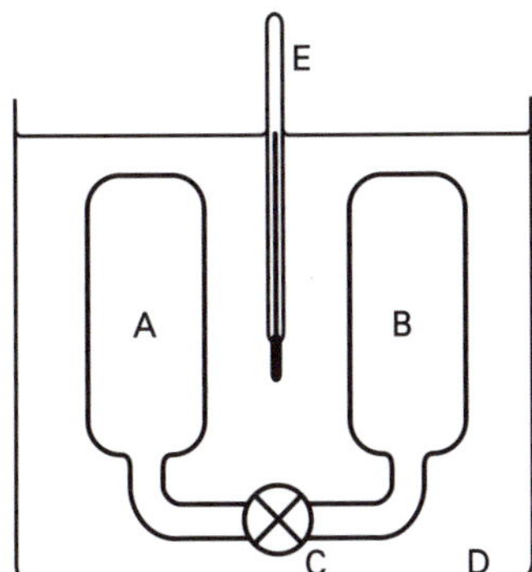

Fig. 5.9 Joule's law apparatus

No work transfer occurred because this was a free expansion into a vacuum, so $W = 0$.

And no heat was transferred during the expansion, so $Q = 0$.

Applying this to the non-flow energy equation for the expansion of a gas

$$Q = \Delta U + W \qquad [1]$$

it follows that for this experiment

$$0 = \Delta U + 0 \qquad [2]$$

or

$$\Delta U = 0 \qquad [3]$$

Both the pressure and the volume changed during the experiment. But the temperature did not change, nor did the internal energy, so it seemed reasonable to assume that the internal energy of a gas was a function of temperature only.

Later experiments, carried out rather more accurately by Joule in conjunction with William Thomson (Lord Kelvin), showed that there was a very small change in temperature during such an expansion of a gas, but this temperature change is small enough to be neglected. Joule's law is assumed to hold in all normal practical cases.

The problem now is to develop an expression which will give the magnitude of the change in internal energy of a gas and, by Joule's law, this expression must be a function of temperature only. Before this can be accomplished, it is necessary to discuss the specific heat capacities of a gas.

5.6 *The specific heat capacities of a gas*

The specific heat capacity of a substance may be defined as the amount of heat transfer required to raise unit mass of a substance through 1 degree difference in temperature (see section 1.25). Apply this statement to a gas and, at first sight, this definition may seem reasonable. But consider Fig. 5.10.

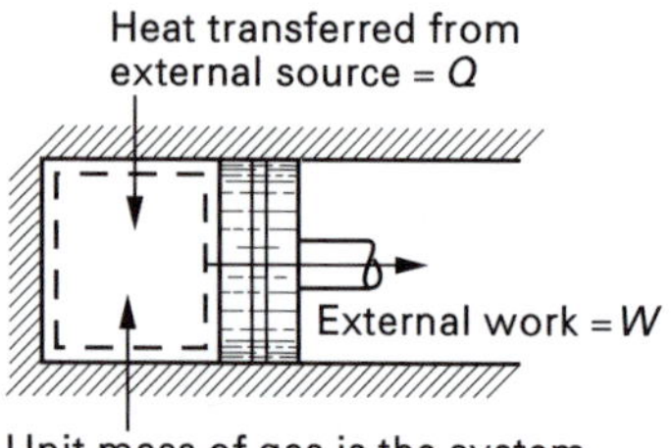

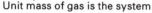

Fig. 5.10 Using a piston to determine specific heat capacity of a gas

The figure shows a piston enclosing unit mass of gas, the system, in a cylinder. This unit mass of gas could be heated from some outside source such that the temperature of the gas is raised by 1 degree. The amount of heat transfer to accomplish this 1 degree rise in temperature will depend upon what happens to the piston. For example, the piston could be fixed, then the gas would be heated at constant volume and a certain quantity of heat would bring about the 1 degree rise in temperature. Or the gas in the cylinder could be allowed to expand, moving the piston and doing external work W. The extent to which the piston is allowed to move is one of an infinite number of possible arrangements. The amount of heat transfer will depend upon the piston movement, so there are infinitely many heat supply quantities, each able to produce a 1 degree rise in temperature. It appears there are an infinite number of possible specific heat capacities for a gas. If the specific heat capacity of a gas is quoted, therefore, it is necessary to define the conditions under which the specific heat capacity was measured. Two important cases are called the **principal specific heat capacities**.

- *The specific heat capacity at constant volume* This is defined as the amount of heat which transfers to or from unit mass of gas while the temperature changes by 1 degree and the volume remains constant. It is written c_v.
- *The specific heat capacity at constant pressure* This is defined as the amount of heat which transfers to or from unit mass of gas while the temperature changes by 1 degree and the pressure remains constant. It is written c_p.

The specific heat capacities at constant volume and constant pressure rise in value with temperature. For calculations, it is usual to assume an average value of specific heat capacity within the temperature range being considered.

Table of average specific heat capacities of gases

Gas	c_p(kJ/kg K)	c_v(kJ/kg K)
Air	1.006	0.718
Carbon dioxide	0.87	0.67
Carbon monoxide	1.04	0.74
Hydrogen	14.4	10.2
Nitrogen	1.04	0.74
Oxygen	0.92	0.66
Methane	2.29	1.74
Sulphur dioxide	0.65	0.52

5.7 The constant volume heating of a gas

Let a mass of gas m be heated at constant volume such that its temperature rises from T_1 to T_2 and its pressure rises from P_1 to P_2, then

Heat received by the gas = mass × specific heat capacity at constant volume × rise in temperature

$$= mc_v(T_2 - T_1) \qquad [1]$$

Constant volume heating is a particular case of a non-flow process carried out on a gas. Consider the non-flow energy equation applied to constant volume heating.

$$Q = \Delta U + W \qquad [2]$$

No external work is done during constant volume heating. This can be seen by inspecting Fig. 5.11, in which pressure is plotted against volume. The process appears as a vertical straight line. There is no area beneath this line, so no external work is done.

$$\therefore \quad W = 0$$

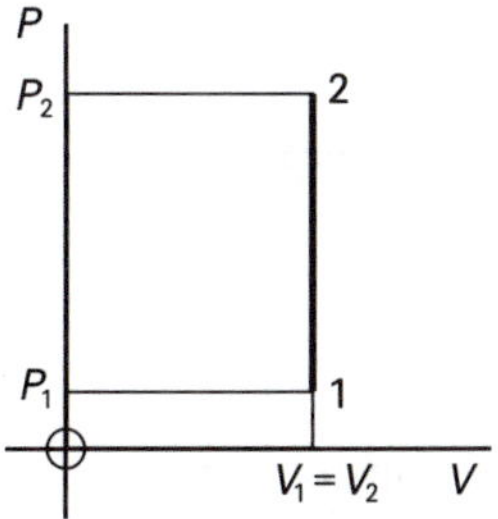

Fig. 5.11 Constant volume process on *P–V* diagram

Hence, equation [2] becomes

$$Q = \Delta U$$

or

$$mc_v(T_2 - T_1) = U_2 - U_1 \qquad [3]$$

which can be written

$$mc_v\Delta T = \Delta U \quad [4]$$

It follows, therefore, that all the heat added during constant volume heating goes completely into increasing the stock of internal energy of the gas. Conversely, if a gas is cooled at constant volume, the heat rejected will be at the expense of the stock of internal energy of the gas. If the new pressure is required, it may be found by the application of the characteristic equation of a perfect gas.

$$\frac{P_1V_1}{T_1} = \frac{P_2V_2}{T_2} \quad [5]$$

and for this case, $V_1 = V_2$.

$$\therefore \quad \frac{P_1}{T_1} = \frac{P_2}{T_2} \quad [6]$$

or

$$P_2 = P_1\frac{T_2}{T_1} \quad [7]$$

If unit mass of gas is considered, equation [3] can be written

$$c_v(T_2 - T_1) = u_2 - u_1 \quad [8]$$

from which

$$c_v = \frac{u_2 - u_1}{T_2 - T_1} = \frac{\Delta u}{\Delta T} \quad [9]$$

At any particular absolute temperature T, therefore, and as $\Delta T \to 0$, equation [9] can be written

$$c_v = \left(\frac{\mathrm{d}u}{\mathrm{d}T}\right)_v \quad [10]$$

The notation $(\,)_v$ means that the process is being considered while holding the volume constant.

Equation [10] is often used mathematically to define the specific heat capacity of a gas at constant volume.

From equation [10]

$$\mathrm{d}u = c_v\mathrm{d}T \quad [11]$$

A further very important point arises out of equation [3]. Remember how Joule's law states that the internal energy of a gas is a function of temperature only. An inspection of equation [3] will show that this expression does, in fact, give the change of internal energy as a function of temperature only since m is constant and c_v is assumed constant, being given an average value within the temperature range T_1 to T_2. The expression $U_2 - U_1 = mc_v(T_2 - T_1)$ will therefore give the change of internal energy during any process where the temperature changes from original temperature T_1 to final temperature T_2. It will be noted that, from this expression for the change of internal energy, at any absolute temperature T, it appears that the total internal energy of a gas, reckoned from the absolute zero of temperature, is given by $U = mc_vT$. This seems to indicate that, at the absolute zero of temperature, a gas possesses no internal energy, a point which is probably true. But the actual value of the internal energy, $U = mc_vT$, is probably not true, since this expression assumes that the gas remains as a gas all the way down to the absolute zero of temperature. This is actually not the case because a gas will liquefy and eventually solidify before absolute zero is reached, thus its gaseous properties will be lost. But in general, only the change of internal energy is required, so there is no need to enquire into its actual value for a gas at any absolute temperature T; the expression for the change of internal energy, $U_2 - U_1 = mc_v(T_2 - T_1)$, will give all that is required.

A further point about the internal energy of a gas concerns its graphical representation. It has been suggested that the change of internal energy is all that is generally required. Hence any graphical representation need only plot its change, not its absolute value. In order to plot values of the change of internal energy, a common origin must be chosen from which to reckon all changes. It is usual to choose the origin as 0 °C at which temperature the internal energy is suggested as being zero. This is really not true, but the overall change of internal energy from one temperature to another is not affected by the choice of an origin for the graph, an arbitrary choice, in any case. But with this choice of origin, all values of internal energy at temperatures below 0 °C will appear as negative.

Consider the equation for the change of internal energy

$$U_2 - U_1 = mc_v(T_2 = T_1)$$

If the origin is chosen as suggested, $U_1 = 0$ when $T_1 = 273.15$ K; substituting in the above equation gives

$$U = mc_v(T - 273.15)$$

where U (written U_2 above) = the value of the internal energy of the gas at temperature T (written T_2 above)

This equation allows determination of the internal energy of a gas, from which a graph may be plotted. It is also possible to produce a set of tables for the internal energy of a gas by using the same method.

Example 5.7 *2 kg of gas, occupying 0.7 m³, had an original temperature of 15 °C. It was then heated at constant volume until its temperature became 135 °C. Determine the heat transferred to the gas and its final pressure. Take* $c_v = 0.72$ *kJ/kg K and* $R = 0.29$ *kJ/kg K.*

SOLUTION

$$\begin{aligned}\text{Heat transferred at constant volume} &= mc_v(T_2 - T_1)\\ &= 2 \times 0.72 \times (135 - 15)\\ &= 2 \times 0.72 \times 120\\ &= \mathbf{172.8\ kJ}\end{aligned}$$

Now $P_1V_1 = mRT_1$ and $T_1 = (15 + 273)\,\text{K} = \mathbf{288\ K}$

$$\therefore \quad P_1 = \frac{mRT_1}{V_1} = \frac{2 \times 0.29 \times 288}{0.7} = \frac{167.04}{0.7} = \mathbf{238.6\ kN/m^2}$$

Since the volume remains constant, then

$$\frac{P_1}{T_1} = \frac{P_2}{T_2} \quad \therefore \quad P_2 = P_1 \frac{T_2}{T_1} \quad \text{and} \quad T_2 = (135 + 273)\ \text{K} = \mathbf{408\ K}$$

$$= 238.6 \times \frac{408}{288} = \mathbf{338.02\ kN/m^2}$$

5.8 The constant pressure heating of a gas

Let a mass of gas m be heated at constant pressure such that its temperature rises from T_1 to T_2 and its volume increases from V_1 to V_2, then

$$\begin{aligned}\text{Heat received by the gas} &= \text{mass} \times \text{specific heat capacity at constant pressure} \times \text{rise in temperature}\\ &= mc_p(T_2 - T_1) \qquad [1]\end{aligned}$$

Constant pressure heating is another case of a non-flow process carried out on a gas. Consider the non-flow energy equation applied to constant pressure heating

$$Q = \Delta U + W \qquad [2]$$

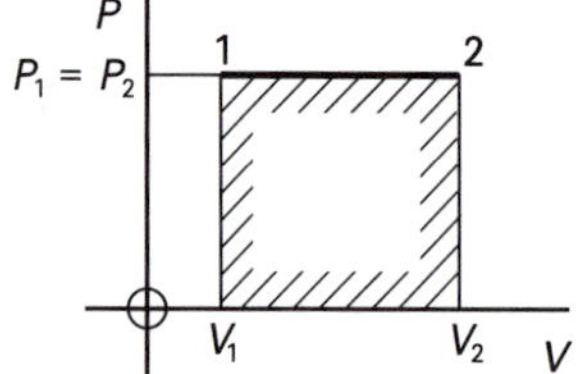

Fig. 5.12 Constant pressure process on *P–V* diagram

In this case external work is done by the gas. Figure 5.12 shows a graph of a constant pressure process plotted on a *P–V* diagram. This graph has a definite area beneath the constant pressure line, which gives the work done, $P(V_2 - V_1)$, where P is the constant pressure ($P = P_1 = P_2$). In this constant pressure case, equation [2] becomes

$$\begin{aligned}mc_p(T_2 - T_1) &= (U_2 - U_1) + P(V_2 - V_1) \qquad [3]\\ &= (U_2 + PV_2) - (U_1 + PV_1)\end{aligned}$$

or

$$mc_p(T_2 - T_1) = H_2 - H_1 \qquad [4]$$

i.e.

Heat transferred at constant pressure = Change of enthalpy

This was shown to be the case when dealing with the constant pressure formation of steam.

Equation [4] could be written

$$mc_p\Delta T = \Delta H \qquad [5]$$

Also, if unit mass of gas is considered, equation [4] becomes

$$c_p(T_2 - T_1) = h_2 - h_1 \qquad [6]$$

from which

$$c_p = \frac{h_2 - h_1}{T_2 - T_1} = \frac{\Delta h}{\Delta T} \qquad [7]$$

At any particular absolute temperature T, therefore, and as $\Delta T \to 0$, equation [7] can be written

$$c_p = \left(\frac{\mathrm{d}h}{\mathrm{d}T}\right)_P \qquad [8]$$

The notation $(\ )_P$ means that the process is being considered while holding the pressure constant.

From equation [8]

$$\mathrm{d}h = c_p\mathrm{d}T \qquad [9]$$

From equation [3]

$$U_2 - U_1 = mc_p(T_2 - T_1) - P(V_2 - V_1)$$

or

$$U_2 - U_1 = mc_p(T_2 - T_1) - mR(T_2 - T_1) \qquad [10]$$

since

$$PV = mRT$$

This again gives an expression for the change of internal energy of a gas in terms of temperature only, so this can also be used as a method for the determination of the change of internal energy during any process when a temperature change from T_1 to T_2 occurs. The expression determined during the constant volume analysis, $U_2 - U_1 = mc_v(T_2 - T_1)$, is of a simpler form and is usually used instead of equation [10].

It should be noted that enthalpy tables could be made up for a gas, using equation [4], in much the same way as indicated for internal energy tables discussed during the constant volume analysis.

If the new volume is required after a constant pressure process, this too may be obtained by using the characteristic equation of a perfect gas.

$$\frac{P_1V_1}{T_1} = \frac{P_2V_2}{T_2} \qquad [11]$$

and for this case,

$$P_1 = P_2 \qquad [12]$$

$$\therefore \quad \frac{V_1}{T_1} = \frac{V_2}{T_2} \qquad [13]$$

or

$$V_2 = V_1 \frac{T_2}{T_1} \qquad [14]$$

Example 5.8 *A gas whose pressure, volume and temperature are 275 kN/m², 0.09 m³ and 185 °C, respectively, has its state changed at constant pressure until its temperature becomes 15 °C. Determine the heat transferred from the gas and the work done on the gas during the process. Take* R = *0.29 kJ/kg K,* c_p = *1.005 kJ/kg K.*

SOLUTION

First determine the mass of gas used.

Now $P_1V_1 = mRT_1$ and $T_1 = (185 + 273)\text{ K} = \mathbf{458\ K}$

$$\therefore \quad m = \frac{P_1V_1}{RT_1} = \frac{275 \times 10^3 \times 0.09}{0.29 \times 10^3 \times 458} = \mathbf{0.186\ kg}$$

Heat transferred $= mc_p(T_2 - T_1)$ and $T_2 = (15 + 273)\text{ K} = \mathbf{288\ K}$

$$\begin{aligned}\therefore \quad \text{Heat transferred} &= 0.186 \times 1.005 \times (288 - 458)\\ &= 0.186 \times 1.005 \times (-170)\\ &= \mathbf{-31.78\ kJ}\end{aligned}$$

Notice the negative sign, indicating that the heat has been transferred from the gas.

Since the pressure remains constant, then

$$\frac{V_1}{T_1} = \frac{V_2}{T_2}$$

$$\therefore \quad V_2 = V_1\frac{T_2}{T_1} = 0.09 \times \frac{288}{458} = \mathbf{0.056\ 6\ m^3}$$

$$\begin{aligned}\text{Work done} &= P(V_2 - V_1)\\ &= 275 \times (0.056\ 6 - 0.009)\\ &= 275 \times (-0.0334)\\ &= \mathbf{-9.19\ kJ}\end{aligned}$$

5.9 *The difference of the specific heat capacities of a gas*

It has been shown that if a mass of gas m has its temperature changed from T_1 to T_2 then the change of internal energy can be determined by the expressions

$$U_2 - U_1 = mc_v(T_2 - T_1) \qquad [1]$$

and

$$U_2 - U_1 = mc_p(T_1 - T_1) - mR(T_2 - T_1) \qquad [2]$$

If the temperature change is the same for both expressions then it follows that equation [1] equals equation [2] because the change of internal energy is a function of temperature only, by Joule's law.

$$\therefore \quad mc_v(T_2 - T_1) = mc_p(T_2 - T_1) - mR(T_2 - T_1)$$

from which

$$c_v = c_p - R$$

since $m(T_2 - T_1)$ is common throughout.

$$\therefore \quad c_p - c_v = R \qquad [3]$$

$$= \frac{Pv}{T}\left(\text{since } \frac{Pv}{T} = R\right) \qquad [4]$$

5.10 *The polytropic process and a gas*

Section 1.20 discussed the general concept of the polytropic process. A gas is no exception to this concept; if a mass of gas is expanded or compressed, the general law of expansion or compression has the polytropic form

$$PV^n = C \qquad [1]$$

For two state points 1 and 2

$$P_1V_1^n = P_2V_2^n \qquad [2]$$

Furthermore

$$\text{Work done} = \frac{P_1V_1 - P_2V_2}{n-1} \qquad [3]$$

This was shown in section 1.21.

By the characteristic equation

$$PV = mRT \qquad [4]$$

Substituting [4] in [3]

$$\text{Work done} = \frac{mR(T_1 - T_2)}{n-1} \qquad [5]$$

Apply the non-flow energy equation

$$Q = \Delta U + W$$

$$= (U_2 - U_1) + W$$

$$= mc_v(T_2 - T_1) + \frac{P_1V_1 - P_2V_2}{n-1} \qquad [6]$$

$$= mc_v(T_2 - T_1) + \frac{mR(T_1 - T_2)}{n-1} \qquad [7]$$

5.11 The combination of the polytropic law $PV^n = C$ and the characteristic equation of a perfect gas

The law $PV^n = C$ will enable calculations to be made of the changes in pressure and volume which occur during a polytropic process. Combining this with the characteristic equation of a perfect gas will enable variations in temperature to be determined.

Consider a polytropic process in which the state of a gas changes from P_1, V_1, T_1 to P_2, V_2, T_2.

By the polytropic law

$$P_1 V_1^n = P_2 V_2^n \qquad [1]$$

By the characteristic equation

$$\frac{P_1 V_1}{T_1} = \frac{P_2 V_2}{T_2} \qquad [2]$$

From equation [2]

$$\frac{T_1}{T_2} = \frac{P_1 V_1}{P_2 V_2} \qquad [3]$$

From equation [1]

$$\frac{P_1}{P_2} = \left(\frac{V_2}{V_1}\right)^n \qquad [4]$$

Substituting equation [4] in equation [3]

$$\frac{T_1}{T_2} = \left(\frac{V_2}{V_1}\right)^n \frac{V_2}{V_1} = \left(\frac{V_2}{V_1}\right)^n \left(\frac{V_2}{V_1}\right)^{-1}$$

or

$$\frac{T_1}{T_2} = \left(\frac{V_2}{V_1}\right)^{(n-1)} \qquad [5]$$

Also, from equation [4]

$$\frac{V_2}{V_1} = \left(\frac{P_1}{P_2}\right)^{1/n} \quad \text{or} \quad \frac{V_1}{V_2} = \left(\frac{P_2}{P_1}\right)^{1/n} \qquad [6]$$

Substituting equation [6] in equation [3]

$$\frac{T_1}{T_2} = \frac{P_1}{P_2}\left(\frac{P_2}{P_1}\right)^{1/n} = \frac{P_1}{P_2}\left(\frac{P_1}{P_2}\right)^{-1/n}$$

or

$$\frac{T_1}{T_2} = \left(\frac{P_1}{P_2}\right)^{(n-1)/n} \qquad [7]$$

Combining equations [5] and [7]

$$\frac{T_1}{T_2} = \left(\frac{P_1}{P_2}\right)^{(n-1)/n} = \left(\frac{V_2}{V_1}\right)^{(n-1)} \qquad [8]$$

or in other words

$$\text{Ratio of absolute temperatures} = (\text{Ratio of pressures})^{(n-1)/n} = (\text{Inverse ratio of volumes})^{(n-1)}$$

This expression gives the relationship between pressure, volume and temperature when a gas state changes according to the law $PV^n = C$.

From equation [8], by raising each term to the power $n/(n-1)$ it follows that

$$\left(\frac{T_1}{T_2}\right)^{n/(n-1)} = \left(\frac{P_1}{P_2}\right)^{(n-1)/n \cdot n/(n-1)} = \left(\frac{V_2}{V_1}\right)^{(n-1) \cdot n/(n-1)}$$

$$\frac{P_1}{P_2} = \left(\frac{V_2}{V_1}\right)^n = \left(\frac{T_1}{T_2}\right)^{n/(n-1)} \qquad [9]$$

Also, by raising each term in equation [8] to the power $1/(n-1)$ it follows that

$$\left(\frac{T_1}{T_2}\right)^{1/(n-1)} = \left(\frac{P_1}{P_2}\right)^{(n-1)/n \cdot 1/(n-1)} = \left(\frac{V_2}{V_1}\right)^{(n-1) \cdot 1/(n-1)}$$

or

$$\frac{V_2}{V_1} = \left(\frac{T_1}{T_2}\right)^{1/(n-1)} = \left(\frac{P_1}{P_2}\right)^{1/n} \qquad [10]$$

Having developed these expressions, it might be useful to note that, initially, it may be difficult to know which one to use for the solution of a particular problem. There is no hard and fast rule, but notice that

$$\frac{P_1 V_1}{T_1} = \frac{P_2 V_2}{T_2}$$

requires five conditions of state to be known before solving for the sixth. If five conditions are not known, another expression may be more appropriate, such as

$$\frac{P_1}{P_2} = \left(\frac{V_2}{V_1}\right)^n \quad \text{or} \quad \frac{T_1}{T_2} = \left(\frac{P_1}{P_2}\right)^{(n-1)/n}$$

Example 5.9 *A gas whose original pressure and temperature were 300 kN/m² and 25 °C, respectively, is compressed according to the law* $PV^{1.4} = C$ *until its temperature becomes 180 °C. Determine the new pressure of the gas.*

SOLUTION

It has been shown that for a polytropic compression, the relationship between pressure and temperature is

$$\frac{T_1}{T_2} = \left(\frac{P_1}{P_2}\right)^{(n-1)/n}$$

From this

$$\frac{P_1}{P_2} = \left(\frac{T_1}{T_2}\right)^{n/(n-1)}$$

$$\therefore \quad P_2 = P_1\left(\frac{T_2}{T_1}\right)^{n/(n-1)}$$

Now

$$T_1 = (25 + 273)\ \text{K} = \mathbf{298\ K}$$

and

$$T_2 = (180 + 273)\ \text{K} = \mathbf{453\ K}$$

Hence

$$P_2 = 300 \times \left(\frac{453}{298}\right)^{1.4/0.4}$$

$$= 300 \times 1.52^{3.5}$$

$$= 300 \times 4.33$$

$$= \mathbf{1299\ kN/m^2} \quad \text{or} \quad \mathbf{1.299\ MN/m^2}$$

Example 5.10 *A gas whose original volume and temperature were 0.015 m³ and 285 °C, respectively, is expanded according to the law* $PV^{1.35} = C$ *until its volume is 0.09 m³. Determine the new temperature of the gas.*

SOLUTION

The relationship between volume and temperature during a polytropic expansion of a gas is

$$\frac{T_1}{T_2} = \left(\frac{V_2}{V_1}\right)^{(n-1)} \quad \text{and} \quad T_1 = (285 + 273)\ \text{K} = \mathbf{558\ K}$$

$$\therefore \quad T_2 = T_1\left(\frac{V_1}{V_2}\right)^{(n-1)} = 558 \times \left(\frac{0.015}{0.09}\right)^{(1.35-1)}$$

$$= \frac{558}{6^{0.35}}$$

$$= \frac{558}{1.87}$$

$$= \mathbf{298.4\ K}$$

$$t_2 = 298.4 - 273 = \mathbf{25.4\ °C}$$

Example 5.11 *0.675 kg of gas at 1.4 MN/m² and 280 °C is expanded to four times the original volume according to the law* $PV^{1.3} = C$. *Determine*
(a) the original and final volume of the gas
(b) the final pressure of the gas
(c) the final temperature of the gas
Take R = 0.287 *kJ/kg K.*

(a)
Now $P_1V_1 = mRT_1$ and $T_1 = (280 + 273)\text{ K} = \mathbf{553\ K}$

$$\therefore \quad V_1 = \frac{mRT_1}{P_1} = \frac{0.675 \times 0.287 \times 10^3 \times 553}{1.4 \times 10^6}$$

$$= \mathbf{0.076\,5\ m^3}$$

The original volume is 0.076 5 m³.
Since the gas is expanded to four times its original volume, then

$$V_2 = 4V_1 = 4 \times 0.076\,5 = \mathbf{0.306\ m^3}$$

The final volume is 0.306 m³.

(b)

$$P_1V_1^n = P_2V_2^n$$

$$\therefore \quad P_2 = P_1\left(\frac{V_1}{V_2}\right)^n = 1.4\left(\frac{1}{4}\right)^{1.3}$$

$$= \frac{1.4}{4^{1.3}}$$

$$= \frac{1.4}{6.06}$$

$$= \mathbf{0.231\ MN/m^2}$$
$$= \mathbf{231\ kN/m^2}$$

The final pressure is 231 kN/m².

(c)

$$\frac{P_1V_1}{T_1} = \frac{P_2V_2}{T_2}$$

$$\therefore \quad T_2 = \frac{P_2}{P_1}\frac{V_2}{V_1}T_1 = \frac{0.231}{1.4} \times 4 \times 553 = 365\text{ K}$$

$$t_2 = 365 - 273 = \mathbf{92\ °C}$$

The final temperature is 92 °C.

Example 5.12 *0.25 kg of air at a pressure of 140 kN/m^2 occupies 0.15 m^3 and from this condition it is compressed to 1.4 MN/m^2 according to the law* $PV^{1.25} = C$. *Determine*
(a) the change of internal energy of the air
(b) the work done on or by the air
(c) the heat received or rejected by the air
Take $c_p = 1.005$ *kJ/kg K*, $c_v = 0.718$ *kJ/kg K*

(a)

Now $c_p - c_v = R$

$$\therefore \quad R = 1.005 - 0.718 = \mathbf{0.287\ kJ/kg\ K}$$

Also $P_1 V_1 = mRT_1$

$$\therefore \quad T_1 = \frac{P_1 V_1}{mR} = \frac{140 \times 10^3 \times 0.15}{0.25 \times 0.287 \times 10^3} = \mathbf{292.7\ K}$$

Also $$\frac{T_1}{T_2} = \left(\frac{P_1}{P_2}\right)^{(n-1)/n}$$

$$\therefore \quad T_2 = T_1\left(\frac{P_2}{P_1}\right)^{(n-1)/n} = 292.7 \times \left(\frac{1.4 \times 10^6}{140 \times 10^3}\right)^{0.25/1.25}$$

$$= 292.7 \times 10^{1/5} = 292.7 \times \sqrt[5]{10}$$
$$= 292.7 \times 1.585$$
$$= \mathbf{463.9\ K}$$

Change of internal energy

$$\Delta U = U_2 - U_1 = mc_v(T_2 - T_1)$$
$$= 0.25 \times 0.718 \times (463.9 - 292.7)$$
$$= 0.25 \times 0.718 \times 171.2$$
$$= \mathbf{30.73\ kJ}$$

This is positive, so it is a gain of internal energy to the air.

(b)

$$\text{Work done, } W = \frac{mR(T_1 - T_2)}{n - 1} = \frac{0.25 \times 0.287 \times (292.7 - 463.9)}{1.25 - 1}$$

$$= \frac{0.25 \times 0.287 \times (-171.2)}{0.25}$$

$$= \mathbf{-49.1\ kJ}$$

This is negative, so the work is done on the air.

(c)

$$Q = \Delta U + W$$
$$\therefore \quad Q = 30.73 - 49.1 = \mathbf{-18.37\ kJ}$$

This is negative, so the heat is rejected by the air.

5.12 *The adiabatic process and a gas*

When dealing with the general case of a polytropic expansion or compression, it was stated that this process followed a law of the form $PV^n = C$. Now the adiabatic process can be a particular case of the polytropic process in which no heat is allowed to enter or leave during the progress of the process. From this it appears there should be a particular value of the index n which will satisfy this condition. An investigation is therefore necessary to see if this is the case.

Consider an adiabatic expansion or compression in which a change of state occurs from P_1, V_1, T_1 to P_2, V_2, T_2. Then

$$\text{Change of internal energy} = mc_v(T_2 - T_1) \qquad [1]$$

Also

$$\text{Work done during the process} = \frac{P_1V_1 - P_2V_2}{(\gamma - 1)} \qquad [2]$$

$$= \frac{mR(T_1 - T_2)}{(\gamma - 1)} \qquad [3]$$

where γ (gamma) is the particular index which will satisfy the case of an adiabatic process (sometimes the adiabatic index is written k).

From the polytropic law, if γ is the adiabatic index

$$P_1V_1^\gamma = P_2V_2^\gamma \qquad [4]$$

Also from the polytropic law

$$\frac{T_1}{T_2} = \left(\frac{P_1}{P_2}\right)^{(\gamma-1)/\gamma} = \left(\frac{V_2}{V_1}\right)^{(\gamma-1)} \qquad [5]$$

and by the characteristic equation

$$\frac{P_1V_1}{T_1} = \frac{P_2V_2}{T_2} \qquad [6]$$

Applying the non-flow energy equation

$$Q = \Delta U + W$$

For an adiabatic process $Q = 0$ (see section 1.27)

$$\therefore \quad 0 = \Delta U + W$$

or

$$W = -\Delta U \qquad [7]$$

Work is done at the expense of internal energy during an adiabatic expansion. Internal energy increases at the expense of work during an adiabatic compression.

Substituting equations [1] and [3] in equation [7]

$$\frac{mR(T_1 - T_2)}{(\gamma - 1)} = -mc_v(T_2 = T_1)$$

$$\therefore \quad \frac{mR(T_1 - T_2)}{(\gamma - 1)} = mc_v(T_1 - T_2)$$

from which

$$\frac{R}{(\gamma - 1)} = c_v \qquad [8]$$

since $m(T_1 - T_2)$ is a common term on both sides.

From this

$$\frac{R}{c_v} = (\gamma - 1) \qquad [9]$$

or

$$\gamma = \frac{R}{c_v} + 1 = \frac{R + c_v}{c_v} \qquad [10]$$

Now $R = c_p - c_v$; substituting in equation [8]

$$\gamma = \frac{c_p - c_v + c_v}{c_v}$$

or

$$\gamma = \frac{c_p}{c_v} \qquad [11]$$

From this, then, the law for an adiabatic expansion or compression of a gas is $PV^\gamma = C$, where $\gamma = c_p/c_v$, the ratio of the specific heat capacities at constant pressure and constant volume. The theoretical adiabatic process is sometimes said to be a **frictionless** adiabatic process. The reason for this is perhaps best understood by attempting to suggest a practical way of carrying out an adiabatic process. If a piece of apparatus for carrying out an expansion or compression could be constructed of a perfect heat insulating material then an adiabatic process would be quite possible. But no perfect heat insulator exists, so perhaps the nearest approach to an adiabatic process is to complete the process very rapidly, in which case there is very little time for heat exchange between the gas and its surroundings.

But, when such a process is carried out, it is found that with both the compression and the expansion, the final temperature is slightly higher than the calculated value. Now since the process is very rapid, the heat transfer required to increase the temperature above the adiabatic temperature could not have transferred from the outside. The answer to this is friction, turbulence and shock within the gas itself. Energy is required to overcome these effects and it will appear as a slightly increased temperature of the gas above its theoretical value. If these effects are neglected then the adiabatic process is said to be **frictionless**.

The average value of γ, the adiabatic index, for air is of the order of 1.4.

Example 5.13 *A gas expands adiabatically from a pressure and volume of 700 kN/m² and 0.015 m³, respectively, to a pressure of 140 kN/m². Determine the final volume and the work done by the gas. Determine, also, the change of internal energy in this case. Take,* $c_p = 1.046$ *kJ/kg K,* $c_v = 0.752$ *kJ/kg K.*

SOLUTION

Adiabatic index $= \gamma = c_p/c_v = 1.046/0.752 = \mathbf{1.39}$

For an adiabatic expansion

$$P_1V_1^\gamma = P_2V_2^\gamma$$

$$\therefore \quad \left(\frac{V_2}{V_1}\right)^\gamma = \frac{P_1}{P_2} \quad \text{or} \quad \frac{V_2}{V_1} = \left(\frac{P_1}{P_2}\right)^{1/\gamma}$$

from which

$$V_2 = V_1\left(\frac{P_1}{P_2}\right)^{1/\gamma} = 0.015 \times \left(\frac{700}{140}\right)^{1.39}$$

$$= 0.015 \times 5^{1/1.39} = 0.015 \times \sqrt[1.39]{5}$$

$$= 0.015 \times 3.18$$

$$= \mathbf{0.048\ m^3}$$

The final volume is 0.048 m³.

$$\text{Work done} = \frac{P_1V_1 - P_2V_2}{\gamma - 1}$$

$$= \frac{(700 \times 0.015) - (140 \times 0.048)}{1.39 - 1}$$

$$= \frac{10.5 - 6.72}{0.39} = \frac{3.78}{0.39}$$

$$= \mathbf{9.69\ kJ}$$

For an adiabatic process

$$W = -\Delta U$$

or

$$\Delta U = -W$$

$\therefore$ Change of internal energy $= \mathbf{-9.69\ kJ}$

This is a loss of internal energy from the gas.

5.13 The isothermal process and a gas

An isothermal process is defined as a process carried out such that the temperature remains constant throughout the process. This is evidently the same as a process carried out according to Boyle's law. The law for an isothermal expansion or compression of a gas is therefore

$$PV = C, \text{ a constant} \quad [1]$$

Thus, for a change of state from 1 to 2

$$P_1V_1 = P_2V_2 \quad [2]$$

$$T_1 = T_2 = T = \text{constant temperature} \quad [3]$$

Now the law $PV = C$ is that of a rectangular hyperbola.
In section 1.22 it was shown that

$$\text{Work done} = PV \ln \frac{V_2}{V_1} \quad [4]$$

This, therefore, is the expression which will give the work done during an isothermal process on a gas.

From the characteristic equation

$$PV = mRT \quad [5]$$

Substituting equation [5] in [4]

$$\text{Work done} = mRT \ln \frac{V_2}{V_1} \quad [6]$$

Applying the non-flow energy equation

$$Q = \Delta U + W.$$

For an isothermal process $T =$ constant, and by Joule's law, the internal energy of a gas is a function of temperature only, so if $T =$ constant, there is no change of internal energy.

Hence, for an isothermal process

$$\Delta U = 0$$

So the energy equation becomes

$$Q = W \quad [7]$$

$$= PV \ln \frac{V_2}{V_1} = mRT \ln \frac{V_2}{V_1} \quad [8]$$

It follows that, during an isothermal expansion, all the heat transferred is converted into external work. Conversely, during an isothermal compression, all the work done on the gas is rejected by the gas as heat transfer. This sometimes seems a little odd at first, but remember the temperature must remain constant throughout, so the internal energy before the process will be the same as the internal energy after the process. For an expansion, external work is performed by the gas. During an isothermal process the internal energy content of the gas must remain constant, so it appears that any heat transferred to the gas must immediately be dissipated in carrying out the external work. A similar analysis holds for the case of an isothermal compression. The energy input is in the form of work done on the gas and is immediately rejected as heat transfer.

Example 5.14 *A quantity of gas occupies a volume of 0.4 m^3 at a pressure of 100 kN/m^2 and a temperature of 20 °C. The gas is compressed isothermally to a pressure of 450 kN/m^2 then expanded adiabatically to its initial volume. For this quantity of gas determine*
(a) the heat transferred during the compression
(b) the change of internal energy during the expansion
(c) the mass of gas
Assume that, for the gas, $\gamma = 1.4$, $c_p = 1.0$ kJ/kg K.

(a)
For the isothermal compression

$$P_1V_1 = P_2V_2$$

$$\therefore \quad V_2 = V_1\frac{P_1}{P_2} = 0.4 \times \frac{100}{450} = \mathbf{0.089\ m^3}$$

Now $Q = \Delta U + W$ and for an isothermal process on a gas $\Delta U = 0$

$$\therefore \quad Q = W = PV \ln r = PV \ln \frac{P_1}{P_2}$$

$$= 100 \times 0.4 \times \ln \frac{100}{450}$$

$$= -100 \times 0.4 \times \ln \frac{450}{100}$$

$$= -40 \times \ln 4.5 = -40 \times 1.5$$

$$= \mathbf{-600\ kJ}$$

This is heat rejected.

(b)
For the adiabatic expansion

$$P_2V_2^\gamma = P_2V_3^\gamma$$

$$\therefore \quad P_3 = P_2\left(\frac{V_2}{V_3}\right)^\gamma = 450 \times \left(\frac{0.089}{0.4}\right)^{1.4}$$

$$= \frac{450}{4.5^{1.4}}$$

$$= \frac{450}{8.21}$$

$$= \mathbf{54.8\ kN/m^2}$$

Now $Q = \Delta U + W$ and for an adiabatic process $Q = 0$

$$\therefore \quad 0 = \Delta U + W$$

or

$$\Delta U = -W = \frac{-(P_2V_2 - P_3V_3)}{\gamma - 1}$$

$$= \frac{-(450 \times 0.089 - 54.8 \times 0.4)}{1.4 - 1}$$

$$= \frac{-(40.05 - 21.9)}{0.4}$$

$$= \frac{-18.15}{0.4}$$

$$= \mathbf{-45.4\ kJ}$$

This is a loss of internal energy.

(c)

$$c_p - c_v = R \quad \text{and} \quad c_p/c_v = \gamma, \quad \text{so} \quad c_v = c_p/\gamma$$

$$\therefore \quad (c_p - c_p/\gamma) = R = c_p\left(1 - \frac{1}{\gamma}\right)$$

$$\therefore \quad R = 1.0 \times \left(1 - \frac{1}{1.4}\right) = 1.0 \times (1 - 0.714) = \mathbf{0.286\ kJ/kg\ K}$$

$$P_1V_1 = mRT_1 \quad \text{and} \quad T_1 = (20 + 273)\ \text{K} = \mathbf{293\ K}$$

$$\therefore \quad m = \frac{P_1V_1}{RT_1} = \frac{100 \times 10^3 \times 0.4}{0.286 \times 10^3 \times 293} = 0.477\ kg$$

5.14 The non-flow energy equation and the polytropic law PV^n = C

Consider the expansion or compression of a gas according to the law $PV^n = C$ in which the state changes from P_1, V_1, T_1, to P_2, V_2, T_2.

It has been shown that the change of internal energy is

$$\Delta U = mc_v(T_2 - T_1) \qquad [1]$$

Also, the work done during the change is

$$W = \frac{mR(T_1 - T_2)}{n - 1} \qquad [2]$$

Substituting equations [1] and [2] into the non-flow energy equation

$$Q = \Delta U + W$$

then

$$Q = mc_v(T_2 - T_1) + \frac{mR(T_1 - T_2)}{n - 1} \qquad [3]$$

Now

$$c_p - c_v = R \qquad [4]$$

and

$$\frac{c_p}{c_v} = \gamma \qquad [5]$$

From equation [5] $c_p = \gamma c_v$; substituting into equation [4] gives

$$\gamma c_v - c_v = R \quad \text{or} \quad c_v(\gamma - 1) = R \qquad [6]$$

$$\therefore \quad c_v = \frac{R}{(\gamma - 1)} \qquad [7]$$

Substituting equation [7] into equation [3]

$$Q = m\frac{R}{(\gamma - 1)}(T_2 - T_1) + m\frac{R}{(n - 1)}(T_1 - T_2)$$

$$= m\frac{R}{(n - 1)}(T_1 - T_2) - m\frac{R}{(\gamma - 1)}(T_1 - T_2)$$

$$= \left[\frac{1}{(n - 1)} - \frac{1}{(\gamma - 1)}\right] mR(T_1 - T_2)$$

$$= \left[\frac{(\gamma - 1) - (n - 1)}{(n - 1)(\gamma - 1)}\right] mR(T_1 - T_2)$$

$$= \frac{(\gamma - n)}{(\gamma - 1)} \frac{mR(T_1 - T_2)}{(n - 1)} \qquad [8]$$

or

$$Q = \frac{\gamma - n}{\gamma - 1} \times \text{Polytropic work} \qquad [9]$$

From this equation it is possible to examine what happens to the heat received or rejected during an expansion or compression of a gas if the value of the index n is varied. For a compression the work done is negative. In this case

If $n > \gamma$ then $\dfrac{\gamma - n}{\gamma - 1}$ is negative, so Q is positive, i.e. heat is received.

If $n < \gamma$ then $\dfrac{\gamma - n}{\gamma - 1}$ is positive, so Q is negative, i.e. heat is rejected.

For an expansion the work done is positive. In this case

If $n > \gamma$ then $\dfrac{\gamma - n}{\gamma - 1}$ is negative, so Q is negative, i.e. heat is rejected.

If $n < \gamma$ then $\dfrac{\gamma - n}{\gamma - 1}$ is positive, so Q is positive, i.e. heat is received.

Note that

If $n = \gamma$ then $\dfrac{\gamma - n}{\gamma - 1} = 0$, so $Q = 0$, i.e. this is the adiabatic case.

If $n = 1$ then $\dfrac{\gamma - n}{\gamma - 1} = 1$, so $Q =$ work done, i.e. this is the isothermal case.

Note that this analysis has shown how to control the value of the index n. The control of the index n is obtained by the extent to which heat is allowed, or not allowed, to pass out of, or into, the gas during the compression or expansion.

Substituting equation [6] in equation [8]

$$Q = \frac{(\gamma - n)}{(\gamma - 1)} mc_v(\gamma - 1) \frac{(T_1 - T_2)}{(n - 1)}$$

$$= mc_v \frac{(\gamma - n)}{(\gamma - 1)} (T_1 - T_2) \qquad [10]$$

$$= mc_n(T_1 - T_2) \qquad [11]$$

where

$$c_n = c_v \frac{(\gamma - n)}{(n - 1)} \qquad [12]$$

c_n is called the **polytropic specific heat capacity**.

Example 5.15 *A gas expands according to the law* $PV^{1.3} = C$ *from a pressure of 1 MN/m*^{2} *and a volume 0.003 m*^{3} *to a pressure of 0.1 MN/m*^{2}*. Determine the heat received or rejected by the gas during this process. Determine the polytropic specific heat capacity. Take* $\gamma = 1.4$*,* $c_v = 0.718$ *kJ/kg K*

SOLUTION

Now $P_1 V_1^n = P_2 V_2^n$

$$\therefore \quad V_2 = V_1 \left(\frac{P_1}{P_2}\right)^{1/n} = 0.003 \times \left(\frac{1}{0.1}\right)^{1/1.3}$$

$$= 0.003 \times 10^{1/1.3}$$
$$= 0.003 \times 5.88$$
$$= \mathbf{0.017\,6\ m^3}$$

Heat received or rejected is

$$Q = \frac{(\gamma - n)}{(\gamma - 1)} \times \text{work done}$$

$$= \frac{(\gamma - n)}{(\gamma - 1)} \times \frac{(P_1 V_1 - P_2 V_2)}{n - 1}$$

$$= \frac{(1.4 - 1.3)}{(1.4 - 1)} \times \frac{(1 \times 0.003 - 0.1 \times 0.017\,6)}{1.3 - 1}$$

$$= \frac{0.1}{0.4} \times \frac{(0.003 - 0.001\,76)}{0.3}$$

$$= \frac{1}{4} \times \frac{0.001\,24}{0.3}$$

$$= \frac{0.001\,24}{1.2}$$

$$= 0.001\,03 \text{ MJ}$$
$$= \mathbf{1.03\ kJ}$$

This is positive, so heat is received by the gas.

$$c_n = c_v \frac{(\gamma - n)}{(n - 1)} = 0.718 \times \frac{(1.4 - 1.3)}{(1.3 - 1)}$$

$$= 0.718 \times \frac{0.1}{0.3}$$

$$= \mathbf{0.239\ kJ/kg\ K}$$

5.15 *Air–steam mixtures (see also Chapter 19)*

The problem of air–steam mixtures is solved by the use of **Dalton's law of partial pressures**. This states that the pressure exerted by a mixture of gases or a mixture of vapours or both is equal to the sum of the pressures of each gas or vapour taken separately if the quantity of gas or vapour in the mixture occupied alone the same volume as that of the mixture and at the same temperature. Thus each constituent behaves as if it occupied the volume alone and is independent of the presence of the other constituents. The pressure of each constituent taken alone is known as its **partial pressure**. Thus, if there is a mixture of air and steam, then

Pressure of mixture = Partial pressure of steam + Partial pressure of air

---

Example 5.16 *A container is filled with a mixture of air and wet steam at a temperature of 39 °C and a pressure of 100 kN/m² (1 bar). The temperature is then raised to 120.2 °C, the steam remaining wet. Determine*

(a) the initial partial pressures of the steam and air

(b) the final partial pressures of the steam and air

(c) the total pressures in the container after heating

(a)

From the steam tables, the pressure of wet steam at 39 °C is 7 kN/m².

Initial partial pressure of the steam = **7 kN/m²**

From this, by Dalton's law of partial pressures, initial partial pressure of the air is

$100 - 7 = \mathbf{93\ kN/m^2}$

(b)

At 120 °C, the pressure of wet steam is 200 kN/m².

Final partial pressure of the steam = **200 kN/m²**

For air, $P_1/T_1 = P_2/T_2$ for a constant volume process.

$$\therefore \quad P_2 = P_1 \frac{T_1}{T_2} = 93 \times \frac{393.2}{312} = \mathbf{117.2\ kN/m^2}$$

Final partial pressure of the air = **117.2 kN/m²**

(c)
Total pressure after heating is

$$200 + 117.2 = \mathbf{317.2\ kN/m^2}$$

Example 5.17 *A vacuum gauge on a condenser reads 660 mm Hg and the barometer height is 765 mm Hg. Steam enters the condenser with a dryness fraction of 0.8 and has a temperature of 41.5 °C. Determine the partial pressures of the air and steam in the condenser. If the steam is condensed at the rate of 1500 kg/h, determine the mass of air which will be associated with this steam. Take* R *for air as 0.29 kJ/kg K.*

SOLUTION

$$\begin{aligned}\text{Absolute pressure in condenser} &= 765 - 660\\ &= \mathbf{105\ mm\ Hg}\\ &= 0.01334 \times 105\\ &= \mathbf{14\ kN/m^2}\end{aligned}$$

At 41.5 °C, from steam tables

$$\text{Partial pressure of the steam} = \mathbf{8\ kN/m^2}$$

$$\therefore \quad \text{Partial pressure of the air} = 14 - 8 = \mathbf{6\ kN/m^2}$$

At 41.5 °C and with a dryness fraction 0.8, specific volume of the steam is

$$xv_g = 0.8 \times 18.1 = \mathbf{14.48\ m^3/kg}$$

The air associated with 1 kg of the steam will occupy this same volume.

$$\text{For air, } PV = mRT \quad \text{and} \quad T = (41.5 + 273)\,\text{K} = \mathbf{314.5\ K}$$

$$\therefore \quad m = \frac{PV}{RT} = \frac{6 \times 14.48}{0.29 \times 314.5} = \frac{86.88}{91.2}$$

$$= \mathbf{0.953\ kg/kg\ steam}$$

$$\therefore \quad \text{Mass of air/h} = 0.953 \times 1500 = \mathbf{1429.5\ kg}$$

Example 5.18 *A cylinder contains a mixture of air and wet steam at a pressure of 130 kN/m^2 and a temperature of 75.9 °C. The dryness fraction of the steam is 0.92. The air–steam mixture is then compressed to one-fifth of its original volume, the final temperature being 120.2 °C. Determine*
(a) the final pressure in the cylinder
(b) the final dryness fraction of the steam

(a)
At 75.9 °C

$$\text{Partial pressure of the wet steam} = \mathbf{40\ kN/m^2}$$

$$\therefore \quad \text{Partial pressure of the air} = 130 - 40 = \mathbf{90\ kN/m^2}$$

Specific volume of the wet steam at 75.9 °C and 0.92 dry is

$$0.92 \times 3.99 = \mathbf{3.67\ m^2/kg}$$

For air

$$\frac{P_1 V_1}{T_1} = \frac{P_2 V_2}{T_2}$$

$$\frac{P_1 V_1}{T_1} = \frac{P_2 V_2}{T_2} \quad \text{and} \quad T_1 = (75.9 + 273)\ \text{K} = \mathbf{348.9\ K}$$

$$T_2 = (120.2 + 273)\ \text{K} = \mathbf{393.2\ K}$$

$$\therefore \quad P_2 = P_1 \frac{V_1}{V_2}\frac{T_2}{T_1} = 90 \times 5 \times \frac{393.2}{348.9} = \mathbf{507.1\ kN/m^2}$$

The final pressure of the air is 507.1 kN/m^2.
For the steam at 120.2 °C

Pressure = 200 kN/m^2

$$\therefore \quad \text{Final pressure in the cylinder} = 507.1 + 200 = \mathbf{707.1\ kN/m^2}$$

(b)
Final specific volume of steam in the cylinder is

$$\frac{3.67}{5} = \mathbf{0.734\ m^3/kg}$$

At 200 kN/m^2

$$v_g = 0.885\ \text{m}^3/\text{kg}$$

$$\therefore \quad \text{Final dryness fraction of the steam} = \frac{0.734}{0.885} = \mathbf{0.83}$$

5.16 General examples

Example 5.19 *A gas at a pressure of 1.4 MN/m^2 and temperature of 360 °C is expanded adiabatically to a pressure of 100 kN/m^2. The gas is then heated at constant volume until it again attains 360 °C, when its pressure is found to be 220 kN/m^2, and finally it is compressed isothermally until the original pressure of 1.4 MN/m^2 is attained. Sketch the* P–V *diagram for these processes and, if the gas has a mass of 0.23 kg, determine*
(a) the value of the adiabatic index γ
(b) the change in internal energy during the adiabatic expansion
Take c_p for the gas as 1.005 kJ/kg K.

SOLUTION

Figure 5.13 is a sketch of the P–V diagram.

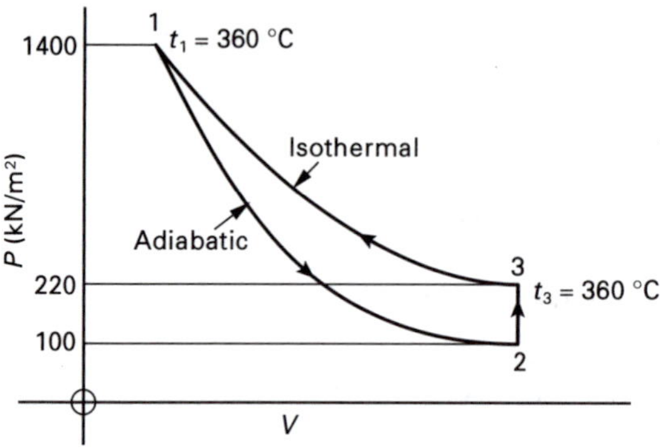

Fig. 5.13 Diagram for Example 5.19

(a)

For the isothermal process, $P_1V_1 = P_3V_3$

$$\therefore \quad \frac{P_1}{P_3} = \frac{V_3}{V_1} = \frac{1400}{220} = 6.36$$

For the adiabatic process, $P_1V_1^\gamma = P_2V_2^\gamma$

$$\therefore \quad \frac{P_1}{P_2} = \left(\frac{V_2}{V_1}\right)^\gamma \quad \text{and} \quad \frac{V_2}{V_1} = \frac{V_3}{V_1} \quad \text{since} \quad V_3 = V_2$$

$$\therefore \quad \frac{1400}{100} = 6.36^\gamma \quad \text{or} \quad 14 = 6.36^\gamma$$

From which

$$\gamma = \mathbf{1.426}$$

(b)

Now $c_p/c_v = \gamma$

$$\therefore \quad c_v = \frac{c_p}{\gamma} = \frac{1.005}{1.426} = \mathbf{0.705\ kJ/kg\ K}$$

For the constant volume process, $P_3/T_3 = P_2/T_2$

$$T_3 = T_1 = (360 + 273)\,\text{K}$$
$$= \mathbf{633\ K}$$

$$\therefore \quad T_2 = \frac{P_2}{P_3}T_3 = \frac{100}{220} \times 633 = 287.7\ \text{K}$$

$$t_2 = 287.7 - 273 = \mathbf{14.7\ °C}$$

Change of internal energy during the adiabatic change is

$$\begin{aligned} U_2 - U_1 &= mc_v(T_2 - T_1) \\ &= 0.23 \times 0.705 \times (287.7 - 633) \\ &= -(0.23 \times 0.705 \times 345.3) \\ &= \mathbf{-55.99\ kJ} \end{aligned}$$

This is a loss of internal energy.

Example 5.20 *An oxygen cylinder has a capacity of 300 litres and contains oxygen at a pressure of 3.1 MN/m² and temperature 18 °C. The stop valve is opened and some gas is used. If the pressure and temperature of the oxygen left in the cylinder fall to 1.7 MN/m² and 15 °C, respectively, determine the mass of oxygen used.*

If after the stop valve is closed the oxygen remaining in the cylinder gradually attains its initial temperature of 18 °C, determine the amount of heat transferred through the cylinder wall from the atmosphere. The density of oxygen at 0 °C and 0.101 325 MN/m² may be taken as 1.429 kg/m³ and γ for oxygen as 1.4.

SOLUTION

Consider the density condition given

$$PV = mRT$$

$$\therefore \quad R = \frac{PV}{mT} = \frac{0.101\,325 \times 10^3 \times 1}{1.429 \times 273} = \mathbf{0.26\ kJ/kg\ K}$$

For the initial conditions in the cylinder

$$P_1V_1 = m_1RT_1$$

$$\therefore \quad m_1 = \frac{P_1V_1}{RT_1} = \frac{3.1 \times 10^6 \times 300 \times 10^{-3}}{0.26 \times 10^3 \times 291} = \frac{3.1 \times 300}{0.26 \times 291} = \mathbf{12.3\ kg}$$

After some of the gas is used

$$m_2 = \frac{P_2V_2}{RT_2} = \frac{1.7 \times 10^6 \times 300 \times 10^{-3}}{0.26 \times 10^3 \times 288} = \frac{1.7 \times 300}{0.26 \times 288} = \mathbf{6.8\ kg}$$

The mass of oxygen remaining in cylinder is 6.8 kg.

$$\therefore \quad \text{Mass of oxygen used} = 12.3 - 6.8 = \mathbf{5.5\ kg}$$

For a non-flow process

$$Q = \Delta U + W$$

and since the volume of the oxygen remains constant, no external work is done.

$$\therefore \quad W = 0 \quad \text{hence} \quad Q = \Delta U$$

This means that the heat transferred through the cylinder wall is equal to the gain of internal energy of the oxygen, $m_2c_v(T_1 - T_2)$. Now

$$c_v = \frac{R}{(\gamma - 1)} = \frac{0.26}{0.4} = \mathbf{0.65\ kJ/kg\ K}$$

$$\begin{aligned} \therefore \quad \text{Heat transferred} &= 6.8 \times 0.65 \times (291 - 288) \\ &= 6.8 \times 0.65 \times 3 \\ &= \mathbf{13.26\ kJ} \end{aligned}$$

Example 5.21 *0.1 m^3 of gas is compressed from a pressure of 120 kN/m^2 and temperature 25 °C to a pressure of 1.2 MN/m^2 according to the law* $PV^{1.2}$ *= constant. Determine*
(a) the work transferred during the compression
(b) the change in internal energy
(c) the heat transferred during the compression
Assume c_v *= 0.72 kJ/kg K and* R *= 0.285 kJ/kg K.*

(a)

$$P_1 V_1^n = P_2 V_2^n$$

$$\therefore \quad \frac{P_1}{P_2} = \left(\frac{V_2}{V_1}\right)^n \quad \text{or} \quad V_2 = V_1 \left(\frac{P_1}{P_2}\right)^{1/n}$$

$$= 0.1 \times \left(\frac{120}{1\,200}\right)^{1/1.2}$$

$$= \frac{0.1}{10^{1/1.2}}$$

$$= \frac{0.1}{6.8}$$

$$= \mathbf{0.014\,7\ m^3}$$

$$\text{Work done} = W = \frac{P_1 V_1 - P_2 V_2}{n-1} = \frac{(120 \times 0.1 - 1200 \times 0.014\,7)}{0.2}$$

$$= \frac{(12 - 17.64)}{0.2}$$

$$= \frac{(-5.64)}{0.2}$$

$$= \mathbf{-28.2\ kJ}$$

(b)

$$\frac{P_1 V_1}{T_1} = \frac{P_2 V_2}{T_2} \qquad \therefore \quad T_2 = \frac{P_2 V_2 T_1}{P_1 V_1}$$

$$= \frac{1200}{120} \times \frac{0.014\,7}{0.1} \times 298$$

$$= \mathbf{438\ K}$$

$$t_2 = 438 - 273 = \mathbf{165\ °C}$$

Now $PV = mRT$, and considering the initial conditions

$$m = \frac{P_1 V_1}{RT_1} = \frac{120 \times 0.1}{0.285 \times 298} = \mathbf{0.141\ kg}$$

Change of internal energy is

$$mc_v(T_2 - T_1) = 0.141 \times 0.72 \times (438 - 298)$$
$$= 0.141 \times 0.72 \times 140$$
$$= \mathbf{14.2\ kJ}$$

(c)

$$Q = \Delta U + W$$
$$= 14.2 - 28.2$$
$$= \mathbf{-14.0\ kJ}$$

This is heat transferred from the gas.

Example 5.22 *An air receiver has a capacity of 0.85 m^3 and contains air at a temperature of 15 °C and a pressure of 275 kN/m^2. An additional mass of 1.7 kg is pumped into the receiver. It is then left until the temperature becomes 15 °C once again. Determine*

(a) the new pressure of the air in the receiver

(b) the specific enthalpy of the air at 15 °C if it is assumed that the specific enthalpy of the air is zero at 0 °C.

Take $c_p = 1.005$ *kJ/kg,* $c_v = 0.715$ *kJ/kg K.*

(a)

If m_1 = original mass of air, then

$$m_1 = \frac{P_1 V_1}{R T_1} \quad \text{and} \quad R = c_p - c_v = 1.005 - 0.715 = \mathbf{0.29\ kJ/kg\ K}$$

$$\therefore \quad m_1 = \frac{275 \times 0.85}{0.29 \times 288} = \mathbf{2.8\ kg}$$

After 1.7 kg of air are pumped in, the final mass of air is

$$m_2 = 2.8 + 1.7 = \mathbf{4.5\ kg}$$

Now $PV = mRT$, so

$$P = \frac{mRT}{V}$$

$$\therefore \quad \frac{P_2}{P_1} = \frac{m_2 R T_2 / V_2}{m_1 R T_1 / V_1} = \frac{m_2}{m_1}$$

since R, T and V are the same in both cases.

$$\therefore \quad P_2 = P_1 \frac{m_2}{m_1} = 275 \times \frac{4.5}{2.8} = \mathbf{442\ kN/m^2}$$

(b)

For 1 kg of air, the change of enthalpy is

$$h_2 - h_1 = c_p(T_2 - T_1)$$

If 0 °C is chosen as the zero of enthalpy, then

$$h = c_p(T - 273) = 1.005 \times (288 - 273)$$
$$= 1.005 \times 15$$
$$= \mathbf{15.075\ kJ/kg}$$

Example 5.23 *A gas has a density of 1.875 kg/m³ at a pressure of 1 bar and with a temperature of 15 °C. A mass of 0.9 kg of the gas requires a heat transfer of 175 kJ to raise its temperature from 15 °C to 250 °C while the pressure of the gas remains constant. Determine*
(a) the characteristic gas constant of the gas
(b) the specific heat capacity of the gas at constant pressure
(c) the specific heat capacity of the gas at constant volume
(d) the change of internal energy
(e) the work transfer

(a)
For a gas, $PV = mRT$

$$\therefore \quad R = \frac{PV}{mT} = \frac{100 \times 1}{1.875 \times 288} = \mathbf{0.185\ kJ/kg\ K}$$

(1 bar = 100 kN/m²)

(b)
For constant pressure heating

$$\text{Heat transfer} = mc_p(t_2 - t_1)$$
$$\therefore \quad 175 = 0.9 \times c_p(250 - 15)$$

from which

$$c_p = \frac{175}{0.9 \times 235} = \mathbf{0.828\ kJ/kg\ K}$$

(c)
Now $c_p - c_v = R$

$$\therefore \quad c_v = c_p - R = 0.828 - 0.185 = \mathbf{0.643\ kJ/kg\ K}$$

(d)
Change of internal energy is

$$mc_v(t_2 - t_1) = 0.9 \times 0.643 \times (250 - 15)$$
$$= 0.9 \times 0.643 \times 235$$
$$= \mathbf{136\ kJ}$$

(e)
Now $Q = \triangle U + W$

$$W = Q - \triangle U = 175 - 136 = \mathbf{39\ kJ}$$

Example 5.24 *A quantity of air has a volume of 0.15 m³ at a temperature of 120 °C and a pressure of 1.2 MN/m². The air expands to a pressure of 200 kN/m² according to the law* $PV^{1.32} = C$. *Determine*
(a) the work transfer
(b) the change of internal energy
(c) the heat transfer
Take c_p = *1.006 kJ/kg K,* c_v = *0.717 kJ/kg K.*

(a)

$$P_1V_1^{1.32} = P_2V_2^{1.32}$$

$$\therefore \quad V_2 = V_1\left(\frac{P_1}{P_2}\right)^{1/1.32}$$

$$= 0.15 \times \left(\frac{1200}{200}\right)^{1/1.32}$$

$$= 0.15 \times 6^{1/1.32}$$
$$= 0.15 \times 3.89$$
$$= \mathbf{0.584\ m^3}$$

Now

$$W = \frac{P_1V_1 - P_2V_2}{n-1}$$

$$= \frac{(1200 \times 0.15) - (200 \times 0.584)}{(1.32 - 1)}$$

$$= \frac{(180 - 116.8)}{0.32}$$

$$= \frac{63.2}{0.32}$$

$$= \mathbf{197.5\ kJ}$$

(b)

$$P_1V_1 = mRT_1$$

$$\therefore \quad m = \frac{P_1V_1}{RT_1} \quad \text{and} \quad R = c_p - c_v = 1.006 - 0.717 = \mathbf{0.289\ kJ/kg\ K}$$

$$\text{also} \quad T_1 = (120 + 273)\ \text{K} = \mathbf{393\ K}$$

Hence

$$m = \frac{1200 \times 0.15}{0.289 \times 393} = \mathbf{1.58\ kg}$$

also

$$\frac{T_1}{T_2} = \left(\frac{P_1}{P_2}\right)^{(n-1)/n}$$

from which

$$T_2 = T_1\left(\frac{P_2}{P_1}\right)^{(n-1)/n} = 393 \times \left(\frac{200}{1200}\right)^{(1.32-1)/1.32}$$

$$= 393 \times \left(\frac{1}{6}\right)^{0.32/1.32}$$

$$= \frac{393}{6^{1/4.125}} = \frac{393}{1.544}$$

$$= \mathbf{254.5\ K}$$

Now, for a gas

$$\Delta U = mc_v\Delta T \quad \text{(see section 5.7)}$$

or

$$\begin{aligned} U_2 - U_1 &= mc_v(T_2 - T_1) \\ &= 1.58 \times 0.717 \times (254.5 - 393) \\ &= 1.58 \times 0.717 \times (-138.5) \\ &= \mathbf{-156.9\ kJ} \end{aligned}$$

Note also that

$$c_v = \frac{R}{\gamma - 1} \quad \text{(see section 5.12)}$$

hence

$$U_2 - U_1 = m\frac{R}{\gamma - 1}(T_2 - T_1) = \frac{P_2V_2 - P_1V_1}{\gamma - 1}$$

since $PV = mRT$ (see section 5.4)

Now

$$\gamma = \frac{c_p}{c_v} = \frac{1.006}{0.717} = \mathbf{1.403}$$

$$\therefore \quad U_2 - U_1 = \frac{(200 \times 0.584) - (1200 \times 0.15)}{(1.403 - 1)} = \frac{-63.2}{0.403}$$

$$= \mathbf{-156.9\ kJ}$$

(c)

$$\begin{aligned} Q &= \Delta U + W \\ &= -156.9 + 197.5 \\ &= \mathbf{40.6\ kJ} \end{aligned}$$

Note also that

$$Q = mc_n(T_1 - T_2) = mc_v\frac{(\gamma - n)}{(n - 1)}(T_1 - T_2) \quad \text{(see section 5.14)}$$

$$= 1.58 \times 0.717 \times \frac{(1.403 - 1.32)}{(1.32 - 1)}(393 - 254.5)$$

$$= 1.58 \times 0.717 \times \frac{0.083}{0.32} \times 138.5$$

$$= \mathbf{40.6\ kJ}$$

Example 5.25 *A pressure vessel contains a gas at an initial pressure of 3.5 MN/m*^{2} *and at a temperature of 60 °C. It is connected through a valve to a vertical cylinder in which there is a piston. Initially there is no gas under the piston. The valve is opened, gas enters the vertical cylinder, and work is done in lifting the piston. The valve is closed and the pressure and temperature of the remaining gas in the cylinder are 1.7 MN/m*^{2} *and 25 °C, respectively. Determine the temperature of the gas in the vertical cylinder if the process is assumed to be adiabatic. Take* $\gamma = 1.4$.

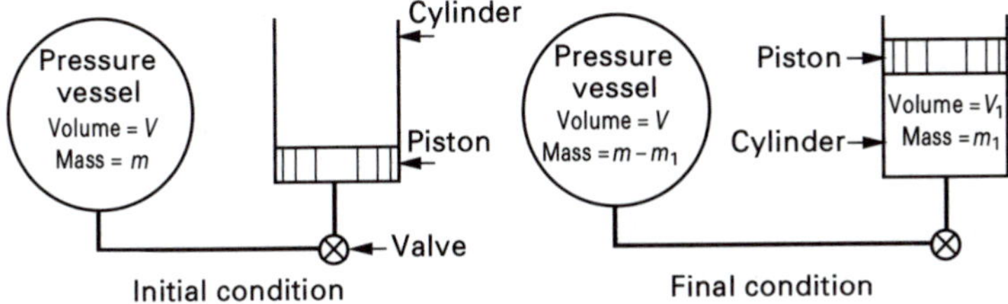

Fig. 5.14 Diagram for Example 5.25

SOLUTION

Firstly, draw a diagram (Fig. 5.14).

Let

Volume of pressure vessel = V
Initial mass of gas in pressure vessel = m
Initial temperature of gas = T
Initial pressure of gas = P

Then, for initial condition in pressure vessel

$$PV = mRT$$

from which

$$R = \frac{PV}{mT} = \frac{3.5 \times 10^6 \times V}{m \times 333} \qquad [1]$$

For final condition in pressure vessel, let

Final pressure = P_2
Final temperature = T_2

then

$$P_2V = (m - m_1)RT_2$$

where m_1 = mass of gas bled from vessel into cylinder

From this

$$R = \frac{P_2V}{(m - m_1)T_2} = \frac{1.7 \times 10^6 \times V}{(m - m_1) \times 298} \qquad [2]$$

Equating equations [1] and [2]

$$\frac{3.5 \times 10^6 \times V}{m \times 333} = \frac{1.7 \times 10^6 \times V}{(m - m_1) \times 298}$$

$$\therefore \quad \frac{3.5}{333m} = \frac{1.7}{298(m - m_1)}$$

From which

$$3.5 \times 298(m - m_1) = 1.7 \times 333m$$
$$1043m - 1043m_1 = 566.1m$$
$$(1043 - 566.1)m = 1043m_1$$

$$476.9m = 1043m_1$$

$$m = \frac{1043}{476.9}m_1$$

$$= 2.19m_1 \qquad [3]$$

The mass of gas in the system remains constant.
Hence, using the non-flow energy equation

$$Q = \Delta U + W$$

$Q = 0$ since the process is adiabatic.

$$\therefore \quad 0 = \Delta U + W$$

or

$$W = -\Delta U = -(U_1 - U)$$
$$W = U - U_1 \qquad [4]$$

Let

Final pressure in cylinder $= P_1$
Final temperature in cylinder $= T_1$
Final volume in cylinder $= V_1$

Then

$$W = P_1 V_1 = m_1 R T_1 = \text{work required to lift piston} \qquad [5]$$

Note that this assumes the pressure of the gas remains constant in the vertical cylinder as the piston is lifted.
Now

$$U = mc_v T \qquad [6]$$

and

$$U_1 = (m - m_1)\, c_v T_2 + m_1 c_v T_1 \qquad [7]$$

So from equations [4], [5], [6] and [7]

$$m_1 R T_1 = mc_v T - [(m - m_1)\, c_v T_2 + m_1 c_v T_1] \qquad [8]$$

But $R = c_p - c_v$; substituting into equation [8]

$$m_1(c_p - c_v)T_1 = c_v [mT - (m - m_1)T_2 - m_1 T_1]$$
$$\therefore \quad m_1(\gamma - 1)T_1 = mT - (m - m_1)T_2 - m_1 T_1 \qquad [9]$$

Substituting equation [3] in equation [9] and $\gamma = 1.4$

$$0.4m_1 T_1 = 2.19m_1 \times 333 - (2.19m_1 - m_1) \times 298 - m_1 T_1$$
$$0.4T_1 = 729.3 - 652.6 + 298 - T_1$$
$$1.4T_1 = 374.7$$
$$T_1 = 374.7/1.4$$
$$= 267.6 \text{ K}$$
$$t_1 = 267.6 - 273$$
$$= \mathbf{-5.4\ °C}$$

Example 5.26 *A pressure vessel is connected, via a valve, to a gas main in which a gas is maintained at a constant pressure and temperature of 1.4 MN/m² and 85 °C, respectively. The pressure vessel is initially evacuated. The valve is opened and a mass of 2.7 kg of gas passes into the pressure vessel. The valve is closed and the pressure and temperature of the gas in the pressure vessel are then 700 kN/m² and 60 °C, respectively. Determine the heat transfer to or from the gas in the vessel. Determine the volume of the pressure vessel and the volume of the gas before transfer.*

For the gas, take $c_p = 0.88$ *kJ/kg K,* $c_v = 0.67$ *kJ/kg K. Neglect the velocity of the gas in the main.*

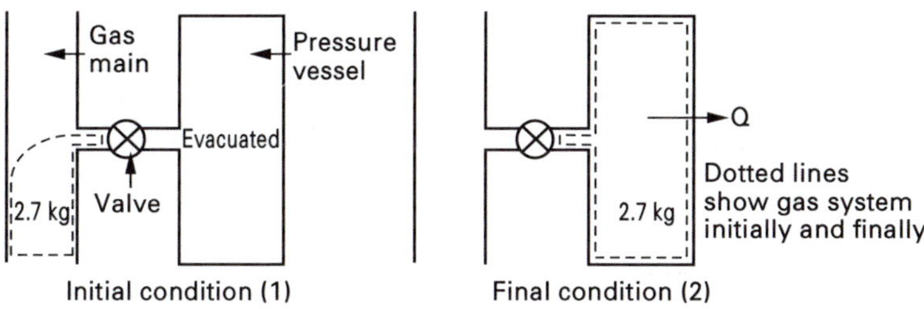

Fig. 5.15 Diagram for Example 5.26

SOLUTION

Firstly, draw a diagram (Fig. 5.15). This problem can be analysed by the use of the steady-flow energy equation.

$$u_1 + P_1 v_1 + \frac{C_1^2}{2} + Q = u_2 P_2 v_2 + \frac{C_2^2}{2} + W \qquad [1]$$

This is neglecting the potential energy terms.

There is no external work done, hence $W = 0$. Also there is no change in kinetic energy, hence the terms $C^2/2$ can be neglected.

There is no energy required for displacement in the pressure vessel since it is initially evacuated. Hence the term $P_2 v_2 = 0$.

Thus, for unit mass of gas, equation [1] becomes

$$u_1 + P_1 v_1 + Q = u_2 \qquad [2]$$

$$\therefore \quad Q = (u_2 - u_1) - P_1 v_1$$
$$= c_v (T_2 - T_1) = P_1 v_1 \qquad [3]$$

For unit mass of gas

$$P_1 v_1 = R T_1 = (c_p - c_v) T_1 \qquad [4]$$

Substituting equation [4] into equation [3]

$$Q = c_v (T_2 - T_1) - (c_p - c_v) T_1$$
$$= 0.67 \times (333 - 358) - 358 \times (0.88 - 0.67)$$
$$= 0.67 \times (-25) - 358 \times 0.21$$
$$= -16.75 - 75.18$$
$$= \mathbf{-91.93\ kJ/kg}$$

This is heat transferred from the vessel. So for 2.7 kg of gas

Heat transferred $= 2.7 \times 91.93 = \mathbf{-248.2\ kJ}$

From equation [4]

$$v_1 = (c_p - c_v)\frac{T_1}{P_1} = \frac{0.21 \times 358}{1400} = \mathbf{0.053\,7\ m^3/kg}$$

$$\therefore \quad V_1 = 2.7 \times 0.053\,7 = \mathbf{0.145\ m^3}$$

The volume of gas before transfer is 0.145 m^3.

$$\text{Now } \frac{P_1 V_1}{T_1} = \frac{P_2 V_2}{T_2}$$

$$\therefore \quad V_2 = \frac{P_1}{P_2}\frac{T_2}{T_1}V_1 = \frac{1400}{700} \times \frac{333}{358} \times 0.145$$

$$= \mathbf{0.27\ m^3}$$

The volume of the pressure vessel is 0.27 m^3.

Questions

1. A quantity of gas has an initial pressure of 140 kN/m^2 and volume 0.14 m^3. It is then compressed to a pressure of 700 kN/m^2 while the temperature remains constant. Determine the final volume of the gas.

   [0.028 m^3]

2. A quantity of gas has an initial volume of 0.06 m^3 and a temperature of 15 °C. It is expanded to a volume of 0.12 m^3 while the pressure remains constant. Determine the final temperature of the gas.

   [303 °C]

3. A mass of gas has an initial pressure of 1 bar and a temperature of 20 °C. The temperature of the gas is now increased to 550 °C while the volume remains constant. Determine the final pressure of the gas.

   [2.81 bar]

4. A mass of air has an initial pressure of 1.3 MN/m^2, volume 0.014 m^3 and temperature 135 °C. It is expanded until its final pressure is 275 kN/m^2 and its volume becomes 0.056 m^3. Determine
   (a) the mass of air
   (b) the final temperature
   Take $R = 0.287$ kJ/kg K.

   [(a) 0.155 kg; (b) 72 °C]

5. A quantity of gas has an initial pressure and volume of 0.1 MN/m^2 and 0.1 m^3, respectively. It is compressed to a final pressure of 1.4 MN/m^2 according to the law $PV^{1.26} = \text{constant}$. Determine the final volume of the gas.

   [0.0123 m^3 (= 12.35 litres)]

6. A quantity of gas has an initial volume and temperature of 1.2 litres and 150 °C, respectively. It is expanded to a volume of 3.6 litres according to the law $PV^{1.4} = \text{constant}$. Determine the final volume of the gas.

   [0 °C]

7. A mass of gas has an initial pressure and temperature of 0.11 MN/m^2 and 15 °C, respectively. It is compressed according to the law $PV^{1.3} = \text{constant}$ until the temperature becomes 90 °C. Determine the final pressure of the gas.

   [0.299 MN/m^2]

8. 0.23 kg of air has an initial pressure of 1.7 MN/m^2 and a temperature of 200 °C. It is expanded to a pressure of 0.34 MN/m^2 according to the law $PV^{1.35} = \text{constant}$. Determine the work transferred during the expansion. Take $R = 0.29$ kJ/kg K.

[30.72 kJ]

9. 0.1 kg of gas is heated by means of an electric heater for a period of 10 min, during which time the pressure of the gas remains constant. The temperature of the gas is increased from 16 °C to 78 °C. The power used by the heater is 20 watts. Assuming no losses, determine
   (a) the specific heat capacity of the gas at constant pressure
   (b) the specific heat capacity of the gas at constant volume
   (c) the characteristic gas constant
   (d) the density of the gas at a temperature of 16 °C and with a pressure of 0.12 MN/m^2

   For the gas, take $\gamma = 1.38$.

[(a) 1.935 kJ/kg K; (b) 1.402 kJ/kg K; (c) 0.533 kJ/kg K; (d) 0.78 kg/m^3]

10. An engine has a swept volume of 15 litres and a volume ratio of compression of 14:1. The air in the engine at the beginning of compression has a temperature and pressure of 30 °C and 95 kN/m^2, respectively. The air is compressed according to the law $PV^{1.34} = C$. At the end of the compression the air is heated at constant volume through a pressure ratio of 1.6:1. Determine
    (a) the temperature and pressure of the air at the end of the compression
    (b) the temperature and pressure of the air at the end of the constant volume process
    (c) the heat transfer required to carry out the constant volume process

    For the air, take $c_p = 1.005$ kJ/kg K, $R = 0.24$ kJ/kg K.

[(a) 469.4 °C, 3262.3 kN/m^2; (b) 915 °C, 5219.7 kN/m^2; (c) 7.16 kJ]

11. One kilogram of gas at an initial pressure of 0.11 MN/m^2 and a temperature of 15 °C. It is compressed isothermally until the volume becomes 0.1 m^3. Determine
    (a) the final pressure
    (b) the final temperature
    (c) the heat transfer

    If the compression had been adiabatic, determine
    (d) the final pressure
    (e) the final temperature
    (f) the work transfer

    For the gas, take $c_p = 0.92$ kJ/kg K, $c_v = 0.66$ kJ/kg K.

[(a) 0.748 MN/m^2; (b) 15 °C; (c) −143.8 kJ; (d) 1.591 MN/m^2; (e) 339.9 °C; (f) −214.4 kJ]

12. A gas has an initial pressure, volume and temperature of 140 kN/m^2, 0.012 m^3 and 100 °C, respectively. The gas is compressed to a final pressure of 2.8 MN/m^2 and volume of 0.001 2 m^2. Determine
    (a) the index of compression if the compression is assumed to follow the law $PV^n = C$
    (b) the final temperature of the gas
    (c) the work transfer
    (d) the change of internal energy of the gas

    For the gas, take $R = 0.287$ kJ/kg K, $c_v = 0.717$ kJ/kg K.

[(a) $n = 1.3$; (b) 471 °C; (c) −5.6 kJ; (d) 4.2 kJ]

13. A gas has a density of 0.09 kg/m^3 at a temperature of 0 °C and a pressure of 1.013 bar. Determine
    (a) the characteristic gas constant
    (b) the specific volume of the gas at a temperature of 70 °C and a pressure of 2.07 bar

    If a volume of 5.6 m^3 of the gas at an initial pressure of 1.02 bar and temperature 0 °C is heated at constant pressure to a final temperature of 50 °C, determine
    (c) the heat transfer
    (d) the change of internal energy of the gas
    (e) the work transfer

    For the gas, take $c_v = 10.08$ kJ/kg K.

    [(a) 4.13 kJ/kg K; (b) 6.84 m^3/kg; (c) 360.2 kJ; (d) 255.5 kJ; (e) 104.7 kJ]

14. A gas has an initial pressure, volume and temperature of 95 kN/m^2, 14 litres and 100 °C, respectively. The gas is compressed according to the law $PV^{1.3} = C$ through a volume ratio of 14:1. Determine
    (a) the work transfer
    (b) the change of internal energy
    (c) the heat transfer

    For the gas, take $R = 0.29$ kJ/kg K, $c_v = 0.72$ kJ/kg K.

    [(a) −5.35 kJ; (b) 4.015 kJ; (c) −1.335 kJ]

15. A gas at an initial pressure of 690 kN/m^2 and temperature of 185 °C has a mass of 0.45 kg. The gas is expanded adiabatically to a final pressure of 138 kN/m^2 with a fall of temperature of 165 °C. The work transfer during the expansion is 53 kJ. For the gas, determine
    (a) the specific heat capacity at constant volume
    (b) the adiabatic index
    (c) the specific heat capacity at constant pressure

    [(a) 0.714 kJ/kg K; (b) 1.385; (c) 0.989 kJ/kg K]

16. An air receiver has a volume of 4.25 m^3 and contains air at a pressure of 650 kN/m^2 and a temperature of 120 °C. The air is cooled to a temperature of 40 °C. Determine
    (a) the final pressure of the air
    (b) the change of internal energy of the air

    For the air, take $R = 0.29$ kJ/kg K, $c_v = 0.717$ kJ/kg K.

    [(a) 517.7 kN/m^2; (b) −1390 kJ]

17. An internal combustion engine has a cylinder bore of 165 mm and a piston stroke of 300 mm. The volume ratio of compression is 8:1. At the commencement of the working stroke the pressure of the gas in the clearance volume is 4.5 MN/m^2 and the temperature is 400 °C. The gas expands at constant pressure while the piston moves a distance of 45 mm down the cylinder. Determine
    (a) the temperature of the gas at the end of the piston movement
    (b) the work transfer during the piston movement
    (c) the heat transfer during the piston movement

    For the gas, take $R = 0.29$ kJ/kg K, $c_p = 1.005$ kJ/kg K.

    [(a) 1102 °C; (b) 4.32 kJ; (c) 14.82 kJ]

18. An engine has a volume ratio of compression of 12:1. At the beginning of compression the gas in the cylinder has a pressure, volume and temperature of 110 kN/m^2, 0.28 m^3 and 80 °C, respectively. The gas is compressed according to the law $PV^{1.28} = C$. Determine
    (a) the pressure of the gas after compression

(b) the temperature of the gas after compression
(c) the work transfer during compression
(d) the heat transfer during compression

For the gas, take $c_p = 1.0$ kJ/kg K, $c_v = 0.71$ kJ/kg K.

[(a) 2.647 MN/m^2; (b) 433 °C; (c) −107.5 kJ; (d) −34.9 kJ]

19. A quantity of gas has an initial pressure, volume and temperature of 1.4 MN/m^2, 0.14 m^3 and 300 °C, respectively. The gas is expanded adiabatically to a pressure of 280 kN/m^2. Determine
(a) the mass of gas
(b) the temperature of the gas after the expansion
(c) the work transfer
(d) the change of internal energy of the gas

For the gas, take $c_p = 1.04$ kJ/kg K, $c_v = 0.74$ kJ/kg K.

[(a) 1.14 kg; (b) 87.4 °C; (c) 179.5 kJ; (d) −179.5 kJ]

20. 0.45 kg of gas is expanded adiabatically until the pressure is halved and the temperature of the gas falls from 220 °C to 130 °C. During the expansion there is a work transfer from 27 kJ. Determine
(a) the adiabatic index of the gas
(b) the characteristic gas constant

[(a) 1.408; (b) 0.272 kJ/kg K]

21. A quantity of gas has a mass of 0.2 kg and an initial temperature of 15 °C. It is compressed adiabatically through a volume ratio of 4:1. The final temperature after compression is 237 °C. The work transfer during compression is 33 kJ. For the gas, determine
(a) the specific heat capacity at constant volume
(b) the adiabatic index
(c) the specific heat capacity at constant pressure
(d) the characteristic gas constant

[(a) 0.743 kJ/kg K; (b) 1.412; (c) 1.049 kJ/kg K; (d) 0.306 kJ/kg K]

22. The cylinder of an engine has a stroke of 300 mm and a bore of 250 mm. The volume ratio of compression is 14:1. Air in the cylinder at the beginning of compression has a pressure of 96 kN/m^2 and a temperature of 93 °C. The air is compressed for the full stroke according to the law $PV^{1.3} = C$. Determine
(a) the mass of air
(b) the work transfer
(c) the heat transfer

For the air, take $\gamma = 1.4$, $c_p = 1.006$ kJ/kg K.

[(a) 0.0145 kg; (b) −6.15 kJ; (c) −1.54 kJ]

23. A quantity of air has a pressure, volume and temperature of 104 kN/m^2, 30 litres and 38 °C, respectively. The temperature of the air is raised (i) by heating while the volume remains constant until the pressure becomes 208 kN/m^2; and (ii) by adiabatic compression to a volume of 6 litres.

For both cases, determine
(a) the final temperature
(b) the work transfer
(c) the change of internal energy
(d) the heat transfer

For the air, take $R = 0.29$ kJ/kg K, $\gamma - 1.4$.

[(i) (a) 349 °C; (b) 0; (c) 7.8 kJ/kg; (d) 7.8 kJ
(ii) (a) 319 °C; (b) −7.05 kJ; (c) 7.05 kJ; (d) 0]

24. A mass of gas has a pressure, volume and temperature of 100 kN/m^2, 0.56 m^3 and 20 °C, respectively. It is compressed to a volume of 0.15 m^3 according to the law $PV^{1.36} = C$. The gas is then cooled at constant pressure until the volume becomes 0.1 m^3. Determine
    (a) the final pressure, the final temperature and the work transfer for the compression
    (b) the final temperature and the heat transfer for the constant pressure process

    For the gas, take $c_p = 1.006$ kJ/kg K, $R = 0.287$ kJ/kg K.

    [(a) 600 kN/m^2, 197.6 °C, −94.4 kJ; (b) 40.7 °C, −105.1 kJ]

25. An air main is connected to a cylinder through a valve. A piston slides in the cylinder. The air in the main is maintained at a constant pressure and temperature of 1 MN/m^2 and 40 °C, respectively. The initial pressure and volume of air in the cylinder are 140 kN/m^2 and 3 litres, respectively. The valve is opened, 0.11 kg of air enters the cylinder then the valve is closed. As a result of this mass transfer, the pressure in the cylinder becomes 700 kN/m^2 and the volume becomes 15 litres. Assuming the process to be adiabatic, determine the work done on the piston.

    For air, take $c_p = 1.006$ kJ/kg K, $c_v = 0.717$ kJ/kg K.

    [7.8 kJ]

26. A cylinder contains a mixture of air and wet steam at a pressure of 130 kN/m^2 and a temperature of 69.1 °C. The temperature is then raised to 151.8 °C. Determine the final pressure in the cylinder if the steam remains wet.

    [624 kN/m^2]

27. A condenser deals with 900 kg of steam per hour with a dryness fraction of 0.9 and temperature 45.8 °C. The air associated with this steam in the condenser is 225 kg/h. The barometric height is 766 mm Hg. Determine the vacuum gauge reading in mm Hg.

    For air, take $R = 0.29$ kJ/kg K.

    [678 mm Hg]

Chapter 6

Thermodynamic reversibility

6.1 General introduction

If a substance passes through the stages of a process in such a manner that, after the process, the substance can be taken back through all the stages in reverse order until it finally reaches its original state, then the process is said to be reversible. After carrying out a reversible process, there would be no evidence anywhere that the process had ever taken place. No such process exists in practice.

Within the substance during any process it is probable that eddies will be set up. Also, due to the viscosity of the substance, however slight, there will be some internal friction. It is also very likely that there will be some small irregularity with regard to the distribution of temperature throughout the substance. The degree to which these occur must have some bearing on the final state of the substance after the process. From here, however, it is unreasonable to assume that these various internal phenomena can be repeated in an exactly reversed sequence in order that the reversed process will return the substance to its original state. For these reasons alone, no actual process can be considered as being truly reversible.

A point to raise here, however, is that the effect of these internal phenomena is not likely to be great, and from a theoretical standpoint, it is possible to neglect them. This being so, it is necessary to consider what else will affect the concept of reversibility. As an example, consider the expansion of a gas.

During any expansion of a gas there will be heat transfer into or out of the gas, with the exception of the adiabatic case in which, by definition, there is no heat transfer (see section 1.27). Now the second law of thermodynamics states that heat transfer will only occur down a temperature gradient as a natural occurrence.

During the expansion, assume that there is some heat transfer from the gas to the surroundings. If this is the case, then by the second law of thermodynamics, the surroundings are at a lower temperature than that of the gas. What if an attempt is now made to reverse the process?

This now means compressing the gas, which is easy enough. However, it is not possible to return the energy lost by heat transfer to the surroundings because the gas is at a higher temperature than the surroundings. The original pressure and temperature could not be attained, however, because of the energy loss by heat transfer to the surroundings during the original expansion.

This energy loss cannot be returned because of the limitations imposed by the second law of thermodynamics. Notice that there would also be some heat transfer loss during the reversed process of a compression because, again, the temperature of the gas would be above that of the surroundings.

Another point to consider is the effect of pressure imbalance. If a substance at a high pressure expands into surroundings, which are at a lower pressure, then the reversed process of the low-pressure surroundings returning the substance to its original high pressure is impossible without the aid of external energy. Thus if pressure imbalance occurs, the process cannot be reversible.

As an extreme example of pressure imbalance, consider the free expansion of air in the Joule's law experiment, discussed in section 5.5.

In this case, the compressed air expanded into a vacuum, so no work was transferred. The reverse of this process would be impossible without the aid of external energy. Hence, the original free expansion process is irreversible. It follows that practical thermodynamic processes are irreversible.

Now irreversibility evidently involves loss, so it appears that reversibility is bound up with efficiency. A truly reversible process involves no loss and is therefore the most efficient thermodynamic process possible.

No external energy is required to return a substance to its original state in a truly reversible process. It is important, therefore, to investigate whether any processes may be considered to be theoretically reversible.

6.2 The adiabatic process

No heat is transferred during an adiabatic process. Thus the effect of the second law of thermodynamics between the substance and its surroundings is eliminated.

If the effects of pressure imbalance, internal friction, non-uniform temperature distribution, etc., are neglected, it follows that the adiabatic process is theoretically reversible.

For an adiabatic non-flow process it is shown that $W = -\Delta U$. Thus, during an adiabatic expansion, external work is done which equals the decrease in internal energy. If this same amount of work is done on the working substance during the reversed process of adiabatic compression, the work will appear as an increase of internal energy of the substance, and this increase will just equal the loss which occurred during the expansion. Thus the substance will be returned exactly to its original state.

The adiabatic process is therefore theoretically reversible.

6.3 The isothermal process

An isothermal process is carried out at constant temperature. For an isothermal, non-flow process, it was shown that the necessary energy exchange was that $Q = W$. This means that, during an isothermal expansion, the working substance must receive an amount of heat equal to the external work done. It follows that if, during the reversed process of compression, an amount of heat equal to the work done on the substance is rejected then, neglecting the effects of pressure imbalance, internal friction, non-uniform temperature distribution, etc., the isothermal process is theoretically reversible.

With regard to the transfer of heat into or out of the substance, it must be remembered that an isothermal process is carried out at constant temperature. Assuming that the external surroundings are at this temperature, heat transfer is equally possible in either direction, namely, into or out of the substance. Actually, by the second law of thermodynamics, a temperature difference is required in order to promote a natural heat transfer. Therefore, during an isothermal expansion, it could be considered that the surroundings are at a slightly elevated temperature above the substance, so the necessary condition that the substance shall receive heat would be met.

Similarly, it could be considered that the substance has a slightly elevated temperature above the surroundings during an isothermal compression. In this case the necessary condition that the substance should reject heat would be met. Since the temperature difference in each case would be small, the rate of heat transfer would be very slow, so from a practical point of view, the isothermal process is very slow. Actually it is all but impossible as a practical process. However, theoretically it exists and, neglecting the effect of internal friction, etc., it is theoretically reversible.

6.4 *The polytropic, constant volume and constant pressure processes*

In all these cases heat is received or rejected by the substance during the progress of the process, and the temperature changes continuously throughout the process. If the temperature of the surroundings remains constant then, by the second law of thermodynamics, heat transfer between the substance and surroundings is unidirectional, being a function of whether the substance is at a higher or lower temperature than the surroundings.

Also, apart from the constant pressure process, in which the pressure of the substance could be made the same as the pressure of the surroundings, in these processes where pressure interaction occurs between the substance and the surroundings, there is pressure imbalance. In these cases, in which there is pressure and temperature difference between the substance and surroundings, the processes are irreversible.

These processes could be considered reversible if the temperature and pressure of the surroundings could be made to vary in the same way as the temperature and pressure of the substance. In this way, in a similar manner to the isothermal case, mutual heat transfer in either direction would be possible, and there would be no pressure imbalance. The processes could then be considered reversible. In the main, these conditions are impossible to achieve, although the constant pressure process can be made to approach the reversible condition in a contraflow heat exchanger.

Figure 6.1 illustrates the principle of the contraflow heat exchanger. A hot fluid enters at temperature t_1 and is progressively cooled at constant pressure until it leaves the heat exchanger at lower temperature t_2. The cool coolant is made to enter at the same end of the heat exchanger at which the cool fluid leaves and passes through the heat exchanger in an opposite direction from the fluid being cooled. The hot coolant thus leaves at the same end at which the hot fluid is entering. The opposite flows account for the name *contraflow*.

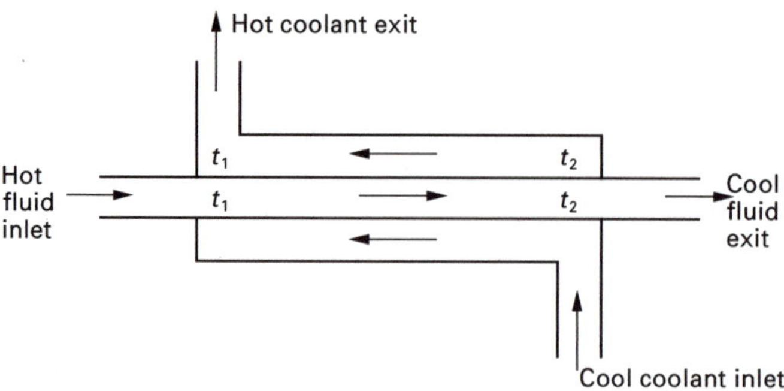

Fig. 6.1 Contraflow heat exchanger

If the fluid and coolant are considered as being at the same temperature at each section of the heat exchanger, then reversible heat transfer is possible; the process could be considered reversible. Thus, if the fluid and coolant flows were reversed, and neglecting losses, the cool fluid at temperature t_2 would be returned to its original state as hot fluid at temperature t_1, and the hot coolant at temperature t_1 would be returned to its original state as cool coolant at temperature t_2.

This ideal of keeping the temperature of the fluid and surroundings the same cannot be achieved in practice. But it illustrates the necessary conditions in order to convert an irreversible process into a reversible process.

6.5 *The non-flow energy equation and reversibility*

The non-flow energy equation connecting the initial and final states of a substance has the form

$$Q = (U_2 - U_1) + W \qquad [1]$$

This equation can be used for a reversible process or an irreversible process, as long as the initial and final states are in equilibrium, i.e. there is pressure and temperature equilibrium throughout the substance and no random internal energies exist due to such things as turbulence. The true amount of external work done must also be known exactly.

If, however, the differential form of this equation is used

$$dQ = dU + dW \qquad [2]$$

then this only applies to a reversible process. To explain this, firstly consider the work done.

For calculation, the work done has been determined by integrating the expression

$$dW = P\,dV \qquad [3]$$

To use this expression it must be assumed that the pressure P is always resisted by an opposing pressure equal to P. Consider a cylinder–piston arrangement. If the substance pressure is higher than the opposing pressure, produced by an external load upon which work is being done, then there will be less work done than $\int P\,dV$ would indicate. In the extreme case when the expansion is free, such as when a substance expands into a vacuum as in Joule's internal energy experiment, no external work is done, even though a change in pressure and volume occurs. In any event, considering the reverse process in which external work is required to compress the substance, a lower external pressure cannot be made to compress a substance already at a higher pressure. If, however, the internal pressure and the external pressure are the same, the compression is theoretically reversible and $\int P\,dV$ gives the external work done exactly.

With regard to the change of internal energy, given by $\int dU$, this is only valid if the substance has passed through a series of equilibrium states, thus there have been no random internal energies such as turbulence. This is only possible in a reversible case.

The heat transfer during a non-flow process is dependent upon the change of internal energy and the external work done, and to use the expression

$$dQ = dU + dW$$

dU and dW must be reversible, so it follows that dQ must also be reversible. This implies that the external surrounding temperature must vary exactly with the substance temperature, as explained for the isothermal case.

And since the heat transfer during a non-flow process is usually calculated using

$$Q = \int dU + \int dW \qquad [4]$$
$$= \int dU + \int P\,dV \qquad [5]$$

this calculated heat transfer is for a reversible case only.

To indicate this, the heat transfer is written Q_{rev}. Thus

$$Q_{rev} = \int dU + \int P\,dV \qquad [6]$$

If, for reversible heat transfer, the external temperature must vary exactly with the substance temperature, then as explained in the isothermal case, the process would be very slow. In fact, to operate at all, it must be a theoretical process of infinite slowness. All practical processes take a finite time; this alone makes them irreversible.

From this discussion, it follows that calculations of heat transfer, change of internal energy and external work done are close approximations only; the calculated results apply only to ideal reversible cases.

6.6 Carnot's principle

It has already been suggested that a reversible process is an efficient process, since it involves no loss. Reversibility, as it applies to the thermodynamic engine, was discussed by a Frenchman, Sadi Carnot, in a paper entitled 'Reflections on the motive power of heat' which was published in 1824. Carnot conceived of an engine working on thermodynamically reversible processes, and from this concept deduced what has since been called **Carnot's principle**:

> No engine can be more efficient than a reversible engine working between the same limits of temperature.

Section 1.30 showed that the principle of the thermodynamic engine is that it receives heat at some high temperature from a heat source. The engine then converts some of this heat into work and rejects the remainder into a sink.

Consider a thermodynamically reversible engine R working between the temperature limits of source T_1 and sink T_2. In some period of time let this engine receive Q units of heat from the source at temperature T_1. It will convert W_R units of this heat into work and then reject $(Q - W_R)$ units of heat into the sink at lower temperature T_2. This is shown in Fig. 6.2(a).

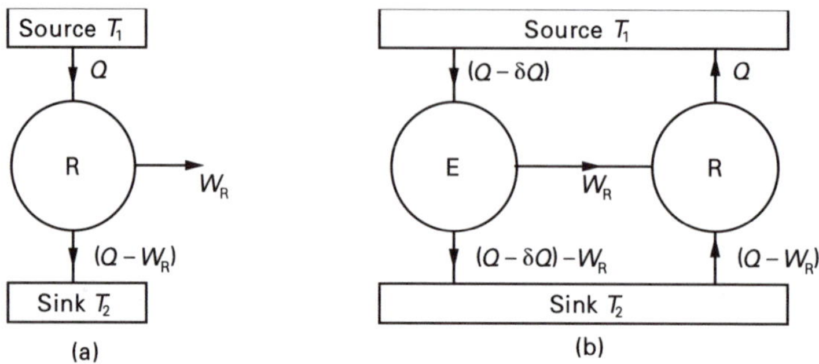

Fig. 6.2 Energy changes for (a) a reversible engine R and (b) when R is driven by a hypothetically more efficient engine E

Now assume that some other engine E can be found which is more efficient than the reversible engine R. Since it is more efficient, engine E will require less heat to perform the same amount of work W_R. Let this engine E drive engine R reversed and let them both work between the same source and sink. This is shown in Fig. 6.2(b).

Engine E, being more efficient, will require $(Q - \delta Q)$ units of heat supplied from the source at temperature T_1. It will convert W_R of this into work and it will reject $(Q - \delta Q) - W_R$ units of heat into the sink at temperature T_2.

Now the work W_R will drive engine R reversed, which now becomes a heat pump. Thus it will take up $(Q - W_R)$ units of heat from the sink at lower temperature T_2. It will convert W_R units of work into heat then reject $(Q - W_R) + W_R = Q$ units of heat into the source at higher temperature T_1.

Investigation of this system will show that, during the time period considered, there has been a gain of heat to the source of $Q - (Q - \delta Q) = \delta Q$ units of heat. Also the sink has lost $(Q - W) - [(Q - \delta Q) - W_R] = \delta Q$ units of heat. This means that the source at higher temperature T_1 is receiving heat from the sink at lower temperature T_2.

Now this arrangement is self-acting and has apparently managed to make more heat transfer up the gradient of temperature than has moved down. This would mean that eventually all the heat would be transferred to the source at temperature T_1 while the sink at lower temperature T_2 would have its energy content reduced to zero! This is contrary to the second law of thermodynamics, so the system is impossible.

If, however, the engine E has the same efficiency as the reversible engine R, then Fig. 6.3 shows that the thermodynamic system balances, in which case both the source and sink gain as much heat as they lose. This means that the energy level of both the source and the sink would remain constant. Once started, this system would continue to run indefinitely, so it would have perpetual motion! (See sections 3.3 and 3.4.)

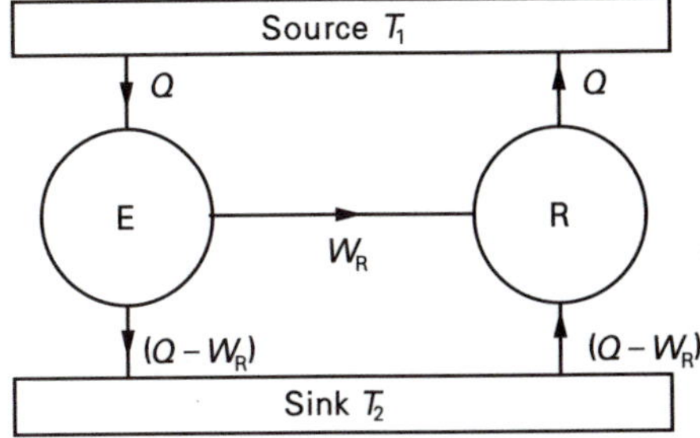

Fig. 6.3 Energy changes when E drives R with E assumed to have the same efficiency as R, i.e. $\delta Q = 0$

No such system exists because no engine can be made to have the same efficiency as a reversible engine. In any case, the above system has assumed no loss, also impossible. However, the important criterion which has been established is that the thermodynamically reversible engine has the maximum efficiency possible.

An alternative analysis showing that the thermodynamically reversible engine has the maximum efficiency possible is as follows. Again, consider a reversible engine R and assume that there exists an engine E which has a higher efficiency than reversible engine R.

Consider Fig. 6.4(a). In this case let engine E receive heat Q from the source at temperature T_1. It will produce work W_E which is greater than W_R because it is more efficient. It will reject heat $(Q - W_E)$ to the sink at temperature T_2.

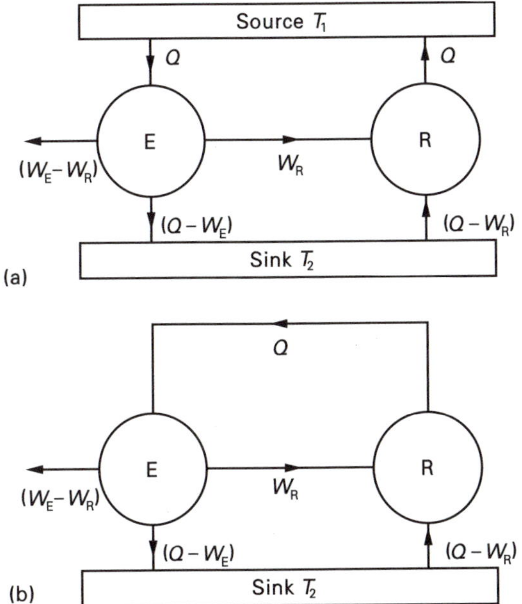

Fig. 6.4 Alternative analysis of Carnot's principle: the source in (a) can be omitted in (b) but this contradicts the second law of thermodynamics

Engine E is arranged to drive engine R reversed. Engine R will require work $W_R (W_R < W_E)$ to drive it and will take up heat $(Q - W_R)$ from the sink at temperature T_2 while it rejects heat Q into the source at temperature T_1.

Now since the reversible engine R rejects heat Q, which is exactly the requirement of engine E, theoretically it would be possible to dispense with the heat source at temperature T_1. This is shown in Fig. 6.4(b). Consider this new arrangement. Here, apparently, is a self-acting machine which is producing net work output $(W_E - W_R)$, while exchanging heat with only a single reservoir, the sink at temperature T_2. This is contrary to the second law of thermodynamics as stated in the combined concept of Kelvin–Planck (see section 3.4).

This analysis once again shows that the thermodynamically reversible engine has the maximum efficiency possible within the given limits of temperature.

For the given temperature limits of source T_1 and sink T_2, it is now necessary to determine this maximum efficiency. To establish it, all processes must be thermodynamically reversible and the arrangement of the process must be such that they are capable of cyclic repetition in order that, having been started, an engine may continue to run by the process of repetition of the cycle (see section 1.13).

Carnot conceived of a cycle made up of thermodynamically reversible processes. By determining the thermal efficiency of this cycle it is possible to establish the maximum possible efficiency between the temperature limits of the cycle. The Carnot cycle will be analysed as it applies to steam plant in Chapter 10, and as it applies to gases in Chapter 15.

Entropy

7.1 General introduction

During many processes it is necessary to investigate the degree to which heat is transferred during the process. The heat transferred during a process can affect any work transfer which may occur during the process and also the end state after the process. Furthermore, it has been shown that the theoretical amount of heat transferred, determined by calculation, is transferred reversibly.

Suppose a graph could be developed such that the area underneath a process plotted on the graph gave the amount of heat transferred reversibly during the progress of the process. Such a graph might perform a useful function. The idea is analogous to the area of the pressure–volume graph, which gives the work done during a reversible process.

The problem now is to decide upon the axes of the graph. Let one axis be absolute temperature T and the other some new function s as shown in Fig. 7.1(a). Absolute temperature is chosen as one axis because it has a very close relationship with the energy level of a substance, notably internal energy and enthalpy.

Also, when a substance is at the absolute zero of temperature, it is assumed that its internal energy content is also zero. And remember that heat is defined as energy transfer which will occur as the result of a temperature difference.

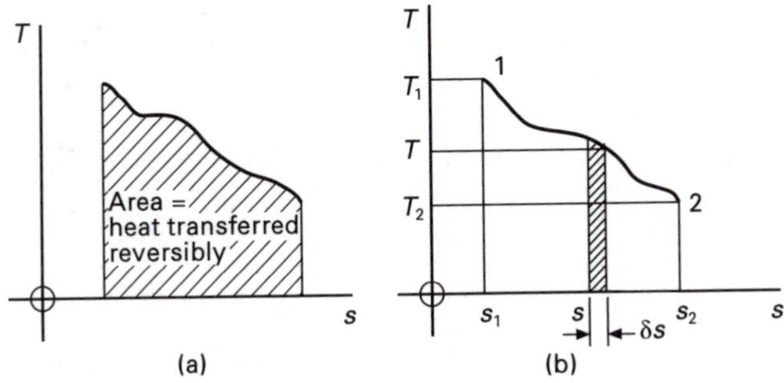

Fig. 7.1 Entropy: (a) absolute temperature plotted against another function so that area = heat transferred reversibly; (b) a small change in this new function during a process

Now consider Fig. 7.1(b) which shows a process plotted on the graph where a change has occurred from state 1 to state 2. Consider some point on this graph where the coordinates are T and s. Let the state change from this point such that there is a small change in s, δs, then

$$\begin{aligned}\text{Heat transferred reversibly} &= \text{Area swept out by this small change}\\ &= T\delta s \text{ (very nearly)}\end{aligned}$$

From this, heat transferred reversibly from 1 to 2 is equal to

$$\text{Total area under graph from 1 to 2} = \sum_{s=s_1}^{s=s_2} T\,\delta s$$

In the limit as $\delta s \to 0$, heat transferred reversibly from 1 to 2 is

$$\int_{s_1}^{s_2} T\,\mathrm{d}s = Q_{\mathrm{rev}} \qquad [1]$$

Differentiating equation [1]

$$\mathrm{d}Q_{\mathrm{rev}} = T\,\mathrm{d}s$$

or

$$\mathrm{d}s = \frac{\mathrm{d}Q_{\mathrm{rev}}}{T} \qquad [2]$$

This equation gives the relationship which must exist between s, T and Q_{rev} in order that the area of the graph shall be heat transferred reversibly.

Now it has already been shown that it is possible to calculate the amount of heat transferred reversibly during a non-flow process. Thus, by using equation [2], changes in s can be determined. It is this function s which is called **entropy**.

Inspection of equation [2] will show that if heat is received, which makes Q_{rev} positive, then the entropy of the receiving substance has increased.

Conversely, if heat is rejected, which makes Q_{rev} negative, then the entropy of the rejecting substance is decreased.

Thus, positive and negative changes of entropy show whether heat has been received or rejected during the process considered.

Now consider an isolated system in which an amount of heat energy Q is transferred from a hot source at temperature T_1 into a cooler sink at temperature T_2.

The loss of entropy from the hot source is Q/T_1 whereas the gain of entropy to the cooler sink is Q/T_2.

The amount of heat transferred is the same for both the source and the sink but, since $T_1 > T_2$, it follows that the gain of entropy to the cooler sink is greater than the loss of entropy from the hot source. This is because

$$\frac{Q}{T_2} > \frac{Q}{T_1} \qquad [3]$$

Now, by the second law of thermodynamics, heat transfer will only occur down a temperature gradient as a natural occurrence. This makes the natural transfer of heat an irreversible process.

Thus a process occurring in an isolated system such that there is an increase in entropy appears to be irreversible.

This leads to the statement of the **principle of increase of entropy**:

An isolated system can only change to states of equal or greater entropy.

Expressed mathematically

$$\Delta s \geqslant 0 \qquad [4]$$

where Δs = change of entropy

The higher the temperature of a system above its surroundings, the greater becomes the availability of the energy obtained by heat transfer.

Since entropy is a function of temperature (see equation [2]), it follows that entropy is associated with the usefulness of energy.

The method of calculation of changes of entropy will now be investigated. The expressions for change of entropy are developed assuming the processes to be reversible. It must be remembered, however, that a change is dependent only upon the end states; it does not matter how the change occurred. Thus the expressions for change of entropy can also be used for irreversible processes.

Note, from equation [2], if unit mass of substance is considered, the unit of specific entropy becomes J/kg K or, very often, kJ/kg K.

7.2 The entropy of vapours (two-phase systems)

Like enthalpy, entropy is treated separately in the three stages of the formation of a vapour from a liquid.

7.3 Entropy of liquid

Consider unit mass of a liquid which will ultimately be raised to unit mass of vapour at constant pressure. For the unit mass of liquid

$$dQ = c_{p_L} dT \qquad [1]$$

where c_{p_L} = specific heat capacity of the liquid at constant pressure.

Dividing equation [1] throughout by T gives

$$\frac{dQ}{T} = c_{p_L}\frac{dT}{T} = ds \qquad [2]$$

since $ds = dQ/T$, for this case

$$ds = c_{p_L}\frac{dT}{T} \qquad [3]$$

Integrating this equation from initial state 1 to final state 2

$$\int_{s_1}^{s_2} ds = c_{p_L}\int_{T_1}^{T_2}\frac{dT}{T}$$

or

$$\Big[s\Big]_{s_1}^{s_2} + c_{p\mathrm{L}}\Big[\ln T\Big]_{T_1}^{T_2}$$

from which

$$s_2 - s_1 = c_{p\mathrm{L}}[\ln T_2 - \ln T_1]$$

or

$$s_2 - s_1 = c_{p\mathrm{L}} \ln \frac{T_2}{T_1} \qquad [4]$$

In the same way as the zero of internal energy and enthalpy are arbitrarily chosen, so the zero of entropy is arbitrarily chosen.

In the case of steam, this arbitrary zero is chosen at the triple point (section 4.4), whose temperature is 273.16 K (273 K is accurate enough for most calculations).

Hence, in equation [4] let

$$s_1 = 0 \quad \text{when} \quad T_1 = 273.16 \text{ K}$$

then

$$s_2 - 0 = c_{p\mathrm{L}} \ln \frac{T_2}{273.16}$$

Dropping the subscript 2

$$s = c_{p\mathrm{L}} \ln \frac{T}{273.16} \qquad [5]$$

If it is assumed that for water

$$c_{p\mathrm{L}} = 4.187 \text{ kJ/kg K}$$

equation [5] becomes

$$s = 4.187 \ln \frac{T}{273.16} \qquad [6]$$

Actually, $c_{p\mathrm{L}}$ for water varies with temperature, but 4.187 kJ/kg K is an average value at normal low-range temperatures.

Maximum liquid entropy is reached at saturation temperature for the particular pressure under consideration. The specific entropy is then written s_f.

Thus, at saturation temperature for water

$$s_\mathrm{f} = c_{p\mathrm{L}} \ln \frac{T_\mathrm{f}}{273.16} \qquad [7]$$

or, if $c_{p\mathrm{L}}$ is assumed to be 4.187 kJ/kg K

$$s_\mathrm{f} = 4.187 \ln \frac{T_\mathrm{f}}{273.16} \qquad [8]$$

Its units will be kJ/kg K.

Example 7.1 *Determine the value of the specific entropy of water at 100 °C.*

SOLUTION

From equation [8]

$$s_f = 4.187 \ln \frac{T_f}{273.16} = 4.187 \ln \frac{373}{273.16}$$

$$= 4.187 \ln 1.366$$

$$= 4.187 \times 0.312$$

$$= \mathbf{1.306\ kJ/kg\ K}$$

From tables, the accurate value of s_f in this case is **1.307 kJ/kg K**.

7.4 Entropy of evaporation

Consider Fig. 7.2. It shows a temperature–entropy (T–s) diagram of the formation of vapour at constant pressure. Curve ab represents the introduction of the liquid enthalpy to the water. At b the water reaches saturation temperature t_f. Horizontal line bc represents the introduction of the enthalpy of evaporation at constant temperature t_f. There is all liquid at b and all dry saturated vapour at c. Curve cd represents the introduction of the superheat. Note that line abcd has a similar appearance to that obtained in the temperature–enthalpy diagram.

Now it has been shown that the area under a T–s diagram gives heat transferred (reversibly). Also, at constant pressure

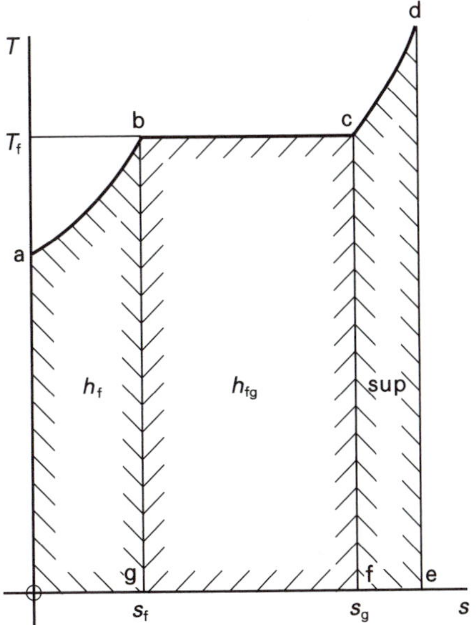

Fig. 7.2 T–s diagram for a vapour at constant pressure

Heat transferred = Change of enthalpy

Hence from Fig. 7.2

Area abgo = h_f = specific liquid enthalpy
Area bcfg = h_{fg} = specific enthalpy of evaporation
Area cdef = sup = specific superheat

From the diagram

Area bcfg = $h_{fg} = T_f(s_g - s_f)$

where s_g = specific entropy of dry saturated steam

Hence

$$s_g - s_f = \frac{h_{fg}}{T_f} = \text{specific entropy of evaporation} = s_{fg} \qquad [1]$$

Also, from equation [1]

$$s_g = s_f + s_{fg} \qquad [2]$$

$$= s_f + (s_g - s_f) \qquad [3]$$

From equation [2], in the case of steam

$$s_g = c_{p_L} \ln \frac{T_f}{273.16} + \frac{h_{fg}}{T_f} \qquad [4]$$

Accurate values of s_g and s_{fg} are given in tables.

If the vapour formed is wet, having a dryness fraction x, the specific enthalpy of evaporation introduced is xh_{fg}.

If the specific entropy of the wet vapour is s, then

$$s = s_f + x\frac{h_{fg}}{T_f} = s_f + xs_{fg} \qquad [5]$$

$$= s_f + x(s_g - s_f) \qquad [6]$$

***Example** 7.2 Determine the value of the specific entropy of wet steam at a pressure of 2 MN/m² (2 MPa) and 0.8 dry:*
(a) by calculation
(b) by using values of entropy from steam tables

(a)

$$s = c_{p_L} \ln \frac{T_f}{273.16} + x\frac{h_{fg}}{T_f}$$

$$= 4.187 \ln \frac{485.4}{273.16} + \left(0.8 \times \frac{1888.6}{485.4}\right)$$

$$= (4.187 \times \ln 1.78) + 3.11$$
$$= (4.187 \times 0.577) + 3.11$$
$$= 2.42 + 3.11$$
$$= \mathbf{5.53\ kJ/kg\ K}$$

(b)

$$
\begin{aligned}
s &= s_f + x s_{fg} \\
&= 2.447 + (0.8 \times 3.89) \\
&= 2.447 + 3.11 \\
&= \mathbf{5.557\ kJ/kg\ K}
\end{aligned}
$$

This is the accurate value.

7.5 Entropy of superheated vapour

Let $c_{p\mathrm{V}}$ = specific heat capacity of superheated vapour at constant pressure. Heat received in the superheat region is

$$c_{p\mathrm{V}}\,\mathrm{d}T = \text{area cdef in Fig. 7.2}$$

Hence, for the superheated vapour

$$\mathrm{d}s = c_{p\mathrm{V}} \frac{\mathrm{d}T}{T}$$

Integrating this equation from saturation temperature T_f to superheated steam temperature T, then

$$\int_{s_g}^{s} \mathrm{d}s = c_{p\mathrm{V}} \int_{T_f}^{T} \frac{\mathrm{d}T}{T}$$

or

$$\Big[s\Big]_{s_g}^{s} = c_{p\mathrm{V}} \Big[\ln T\Big]_{T_f}^{T}$$

hence

$$s - s_g = c_{p\mathrm{V}} [\ln T - \ln T_f]$$

$$\therefore \quad s - s_g = c_{p\mathrm{V}} \ln \frac{T}{T_f}$$

from which

$$s = s_g + c_{p\mathrm{V}} \ln \frac{T}{T_f} \qquad [1]$$

or for superheated steam

$$s = c_{p\mathrm{L}} \ln \frac{T_f}{273.16} + \frac{h_{fg}}{T_f} + c_{p\mathrm{V}} \ln \frac{T}{T_f} \qquad [2]$$

Accurate values of s for superheated vapours are given in tables.

Example 7.3 *Determine the value of specific entropy of steam at 1.5 MN/m² (1.5 MPa) with a temperature of 300 °C:*

(a) by calculation

(b) from steam tables

(a)

$$s = 4.187 \ln \frac{T_f}{273.16} + \frac{h_{fg}}{T_f} + c_{pv} \ln \frac{T}{T_f}$$

$$= 4.187 \ln \frac{471.3}{273.16} + \frac{1946}{471.3} + 2.093\,4 \ln \frac{573}{471.3}$$

$$= 4.187 \ln 1.73 + 4.127 + 2.093\,4 \ln 1.216$$
$$= (4.187 \times 0.548) + 4.127 + (2.093\,4 \times 0.196)$$
$$= 2.294 + 4.127 + 0.41$$
$$= \mathbf{6.831\ kJ/kg\ K}$$

(b)

From tables, $s =$ **6.919 kJ/kg K**
This is the accurate value.

7.6 The temperature–entropy chart for vapours

Extending Fig. 7.2 to take into account a wide pressure range in the formation of a vapour, the complete temperature–entropy chart for the vapour has the appearance shown in Fig. 7.3. Note its similarity with the temperature–enthalpy diagram shown in Fig. 4.4.

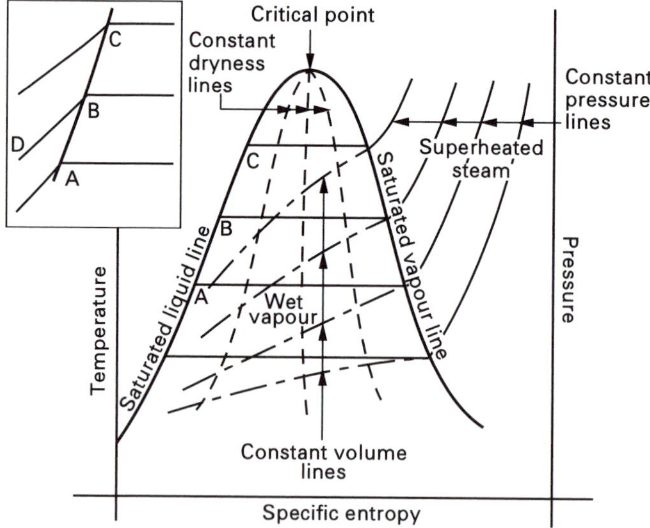

Fig. 7.3 *T–s* diagram for steam

Once again the liquid line and the saturated vapour line are drawn, which together enclose the wet vapour region. Again they join at the top of the critical point. Note, however, that the dry saturated vapour line is concave; in the case of the temperature–enthalpy diagram it was convex. On this chart, lines of constant dryness and of constant volume are often plotted in the wet region. The chart may be further extended by the introduction of a pressure axis, as shown. An important point is that this pressure axis will only correspond to pressures in the wet region. This will become clear if reference is made to Fig. 7.2, where a constant pressure line is shown.

Note that the line rises up to the transformation stage, is horizontal in the transformation stage then rises from this level into the superheat region. All along this line the pressure is constant. Because the section in the transformation stage is horizontal, a vertical scale of pressure can be arranged to correspond with the pressures associated with these horizontal lines. To determine pressures in the superheat region and along the liquid line, the lines concerned must be traced along to the transformation stage; then the corresponding pressure is determined from the vertical scale.

The large area in the lower-temperature region below the triple point is not usually shown in the temperature–entropy chart for steam since it has little value in the solution of problems concerning steam.

Figure 7.3 inset shows an exaggerated section of the saturated liquid line, A, B, C. The saturated liquid line appears as a locus through points A, B and C, whereas the liquid sections of the constant pressure lines below saturation temperature, such as DB, appear as separate lines. To normal scales, however, these separate liquid lines would very nearly merge into a single line, the saturated liquid line in the main diagram of Fig. 7.3.

7.7 The isothermal process on the temperature–entropy chart

By definition, an isothermal process is a process carried out at constant temperature. Consider the vapour at state A in Fig. 7.4. Here the vapour is wet. Expansion from A will proceed as illustrated; the temperature will remain constant and, since it is in the wet region, the pressure will also remain constant until point B is reached, where the vapour becomes dry saturated. Further expansion takes the vapour into the superheat region shown as point C. But in the superheat region the pressure drops. The pressure–volume diagram of the process is shown inset. The vapour in the superheat region can be approximately considered to behave as a gas, so the isothermal expansion BC could be considered as expanding according to the law $PV = C$, approximately.

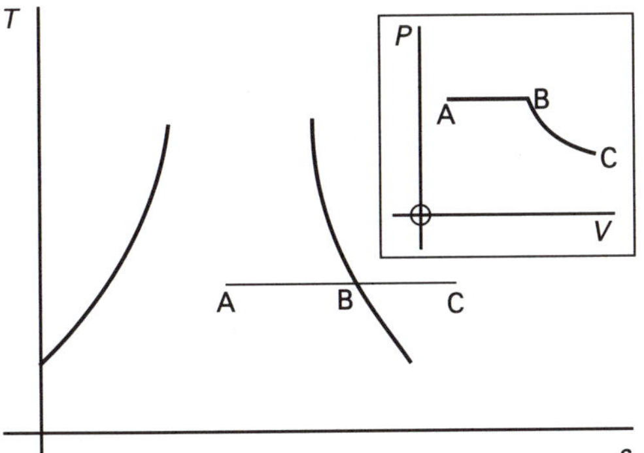

Fig. 7.4 Isothermal process on *T–s* diagram

7.8 The frictionless adiabatic process on the temperature–entropy chart

An adiabatic process is defined as a process carried out such that there is no heat transferred during the process. The adiabatic process, neglecting friction, shock, etc., must therefore have no area underneath it when plotted on a temperature–entropy diagram. It appears as a vertical line showing it to have constant entropy. A line of constant entropy is called an **isentrope**. The frictionless adiabatic process is therefore an **isentropic process**.

Process AB in Fig. 7.5 shows the frictionless adiabatic expansion of very wet vapour. Note that at B the vapour has a higher dryness fraction than at A.

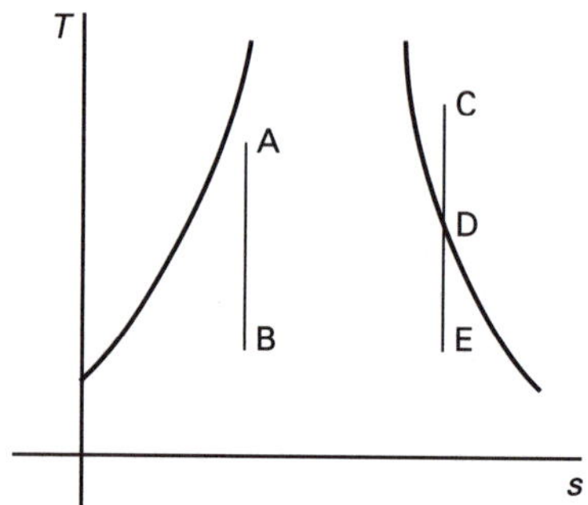

Fig. 7.5 Frictionless adiabatic (isentropic) process on *T–s* diagram

Process CDE shows the frictionless adiabatic expansion of superheated vapour. It illustrates the fact that superheated vapour, if adiabatically expanded, will become dry saturated, as at D, and eventually wet, as at E.

Since the frictionless adiabatic process is isentropic, then

$$\text{Entropy before expansion} = \text{Entropy after expansion}$$

or from Fig. 7.5

$$s_A = s_B$$

and

$$s_C = s_D = s_E$$

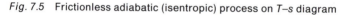

Example 7.4 *Steam at 2 MN/m² and with a temperature of 350 °C is expanded adiabatically until its pressure becomes 0.28 MN/m². Determine the final dryness fraction of the steam. The adiabatic expansion may be assumed to be isentropic.*

SOLUTION

At 2 MN/m^2 and 350 °C the steam is superheated because the saturation temperature at 2 MN/m^2 is 212.4 °C.

From tables, at 2 MN/m^2 and 350 °C

$$s_1 = \mathbf{6.957\ kJ/kg\ K}$$

For an isentropic process

$$s_1 = s_2$$

and

$$s_2 = s_{f2} + x_2 s_{fg2}$$

$$\therefore \quad 6.957 = 1.647 + 5.368\,x_2$$

$$x_2 = \frac{6.957 - 1.647}{5.368}$$

$$= \frac{5.310}{5.368}$$

$$= \mathbf{0.989}$$

Example 7.5 *Steam at 2.0 MN/m² and 250 °C is expanded isentropically to 0.36 MN/m² and it is then further expanded hyperbolically to 0.06 MN/m². Using steam tables, determine*
(a) the final condition of the steam
(b) the change in specific entropy during the hyperbolic process

(a)

$s_1 = s_2$ and $s_1 = \mathbf{6.545\ kJ/kg\ K}$
At 0.36 MN/m², $s_g = \mathbf{6.930\ kJ/kg\ K}$

So after isentropic expansion, steam is wet.
Hence

$$s_1 = s_{f2} + x_2 s_{fg2}$$

$$6.545 = 1.738 + 5.192\,x_2$$

$$x_2 = \frac{6.545 - 1.738}{5.192} = \frac{4.807}{5.192} = \mathbf{0.927}$$

$$v_2 = x_2 v_{g2} = 0.927 \times 0.510 = \mathbf{0.473\ m^3/kg}$$

For the hyperbolic process

$$P_2 v_2 = P_3 v_3$$

$$\therefore \quad v_3 = \frac{P_2 v_2}{P_3} = \frac{0.36}{0.06} \times 0.473 = \mathbf{2.84\ m^3/kg}$$

From tables, at 0.06 MN/m² and with a specific volume of 2.84 m³/kg, steam is superheated and has a temperature of 100 °C.

(b)
At this condition, $s_3 = \mathbf{7.609\ kJ/kg\ K}$.

$$\therefore \quad s_3 - s_2 = 7.609 - 6.930 = \mathbf{0.679\ kJ/kg\ K}$$

Example 7.6 *4.5 kg of steam has an initial pressure of 3 MN/m² and temperature of 300 °C. The steam then expands reversibly to a new pressure of 0.1 MN/m² with a dryness fraction of 0.96. The expansion of the steam appears as a straight line when plotted on a temperature–entropy chart. Determine*
(a) the heat transfer during the expansion
(b) the work done during the expansion

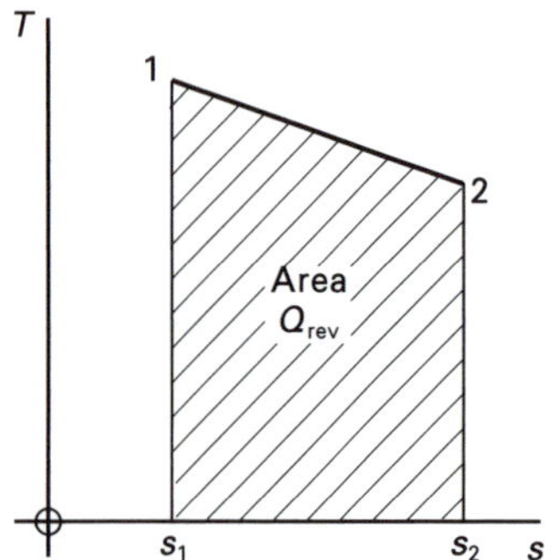

Fig. 7.6 Diagram for Example 7.6

SOLUTION
First draw a diagram (Fig. 7.6)
For state point 1

$$s_1 = 6.541 \text{ kJ/kg K}$$
$$T_1 = (300 + 273) \text{ K} = 573 \text{ K}$$
$$u_1 = 2751 \text{ kJ/kg}$$

For state point 2

$$\begin{aligned} s_2 = s_{f2} + x_2 s_{fg2} &= 1.303 + (0.96 \times 6.056) \\ &= 1.303 + 5.814 \\ &= 7.117 \text{ kJ/kg K} \\ T_2 = (99.6 + 273) \text{ K} &= 372.6 \text{ K} \\ h_2 = h_{f2} + x_2 h_{fg2} &= 417 + (0.96 \times 2258) \\ &= 417 + 2167.7 \\ &= 2584.7 \text{ kJ/kg} \\ u_2 = h_2 - P_2 x_2 v_{g2} &= 2584.7 - 100 \times (0.96 \times 1.694) \\ &= 2584.7 - 162.6 \\ &= 2422.1 \text{ kJ/kg} \end{aligned}$$

(a)

$$\begin{aligned} Q_{rev} &= \text{area of } T\text{–}s \text{ diagram} \\ &= \left(\frac{T_1 + T_2}{2}\right)(s_2 - s_1) \\ &= \left(\frac{573 + 372.6}{2}\right) \times (7.117 - 6.541) \end{aligned}$$

$$= \frac{945.6}{2} \times 0.576$$

$$= 472.8 \times 0.576$$
$$= \mathbf{272.3\ kJ/kg}$$

$\therefore$ Heat transfer for 4.5 kg is

$372.3 \times 4.5 =$ **1225.4 kJ (received)**

(b)

$Q_{\text{rev}} = \Delta u + W$

$$\therefore \quad W = Q_{\text{rev}} - \Delta u$$
$$= Q_{\text{rev}} - (u_2 - u_1)$$
$$= 272.3 - (2422.1 - 2751)$$
$$= 272.3 - (-328.9)$$
$$= 272.3 + 328.9$$
$$= \mathbf{601.2\ kJ/kg}$$

$\therefore$ Work done by 4.5 kg is

$601.2 \times 4.5 =$ **2705.4 kJ**

7.9 The enthalpy–entropy chart for vapours

A commonly used vapour chart is the **enthalpy–entropy chart** as illustrated in Fig. 7.7. Here the axes of enthalpy and entropy are used. The saturated liquid line and the saturated vapour line appear, meeting at the critical point. Lines of constant pressure cross the chart and lines of constant dryness fraction appear in the wet vapour region. In the superheated vapour region are drawn lines of constant temperature and lines of constant degree of superheat.

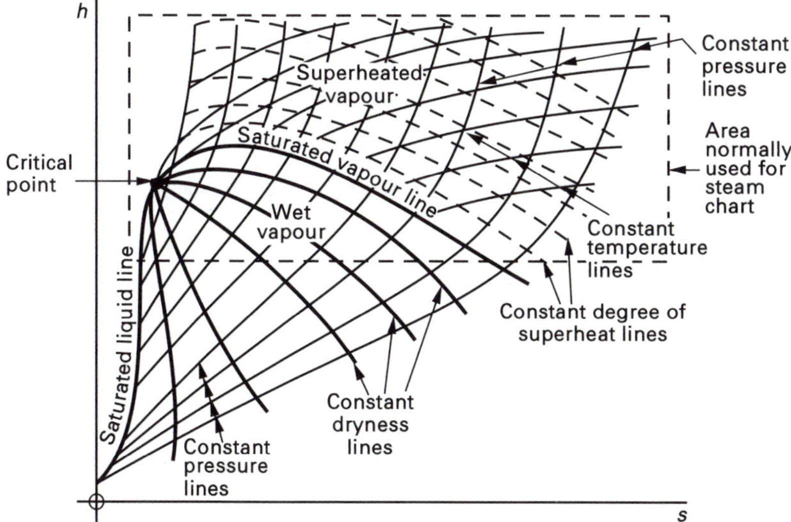

Fig. 7.7 *h–s* chart for a vapour

This chart is very useful for the determination of property changes, such as change of enthalpy during an isentropic process, illustrated as a vertical line on this chart. A throttling process becomes a horizontal line on this chart. Isothermal and constant pressure processes can be traced on the chart.

In the case of the enthalpy–entropy chart for steam, only the dotted area of Fig. 7.7 is usually illustrated. This is because a steam plant, e.g. the steam turbine, it is only steam of good quality (high dryness fraction or superheated) which is useful, so the expansions which occur can be easily plotted within the chart area illustrated.

This chart is sometimes called a Mollier chart, after Richard Mollier (1863–1935), a German scientist who introduced it in about 1904.

7.10 The pressure–enthalpy chart for vapours

A chart which is often used for substances, called refrigerants, required in the process of refrigeration is the **pressure–enthalpy chart**. Again, the saturated liquid and the saturated vapour lines appear and they join at the critical point as shown in Fig. 7.8. Lines of constant temperature (isotherms), constant entropy (isentropes), constant volume (isochors) and constant dryness are drawn on the chart.

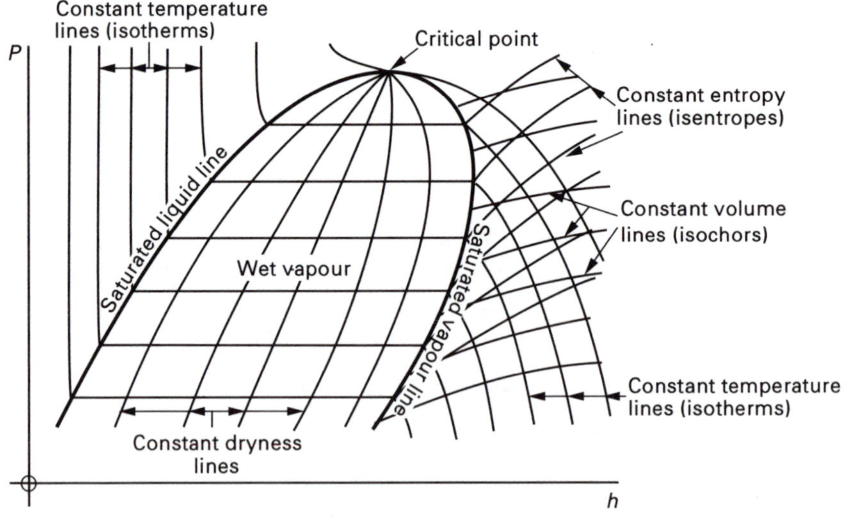

Fig. 7.8 *P–s* chart for a vapour

This chart is particularly useful in the determination of property changes during constant pressure (isobaric) processes. The constant pressure process commonly occurs in the process of refrigeration, and change of enthalpy during the constant pressure process is often required. This accounts for the choice of pressure and specific enthalpy as axes on a chart for refrigerants.

7.11 Changes of entropy

7.11.1 The change of entropy for a gas (single-phase systems)

For a gas, the heat transferred during a non-flow process can be determined by using the non-flow energy equation

$$dQ = dU + dW \qquad [1]$$

Consider unit mass of gas and let its state change from pressure P_1, specific volume v_1 and temperature T_1 to new state P_2, v_2, T_2.

From equation [1]

$$dQ = c_v\,dT + P\,dv \qquad [2]$$

Divide throughout by T, then

$$\frac{dQ}{T} = \frac{c_v\,dT}{T} + \frac{P\,dv}{T} \qquad [3]$$

But $dQ/T = ds$, so equation [3] now becomes

$$ds = \frac{c_v\,dT}{T} + \frac{P\,dv}{T} \quad \text{(see section 7.1)} \qquad [4]$$

Now to obtain the change of entropy, equation [4] must be integrated. However, it cannot be integrated as it stands because the expression $P\,dv/T$ contains too many variables. But this can be changed by using the characteristic equation.

For unit mass of gas

$$Pv = RT \quad \text{or} \quad \frac{P}{T} = \frac{R}{v}$$

and substituting this in equation [4]

$$ds = \frac{c_v\,dT}{T} + R\frac{dv}{v} \qquad [5]$$

Hence

$$\int_{s_1}^{s_2} ds = c_v \int_{T_1}^{T_2} \frac{dT}{T} + R\int_{v_1}^{v_2} \frac{dv}{v}$$

Integrating

$$\Big[s\Big]_{s_1}^{s_2} = c_v\Big[\ln T\Big]_{T_1}^{T_2} + R\Big[\ln v\Big]_{v_1}^{v_2}$$

or

$$s_2 - s_1 = c_v[\ln T_2 - \ln T_1] + R[\ln v_2 - \ln v_1]$$

or

$$s_2 - s_1 = c_v \ln\frac{T_2}{T_1} + R\ln\frac{v_2}{v_1} \qquad [6]$$

Equation [6] determines the change of specific entropy of a gas from a knowledge of temperature and volumes.

Now $c_p - c_v = R$; substituting this into equation [6] gives

$$s_2 - s_1 = c_v \ln\frac{T_2}{T_1} + c_p \ln\frac{v_2}{v_1} - c_v \ln\frac{v_2}{v_1}$$

$$= c_p \ln\frac{v_2}{v_1} + c_v\left(\ln\frac{T_2}{T_1} - \ln\frac{v_2}{v_1}\right)$$

$$= c_p \ln\frac{v_2}{v_1} + c_v \ln\frac{T_2}{T_1}\frac{v_1}{v_2} \qquad [7]$$

Now from the characteristic equation

$$\frac{P_1 v_1}{T_1} = \frac{P_2 v_2}{T_2}$$

from which

$$\frac{T_2}{T_1}\frac{v_1}{v_2} = \frac{P_2}{P_1} \qquad [8]$$

Substituting equation [8] into equation [7]

$$s_2 - s_1 = c_p \ln\frac{v_2}{v_1} + c_v \ln\frac{P_2}{P_1} \qquad [9]$$

Equation [9] determines the change of specific entropy of a gas from a knowledge of volumes and pressures.

Again, from $c_p - c_v = R$, it follows that $c_v = c_p - R$ and substituting this into equation [6] gives

$$s_2 - s_1 = c_p \ln\frac{T_2}{T_1} - R\ln\frac{T_2}{T_1} + R\ln\frac{v_2}{v_1}$$

$$= c_p \ln\frac{T_2}{T_1} - R\left(\ln\frac{T_2}{T_1} - \ln\frac{v_2}{v_1}\right)$$

$$= c_p \ln\frac{T_2}{T_1} - R\ln\frac{T_2}{T_1}\frac{v_1}{v_2} \qquad [10]$$

Substituting equation [8] into equation [10]

$$s_2 - s_1 = c_p \ln\frac{T_2}{T_1} - R\ln\frac{P_2}{P_1} \qquad [11]$$

Equation [11] determines the change of specific entropy of a gas from a knowledge of temperatures and pressures.

Note that for any given change of state from P_1, v_1, T_1 to P_2, v_2, T_2 each of the equations [6], [9] and [11] will give the same result. The choice of equation is a matter of convenience only.

7.11.2 The change of entropy during a constant temperature (isothermal) process for a gas

Here $T_1 = T_2$ and hence $\ln T_2/T_1 = \ln 1 = 0$; so from equation [6]

$$s_2 - s_1 = R\ln\frac{v_2}{v_1} \qquad [12]$$

and from equation [11]

$$s_2 - s_1 = -R \ln \frac{P_2}{P_1}$$

or

$$s_2 - s_1 = R \ln \frac{P_1}{P_2} \qquad [13]$$

From equation [9]

$$s_2 - s_1 = c_p \ln \frac{v_2}{v_1} + c_v \ln \frac{P_2}{P_1} \quad \text{(no change)}$$

7.11.3 The change of entropy during a constant volume (isochoric) process for a gas

Here $v_1 = v_2$ and hence $\ln v_2/v_1 = \ln 1 = 0$; so from equation [6]

$$s_2 - s_1 = c_v \ln \frac{T_2}{T_1} \qquad [14]$$

and from equation [9]

$$s_2 - s_1 = c_v \ln \frac{P_2}{P_1} \qquad [15]$$

From equation [11]

$$s_2 - s_1 = c_p \ln \frac{T_2}{T_1} - R \ln \frac{P_2}{P_1} \quad \text{(no change)}$$

7.11.4 The change of entropy during a constant pressure (isobaric) process for a gas

Here $P_1 = P_2$ and hence $\ln P_2/P_1 = \ln 1 = 0$; so from equation [4]

$$s_2 - s_1 = c_p \ln \frac{v_2}{v_1} \qquad [16]$$

and from equation [11]

$$s_2 - s_1 = c_p \ln \frac{T_2}{T_1} \qquad [17]$$

From equation [6]

$$s_2 - s_1 = c_v \ln \frac{T_2}{T_1} + R \ln \frac{v_2}{v_1} \quad \text{(no change)}$$

7.11.5 The change of entropy during a polytropic process ($PV^n = C$) for a gas

For the polytropic process it has been shown in section 5.14 that

$$\text{Heat transferred} = \frac{(\gamma - n)}{(\gamma - 1)} \times \text{Polytropic work}$$

or for unit mass of gas

$$dQ = \frac{(\gamma - n)}{(\gamma - 1)} P dv \tag{18}$$

Divide equation [18] throughout by T, then

$$\frac{dQ}{T} = ds = \frac{(\gamma - n)}{(\gamma - 1)} \frac{P dv}{T} \tag{19}$$

Again, for unit mass of gas

$$Pv = RT \quad \text{or} \quad \frac{P}{T} = \frac{R}{v}$$

and substituting into equation [19]

$$ds = \frac{(\gamma - n)}{(\gamma - 1)} R \frac{dv}{v} \tag{20}$$

from this

$$\int_{s_1}^{s_2} ds = \frac{(\gamma - n)}{(\gamma - 1)} R \int_{v_1}^{v_2} \frac{dv}{v}$$

Integrating

$$\Big[s\Big]_{s_1}^{s_2} = \frac{(\gamma - n)}{(\gamma - 1)} R \Big[\ln v\Big]_{v_1}^{v_2}$$

or

$$s_2 - s_1 = \frac{(\gamma - n)}{(\gamma - 1)} R [\ln v_2 - \ln v_1]$$

or

$$s_2 - s_1 = \frac{(\gamma - n)}{(\gamma - 1)} R \ln \frac{v_2}{v_1} \tag{21}$$

Now $c_p - c_v = R$ and $c_p/c_v = \gamma$ or $c_p = \gamma c_v$; so $\gamma c_v - c_v = R$ or $c_v(\gamma - 1) = R$. Substituting this into equation [21] gives

$$s_2 - s_1 = \frac{(\gamma - n)}{(\gamma - 1)} c_v (\gamma - 1) \ln \frac{v_2}{v_1}$$

or

$$s_2 - s_1 = c_v (\gamma - n) \ln \frac{v_2}{v_1} \tag{22}$$

This expression gives the change of entropy in terms of volumes.

Now

$$\frac{v_2}{v_1} = \left(\frac{T_1}{T_2}\right)^{1/(n-1)} = \left(\frac{P_1}{P_2}\right)^{1/n} \tag{23}$$

Substituting the temperature–volume relationship into equation [22]

$$s_2 - s_1 = c_v(\gamma - n)\ln\left(\frac{T_1}{T_2}\right)^{1/(n-1)}$$

or

$$s_2 - s_1 = c_v\frac{(\gamma - n)}{(n-1)}\ln\frac{T_1}{T_2} \qquad [24]$$

This expression gives the change of entropy in terms of temperatures.

Also, substituting the pressure–volume relationship from equation [23] into equation [22] gives

$$s_2 - s_1 = c_v(\gamma - n)\ln\left(\frac{P_1}{P_2}\right)^{1/n}$$

or

$$s_2 - s_1 = c_v\frac{(\gamma - n)}{n}\ln\frac{P_1}{P_2} \qquad [25]$$

This expression gives the change of entropy in terms of pressures.

It should again be noted that for a given polytropic change from P_1, v_1, T_1 to P_2, v_2, T_2, each of the equations [21], [22], [24] and [25] will give the same result. The choice is simply a matter of convenience.

It is important to remember that all the above expressions for the change of entropy are for unit mass of gas only. In any particular problem these expressions must be multiplied by the actual mass of gas being considered in order to determine the actual change of entropy which has occurred.

And note that in the above expressions, specific volume v has been used. Now for a mass m of the gas, its volume $V = mv$. The volume in the above expressions for the change of entropy appears in the ratio v_2/v_1.

Now

$$v_2/v_1 = mv_2/mv_1 = V_2/V_1$$

Thus, when calculating changes of entropy, it is not always necessary to find the ratio of the specific volumes. The ratio of the actual volumes will give the same result.

7.12 The entropy chart for a gas

In order to make up a chart, numerical values of entropy must be known so that points can be plotted on the chart. The entropy equations already determined are for changes of entropy and therefore do not give individual values at any particular state. The absolute value of entropy for a gas is not known because of the discontinuity of state as the gas is cooled. The gas liquefies then solidifies before absolute zero of temperature is reached. However, it is the change of entropy which is important when discussing a process with a gas. This change of entropy is quite independent of any zero which may exist, so when making up charts, it is usual to choose an arbitrary zero for entropy. Thus, for air, it is commonly arranged that its entropy is considered zero when its pressure is 0.101 MN/m^2 and its temperature is 0 °C.

Consider the effect of this choice of zero on the equations for the change of entropy already determined.

Take equation [11], section 7.11, for example.

$$s_2 - s_1 = c_p \ln \frac{T_2}{T_1} - R \ln \frac{P_2}{P_1}$$

Assume that $s_1 = 0$, then by the choice of the arbitrary zero $T_1 = 273.15$ K (say 273 K) and $P_1 = 0.101$ MN/m^2.

Since $s_1 = 0$, this equation becomes

$$s_2 = c_p \ln \frac{T_2}{273} - R \ln \frac{P_2}{0.101}$$

or, more generally, neglecting the subscript 2

$$s_2 = c_p \ln \frac{T}{273} - R \ln \frac{P}{0.101} \qquad [1]$$

Thus, from equation [1] it is possible to determine the value of specific entropy s for a gas at any absolute temperature T and corresponding pressure P.

A similar arrangement is made for the other expressions. In the case of equations containing volume, the volume at the arbitrary zero condition is calculated, v_0 say, and this is substituted in place of v_1. Thus in the case of equation [6], section 7.11, the value of entropy becomes

$$s = c_v \ln \frac{T}{273} + R \ln \frac{v}{v_0} \qquad [2]$$

Also, from equation [9]

$$s = c_p \ln \frac{v}{v_0} + c_v \ln \frac{P}{0.101} \qquad [3]$$

Using these equations for the value of specific entropy, it is now possible to make up a temperature–entropy chart, as shown in Fig. 7.9.

The chart usually has the axes of absolute temperature vertical and specific entropy horizontal. On the chart is drawn a network of constant volume and constant pressure lines, as shown. These lines can be drawn by making use of the expressions already determined. Using equation [2] the constant volume lines can be drawn. By selecting a suitable constant volume and substituting into equation [2] the part of the expression $\ln(v/v_0)$ remains constant. Thus, by selecting a number of values for the absolute temperature T, a series of values of s can be determined for the constant volume v. These values can then be plotted to form a constant volume line. Taking various new values of v, a series of constant volume lines can be drawn.

A similar procedure is adopted when drawing the constant pressure lines. Here equation [3] can be used. On selecting a constant pressure P, the part of the expression $\ln(P/0.101)$ now becomes constant. The procedure then follows that adopted for the constant volume lines. The chart can be made more useful if two vertical axes of specific internal energy and specific enthalpy are introduced.

When dealing with gas laws, it was shown that the internal energy of a gas is a function of temperature only (see section 5.5). Thus the internal energy axis is parallel

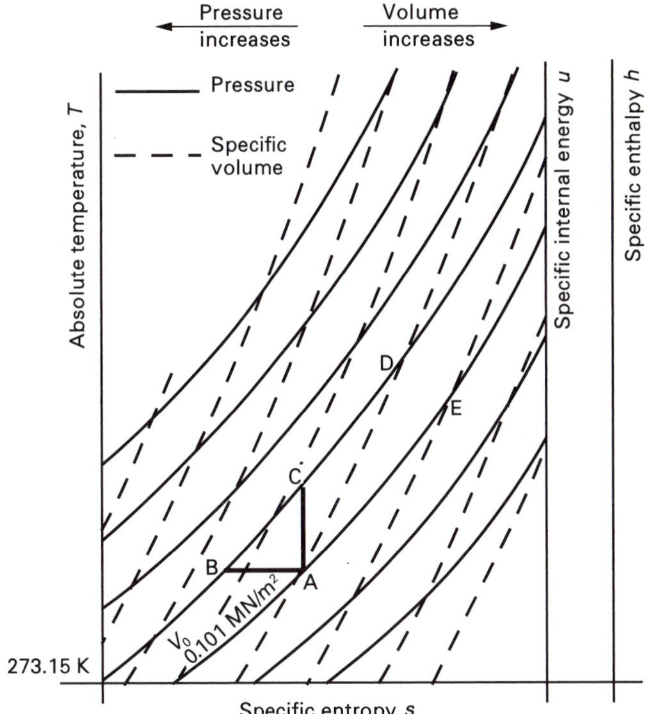

Fig. 7.9 Entropy chart for a gas

to the temperature axis. Now the change of specific internal energy for a gas is $c_v(T_2 - T_1)$. Assuming an arbitrary zero of internal energy at 0 °C, the value of internal energy above 0 °C can be obtained from

$$u = c_v(T - 273)$$

Using this equation, the scale of specific internal energy can be introduced.

With regard to enthalpy, it was shown in the chapter on gases that the change of specific enthalpy for a gas is $c_p(T_2 - T_1)$. Using a similar analogy as that used for the internal energy of a gas, the value of the enthalpy above 0 °C can be obtained from

$$h = c_p(T - 273)$$

Thus the scale of specific enthalpy can now be introduced.

Looking at Fig. 7.9 and selecting, say, point A, it will be observed that the constant volume and constant pressure lines cross, thus identifying the volume and pressure of the gas at this point. Further from the axes of the graph, it is possible to identify the temperature, specific internal energy, specific enthalpy and specific entropy of the gas for the point A. Thus a complete knowledge of the state of the gas is known at point A when it has been identified on the chart.

Again, reference to Fig. 7.9 will show that various processes, and the changes which occur throughout, are readily determined on the chart. An isothermal compression AB is shown as a horizontal straight line. It will be observed that line AB finishes on a constant pressure line and the new pressure can be read from the value on the line. Point B does not coincide with a given constant volume line. The point B does, however, lie between two given constant volume lines. The volume at B is thus interpolated from these two given values. The change of specific entropy during the process can be determined from the specific entropy axis. The chart also shows that, during the isothermal process, no change in internal energy or enthalpy occurs because the temperature remains constant.

In the case of the frictionless adiabatic process, it must be remembered that no heat is transferred during the process. Now the area under a process plotted on a temperature–entropy chart gives the heat transferred during the progress of the process. No heat is transferred during the frictionless adiabatic process, so there must be no area under the graph of the process when plotted on the temperature–entropy chart. For this to be so, the graph will appear as a vertical straight line, which shows that the frictionless adiabatic process is carried out at constant entropy. A line of constant entropy is said to be an **isentropic** line, thus the frictionless adiabatic process is an **isentropic process**. Such a process is shown as AC in Fig. 7.9. The process starts at point A; the new pressure, volume and temperature can be determined from point C. The change of specific internal energy and specific enthalpy can be determined from the axes of the chart.

A constant volume process from point A will appear as AD. The new pressure and temperature can be determined from the point D. Also, the change in specific internal energy, specific enthalpy and specific entropy can be determined from the axes.

A constant pressure process from point A will appear as AE. The new specific volume and temperature can be determined from the point E. Also, the change in specific internal energy, specific enthalpy and specific entropy can be determined from the axes.

Note that the chart is usually made out for unit mass of gas, usually 1 kg. Thus, in any particular problem, the values obtained from the chart for the specific volume and changes in specific internal energy, specific enthalpy and specific entropy must be multiplied by the actual mass of gas being used.

***Example** 7.7* *A quantity of gas has an initial pressure, volume and temperature of 140 kN/m^2, 0.14 m^3 and 25 °C, respectively. It is compressed to a pressure of 1.4 MN/m^2 according to the law* $PV^{1.25}$ = *constant. Determine*

(a) the change of entropy

(b) the approximate change of entropy obtained by dividing the heat transferred by the gas by the mean absolute temperature during the compression

Take c_p = *1.041 kJ/kg K,* c_v = *0.743 kJ/kg K.*

(a)

$$R = c_p - c_v = 1.041 - 0.743 = \mathbf{0.298\ kJ/kg\ K}$$

$$P_1V_1 = mRT_1 \quad \text{and} \quad T_1 = (25 + 273)\ \text{K} = \mathbf{298\ K}$$

$$\therefore \quad m = \frac{P_1V_1}{RT_1} = \frac{140 \times 0.14}{0.298 \times 298} = \mathbf{0.221\ kg}$$

For 1 kg of gas

$$s_2 - s_1 = c_p \ln\frac{V_2}{V_1} + c_v \ln\frac{P_2}{P_1}$$

Also, $P_1 V_1^n = P_2 V_2^n$

$$\therefore \quad V_2 = V_1 \left(\frac{P_1}{P_2}\right)^{1/n} = 0.14 \times \left(\frac{140}{1400}\right)^{1/1.25}$$

$$= \frac{0.14}{10^{1/1.25}}$$

$$= \frac{0.14}{6.31}$$

$$= \mathbf{0.022\ 2\ m^3}$$

$$\therefore \quad s_2 - s_1 = 1.041 \ln \frac{0.022\ 2}{0.14} + 0.743 \ln \frac{1400}{140}$$

$$= -1.041 \ln \frac{0.14}{0.022\ 2} + 0.743 \ln \frac{1400}{140}$$

$$= -1.041 \ln 6.31 + 0.743 \ln 10$$
$$= -1.041 \times 1.842 + 0.743 \times 2.303$$
$$= -1.918 + 1.711$$
$$= \mathbf{-0.207\ kJ/kg\ K}$$

But there is 0.221 kg of gas.

$\therefore$ Change of entropy $= -0.207 \times 0.221$
$= \mathbf{-0.045\ 7\ kJ/K}$ (a decrease)

(b)

$$\text{Polytropic work} = \frac{P_1 V_1 - P_2 V_2}{n - 1}$$

$$= \frac{(140 \times 0.14) - (1400 \times 0.022\ 2)}{1.25 - 1}$$

$$= \frac{19.6 - 31.08}{0.25} = \frac{11.48}{0.25}$$

$= \mathbf{-45.92\ kJ}$ (work transfer to the gas)

Heat transferred $= \dfrac{(\gamma - n)}{(\gamma - 1)} \times$ polytropic work, and

$\gamma = c_p/c_v = 1.041/0.743 = \mathbf{1.401}$

$$\therefore \quad \text{Heat transferred} = \frac{(1.401 - 1.25)}{(1.401 - 1)} \times (-45.92)$$

$$= \frac{0.151}{0.401} \times -45.92$$

$= \mathbf{-17.29\ kJ}$ (heat transfer from the gas)

$$\frac{T_1}{T_2} = \left(\frac{V_2}{V_1}\right)^{(n-1)}$$

$$\therefore \quad T_2 = T_1\left(\frac{V_1}{V_2}\right)^{(n-1)} = 298 \times 6.31^{(1.25-1)}$$

$$= 298 \times 6.31^{0.25}$$
$$= 298 \times 1.585$$
$$= \mathbf{472.3\ K}$$

$$\therefore \quad \text{Mean absolute temperature} = \frac{472.3 + 298}{2} = \frac{770.3}{2}$$

$$= \mathbf{385.15\ K}$$

Hence approximate change of entropy is

$$-\frac{17.29}{385.15} = \mathbf{-0.044\,9\ kJ/K}$$

Note that in this problem the other expressions could have been used for the determination of the change of entropy. For example

$$s_2 - s_1 = c_p \ln\frac{T_2}{T_1} - R\ln\frac{P_2}{P_1}$$

$$= 1.041 \ln\frac{472.3}{298} - 0.298 \ln\frac{1400}{40}$$

$$= 1.041 \ln 1.585 - 0.298 \ln 10$$
$$= (1.041 \times 0.46) - (0.298 \times 2.303)$$
$$= 0.479 - 0.686$$
$$= \mathbf{-0.207\ kJ/kg\ K} \text{ (as before)}$$

Alternatively

$$s_2 - s_1 = c_v \ln\frac{T_2}{T_1} + R\ln\frac{V_2}{V_1}$$

$$= 0.743 \ln\frac{472.3}{298} + 0.298 \ln\frac{0.022\,2}{0.14}$$

$$= (0.743 \times 0.46) - (0.298 \times 1.842)$$
$$= \mathbf{-0.207\ kJ/kg\ K} \text{ (as before)}$$

Alternatively

$$s_2 - s_1 = c_v(\gamma - n)\ln\frac{V_2}{V_1}$$

$$= 0.743 \times (1.401 - 1.25)\ln\frac{0.022\,2}{0.14}$$

$$= 0.743 \times 0.151 \times (-1.842)$$
$$= \mathbf{-0.207\ kJ/kg\ K} \text{ (as before)}$$

Alternatively

$$s_2 - s_1 = c_v \frac{(\gamma - n)}{(n - 1)} \ln \frac{T_1}{T_2}$$

$$= 0.743 \times \frac{(1.401 - 1.25)}{(1.25 - 1)} \ln \frac{298}{472.3}$$

$$= 0.743 \times \frac{0.151}{0.25} \times -0.46$$

$$= \mathbf{-0.207\ kJ/kg\ K} \text{ (as before)}$$

Alternatively

$$s_2 - s_1 = c_v \frac{(\gamma - n)}{n} \ln \frac{P_1}{P_2}$$

$$= 0.743 \times \frac{(1.401 - 1.25)}{1.25} \ln \frac{140}{1400}$$

$$= 0.743 \times \frac{0.151}{1.25} \times -2.303$$

$$= \mathbf{-0.207\ kJ/kg\ K} \text{ (as before)}$$

Example 7.8 *0.3 kg of air at a pressure of 350 kN/m^2 and a temperature of 35 °C receives heat energy at constant volume until its pressure becomes 700 kN/m^2. It then receives heat energy at constant pressure until its volume becomes 0.228 9 m^3. Determine the change of entropy during each process. Take c_p = 1.006 kJ/kg K, c_v = 0.717 kJ/kg K.*

CONSTANT VOLUME PROCESS

$$R = c_p - c_v = 1.006 - 0.717 = \mathbf{0.289\ kJ/kg\ K}$$
$$P_1 V_1 = mRT_1$$

$$\therefore \quad V_1 = \frac{mRT_1}{P_1} = \frac{0.3 \times 0.289 \times 308}{350} = \mathbf{0.076\ 3\ m^3}$$

For a constant volume process, $P_2/T_2 = P_1/T_1$

$$\therefore \quad T_2 = T_1 \frac{P_2}{P_1} = 308 \times \frac{700}{350} = 308 \times 2 = \mathbf{616\ K}$$

Also $s_2 - s_1 = c_v \ln \frac{P_2}{P_1} = 0.717 \ln \frac{700}{350}$

$$= 0.717 \ln 2$$
$$= 0.717 \times 0.693$$
$$= \mathbf{0.497\ kJ/kg\ K}$$

$\therefore$ for 0.3 kg

Change of entropy $= 0.497 \times 3$
$= \mathbf{1.491\ kJ/K}$ (an increase)

Alternatively

$$s_2 - s_1 = c_v \ln\frac{T_2}{T_1} = 0.717 \ln\frac{616}{308}$$

$$= 0.717 \ln 2$$
$$= \mathbf{0.496\ kJ/kg\ K}\ \text{(as before)}$$

Alternatively

$$s_2 - s_1 = c_p \ln\frac{T_2}{T_1} - R \ln\frac{P_2}{P_1}$$

$$= 1.006 \ln\frac{616}{308} - 0.289 \ln\frac{700}{350}$$

$$= 1.006 \ln 2 - 0.289 \ln 2$$
$$= (1.006 - 0.289) \ln 2$$
$$= 0.717 \ln 2$$
$$= \mathbf{0.497\ kJ/kg\ K}\ \text{(as before)}$$

CONSTANT PRESSURE PROCESS

For a constant pressure process, $V_3/T_3 = V_2/T_2$

$$\therefore \quad T_3 = T_2 \frac{V_3}{V_2} = 616 \times \frac{0.228\,9}{0.076\,3} = 616 \times 3 = \mathbf{1848\ K}$$

$$\text{Also} \quad s_3 - s_2 = c_p \ln\frac{V_3}{V_2} = 1.006 \ln\frac{0.228\,9}{0.076\,3}$$

$$= 1.006 \ln 3$$
$$= 1.006 \times 1.099$$
$$= \mathbf{1.105\,6\ kJ/kg\ K}$$

$\therefore$ for 0.3 kg

Change of entropy $= 1.105\,6 \times 0.3$
$= \mathbf{0.332\ kJ/K}$

Alternatively

$$s_3 - s_2 = c_p \ln\frac{T_3}{T_2} = 1.006 \ln\frac{1848}{616}$$

$$= 1.006 \ln 3$$
$$= \mathbf{1.105\,6\ kJ/kg\ K}\ \text{(as before)}$$

Alternatively

$$s_3 - s_2 = c_v \ln\frac{T_3}{T_2} + R \ln\frac{V_2}{V_1}$$

$$= 0.717 \ln 3 + 0.289 \ln 3$$
$$= (0.717 + 0.289) \ln 3$$
$$= 1.006 \ln 3$$
$$= \mathbf{1.105\,6\ kJ/kg\ K}\ \text{(as before)}$$

Example 7.9 *A quantity of gas has a pressure of 700 kN/m² and it occupies a volume of 0.014 m³ at a temperature of 150 °C. The gas is expanded isothermally to a volume of 0.084 m³. Determine the change of entropy.*

SOLUTION

This problem is solved by going back to the original theory that the area under a process plotted on a T–s diagram is equal to the heat transferred reversibly during that process (see section 7.1).

Also

$$\text{Heat transfer} = PV \ln \frac{V_2}{V_1} \quad \text{(see section 5.13)}$$

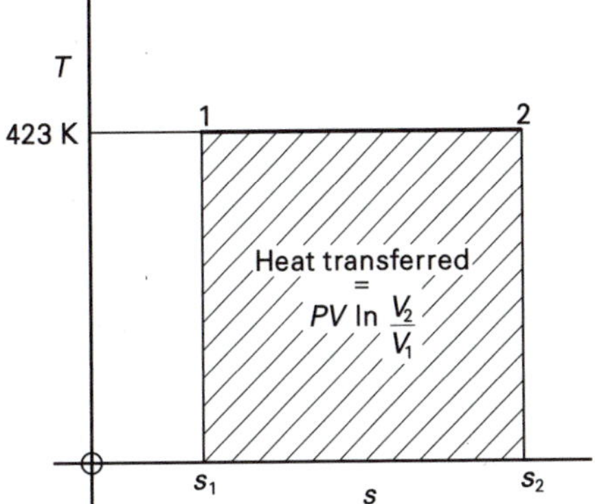

Fig. 7.10 Diagram for Example 7.9

Isotherm 1–2 in Fig. 7.10 is a horizontal line of length $s_2 - s_1$, and it occurs at a temperature of 423 K.

$$\text{Area under this line} = 423(s_2 - s_1)\ \text{kJ}$$

$$\therefore \quad 423(s_2 - s_1) = PV \ln \frac{V_2}{V_1} \quad (PV = C \text{ for an isothermal process on a gas})$$

or

$$s_2 - s_1 = \frac{PV}{423} \ln \frac{V_2}{V_1}$$

$$= \frac{700 \times 0.014}{423} \ln \frac{0.084}{0.014}$$

$$= \frac{700 \times 0.014}{423} \ln 6$$

$$= \frac{700 \times 0.014 \times 1.792}{423}$$

$$= \mathbf{0.041\,5\ kJ/K}$$

Questions

1. Steam initially at a pressure of 700 kN/m^2 with a dryness fraction of 0.65 passes through a reversible process which appears as a semicircle when plotted on a temperature–entropy chart. The final pressure is 100 kN/m^2 and the final temperature is the same as the initial steam temperature at the pressure of 700 kN/m^2. During the process the maximum temperature is 230 °C. For this process, determine
   (a) the heat transfer per kilogram of steam
   (b) the work done per kilogram of steam.

   [(a) 1282 kJ/kg; (b) 591.7 kJ/kg]

2. Steam at a pressure of 2 MN/m^2 and temperature 250 °C is expanded to a pressure of 0.32 MN/m^2 according to the law $PV^{1.25}$ = constant. For this expansion, determine
   (a) the final condition of the steam
   (b) the specific heat transfer
   (c) the change of specific entropy

   [(a) 0.847; (b) −163.9 kJ/kg; (c) −0.383 kJ/kg K]

3. Steam at a pressure of 1.9 MN/m^2 and with a temperature of 225 °C is expanded isentropically to a pressure of 0.3 MN/m^2. It is then further expanded hyperbolically to a pressure of 0.12 MN/m^2. Using steam tables, determine
   (a) the final condition of the steam
   (b) the change of specific entropy during the hyperbolic process

   [(a) 0.954; (b) 0.572 7 kJ/kg K, an increase]

4. Steam initially at a pressure of 1.4 MPa and having a specific volume of 0.12 m^3/kg is throttled to a pressure of 0.7 MPa then expanded hyperbolically to a pressure of 0.11 MPa. For each process, determine the change in specific entropy.

   [0.256 kJ/kg K; 1.09 kJ/kg K]

5. Steam at a pressure of 6.8 MN/m^3 and at a temperature of 375 °C is isentropically expanded to a pressure of 1.0 MN/m^2. It is then reheated at constant pressure until it has a temperature of 300 °C. It is then further isentropically expanded to a pressure of 0.14 MN/m^2. Using steam tables, determine
   (a) the condition of the steam after both isentropic expansions
   (b) the heat transfer per kilogram of steam to carry out the constant pressure reheat

   Sketch the *T*–*s* and *h*–*s* diagrams for the processes.

   [(a) 0.953, 0.979; (b) 370.4 kJ/kg]

6. A quantity of gas has an initial pressure, volume and temperature of 130 kN/m^2, 0.224 m^3 and 21 °C, respectively. It is compressed to a volume of 0.028 m^3 according to the law $PV^{1.3}$ = constant. Determine the change of entropy and state whether it is an increase or decrease.

   Take $c_v = 0.717$ kJ/kg K, $R = 0.287$ kJ/kg K.

   [−0.051 8 kJ/K, a decrease]

7. 1 kg of air has a volume of 56 litres and a temperature of 190 °C. The air then receives heat at constant pressure until its temperature becomes 500 °C. From this state the air rejects heat at constant volume until its pressure is reduced to 700 kN/m^2. Determine the change of entropy during each process, stating whether it is an increase or decrease.

   Take $c_p = 1.006$ kJ/kg K, $c_v = 0.717$ kJ/kg K.

   [0.516 kJ/kg K, an increase; −0.88 kJ/kg K, a decrease]

8. A quantity of air has an initial pressure, volume and temperature of 2.72 MPa, 3 litres and 260 °C, respectively. It is expanded to a pressure of 0.34 MPa according to the law $PV^{1.25}$ = constant. Determine the change of entropy and state whether it is an increase or decrease.

   Take $c_p = 1.005$ kJ/kg K, $c_v = 0.715$ kJ/kg K.

   [0.009 77 kJ/K, an increase]

9. A quantity of gas has an initial pressure, volume and temperature of 1.1 bar, 0.16 m^3 and 18 °C, respectively. It is compressed isothermally to a pressure of 6.9 bar. Determine the change of entropy.

   Take $R = 0.3$ kJ/kg K.

   [−0.111 kJ/K, a decrease]

10. 0.5 kg of air is expanded according to the law $PV^{1.2}$ = constant from initial conditions of a pressure of 1.4 MN/m^2 and a volume of 0.06 m^3 to a final pressure of 0.35 MN/m^2. Determine the heat transferred by the air during the expansion. Show that this amount of heat transfer is approximately equal to the change of entropy multiplied by the mean absolute temperature.

    Take $v_p = 1.006$ kJ/kg K, $c_v = 0.717$ kJ/kg K.

    [43.6 kJ, 43.79 kJ]

11. 0.3 m^3 of air at a temperature of 15 °C and pressure 100 kN/m^2 is compressed to a volume of 0.03 m^3 according to the law $PV^{1.3}$ = constant. Determine the change of entropy.

    Take $\gamma = 1.41$, $c_v = 0.72$ kJ/kg K.

    [−0.058 6 kJ/K, a decrease]

Chapter 8

Combustion

8.1 General introduction

Until the advent of atomic energy, all thermodynamic engine processes derived their energy supply either directly or indirectly from the combustion of **fuel**. Combustion means burning and a fuel is a substance which, when burnt, liberates a large amount of energy for a given bulk. A fuel can be a solid, a liquid or a gas. It should be obtainable in abundant quantity, it should be relatively easy to handle and its combustion should allow satisfactory control.

The vast majority of fuels are based on carbon, hydrogen or some combination of carbon and hydrogen. Called **hydrocarbons**, these *C–H* combinations occur as solids, liquids and gases. Coal, oils and natural gas are natural fuels that were laid down many millions of years ago, so they are often called **fossil fuels**.

The acquisition of fossil fuels presents practical difficulties, such as mining and drilling. However, oil and gas have been, and are, the source of much international tension. Wars have been fought to obtain them because industrial society largely depends upon the use of fuel as its fundamental energy base.

The increasing use of fossil fuels also produces immense amounts of atmospheric pollution as the products of combustion are dispersed into the earth's atmosphere in ever increasing quantities. This is having an adverse effect on the planet's flora and fauna.

Coal has largely been displaced by oil for political and economic reasons and because oil is easier to extract. Both coal and oil exist in finite quantities, so perhaps at some time in the future, as the oil supply diminishes, coal will probably be brought back on stream, and used in a more refined form than at present.

In the vast majority of engines, the burning of fuel is the prerequisite of thermodynamic processes, so it is important to understand the chemical reactions which take place during combustion.

8.2 Exothermic and endothermic reactions

In many chemical reactions energy is liberated; in others energy is absorbed. A chemical reaction in which energy is liberated is called an **exothermic** reaction. Thus, fuel-burning reactions are exothermic reactions. A chemical reaction in which energy is absorbed is called an **endothermic** reaction. For example, in the manufacture of coal gas there is a reaction between carbon and carbon dioxide which produces carbon monoxide. Energy is required to carry out this reaction, so it is an endothermic reaction.

8.3 Elements, compounds and mixtures

The basic materials of matter are called **elements**. Elements cannot be subdivided into any other substances. Oxygen and hydrogen cannot be subdivided into any other substances, so they are elements.

Substances which are made from elements are called **compounds**. The elements in compounds lose their original identity; they are in chemical combination with each other. The properties of a compound may be quite different from the original properties of its constituent elements. Water is a compound which occurs naturally as a liquid. Its constituent elements, hydrogen and oxygen, occur naturally not as liquids but as gases.

A **mixture** occurs when several elements or compounds, or both, are mixed together but no new substance is formed. Air is a mixture of gases, mainly oxygen and nitrogen.

8.4 The atom and relative atomic mass (atomic weight)

Matter is currently thought to consist of extremely small particles. These particles are, as it were, the bricks from which the greater bulk of matter is made. The small particles of elements are called **atoms**. The atoms of each element have different characteristics which make them what they are. Apart from an obvious difference of substance, another characteristic is that their masses are different. To identify this difference, a relative scale of masses has been compiled. The reference of an element to this relative scale of mass is commonly called its **atomic weight**. It is now recommended that it be called the **relative atomic mass**.

The lightest of all elements is hydrogen and, for the relative atomic mass scale, its mass was originally taken as unity. The other elements were then compared for mass against the relative atomic mass of hydrogen. An oxygen atom was found to be 16 times the mass of the hydrogen atom, so the relative atomic mass of oxygen was 16. A carbon atom was found to be 12 times the mass of the hydrogen atom, so the relative atomic mass of carbon was 12.

More accurate work on relative atomic masses, assuming the relative atomic mass of oxygen to be 16, has shown that small corrections are necessary. For example, with the relative atomic mass of oxygen as 16, the relative atomic mass of hydrogen is 1.008.

More recently it has been suggested that the relative atomic mass scale be based on an atomic mass unit (a.m.u.) which is equal to 1/12 the mass of the neutral carbon atom ^{12}C. For the general analysis of the combustion of fuels in thermodynamic engines, this degree of accuracy is not required.

8.5 The molecule and relative molecular mass (molecular weight)

Some elements do not normally exist as structures of single atoms but form themselves into structures of minute particles containing two atoms; examples are oxygen, hydrogen and nitrogen.

In a similar way, when compounds are formed, they are again made up of minute particles, and each of these particles is made up of two or more atoms. For example, a particle of carbon dioxide is made up of one atom of carbon in chemical combination with two atoms of oxygen. Such minute particles, which contain more than one atom, are called **molecules**.

A molecule is made up of atoms, so its mass relative to the relative atomic mass scale can be established. The mass of a molecule so determined is commonly called its **molecular weight**. It is now recommended that it be called the **relative molecular mass**.

Take the case of carbon dioxide.

Relative mass of one atom of carbon $= 12$

Relative mass of two atoms of oxygen $= 2 \times 16 = 32$

$\therefore$ Relative molecular mass of carbon dioxide $= 12 + 32 = 44$

Molecules containing two atoms, such as oxygen, hydrogen and carbon monoxide, are **diatomic**. Molecules containing three atoms, such as carbon dioxide and water, are **triatomic**. Molecules containing more than three atoms are **polyatomic**.

8.6 Chemical symbols

To save writing them out in full, chemical names are abbreviated using conventional symbols.

Carbon is written C

Sulphur is written S

To indicate the structure of a molecule, subscripts may be written on one or more of its symbols. Single atoms, such as carbon and sulphur (above), take no subscripts. The figure '1' should be there but is never written.

The element symbols for oxygen, hydrogen and nitrogen are O, H and N respectively. But these gases form diatomic molecules, so

Oxygen is written O_2

Hydrogen is written H_2

Nitrogen is written N_2

In a similar way, compounds have subscripts given to their constituents indicating the number of atoms of each constituent per molecule.

Carbon dioxide is written CO_2

indicating that there is one atom of carbon combined with two atoms of oxygen to produce one molecule of carbon dioxide.

Water is written H_2O

indicating that two atoms of hydrogen combined with one atom of oxygen produce one molecule of water. In the water molecule a single atom of oxygen can exist in equilibrium with two atoms of hydrogen, but on its own, oxygen forms into molecules of two atoms.

The number of molecules under discussion is written in front of the chemical formula.

$2H_2O$ means two molecules of water
$4CO_2$ means four molecules of carbon dioxide

Table of some relative atomic and molecular masses

Substance	Symbol	Relative atomic mass	Relative molecular mass
Elements			
Carbon	C	12	–
Hydrogen	H_2	1	2
Oxygen	O_2	16	32
Nitrogen	N_2	14	28
Sulphur	S	32	–
Compounds			
Carbon monoxide	CO	–	28
Carbon dioxide	CO_2	–	44
Water	H_2O	–	18

8.7 Air

During combustion a fuel always reacts with oxygen and liberates energy. The supply of oxygen is nearly always obtained from air. A knowledge of the constituents of air is therefore required to find the amount of oxygen in a given bulk.

Now air contains many gases as well as oxygen; nitrogen, argon, helium, neon, krypton, xenon and carbon dioxide are also present, together with some water vapour. Of all these constituents, oxygen and nitrogen make up the main bulk. So much so, in fact, that for most combustion purposes it is usual to assume the air consists entirely of oxygen and nitrogen.

Oxygen is the reacting agent during the combustion of a fuel. This oxygen cannot be obtained without it being accompanied by nitrogen. Now nitrogen is an inert gas, meaning that it does not take part in the combustion reaction. But it will slow down the combustion reaction because it will interfere with the necessary contact between the oxygen and the fuel. Nitrogen will also absorb some of the energy liberated by the combustion and will therefore be responsible for a lower combustion temperature. But the presence of nitrogen is not exclusively detrimental. Combustion with pure oxygen would be very rapid and would produce extremely high temperatures. Very rapid combustion would probably produce control difficulties and excessive combustion temperatures would produce damage to materials.

A gas has both mass and volume. The constituents of air are therefore given as percentage composition by mass and by volume. Neglecting all other gases but oxygen and nitrogen, the composition is given in the table.

	By mass	By volume
Oxygen	23.2%	21%
Nitrogen	76.8%	79%

But it is very common in problems to be given a mass analysis

Oxygen	23%
Nitrogen	77%

8.8 Combustion equations – stoichiometry

The chemical process of combustion is written out in the form of equations. Consider the combustion of carbon, C, with oxygen, O_2, which results in the production of carbon dioxide, CO_2, if the combustion is complete. Begin the combustion equation by writing

$$C + O_2$$

This means that oxygen is reacted with carbon. The product of combustion is introduced by writing

$$C + O_2 = CO_2$$

This means that oxygen is reacted with carbon to produce carbon dioxide. But now a check is made to see whether the numbers of carbon and oxygen atoms on the left-hand side are the same as the numbers of carbon and oxygen atoms on the right-hand side. A check is made because the amount of matter present must always be the same.

In the equation

$$C + O_2 = CO_2$$

there is one atom of carbon on each side and two atoms of oxygen on each side. This equation balances so it is the complete equation for the combustion of carbon with oxygen to form carbon dioxide.

A balanced chemical equation is called a **stoichiometric** equation; the carbon combustion equation represents the correct balance between the amount of fuel provided and the amount of oxygen supplied. Nitrogen will also appear in the stoichiometric equation when the combustion of a fuel in air is considered.

Now take the case of the combustion of hydrogen, H_2, with oxygen, O_2, to form water, H_2O. Begin by writing

$$H_2 + O_2 = H_2O$$

This means *react oxygen with hydrogen to form water*. Note that hydrogen is written H_2 and oxygen is written O_2 because they are diatomic molecules. In their free state, it is impossible to have single atoms of hydrogen or oxygen.

Now check the balance of atoms on both sides of the equation. On the left-hand side there are two atoms of hydrogen. On the right-hand side there are two atoms of hydrogen. The hydrogen balances.

On the left-hand side there are two atoms of oxygen. On the right-hand side there is only one atom of oxygen in combination with the hydrogen. The oxygen does not balance. The oxygen molecule contains two oxygen atoms whereas the water molecule contains only one oxygen atom, so the oxygen molecule will form two water molecules.

To accommodate this, the combustion equation may be extended to

$$H_2 + O_2 = 2H_2O$$

The oxygen atoms balance but now there are four hydrogen atoms on the right-hand side, which does not balance on the left-hand side. This may be corrected by writing

$$2H_2 + O_2 = 2H_2O$$

Now the atoms on both sides completely balance so this is the stoichiometric combustion equation of oxygen with hydrogen to produce water.

Now consider a hydrocarbon such as methane, a gas, whose chemical symbol is CH_4. Sometimes called marsh gas, it contains both carbon and hydrogen. For the complete combustion with oxygen, the carbon will form carbon dioxide and the hydrogen will form water. Begin by writing

$$CH_4 + O_2 = CO_2 + H_2O$$

Check the balance of atoms. In the molecule of CH_4 there is one atom of carbon; this will produce one molecule of CO_2 because there is one atom of carbon in a molecule of CO_2. One molecule of CH_4 also contains four atoms of hydrogen. In a molecule of water there are two atoms of hydrogen, so the four atoms of hydrogen will form two molecules of water. Thus the combustion equation may be modified as

$$CH_4 + O_2 = CO_2 + 2H_2O$$

Now to balance the oxygen. On the right-hand side there are four atoms of oxygen which may be balanced on the left-hand side by writing $2O_2$. Hence the stoichiometric combustion equation of methane with oxygen is written

$$CH_4 + 2O_2 = CO_2 + 2H_2O$$

Methane, CH_4, commonly forms the larger part of natural gas piped to domestic and industrial users.

As a further example, consider the combustion of hexane, C_6H_{14}, a paraffin, with oxygen. Again this may be started by writing

$$C_6H_{14} + O_2 = CO_2 + 2H_2O$$

indicating that the carbon will form CO_2 and the hydrogen will form H_2O.

Now C_6H_{14} contains six carbon atoms which will form $6CO_2$; it also contains 14 hydrogen atoms which will form $7H_2O$. Hence the combustion equation may be extended to

$$C_6H_{14} + O_2 = 6CO_2 + 7H_2O$$

Now to balance up the oxygen from that now shown on the right-hand side.

With the CO_2 there are $6O_2$
With the H_2O there are $3.5O_2$

Hence the total oxygen is $9.5O_2$. The combustion equation may now be written

$$C_6H_{14} + 9.5O_2 = 6CO_2 + 7H_2O$$

Although the atoms on both sides are balanced, it is impossible to have half a molecule as suggested in $9.5O_2$. This can be corrected by multiplying throughout by 2. Hence the equation becomes

$$2C_6H_{14} + 19O_2 = 12CO_2 + 14H_2O$$

This equation is now correct; it is the stoichiometric combustion equation of hexane with oxygen.

It must be remembered that chemical reactions always follow definite natural laws, consequently, definite natural substances are formed. For example, the reaction of oxygen with carbon could not be considered as forming C_3O_{16}! The reaction of oxygen with carbon either forms carbon monoxide, CO, if incomplete, or carbon dioxide, CO_2, if complete. The reactions are fixed by natural laws and cannot be altered.

8.9 Combustion analysis by mass and by volume

A primary objective of combustion analysis is to determine the amount of air required to burn a fuel and then to determine how much of each combustion product has been formed. Some analyses are made by mass, some by volume and some by both.

8.9.1 Analysis by mass – gravimetric analysis

All chemical substances are given a relative atomic mass or relative molecular mass. These relative masses are used in determining the combining masses and product masses associated with combustion.

For example, consider the combustion of hydrogen with oxygen to form water. This has the combustion equation

$$2H_2 + O_2 = 2H_2O$$

The relative atomic mass of hydrogen is 1. Thus the relative mass of the hydrogen present here is

$$2 \times 2 = 4$$

Similarly, the relative mass of the oxygen present here is

$$2 \times 16 = 32$$

The relative mass of the water is

$$2 \times (2 + 16) = 2 \times 18 = 36$$

Thus, it appears that

$$4 \text{ masses } H_2 \text{ combined with } 32 \text{ masses } O_2 = 36 \text{ masses } H_2O$$

Dividing throughout by 4, then

$$1 \text{ mass } H_2 + 8 \text{ masses } O_2 = 9 \text{ masses } H_2O$$

Now the mass chosen can be on any mass scale. Assume the mass to be in kilogrammes, then

$$1 \text{ kg } H_2 + 8 \text{ kg } O_2 = 9 \text{ kg } H_2O$$

These are the necessary combining masses. Note that the mass of the product formed, in this case H_2O, is the same as the combined masses of the constituents.

Oxygen is contained in air and is 23.2 per cent by mass, so the 8 kg O_2 above is 23.2 per cent of the mass of air theoretically necessary for the complete combustion of the H_2.

Hence the mass of air required to completely burn 1 kg H_2 is

$$8/0.232 = 34.5 \text{ kg}$$

Of this 34.5 kg, there are 8 kg O_2 so the nitrogen, N_2, present is

$$34.5 - 8 = 26.5 \text{ kg}$$

But this N_2 is inert; it does not take any part in the combustion process itself. There is as much N_2 after combustion as there was before. The above analysis shows

> If 1 kg H_2 is burnt in 34.5 kg of air then the products of combustion will be 9 kg H_2O together with 26.5 kg N_2.

The mass of oxygen or air calculated above is called the **stoichiometric mass** because it is the minimum mass required to completely burn 1 kg H_2. It is sometimes called the **theoretical mass**, the **correct mass** or the **minimum mass**.

8.9.2 Analysis by volume

Before the analysis by volume can be carried out, it is necessary to investigate **Avogadro's hypothesis**; this states:

> Equal volumes of different gases at the same pressure and temperature contain the same number of molecules.

From this it follows that proportions by numbers of molecules are also proportions by volumes. Thus, if there are twice the number of molecules present for gas 1 than for gas 2, gas 1 has twice the volume of gas 2, at the same pressure and temperature. This hypothesis can be used in the solution of combustion volumetric analysis.

Consider the complete combustion of hydrogen. The combustion equation is

$$2H_2 + O_2 = 2H_2O$$

The numbers in front of the chemical symbols are the numbers of molecules of those chemicals. Thus

$2H_2$ means two molecules of H_2
O_2 means one molecule of O_2
$2H_2O$ means two molecules of H_2O

From Avogadro's hypothesis, proportions by molecules are also proportions by volume, so the above combustion equation suggests that

2 volumes H_2 combine with 1 volume O_2 to produce 2 volumes of H_2O

Dividing throughout by 2

$$1 \text{ vol } H_2 + 0.5 \text{ vol } O_2 = 1 \text{ vol } H_2O$$

Assume that volume is measured in cubic metres, then

$$1 \text{ m}^3\ H_2 + 0.5 \text{ m}^3\ O_2 = 1 \text{ m}^3\ H_2O$$

Note that there has been an overall volumetric contraction of 0.5 m^3 from 1.5 m^3 of gas before combustion to 1 m^3 of water vapour after combustion. It is important to realise that after combustion the substance is water vapour and this phase has a volume of only 1 m^3. If it condenses then its volume is negligible compared with its volume as a vapour; the overall volumetric contraction is almost 1.5 m^3.

The reason for this contraction in volume after combustion is due to the recombination of the atoms from their original form as diatomic molecules, H_2 and O_2, into the new triatomic form of H_2O. There are the same numbers of atoms before and after combustion, so there will be fewer triatomic molecules formed than the original diatomic molecules. From Avogadro's hypothesis, proportions by molecules are proportions by volume; in this case there are fewer molecules after combustion than before, so the volume after combustion is smaller than before combustion.

Once again, an analysis of the air and its associated nitrogen can be made, this time by volume. Air contains 21 per cent O_2 by volume. Hence 0.5 m^3 O_2 is contained in

$$0.5/0.21 = 2.38 \text{ m}^3 \text{ air}$$

In this air there will be

$$2.38 - 0.5 = 1.88 \text{ m}^3 \text{ N}_2$$

Hence

$$1 \text{ m}^3 \text{ H}_2 + 2.38 \text{ m}^3 \text{ air} = 1 \text{ m}^3 \text{ H}_2\text{O} + 1.88 \text{ m}^3 \text{ N}_2$$

And once again, the N_2 plays no part in the combustion; it has the same volume before and after.

The volume of oxygen or air calculated above is called the **stoichiometric volume** because it is the minimum volume required to completely burn 1 m^3 H_2. It is sometimes called the **theoretical volume**, the **correct volume** or the **minimum volume**.

In a later section it will be shown that combustion analysis can be made using a quantity known as the **mole (mol)**. For now the present analysis technique will be adopted. The solution in either case will be the same.

8.10 Complete combustion of carbon to carbon dioxide

8.10.1 Analysis by mass

$$\begin{aligned}
&\text{Combustion equation: } \text{C} + \text{O}_2 &&= \text{CO}_2\\
&\text{Proportion by mass: } 12 + (2 \times 16) &&= 12 + (2 \times 16)\\
&\qquad 12 + 32 &&= 44\\
&\text{Divide through by 12: } 1 + 2\tfrac{2}{3} &&= 3\tfrac{2}{3}
\end{aligned}$$

or

$$1 \text{ kg C} + 2\tfrac{2}{3} \text{ kg O}_2 = 3\tfrac{2}{3} \text{ kg CO}_2$$

Hence

$$2\tfrac{2}{3} \text{ kg O}_2 = \text{stoichiometric mass of O}_2$$

Now $2\frac{2}{3}$ kg O_2 are contained in

$$2\tfrac{2}{3}/0.232 = 11.5 \text{ kg air}$$
$$= \text{stoichiometric mass of air}$$

This air will contain

$$11.5 - 2.66 = 8.84 \text{ kg } N_2$$

Hence

$$1 \text{ kg C} + 11.5 \text{ kg air} = 3\tfrac{2}{3} \text{ kg } CO_2 + 8.84 \text{ kg } N_2$$

8.10.2 Analysis by volume

Combustion equation: $C + O_2 = CO_2$

Now carbon, in its natural state, is a solid, but Avogadro's hypothesis is concerned with gases, so it will not apply to carbon. On the other hand, O_2 and CO_2 are gases, so they can be analysed.

In each case the combustion equation shows one molecule of each of these gases, so

1 m^3 O_2 when reacted with C, during complete combustion, will produce 1 m^3 CO_2

1 m^3 CO_2 is contained in

$$\frac{1}{0.21} = 4.76 \text{ m}^3 \text{ air}$$

This air will contain

$$4.76 - 1 = 3.76 \text{ m}^3 \text{ } N_2$$

Hence

4.76 m^3 air when reacted with C, during complete combustion, will produce 1 m^3 CO_2 together with 3.76 m^3 N_2

8.11 Incomplete combustion of carbon to carbon monoxide

The combustion of carbon to carbon monoxide is **incomplete**. This is because carbon monoxide is a fuel, so it is capable of being burnt to produce carbon dioxide. When carbon is burned to carbon monoxide, only part of the available energy is obtained, so the combustion is said to be incomplete. It should be noted that carbon monoxide, CO, is a very toxic gas that should not be inhaled and should be treated with care and caution.

8.11.1 Analysis by mass

Combustion equation:	$2C + O_2$	$= 2CO$
Proportion by mass:	$(2 \times 12) + (2 \times 16)$	$= 2(12 + 16)$
	$24 + 32$	$= 56$
Divide through by 24:	$1 + 1\frac{1}{3}$	$= 2\frac{1}{3}$

or

$$1 \text{ kg C} + 1\tfrac{1}{3} \text{ kg } O_2 = 2\tfrac{1}{3} \text{ kg CO}$$

Hence

$$1\tfrac{1}{3}\ \text{kg O}_2 = \text{stoichiometric mass of O}_2$$

Now $1\frac{1}{3}$ kg O_2 are contained in

$$1\tfrac{1}{3}/0.232 = 5.75\ \text{kg air}$$
$$= \text{stoichiometric mass of air}$$

This air will contain

$$5.75 - 1.33 = 4.43\ \text{kg N}_2$$

Hence

$$1\ \text{kg C} + 5.75\ \text{kg air} = 2\tfrac{1}{3}\ \text{kg CO} + 4.42\ \text{kg N}_2$$

8.11.2 Analysis by volume

Here

1 m^3 O_2 when reacted with C, during incomplete combustion, will produce 2 m^3 CO

1 m^3 O_2 is contained in

$$\frac{1}{0.21} = 4.76\ \text{m}^3\ \text{air}$$

This will contain

$$4.76 - 1 = 3.76\ \text{m}^3\ \text{N}_2$$

Hence

4.76 m^3 air when reacted with C, during incomplete combustion, will produce 2 m^3 CO together with 3.76 m^3 N_2

8.12 Complete combustion of carbon monoxide to carbon dioxide

8.12.1 Analysis by mass

$$\begin{aligned}
&\text{Combustion equation: } 2\text{C} + \text{O}_2 &&= 2\text{CO}_2\\
&\text{Proportion by mass: } 2(12 + 16) + (2 \times 16) &&= 2[12 + (2 \times 16)]\\
&\qquad\qquad 56 + 32 &&= 88\\
&\text{Divide through by 56: } 1 + \tfrac{4}{7} &&= 1\tfrac{4}{7}
\end{aligned}$$

or

$$1\ \text{kg CO} + \tfrac{4}{7}\ \text{kg O}_2 = 1\tfrac{4}{7}\ \text{kg CO}_2$$

Hence

$$\tfrac{4}{7}\ \text{kg O}_2 = \text{stoichiometric mass of O}_2$$

Now $\frac{4}{7}$ kg O_2 is contained in

$$\frac{4}{7 \times 0.232} = 2.46\ \text{kg air}$$
$$= \text{stoichiometric mass of air}$$

This air will contain

$$2.46 - \tfrac{4}{7} = 2.46 - 0.57$$
$$= 1.89 \text{ kg } N_2$$

Hence

$$1 \text{ kg CO} + 2.46 \text{ kg air} = 1.57 \text{ kg } CO_2 + 1.89 \text{ kg } N_2$$

8.12.2 Analysis by volume

Carbon monoxide is a gas so a complete volumetric analysis is possible.

Combustion equation: $2CO + O_2 = 2CO_2$
Proportion by volume: $2 + 1 = 2$
Divide through by 2: $1 + 0.5 = 1$

or

$$1 \text{ m}^3 \text{ CO} + 0.5 \text{ m}^3 \text{ } O_2 = 1 \text{ m}^3 \text{ } CO_2$$

Hence

$$0.5 \text{ m}^3 \text{ } O_2 = \text{stoichiometric volume of } O_2$$

Note that there is a volumetric contraction of 0.5 m^3 after combustion. Now 0.5 m^3 is contained in

$$\frac{0.5}{0.21} = 2.38 \text{ m}^3 \text{ air}$$

$$= \text{stoichiometric volume of air}$$

This air will contain

$$2.38 - 0.5 = 1.88 \text{ m}^3 \text{ } N_2$$

Hence

$$1 \text{ m}^3 \text{ CO} + 2.38 \text{ m}^3 \text{ air} = 1 \text{ m}^3 \text{ } CO_2 + 1.88 \text{ m}^3 \text{ } N_2$$

8.13 Complete combustion of methane

Methane, CH_4, is the major constituent of Natural Gas obtained from oilfields.

8.13.1 Analysis by mass

Combustion equation: $CH_4 + 2O_2 = CO_2 + 2H_2O$
Proportion by mass: $(12 + 4) + 2(2 \times 16) = (12 + 32) + 2(2 + 16)$
$16 + 64 = 44 + 36$
Divide through by 16: $1 + 4 = 2\tfrac{3}{4} + 2\tfrac{1}{4}$

or

$$1\ \text{kg CH}_4 + 4\ \text{kg O}_2 = 2\tfrac{3}{4}\ \text{kg CO}_2 + 2\tfrac{1}{4}\ \text{kg H}_2\text{O}$$

Hence

$$4\ \text{kg O}_2 = \text{stoichiometric mass of O}_2$$

Now 4 kg O_2 is contained in

$$\frac{4}{0.232} = 17.24\ \text{kg air}$$
$$= \text{stoichiometric mass of air}$$

This air will contain

$$17.24 - 4 = 13.24\ \text{kg N}_2$$

Hence

$$\begin{aligned} 1\ \text{kg CH}_4 + 17.24\ \text{kg air} &= 2\tfrac{3}{4}\ \text{kg CO}_2 + 2\tfrac{1}{4}\ \text{kg H}_2\text{O} + 13.24\ \text{kg N}_2 \\ &= 2.75\ \text{kg CO}_2 + 2.25\ \text{kg H}_2\text{O} + 13.24\ \text{kg N}_2 \end{aligned}$$

8.13.2 *Analysis by volume*

Methane is a gas so a complete volumetric analysis is possible.

$$\text{Combustion equation: } CH_4 + 2O_2 = CO_2 + 2H_2O$$
$$\text{Proportion by volume: } 1 + 2 \qquad = 1 + 2$$

or

$$1\ \text{m}^3\ \text{CH}_4 + 2\ \text{m}^3\ \text{O}_2 = 1\ \text{m}^3\ \text{CO}_2 + 2\ \text{m}^3\ \text{H}_2\text{O}$$

Hence

$$2\ \text{m}^3\ \text{O}_2 = \text{stoichiometric volume of O}_2$$

Note that there is no volumetric contraction. Now 2 m^3 O_2 is contained in

$$\frac{2}{0.21} = 9.52\ \text{m}^3\ \text{air}$$
$$= \text{stoichiometric volume of air}$$

This air will contain

$$9.52 - 2 = 7.52\ \text{m}^3\ \text{N}_2$$

Hence

$$1\ \text{m}^3\ \text{CH}_4 + 9.52\ \text{m}^3\ \text{air} = 1\ \text{m}^3\ \text{CO}_2 + 2\ \text{m}^3\ \text{H}_2\text{O} + 7.52\ \text{m}^3\ \text{N}_2$$

In this volumetric analysis, it is again assumed that the H_2O formed remains as a vapour.

8.14 Complete combustion of sulphur to sulphur dioxide

It should be noted that sulphur dioxide is a toxic gas; it is a serious atmospheric pollutant in some industrial areas and places of high motor traffic density. Over long periods it can produce extensive surface damage to building material.

8.14.1 Analysis by mass

Combustion equation: $S + O_2 = SO_2$

Proportion by mass: $32 + 32 = 64$

Divide through by 32: $1 + 1 = 2$

or

$$1\ \text{kg S} + 1\ \text{kg O}_2 = 2\ \text{kg SO}_2$$

Hence

$$1\ \text{kg O}_2 = \text{stoichiometric mass of O}_2$$

Now 1 kg O_2 is contained in

$$\frac{1}{0.232} = 4.3\ \text{kg air}$$

$$= \text{stoichiometric mass of air}$$

This air will contain

$$4.3 - 1 = 3.3\ \text{kg N}_2$$

Hence

$$1\ \text{kg S} + 4.3\ \text{kg air} = 2\ \text{kg SO}_2 + 3.3\ \text{kg N}_2$$

8.14.2 Analysis by volume

Here again sulphur is a solid, so

1 m^3 O_2 reacted with S will produce 1 m^3 SO_2

1 m^3 O_2 is contained in

$$\frac{1}{0.21} = 4.76\ \text{m}^3\ \text{air}$$

This air will contain

$$4.76 - 1 = 3.76\ \text{m}^3\ \text{N}_2$$

Hence

4.76 m^3 air when reacted with sulphur will produce 2 m^3 SO_2 together with 3.76 m^3 N_2.

8.15 Stoichiometric mass of air for the complete combustion of a fuel

If the analysis of a fuel is given by mass, then proceed as follows:

- Determine the mass of oxygen required for each constituent. From this, find the total mass of oxygen by adding all the separate masses required.
- Subtract any oxygen which may be in the fuel because this does not have to be supplied.
- Stoichiometric mass of air = O_2 required/0.232.

Example 8.1 *A fuel consists of 72 per cent carbon, 20 per cent hydrogen and 8 per cent oxygen by mass. Determine the stoichiometric mass of air required to completely burn 1 kg of this fuel.*

SOLUTION

A convenient form of solution here is to tabulate results.

Constituent	Mass of constit. (kg/kg fuel)	O_2 required (kg/kg constit.)	O_2 required (kg/kg fuel)
C	0.72	$2\frac{2}{3}$	$0.72 \times 2\frac{2}{3} = 1.92$
H_2	0.2	8	$0.2 \times 8 = 1.6$
O_2	0.08	–	-0.08

$$O_2 \text{ required} = 1.92 + 1.6 - 0.08 = \mathbf{3.44\ kg/kg\ fuel}$$

$$\therefore \quad \text{Stoichiometric air required} = \frac{3.44}{0.232} = \mathbf{14.8\ kg/kg\ fuel}$$

Suppose the fuel contained carbon and hydrogen only. As 1 kg carbon requires 11.5 kg air and 1 kg hydrogen requires 34.5 kg air, it follows that

$$\text{Stoichiometric mass of air} = (11.5\text{C} + 34.5\text{H}_2)\ \text{kg}$$

where C and H_2 are the masses of carbon and hydrogen per kilogram of fuel.

Example 8.2 *Determine the stoichiometric mass of air required to completely burn 1 kg of heptane,* C_7H_{16}.

SOLUTION

The first part of the problem is to determine the masses of carbon and hydrogen per kilogram of fuel. This is achieved by an analysis of its relative molecular mass.

$$\begin{aligned}\text{Relative molecular mass of } C_7H_{16} &= (7 \times 12) + (1 \times 16)\\ &= 84 + 16\\ &= 100\end{aligned}$$

From this it appears that the fuel is made up of

$$\frac{84}{100} \text{ parts carbon} = 84\% \text{ by mass}$$

and

$$\frac{16}{100} \text{ parts hydrogen} = 16\% \text{ by mass}$$

Hence in every 1 kg of fuel there is 0.84 kg C and 0.16 kg H_2.

Thus

$$\begin{aligned}\text{Stoichiometric mass of air} &= (11.5 \times 0.84) + (34.5 \times 0.16)\\ &= 9.66 + 5.52\\ &= \mathbf{15.18\ kg/kg\ fuel}\end{aligned}$$

This problem could also have been solved by using the combustion equation.

Combustion equation: $C_7H_{16} + 11O_2 = 7CO_2 + 8H_2O$
Proportion by mass: $100 + (11 \times 32)$
$100 + 352$
Divide through by 100: $1 + 3.52$

Hence

1 kg C_7H_{16} requires 3.52 kg O_2

$$\begin{aligned}\text{Stoichiometric mass of air} &= \frac{3.52}{0.232}\\ &= \mathbf{15.18\ kg/kg\ fuel}\end{aligned}$$

8.16 The products of combustion by mass

After burning, a fuel will produce certain quantities of combustion products. The products of combustion of individual constituents such as carbon and hydrogen have already been dealt with. The combustion products of a fuel are best illustrated by an example.

Example 8.3 *A fuel oil consists of the following percentage analysis by mass:*

82%C, 12%H_2, 2%O_2, 1%S, 3%N_2

Determine the stoichiometric mass of air required to completely burn 1 kg of this fuel and determine the products of combustion both by mass and as a percentage.

SOLUTION

Again, the tabular form is suitable for this calculation.

Constituent	Mass of constit. (kg/kg fuel)	O_2 required (kg/kg fuel)	Products of combustion (kg/kg fuel)			
			CO_2	H_2O	SO_2	N_2
C	0.82	$0.82 \times 2\frac{2}{3} = 2.19$	$0.82 \times 3\frac{2}{3} = 3.01$	–	–	$0.82 \times 8.84 = 7.25$
H_2	0.12	$0.12 \times 8 = 0.96$	–	$0.12 \times 9 = 1.08$		$0.12 \times 26.5 = 3.18$
O_2	0.02	−0.02	–	–	–	$-0.02 \times \frac{0.768}{0.232} = -0.066$
S	0.01	$0.01 \times 1 = 0.01$	–	–	$0.01 \times 2 = 0.02$	$0.01 \times 3.3 = 0.033$
N_2	0.03	–	–	–	–	0.03

From the table

$$\text{Total } O_2 \text{ required} = 2.19 + 0.96 + 0.01 - 0.02$$
$$= 3.16 - 0.02$$
$$= \mathbf{3.14\ kg/kg\ fuel}$$

$$\therefore \quad \text{Stoichiometric air} = \frac{3.14}{0.232} = \mathbf{13.54\ kg/kg\ fuel}$$

The products of combustion need a little explanation. Consider the carbon. It has been shown that when 1 kg C is burnt with the stoichiometric air, the products of combustion will be $3\frac{2}{3}$ kg CO_2 and 8.84 kg N_2.

Hence with 0.82 kg C, the products will be

$$0.82 \times 3\tfrac{2}{3} = \mathbf{3.01\ kg\ CO_2}$$

and

$$0.82 \times 8.84 = \mathbf{7.25\ kg\ N_2}$$

The same procedure has been adopted in the case of the H_2 and the S. Note that the N_2 originally in the fuel will appear as the same quantity in the products. Note also that since there is 0.02 kg O_2 present per kilogram of fuel, so the N_2 content in the products of combustion must be reduced by the amount of N_2 associated with this 0.02 kg O_2 because it was not supplied by the air.

Now the air associated with 0.02 kg O_2 = 0.02/0.232 kg. But 1 kg air contains 0.768 kg N_2, hence

$$N_2 \text{ associated with } 0.02 \text{ kg } O_2 = 0.02 \times \frac{0.768}{0.232}$$
$$= \mathbf{0.066\ kg}$$

This is the quantity of N_2 which must be subtracted from the total N_2 that assumes all the O_2 was supplied from the air.

Hence, actual N_2 from the air is

$$7.25 + 3.18 + 0.033 + 0.03 - 0.066 = 10.493 - 0.066$$
$$= \mathbf{10.427\ kg/kg\ fuel}$$

Hence the products of combustion per kilogram of fuel are

3.01 kg CO_2, 1.08 kg H_2O, 0.02 kg SO_2, 10.427 kg N_2

The total mass of combustion products per kilogram of fuel is

$$3.01 + 1.08 + 0.02 + 10.427 = \mathbf{14.537\ kg}$$

Hence, percentage analysis of products by mass is

$$CO_2 = \frac{3.01}{14.537} \times 100 = 20.7\%$$

$$H_2O = \frac{1.08}{14.537} \times 100 = 7.43\%$$

$$SO_2 = \frac{0.02}{14.537} \times 100 = 0.14\%$$

$$N_2 = \frac{10.427}{14.537} \times 100 = 71.73\%$$

Note that from the total mass of products per kilogram of fuel it is possible to obtain the mass of air per kilogram of fuel; it is

$$14.537 - 1 = \mathbf{13.537\ kg}$$

This is so because

Mass of fuel + Mass of air = Mass of products

Note also that the N_2 in the products could also have been obtained from the fact that 13.54 kg of air is required per kilogram of fuel.

Now air contains 76.8 per cent N_2 by mass

$$\therefore \quad N_2 \text{ in } 13.54 \text{ kg air} = 13.54 \times 0.768 = \mathbf{10.397\ kg}$$

But there is 0.03 kg N_2 per kilogram of fuel

$$\therefore \quad N_2 \text{ in products} = 10.397 + 0.03 = \mathbf{10.427\ kg}$$

This is as before.

8.17 Stoichiometric volume of air for the complete combustion of a fuel

If the analysis of a fuel is given by volume, then proceed as follows:

- Determine the volume of oxygen required for each constituent. From this, find the total volume of oxygen by adding all the separate volumes required.
- Subtract any oxygen which may be in the fuel because this does not have to be supplied.
- Stoichiometric volume of air = O_2 required/0.21.

Example 8.4 *The analysis by volume of a producer gas is as follows:*

14% H_2, 2% CH_4, 22% CO, 5% CO_2, 2% O_2, 55% N_2

Find the stoichiometric volume of air required for the complete combustion of 1 m^3 of this gas.

SOLUTION

Again, it is convenient to draw up a table.

Constituent	Volume of constit. (m^3/m^3 fuel)	O_2 required (m^3/m^3 constit.)	O_2 required (m^3/m^3 fuel)
H_2	0.14	0.5	$0.14 \times 0.5 = 0.07$
CH_4	0.02	2.0	$0.02 \times 2.0 = 0.04$
CO	0.22	0.5	$0.22 \times 0.5 = 0.11$
CO_2	0.05	–	–
O_2	0.02	–	−0.02
N_2	0.55	–	–

$$\begin{aligned} O_2 \text{ required} &= 0.07 + 0.04 + 0.11 - 0.02 \\ &= 0.22 - 0.02 \\ &= \mathbf{0.2\ m^3/m^3\ fuel} \end{aligned}$$

$$\therefore \quad \text{Stoichiometric air required} = \frac{0.02}{0.21} = \mathbf{0.952\ m^3/m^3\ fuel}$$

Suppose the fuel had contained only H_2, CO and CH_4. Since 1 m^3 of H_2 and CO require 2.38 m^3 air each and 1 m^3 CH_4 requires 9.52 m^3 air for complete combustion, it follows that

$$\text{Stoichiometric volume of air} = [2.38(H_2 + CO) + 9.52CH_4]\ m^3$$

where H_2, CO and CH_4 are the volumes of hydrogen, carbon monoxide and methane per cubic metre of fuel.

Example 8.5 *A fuel contains 45 per cent H_2, 40 per cent CO and 15 per cent CH_4 by volume. Determine the volume of air required to burn 1 m^3 of this fuel.*

SOLUTION

$$\begin{aligned}\text{Stoichiometric volume of air} &= 2.38 \times (0.45 + 0.4) + (9.52 \times 0.15)\\ &= (2.38 \times 0.85) + (9.52 \times 0.15)\\ &= 2.023 + 1.428\\ &= \mathbf{2.451\ m^3/m^3\ fuel}\end{aligned}$$

8.18 The products of combustion by volume

The products of combustion are always gaseous so they allow volumetric analysis to be performed. The combustion products of carbon and hydrogen have already been analysed; a fuel is best dealt with by an example.

Example 8.6 *A gas consists of*

14.2% CH_4, 5.9% CO_2, 36% CO, 40.5% H_2, 0.5% O_2, 2.9% N_2

Determine the stoichiometric volume of air for the complete combustion of 1 m^3 of this gas and also the products of combustion both in m^3/m^3 of gas and as a percentage.

SOLUTION

A table is again a useful way of setting out this problem.
From the table

$$\begin{aligned}\text{Stoichiometric volume of } O_2 \text{ required} &= 0.284 + 0.18 + 0.202\,5 - 0.005\\ &= 0.666\,5 - 0.005\\ &= \mathbf{0.661\,5\ m^3/m^3\ fuel}\end{aligned}$$

$$\begin{aligned}\therefore \quad \text{Stoichiometric volume of air required} &= \frac{0.661\,5}{0.21}\\ &= \mathbf{3.15\ m^3/m^3\ fuel}\end{aligned}$$

The products of combustion are dealt with as follows. Take the CH_4 for example. When 1 m^3 of CH_4 is burnt with the stoichiometric air, it will produce 1 m^3 CO_2, 2 m^3 H_2O and 7.52 m^3 N_2. Hence for 0.142 m^3 CH_4, the products will be

Constituent	Volume of constit. (m^3/m^3 fuel)	O_2 required (m^3/m^3 fuel)	Products of combustion (m^3/m^3 fuel)		
			CO_2	H_2O	N_2
CH_4	0.142	$0.142 \times 2 = 0.284$	$0.142 \times 1 = 0.142$	$0.142 \times 2 = 0.284$	$0.142 \times 7.52 = 1.068$
CO_2	0.059	–	0.059	–	–
CO	0.36	$0.36 \times 0.5 = 0.18$	$0.36 \times 1 = 0.36$	–	$0.36 \times 1.88 = 0.676$
H_2	0.405	$0.405 \times 0.5 = 0.202\,5$	–	$0.405 \times 1 = 0.405$	$0.405 \times 1.88 = 0.762$
O_2	0.005	-0.005	–	–	$-0.005 \times (0.79/0.21) = -0.019$
N_2	0.029	–	–	–	0.029

$0.142 \times 1 \quad = 0.142 \text{ m}^3 \text{ CO}_2$
$0.142 \times 2 \quad = 0.284 \text{ m}^3 \text{ H}_2\text{O}$
$0.142 \times 7.52 = 1.068 \text{ m}^3 \text{ N}_2$

The products for the CO and H_2 are dealt with in a similar manner. The CO_2 and the N_2 in the fuel appear as the same quantity in the products.

Note that there is O_2 in the fuel, so the N_2 in the products of combustion must be reduced by an amount proportional to this O_2 because it is not supplied by the air.

$$\text{Air associated with } 0.005 \text{ m}^3 \text{ O}_2 = \frac{0.005}{0.21} \text{ m}^3$$

$$\text{N}_2 \text{ associated with this air} = 0.005 \times \frac{0.79}{0.21} = \mathbf{0.019 \text{ m}^3}$$

This N_2 must be subtracted from the total N_2 that assumes all the O_2 was supplied by the air. Hence

$$\begin{aligned}\text{N}_2 \text{ in products} &= 1.068 + 0.676 + 0.762 + 0.029 - 0.019 \\ &= 2.535 - 0.019 \\ &= \mathbf{2.516 \text{ m}^3/\text{m}^3 \text{ fuel}}\end{aligned}$$

$$\begin{aligned}\text{CO}_2 \text{ in products} &= 0.142 + 0.059 + 0.36 \\ &= \mathbf{0.561 \text{ m}^3/\text{m}^3 \text{ fuel}}\end{aligned}$$

$$\begin{aligned}\text{H}_2\text{O in products} &= 0.284 + 0.405 \\ &= \mathbf{0.689 \text{ m}^3/\text{m}^3 \text{ fuel}}\end{aligned}$$

$$\begin{aligned}\text{Total volume of products} &= 2.516 + 0.561 + 0.689 \\ &= \mathbf{3.766 \text{ m}^3/\text{m}^3 \text{ fuel}}\end{aligned}$$

The percentage analysis of the products by volume is

$$\text{CO}_2 = \frac{0.561}{3.766} \times 100 = 14.9\%$$

$$\text{H}_2\text{O} = \frac{0.689}{3.766} \times 100 = 18.3\%$$

$$\text{N}_2 = \frac{2.516}{3.766} \times 100 = 66.8\%$$

8.19 Conversion of volumetric to gravimetric (mass) analysis

From Avogadro's hypothesis, proportions by molecules are also proportions by volume. And equal volumes of different gases at the same pressure and temperature contain the same number of molecules, so the mass of equal volumes will be proportional to the relative molecular masses of the gases. It follows that for a gas

Proportion by mass = Proportion by volume × Relative molecular mass (RMM)

Using this relationship, conversion of volumetric analysis to gravimetric analysis proceeds as follows, assuming the volumetric analysis is given as a percentage analysis. Multiply the percentage by volume of each constituent by its relative molecular mass. Then determine the sum of all the products, and hence determine the percentage analysis by mass.

Example 8.7 *A gas consists of 20 per cent CO_2, 70 per cent N_2 and 10 per cent O_2 by volume. Determine the percentage analysis of the gas by mass.*

SOLUTION
The conversion is conveniently displayed in tabular form.

Constituent	Percentage by volume	Percentage by volume × RMM	Percentage by mass
CO_2	20	20 × 44 = 880	$\frac{880}{3160} \times 100 =$ 27.9
N_2	70	70 × 28 = 1 960	$\frac{1960}{3160} \times 100 =$ 62.0
O_2	10	10 × 32 = 320	$\frac{320}{3160} \times 100 =$ 10.1
Total	100	3 160	100

8.20 Conversion of gravimetric (mass) to volumetric analysis

This is the reverse of conversion from volumetric to gravimetric analysis. Remember that it applies only to gaseous fuels or products.

To convert mass to volumetric analysis, divide the percentage mass by the relative molecular mass and sum the quotients, hence obtain the percentage by volume.

Example 8.8 *A gas consists of the following percentage analysis by mass:*

30% CO, 20% N_2, 15% CH_4, 25% H_2, 10% O_2

Determine the percentage composition of the gas by volume.

SOLUTION
The conversion is conveniently displayed in tabular form.

Constituent	Percentage by mass	$\frac{\text{Percentage by mass}}{\text{RMM}}$	Percentage by volume
CO	30	$\frac{30}{28} =$ 1.071	$\frac{1.071}{15.536} \times 100 =$ 6.9
N_2	20	$\frac{20}{28} =$ 0.714	$\frac{0.714}{15.536} \times 100 =$ 4.6
CH_4	15	$\frac{15}{16} =$ 0.938	$\frac{0.938}{15.536} \times 100 =$ 6.0
H_2	25	$\frac{25}{2} =$ 12.5	$\frac{12.5}{15.536} \times 100 =$ 80.5
O_2	10	$\frac{10}{32} =$ 0.313	$\frac{0.313}{15.536} \times 100 =$ 2.0
Total	100	15.536	100

8.21 *Dry flue or exhaust gas*

The analysis of flue or exhaust gas is required to determine whether or not a fuel is being burnt efficiently. If a fuel contains hydrogen, H_2O will be formed and this will appear initially as steam in the combustion products. If a fuel contains carbon, this can appear in the combustion products as CO_2, CO or both.

The analysis of the combustion products will depend upon the amount of air being supplied to the fuel and the type of fuel. If too little air is supplied, the initial fuel–air mixture is rich, meaning rich in fuel. Most fuels are hydrocarbons and all the hydrogen is nearly always burnt off to H_2O. Hence for a rich mixture, this leaves the carbon deficient in air. Although some carbon will be burnt off to CO_2, the remainder will form CO, showing incomplete combustion. The N_2 from the air will also appear in the products. There may also be a small amount of oxides of nitrogen (NO_x) in the products.

If too much air is supplied, the initial fuel–air mixture will be weak or lean, meaning weak or lean in fuel. For a hydrocarbon fuel, the hydrogen will burn off to H_2O, the carbon will all burn off to CO_2 and there will be some O_2 left, which will appear in the products together with the N_2 associated with the air.

If a percentage analysis of dry flue or exhaust gas is made from very weak to very rich mixtures of fuel and air, the graph obtained will have the general shape of Fig. 8.1. This shows that from a weak mixture the percentage CO_2 increases while the O_2 decreases. The CO_2 is at a maximum at the stoichiometric fuel–air ratio, where the O_2 has theoretically reduced to zero. Over to the rich side shows the CO_2 percentage decreasing and CO begins to appear; it increases as the richness increases. In practice, the O_2 and CO lines generally overlap slightly.

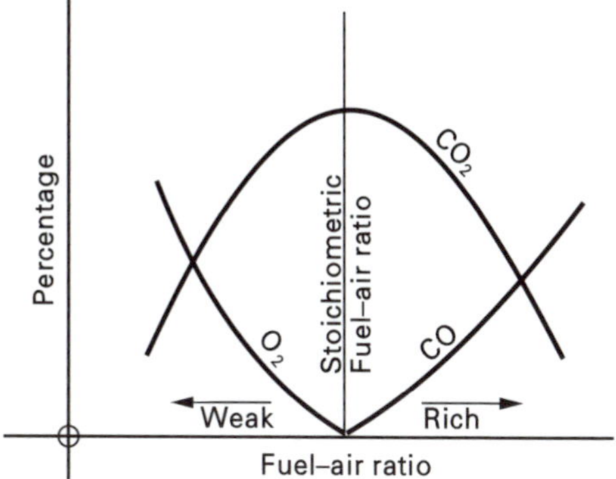

Fig. 8.1 Dry flue or exhaust gas analysis

It appears that, since all the hydrogen is nearly always burnt off to H_2O, the efficiency of combustion is generally a function of the degree to which CO_2, CO and O_2 appear in the products of combustion. As a practical point, no combustion in a furnace or engine is perfect. This is mainly due to the presence of N_2 in air, which limits the degree of intimacy between the fuel and oxygen. The mixing of the fuel and air, to improve intimacy, is achieved by the introduction of some turbulence to the air and atomisation or pulverisation of the fuel. Nevertheless, it is always possible that some oxygen will get through to the flue or exhaust uncombined with the fuel and will probably produce a small amount of CO. But this effect is generally small compared with the effect of deficient or excess air. Very often the CO content of the products is taken as a measure of the combustion efficiency. If the CO is high, there is a deficiency of air. The air is therefore increased until the CO is eliminated or all but eliminated. The introduction of too much air can be detected by the appearance of O_2 in the products.

Now because CO_2, CO and O_2 in the product show up the efficiency of combustion, very often they are the only gases to be considered in the flue or exhaust gas analysis. There is no H_2O, hence it is called **dry flue analysis** or **dry exhaust gas analysis**.

8.22 *Excess air*

If O_2 appears in a dry flue gas analysis it means that air has been supplied to the fuel beyond that required for complete combustion. From the dry flue gas analysis it is possible to determine the mass of excess air supplied.

The first procedure is to determine the mass of dry flue gas formed per kilogram of fuel burnt. First the volumetric analysis of the dry flue gas is converted into a mass analysis. From this an analysis is made of the mass of carbon per kilogram of flue gas; this is compared with the mass of carbon per kilogram of fuel. The comparison is relevant because, the carbon in the fuel ultimately appears in the flue gas combined with O_2 to form either CO_2 or CO, so it must be the same quantity.

Now the relative molecular mass of $CO_2 = 44$. This is made up of 12 parts C and 32 parts O_2. Hence in every 1 kg CO_2 there is $12/44 = 3/11$ kg C. Similarly the relative molecular mass of $CO = 28$. This is made up of 12 parts C and 16 parts O_2. Hence in every 1 kg CO_2 there is $12/28 = 3/7$ kg C. Thus in every 1 kg dry flue gas, if m_{CO_2} and m_{CO} are the masses of carbon dioxide and carbon monoxide, respectively, then

$$\text{Mass of C/kg flue gas} = (3/11 m_{CO_2} + 3/7 m_{CO_2}) \text{ kg}$$

If m = mass of C/kg fuel, then

$$\text{Mass of dry flue gas/kg fuel} = \frac{m}{(3/11 m_{CO} + 3/7 m_{CO})} \text{ kg}$$

Now from the mass analysis of the dry flue gas, the amount of O_2 per kilogram of dry flue gas is known, hence

$$\text{Mass of excess } O_2\text{/kg fuel} = \text{Mass of } O_2\text{/kg dry flue gas} \times \text{Mass of dry flue gas/kg fuel}$$

From this

$$\text{Mass of excess air/kg fuel} = \frac{\text{Mass of excess } O_2\text{/kg fuel}}{0.232}$$

If the stoichiometric air is calculated for the fuel in the usual manner, then

$$\text{Total mass of air/kg fuel} = \text{Stoichiometric air} + \text{Excess mass of air}$$

Example 8.9 *The dry exhaust gas from an oil engine had the following percentage composition by volume:*

8.85% CO_2, 1.2% CO, 6.8% O_2, 83.15% N_2

The fuel oil had a percentage composition by mass as follows:

84% C, 14% H_2, 2% O_2

Determine the mass of air supplied per kilogram of fuel burnt. Air contains 23 per cent O_2 by mass.

SOLUTION
First draw up a table.

Constituent	Mass of constit. (kg/kg fuel)	O_2 required (kg/kg constit.)	O_2 required (kg/kg fuel)
C	0.84	$2\frac{2}{3}$	$0.84 \times 2\frac{2}{3} = 2.24$
H_2	0.14	8	$0.14 \times 8 = 1.12$
O_2	0.02	–	-0.02

$$\begin{aligned}\text{Theoretical total } O_2 \text{ required} &= 2.24 + 1.12 - 0.02\\ &= 3.36 - 0.02\\ &= \mathbf{3.34\ kg/kg\ fuel}\end{aligned}$$

$$\therefore \quad \text{Theoretical mass of air} = 3.34/0.23 = \mathbf{14.5\ kg/kg\ fuel}$$

Using a new table, convert the volumetric analysis of the flue gas into a mass analysis.

Constituent	Percentage by volume	Percentage by volume × RMM	Percentage by mass
CO_2	8.85	$8.85 \times 44 = 389.4$	$\frac{389.4}{2968.8} \times 100 = 13.12$
CO	1.2	$1.2 \times 28 = 33.6$	$\frac{33.6}{2968.8} \times 100 = 1.13$
O_2	6.8	$6.8 \times 32 = 217.6$	$\frac{217.6}{2968.8} \times 100 = 7.33$
N_2	83.15	$83.15 \times 28 = 2\,328.2$	$\frac{2328.2}{2968.8} \times 100 = 78.42$
		Total = 2 968.8	

$$\begin{aligned}\text{Mass of C/kg of dry flue gas} &= (0.131\,2 \times 12/44) + (0.011\,3 \times 12/28)\\ &= 0.035\,78 + 0.004\,84\\ &= \mathbf{0.040\,62\ kg}\end{aligned}$$

Mass of C/kg fuel = 0.84 kg

$$\therefore \quad \text{Mass of dry flue gas/kg fuel} = \frac{0.84}{0.040\,62} = \mathbf{20.68\ kg}$$

Hence

$$\text{Mass of excess } O_2/\text{kg fuel} = 20.68 \times 0.073\,3 = \mathbf{1.516\ kg}$$

$$\therefore \quad \text{Mass of excess air} = \frac{1.516}{0.23} = \mathbf{6.6\ kg/kg\ fuel}$$

From this

$$\text{Mass of air supplied/kg fuel} = 14.5 + 6.6 = \mathbf{21.2\ kg}$$

Alternatively

$$N_2 \text{ in the dry flue gas} = \text{Total } N_2 \text{ from the air}$$

$$\therefore \quad \text{Mass of } N_2/\text{kg fuel} = 20.68 \times 0.784\,6 = \mathbf{16.22\ kg}$$

$$\therefore \quad \text{Mass of air/kg fuel} = \frac{16.22}{0.77} = \mathbf{21.1\ kg}$$

Example 8.10 *A certain petrol consists of 86 per cent carbon and 14 per cent hydrogen by mass. If the fuel is burnt with 20 per cent excess of air and the combustion is complete, estimate the volumetric composition of the products of combustion, including the water vapour formed. Assume air contains 23 per cent of oxygen by mass.*

SOLUTION

$$\text{Stoichiometric } O_2 \text{ required} = \underbrace{(0.86 \times 2\tfrac{2}{3})}_{\text{for } CO_2} + \underbrace{(0.14 \times 8)}_{\text{for } H_2}$$

$$= 2.293 + 1.12$$

$$= \mathbf{3.413\ kg/kg\ petrol}$$

$$\therefore \quad \text{Stoichiometric air required} = \frac{3.413}{0.23} = \mathbf{14.84\ kg/kg\ petrol}$$

Products of combustion by mass per kilogram of petrol are

$$\begin{aligned} CO_2 &= 0.86 \times 3\tfrac{2}{3} &&= 3.153\ \text{kg} \\ H_2O &= 0.14 \times 9 &&= 1.26\ \text{kg} \\ \text{Excess } O_2 &= 3.413 \times 0.2 &&= 0.683\ \text{kg} \\ N_2 &= 14.84 \times 1.2 \times 0.77 &&= 13.712\ \text{kg} \\ & & \text{Total} &= 18.808\ \text{kg} \end{aligned}$$

Therefore percentage analysis of products by mass is

$$CO_2 = \frac{3.153}{18.808} \times 100 = 16.76\%$$

$$H_2O = \frac{1.26}{18.808} \times 100 = 6.7\%$$

$$O_2 = \frac{0.683}{18.808} \times 100 = 3.63\%$$

$$N_2 = \frac{13.712}{18.808} \times 100 = 72.9\%$$

Using a table, convert products by volume.

Constituent	Percentage by mass	$\dfrac{\text{Percentage by mass}}{\text{RMM}}$	Percentage by volume
CO	16.76	$\dfrac{16.76}{44} = 0.381$	$\dfrac{0.381}{3.47} \times 100 =$ 10.98
H_2O	6.7	$\dfrac{6.7}{18} = 0.372$	$\dfrac{0.372}{3.47} \times 100 =$ 10.72
O_2	3.63	$\dfrac{3.63}{32} = 0.113$	$\dfrac{0.113}{3.47} \times 100 =$ 3.26
N_2	72.9	$\dfrac{72.9}{28} = 2.604$	$\dfrac{2.604}{3.47} \times 100 =$ 75.04
Total	100	3.47	100

Therefore the composition of the combustion products by volume is

10.98% CO_2, 10.72% H_2O, 3.26% O_2, 75.04% N_2

Example 8.11 *The fuel supplied to a boiler contains 78 per cent carbon, 6 per cent hydrogen, 9 per cent oxygen, 7 per cent ash by mass as fired. The air supplied is 50 per cent in excess of that required for theoretically correct combustion. If the boiler house temperature is 20 °C and the flue gas temperature is 320 °C estimate the energy carried away by the dry flue gas per kilogram of fuel burned.*

Assume air contains 23 per cent oxygen by mass and the mean specific heat capacity of the dry flue gas is 1.006 kJ/kg K.

SOLUTION

For 1 kg of fuel

$$\begin{aligned} 0.78\text{ kg carbon requires } 0.78 \times 2\tfrac{2}{3} &= 2.08\text{ kg }O_2 \\ 0.06\text{ kg hydrogen requires } 0.06 \times 8 &= 0.48\text{ kg }O_2 \\ \text{Stoichiometric }O_2\text{ required} &= 2.08 + 0.48 \\ &= \mathbf{2.56\ kg/kg\ fuel} \end{aligned}$$

But there is 0.09 kg O_2 in fuel

$$\begin{aligned} \therefore \quad \text{Stoichiometric }O_2\text{ required from air} &= 2.56 - 0.09 \\ &= \mathbf{2.47\ kg/kg\ fuel} \\ \text{Stoichiometric air required} &= 2.47/0.23 \\ &= \mathbf{10.74\ kg/kg\ fuel} \\ \text{Actual air supplied/kg fuel} &= 10.74 \times 1.5 \\ &= \mathbf{16.11\ kg} \end{aligned}$$

From this, the total mass of flue gas per kilogram of fuel is

$16.11 + 1 = \mathbf{17.11\ kg}$

Now 0.06 kg H_2 will produce

$$0.06 \times 9 = \mathbf{0.54\ kg\ H_2O}$$

So the mass of dry flue gas per kilogram of coal is

$$17.11 - 0.54 = \mathbf{16.57\ kg}$$

Hence, the energy per kilogram of coal carried away by the dry flue gas is

$$\begin{aligned} 16.57 \times 1.006 \times (320 - 20) &= 16.57 \times 1.006 \times 300 \\ &= \mathbf{5000\ kJ} \end{aligned}$$

Example 8.12 *A natural gas consists of the following volumetric composition*

Nitrogen (N_2) 1.8%
Methane (CH_4) 94%
Ethane (C_2H_6) 3.5%
Propane (C_3H_8) 0.7%

For this gas, determine
(a) the stoichiometric volume of air for the complete combustion of 1 m^3
(b) the percentage volumetric analysis of the products of combustion

(a)
For the methane

$$CH_4 + 2O_2 = CO_2 + 2H_2O$$

Proportions by volume

$$1 + 4 = 1 + 2$$

Thus

$$1\ m^3\ CH_4 + 2\ m^3\ O_2 = 1\ m^3\ CO_2 + 2\ m^3\ H_2O\ \text{(vapour)}$$

$$\text{Stoichiometric volume air} = \frac{2}{0.21} = \mathbf{9.52\ m^3/m^3\ CH_4}$$

This air will contain $9.52 \times 0.79 = \mathbf{7.52\ m^3\ N_2}$
For the ethane

$$2C_2H_6 + 7O_2 = 4CO_2 + 6H_2O$$

Proportions by volume

$$2 + 7 = 4 + 6$$

Thus

$$2\ m^3\ C_2H_6 + 7\ m^3\ O_2 = 4\ m^3\ CO_2 + 6\ m^3\ H_2O\ \text{(vapour)}$$

$$\text{Stoichiometric volume air} = \frac{7}{2 \times 0.21} = \mathbf{16.7\ m^3/m^3\ C_2H_6}$$

This air will contain 16.7 × 0.79 = **13.2 m³ N_2**
For the propane

$$C_3H_8 + 5O_2 = 3CO_2 + 4H_2O$$

Proportions by volume

$$1 + 5 = 3 + 4$$

Thus

$$1\ m^3\ C_3H_8 + 5\ m^3\ O_2 = 3\ m^3\ CO_2 + 4\ m^3\ H_2O$$

$$\text{Stoichiometric volume air} = \frac{5}{0.21} = \mathbf{23.8\ m^3\ C_3H_8}$$

This air will contain 23.8 × 0.79 = **18.8 m³ N_2**
From this

$$\begin{aligned}\text{Stoichiometric volume of air required} &= (0.94 \times 9.52) + (0.035 \times 16.7) + (0.007 \times 23.8)\\ &= 8.949 + 0.585 + 0.167\\ &= \mathbf{9.701\ m^3/m^3\ gas}\end{aligned}$$

(b)
The products of combustion can be tabulated.

Constituent	Volume of constit. (m^3/m^3 fuel)	Products of combustion (m^3/m^3 fuel)		
		CO_2	H_2O	N_2
CH_4	0.94	0.94 × 1 = 0.94	0.94 × 2 = 1.88	0.94 × 7.52 = 7.07
C_2H_6	0.035	0.035 × 2 = 0.07	0.035 × 3 = 0.105	0.035 × 13.2 = 0.462
C_3H_8	0.007	0.007 × 3 = 0.021	0.007 × 4 = 0.028	0.007 × 18.8 = 0.132
N_2	0.018	–	–	0.018

Thus, for the combustion of 1 m³ of the natural gas, the products will be

$$\begin{aligned}CO_2 &= 0.94 + 0.07 + 0.021 &&= 1.031\ m^3\\ H_2O &= 1.88 + 0.105 + 0.028 &&= 2.013\ m^3\\ N_2 &= 7.07 + 0.462 + 0.132 + 0.018 &&= 7.682\ m^3\\ & &\text{Total} &= 10.726\end{aligned}$$

Hence, as percentages

$$CO_2 = \frac{1.031}{10.726} \times 100 = 9.6\%$$

$$H_2O = \frac{2.013}{10.726} \times 100 = 18.8\%$$

$$N_2 = \frac{7.687}{10.726} \times 100 = 71.6\%$$

Example 8.13 *The analysis, by mass, of the coal fired to a boiler is,*

82% carbon, 8% hydrogen, 3% oxygen, 7% ash

The boiler uses 0.19 kg/s and air is supplied to the furnace by a fan; the air supply is 30 per cent in excess of that required for theoretically correct combustion. Calculate

*(a) the volume of air taken in by the fan per second when the pressure and temperature of the air at the fan intake are 100 kN/m² and 18 °C, respectively (*R *for air = 0.287 kJ/kg K)*

(b) the percentage composition by mass of the dry flue gases

Air contains 23 per cent oxygen by mass. Relative atomic masses: hydrogen, 1; carbon, 12; oxygen, 16.

(a)

For 1 kg of coal

$$0.82 \text{ kg carbon require } 0.82 \times 2\tfrac{2}{3} = 2.19 \text{ kg } O_2$$
$$0.08 \text{ kg hydrogen require } 0.08 \times 8 = 0.64 \text{ kg } O_2$$

$\therefore$ Stoichiometric O_2 required = **2.83 kg/kg coal**

But there are 0.03 kg O_2 in the coal

$\therefore$ O_2 actually required = 2.83 − 0.03 = **2.8 kg/kg coal**

Hence

$$\text{Air theoretically required} = \frac{2.8}{0.23} = \mathbf{12.17\ kg/kg\ coal}$$

But 30 per cent excess air is supplied

$\therefore$ Actual air supplied = 12.17 × 1.3 = **15.82 kg/kg coal**

Air supplied/s = 0.19 × 15.82
= **3.01 kg**

Now $PV = mRT$ and $T = 18 + 273 = 291$ K

$$\therefore \quad V = \frac{mRT}{P} = \frac{3.01 \times 0.287 \times 291}{100}$$

$$= \mathbf{2.51\ m^3/s}$$

(b)

0.82 kg carbon will produce

$0.82 \times 3\tfrac{2}{3} =$ **3.01 kg CO_2**

Also there will be

$2.8 \times 0.3 =$ **0.84 kg O_2**

together with

$15.82 \times 0.77 =$ **12.18 kg N_2**

So the dry flue gas per kilogram of coal will be

CO_2 = 3.01 kg = 18.78%
O_2 = 0.84 kg = 5.24%
N_2 = 12.18 kg = 75.98%
Total = 16.03 kg

Example 8.14 *A single-cylinder, four-stroke, compression-ignition oil engine gives 15 kW at 5 rev/s and uses fuel having the composition by mass: carbon, 84 per cent; hydrogen 16 per cent. The air supply is 100 per cent in excess of that required for perfect combustion. The fuel has a calorific value of 45 000 kJ/kg and the brake thermal efficiency of the engine is 30 per cent. Calculate*

(a) the mass of fuel used per cycle

(b) the actual mass of air taken in per cycle

(c) the volume of air taken in per cycle if the pressure and temperature of the air are 100 kN/m^2 and 15 °C, respectively

Take R = *0.29 kJ/kg K*

(a)

Work done/s = **15 kJ** (1 W = 1 J/s)

$$\text{Work done/cycle} = 15 \times \frac{2}{5} = \mathbf{6\ kJ}$$

$$\text{Energy supplied/cycle} = \frac{6}{0.3} = \mathbf{20\ kJ}$$

$$\therefore \quad \text{Mass of fuel used/cycle} = \frac{20}{45\ 000} = 0.000\ 44\ \text{kg}$$

$$= \mathbf{0.444\ g}$$

(b)

1 kg C requires $2\frac{2}{3}$ kg O_2

$\therefore$ 0.84 kg C requires $2\frac{2}{3} \times 0.84 =$ **2.24 kg O_2**

1 kg H_2 requires 8 kg O_2

$\therefore$ 0.16 kg H_2 requires $8 \times 0.16 =$ **1.28 kg H_2**

$\therefore$ Total O_2/kg fuel = 2.24 + 1.28 = **3.52 kg**

Stoichiometric air required = 3.52/0.23 = **15.3 kg/kg fuel**
Actual mass of air supplied = 15.3 × 2 = **30.6 kg/kg fuel**

$\therefore$ Mass of air supplied/cycle = 30.6 × 0.000 444 = **0.013 6 kg**

(c)

$PV = mRT$ and $T = 15 + 273 = 288$ K

$$\therefore \quad V = \frac{mRT}{P} = \frac{0.013\ 6 \times 0.29 \times 288}{100}$$

$$= \mathbf{0.011\ 4\ m^3}$$

The volume of air taken in per cycle is 0.0114 m^3.

Example 8.15 *A boiler generates 900 kg of steam per hour, 0.96 dry at a pressure of 1.4 MN/m^2 from feedwater at 52 °C; the boiler efficiency is 71 per cent and the calorific value of the coal is 33 000 kJ/kg. The coal has a composition by mass: carbon, 83 per cent; hydrogen, 5 per cent; oxygen, 3 per cent; ash, 9 per cent; and the air supply is 22 per cent in excess of that required for perfect combustion. Determine*

(a) the mass of coal used per hour

(b) the mass of air used per hour

(c) the percentage analysis of the flue gases by mass

Air contains 23 per cent O_2 by mass. Specific heat capacity of water is 4.187 kJ/kg K.

Extract from steam tables

Pressure (MN/m^2)	Sat. temp. t_f (°C)	Specific enthalpy (kJ/kg)			Spec. vol. v_g (m^3/kg)
		h_f	h_{fg}	h_g	
1.4	195	830.1	1957.1	2787.8	0.1407

(a)

Specific enthalpy of steam generated by boiler

$$\begin{aligned} h_f + xh_{fg} &= 830.1 + (0.96 \times 1957.7) \\ &= 830.1 \times 1879 \\ &= \mathbf{2709.1\ kJ/kg} \end{aligned}$$

$$\begin{aligned} \text{Specific enthalpy of feedwater} &= 4.187 \times 52 \\ &= \mathbf{218\ kJ/kg} \end{aligned}$$

$$\text{Energy to steam generated} = 2709.1 - 218 = \mathbf{2491.1\ kJ/kg}$$

$$\text{Energy to steam/h} = (900 \times 2491.1)\ \text{kJ}$$

$$\text{Energy required from fuel} = \left(\frac{900 \times 2491.1}{0.71}\right)\text{kJ}$$

$$\therefore\quad \text{Mass of coal/h} = \frac{900 \times 2491.1}{0.71 \times 33\,000}$$

$$= \mathbf{95.7\ kg}$$

(b)

Consider 1 kg coal

$$\begin{aligned} 0.83\ \text{kg C requires } 0.83 \times 2\tfrac{2}{3} &= 2.21\ \text{kg } O_2 \\ 0.05\ \text{kg } H_2 \text{ requires } 0.05 \times 8 &= 0.4\ \text{kg } O_2 \\ \text{Total } O_2 &= 2.61\ \text{kg} \end{aligned}$$

But there is 0.03 kg O_2 in coal

$$\begin{aligned} \therefore\quad \text{Actual } O_2 \text{ supplied} &= 2.61 - 0.03 \\ &= \mathbf{2.58\ kg/kg\ coal} \end{aligned}$$

$$\text{Theoretical air} = \frac{2.58}{0.23} = \mathbf{11.22\ kg/kg\ coal}$$

But there is 22 per cent excess air supplied

$$\begin{aligned} \therefore\quad \text{Actual air supplied} &= 11.22 \times 1.22 \\ &= \mathbf{13.69\ kg/kg\ coal} \end{aligned}$$

Hence, the mass of air supplied per hour = 13.69 × 116 = **1588 kg**

(c)
Consider 1 kg of coal

$$
\begin{aligned}
0.83 \text{ kg C produces } 0.83 \times 3\tfrac{2}{3} &= \mathbf{3.043\ kg\ CO_2} \\
0.05 \text{ kg } H_2 \text{ produces } 0.05 \times 9 &= \mathbf{0.45\ kg\ H_2O} \\
\text{Mass of excess } O_2 \text{ in flue gas} &= 2.58 \times 0.22 \\
&= \mathbf{0.568\ kg} \\
\text{Mass of } N_2 \text{ in flue gas} &= 13.69 \times 0.77 \\
&= \mathbf{10.54\ kg}
\end{aligned}
$$

So the flue gas composition is

$$
\begin{aligned}
CO_2 &= 3.043 \text{ kg} = 20.84\% \\
H_2O &= 0.45 \text{ kg} = 3.08\% \\
O_2 &= 0.568 \text{ kg} = 3.89\% \\
N_2 &= 10.54 \text{ kg} = 72.19\% \\
& \overline{14.601 \text{ kg}}
\end{aligned}
$$

8.23 The universal gas constant

The following work concerns gases and gas mixtures together with some combustion analysis. Consider two gases A and B both with the same pressure P, volume V and temperature T. Let their masses be m_A and m_B, respectively.

By the characteristic equation of a perfect gas

$$\frac{PV}{T} = m_A R_A = m_B R_B \qquad [1]$$

where R = characteristic gas constant

Let the relative molecular masses of the gases be M_A and M_B, respectively. By Avogadro's hypothesis, the masses of equal volumes of different gases at the same pressure and temperature are proportional to their relative molecular masses. Equation [1] requires

$$m_A R_A = m_B R_B$$

so it follows that

$$M_A R_A = M_B R_B \qquad [2]$$

It appears that the product of the relative molecular mass and the characteristic gas constant of all gases is always a constant.

This constant is called the **universal gas constant** or **molar gas constant**; it has the symbol $\tilde{R}$.

Hence

$$\tilde{M} R = \tilde{R} \qquad [3]$$

Thus for air, the values $R = 0.287$ kJ/kg K and $\tilde{M} = 28.95$ kg (which is 1 kmol of air, see next section) give

$$
\begin{aligned}
\tilde{R} &= 0.287 \times 28.95 \\
&= 8.309 \text{ kJ/kmol K}
\end{aligned}
$$

The accurate value of $\tilde{R}$ is 8.314 3 kJ/kmol K.

8.24 The mole

The mole is used in physical chemistry and molecular physics; it is the unit for '**amount of substance**'. Its abbreviation is **mol**. One mole is described here as an amount of substance of a system which contains as many elementary entities as there are carbon atoms in 0.012 kg of ^{12}C (carbon-12). The elementary unit must be specified and may be an atom, a molecule, an ion, an electron, etc., or a specified group of such particles. A mole of substance is defined as the mass of the substance equivalent to its relative molecular mass and is called the **molar mass**, $\tilde{M}$.

If the unit of mass is taken as the kilogram, then

1 kmol O_2 = 32 kg O_2	1 kmol C = 12 kg C
1 kmol H_2 = 2 kg H_2	1 kmol S = 32 kg S
1 kmol CO_2 = 44 kg CO_2	

Since the kilogram mass has been used, the amount of substance is in kilomoles, written kmol.

To determine the number of moles of gas, for example, it is necessary to divide the mass of gas by its molar mass. Thus

$$n = m/\tilde{M} \qquad [1]$$

where n = number of moles
m = mass of gas, kg
$\tilde{M}$ = molar mass of gas, kg/kmol

From this,

$$m = n\tilde{M} \qquad [2]$$

Also, since

$$\tilde{M}R = \tilde{R}$$

then

$$R = \tilde{R}/\tilde{M} \qquad [3]$$

Now from the characteristic equation of a perfect gas

$$PV = mRT \qquad [4]$$

Substituting equations [2] and [3] into equation [4]

$$PV = n\tilde{M}\frac{\tilde{R}T}{\tilde{M}}$$

or

$$PV = n\tilde{R}T \qquad [5]$$

8.25 Volume of one mole of gas

Since

$$PV = n\tilde{R}T$$

then

$$V = \frac{n\tilde{R}T}{P}$$

which for 1 mol of gas becomes

$$V = \frac{\tilde{R}T}{P} \text{ since } n = 1$$

This equation is independent of any particular gas, so it shows that the volume of one mole of any gas at the same pressure and temperature is constant.

Thus, for any gas at a pressure of 101.325 kN/m^2 and a temperature of 0 °C, the volume of 1 kmol is

$$V = \frac{8.314\,3 \times 273.15}{101.315}$$

$$= \mathbf{22.41\ m^3}$$

8.26 Average molar mass of a gas mixture

It has been shown that

$$\text{Number of moles of gas} = \frac{\text{Mass of gas}}{\text{Molar mass of gas}} \qquad [1]$$

or

$$n = \frac{m}{\tilde{M}}$$

From this

$$\text{Molar mass of gas} = \frac{\text{Mass of gas}}{\text{Number of moles of gas}}$$

or

$$\tilde{M} = \frac{m}{n} \qquad [2]$$

For a gas mixture, equation [2] can be used to determine the average molar mass of the gas mixture. The mass of a gas mixture is made up of the masses of the individual amounts of gas present.

From equation [1]

$$m = n\tilde{M} \qquad [3]$$

Thus the mass of a gas mixture is given by

$$n_1\tilde{M}_1 + n_2\tilde{M}_2 + n_3\tilde{M}_3 + \ldots \qquad [4]$$

where 1, 2, 3, etc., represent the different gases in the mixture.

Also, the number of moles of gas mixture is given by

$$n_1 + n_2 + n_3 + \ldots \qquad [5]$$

Hence, from equation [2] the average molar mass of a gas mixture, $\tilde{M}_{av}$, is given by

$$\tilde{M}_{av} = \frac{n_1\tilde{M}_1 + n_2\tilde{M}_2 + n_3\tilde{M}_3 + \ldots}{n_1 + n_2 + n_3 + \ldots}$$

or

$$\tilde{M}_{av} = \frac{\Sigma\, n\tilde{M}}{\Sigma\, n} \qquad [6]$$

In using this equation it is not essential to determine the number of moles of each gas present in any given gas mixture. The solution can also be determined from a percentage analysis of the gas by volume. One mole of any gas at the same pressure and temperature occupies the same volume, so it follows that a percentage analysis by volume is also a percentage analysis by moles. If a percentage analysis is used then $\Sigma n = 100$. Thus, for air, the volumetric analysis is

21% O_2, 78.05% N_2, 0.95% other gases

The relative molecular masses are

$O_2 = 32$, $N_2 = 28$, other gases $= 39.9$

For air, then

$$\tilde{M}_{av} = \frac{\Sigma\, n\tilde{M}}{\Sigma\, n} = \frac{(21\times 32)+(78.05\times 28)+(0.95\times 39.9)}{100}$$

$$= \mathbf{28.95\ kg/kmol}$$

8.27 *The density of a gas mixture*

From the volumetric analysis of a gas mixture, the average molar mass is

$$\tilde{M}_{av} = \frac{\Sigma\, n\tilde{M}}{\Sigma\, n} \qquad [1]$$

By definition, one mole of the gas mixture will have a mass of $\tilde{M}_{av}$ kg.

The volume of one mole of the gas mixture at its pressure and temperature can be calculated from

$$V = \frac{\tilde{R}T}{P} \qquad [2]$$

From equations [1] and [2]

$$\text{Density} = \rho = \frac{\tilde{M}_{av}}{V}\ \text{kg/m}^3 \qquad [3]$$

From this

$$\text{Specific volume} = v = \frac{V}{\tilde{M}_{av}}\ \text{m}^3/\text{kg} \qquad [4]$$

8.28 The average value of the characteristic gas constant for a gas mixture

The average molar mass of the gas mixture is

$$\tilde{M}_{\text{av}} = \frac{\Sigma\, n\tilde{M}}{\Sigma\, n} \qquad [1]$$

For a gas, it has been shown that

$$\tilde{M}R = \tilde{R} \qquad [2]$$

From equations [1] and [2]

$$R_{\text{av}} = \frac{\tilde{R}}{\tilde{M}_{\text{av}}} \qquad [3]$$

8.29 Molar heat capacity

The definition of molar heat capacity is similar to the definition of specific heat capacity, except that molar heat capacity considers one mole of gas instead of unit mass of gas. Thus

> Molar heat capacity at constant volume, C_v, is defined as the amount of heat which transfers to or from one mole of gas while the temperature changes by one degree and the volume remains constant.

Also

> Molar heat capacity at constant pressure, C_p, is defined as the amount of heat which transfers to or from one mole of gas while the temperature changes by one degree and the pressure remains constant.

In both cases the units are J/kmol K

Since there are $\tilde{M}$ kg in one mole, it follows that

$$Mc_v = \tilde{C}_v \qquad [1]$$

and

$$Mc_p = \tilde{C}_p \qquad [2]$$

Now it has been shown for a gas that

$$c_p - c_v = R \qquad [3]$$

Multiplying equation [3] throughout by $\tilde{M}$ gives

$$\tilde{M}c_p - \tilde{M}c_v = \tilde{M}R$$

which from equations [1] and [2] becomes,

$$\tilde{C}_p - \tilde{C}_v = \tilde{M}R \qquad [4]$$

Now $\tilde{M}R - \tilde{R} = 8.314\,3$ kJ/kmol K; substituting this into equation [4] gives

$$\tilde{C}_p - \tilde{C}_v = 8.314\,3 \qquad [5]$$

And $c_p/c_v = \gamma$, the adiabatic index. Multiplying top and bottom by $\tilde{M}$ gives

$$\frac{\tilde{M}c_p}{\tilde{M}c_v} = \gamma$$

From equations [1] and [2]

$$\frac{\tilde{C}_p}{\tilde{C}_v} = \gamma \qquad [6]$$

From equation [5]

$$\tilde{C}_p = \tilde{C}_v + 8.314\ 3$$

and substituting in equation [6]

$$\gamma = \frac{\tilde{C}_v + 8.314\ 3}{\tilde{C}_v} \qquad [7]$$

Also from equation [5]

$$\tilde{C}_v = \tilde{C}_p + 8.314\ 3$$

and substituting in equation [6]

$$\gamma = \frac{\tilde{C}_p}{C_p - 8.314\ 3} \qquad [8]$$

8.30 Average molar heat capacity of a gas mixture

For any gas mixture, let

Total number of moles of gas mixture $= n_T$
Average molar heat capacity at constant volume $= \tilde{C}_{v\,av}$
Number of moles of each individual gas in the mixture $= n_1, n_2, n_3, \ldots$
Molar heat capacities at constant volume of each individual gas in the mixture $= \tilde{C}_{v1}, \tilde{C}_{v2}, \tilde{C}_{v3}, \ldots$

Now the amount of heat required to raise n moles of gas through one degree rise of temperature while the volume remains constant is nC_v.

Hence, for the gas mixture, the amount of heat required to raise the temperature through one degree rise of temperature is

$$n_1\tilde{C}_{v1} + n_2\tilde{C}_{v2} + n_3\tilde{C}_{v3} + \ldots = n_1\tilde{C}_{v\,av}$$
$$= (n_1 + n_2 + n_3 + \ldots)\,\tilde{C}_{v\,av}$$

From this

$$\tilde{C}_{v\,av} = \frac{n_1\tilde{C}_{v1} + n_2\tilde{C}_{v2} + n_3\tilde{C}_{v3} + \ldots}{n_1 + n_2 + n_3 + \ldots}$$

or

$$\tilde{C}_{v\,av} = \frac{\Sigma n\tilde{C}_v}{\Sigma n} \qquad [1]$$

By similar analysis, it can be shown that the average molar heat capacity at constant pressure is given by

$$\tilde{C}_{p\,\text{av}} = \frac{\Sigma n\tilde{C}_p}{\Sigma n} \qquad [2]$$

Example 8.16

(a) Given that standard pressure and temperature (STP) may be taken as 101.32 kN/m² and 0 °C, respectively, calculate the volume of 1 kmol of a perfect gas as STP.

(b) Assuming that air contains 21 per cent oxygen by volume, the rest nitrogen, determine

(i) the average molecular mass of air

(ii) the value of R, *the characteristic gas constant, in kJ/kg K*

(iii) the mass of one cubic metre of air at STP

Molar gas constant $\tilde{R} = 8.314\ 3$ kJ/kmol K

(a)

$PV = n\tilde{R}T$ and for 1 kmol $PV = \tilde{R}T$

$$\therefore \quad V = \frac{\tilde{R}T}{P} = \frac{8.314\ 3 \times 273}{101.32}$$

$$= \mathbf{22.4\ m^3}$$

(b)

(i)
$$\tilde{M}_{\text{av}} = \frac{\Sigma n\tilde{M}}{\Sigma n} = \frac{(21 \times 32) + (79 \times 28)}{100}$$

$$= \frac{672 \times 2212}{100}$$

$$= \frac{2884}{100}$$

$$= \mathbf{28.84}$$

(ii)
$$R_{\text{av}} = \frac{\tilde{R}}{\tilde{M}_{\text{av}}} = \frac{8.314\ 3}{28.84} = \mathbf{0.288\ kg/kg\ K}$$

(iii)
$$\rho = \frac{\tilde{M}_{\text{av}}}{V} = \frac{28.84}{22.4} = \mathbf{1.288\ kg/m^3}$$

Example 8.17 *A vessel contains 8 kg of oxygen, 7 kg of nitrogen and 22 kg of carbon dioxide. The total pressure in the vessel is 416 kN/m² and the temperature is 60 °C. Calculate*

(a) the partial pressure of each gas in the vessel

(b) the volume of the vessel

(c) the total pressure in the vessel when the temperature is raised to 228 °C

Molar gas constant R *= 8.314 3 kJ/kmol K.*

(a)

$$\text{Number of moles}\quad O_2 = \frac{8}{32} = 0.25 = n_1$$

$$\text{Number of moles}\quad N_2 = \frac{7}{28} = 0.25 = n_2$$

$$\text{Number of moles}\ CO_2 = \frac{2}{44} = 0.5 = n_3$$

Total number of moles of gas in vessel, n, is given by

$$n = 0.25 + 0.25 + 0.5 = 1$$

Now

$$PV = n\tilde{R}T$$

$$\therefore \quad P = \frac{n\tilde{R}T}{V} \qquad [1]$$

The partial pressure of each gas is as if the gas occupied the complete volume of the vessel. Also, each gas is at the same temperature so $\tilde{R}T/V$ is constant for each gas. From equation [1]

$$P \propto n$$

Let P_1 = partial pressure of O_2
P_2 = partial pressure of N_2
P_3 = partial pressure of CO_2
P = total pressure

then

$$\frac{P_1}{P} = \frac{n_1}{n}$$

or

$$P_1 = \frac{n_1 P}{n} = \frac{(0.25 \times 416)}{1} = \mathbf{104\ kN/m^2}$$

Similarly

$$P_2 = \frac{n_2 P}{n} = \frac{(0.25 \times 416)}{1} = \mathbf{104\ kN/m^2}$$

$$P_3 = \frac{n_3 P}{n} = \frac{(0.5 \times 416)}{1} = \mathbf{208\ kN/m^2}$$

(b)

$$PV = n\tilde{R}T \text{ and } T = 60 + 273 = 333\ \text{K}$$

$$\therefore \quad V\frac{n\tilde{R}T}{P} = \frac{1 \times 8.314\,3 \times 333}{416}$$

$$= \mathbf{6.655\ m^3}$$

The volume of the vessel is 6.655 m^3.

(c)
For a fixed mass of gas at constant volume $P/T = \text{constant}$.

$$\therefore \quad \frac{P_1}{T_1} = \frac{P_2}{T_2} \quad \text{and} \quad T_2 = 228 + 273 = \mathbf{501\ K}$$

$$\therefore \quad P_2 = \frac{P_1 T_2}{T_1} = 416 \times \frac{501}{333}$$

$$= \mathbf{626\ kN/m^2}$$

Example 8.18 *The coal supplied to a boiler has the following composition by mass*

carbon: 0.84, hydrogen: 0.04, oxygen: 0.05, remainder: ash

When 635 kg of coal are burned per hour, the air supplied is 25 per cent in excess of the theoretical minimum required for complete combustion.

Calculate the actual mass of air supplied per kilogram of coal and, assuming complete combustion, estimate the velocity of the flue gas in m/s if it is at a pressure of 100 kN/m² and temperature 344 °C at a section in the flue where the cross-sectional area is 1.1 m².

The volume of $\tilde{M}$ kg of any gas at 101.325 kN/m² and 0 °C *can be assumed to be 22.4 m², where* $\tilde{M}$ *is the molar mass of the gas. Use the following relative molecular masses:* $C = 12$, $H_2 = 2$, $O_2 = 32$, $N_2 = 28$. *Air contains 23 per cent oxygen and 77 per cent nitrogen by mass.*

SOLUTION
First draw up a table.

Constituent	Mass of constit. (kg/kg fuel)	O_2 required (kg/kg fuel)
C	0.84	$0.84 \times 2\frac{2}{3} = 2.24$
H_2	0.04	$0.04 \times 8 = 0.32$
O_2	0.05	-0.05
Ash	0.07	–

$$\text{Total theoretical } O_2 \text{ required} = 2.24 + 0.32 - 0.05 = 2.56 - 0.05 = \mathbf{2.51\ kg/kg\ coal}$$

$$\text{Stoichiometric mass of air} = \frac{2.51}{0.23} = \mathbf{10.91\ kg/kg\ coal}$$

But 25 per cent excess air is supplied.

$$\therefore \quad \text{Actual mass of air supplied} = 10.91 \times 1.25 = \mathbf{13.64\ kg/kg\ coal}$$

Products of combustion by mass per kilogram of coal

$$\begin{aligned} CO_2 &= 0.84 \times 3\tfrac{2}{3} = 3.08 \\ H_2O &= 0.04 \times 9 = 0.36 \\ O_2 &= 2.51 \times 0.25 = 0.628 \\ N_2 &= 13.64 \times 0.77 = 10.5 \end{aligned}$$

$$\text{Total} = 14.568 \text{ kg}$$

Percentage analysis of products by mass

$CO_2 = 21.14\%$
$H_2O = 2.47\%$
$O_2 = 4.31\%$
$N_2 = 72.08\%$

Using a table, convert the analysis of the products of combustion from mass to volume.

Constituent	Percentage by mass	$\frac{\text{Percentage by mass}}{\text{RMM}}$	Percentage by volume
CO_2	21.14	$\frac{21.14}{44} = 0.481$	14.46
H_2O	2.47	$\frac{2.47}{18} = 0.137$	4.12
O_2	4.31	$\frac{4.31}{32} = 0.135$	4.06
N_2	72.08	$\frac{72.08}{28} = 2.574$	77.36
Total	100	3.327	100

Average molar mass of the products is given by

$$\tilde{M}_{av} = \frac{\sum n\tilde{M}}{\sum n} = \frac{(14.46 \times 44) + (4.12 \times 18) + (4.06 \times 32) + (77.36 \times 28)}{100}$$

$$= \frac{636.2 + 74.2 + 129.9 + 2166.1}{100}$$

$$= \frac{3006.4}{100}$$

$$= \mathbf{30.064\ kg/kmol}$$

Volume of $\tilde{M}_{av}$ kg of products at 100 kN/m^2 and 344 °C is given by

$$\frac{P_1 V_1}{T_1} = \frac{P_2 V_2}{T_2}$$

$$V_2 = \frac{P_1 V_1 T_2}{P_2 T_1} = \frac{101.325 \times 22.4 \times 617}{100 \times 273} = \mathbf{51.3\ m^3}$$

$$\text{Mass of coal used/s} = \frac{635}{3600}\ \text{kg}$$

$$\text{Mass of products of combustion/s} = \frac{635 \times 14.568}{3600}$$

$$= \mathbf{2.57\ kg}$$

But

Volume of 30.064 kg of products = 51.3 m^3

$\therefore$ Volume of 2.57 kg of products $= \dfrac{51.3}{30.064} \times 2.57 = \mathbf{4.39\ m^3/s}$

Volume of gas flowing/s = area × velocity/s

$\therefore$ Velocity = volume flowing/area

$$= \frac{4.39}{1.1}$$

$$= \mathbf{3.99\ m/s}$$

Example 8.19 *A gas mixture has the following composition by volume*

26% CO, 16% H_2, 7% CH_4, 51% N_2

This gas is mixed with air in the ratio 1 volume of gas to 2 volumes of air. The mixture is enclosed in a cylinder at a pressure of 103 kN/m^2 and a temperature of 21 °C. It is then compressed adiabatically through a volume compression ratio of 7:1. Determine
(a) the temperature of the gas after compression
(b) the density of the air–gas mixture at the initial condition of 103 kN/m^2 and 21 °C
Use the following molar specific heat capacities at constant volume: diatomics = 21 kJ/kmol K, CH_4 = 36 kJ/kmol K. Air contains 21 per cent O_2 by volume.

(a)

$$C_{v\,\mathrm{av}} = \frac{\sum n\tilde{C}_v}{\sum n} = \frac{\overbrace{(26 \times 21) + (16 \times 21) + (7 \times 36) + (51 \times 21)}^{\text{Gas}} + \overbrace{(200 \times 21)}^{\text{Air}}}{\underbrace{100}_{\text{Gas}} + \underbrace{200}_{\text{Air}}}$$

$$= \frac{(293 \times 21) + (7 \times 36)}{300}$$

$$= \frac{6153 + 252}{300}$$

$$= \frac{6405}{300}$$

$$= \mathbf{21.35\ kJ/kmol\ K}$$

$$\gamma = \frac{\tilde{C}_v + 8.314\,3}{\tilde{C}_v} = \frac{21.35 + 8.314\,3}{21.35}$$

$$= \frac{29.664\,5}{21.35}$$

$$= \mathbf{1.389}$$

$T_1/T_2 = (V_2/V_1)^{(\gamma - 1)}$ and $T_1 = 21 + 273 = \mathbf{294\ K}$

$$\therefore \quad T_2 = T_1 \left(\frac{V_1}{V_2}\right)^{(\gamma - 1)} = 294 \times 7^{0.389}$$
$$= 294 \times 2.132$$
$$= \mathbf{626.8\ K}$$
$$t_2 = 626.8 - 273$$
$$= \mathbf{353.8\ °C}$$

(b)

$$\tilde{M}_{av} = \frac{\sum n\tilde{M}}{\sum n} = \frac{\overbrace{(26 \times 28) + (16 \times 2) + (7 \times 16) + (51 \times 28)}^{\text{Gas}} + \overbrace{(42 \times 32) + (158 \times 28)}^{\text{Air}}}{300}$$

$$= \frac{728 + 32 + 112 + 1428 + 1344 + 4424}{300}$$

$$= \frac{8068}{300}$$

$$= \mathbf{26.89}$$

For 1 kmol of gas

$$V = \frac{\tilde{R}T}{P}$$

$$= \frac{8.314\,3 \times 294}{103}$$

$$= \mathbf{23.73\ m^3}$$

$$\therefore \quad \text{Density} = \rho = \frac{M_{av}}{V} = \frac{26.89}{23.73} = \mathbf{1.133\ kg/m^3}$$

8.31 Solution of combustion problems by the mole method

Now that the concept of the mole has been introduced, a second method of solving combustion problems will be discussed. It consists of writing out the combustion equation, this time in terms of numbers of moles, then equating the number of moles of each constituent before combustion to the number of moles of products after combustion. The solution often involves the use of simultaneous algebraic equations. A few examples will best illustrate the method. First, consider the formation of two stoichiometric equations from section 8.8.

Example 8.20 *Form the stoichiometric equation for the combustion of hydrogen.*

SOLUTION

The equation is firstly written out in the following form.

$$H_2 + a\,O_2 + \frac{79}{21}a\,N_2 = b\,H_2O + \frac{79}{21}a\,N_2$$

The equation must now be balanced for the number of moles of each constituent.

Equating H_2 coefficients: $b = 1\,\text{mol}$
Equating O_2 coefficients: $a = b/2 = 0.5\,\text{mol}$

So the equation becomes

$$H_2 + \frac{1}{2}O_2 + \frac{79}{21} \times \frac{1}{2}N_2 = H_2O + \frac{79}{21} \times \frac{1}{2}N_2$$

Bring the numbers of moles to whole numbers by multiplying throughout by two. The equation now becomes

$$2H_2 + O_2 + 3.76N_2 = 2H_2O + 3.76N_2$$

Example 8.21 *Form the stoichiometric equation for the combustion of hexane, C_6H_{14}.*

SOLUTION
Write out the equation using algebraic coefficients

$$C_6H_{14} + a\,O_2 + \frac{79}{21}a\,N_2 = b\,CO_2 + d\,H_2O + \frac{79}{21}a\,N_2$$

Equating carbon coefficients: $b = 6\,\text{mol}$

Equating hydrogen coefficients: $d = 14/2 = 7\,\text{mol}$

Equating oxygen coefficients: $a = b + d/2 = 6 + 3.5 = 9.5\,\text{mol}$

So the equation becomes

$$C_6H_{14} + 9.5O_2 + (3.76 \times 9.5)N_2 = 6CO_2 + 7H_2O + 35.7N_2$$

Multiplying throughout by 2

$$2C_6H_{14} + 19O_2 + 71.4N_2 = 12CO_2 + 14H_2O + 71.4N_2$$

Example 8.22 *A gaseous fuel has the percentage analysis by volume:*

12% CO, 41% H_2, 27% CH_4, 2% O_2, 3% CO_2, 15% N_2

Determine the percentage gravimetric analysis of the total products of combustion when the air supplied is 15 per cent in excess of the minimum theoretically required for complete combustion. Air contains 21 per cent O_2 and 79 per cent N_2 by volume.

SOLUTION
Write out the combustion equation by numbers of moles

$$12CO + 41H_2 + 27CH_4 + 2O_2 + 3CO_2 + 15N_2 + aO_2 + \frac{79}{21}aN_2$$

$$= bCO_2 + d\,H_2O + e\,O_2 + 15N_2 + \frac{79}{21}aN_2$$

Equating C coefficients: $b = 12 + 27 + 3 = 42$ mol
Equating H_2 coefficients: $d = 14 + (27 \times 2) = 41 + 54 = 95$ mol

Now O_2 is 15 per cent in excess of minimum

$$\therefore \quad \frac{e}{a - e} = 0.15$$

$$\therefore \quad e = 0.15a - 0.15e$$

$$1.15e = 0.15a$$

$$\therefore \quad e = \frac{0.15}{1.15}a = 0.13a$$

Equating O_2 coefficients

$$2 + 3 + a = b + \frac{d}{2} + e$$

$$5 + a = 42 + \frac{95}{2} + 0.13a$$

$$\therefore \quad a - 0.13a = 42 + 47.5 - 5 = 84.5$$

$$0.87a = 84.5$$

$$a = \frac{84.5}{0.87} = \mathbf{97.1\ mol}$$

$$\therefore \quad e = 0.13 \times 97.1 = \mathbf{12.62\ mol}$$

Products of combustion in moles per 100 moles of fuel

$$= 42CO_2 + 95H_2O + 12.62O_2 + 15N_2 + (3.76 \times 97.1)N_2$$

$$= 42CO_2 + 95H_2O + 12.62O_2 + 15N_2 + 365N_2$$

$$= 42CO_2 + 95H_2O + 12.62O_2 + 380N_2$$

Gravimetric analysis of products

$$= (42 \times 44)CO_2 + (95 \times 18)H_2O + (12.62 \times 32)O_2 + (380 \times 28)N_2$$

$$= 1848CO_2 + 1710H_2O + 403.8O_2 + 10\,640N_2$$

$$\text{Total} = 1848 + 1710 + 403.8 + 10\,640$$

$$= \mathbf{14\,601.8\ kg}$$

And the percentage analysis is

$$CO_2 = \frac{1848}{14\,601.8} \times 100 = 12.65\%$$

$$H_2O = \frac{1710}{14\,601.8} \times 100 = 11.71\%$$

$$O_2 = \frac{403.8}{14\,601.8} \times 100 = 2.77\%$$

$$N_2 = \frac{10\,640}{14\,601.8} \times 100 = 72.87\%$$

Example 8.23 *A single-cylinder, single-acting oil engine operates on the four-stroke cycle; it has a bore of 300 mm, a stroke of 460 mm and runs at 200 rev/min. The fuel oil has a composition by mass of 87 per cent C, 13 per cent H; it is consumed at a rate of 6.75 kg/h. The volumetric composition of the dry exhaust gases is 7 per cent CO_2, 10.5 per cent O_2, 82.5 per cent N_2. Atmospheric temperature and pressure are 17 °C and 100 kN/m^2, respectively. Determine*

(a) the actual quantity of air supplied per kg of fuel

(b) the volumetric efficiency of the engine

Take R *for air as 0.287 kJ/kg K*

(a)

Write out the combustion equation in terms of moles

$$\frac{87}{12}\text{C} + \frac{13}{2}\text{H}_2 + a\,\text{O}_2 + \frac{79}{21}a\text{N}_2 = b\,\text{CO}_2 + d\text{H}_2\text{O} + e\text{O}_2 + f\,\text{N}_2$$

Equating carbons

$$b = \frac{87}{12} = 7.25 \text{ mol}$$

Consider the volumetric analysis of the dry exhaust gas

$$\text{CO}_2\text{:}\ \frac{b}{b+e+f} \times 100 = 7 \qquad [1]$$

$$\text{O}_2\text{:}\ \frac{e}{b+e+f} \times 100 = 10.5 \qquad [2]$$

$$\text{N}_2\text{:}\ \frac{f}{b+e+f} \times 100 = 82.5 \qquad [3]$$

Divide [2] by [1]

$$\frac{e}{b} = \frac{10.5}{7} = 1.5$$

$$\therefore \quad e = 1.5b = 1.5 \times 7.25 = 10.875 \text{ mol}$$

Divide [3] by [1]

$$\frac{f}{b} = \frac{82.5}{7} = 11.8$$

$$\therefore \quad f = 11.8b = 11.8 \times 7.25 = 85.5 \text{ mol}$$

Equating N_2

$$\frac{79}{21}a = 85.5$$

$$\therefore \quad a = 85.5 \times \frac{21}{79} = \mathbf{22.7\ mol}$$

So the fuel side of the combustion equation becomes

$$\frac{87}{12}\text{C} + \frac{13}{2}\text{H}_2 + 22.7\text{O}_2 + 85.5\text{N}_2$$

Convert to gravimetric analysis

$$\left(\frac{87}{12}\times 12\right)C+\left(\frac{13}{2}\times 2\right)H_2+(22.7\times 32)O_2+(85.5\times 28)N_2$$

$$\overbrace{87C+13H_2}^{\text{Fuel}}+\overbrace{726O_2+2\,394N_2}^{\text{Air}}$$

$\therefore$ Air/100 kg fuel = 726 + 2394 = **3120 kg**

$\therefore$ Air/kg fuel = 3120/100 = **31.20 kg**

(b)

$$\text{Mass of air/min} = \frac{31.20\times 6.75}{60} = \mathbf{3.51\ kg}$$

$$V = \frac{mRT}{P} = \frac{351\times 0.287\times 290}{100} = \mathbf{2.92\ m^3/min}$$

where $T = 17 + 273 = \mathbf{290\ K}$

$$\text{Swept volume/min} = \pi\times\frac{0.3^2}{4}\times 0.46\times 100 = \mathbf{3.25\ m^3}$$

$$\therefore\quad \text{Volumetric } \eta = \frac{2.91}{3.25}\times 100 = \mathbf{89.8\%}$$

Example 8.24 *This is the same as Example 8.9 but now it is solved by the mole method.*
The dry exhaust gas from an oil engine had the following percentage composition by volume:

8.85% CO_2, 1.2% CO, 6.8% O_2, 83.15% N_2

The fuel oil had a percentage composition by mass as follows:

84% C, 14% H_2, 2% O_2

Determine the mass of air supplied per kilogram of fuel burnt. Air contains 23 per cent O_2 by mass.

SOLUTION

Convert the air mass analysis to a volumetric analysis.

$$\text{Moles } O_2 = \frac{23}{32} = 0.718$$

$$\text{Moles } N_2 = \frac{77}{28} = 2.75$$

Total moles = 0.718 + 2.75 = **3.468**

So the volumetric analysis is

$$O_2 = \frac{0.718}{3.468}\times 100 = \mathbf{20.7\%}$$

$$N_2 = \frac{2.75}{3.468} = \mathbf{79.3\%}$$

Write out the combustion equation for the fuel in terms of moles.

$$\frac{84}{12}C + \frac{14}{2}H_2 + \frac{2}{32}O_2 + a\,O_2 + \frac{79.3}{20.7}aN_2 = bCO_2 + d\,CO_2 + e\,O_2 + fN_2 + gH_2O$$

Equating carbon coefficients

$$b + d = \frac{84}{12} = 7 \text{ mol} \qquad [1]$$

From the volumetric analysis of the dry flue gas

$$CO_2\text{:}\ \frac{b}{b+d+e+f} \times 100 = 8.85 \qquad [2]$$

$$CO\text{:}\ \frac{d}{b+d+e+f} \times 100 = 1.2 \qquad [3]$$

$$O_2\text{:}\ \frac{e}{b+d+e+f} \times 100 = 6.8 \qquad [4]$$

$$N_2\text{:}\ \frac{f}{b+d+e+f} \times 100 = 83.15 \qquad [5]$$

Divide [2] by [3]

$$\frac{b}{d} = \frac{8.85}{1.2} = 7.37$$

$$\therefore \quad b = 7.37d \qquad [6]$$

Substitute [6] in [1]

$$7.37d + d = 7$$

$$\therefore \quad 8.37d = 7$$

$$d = \frac{7}{8.37} = 0.836 \text{ mol}$$

$$\therefore \quad b = 7.37 \times d = 7.37 \times 0.836 = 6.16 \text{ mol}$$

Divide [5] by [3]

$$\frac{f}{d} = \frac{83.15}{1.2} = 69.3$$

$$\therefore \quad f = 69.3 \times d = 69.3 \times 0.836 = 58 \text{ mol}$$

Equating N_2 coefficients

$$\frac{79.3}{20.7}a = 3.83a = 58$$

$$\therefore \quad a = \frac{58}{3.83} = 15.14 \text{ mol}$$

So the fuel side of the combustion equation becomes

$$\frac{84}{12}C + \frac{14}{2}H_2 + \frac{2}{32}O_2 + 15.14O_2 + 58N_2$$

Convert to gravimetric analysis

$$\left(\frac{84}{12}\times 12\right)\text{C}\times\left(\frac{14}{2}\times 2\right)\text{H}_2+\left(\frac{2}{32}\times 32\right)\text{O}_2+(15.14\times 32)\,\text{O}_2+(58\times 28)\,\text{N}_2$$

$$\overbrace{84\text{C}+14\text{H}_2+2\text{O}_2}^{\text{Fuel}}+\overbrace{484\text{O}_2+1\,624\text{N}_2}^{\text{Air}}$$

$$\therefore\quad \text{Air/100 kg fuel} = 484 + 1\,624 = \mathbf{2108\ kg}$$

$$\therefore\quad \text{Air/kg fuel} = \frac{2108}{100} = \mathbf{21.08\ kg}$$

8.32 Heat energy release by the complete combustion of a fuel

The heat energy released by the complete combustion of unit quantity of fuel is called the **calorific value** of the fuel. The conditions under which the combustion is carried out must be specified and the calorific value so determined is for the said conditions. *Calorific value* is used in the gas and solid fuel industries as a contractual and legal term. An alternative term in the oil industry is **heat of combustion**. Calorific value is also called **specific energy**.

The determination of the calorific value of fuels is carried out in specially designed calorimeters. The type of calorimeter used will depend upon the form which the fuel takes. In the case of solid and some liquid fuels the calorific value is usually determined in a **bomb calorimeter**. In the case of gaseous and some liquid fuels the calorific value is determined in a **gas calorimeter**.

The definition of calorific value will depend upon whether the fuel is tested at constant volume or constant pressure.

8.33 Definitions of calorific value

8.33.1 Gross calorific value at constant volume ($Q_{gr,v}$)

Gross calorific value at constant volume is the amount of heat liberated per unit quantity of fuel when burned in oxygen (O_2) in a bomb calorimeter under standard conditions and the products of combustion are cooled to the original standard conditions. The international thermochemical reference temperature is 25 °C.

Correction should be made for products of combustion such as oxygen (O_2), carbon dioxide (CO_2), sulphur dioxide (SO_2), nitrogen (N_2), liquid water (H_2O), hydrogen chloride (HCl), oxides of nitrogen (NO_x) and acids of sulphur.

8.33.2 Net calorific value at constant volume ($Q_{net,v}$)

Net calorific value at constant volume is the amount of heat liberated per unit quantity of fuel when burned in oxygen (O_2) in a bomb calorimeter under standard conditions and the products of combustion are cooled to the original standard conditions. Corrections are made for the products of combustion but the water (H_2O) formed remains as vapour (i.e. it remains uncondensed).

8.33.3 Gross calorific value at constant pressure ($Q_{gr,p}$)

Gross calorific value at constant pressure is used mostly in the case of gaseous fuels and is the amount of heat liberated per unit volume of fuel when burned in oxygen (O_2) at constant pressure.

The products of combustion are carbon dioxide (CO_2), sulphur dioxide (SO_2), oxygen (O_2), nitrogen (N_2) and liquid water (H_2O).

The standard conditions for measurement are 15 °C and 101.325 kPa. In the United Kingdom this is set by the UK Gas Act 1986.

8.33.4 Net calorific value at constant pressure ($Q_{net,p}$)

Net calorific value at constant pressure is used mostly in the case of gaseous fuels and is the amount of heat liberated per unit volume of fuel when burned in oxygen (O_2) at constant pressure.

The products of combustion are CO_2, SO_2, O_2, N_2 and H_2O; in this case the H_2O remains as vapour (i.e., it remains uncondensed). The standard conditions for measurement are 15 °C and 101.325 kPa.

8.34 Determination of net calorific value (Q_{net})

Most fuels contain some hydrogen and maybe some moisture. When burnt, the hydrogen will form H_2O and this, together with any moisture in the fuel, will appear as steam in the exhaust or flue. Now, in general, it is not convenient to cool the exhaust products sufficiently, so the H_2O leaves as steam. The steam leaves without giving up its enthalpy of evaporation to be used in the plant. For this reason, the lower or net calorific value of a fuel has been introduced. It is determined by reducing the higher calorific value by the amount of the enthalpy of evaporation leaving in the H_2O in the products.

In the case of solid and liquid fuels, the mass of H_2O in the products of combustion per kilogram of fuel burned is

$$(m + 9m_{H_2}) \text{ kg}$$

where m = mass of moisture per kilogram of fuel

m_{H_2} = mass of H_2 per kilogram of fuel

The specific enthalpy of evaporation steam which leaves with the products of combustion is taken as 2442 kJ/kg. This is the specific enthalpy of evaporation of steam at 25 °C (see section 8.33.1). From this

$$Q_{net,v} = [Q_{gr,v} - 2442(m + 9m_{H_2})] \text{ kJ/kg} \qquad [1]$$

For gaseous fuels, let

$\dot{V}_S$ = volume of gas used, converted to standard reference conditions 15 °C and 1.013 25 kPa (760 mm Hg)

$\dot{m}$ = mass of condensate (H_2O) collected from combustion products in same time as gas, kg

then

$$\frac{\dot{m}}{\dot{V}_S} = \text{mass of condensate/standard m}^3 \text{ gas}$$

The specific enthalpy of steam at the standard reference condition is taken as 2466 kJ/kg (see section 8.33.3). From this

$$Q_{net,p} = \left(Q_{gr,p} - 2466\frac{\dot{m}}{\dot{V}_S}\right) \text{ kJ/standard m}^3 \qquad [2]$$

8.35 Practical determination of calorific value

8.35.1 The bomb calorimeter

In the bomb calorimeter a small quantity of the fuel under test is burnt at constant volume in a high-pressure container, hence its name. It is sometimes called the oxygen bomb calorimeter. Oxygen is admitted to the bomb under pressure in order that the fuel may be burnt. The energy liberated is measured and the calorific value of the fuel determined.

A Schole's stainless steel bomb calorimeter is illustrated in Fig. 8.2. Named after its designer, it consists of a thick wall, stainless steel, high-pressure cylinder A into the top of which is introduced a non-return, oxygen admission valve B and a pressure release valve C. On to the bottom of this cylinder is screwed a base cap D and the seal between this and the cylinder is made by a U-section washer E made of rubber or plastic. Set into base cap D and protruding up into the bomb are two pillars F and G. Pillar F is set directly into the base whereas pillar G is electrically insulated from the base and is connected to an electrical contact which protrudes from the base cap at the bottom. On to pillar F is fitted a support ring H which can be slid up or down the pillar. Both pillars have a slot at the top into which a piece of fuse wire J can be wedged. A crucible K is mounted in the support ring. Crucibles may be made from quartz, Inconel, stainless steel and platinum.

In use the calorimeter is mounted in a water-container can L, into the bottom of which is set a mount M for the calorimeter. In this mount there are registers for the feet of the bomb and also electrical contacts which mate with the corresponding contacts in the base of the bomb. Electrical power is fed to the mount via lead-covered wire N from a plug and socket O. The can L is itself mounted in another outer can P which serves as a heat insulator. Into the base of outer can P is set a wooden platform Q which serves as a register for the can L.

The apparatus is covered by cover plate R into which is set a stirrer S. This stirrer is designed to be spring-loaded and is horseshoe-shaped round the bomb. Other stirrer designs may incorporate an electrically driven mechanism. The cover plate is also pierced by a hole through which is inserted a thermometer T, usually a Beckmann. The thermometer must have a high degree of accuracy. Recommended devices include the platinum resistance thermometer, the mercury-in-glass thermometer (probably Beckmann) or Thermistor, and the quartz crystal resonator.

The electrical circuit for firing the fuse J is shown in Fig. 8.2(b). The fuse is in series with an ammeter, a power supply (battery or mains), a rheostat and a switch. A voltmeter is in parallel, across the circuit.

Apparatus is also required to prepare specimens for the bomb calorimeter: a pestle and mortar (for coal), a small press (for compressing powdered coal into a small pellet) and a chemical balance. A high-pressure oxygen supply, usually from a high-pressure cylinder, is also required along with a stop-clock.

8.35.2 Using the bomb calorimeter – solid fuel

Assume that it is coal under test. The coal is first crushed in a pestle and mortar until it is a fine powder. According to the relevant British Standard (BS 1016: Part 5), the coal should be ground to pass through a 210 μm (72 mesh) sieve. The crucible from the calorimeter is then weighed. The powdered coal is pressed into the small press to

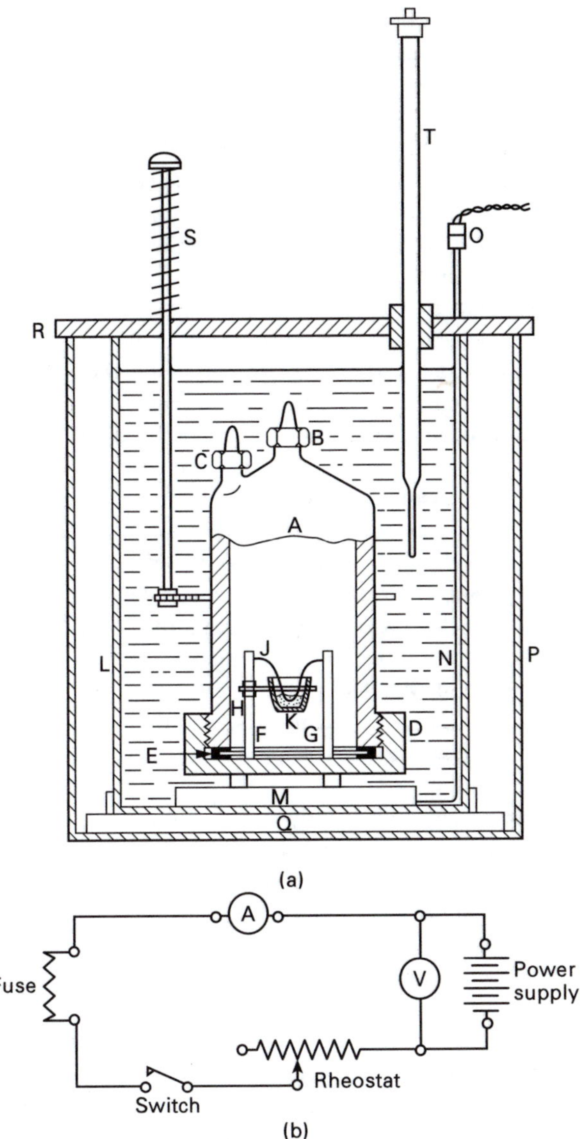

Fig. 8.2 Bomb calorimeter: (a) apparatus and (b) fuse-firing circuit

form a pellet then approximately 1 g is placed in the crucible. The crucible plus coal is weighed; the weight of coal is determined by subtracting the known weight of the crucible.

The top and bottom of the bomb are then separated and the crucible is set centrally in the base of the bomb in the support ring. A fuse is set across the support pillars and adjusted so it touches the coal in the crucible. Sometimes the fuse wire is set tightly across the pillars then a piece of cotton is tied at the centre of the wire, its free end hanging down into the coal.

A small measured quantity of distilled water (1 ml) is poured into the base, which is then reassembled with the top and screwed home tightly by hand. Care is needed to keep the bomb upright so as not to upset its contents. The bomb is then charged with oxygen to a pressure of 3 MN/m^2 without displacing the original air content. After charging, the valves at the top of the bomb are carefully sealed. The charged bomb is now inserted into the water-container can, making sure that the bomb feet and the electrodes register correctly in the base of the can.

A measured quantity of water is now poured into the can, such that it just covers the bomb. The temperature of the water should be such that the final temperature reached is within a degree of the standard reference temperature, 25 °C. The cover and its stirrer are now assembled and the thermometer inserted through the cover. If the thermometer is a Beckmann, it is adjusted so that the mercury in the thermometer is just on the lower part of the thermometer scale. Electrical connections are then made, making sure that the switch is open circuit. The rheostat in the circuit is then adjusted for a moderate current and the experiment is then ready to proceed.

Stirring is begun at a moderate pace and continued throughout the experiment. After a period of not less than 5 min the thermometer is read to an accuracy of 0.001 °C and at the same time the stop-clock is started. The thermometer is then read at intervals of 1 min for the next 5 min. After exactly 5 min the switch is closed to fire the bomb. The ammeter may be quickly observed here to see if the bomb fuse has fired. This is detected by the meter swinging up to a maximum value then quickly returning to zero, showing that the fuse wire has broken. If the meter stays at a maximum value, the current must be quickly increased using the rheostat until the fuse does fire. This marks the end of the **preliminary period** and the beginning of the **chief period**.

The thermometer will now commence to rise fairly rapidly, showing successful firing of the bomb. The thermometer is read every minute to a minimum accuracy of 0.01 °C and this continues for the remainder of the experiment. When the maximum temperature has been reached this marks the end of the chief period and the beginning of the **after period**. The temperature fall in the after period may, at first, be very slow. When the temperature fall shows a steady rate, temperature recordings are made for a period of not less than 5 min.

After this the bomb is removed from the apparatus and the pressure release valve is opened to reduce the pressure in the bomb to atmospheric. The bomb is then opened and the crucible is inspected to check for a complete absence of sooty deposit, showing that the coal has been completely fired. If this is so, the calorific value of the coal can be determined from the experimental results.

8.35.3 Results and calculations

After the bomb has been fired, the temperature inside the apparatus will rise, so some energy will be lost to the surroundings. A cooling correction must therefore be applied. BS 1016: Part 5 recommends the Regnault–Pfaundler cooling correction. This cooling correction takes the following form:

$$\text{Correction} = nv' + \frac{v'' - v'}{t'' - t'}\left[\sum_{1}^{n-1}(t) + \tfrac{1}{2}(t_0 + t_n) - nt'\right]$$

$$= nv + kS$$

where S = expression within the brackets

n = number of minutes within chief period

v' = rate of fall of temperature per minute within the preliminary period; if the temperature rises during this period then v is negative

v'' = rate of fall of temperature per minute in the after period

t' and t'' = average temperature during the preliminary and after periods respectively

$\sum_{1}^{n-1}(t)$ = sum of the temperature readings during the chief period

$\frac{1}{2}(t_o + t_n)$ = mean of the firing temperature t_0 and the first temperature t_n after which the rate of change is constant

$k = \dfrac{v'' - v'}{t'' - t'}$ = cooling constant of the calorimeter (if the total water equivalent is not less than 2500 g and with adequate heat insulation, k should not exceed 0.002 5)

The cooling correction can be illustrated using a table of typical results.

Preliminary period		Chief period		After period	
Time	Temp. (°C)	Time	Temp. (°C)	Time	Temp. (°C)
0	1.249	6	3.45	$10 = t_n$	3.868
1	1.251	7	3.86	11	3.864
2	1.253	8	3.871	12	3.860
3	1.256	9	3.811	13	3.857
4	1.258			14	3.853
$5 = t_0$	1.260			15	3.849
$t' = 1.255$		$\sum_{1}^{n-1}(t) = 14.992$		$t'' = 3.858$	
$v' = -0.002\,2$		$\frac{1}{2}(t_0 - t_n) = 2.56$		$v'' = 0.003\,8$	
$n = 5$		$-nt' = -6.275$		$k = \dfrac{v'' - v'}{t'' - t'} = 0.002\,3$	
$nv' = -0.011$		$S = 11.277$			

From the table

$$\begin{aligned}\text{Cooling correction} &= nv + kS \\ &= -0.011 + 0.026 \\ &= \mathbf{0.015\ ^\circ C}\end{aligned}$$

$$\begin{aligned}\text{Uncorrected temperature rise} &= t_n - t_0 \\ &= 3.868 - 1.26 \\ &= \mathbf{2.608\ ^\circ C}\end{aligned}$$

$$\begin{aligned}\text{Corrected temperature rise} &= 2.608 + 0.015 \\ &= \mathbf{2.623\ ^\circ C}\end{aligned}$$

Assume that the water equivalent of the apparatus is 2600 = 2.6 kg, then

$$\text{Energy liberated by coal} = (2.6 \times 2.623 \times 4.1868)\ \text{kJ}$$

Assume that there was 0.98 g = 0.98×10^{-3} kg of coal

$$\text{Calorific value of coal} = \frac{2.6 \times 2.623 \times 4.1868}{0.98 \times 10^{-3}}$$

$$= \mathbf{29\,136\ kJ/kg}$$

According to BS 1016: Part 5 the difference between duplicate determinations should not exceed 125 kJ/kg.

In the above determination, no corrections have been applied as a result of acid formation, etc., during combustion. Accurate calorific value determination includes this. An analysis of the solution in the base of the bomb is made in this case. This started as the small quantity of distilled water originally placed in the bomb on assembly.

The water equivalent (sometimes called the mean effective heat capacity) of the bomb may be determined by burning a known weight of pure benzoic acid in the bomb in the same way as described above. The calorific value of benzoic acid is 26 500 kJ/kg. The only unknown in the experiment with benzoic acid is the water equivalent of the bomb which may thus be determined.

To determine the calorific value of a volatile fuel, such as petrol, a small quantity of the fuel is sealed in a crucible. In this condition it is introduced into the bomb and the cotton hanging from the fuse wire passes through the seal into the petrol. The seal prevents loss due to evaporation. The seal at the top of the crucible is usually made using adhesive tape. The liquid fuel can be introduced through the adhesive tape using a hypodermic syringe.

The calorimeter arrangement described above is as an **isothermal jacket calorimeter**. The air gap is recommended as being 10 mm all round the calorimeter. This air gap should remain at a uniform temperature with a maximum variation of 0.05 °C.

Another arrangement is the **adiabatic calorimeter**. This is capable of circulating water continually through both jacket and lid. The system is fitted with heating and cooling devices so that the jacket water temperature is rapidly adjusted to that of the water in the calorimeter.

8.36 The gas calorimeter

Figure 8.3 shows the fundamental principle behind the construction of one kind of gas calorimeter used for the determination of the calorific value of a gaseous fuel. The basic elements are a central funnel surrounding a burner at its base. A cooling coil is fitted around the outside of the funnel: in some calorimeters the coil is replaced by a nest of tubes. Underneath the coil and arranged round the funnel is a condensate trap. Round the outside is fitted a cover, and the whole apparatus is heat insulated by lagging. The cooling water inlet passes through the cover at the bottom; the outlet is at the top. Thermometers are installed to measure the cooling water inlet and outlet temperatures.

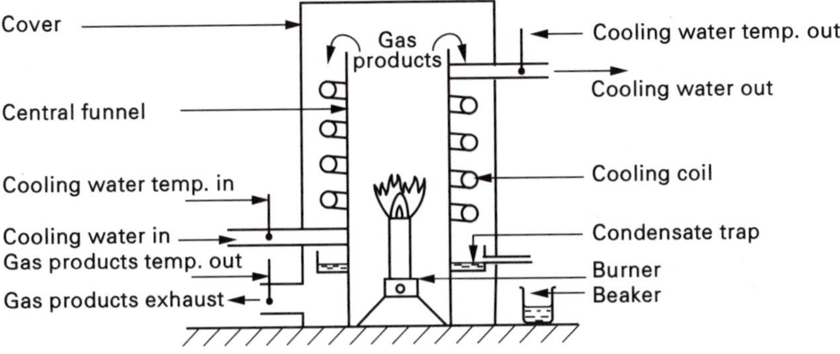

Fig. 8.3 Fundamental type of gas calorimeter

The gas products are arranged to pass out at the bottom through an exhaust. The exhaust is fitted with a thermometer to measure its temperature. Most gases under test contain H_2; during combustion this will form H_2O, which will condense as it passes over the cooling coils. The coils are arranged so that H_2O will drip from the coils into the condensate trap. A small tube is fed out from this trap through the cover and a beaker is placed underneath to collect the H_2O as it leaves the apparatus.

The calorimeter uses a constant head water-tower to maintain a constant flow of water. There is also a gas governor to maintain a constant gas pressure, and a water manometer to measure the pressure of the gas as it enters the apparatus. The energy gain by the cooling water is the energy liberated by burning the gas. Using this information it is possible to find the calorific value of the gas. A further point concerns the density of the gas under test. The density of a gas depends upon its pressure and temperature. Now the calorific value of a gaseous fuel is given per unit volume. The volume taken is the standard cubic metre. This is 1 m^3 of gas when at a temperature of 15 °C and a pressure of 760 mm Hg, sometimes called standard reference conditions (SRC), see section 8.33.3.

Here is an example calculation of the calorific value of a gas.

Volume of gas used, $\dot{V}$	$= 0.008\,5\ m^3 = 8.5$ litres
Temp. of gas	$= 18$ °C
Gas manometer height	$= 100$ mm H_2O
Barometric height	$= 755$ mm Hg
Mass of cooling water used in the same time as gas, $\dot{M}$	$= 4.26$ kg
Mass of conductance in same time as gas, $\dot{m}$	$= 0.013$ kg $= 13$ g
Temp. cooling water in, t_1	$= 14$ °C
Temp. cooling water out, t_2	$= 32$ °C

$$\text{Absolute pressure of gas} = 755 + \frac{100}{13.6}$$

$$= 755 + 7.35$$

$$= \mathbf{762.35\ mm\ Hg}$$

The volume of gas used must be converted to the volume at SRC (15 °C and 760 mm Hg). Let this volume be $\dot{V}_S$

Then

$$\dot{V}_S = \frac{P}{760} \times \frac{288}{T} \times \dot{V} \text{ (since } PV/T = \text{constant)}$$

$$= \frac{762.35}{760} \times \frac{288}{291} \times 0.008\,5$$

$$= \mathbf{0.008\,44\ m^3}$$

Now

Energy transferred by gas = Energy gained by water

$$\dot{V}_S \times Q_{gr,p} = \dot{M}(h_2 - h_1)$$

$$\therefore \quad Q_{gr,p} = \frac{\dot{M}(h_2 - h_1)}{\dot{V}_S}$$

$$= \frac{4.26 \times (134 - 58.8)}{0.008\,44} \quad \text{(Values of } h_1 \text{ and } h_2 \text{ from steam tables)}$$

$$= \frac{4.26 \times 75.2}{0.008\,44}$$

$$= \mathbf{37\,456\ kJ/standard\ m^3}$$

Now $\frac{\dot{m}}{\dot{V}_S}$ = Mass of condensate/standard m^3

From this $Q_{net,p}$ is given by

$$Q_{net,p} = \left(Q_{gr,p} - \frac{\dot{m}}{\dot{V}_S} \times 2466\right)$$

$$= 37\,456 - \frac{0.013}{0.008\,44} \times 2466$$

$$= 37\,456 - 3798$$

$$= \mathbf{33\,658\ kJ/standard\ m^3}$$

The British Standard Specification is BS 3804: Part 1, The Determination of Calorific Value of Fuel Gas. The recommended type of calorimeter is the Boys gas calorimeter. This method is now largely obsolete, especially commercially (section 8.37 contains a list of specifications).

The gas calorimeter can also be used for the determination of the calorific value of some fuel oils. A suitable burner is placed in the calorimeter and a pipe feeds the oil from a container outside the calorimeter. The oil is usually fed under pressure in a similar manner to a Primus stove. The calorific value of the fuel can be determined from a knowledge of the quantity of fuel burnt and the readings on the calorimeter.

In the gas calorimeter the fuel is burnt at constant pressure whereas in the bomb calorimeter it is burnt at constant volume. Some fuels can be tested in both calorimeters. If a fuel is burnt at constant volume the energy evolved will all go into the stock of internal energy of the products of combustion; this is because no external work is done. So this calorific value is sometimes called the **internal energy of combustion** or the **internal energy of reaction**.

If a fuel is burnt at constant pressure, the calorific value will depend on whether the volume increases, decreases or remains the same. If there is an increase in volume then some energy will be used up in performing the work of expansion; the calorific value will be decreased. If there is no change of volume, the calorific value remains unchanged. If there is a volumetric contraction, the calorific value is increased by an amount equal to the contraction work. Whichever the case, the process occurs at constant pressure and the energy evolved will appear as a change of enthalpy. So this calorific value is sometimes called the **enthalpy of combustion** or the **enthalpy of reaction**.

If a gas consists mainly of methane (CH_4), as is the case with natural gas obtained and refined from oil wells, it is considered reasonable to estimate the net calorific value as being 10 per cent less than the gross calorific value. In this case

$$Q_{\text{net},p} = 0.9Q_{\text{gr},p} \qquad [1]$$

And if the chemical composition of a fuel is known then the higher and lower calorific values can be estimated (see specifications in section 8.37).

8.37 Fuel testing specifications

BS 7420 : 1991	British Standard guide for the determination of the calorific value of solid, liquid and gaseous fuels (a general publication, it includes definitions and useful references)
BS 1016 : Part 5	Methods of analysis and testing of coal and coke: gross calorific value of coal and coke
BS 2000 : Part 12	Methods of test for petroleum and its products: heat of combustion of liquid hydrocarbon fuels
BS 3804 : Part 1	Methods for the determination of the calorific value of fuel gases: non-recording methods (largely obsolete industrially)
ASTM D 1826–88	American Society for Testing and Materials test method for the calorific value of gases in natural gas range by the continuous recording calorimeter
ASTM D 4891–89	Test method for heating values of gases in natural gas range by stoichiometric combustion

8.38 Principal solid fuels

8.38.1 Wood

Wood is generally obtained from trees. It is considered a low-grade fuel, having a relatively low calorific value. It has the advantage of being renewable by the cultivation of new trees.

8.38.2 Peat

Peat is formed by the partial carbonisation of vegetable matter in boggy areas. It is considered a low-grade fuel.

8.38.3 Lignite (brown coal)

Lignite is a fossil fuel somewhere between peat and bituminous coal. It can display some characteristics of the wood from which it was originally laid down.

8.38.4 Coal

Coal is a fossil fuel laid down from moist vegetable matter and compacted under pressure and temperature within the surface of the earth. It is the most widely used of the commercial solid fuels.

There are various grades of coal. They vary from anthracite (hard coal), composed mostly of carbon, through the bituminous coals (soft coals), containing medium quantities of volatile matter, to cannel coals which contain higher amounts of volatile matter.

All coals have an inherent moisture (H_2O) content. The coal can be air-dried to reduce the quantity of moisture. Bituminous and cannel coals contain volatile matter. This takes the form of a black mixture of hydrocarbons which distils off and burns, usually brightly, as the coal is heated. Unless controlled correctly, this volatile matter can produce smoke, releasing unburnt combustible gas, thus increasing atmospheric pollution.

Table of some approximate characteristics of solid fuels

Fuel	Volatile matter (%)	Calorific value, $Q_{gr,v}$ (kJ/kg)
Wood		15 800
Peat	55–80	20 900–26 000
Lignite	35–60	26 000–29 000
Anthracite	0–8	34 900–38 500
Bituminous coal (average)	30–50	32 000–35 000
Cannel coal	60–80	35 000–38 500

8.39 Solid fuels versus liquid or gaseous fuels

All solid fuels will leave some ash to be disposed of after burning. The use of solid fuel, especially coal, is widespread throughout the world. However, in some industrial societies, the use of oil and natural gas has largely replaced the use of coal. The reasons are mostly political and economic. A caveat should be added here. Oil, natural gas and coal are all fossil fuels. Their supply and reserves may appear large but are progressively diminishing. However, they are in finite quantities, and will runout eventually. Short-term attitudes mostly prevail; more emphasis on long-term management is required. More will be said about the use of fossil fuels later in the text.

Table of properties for some gaseous fuels: single gases

Gas	Molar mass (kg/kmol)	Density at 15 °C and 101.325 kPa (kg/m^3)	Relative density (air = 1)	Calorific value at 15 °C and 103.325 kPa (MJ/m^3)	
				Gross	Net
Carbon monoxide (CO)	28.01	1.19	0.967	11.97	11.97
Hydrogen (H_2)	2.02	0.085	0.07	12.10	10.22
Methane (CH_4)	16.04	0.679	0.55	37.71	33.95
Ethylene (C_2H_4)	28.05	1.19	0.97	59.72	55.96
Ethane (C_2H_6)	30.07	1.27	1.04	66.07	60.43
Propane (C_3H_8)	44.1	1.87	1.52	93.94	86.42
Butane (C_4H_{10})	58.12	2.46	2.01	121.80	112.41
Pentane (C_5H_{12})	72.15	3.05	2.49	149.66	138.39
Benzene (C_6H_6)	78.12	3.3	2.7	139.69	134.05
Hexane (C_6H_{14})	86.16	3.65	2.97	177.55	164.4
Octane (C_8H_{18})	114.23	4.83	3.94	233.29	216.38

Table of average properties for some gaseous fuel mixtures

Gas constituents	Natural gas (vol%)	Liquid petroleum gas (commercial)	
		Propane (vol %)	Butane (vol %)
Nitrogen (N_2)	1.5	–	–
Methane (CH_4)	94.4	–	0.1
Ethane (C_2H_6)	3.0	1.5	0.5
Propane (C_3H_8)	0.5	91.0	7.2
Butane (C_4H_{10})	0.2	2.5	88.0
Pentane (C_5H_{12})	0.1	–	–
Hexane (C_6H_{14})	0.1	–	–
Propylene (C_3H_6)	–	5.0	4.2
Density at 15 °C and 101.325 kPa (kg/m^3	0.72	1.87	2.38
Theoretical air (vol/vol)	9.75	23.76	29.92
Calorific value at 15 °C and 101.325 kPa (MJ/m^3/gas)			
Gross	38.62	93.87	117.75
Net	34.82	86.43	108.69

Table of average properties for some oil fuels

	Diesel	Paraffin (kerosene)	Gas oil	Fuel oil		
				Light	Medium	Heavy
Carbon, C (%)	86.5	85.8	86.1	85.6	85.6	85.4
Hydrogen, H_2 (%)	13.2	14.1	13.2	11.7	11.5	11.4
Sulphur, S (%)	0.3	0.1	0.7	2.5	2.6	2.8
O_2, N_2, ash (%)	–	–	–	0.2	0.3	0.4
Stoichioimetric mass dry air/mass fuel	14.55	14.69	14.44	13.95	13.88	13.84
Volume dry air/mass fuel (m^3/kg)	11.25	11.36	11.17	10.79	10.75	10.70
Calorific value (MJ/kg)						
Gross	45.7	46.5	45.6	43.5	43.1	42.9
Net	42.9	43.5	42.8	41.1	40.8	40.5

Table of average properties for petrol (gasoline)

Composition by mass (%)	
Carbon (C)	85.0–88.5
Hydrogen (H_2)	11.5–15.0
Sulphur (S)	< 0.1
Freezing point (°C)	< -40
Calorific value (MJ/kg)	
Gross	44.8–46.9
Net	41.9–44.0

8.40 Atmospheric and ecological pollution

The combustion of fuels pollutes the atmosphere. And fuels are burned in increasing amounts, so the problems of pollution, both atmospheric and ecological, need serious consideration.

8.40.1 Carbon monoxide (CO)

Carbon monoxide is produced by the incomplete combustion of carbon with oxygen. If breathed in, it reacts with the haemoglobin in the blood and prevents the uptake of oxygen. It impairs thought, alertness and reflexes. It is toxic enough to kill in sufficient dosage.

8.40.2 Carbon dioxide (CO_2)

Carbon dioxide is produced by the complete combustion of carbon with oxygen. Large masses are now emitted into the atmosphere from the various combustion processes. This is especially true of motor car and lorries. It has become possibly the main contributor to the greenhouse effect, producing global warming.

8.40.3 Sulphur dioxide (SO_2)

Sulphur dioxide is produced by the complete combustion of sulphur with oxygen. It is an atmospheric pollutant mostly from diesel engines and flue gas emission in industrial areas. It is a toxic gas and can be a respiratory irritant. It can produce atmospheric acid, a precursor of acid rain.

8.40.4 Oxides of nitrogen (NO_x)

Various oxides of nitrogen can be produced from the nitrogen in the air reacting with the oxygen in the air as the result of high temperatures obtained during the combustion of a fuel. The oxides are toxic and can be respiratory irritants. They can produce atmospheric acids, precursors of acid rain. They also contribute to photochemical smog.

8.40.5 Volatile organic compounds (VOCs)

Volatile organic compounds (VOCs) come from the evaporation of fuels from petrol stations, petrol tanks, leaks of hydrocarbons, etc. They contain cancer-forming compounds, carcinogens, e.g. benzene. They also contribute to photochemical smog.

8.40.6 Photochemical smog

Photochemical smogs occur on sunny, windless days, which perhaps have a temperature inversion in the lower atmosphere (i.e. when atmospheric temperature increases with height instead of falling as normal). If there are high concentrations of NO_x and VOCs, they may react with the ultraviolet light from the sun to produce ozone (O_3) an irritating, bluish gas with a pungent smell. Ozone in the very high atmosphere is formed naturally and protects the earth from some of the sun's ultraviolet radiation. But near to the ground it is photochemical smog.

8.40.7 Particulates

Particulates are small particles which mix with atmospheric air. They are produced during inefficient fuel burning (e.g. black, sooty smoke from some diesel engines, coal and oil burning). Dust particles may be thrown up by moving vehicles or may come from their brakes. Like many pollutants, particulates can be carcinogenic and produce breathing difficulties.

8.40.8 Lead (P_b)

Small quantities of lead compounds, such as tetraethyl lead, $Pb(C_2H_5)_4$, and tetramethyl lead, $Pb(CH_3)$, continue to be added to petrol to improve combustion characteristics (octane rating, section 16.19). After combustion the lead leaves with the exhaust and appears as a toxic pollutant in the atmosphere. Apart from its general toxicity, lead appears to be capable of retarding brain development in children.

The pollutant effect of lead in the atmosphere has prompted the development of unleaded petrol (ULG, unleaded gasoline) or green fuel. Unfortunately, aromatic additives are included to help improve the combustion efficiency of ULG, but additives such as benzene can be toxic (see VOCs). Many cars are now fitted with catalytic converters in the engine exhaust system. The exhaust gases must pass through the catalytic converter, which removes some of the nitrous oxides, unburnt hydrocarbons and carbon monoxide. But it does increase the emission of carbon dioxide. The catalysts are based on rhodium (Rh) and platinum (Pt); they require high temperatures to be effective so the converter must run hot.

8.40.9 Ecological considerations

The aquisition and use of coal and oil present ecological problems. Coal has to be mined, both on the surface and in deep mines. Mining scars the earth's surface, produces subsidance and creates large mounds of earth spoil and coal storage. After burning, coal can produce atmospheric particulates and quantities of ash which need disposal.

In the case of oil, both land and sea rigs are used for drilling and for receiving the acquired oil. Large tanks are used for storage. Inevitably, some spillage occurs; tanker disasters during sea transport cause large oil discharges into the sea and may severely damage the local ecology.

Another important problem is acid rain. If enough acid rain falls into lakes and onto forests, for example, it can cause serious, even disasterous, damage to flora and fauna.

A further problem occurs when an oil or gas field becomes depleted. The rig used for drilling the well and receiving the oil or gas is of no further use and requires careful disposal.

Questions

1. The coal supplied to a boiler has the following composition by mass: hydrogen 4 per cent, carbon 84 per cent, moisture 5 per cent, the remainder ash.

   The air supplied is 40 per cent in excess of that required for complete combustion and coal is burned at the rate of 2000 kg/h. Assuming that the specific volume of the flue gases at entrance to the flue stack is 1.235 m^3/kg and that the maximum permissible flue gas velocity is 7.75 m/s, find the cross-sectional area of the stack entrance.

   Assume that air contains 23 per cent of oxygen by mass.

   [1.47 m^2.]

2. A single-cylinder, four-stroke cycle, petrol engine is supplied with a fuel having an analysis by mass of carbon 84 per cent and hydrogen 16 per cent. Assuming the air supplied is just sufficient to burn completely the whole of the hydrogen to H_2O, 91 per cent of the carbon of CO_2 and the remaining 9 per cent of the carbon to CO., determine per kilogram of fuel

   (a) the mass of each of the actual exhaust gas constituents
   (b) the actual mass of air supplied

   If the engine develops an indicated power of 9 kW at 42 rev/s and uses 0.25 kg fuel/kWh, find the volume of air at 20 °C and 0.1 MN/m^2 drawn into the engine per cycle. Air contains 23 per cent O_2 by mass, R for air is 0.287 kJ/kg K.

   [(a) 1.44 kg H_2O, 2.8 kg CO_2, 0.176 kg CO, 11.45 kg N_2;
   (b) 14.87 kg air/kg fuel; 3.72×10^{-4} m^3]

3. The following results were obtained during an experiment with a continuous flow gas calorimeter. Calculate the higher calorific value of the gas at SRC (i.e. 15 °C and

760 mm Hg). Determine the lower calorific value of the gas, allowing 2466 kJ/kg of steam formed.

Volume of gas burned at 16 °C and 120 mm water pressure	$= 0.007\ m^3$
Volume of cooling water flowing during test	$= 4.12 \times 10^{-3}\ m^3$
Temperature of cooling water at inlet to calorimeter	= 14 °C
Temperature of cooling water at outlet from calorimeter	= 28 °C
Water collected during test	$= 3.5 \times 10^{-6}\ m^3$
Barometer	= 758 mm Hg
Specific gravity of mercury	= 13.6

[$Q_{gr} = 36.34\ MJ/m^3$; $Q_{net} = 34.26\ MJ/m^3$]

4. During a boiler trial, a sample of coal gave the following analysis by mass: carbon 89 per cent, hydrogen 4 per cent, oxygen 3 per cent, sulphur 1 per cent, the remainder being incombustible.

   Determine the stoichiometric mass of air required per kilogram of coal for chemically correct combustion. If 60 per cent air is supplied, estimate the percentage analysis by mass of the dry flue gas.

   Air contains 23 per cent oxygen by mass.

   [11.6 kg air/kg coal; CO_2 17.01%, SO_2 0.1%, O_2 8.35%, N_2 74.54%]

5. The following readings were obtained from a gas calorimeter during a test to determine the calorific value of a gas:

Water collected	$= 4.7 \times 10^{-3}\ m^3$
Inlet temperature of cooling water	= 16 °C
Outlet temperature of cooling water	= 21 °C
Gas consumed	$= 0.002\,8\ m^3$
Gas temperature	= 17 °C
Gas pressure	= 786 mm Hg

   Determine the calorific value of the gas in MJ/standard m^3 (measured at 15 °C and 760 mm Hg).

   [34.13 MJ/m^3]

6. A bomb calorimeter was used to determine the calorific value of a sample of coal and the following results were recorded:

   Mass of coal sample = 1 g
   Mass of water in calorimeter = 2500 g
   Water equivalent of apparatus = 744 g
   Initial temperature of water = 17.48 °C
   Maximum observed temperature of water = 20.07 °C
   Cooling correction = +0.015 °C

   From these results, determine the calorific value of the coal. Take the specific heat capacity of water as 4.18 kJ/kg K.

   [35 324 kJ/kg]

7. The analysis by mass of a coal used in a boiler furnace is C 70 per cent, H_2 10 per cent, O_2 9 per cent and the remainder ash. A boiler using this coal is supplied with 17 kg of air/kg of coal. The temperature of the air entering the furnace and of the gases entering the chimney are 16 °C and 354 °C, respectively. Determine
   (a) the energy carried away by the dry products of combustion per kilogram of coal
   (b) the percentage by mass of the CO_2 in the dry products of combustion

   [(a) 5743 kJ; 15.1%]

8. The analysis by mass of a fuel is carbon 72 per cent, hydrogen 12 per cent, oxygen 8 per cent and the remainder ash.
   (a) Determine the higher and lower calorific values of this fuel assuming that the calorific value of C to CO_2 is 35 000 kJ/kg, the higher calorific value of H_2 to H_2O is 143 000 kJ/kg and the specific enthalpy of evaporation of the moisture formed by combustion is 2442 kJ/kg.
   (b) Calculate, from first principles, the minimum mass of air required for the complete combustion of 1 kg of this fuel.

   Air contains 23 per cent oxygen by mass.

   [(a) Q_{gr} = 42 360 kJ/kg; Q_{net} = 39 723 kJ/kg; (b) 12.17 kg/kg fuel]

9. A fuel has the following percentage composition by mass C 85 per cent, H_2 15 per cent. If air contains 23 per cent oxygen by mass: determine
   (a) the stoichiometric mass of air required for complete combustion of 1 kg of this fuel
   (b) the percentage composition by mass of the dry products of combustion
   (c) the lower calorific value of the fuel, given the following data

   Lower calorific value of H_2 to H_2O = 121 000 kJ/kg
   Calorific value of C to CO_2 = 35 000 kJ/kg

   [(a) 15.1 kg; (b) CO_2 21.15%, N_2 78.85%; (c) 47 900 kJ/kg]

10. A sample of anthracite coal has the following composition by mass: carbon 89 per cent, hydrogen 3.5 per cent, oxygen 3.0 per cent, nitrogen 1.0 per cent, sulphur 0.4 per cent, the remainder ash. Determine the stoichiometric air required for the complete combustion of 1 kg of this fuel. If 16 kg air is supplied, determine
    (a) the percentage excess air
    (b) the masses of the constituents of the wet flue gases per kilogram of fuel burnt

    Air contains 23 per cent oxygen by mass.

    [11.4 kg; (a) 40%; (b) 3.26 kg CO_2, 0.315 kg H_2O, 0.008 kg SO_2, 1.058 kg O_2, 12.33 kg N_2]

11. An oil engine develops 37.5 kW at a specific fuel composition of 0.255 kg/kWh. The air supply to the engine flows through a 75 mm diameter pipe at a velocity of 9 m/s; the air is at a pressure of 96.5 kN/m^2 and a temperature of 15 °C. The fuel composition by mass is carbon 85.8 per cent, hydrogen 14.2 per cent. Determine
    (a) the mass of air supplied per kilogram of fuel
    (b) the percentage excess air
    (c) the percentage analysis by mass of the exhaust gases if combustion is complete

    For air, $R = 0.287$ kJ/kg K. Air contains 23 per cent oxygen by mass.

    [(a) 17.4 kg; (b) 16.8; (c) CO_2 17.1%; H_2O 6.9%, O_2 3.1%, N_2 72.9%]

12. An oil engine uses fuel oil having a composition by mass of C 86 per cent and H_2 14 per cent. The fuel oil is used at the rate of 55 kg/h. The air supply is 20 per cent in excess of the stoichiometric requirement. The air is supplied to the engine at a pressure of 96.5 kN/m^2 and a temperature of 17 °C through a pipe of 150 mm diameter. Determine
    (a) the percentage analysis of the exhaust gas by mass
    (b) the velocity of the air in the supply pipe in m/s

    For air, take $R = 0.287$ kJ/kg K.

    [(a) CO_2 16.7%, H_2O 6.7%, O_2 3.8%, N_2 72.8%; (b) 13.3 m/s]

13. A boiler uses coal at a rate of 910 kg/h. The coal is of the following composition by mass: carbon 82 per cent, hydrogen 6 per cent, oxygen 3 per cent, ash 9 per cent. The air is supplied by a forced draught fan at a pressure of 100 kN/m^2 and a temperature of 13 °C. If the total quantity of air is 25 per cent in excess of that theoretically required for the complete combustion, determine
    (a) the mass and volume of the air taken in by the fan per minute
    (b) the percentage composition of the flue gases by mass
    Assume air contains 23 per cent oxygen by mass. For air, $R = 0.287$ kJ/kg K.
    [(a) 217.3 kg; 178.3 m^3; (b) CO_2 19.8%, H_2O 3.5%, O_2 4.3%, N_2 72.4%]

14. A fuel oil consists of 86 per cent C and 14 per cent H_2 by mass. During a test on an engine using this oil, the dry exhaust gas analysis by volume was 11.25 per cent CO_2, 1.2 per cent O_2, 2.8 per cent CO and the remainder N_2. Estimate the air–fuel ratio by mass being supplied to the engine.
    [15.7]

15. The volumetric analysis of a gas is as follows: CO 30 per cent, H_2 60 per cent, CH_4 10 per cent. This gas is mixed with air in the proportions 1 volume of gas to 1.5 volumes of air. For the air–gas mixture, determine
    (a) its mean relative molecular mass
    (b) its density in kg/m^3 at a temperature of 27 °C and a pressure of 138 kN/m^2
    (c) its adiabatic index
    (d) its characteristic gas constant
    Use the following molar specific heat capacities at constant volume: $CH_4 = 36$ kJ/kmol K, diatomics = 20.9 kJ/kmol K. Air contains 21 per cent oxygen by volume.
    [(a) 21.78; (b) 1.205 kg/m^3; (c) 1.387; (d) 0.382 kJ/kg K]

16. Determine the composition by mass and by volume of the dry products of the combustion mixture from 1 kg of coal and 18 kg of air. Composition of the coal by mass is C 80 per cent, H_2 6 per cent, O_2 8 per cent, ash 6 per cent.
    [Mass: CO_2 15.9%; O_2 8.8%; N_2 75.3%;
    Vol: CO_2 10.8%; O_2 8.3%; N_2 80.9%]

17. Hexane (C_6H_{14}) is mixed with air in the correct stoichiometric proportions. For this mixture, determine:
    (a) its average relative molecular mass before combustion
    (b) its density before combustion in kg/m^3 at a temperature of 15 °C and a pressure of 100 kN/m^2
    (c) the adiabatic index for the products of combustion
    Molar specific heat capacities at constant volume

    $$\text{Diatomics} = 21 \text{ kJ/kmol K}$$
    $$\text{Triatomics} = 31.4 \text{ kJ/kmol K}$$
    $$\tilde{R} = 8.314\,4 \text{ kJ/kmol K}$$

    Air contains 21 per cent O_2, 79 per cent N_2 by volume.
    [(a) 30.1; (b) 1.257 k/m^{3}' (c) 1.349]

Chapter 9

Heat transfer

9.1 General introduction

The transfer of heat can take place by two phenomena known as **conduction** and **radiation**. These phenomena may take place in a given system on their own or they may occur simultaneously.

As the result of conduction and/or radiation occurring into a fluid media then a transport of heat may occur called **convection**.

These various phenomena will be discussed separately.

9.2 Conduction

This is the form of heat transfer which takes place when heat transfers through a material. It is considered that energy transfer occurs due to atomic or molecular impact which results from atomic or molecular vibration in the case of solids, or movement in the case of liquids or gases. It is also probable that there is some free electron drift showing an energy flux in the direction of reducing temperature. So metals, with their more compact molecular structure, will show greater thermal conductivity than liquids, with their greater molecular dispersal, and then gases, with their even greater molecular dispersal.

As an example of heat transfer by conduction, consider the transfer of heat from steam in a pipe to the outside. The heat must pass through material of the pipe. The heat being transferred through the pipe is said to be **conducted** through the pipe.

9.3 Conduction through a flat plate or wall

The transfer of heat by conduction is found to depend upon the following factors:

- The area through which heat transfer takes place.
- The temperature difference of the faces through which the heat is passing.
- The time taken for the heat transfer.
- The thickness of material through which the heat is passing.
- The type of material.

Now the greater the area and temperature difference, the greater will be the heat transfer. The heat transfer is therefore proportional to area and temperature difference.

On the other hand, the greater the thickness of material, the smaller the heat transfer. Heat transfer is therefore inversely proportional to thickness.

Consider a flat plate, or wall, of thickness x and heat transfer area A. Let the temperatures of its faces be t_1 and t_2 respectively. This is shown in Fig. 9.1.

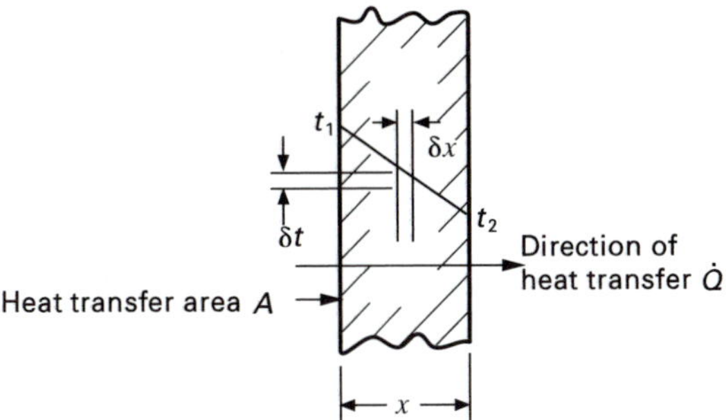

Fig. 9.1 Conduction through flat plate or wall

Consider now an elemental thin slice within the material of thickness δx. Let the temperature fall across this elemental thin slice $= \delta t$.

Then

$$\dot{Q} \propto A\frac{\delta t}{\delta x} \qquad [1]$$

or

$$\dot{Q} = -kA\frac{\delta t}{\delta x} \qquad [2]$$

where $k =$ constant of proportionality

In this case k is called the **coefficient of thermal conductivity**.

The minus sign in equation [2] indicates that the heat transfer occurs in the direction of diminishing temperature $(-\delta t/\delta x)$ or, in other words, this is an association with the second law of thermodynamics that heat will only transfer down a temperature gradient as a natural occurrence.

Assuming the temperature fall to be linear through the material of thickness x, equation [2] can be written

$$\dot{Q} = \frac{-kA(t_2 - t_1)}{x} \qquad [3]$$

$$= \frac{kA(t_1 - t_2)}{x} \qquad [4]$$

Let the transfer of heat be $\dot{Q}$ J/s $= \dot{Q}$ W, then

$$\dot{Q} = \frac{kA(t_1 - t_2)}{x} \qquad [5]$$

This is known as **Fourier's equation**, determined in about 1822. It is named after

Baron Jean-Baptiste-Joseph Fourier (1768–1830), a French physicist, mathematician and politician.

In this equation

k = coefficient of thermal conductivity, W/mK
A = area of transfer, m^2
t_1 = inlet face temperature, °C
t_2 = exit face temperature, °C
x = thickness, m

From the Fourier equation

$$k = \frac{\dot{Q}x}{A(t_1 - t_2)} \qquad [6]$$

from which the units of k become

$$\frac{\text{W m}}{\text{m}^2\text{ K}} = \frac{\text{W}}{\text{m K}} \quad \text{or} \quad \text{W m}^{-1}\text{ K}^{-1}$$

Table of average values of coefficients of thermal conductivity

Substance	Coefficient of thermal conductivity (W/m K)
Air, still, at 15 °C	0.025
Aluminium	206
Brass	104
Brick, common	0.6
Concrete	0.85
Copper	380
Cork, ground	0.043
Diatomaceous earth	0.086
Felt	0.038
Glass	1.0
Glass, fibre	0.04
Iron, cast	70
Magnesia	0.06
Plastic, cellular	0.04
Steel	60
Vermiculite	0.065
Wood	0.15
Wallboard, paper	0.076

9.4 Conduction through a composite wall

Consider the composite wall shown in Fig. 9.2; in this case there are three layers.

If $\dot{Q}$ W are passing through this wall then $\dot{Q}$ W are passing through each layer.

$$\therefore \quad \dot{Q} = \frac{k_1 A(t_1 - t_2)}{x_1} \qquad [1]$$

$$= \frac{k_2 A(t_2 - t_3)}{x_2} \quad [2]$$

$$= \frac{k_3 A(t_3 - t_4)}{x_3} \quad [3]$$

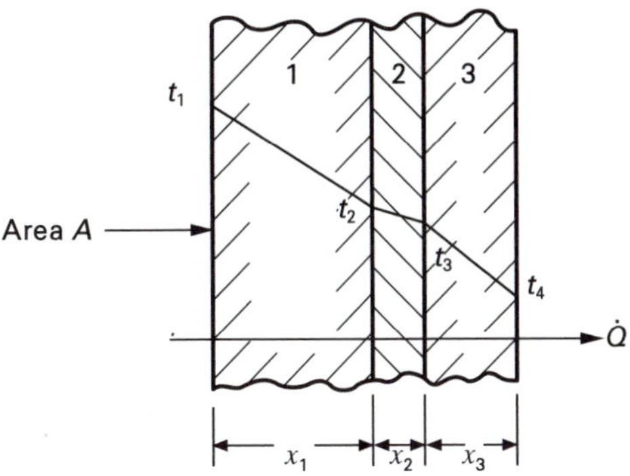

Fig. 9.2 Conduction through composite wall

As it stands, none of these equations are soluble since the interface temperatures t_2 and t_3 are, as yet, unknown.

Transposing equations [1], [2] and [3] gives

$$t_1 - t_2 = \frac{\dot{Q}x_1}{k_1 A} \quad [4]$$

$$t_2 - t_3 = \frac{\dot{Q}x_2}{k_2 A} \quad [5]$$

$$t_3 - t_4 = \frac{\dot{Q}x_3}{k_3 A} \quad [6]$$

Adding equations [4], [5] and [6]

$$t_1 - t_4 = \frac{\dot{Q}}{A}\left(\frac{x_1}{k_1}\ \frac{x_2}{k_2}\ \frac{x_3}{k_3}\right) \quad [7]$$

From equation [7]

$$\dot{Q} = \frac{A(t_1 - t_4)}{x_1/k_1 + x_2/k_2 + x_3/k_3} \quad [8]$$

From this equation, it can be inferred that if there are n layers, then

$$\dot{Q} = \frac{A(t_1 - t_{n+1})}{\sum x/k} \quad [9]$$

Thus, the heat transferred per second can be calculated. When this is known, by substituting back into equations [1], [2] and [3], the interface temperatures can be calculated.

Example 9.1 *A brick wall 250 mm thick is faced with concrete 50 mm thick. The brick has a coefficient of thermal conductivity of 0.69 W/m K and the concrete 0.93 W/m K. If the exposed brick face is at a temperature of 30 °C and the concrete at 5 °C, determine the heat lost per hour through a wall 10 m long and 5 m high. Determine the interface temperature.*

SOLUTION

$$\dot{Q} = \frac{A(t_1 - t_3)}{x_1/k_1 + x_2/k_2} = \frac{(10 \times 5) \times (30 - 5)}{0.25/0.69 + 0.05/0.93}$$

$$= \frac{50 \times 25}{0.362 + 0.054}$$

$$= \frac{1250}{0.416}$$

$$= 3005 \text{ W}$$
$$= 3005 \text{ J/s}$$
$$= \mathbf{3.005\ kJ/s}$$

$$\therefore \quad \text{Heat lost/h} = 3.005 \times 3600 = \mathbf{10\,818\ kJ}$$

For the brick wall

$$\dot{Q} = \frac{k_1 A(t_1 - t_2)}{x_1}$$

$$\therefore \quad t_2 = t_1 - \frac{\dot{Q}x_1}{k_1 A} = 30 - \frac{3005 \times 0.25}{0.69 \times 50}$$

$$= 30 - 21.8$$
$$= \mathbf{8.2\ °C}$$

The interface temperature is 8.2 °C.

Alternatively, for the concrete

$$\dot{Q} = \frac{k_2 A(t_2 - t_3)}{x_2}$$

$$\therefore \quad t_2 = t_3 \frac{+\dot{Q}x_2}{k_2 A} = 5 + \frac{3005 \times 0.05}{0.93 \times 50}$$

$$= 5 + 3.2$$
$$= \mathbf{8.2\ °C}$$

9.5 *Conduction through a thin cylinder*

There is no hard and fast rule as to what constitutes a thick cylinder and what constitutes a thin cylinder. A thin cylinder may be considered as a cylinder whose internal surface area is very nearly the same as its external surface area.

From the viewpoint of heat transfer, if this is the case, then the area through which the heat is passing is always very nearly the same.

For a thin cylinder of radius r (either internal or external or mean, since they are all very nearly the same) and thickness x, the area of heat transfer for a length of cylinder $L = 2\pi rL$.

Hence

$$\dot{Q} = \frac{k \times 2\pi rL \times (t_1 - t_2)}{x}$$

where k = coefficient of thermal conductivity
t_1, t_2 = surface temperatures

9.6 *Conduction through a thick cylinder*

In the thick cylinder the internal surface area is considerably smaller than the external surface area, so the thick cylinder does not admit of the simple treatment given in the case of the thin cylinder. However, part of the solution follows the treatment for the thin cylinder. Figure 9.3 shows a thick cylinder of internal radius r_1 and external radius r_2. The cylinder length is L, its internal surface temperature is t_1 and its external surface temperature is t_2.

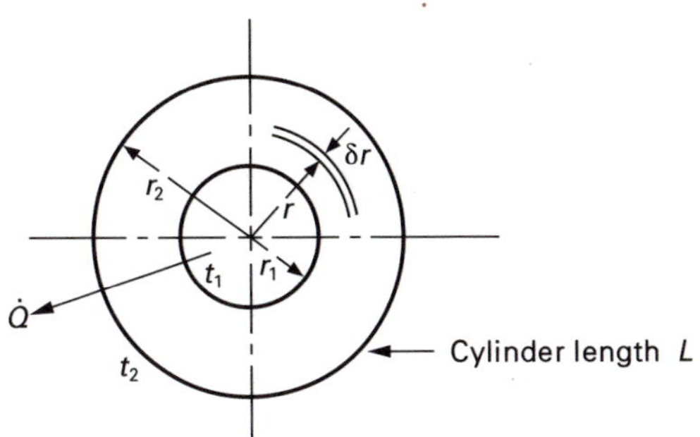

Fig. 9.3 Conduction through thick cylinder

Assume that heat transfer is from inside to outside, in which case $t_1 > t_2$.
Let the heat transfer per second be $\dot{Q}$.
Consider an elemental cylinder within the cylinder at radius r and with thickness δr.
Let the change of temperature across this elemental cylinder be δt.
From work on the thin cylinder

$$\dot{Q} = -k2\pi rL\frac{\delta t}{\delta r} \qquad [1]$$

The sign of $\delta t/\delta r$ is negative because there is a temperature fall across the cylinder.
From equation [1]

$$\delta t = -\frac{\dot{Q}}{k2\pi L}\frac{\delta r}{r}$$

Integrating across the thick cylinder

$$\int_{t_1}^{t_2} \mathrm{d}t = -\frac{\dot{Q}}{k2\pi L}\int_{r_1}^{r_2} \frac{\mathrm{d}r}{r}$$

$$\Big[t\Big]_{t_1}^{t_2} = -\frac{\dot{Q}}{k2\pi L}\Big[\ln r\Big]_{r_1}^{r_2}$$

or

$$(t_2 - t_1) = -\frac{\dot{Q}}{k2\pi L}[\ln r_2 - \ln r_1]$$

multiplying throughout by -1 and using $\ln r_2 - \ln r_1 = \ln(r_2/r_1)$ gives

$$(t_1 - t_2) = \frac{\dot{Q}}{k2\pi L}\ln\frac{r_2}{r_1}$$

From which

$$\dot{Q} = \frac{k2\pi L(t_1 - t_2)}{\ln(r_2/r_1)} \qquad [2]$$

Example 9.2 *A steam pipe is 75 mm external diameter and 80 m long. It conveys 1000 kg of steam per hour at a pressure of 2 MN/m². The steam enters the pipe with a dryness fraction of 0.98 and is to leave the other end of the pipe with a minimum dryness fraction of 0.96. This is to be accomplished by suitably lagging the pipe; the coefficient of thermal conductivity of the lagging is 0.08 W/m K. Assuming that the temperature drop across the steam pipe is negligible, determine the minimum thickness of the lagging required to meet the necessary conditions. Take the temperature of the outside surface of the lagging as 27 °C.*

SOLUTION

At 2 MN/m² the enthalpy of evaporation is

$$h_{fg} = \mathbf{1888.6\ kJ/kg}$$

The heat loss per kilogram of steam passing through the pipe is

$$(0.98 - 0.96) \times 1888.6 = 0.02 \times 1888.6$$
$$= \mathbf{37.77\ kJ}$$

So the heat loss per second through the pipe is

$$\dot{Q} = \frac{1000}{3600} \times 37.77$$
$$= \mathbf{10.5\ kJ}$$

For a thick cylinder

$$\dot{Q} = \frac{k2\pi L(t_1 - t_2)}{\ln(r_2/r_1)}$$

From this

$$\ln(r_2/r_1) = \frac{k2\pi L(t_1 - t_2)}{\dot{Q}}$$

The saturation temperature at 2 MN/m^2 is 212.4 °C

$$\therefore \quad \ln(r_2/r_1) = \frac{0.08 \times 2\pi \times 80 \times (212.4 - 27)}{10.5 \times 10^3}$$

$$= \frac{0.08 \times 2\pi \times 80 \times 185.4}{10.5 \times 10^3}$$

$$= \mathbf{0.71}$$

From this

$$r_2/r_1 = \mathbf{2.03}$$

Hence

$$r_2 = 2.03 r_1$$

$$= 2.03 \times \frac{75}{2}$$

$$= \mathbf{76.1\ mm}$$

$$\therefore \quad \text{Minimum thickness of lagging} = 76.1 - 75/2 = 76.1 - 37.5 = \mathbf{38.6\ mm}$$

9.7 Conduction through a composite thick cylinder

The solution to this arrangement is similar to that used on the composite wall.

If $\dot{Q}$ heat units per second pass through the composite thick cylinder, then $\dot{Q}$ heat units per second pass through each separate thick cylinder. For the composite thick cylinder in Fig. 9.4.

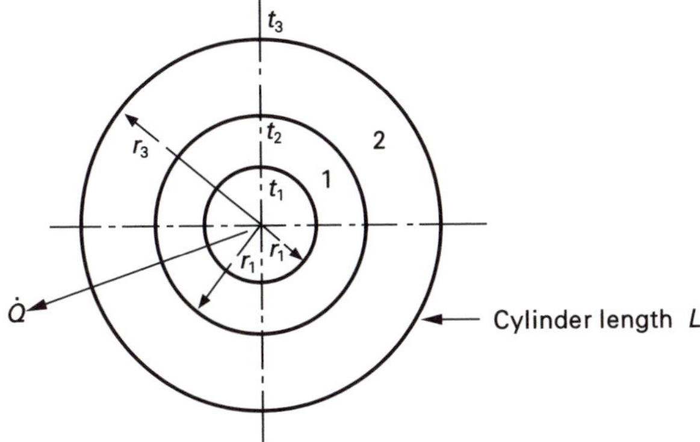

Fig. 9.4 Conduction through composite thick cylinder

$$\dot{Q} = \frac{k_1 2\pi L(t_1 - t_2)}{\ln(r_2/r_1)} \quad [1]$$

$$= \frac{k_2 2\pi L(t_2 - t_3)}{\ln(r_3/r_2)} \quad [2]$$

Again, neither of these equations is soluble due to the lack of knowledge of the interface temperature t_2.

Hence transposing equations [1] and [2]

$$t_1 - t_2 = \frac{\dot{Q}\ln(r_2/r_1)}{k_1 2\pi L} \qquad [3]$$

$$t_2 - t_3 = \frac{\dot{Q}\ln(r_3/r_2)}{k_2 2\pi L} \qquad [4]$$

Adding equations [3] and [4]

$$t_1 - t_3 = \frac{\dot{Q}}{2\pi L}\left[\frac{\ln(r_2/r_1)}{k_1} + \frac{\ln(r_3/r_2)}{k_2}\right]$$

from which

$$\dot{Q} = \frac{2\pi L(t_1 - t_3)}{\dfrac{\ln(r_2/r_1)}{k_1} + \dfrac{\ln(r_3/r_2)}{k_2}} \qquad [5]$$

The solution to a composite thick cylinder with more than two thicknesses follows the same procedure.

Since $\dot{Q}$ is known from equation [5], substitution into either equation [1] or [2] determines the interface temperature t_2.

Example 9.3 *A 100 mm diameter steam main is covered by two layers of lagging. The inside layer is 40 mm thick and has a coefficient of thermal conductivity of 0.07 W/m K. The outside layer is 25 mm thick and has a coefficient of thermal conductivity of 0.1 W/m K. The main conveys steam at a pressure of 1.7 MN/m² with 30 K superheat. The outside temperature of the lagging is 24 °C. If the steam main is 20 m long, determine*

(a) the heat lost per hour

(b) the interface temperature of the lagging

Neglect the temperature drop across the steam main.

(a)

$$\dot{Q} = \frac{2\pi L(t_1 - t_3)}{\dfrac{\ln(r_2/r_1)}{k_1} + \dfrac{\ln r_3/r_2}{k_2}}$$

The saturation temperature at 1.7 MN/m² is 204.3 °C

$$\therefore \quad t_1 = 204.3 + 30 = \mathbf{234.3\ ^\circ C}$$

$$\therefore \quad \dot{Q} = \frac{2\pi \times 20 \times (234.3 - 24)}{\dfrac{\ln 90/50}{0.07} + \dfrac{\ln 115/90}{0.1}}$$

$$= \frac{2\pi \times 20 \times 210.3}{\dfrac{\ln 1.8}{0.07} + \dfrac{\ln 1.28}{0.1}}$$

$$= \frac{2\pi \times 20 \times 210.3}{8.4 + 2.47} = \frac{2\pi \times 20 \times 210.3}{10.87}$$

$$= \mathbf{2431\ W}$$

$$\therefore \quad \text{Heat lost/h} = (2431 \times 3600)\ \text{J}$$

$$= \frac{2431 \times 3600}{10^3}$$

$$= \mathbf{8752\ kJ}$$

(b)

$$\dot{Q} = \frac{2\pi k_1 L(t_1 - t_2)}{\ln r_2/r_1}$$

$$\therefore \quad t_2 = t_1 - \frac{\dot{Q} \ln r_2/r_1}{2\pi k_1 L}$$

$$= 234.3 - \frac{2431 \ln 90/50}{2\pi \times 0.07 \times 20}$$

$$= 234.3 - \frac{2431 \times 0.588}{2\pi \times 0.07 \times 20}$$

$$= 234.3 - 162.5$$
$$= \mathbf{71.8\ ^\circ C}$$

The interface temperature is 71.8 °C. Alternatively, the solution could have been determined using

$$\dot{Q} = \frac{2\pi k_2 L(t_2 - t_3)}{\ln (r_3/r_2)}$$

9.8 Radiation

Radiation is an electromagnetic phenomenon of varying wavelengths closely allied to the transmission of light and radio. It proceeds in straight lines at the speed of light, (2.998×10^8 m/s).

Unlike conduction, radiation requires no transfer medium between the emitting and receiving surfaces. In fact, any material medium between such surfaces could impede radiation transfer of energy.

The classic example of energy transmission by radiation is the sun, which transmits abundant energy to the earth by this means.

Radiation is independent of mass, except for nuclear reactions.

The amount of energy (heat) transferred by conduction largely depends upon temperature difference rather than temperature level. But in radiation it is the temperature of the emitting surface that controls the quantity and the quality of the energy transmitted.

Radiation is very much a surface phenomenon and will leave the transmitting surface through a wide wavelength band.

A surface will emit or absorb radiant energy without a temperature difference, but in order for energy transfer to occur there must be a temperature difference between the exchange surfaces.

Figure 9.5 shows the general shape of the spectromagnetic curve of radiant power against wavelength for a **black body**. A body which is defined as being **black** is a perfect radiator of energy.

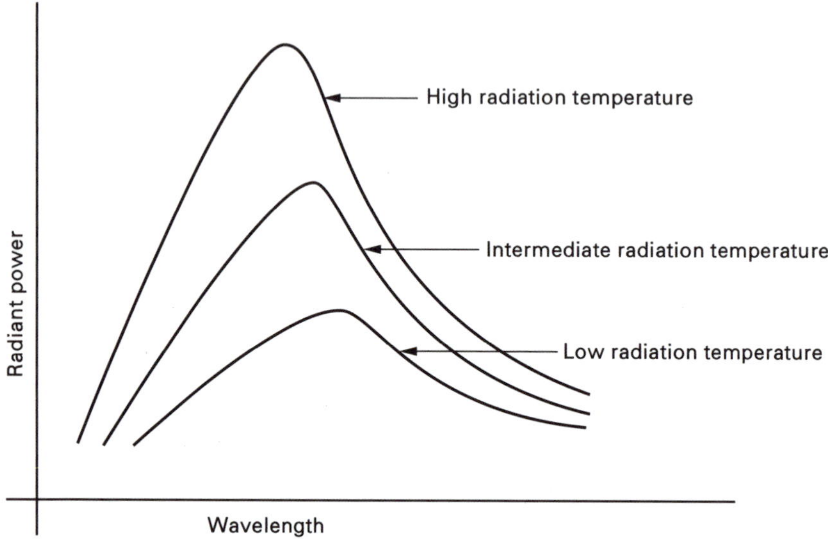

Fig. 9.5 Spectroradiometric curves

In Fig. 9.5 the height of the curve determines the quantity, and the general shape shows the quality of the emitted radiant energy. At the lower radiation temperatures there is a shift towards emissions of lower wavelength.

A radiating surface at a high temperature, perhaps over 800 °C, will emit some wavelengths which are within the visible light spectrum, approximately 10^{-6} to 10^{-7} metres. At lower temperatures, less than 800 °C, the radiation will be in the **infrared** range, approximately 10^{-2} to 10^{-6} metres.

If a body emits energy across the black body wavelength spectrum but only a fraction of the power of the black body, then it is called a **grey body**.

The terms *black* and *grey* do not necessarily refer to the colour of the body; they merely describe its effectiveness as a radiator. A black body is a perfect radiator; a grey body is not a perfect radiator.

9.9 Stefan–Boltzmann law and radiation surface parameters

9.9.1 Stefan–Boltzmann law

Joseph Stefan (1835–1893), an Austrian physicist, showed experimentally that the energy, E, transmitted by thermal radiation is proportional to the fourth power of the absolute temperature

$$E \propto T^4 \qquad [1]$$

Ludwig Boltzmann (1844–1906) was also an Austrian physicist and he, too, worked on the problem of thermal radiation.

The work of these two physicists, resulted in what is now termed the Stefan–Boltzmann law. For a perfect black body this law has been shown empirically to have the form

$$\dot{q} = \sigma T^4 \ (\mathrm{W/m^2}) \quad [2]$$

where σ = Stefan's constant
$= 5.669\,7 \times 10^{-8}\ \mathrm{W/m^2\ K^4}$
$\simeq 5.67 \times 10^{-8}\ \mathrm{W/m^2\ K^4}$
$\dot{q}$ = rate of energy emitted, $\mathrm{W/m^2}$

For surface area A_S m^2

$$\dot{Q} = \sigma T^4 A_S \ (\mathrm{W}) \quad [3]$$

9.9.2 Radiation surface parameters

Surfaces are capable of emitting, absorbing, reflecting or transmitting radiant energy.

It can be shown that the emissivity and absorptivity of any given surface are of equal value.

For any given surface, the radiation received must be reflected, absorbed or transmitted through the material.

For a black surface, therefore, the net sum of the maximum reflected, absorbed and transmitted energy must be equal to unity when used in the Stefan–Boltzmann equation.

$$\rho + \alpha + \tau = 1 \quad [4]$$

where ρ = reflected energy
α = absorbed energy
τ = transmitted energy

For materials that are opaque, the transmissivity is zero, so $\tau = 0$ and equation [4] becomes

$$\rho + \alpha = 1 \quad [5]$$

which becomes

$$\rho + \varepsilon = 1 \quad [6]$$

where ε = emissivity

Because emissivity and absorptivity can be shown to be equal, for a perfect black surface, the reflectivity is also considered to be zero. Thus from equation [6] the emissivity of a perfect black surface, ε_B is unity.

$$\varepsilon_B = 1 \quad [7]$$

For a grey surface the emissivity, ε_S, which is the ratio of the grey surface emission to the black surface emission, must therefore be of the form

$$\varepsilon_S = E_S/E_B < 1 \quad [8]$$

For a grey surface with emissivity ε_S the Stefan–Boltzmann equation (equation [3]) becomes

$$\dot{Q} = \varepsilon_S \sigma T^4 A_S \text{ watts (W)} \qquad [9]$$

Emissivity, and hence absorptivity, vary somewhat with surface temperature and are also a function of surface finish.

Generally, highly polished surfaces have low emissivities whereas rough and oxidised surfaces have much higher emissivities.

Table of average emissivities

Substance	Average emissivity ε_S
Aluminium	0.25
Brick, rough	0.93
Carbon	0.85
Chromium	0.4
Copper, oxidised	0.7
Copper, polished	0.02
Glass	0.94
Iron (steel), oxidised	0.7
Iron (steel), polished	0.15
Paints, all colours	0.85
Water	0.95

9.10 Some further radiation phenomena

If a radiant energy system is completely enclosed then its emissivity is 1, independent of the nature of its surfaces. This occurs in a boiler furnace. The reason for this is that, although the emissivity of each individual furnace wall is less than unity, heat energy will radiate backwards and forwards from wall to wall until they absorb it completely. That is why radiant heat boilers have been developed, because the maximum amount of radiant heat is absorbed under the closed conditions that exist in a boiler furnace. This same phenomenon occurs in any enclosed furnace or oven.

It can be noted that, in very sunny climates, buildings are often painted white in order to reflect some of the energy received by radiation from the sun. Thick walls are also often used in order to attenuate conduction energy penetration, thus leaving the interior of the building relatively cool.

Thick, white, loose-fitting clothing is also worn to reduce radiation energy influx, thus keeping the person wearing the clothing relatively cool. This coolness remains so long as the person is not too physically active.

An important energy radiation level is that of the **infrared** waveband. This occurs at wavelengths between the approximate limits of 10^{-2} and 10^{-6} metres. This is below the visual limit of the colour red, hence its name.

Because it is not detectable by the human eye, special detectors are required to record its presence. This has led to the development of infrared thermography and photography.

In infrared thermography, a detector is aimed at a surface. The infrared radiant energy is received and, through subsequent apparatus, is translated into a visual screening (bright for high temperature through to dark for low temperature). Energy loss by radiation can then be assessed and calibrated. This apparatus is particularly useful in the detection of heat transfer loss from buildings.

Infrared energy received from an object can be translated into a photographic image; infrared cameras can be used at night because no direct visual wavelengths are required from the original energy-transmitting object. Infrared binoculars have also been developed, enabling an observer to see at night.

Example 9.4 *A rectangular surface has a height of 3 m and a width of 4 m. The emissivity of the surface is 0.9 and its surface temperature is 600 °C. Determine the energy emitted from the surface in kilowatts.*

SOLUTION

$$\text{Area of surface} = 3 \times 4 = 12 \text{ m}^2$$

From equation [9]

$$\begin{aligned}\dot{Q} &= \varepsilon_S \sigma T^4 A_S \\ &= 0.9 \times 5.67 \times 10^{-8} \times (600 + 273)^4 \times 12 \\ &= 0.9 \times 5.67 \times 10^{-8} \times (873)^4 \times 12 \\ &= 355\,700 \text{ W} \\ &= \mathbf{355.7 \text{ kW}}\end{aligned}$$

Example 9.5 *A cubical furnace has internal dimensions 1.25 m × 1.25 m × 1.25 m. The internal surface temperature is 800 °C. At the internal centre is placed a sphere of radius 0.2 m which has an emissivity of 0.6 and a surface temperature of 300 °C. Determine the rate of energy transfer between the furnace walls and the sphere; determine its direction.*

SOLUTION

For the furnace

$$\text{Internal surface area} = 1.25 \times 1.25 \times 6 = \mathbf{9.45 \text{ m}^2}$$

The internal furnace can be considered as black.

$$\therefore \quad Q_f = \frac{5.67}{10^8} \times (800 + 273)^4 \times \frac{9.45}{10^3}$$

$$= \mathbf{710.25 \text{ kW}}$$

For the sphere

$$\begin{aligned}\text{Surface area} &= 4\pi r^2 \\ &= 4\pi \times 0.2^2 \\ &= \mathbf{0.503 \text{ m}^2}\end{aligned}$$

Thus the sphere will emit radiation

$$Q_S = 0.6 \times \frac{5.67}{10^8} \times (300 + 273)^4 \times \frac{0.503}{10^3}$$

$$= \mathbf{184.5\ kW}$$

Thus the transfer of energy will be from furnace to sphere at the rate

$$Q_f - Q_s = 710.25 - 184.5$$
$$= \mathbf{525.75\ kW}$$

9.11 Convection

Convection is the heat transfer which occurs as the result of the movement of a fluid. It cannot take place in a solid.

Heat will transfer by conduction through the walls of the fluid container. From the container walls, the heat will transfer by conduction and radiation into the fluid, causing convection currents to be set up in the fluid.

Figure 9.6 shows a vessel with a pipe leading from its top. This pipe bends round to re-enter the vessel at the bottom. Consider the system full of water. If heat is transferred through the base of the vessel, the water in the immediate vicinity of the heat supply will receive heat and will therefore expand. Its density will decrease as a result of the expansion, so it will begin to rise through the main water bulk. Its place at the bottom will at once be taken by cooler water which, upon being heated, will itself begin to rise.

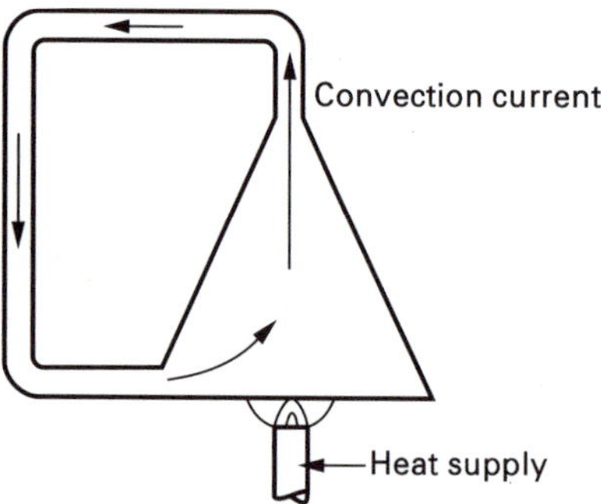

Fig. 9.6 Formation of a convection current

Most of the hot water will eventually pass up into the pipe and, in its passage through the pipe, will lose some heat. It will therefore fall through the pipe as its density increases and will eventually reappear at the bottom of the vessel, once again to receive heat. Thus a circulation of water has been created called a **convection current**.

The pipe was arranged so as to assist the formation of the convection current. This is always the case in any heating device of high efficiency. In domestic heating plant and boilers the pipework is always arranged to give the greatest possible assistance to the formation of convection currents. In this way, rapid heat transfer is made possible.

Convection, as described, is called **natural** or **thermo-syphon** convection. To improve heat transfer the convection currents may be further assisted by using a circulating pump; this is called forced convection. Most motor vehicle engines and heating systems have a pump in the cooling system to help create convection currents; they use forced convection.

It will appear from the following that the solution of convection problems can be quite complex. The transfer of heat from the hot source to the fluid will depend upon many factors.

- the direction of flow of the fluid across the retaining wall surface
- the nature of the fluid
- the nature of the retaining wall
- the velocity of the fluid across the retaining-wall surface
- the degree of turbulence of the fluid
- whether a change of state occurs within the fluid, i.e. from liquid to vapour or from vapour to liquid
- whether the fluid flow is natural or forced

Convection is an energy transport process because the heat transferred into the convecting fluid (liquid or gas) is transported by the convection current to another part of the fluid-containing circuit.

9.12 Heat insulation

Unless precautions are taken, heat losses in any thermal system can be considerable. The system can be heat insulated, or **lagged**, to cut these losses to a minimum. Lagging means the application to such things as pipes, walls and engines of some form of heat insulating material.

A heat insulating material will be a material which has a low thermal conductivity. Apart from its heat resisting properties it must also be able to withstand high or low temperatures, it must be relatively easy to apply, it should have a long life, it should withstand a moderate degree of rough handling and its cost should not be excessive. It should present no fire risk.

Insulating materials are made up from several substances:

- *Asbestos* Use of asbestos (a mined mineral) is now largely deprecated due to its health hazard.
- *Diatomaceous earth* Formed from small algae which are part of marine and freshwater plankton. The earth is composed of the hard shell skeletons of the dead organisms. The skeletons are composed of silica.
- *Glass and rock-wool* Formed by making wool of the fine fibres made from silica minerals.
- *Magnesium carbonate* Made from dolomite limestone.
- *Cork* Bark stripped from the cork-oak tree.
- *Expanded polystyrene* A synthetic plastic material.

These insulating materials, and others, generally owe their low thermal conductivities to their cellular structure. It so happens that air has a low thermal conductivity, which makes it good as a heat insulator except for convection. If the air can be kept static, it is a good thermal insulator. In the insulating materials the air is locked up in their cellular structures, so the property of the materials as thermal insulators really belongs to their static air content rather than the material itself. The heat insulation of clothing is really due to the layer of static air that a person puts on, air trapped by the cellular structure of the clothing; very little is due to the clothing itself.

Insulating materials are generally manufactured as fibres, pastes, sheets or granules. Such items as pipes and boilers often have their lagging applied as paste held in place by wire mesh and fabric coverings.

Alternatively, pipes may be lagged with sectional covering. This is preformed, semicylindrical lagging which is simply fitted round the pipe and held in place by clips on fabric. Large surfaces can be lagged with prefabricated blankets or boards made of various insulating materials.

Yet another form of insulation is to pack rigid spaces with granular insulation such as granular cork or vermiculite.

The rigid spaces may also be filled with glass wool or expanded polystyrene.

For high thermal efficiency of any plant, lagging is an absolute essential. But the **Dewar flask** achieves successful heat insulation without the use of lagging. Named after Sir James Dewar (1842–1923), a Scottish chemist and physicist, it is more commonly known by the trade name Thermos flask.

The Dewar flask is used for maintaining fluids, normally in liquid form, either hot or cold over long periods. It is used domestically as a picnic flask and commercially for the bulk transport of liquid gases. Due to the high volatility of the liquid gas, the commercial flask, which is rather large, remains uncorked. This is also the case with any other highly volatile fluid.

Figure 9.7 shows it to consist of two glass vessels, one sealed inside the other leaving a space between them. The inside surface of the outer vessel and the outside surface of the inner vessel are silvered. The space between the two vessels is pumped to vacuum and then sealed. The Dewar flask accomplishes heat insulation in several ways.

9.12.1 Dewar flask – conduction

Heat loss by conduction can only take place through the glass. For heat loss from inside to out, the heat must pass up through the thin glass section to the top. The reverse is true for a heat gain to the inside from the outside. Now glass has a low thermal conductivity and the area presented for heat transfer is small. Also the path of heat transfer through the glass becomes progressively longer from the top to the bottom of the flask. Thus, heat loss due to conduction is very low.

9.12.2 Dewar flask – convection

The space between the two flasks is evacuated to vacuum, so there is no convecting medium and no convection loss. There may be some convection loss above the surface of a fluid contained in the flask; the convection currents may pass out through the neck of the flask. This neck is often sealed using a cork or stopper, which itself has a low thermal conductivity. Under corked conditions, convection currents will be confined to the interior of the flask.

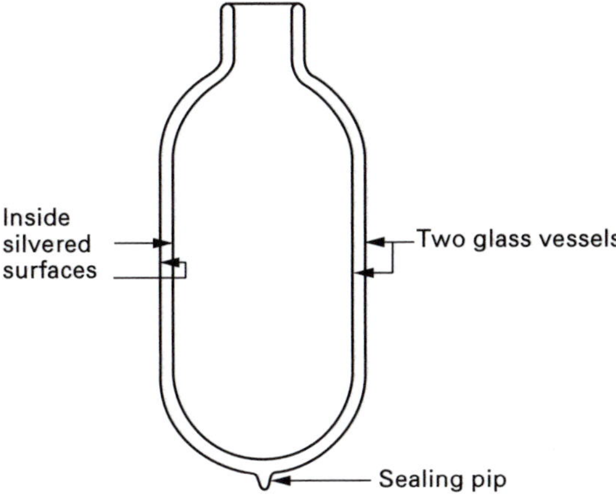

Fig. 9.7 Dewar flask

9.12.3 Dewar flask – radiation

As with light, a mirror surface is a good reflector of radiant heat energy. Thus any radiant energy passing from the internal flask will be nearly all reflected by the mirror surface on the inside of the outer flask. This, in turn, is reflected to a large extent by the mirror surface on the inner flask, and so on. In this way heat loss due to radiation is reduced to a minimum.

9.13 Surface and overall heat transfer

The examples so far have only considered surface temperatures. However, the control temperatures in any heat transfer system will be the surrounding, or ambient, temperatures. Now there must be a temperature difference between a surface and its surroundings if there is to be any heat transfer (second law of thermodynamics). The transfer of heat at a surface can include conduction, radiation and convection and any analysis of this transfer could be very complex.

A simple solution can be obtained by using experimental methods to determine the heat transfer through known and commonly used systems. Many commonly used systems appear in buildings, refrigerators, heat exchangers, etc. These systems tend to be solved experimentally, not analytically.

Consider the composite wall shown in Fig. 9.8.

As before, if heat transfer $\dot{Q}$ passes through the wall, then $\dot{Q}$ passes through each layer of the wall.

Let

t_{a1} = ambient temperature on the inlet side of the wall
t_1 = inlet face temperature
t_2 = interface temperature
t_3 = exit face temperature
t_{a2} = ambient temperature on the exit side of the wall

Now $t_{a1} > t_1$ and $t_3 > t_{a2}$ because there must be a temperature difference in order that heat transfer can take place.

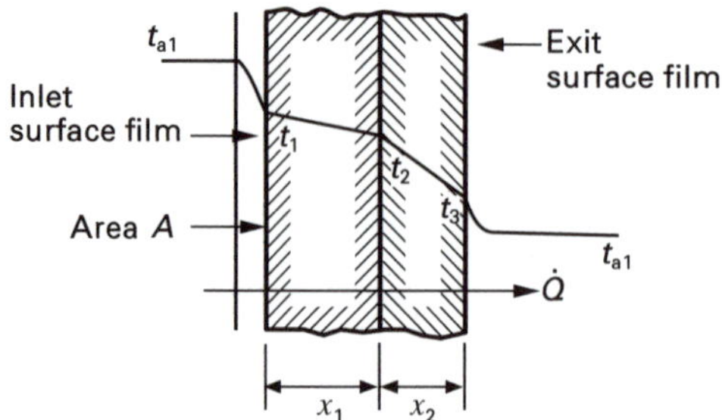

Fig. 9.8 Overall heat transfer through composite wall

The solution at the surfaces is obtained from

$$\dot{Q} = U_{s1}A(t_{s1} - t_1) \quad [1]$$

$$= U_{s2}A(t_3 - t_{a2}) \quad [2]$$

In each case U_s is called the **surface transfer coefficient** and has the units

$$U_s = \frac{\dot{Q}}{A\Delta t} = \frac{\mathrm{W}}{\mathrm{m^2\ K}}$$

U_s is determined experimentally and will include the effects of conduction, radiation and convection at the considered surface.

Again consider the composite wall shown in Fig. 9.8. Two thicknesses are shown but the solution is similar for a single or any multithickness wall.

For this wall

$$\dot{Q} = U_{s1}A(t_{a1} - t_1) \quad [3]$$

$$= \frac{k_1 A(t_1 - t_2)}{x_1} \quad [4]$$

$$= \frac{k_2 A(t_2 - t_3)}{x_2} \quad [5]$$

$$= U_{s2}A(t_3 - t_{a2}) \quad [6]$$

From this

$$(t_{a1} - t_1) = \frac{\dot{Q}}{U_{s1}A} \quad [7]$$

$$(t_1 - t_2) = \frac{\dot{Q}x_1}{k_1 A} \quad [8]$$

$$(t_2 - t_3) = \frac{\dot{Q}x_1}{k_2 A} \quad [9]$$

$$(t_3 - t_{a2}) = \frac{\dot{Q}}{U_{s2}A} \quad [10]$$

Adding equations [7], [8], [9], [10]

$$(t_{a1} - t_{a2}) = \frac{\dot{Q}}{A}\left(\frac{1}{U_{s1}} + \frac{x_1}{k_1} + \frac{x_2}{k_2} + \frac{1}{U_{s2}}\right) \qquad [11]$$

From which

$$\dot{Q} = \frac{A(t_{a1} - t_{a2})}{\left(\frac{1}{U_{s1}} + \frac{x_1}{k_1} + \frac{x_2}{k_2} + \frac{1}{U_{s2}}\right)} \qquad [12]$$

$$= UA(t_{s1} - t_{s2}) \qquad [13]$$

where

$$U = \frac{1}{\left(\frac{1}{U_{s1}} + \frac{x_1}{k_1} + \frac{x_2}{k_2} + \frac{1}{U_{s2}}\right)}$$

U is called the **overall transfer coefficient** and sometimes the **U coefficient** or **U value**. The units of U will be W/m^2 K, as for the surface transfer coefficient. The overall transfer coefficient can be determined directly for known systems by experiment. For building construction a U value of not more than 0.6 W/m^2 K is recommended.

Example 9.6 *A composite wall is made up of an external thickness of brickwork 110 mm thick, inside which is a layer of fibreglass 75 mm thick. The fibreglass is faced internally by an insulating board 25 mm thick. The coefficients of thermal conductivity for the three materials are*

Brickwork	*0.6 W/m K*
Fibreglass	*0.04 W/m K*
Insulating board	*0.06 W/m K*

The surface transfer coefficient of the inside wall is 2.5 W/m² K; that of the outside wall is 3.1 W/m² K.

Determine the overall transfer coefficient for the wall and, using the coefficient, determine the heat lost per hour through such a wall 6 m high and 10 m long. Take the internal ambient temperature as 27 °C and the external ambient temperature as 10 °C.

SOLUTION

$$U = \frac{1}{\left(\frac{1}{U_{s1}} + \frac{x_1}{k_1} + \frac{x_2}{k_2} + \frac{x_3}{k_3} + \frac{1}{U_{s2}}\right)}$$

$$= \frac{1}{\left(\frac{1}{2.5} + \frac{0.025}{0.06} + \frac{0.075}{0.04} + \frac{0.110}{0.6} + \frac{1}{3.1}\right)}$$

$$= \frac{1}{0.4 + 0.417 + 1.875 + 0.183 + 0.322}$$

$$= \frac{1}{3.197}$$

$$= \mathbf{0.313\ W/m^2\ K}$$

$$\dot{Q} = UA(t_{a1} - t_{a2})$$
$$= 0.313 \times 6 \times 10 \times (27 - 10)$$
$$= 0.313 \times 60 \times 17$$
$$= 319 \text{ W}$$
$$= \mathbf{319\ J/s}$$

$$\therefore \quad \text{Heat lost/h} = \frac{319 \times 3600}{10^3} = \mathbf{1148\ kJ}$$

Example 9.7 *A steam pipe, which is 150 mm external diameter, carries wet steam at a pressure of 3.6 MN/m². It is covered with two layers of lagging each 40 mm thick. The coefficients of thermal conductivity for the two layers are 0.07 W/m K for the inner layer and 0.1 W/m K for the outer layer. The surface transfer coefficient for the outer surface is 7.0 W/m² K. Estimate the heat loss per hour for a 50 m length of the lagged pipe. The ambient temperature is 27 °C. What would be the surface temperature of the lagging? Neglect the thickness of the steam pipe and assume that its temperature is constant throughout and, together with the inside surface of the inner layer of lagging, is at the same temperature as the wet steam.*

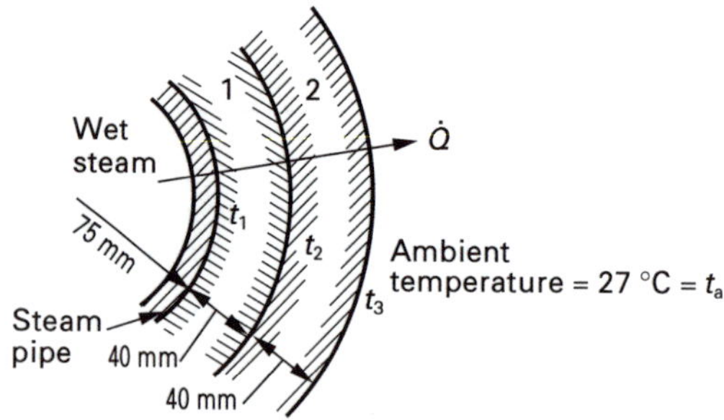

Fig. 9.9 Diagram for Example 9.7

SOLUTION

First draw a diagram (Fig. 9.9).

$$\dot{Q} = \frac{k_1 2\pi L(t_1 - t_2)}{\ln(r_2/r_1)} \qquad [1]$$

$$= \frac{k_2 2\pi L(t_3 - t_2)}{\ln(r_3/r_2)} \qquad [2]$$

$$= U_s 2\pi r_3 L(t_3 - t_a) \qquad [3]$$

From this

$$(t_1 - t_2) = \frac{\dot{Q}}{2\pi L}\frac{\ln(r_2/r_1)}{k_1} \qquad [4]$$

$$(t_3 - t_2) = \frac{\dot{Q}}{2\pi L}\frac{\ln(r_3/r_2)}{k_2} \qquad [5]$$

$$(t_3 - t_a) = \frac{\dot{Q}}{2\pi L}\frac{1}{U_s r_3} \qquad [6]$$

Adding equations [4], [5] and [6]

$$(t_1 - t_a) = \frac{\dot{Q}}{2\pi L}\left[\frac{\ln(r_2/r_1)}{k_1} + \frac{\ln(r_3/r_2)}{k_2} + \frac{1}{U_s r_3}\right]$$

Hence

$$\dot{Q} = \frac{2\pi L(t_1 - t_a)}{\left[\dfrac{\ln(r_2/r_1)}{k_1} + \dfrac{\ln(r_3/r_2)}{k_2} + \dfrac{1}{U_s r_3}\right]} \qquad [7]$$

From steam tables, saturation temperature at 3.6 MN/m^2 is $t_1 = 244.2$ °C and

$r_1 = 75$ mm
$r_2 = 75 + 40 = 115$ mm $= 0.115$ m
$r_3 = 75 + 40 + 40 = 155$ mm $= 0.155$ m

So from equation [7]

$$\dot{Q} = \frac{2\pi \times 50 \times (244.2 - 27)}{\left(\dfrac{\ln 115/75}{0.07} + \dfrac{\ln 155/115}{0.1} + \dfrac{1}{7 \times 0.155}\right)}$$

$$= \frac{100\pi \times 217.2}{\left(\dfrac{\ln 1.533}{0.07} + \dfrac{\ln 1.348}{0.1} + \dfrac{1}{7 \times 0.155}\right)}$$

$$= \frac{100\pi \times 217.2}{\left(\dfrac{0.427}{0.07} + \dfrac{0.3}{0.1} + \dfrac{1}{0.085}\right)}$$

$$= \frac{100\pi \times 217.2}{6.1 + 3 + 0.922}$$

$$= \frac{100\pi \times 217.2}{10.022}$$

$= 6808.6$ W
$=$ **6808.6 J/s**

$\therefore$ Heat loss/h $= \dfrac{6808.6 \times 3600}{10^3} =$ **24 511 kJ**

From equation [3]

$$\dot{Q} = U_s 2\pi r_3 L(t_3 - t_a)$$

$$\therefore \quad t_3 = t_a + \frac{\dot{Q}}{U_s 2\pi r_3 L}$$

$$= 27 + \frac{6808.6}{7 \times 2\pi \times 0.155 \times 50}$$

$= 27 + 19.7$
$=$ **46.97 °C**

The surface temperature of the lagging is 46.97 °C.

Questions

1. A wall is made up of two layers of bricks each 100 mm thick with a 50 mm air space between them. The coefficients of thermal conductivity are

Inside brick	0.6 W/m K
Air	0.025 W/m K
Outside brick	0.8 W/m K

The wall is 6.15 m long and 5.5 m high. Determine the heat loss per hour through the wall if the inside face temperature is 24 °C and the outside face temperature is 7 °C. Determine the interface temperatures.

[902.7 kJ; 22.76 °C; 7.93 °C]

2. A refrigerator room has a wall 6.0 m long and 3.0 m high. The wall is built of 120 mm thick brick, insulated on the inside with an 80 mm layer of cork faced with a thin metal sheet. The coefficient of thermal conductivity of the brick is 0.7 W/m K; that of the cork is 0.043 W/m K. The exterior brick surface temperature of the wall is 21 °C and that of the interior metal-faced surface is −4 °C. The temperature of the metal sheet can be considered as −4 °C throughout. Estimate the heat leakage through the wall in 24 hours and also the temperature of the interface between the cork and the brick.

[19 155 kJ; 18.9 °C]

3. During a test on pipe lagging a 2 m length of pipe, 80 mm external diameter was covered with a 40 mm thickness of lagging. The pipe was then coupled to a steam main which supplied steam at a pressure of 7 bar and 0.9 dry. As a result of heat loss through the lagging, 0.365 kg of condensate was collected from the pipe in 1 hour. The condensate was at saturation temperature for a pressure of 7 bar. The outside surface temperature of the lagging was 38 °C. Assuming that the inside temperature of the lagging was the same as the saturation temperature for the steam, estimate the coefficient of thermal conductivity of the lagging in W/m K.

[0.082 W/m K]

4. A steel pipe 150 mm external diameter conveys steam at a temperature of 260 °C and it is covered by two layers of lagging, each 50 mm thick. The thermal conductivity of the inside layer of the lagging is 0.086 W/m K; that of the outside layer is 0.055 W/m K. The outside surface temperature of the steel pipe can be taken as being the same as the temperature of the steam; the temperature of the outside layer of lagging is 27 °C. Determine
   (a) the heat lost per hour for a pipe length of 30 m
   (b) the interface temperature between the two layers of lagging.

[(a) 13 111 kJ; (b) 145.2 °C]

5. A cold room has internal dimensions 6 m × 6 m × 3 m high. Each containing wall, including the ceiling and floor, consists of an inner layer of 25 mm thick wood, backed up with a 75 mm layer of fibreglass and, on the outside, a 100 mm layer of brick. The coefficients of thermal conductivity are

Wood	0.2 W/m K
Fibreglass	0.04 W/m K
Brick	0.75 W/m K

Surface transfer coefficients are

Wood to still air	2.5 W/m^2 K
Moderately turbulent air to brick	5 W/m^2 K

Average external ambient temperature can be taken as 15 °C. It is required to maintain the interior ambient temperature of the cold room at −15 °C. The enthalpy of evaporation of the refrigerant in the evaporator is 158.7 kJ/kg. The refrigerant is 0.1 dry at entry to the evaporator and is dry saturated at exit. Determine

(a) the overall transfer coefficient in W/m^2 K
(b) the mass flow of refrigerant through the refrigeration plant in kg/h
(c) the temperature of the exposed surfaces and of the interfaces

Neglect corner effects at surface joints.

[(a) 0.364 W/m^2 K; (b) 39.6 kg/h; (c) surfaces 12.82 °C, −10.62 °C; interfaces 11.22 °C, −9.25 °C]

6. Wet steam at a pressure of 2 MN/m^2 flows through a pipe 20 m long. The pipe has an external diameter of 80 mm. The pipe is covered with lagging 35 mm thick which has a coefficient of thermal conductivity of 0.065 W/m K. The surface transfer coefficient is 4.5 W/m^2 K and the ambient temperature is 15 °C. The steam flow rate is 300 kg/h and it enters the pipe with a dryness fraction of 0.97. Assuming there is no temperature drop across the pipe, determine

   (a) the dryness fraction of the steam as it leaves the pipe
   (b) the surface temperature of the lagging

   [(a) 0.958; (b) 61.3 °C]

7. A steam pipe, 100 mm external diameter, is to be lagged with two layers of different lagging material, each 25 mm thick. Material A has coefficient of thermal conductivity 0.052 W/m K; material B has coefficient 0.086 W/m K. Determine which of the two lagging materials must be on the inside to produce the best insulation arrangement. If the internal surface temperature is 320 °C and the external surface temperature is 20 °C, determine the heat loss per hour for a 10 m run of lagged pipe with the best lagging arrangement.

   [Best, A inside; 6037 kJ/h]

8. A composite wall is made up of material A, thickness 3 cm and coefficient of thermal conductivity 0.6 W/m K, and material B, thickness 2 cm and coefficient of thermal conductivity 0.15 W/m K. The external surface temperature of material A is 100 °C. The external surface temperature of material B is 30 °C and its surface has an emissivity of 0.9. The wall has an area of 25 m^2. Neglecting edge effects, determine

   (a) the rate of heat transfer through the wall
   (b) the interface temperature
   (c) the percentage of the heat transfer lost by radiation from the surface of material B

   [(a) 9562.8 W; (b) 81 °C; (c) 87.5%]

9. Wet steam at a pressure of 20 bar flows through a pipe of 0.06 m diameter and 20 m long. Assuming that the surface emissivity of the pipe is 0.72 and that the surface temperature of the pipe is the same as that of the wet steam, determine the rate of energy lost by radiation from the surface of the pipe.

   [8.544 kW]

Chapter 10

Steam plant

10.1 General introduction

The general arrangement of the basic elements of a steam plant are illustrated diagrammatically in Fig. 10.1.

Steam is generated in a **boiler** from which it passes into the steam main. The steam main feeds the steam into a turbine or engine or it may pass into some other plant such as heaters or process machinery. After expanding through the turbine or engine or passing through some other plant, if the plant is working on a 'dead-loss' system, then the exhaust steam passes away to atmosphere. Such is the case with the steam locomotive, which is still in use on some railways in many countries of the world. This system is very inefficient and is rarely adopted in modern plant. It is used in the steam locomotive since, in this case, the plant is mobile and there is not sufficient room for the complex steam recovery equipment which can be installed in a power station or factory. If steam recovery plant is installed, the exhaust steam passes into a **condenser** where it is condensed to water, called **condensate**.

The condensate is extracted from the condenser by the **condensate extraction pump** from which it passes as feedwater into the feedwater main and back to the boiler. Because the boiler is operating at a high pressure the water pressure must be increased in order to get it into the boiler. This is dealt with by means of a pump called the **feed pump**. Thus the water returns to the boiler and, neglecting system losses, a steam recovery plant circulates the same water all the time. Actually, there are losses; they are made up in the condenser by means of a make-up water supply. The advantages of steam recovery plant are primarily as follows.

Firstly, the pressure in the condenser can be operated well below atmospheric pressure. This means that a greater expansion of the steam can be obtained, which results in more work. Secondly, the water in the circuit can be chemically treated to reduce scale formation in the boiler. The formation of scale in the boiler impedes the transfer of heat from the furnace to the water, so it reduces the boiler efficiency. It may also cause local overheating with resultant damage and, if overheating is serious, it may even cause a burst in the vicinity.

The condenser is cooled by circulating cooling water through it. If an abundant supply of water is nearby, such as a river or lake, then this can be used. Trouble may be experienced here due to water pollution. This may take the form of fish or mud

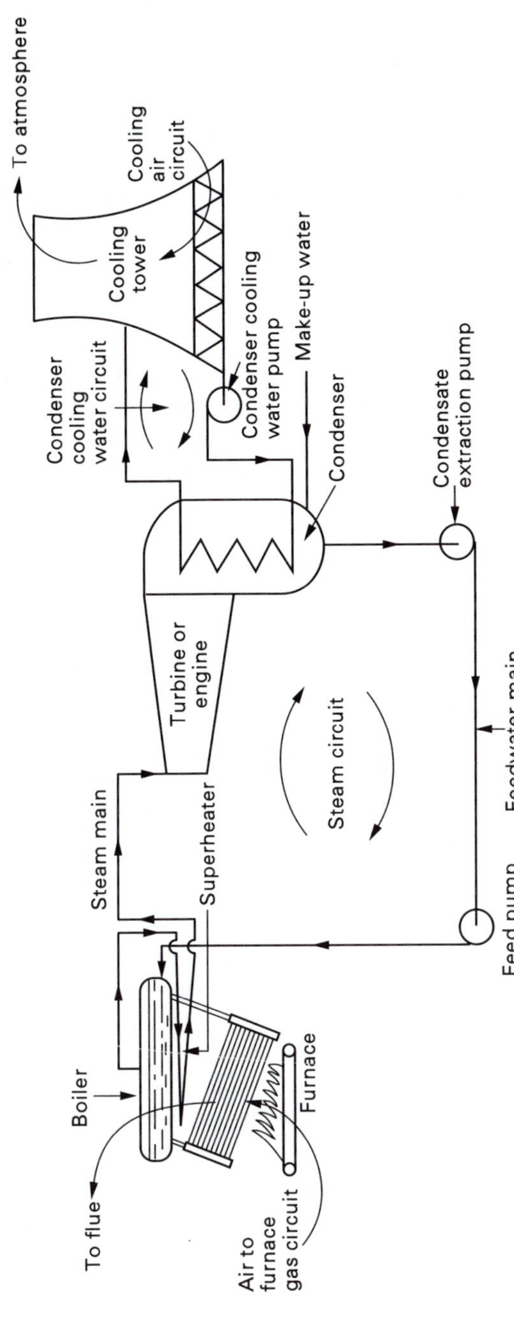

Fig. 10.1 Steam plant

entering the condenser system. Filters are usually installed to cut down pollution, otherwise the condenser cooling water circuit may become blocked. On the other hand, the river or lake may itself become polluted by hot water returning from the condenser. If the amount of hot water is large it could have an effect on the flora and fauna of the river or lake. If a river or lake is not to hand, or the risk of pollution is too high, it is common to install a **cooling tower** of wood or concrete. The hot water from the condenser enters the tower approximately midway up and is sprayed to the bottom. Air circulates into the bottom of the tower and passes up through the water spray. Heat transfer occurs between the water and the air, thus cooling the water. The warmed air passes out at the top of the tower. The cooled water is collected at the bottom of the tower from where it is pumped back to the condenser. The same cooling water is circulated through the condenser. It is only necessary to make up any loss. The entire steam plant contains four separate circuits.

10.1.1 The furnace gas circuit

Air is taken into the furnace from the atmosphere to supply the necessary oxygen for combustion. The combustion products pass through the boiler, transferring heat, then pass out to the atmosphere through the flue. Care should be taken to reduce atmospheric pollution from the combustion products to an absolute minimum. Most furnaces are fired by coal, gas or oil.

10.1.2 The steam circuit

Water is passed into the boiler where it is converted into steam. It passes into the plant where it is expanded, giving up some of its energy. It is then condensed in a condenser and passes as condensate to be pumped back into the boiler.

10.1.3 Condenser cooling water circuit

Cool water passes into the condenser, has heat transferred into it by the condensing steam then, at a higher temperature, passes out to be cooled in a river, lake or cooling tower. Cool water then circulates back to the condenser. Care must be taken to reduce pollution to a minimum if the discharge of condenser cooling water is into a river or lake.

10.1.4 Cooling air circuit

In the case of a cooling tower, cool air passes into the bottom of the tower from the atmosphere and heat is transferred into it from the falling hot water spray. The warm air then passes back to the atmosphere through the top of the tower. In the case of a river or lake the condenser cooling water will mix with river or lake water which will be cooled by heat transfer to the atmosphere.

In some steam plant the condensate from the condenser is passed into a tank, called the hot well, which acts as a reservoir for feedwater. From the hot well, feedwater is pumped through the feed pump back into the boiler. In this case, make-up water could be fed into the hot well.

10.2 Boilers

A boiler is the device in which steam is generated. Generally, it must consist of a water container and some heating device. There are many designs of boiler but they can be divided into two types: **fire-tube boilers** and **water-tube boilers**. Before describing various boiler designs, it will be useful to discuss the formation of steam in a boiler, and the methods employed to improve its thermal efficiency.

Whatever the type of boiler, steam will leave the water at its surface and pass into what is called the **steam space**. This is the space in the water container directly above the water. Steam formed above the surface of water is always wet and will remain wet so long as there is water present. This is because the steam, rising from the surface of the turbulent boiling water, will carry away with it some minute droplets of water (see section 4.2).

The water container must always contain water, so the steam in the steam space is always wet. If wet steam is all that is required, the steam is piped directly from the steam space into the steam main.

But if superheated steam is required, the wet steam is removed from the steam space and piped into a **superheater**. This consists of a long tube or series of tubes which are suspended across the path of the hot gases from the furnace. As the wet steam progresses through the tube or tubes it is gradually dried out and eventually superheated. From the superheater it passes to the steam main. If a control of the degree of superheat is required, as in some of the larger boilers, then an **attemperator** is fitted. The control of the degree of superheat is obtained by the injection of water or steam into the superheated steam. If an attemperator is fitted, the superheater is generally divided into two parts. The first part is called the primary superheater. Then comes the attemperator followed by the second part of the superheater called the secondary superheater.

Now the flue gases will still be hot, having passed through the main boiler then the superheater. The energy in these flue gases can be used to improve the thermal efficiency of the boiler. To achieve this thermal efficiency improvement, the flue gases are firstly passed through an **economiser**. The economiser is really a heat exchanger in which the feedwater being pumped into the boiler is heated. The feedwater thus arrives in the boiler at a higher temperature than would be the case if no economiser were fitted. Hence, less energy is required to raise the steam, or if the same energy is supplied, then more steam is raised. This results in a higher thermal efficiency.

Having passed through the economiser, the flue gases are still moderately hot. Further thermal efficiency improvement can be obtained by passing them through an air heater. This, too, is a heat exchanger in which the air being ducted to the boiler furnace is heated. The air thus arrives at the furnace hotter than if the air heater were not fitted. This results in a higher furnace temperature which thus increases the furnace potential for steam raising. Thermal efficiency improvement results.

Still further improvement of boiler thermal efficiency is obtained by the installation of a **reheater**. The reheater will often appear in the flue gas path before the economiser. In some of the larger steam turbines, e.g. in power stations, steam is removed from the turbine after partial expansion. This steam is fed back to the boiler and then to the reheater. Here it is reheated to a higher temperature and then passed back to the turbine where it completes its expansion in the latter stages.

The object of reheating steam in a turbine plant is to preserve the steam quality in the low-pressure stages of the turbine. If there were no reheat, the steam in the low-pressure stages would become too wet. Wet steam has erosive and corrosive effects on the turbine blades. By returning the steam to the boiler, after partial expansion, the quality of the steam is improved and wet steam in the low-pressure stages is therefore largely avoided. Reheating also gives greater potential work output from the steam in the low-pressure stages. It may also give a slight improvement in thermal efficiency.

In power stations it was usually the practice to cross-connect all boilers and turbines; any boiler could run any turbine. But the installation of reheaters was difficult, so it was rarely adopted. Modern boilers, however, have become much more reliable and new installations have one boiler connected to one turbine, making a single boiler–turbine unit. Reheaters are more easily installed because there is no cross-coupling.

The location of the superheater, reheater, economiser and air heater are illustrated in Fig. 10.2. After the air heater, the flue gas passes to the exhaust chimney.

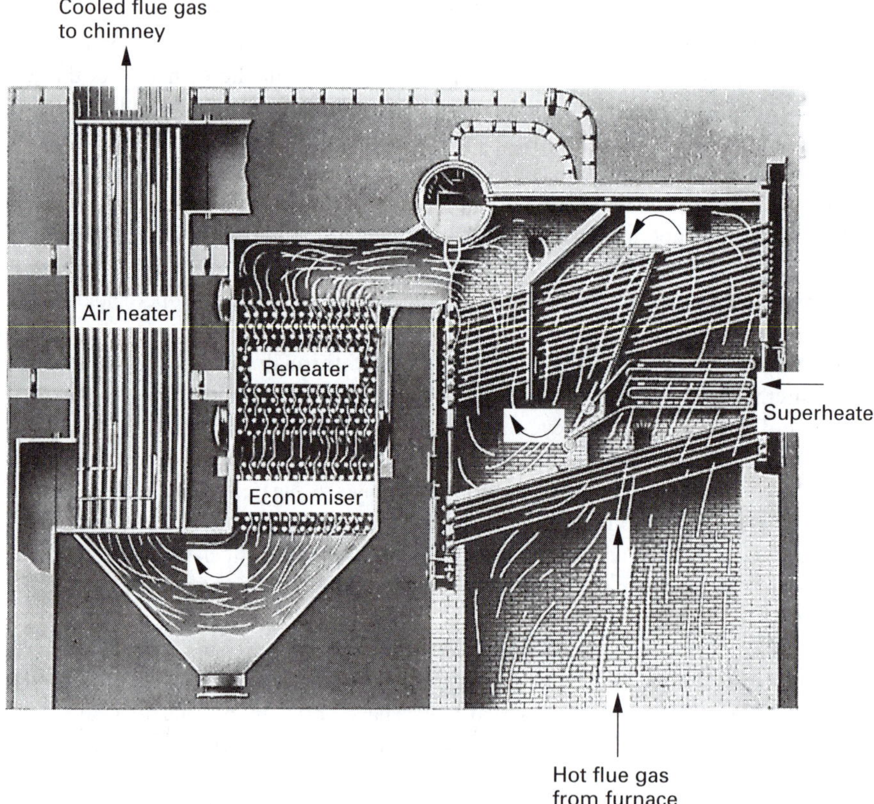

Fig. 10.2 Important parts of a boiler

The following auxiliary equipment is fitted to all boilers:

- *A pressure gauge* This will record the gauge pressure of the saturated steam formed in the steam space.
- *A water gauge glass* This will record the water level in the boiler. Often two are fitted in case one breaks.
- *A pressure relief valve* This is fitted as a safety precaution and is set to blow-off at a particular pressure. Often two are fitted as an added precaution in case one sticks. They are either of the dead-weight or spring-loaded types.

10.3 Fire-tube boilers

A type of fire-tube boiler is illustrated in Fig. 10.3. Sometimes called an **economic** boiler, it has a cylindrical outer shell and contains two large-bore flues into which are set the furnaces. The one illustrated has a mechanical stoker and ash remover. The hot flue gases pass out of the furnace flues at the back of the boiler into a brickwork setting, which deflects them back to pass through a number of small-bore tubes arranged above the large-bore furnace flues. These small-bore tubes break up the water bulk in the boiler and present a large heating surface to the water. The flue gases pass out of the boiler at the front and into an induced-draught fan, which passes them into the chimney.

The general range of sizes of the economic boiler is from small, about 3 m long and 1.6 m diameter, to large, about 6.5 m long and 4 m diameter. Equivalent evaporation ranges from about 900 kg steam per hour to about 14 000 kg steam per hour.

Another type of economic boiler is illustrated in Fig. 10.4. This type is called a **supereconomic boiler**. The boiler illustrated is oil-fired through a central large-bore corrugated flue. This gives the flue gas its first pass. At the rear of the boiler the flue gas is deflected down and back to pass through a number of small-bore tubes set in the bottom of the boiler. This gives the flue gas its second pass. At the front of the boiler the flue gas is deflected up and into the boiler again for its third pass, through another set of small-bore tubes set in the sides of the boiler. At the back of the boiler the flue gas then passes to the chimney.

With the large number of tubes set in this boiler, there is quite a high heating surface area and the boiler will therefore have a high evaporation rate for its size. The boiler is called supereconomic because of its three gas passes as against the economic two passes.

The boiler illustrated is capable of being installed completely as a self-contained working unit, having been assembled at the manufacturer before delivery. It is fully automatic and electronically controlled. The size ranges from an overall length and height of 3.4 m and 2.3 m to an overall length and height of 6.1 m and 3.7 m. The equivalent evaporation ranges from 680 to 8000 kg steam per hour. Such a boiler may also be called a **package boiler**.

It is useful to note that not all fire-tube boilers are horizontal. Some, especially the smaller boilers, stand vertically, with the furnace at the bottom, so they occupy a smaller floor area. This is especially convenient where space is at a premium.

10.4 Water-tube boilers

With the increasing demand for higher power output from steam plant it became necessary to develop boilers with higher pressures and steam outputs than could be handled by the shell-type boilers. This led to the development of the **water-tube boiler**.

Consider the pressure. For a constant thickness of material, a tube of smaller diameter can withstand a higher internal pressure than a tube of larger diameter.

And the steam output. If the water is contained in a large number of tubes, there is a large heating surface area and each tube contains a small water bulk, so the steam output is greatly increased.

By using a large number of tubes, both the pressure and the steam output can be increased.

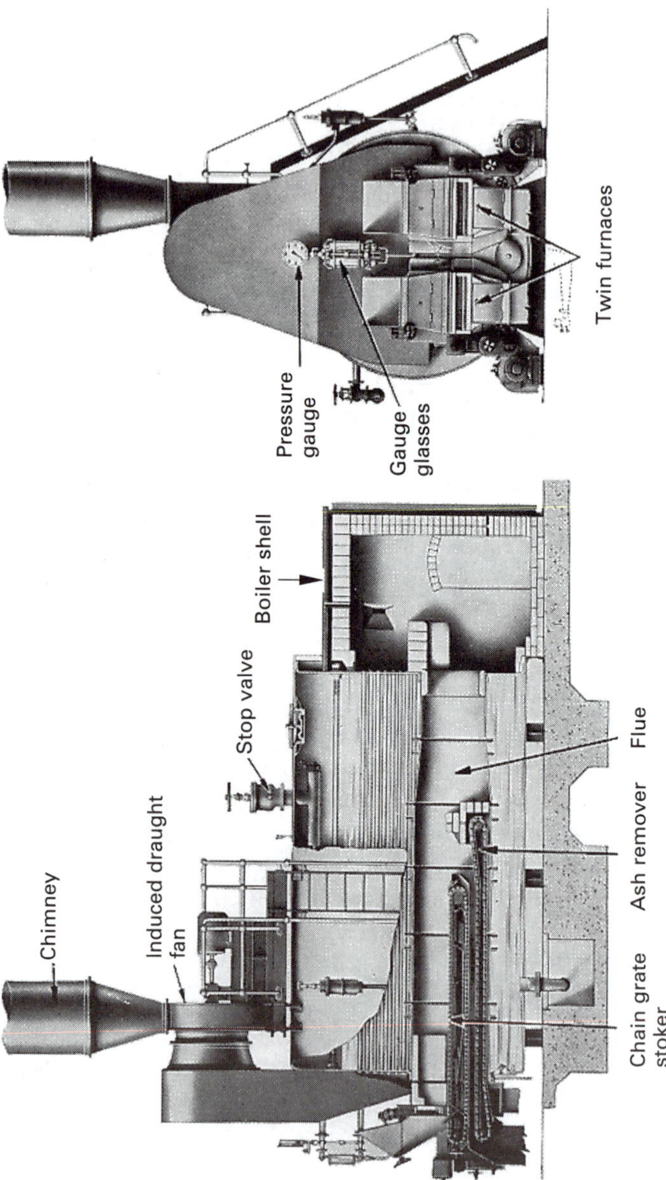

Fig. 10.3 Economic boiler

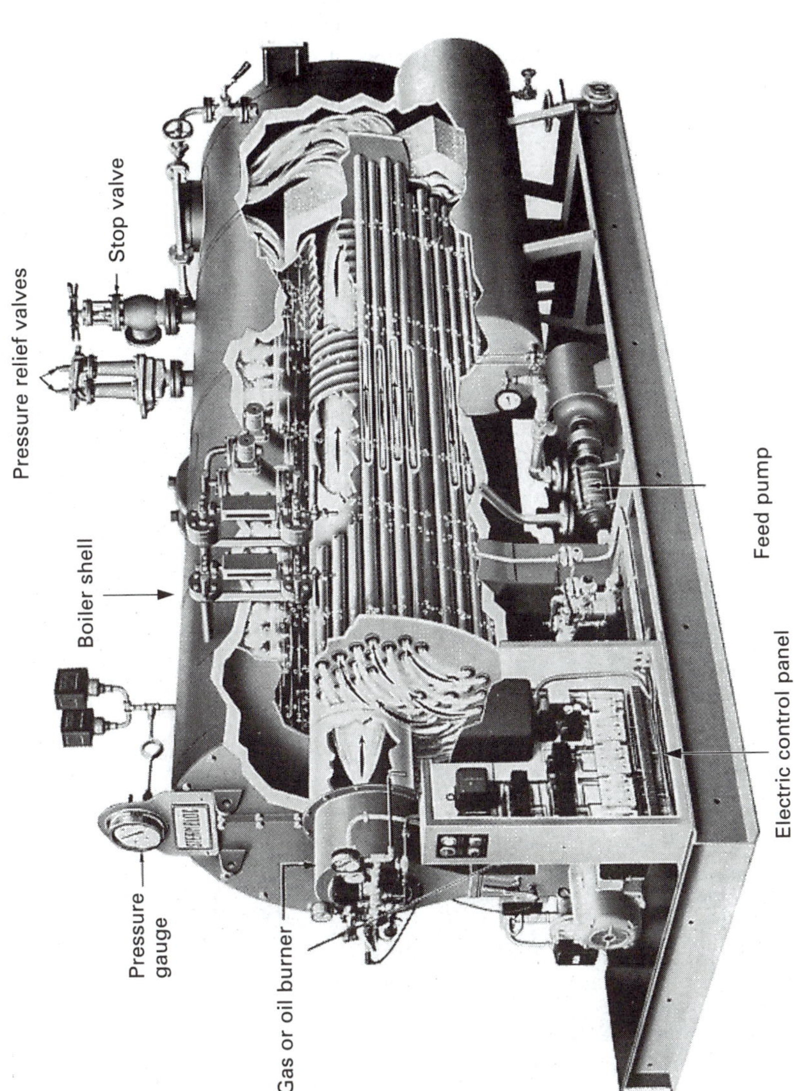

Fig. 10.4 Supereconomic boiler

In most water-tube boilers the water circulation is by natural convection, but a few designs employ forced convection.

A very large water-tube boiler of modern design is illustrated in Fig. 10.5. It is called a **radiant heat boiler**. The boiler is fired with **pulverised coal**. The coal-pulverising mills are shown at the bottom. Coal is fed into the mills where it is crushed to powder. Primary air is fed through the mills where it mixes with the powdered coal. The air–coal mixture is then fed through ducting to the boiler burners. The mills are driven by electric motors. The powdered coal and primary air blow through the centre of the burners. To encourage and control the burning, secondary air is controlled and fed round the delivery of the coal and primary air supply.

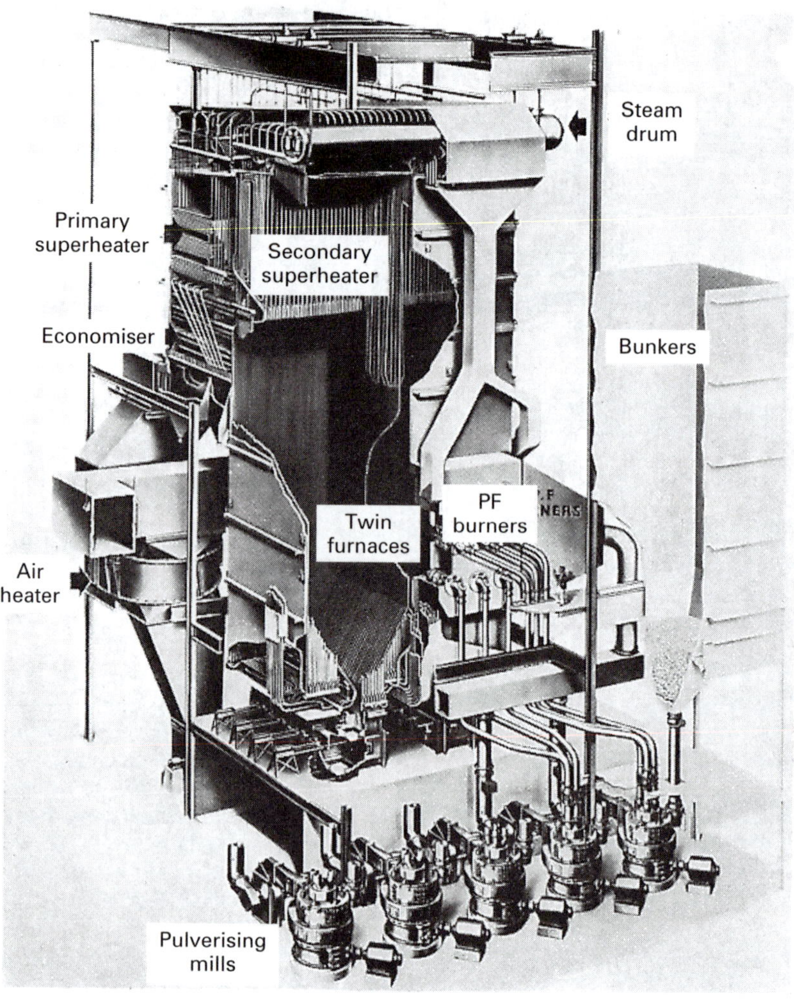

Fig. 10.5 Large water-tube boiler

The boiler shown in Fig. 10.5 has twin furnaces. The furnaces are completely water cooled; in fact they consist of water tubes. Most heat energy in this larger type of boiler is transferred by radiation to the vertical water tubes. By the time the flue gas has passed up the boiler and through the superheater, the temperature is not very much higher than the saturation temperature of the steam drum. Under these circumstances there would be little steaming improvement if a convection surface were provided. Thus, convection water tubes, as used in the other types of water-tube boiler, are not included. The superheater is split into primary and secondary sections so as to provide superheat control by the introduction of an attemperator. And note the provision of an economiser and air heater. This boiler has a steam output of 380 000 kg/h. The superheater outlet pressure is 11 MN/m^2 with a final steam temperature of 570 °C.

A very large, radiant heat, oil-fired boiler can have a steaming capacity of about 1600 tonne/h (1.6×10^6 kg/h) at a pressure of 16.5 MN/m^2 (165 bar) with a delivery steam temperature of 550 °C. It will use about 118 tonne (118×10^3 kg) of fuel oil per hour.

One of the difficulties of a pulverised fuel (PF) burning boiler is that the ash in the coal is also pulverised, so it is blown into the furnace and passes up with the flue gas out of the boiler. This dust ash must be removed so that it does not pollute the atmosphere. The dust ash is removed in a **precipitator**. One type of precipitator gives a vortex motion to the flue gas. The dust is thus flung out of the gas and is collected for disposal. This type of precipitator is called a **cyclone precipitator**.

Another design, the **electrostatic precipitator**, is operated electrically. The precipitator consists of a bank of plates or wires, some positively charged and some negatively charged. The ash-laden flue gas is passed, relatively slowly, through the plates or wires. During passage through the plates, the ash particles become negatively charged; they move over and cling to the positive plates or wires. The ash is removed by rapping the plates or wires with mechanical rappers. It falls into hoppers from which it is removed.

Another effect of the combustion of coal, or the combustion of hydrocarbon fuels such as oil and petrol, is that the combustion can produce undesirable flue and exhaust products which pollute the atmosphere when they are discharged. Such pollutants may have an unfortunate and undesirable effect not only locally but also at some distance from their source. They may be carried long distances in the atmosphere. Two of the main pollutants appear to be the sulphur oxides (written generally as SO_x), and the nitrogen oxides (written generally as NO_x). They can form acids and compounds which have a corrosive effect on surroundings and buildings, and also have a contributory effect on the formation of what is generally called acid rain. The acid rain appears to have a marked effect on some flora and fauna of forests and lakes. These undesirable effects have led to increasing efforts to scrub the flue gases of pollutants before they are discharged into the atmosphere. Chapter 8 gives further comment on atmospheric pollution.

It should be noted that boilers can be fuelled by any substance that will burn. The fuels are mostly coal, fuel oil or gas, which is mostly natural gas. Large water-tube boilers are usually custom-built on site.

10.5 Once-through boilers

In this type of boiler, water is force-circulated in a single passage through the boiler which consists of a number of tubes in parallel. The pressure in the boiler can be above the critical pressure for steam (approximately 22.12 MN/m^2). Thus these boilers will operate at pressures of some 22 MN/m^2 to about 34 MN/m^2. Steam temperature will be of the order of 600 °C. Radiant heat furnaces are employed and general sizes vary from industrial plant up to large power stations. Advantages claimed with this type of boiler are that the welded construction avoids expansion troubles due to starting up and shutting down. Starting up and shutting down can be accomplished more rapidly. The boiler can be operated at any pressure and temperature over its load range. Steam can be supplied at gradually increased superheat temperature which assists turbine starting.

Small, once-through, subcritical boilers are also manufactured.

10.6 Fluidised bed combustion

In the preceding discussion on boilers, the use of coal-burning equipment such as the chain grate and the pulverised fuel burner were indicated and described. A further method of utilising coal is by means of **fluidised bed** combustion. A diagram of a fluidised bed combustor is shown in Fig. 10.6.

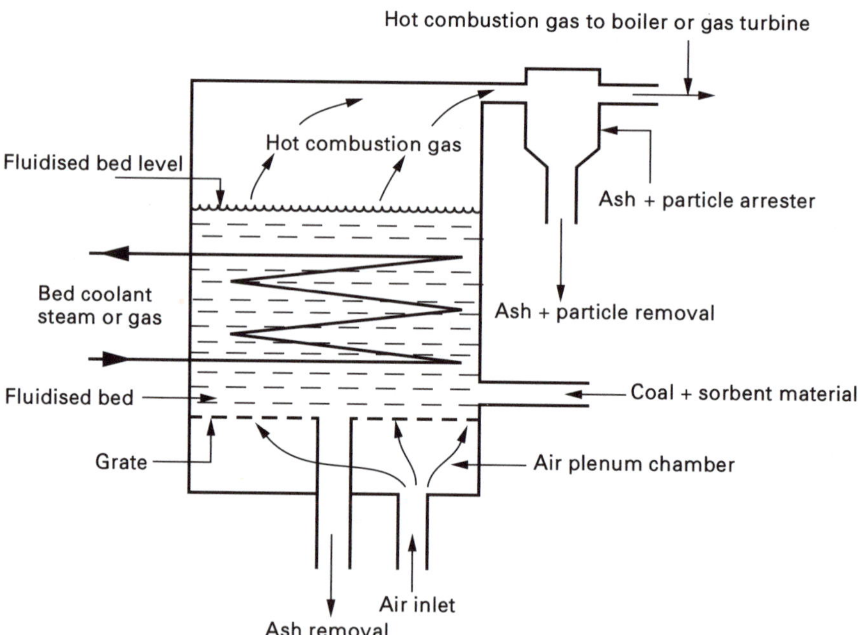

Fig. 10.6 Fluidised bed combustor

The bed consists of a thick layer of fine inert particles of sand, limestone or another natural substance. Air is blow through the bed. The air is evenly distributed through a grate at the base of the bed using a device such as a plenum chamber. At a particular velocity and mass flow of air, the bed will begin to behave like a fluid; in other words, it becomes a fluidised bed. In fact, a hollow vessel, such as a ball, would float on its surface in the same way as it would float on the surface of a fluid.

If particles of coal are added to the fluidised bed, they become well mixed throughout the bed. If the temperature of the bed is high enough, the coal will chemically combine with the oxygen in the airflow and it will burn. The fluid nature of the bed will ensure an even heat transfer through the bed and also to any coolant device immersed in it. Such coolant devices could be required to produce superheated steam in a boiler or hot gas for use in a gas turbine. Hot gas from the combustion process will leave the bed at its top surface. The hot gas can be arranged to proceed through a boiler or a gas turbine or industrial process plant.

A device is required for the removal of the coal ash at the base of the bed through the bed grate. A further device, sometimes called an arrester, is located at the top of the combuster to intercept any small ash particles (fly ash), and any particles of bed material which leave the bed with the heated combustion gas. This leaves a cleaner hot gas to proceed to the plant which follows. Particles collected in the arrester are removed from its base.

Sorbent material, such as limestone or dolomite, can be arranged to be present in the fluidised bed. As mentioned in section 10.4, the combustion of coal can produce undesirable sulphur or nitrogen oxides. These oxides are attracted to the sorbent material in the bed and, as a result, are much reduced in the flue gas, thus reducing their effect on atmospheric pollution.

10.7 Waste heat boilers – combined heat and power (CHP)

Many engineering processes produce large quantities of energy which can be transferred as a heat by-product. For example, in the manufacture of steel the furnaces produce a large quantity of such energy. Ships using large oil engines produce a considerable amount of energy in the exhaust gas, and large quantities of exhaust gas are available over long periods at sea. A similar situation arises with the use of large industrial gas turbines. In these cases, the energy available for transfer as heat would be wasted unless an attempt were made to recover some. The recovery can be accomplished by employing a **waste heat boiler**.

In the waste heat boiler the hot gases from the furnace, engine or gas turbine are employed in raising steam which can be used to run auxiliary plant such as a steam turbine. They can be used for the generation of electrical power for lighting and heating. In this way the overall efficiency of the plant is improved.

The waste heat boiler is usually provided with an auxiliary furnace, normally oil-fired or gas-fired, such that it can be operated when waste energy is not available. Waste heat boilers are part of the concept called **combined heat and power** (CHP), see also Chapter 16.

Much more care is now being taken in the use of energy and the reduction of energy losses to a minimum. In power production plant there is considerable energy loss in the exhaust system (steam in the case of steam turbines, gas in the case of engines). These losses can be between 65 and 85 per cent of the energy available from the fuel supply. Consequently, much effort is being made to design plant in which both power and heat transfer are made available from the plant.

For example, it could be arranged that a gas turbine produces power and its exhaust feeds a boiler which produces steam for a steam turbine to produce additional power. Alternatively, a steam turbine could produce power and the exhaust steam from the turbine could heat water by condensing in its passage through a condenser. The hot water from the condenser could be used for heating purposes.

As a further example, waste heat recovery is made use of in a motor car. Hot cooling water from the engine is circulated through a heat exchanger, which serves to heat air passing through it on its way to warm the passenger compartment.

Such ideas are not new. But with the increasing emphasis on the efficient use of fuel and energy, there is much more interest in the concept of combined heat and power.

10.8 Steam generation by nuclear reaction

In nuclear reaction, atoms of uranium or plutonium are bombarded by neutrons, elementary particles present in the nuclei of atoms. As a result of the neutron bombardment, the atoms of uranium or plutonium undergo fission, which means that they are split. The fission releases heat energy and more neutrons, thus the nuclear reaction continues in a **chain reaction**. This chain reaction is quite unlike the chemical reaction which occurs in fuels such as coal or oil. The device in which a nuclear reaction occurs is called a **reactor**.

There are two principal types of nuclear reactor: **thermal reactors** and **fast reactors**.

Natural uranium consists of two forms of uranium, called **isotopes** of uranium. Most of the natural uranium, about 99.3 per cent, consists of the isotope uranium-238 (U-238). The remaining 0.7 per cent consists of the isotope uranium-235 (U-235).

The isotope U-235 will fission much more easily than the isotope U-238. Uranium fuel for use in thermal reactors is usually arranged to be slightly enriched in U-235 in order to increase the fission potential.

In the thermal reactor, the neutrons need to be slowed down in order to produce a higher probability of fission of the U-235. The neutrons are slowed down using a **moderator**, which can be of either graphite or water.

The graphite-moderated reactors are of two general types. In the **magnox reactor** the fuel is uranium metal clad in a magnesium alloy (Magnox). In the **advanced gas-cooled reactor** (AGR) the fuel is uranium dioxide clad in stainless steel.

Both types of reactor are cooled by carbon dioxide (CO_2) gas, which is heated by passing it over the fuel in the reactor core. The hot gas then passes through a steam generator and transfers heat into water in the generator, thus producing steam. The cooled gas then returns to the reactor. The fuel used in the AGR can operate at a higher temperature than the fuel used in the magnox reactor, so it gives a smaller size of reactor for a given energy output.

In the **pressurised water** reactors (PWRs) high-pressure water is used, not only as the moderator but also as the coolant of the reactor. The fuel used is uranium dioxide (UO_2) clad in zirconium alloy (Zircaloy). The hot, high-pressure water from the reactor is pumped through a steam generator and transfers heat into water in the generator, thus producing steam. The cooled water then passes back to the reactor.

One type of PWR has the general arrangement shown in Fig. 10.7. And Fig. 10.8 illustrates how four steam generators can be fed from a single reactor in a four-loop system.

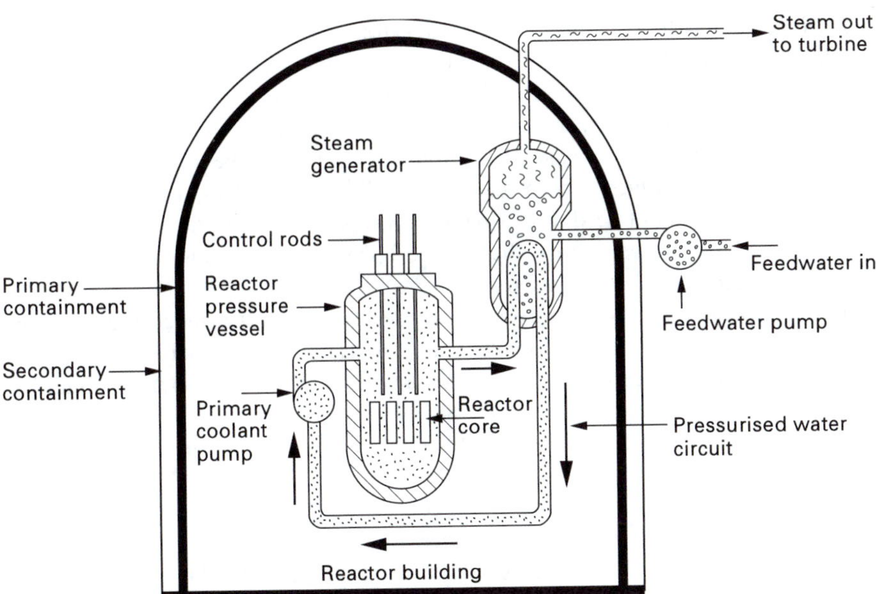

Fig. 10.7 Pressurised water reactor building

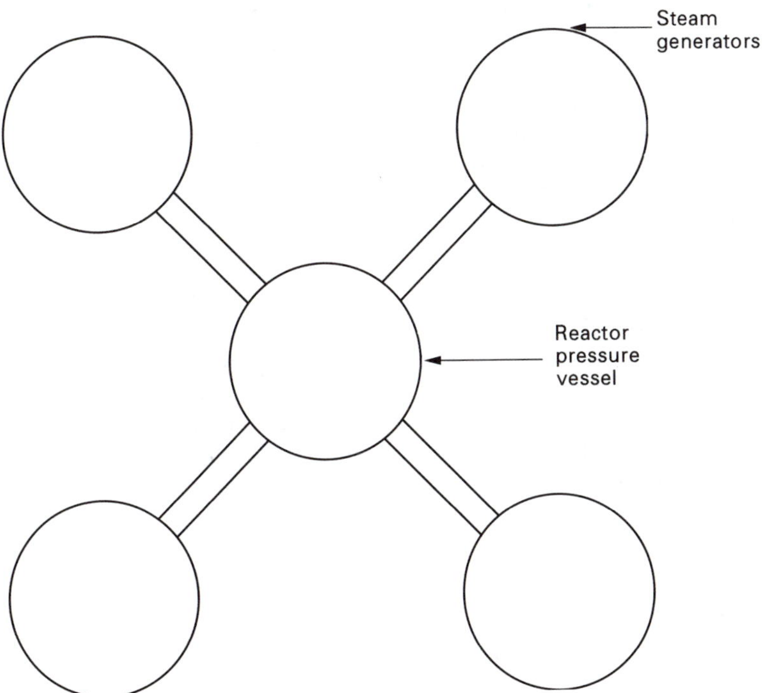

Fig. 10.8 Four-loop PWR

During the nuclear reaction in thermal reactors, the isotope U-238 captures some of the neutrons released in the fission of the isotope U-235 and forms a new radioactive element, plutonium (Pu-239), which will fission. By reprocessing the spent fuel from thermal reactors, a new fuel can be produced, comprising some 20–30 per cent plutonium and the remainder uranium. This fuel can be used in **fast reactors**.

In a fast reactor, which can be relatively small, no moderator is required because the fast neutrons do not need to be slowed down. And if the reactor is surrounded by a blanket of uranium containing the isotope U-238, more plutonium is created and the reactor becomes a **fast breeder reactor**.

The high heat energy output from a fast reactor requires a much more efficient coolant than gas or water. The coolant used is usually liquid sodium. The sodium passing through the reactor core will become radioactive, so two sodium circuits are required. The first is contained within the core of the reactor; it cools the core and transfers heat to a second sodium circuit. The second circuit transfers heat to a steam generator.

By the use of fast breeder reactors, it is considered that uranium stocks can remain useful as an energy source for some considerable time.

There are many nuclear power reactors in use throughout the world. In power production plant the nuclear reactor takes the place of the conventional coal or oil furnace. The whole of a reactor, its equipment and the building in which it is housed, is shielded against radiation and leakage. This is shown diagrammatically in Fig. 10.7. The diagram is not to scale; it is based on the Sizewell B nuclear power station in Suffolk, UK. The dome-topped primary containment building is made of prestressed concrete with a carbon steel liner. It has a wall thickness of 1.3 m with a liner thickness of 6 mm. It contains any leakages or emitted radiation. The containment building houses the reactor, the steam generator and any auxiliary equipment. The primary containment building is surrounded by a steel fabrication, the secondary containment. This is an additional precaution and acts as a leakage collector.

The thermal reactor power of the station is 3411 MW with a gross electrical power of 1245 MW. Turbine speed is 3000 rev/min. Steam is produced at a pressure of 69 bar with a temperature of 285 °C from feedwater at 227 °C. The steam quality is 99.75 per cent dry. Steam is always wet when produced from a water surface. The high dryness fraction in this case is produced by the introduction of swirl-vane moisture separators and a steam dryer at the exit from the steam generator. The primary water coolant is pressurised to 155 bar using a pump, and its outlet temperature from the reactor is 325 °C. This temperature is high enough to produce steam in the steam generator but is below the saturation temperature in the primary coolant circuit. Thus it remains as liquid water in this circuit. The illustration shows just one design of a nuclear power station. Many other designs, with varying outputs, are used throughout the world.

10.9 Boiler calculations

10.9.1 Heat transfer required to form steam

Let h_2 = specific enthalpy of steam formed, kJ/kg
h_1 = specific liquid enthalpy of feedwater, kJ/kg (see Chapter 4)

Then since the steam is formed at constant pressure

$$\text{Heat transfer required to form 1 kg of steam in the boiler} = (h_2 - h_1)\ \text{kJ} \qquad [1]$$

10.9.2 *Energy received from fuel*

This is obtained from a knowledge of the mass of fuel used and its calorific value.

If the mass of fuel used is m kg, then

$$\text{Calorific value of fuel} = CV\ \text{kJ/kg} \quad \text{(see section 1.30 and Chapter 8)}$$

then

$$\text{Energy received from fuel} = (m \times CV)\ \text{kJ} \qquad [2]$$

10.9.3 *Boiler thermal efficiency*

This is given by the ratio of the energy received by the steam to the energy supplied by the fuel to produce the steam. Thus

$$\text{Boiler thermal efficiency} = \frac{\text{Energy to steam}}{\text{Energy from fuel}}$$

And if

$\dot{m}_s$ = mass of steam raised in a given time
$\dot{m}$ = mass of fuel used in the same time

then

$$\text{Boiler thermal efficiency} = \frac{\dot{m}_s(h_2 - h_1)}{(\dot{m} \times CV)} \times 100\% \qquad [3]$$

10.9.4 *Equivalent evaporation of a boiler*

Some boilers can be operated under many different running conditions; for these boilers it is necessary to have some standard upon which to base, and compare, their respective evaporative capacities. The standard commonly adopted is the equivalent evaporation of a boiler from and at 100 °C.

If

$\dot{m}_s$ = mass of steam formed in a given time (or sometimes per kilogram of fuel burnt), kg
h_2 = specific enthalpy of steam formed, kJ/kg
h_1 = specific enthalpy of feedwater, kJ/kg

then

$$\text{Energy received by steam} = \dot{m}_s(h_2 - h_1)\ \text{kJ}$$

From this is determined the amount of water at 100 °C which could be evaporated into dry saturated steam at 100 °C if supplied with this same amount of energy. This is then called the **equivalent evaporation of the boiler**, from and at 100 °C.

At 100 °C it will be the enthalpy of evaporation which is supplied.

Specific enthalpy of evaporation at 100 °C = **2256.9 kJ/kg** [4]

Thus, the equivalent evaporation of a boiler, from and at 100 °C is

$$\frac{\dot{m}_s(h_2 - h_1)}{2256.9}\text{ kg} \quad \text{in the given time or per kilogram of fuel} \qquad [5]$$

Example 10.1 *A boiler working at a pressure of 1.4 MN/m² evaporates 8 kg of water per kilogram of coal fired from feedwater entering at 39 °C. The steam at the stop valve is 0.95 dry. Determine the equivalent evaporation, from and at 100 °C, in kg steam/kg coal.*

SOLUTION

Specific enthalpy of steam at 1.4 MN/m^2 and 0.95 dry is

$$\begin{aligned} h_f + xh_{fg} &= 830.1 + (0.95 \times 1957.7) \\ &= 830.1 + 1859.8 \\ &= \mathbf{2689.9\ kJ/kg} \end{aligned}$$

Specific enthalpy of water at 39 °C is 163.4 kJ/kg. So heat transfer required to raise 1 kg of steam is

$$2698.9 - 163.4 = 2526.5 \text{ kJ/kg}$$

$\therefore$ Heat transfer to 8 kg = (8 × 2526.5) kJ

Equivalent evaporation from and at 100 °C is

$$\frac{8 \times 2526.5}{2256.9} = \mathbf{8.96\ kg\ steam/kg\ coal}$$

Example 10.2 *A boiler generates 5000 kg of steam per hour at 1.8 MN/m². The steam temperature is 325 °C and the feedwater temperature is 49.4 °C. The efficiency of the boiler plant is 80 per cent when using oil of calorific value 45 500 kJ/kg. The steam generated is supplied to a turbine which develops 500 kW and exhausts at 0.18 MN/m²; the dryness fraction of the steam is 0.98. Estimate the mass of oil used per hour and the fraction of the enthalpy drop through the turbine that is converted into useful work.*

If the turbine exhaust is used for process heating, find the heat transfer available per kilogram of exhaust steam above 49.4 °C.

SOLUTION

Specific enthalpy of steam generated is

$$\begin{aligned} 3106 - 0.8 \times (3106 - 3083) &= 3106 - (0.8 \times 23) \\ &= 3106 - 18.4 \\ &= \mathbf{3087.6\ kJ/kg} \end{aligned}$$

Specific enthalpy of feedwater at 49.4 °C is 206.9 kJ/kg

$$\therefore \quad \text{Energy to raise steam} = 3087.6 - 206.9 = \mathbf{2880.7\ kJ/kg}$$

$$\therefore \quad \text{Energy to steam/h} = \mathbf{(2880.7 \times 5000)\ kJ}$$

This is 80 per cent of the energy from the fuel.

$$\therefore \quad \text{Energy from fuel/h} = \frac{2880.7 \times 5000}{0.8}$$

$$\text{Mass of oil/h} = \frac{2880.7 \times 5000}{0.8 \times 45\,500}$$

$$= \mathbf{395.7\ kg}$$

The specific enthalpy of exhaust steam is

$$\begin{aligned} h_f + xh_{fg} &= 490.7 + (0.98 \times 2210.8) \\ &= 490.7 + 2166.6 \\ &= \mathbf{2657.3\ kJ/kg} \end{aligned}$$

And the specific enthalpy drop in the turbine is

$$3087.6 - 2657.3 = \mathbf{430.3\ kJ/kg}$$

The mass of steam per second is 5000/3600 kg.

$$\therefore \quad \text{Specific enthalpy drop in turbine/s} = 430.3 \times \frac{5000}{3600}$$

The energy output from the turbine is 500 kW = 500 kJ/s. So the fraction of enthalpy drop converted into useful work is

$$\frac{500}{430.3 \times \dfrac{5000}{3600}} = \frac{500 \times 3600}{430.3 \times 5000}$$

$$= \mathbf{0.837}$$

Heat transfer available in exhaust steam above 49.4 °C is

$$2657.3 - 206.9 = \mathbf{2450.4\ kJ/kg}$$

Example 10.3 *A boiler delivers 5400 kg of steam per hour at a pressure of 750 kN/m² and with a dryness fraction of 0.98. The feedwater to the boiler is at a temperature of 41.5 °C. The coal used for firing the boiler has a calorific value of 31 000 kJ/kg and is used at the rate of 670 kg/h. Determine (a) the thermal efficiency of the boiler and (b) the equivalent evaporation of the boiler in kg/coal.*

An economiser is fitted to the boiler which raises the feedwater temperature to 100 °C. The thermal efficiency of the boiler is increased by 5 per cent, all other conditions remaining unaltered. Determine (c) the new coal consumption in kg/h and hence the saving in coal in kg/h obtained by fitting the economiser.

(a)

$$\text{Steam raised/kg coal} = \frac{5400}{670} = \mathbf{8.06\ kg}$$

Specific enthalpy of steam raised is

$$\begin{aligned} h_f + xh_{fg} &= 709.3 + (0.98 \times 2055.5) \\ &= 709.3 + 2014.4 \\ &= \mathbf{2723.7\ kJ/kg} \end{aligned}$$

Specific enthalpy of feedwater is 173.9 kJ/kg

$$\begin{aligned} \therefore \quad \text{Efficiency of boiler} &= \frac{8.06 \times (2723.7 - 173.9)}{31\,000} \\ &= \frac{8.06 \times 2549.8}{31\,000} \\ &= 0.663 \\ &= \mathbf{66.3\%} \end{aligned}$$

(b)

Equivalent evaporation from and at 100 °C is

$$\frac{8.06 \times 2549.8}{2256.9} = \mathbf{9.11\ kg/kg\ coal}$$

(c)

Specific enthalpy of feedwater at 100 °C is 419.1 kJ/kg

$$\begin{aligned} \therefore \quad \text{Energy to steam under new conditions} &= 2723.7 - 419.1 \\ &= \mathbf{2304.6\ kJ/kg} \\ \text{Energy to steam/h} &= \mathbf{(2304.6 \times 5400)\ kJ} \end{aligned}$$

This is with a new boiler efficiency of 66.3 + 5 = 71.3%

$$\therefore \quad \text{Energy from coal/h} = \frac{2304.6 \times 5400}{0.713}\ \text{kJ}$$

Mass of coal used per hour under new conditions is

$$\frac{2304.6 \times 5400}{0.713 \times 31\,000} = \mathbf{563\ kg}$$

$$\begin{aligned} \therefore \quad \text{Saving in coal/h} &= 670 - 563 \\ &= \mathbf{107\ kg} \end{aligned}$$

Example 10.4 *A gas-fired boiler operates at a pressure of 100 bar. The feedwater temperature is 256 °C. Steam is produced with a dryness fraction of 0.9 and in this condition it enters a superheater. Superheated steam leaves the superheater at a temperature of 450 °C. The boiler generates 1200 tonne of steam per hour with a thermal efficiency of 92 per cent. The gas used has a calorific value of 38 MJ/m*^{3}*. Determine*

(a) the heat transfer per hour in producing wet steam in the boiler
(b) the heat transfer per hour in producing superheated steam in the superheater
(c) the gas used in m^3/h

(a)
From steam tables

$$\text{Specific enthalpy of feedwater} = \mathbf{1115.4\ kJ/kg}$$

$$\begin{aligned}\text{Specific enthalpy of wet steam} &= 1408 + 0.92 \times (2727.7 - 1408)\\ &= 1408 + (0.92 \times 1319.7)\\ &= 1408 + 1214.12\\ &= \mathbf{2622.12\ kJ/kg}\end{aligned}$$

$$\begin{aligned}\therefore\quad \text{Heat transfer/h for wet steam} &= \frac{1200 \times 10^3}{10^3} \times (2622.12 - 1115.4)\\ &= 1200 \times 1506.72\\ &= \mathbf{1\,808\,064\ MJ}\end{aligned}$$

(b)

$$\text{Specific enthalpy of superheated steam} = \mathbf{3244\ kJ/kg}$$

$$\begin{aligned}\therefore\quad \text{Heat transfer/h in superheater} &= \frac{1200 \times 10^3}{10^3}\,(3244 - 2622.12)\\ &= 1200 \times 621.88\\ &= \mathbf{746\,256\ MJ}\end{aligned}$$

(c)

$$\text{Thermal } \eta = \frac{\text{Heat transfer/h}}{\text{Volume of gas/h} \times CV}$$

$$\begin{aligned}\therefore\quad \text{Volume of gas/h} &= \frac{\text{Heat transfer/h}}{\text{Thermal } \eta \times CV}\\ &= \frac{1\,808\,064 + 746\,256}{0.92 \times 38}\\ &= \frac{2\,554\,320}{0.92 \times 38}\\ &= \mathbf{73\,064\ m^3}\end{aligned}$$

10.10 Condensers

There are two main types of condenser; **surface condensers** and **jet condensers**.

In a surface condenser, the steam to be condensed is usually passed over a large number of tubes through which cooling water is flowing. The steam is condensed on the surface of the tubes as it gives up its enthalpy to the cooling water flowing through the tubes. The condensate and cooling water leave separately.

In a jet condenser, the steam to be condensed comes into direct contact with the cooling water, which is usually introduced in the form of a spray from a jet. The steam gives up its enthalpy to the cooling water spray, is condensed and finally leaves as condensate with the cooling water.

Both types of condenser may be operated as either a wet or dry condenser. In a **wet condenser**, any gas which does not dissolve in the condensate (and cooling water for a jet condenser) is removed by the same pump which is dealing with the condensate. In a **dry condenser**, the free gas and the condensate (and cooling water for a jet condenser) are removed separately.

There is a further subdivision according to the relative directions of flow of the condensing steam and the cooling water. The three possibilities are as follows:

- *Transverse flow* The steam flows across the path of the cooling water; this is only possible in the surface condenser.
- *Parallel flow* The steam flow is in the same direction as the cooling water.
- *Counterflow* Also called contraflow; the steam flows in the opposite direction to the cooling water.

Another subdivision is according to how the condensate in the condenser is removed. Two arrangements are possible: the barometric condenser and the low-level condenser.

10.10.1 The barometric condenser

The barometric condenser (Fig. 10.9) is mounted on a pipe, usually at least 10.34 m long. The long pipe, called the **barometric leg**, acts rather like a Fortin barometer. If water were used in a Fortin barometer, the barometric height would be about 10.34 m. But instead of having a Torricellian vacuum on the top of the water in the pipe, there is a positive pressure, less than atmospheric pressure. The height of the water column will therefore be less than 10.34 m; it will be a function of the degree of vacuum that exists. This is illustrated as h in Fig. 10.9. Using this atmospheric leg it is possible for the condensate to drain away by gravity into the atmospheric tank at the bottom. The atmospheric leg dips deeply into the water in the atmospheric tank. The discharge from the atmospheric tank is from a standpipe, whose entry is high up, maintaining a constant high-level discharge. In this way there is no possibility of breaking the vacuum in the condenser.

10.10.2 The low-level condenser

In the low-level condenser, the condensate is removed using a pump. Low-level condensers are appropriate when there is not enough height available for a barometric condenser.

Figure 10.10 shows a transverse-flow surface condenser. Steam is admitted to the top of the condenser and is removed as condensate from the bottom, having been condensed at the surface of the water tubes. Cooling water flows in at the bottom and out at the top of the condenser. Inlet and exit in this case are both at the same end of the condenser, so the water makes two passes through the condenser. Air extraction from the condenser is from the side, as illustrated. The air extraction point is shrouded using a baffle-plate in order to achieve as much separation of the condensate and air as possible. The condenser illustrated is a dry condenser because air and condensate are extracted separately.

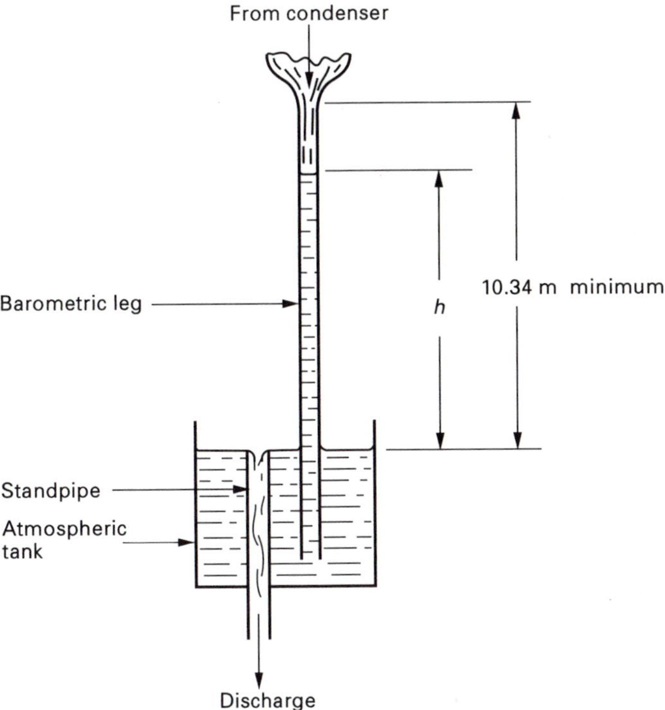

Fig. 10.9 Barometric condenser

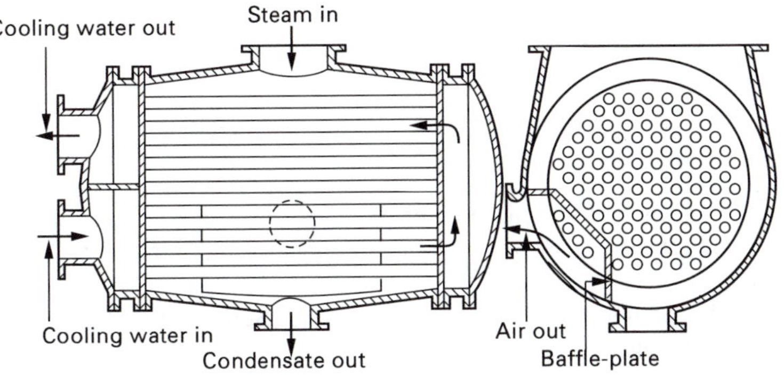

Fig. 10.10 Transverse flow surface condenser

Figure 10.11 shows a simple, but effective, jet condenser. It consists of a tall cylinder into which are introduced perforated baffle-plates. They are fixed alternatively on either side of the cylinder and cover just over half the cross-sectional area of the cylinder. Cooling water spray is introduced at the top of the condenser. The perforated baffle-plates help to maintain this spray throughout the condenser. Steam is introduced at the bottom of the condenser. Due to its low density it will begin to rise up in the condenser through the spray curtain. It will thus be condensed; the spray and condensate will fall together to the bottom of the condenser and will be extracted. Air entering the condenser will be warm, so it will rise to the top and will be extracted.

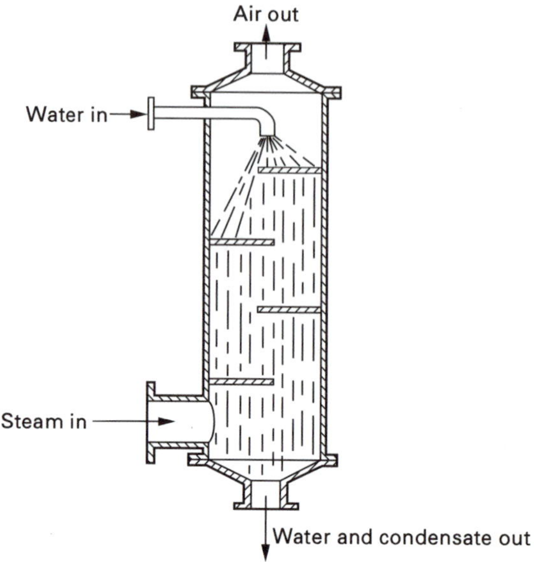

Fig. 10.11 Simple jet condenser

This condenser is also a dry condenser because both air and condensate are extracted separately. The steam and cooling water move in opposite directions through the condenser, so this is a counterflow condenser.

The jet condenser is used where steam recovery is not a fundamental requirement. If steam recovery were required, the cooling water and condensate would have to be kept separate, so a surface condenser would be used. This is the case when treated water is being fed to the boiler.

10.11 Condenser vacuum

The vacuum in a condenser is usually measured by means of a Bourdon pressure gauge which is calibrated to read the pressure in millimetres of mercury below atmospheric pressure.

If the gauge reads 635 mm Hg this means that the pressure in the condenser is 635 mm Hg below atmospheric pressure.

If atmospheric pressure is 760 mm Hg, the absolute pressure in the condenser will be

$$\begin{aligned} 760 - 635 &= 125 \text{ mm Hg} \\ &= 125 \times 0.133\,4 \\ &= \mathbf{16.7\ kN/m^2} \end{aligned}$$

The degree of vacuum in a condenser will be a function of the partial pressures of the steam and the air in the condenser. By Dalton's law of partial pressures the absolute pressure in the condenser will be made up of the sum of the partial pressure of the steam and the partial pressure of any air that is present. The partial pressure of the air will depend upon the amount of air which has been dissolved in the original water and the amount of air which has leaked into the system.

The partial pressure of the steam will be a function of the temperature to which it is cooled. The steam is being condensed so it must be wet steam. Its temperature must therefore be saturation temperature, which will correspond to a particular pressure.

Thus, if the temperature of the air–steam mixture in a condenser is 59 °C, inspection of the steam tables will show that the partial pressure of the steam is 19 kN/m^2.

If the pressure in the condenser is 28 kN/m^2, the partial pressure of the air present is 28 − 19 = 9 kN/m^2.

Note that the degree of vacuum in a condenser will largely depend upon the level to which the temperature in the condenser can be reduced. The cooler the air–steam mixture can be made, the greater will be the vacuum.

Calculations involving air–steam mixtures are dealt with in section 5.15 and Chapter 19.

10.12 Condenser calculations

Consider the surface condenser illustrated in Fig. 10.12.

Let

$\dot{m}_s$ = mass of steam entering condenser, kg/unit time
h_1 = specific enthalpy of steam entering condenser, kJ/kg
h_{f1} = specific enthalpy of condensate leaving condenser, kJ/kg
$\dot{m}_w$ = mass of cooling water entering condenser, kg/unit time
h_{f2} = specific enthalpy of cooling water entering condenser, kJ/kg
h_{f3} = specific enthalpy of cooling water leaving condenser, kJ/kg

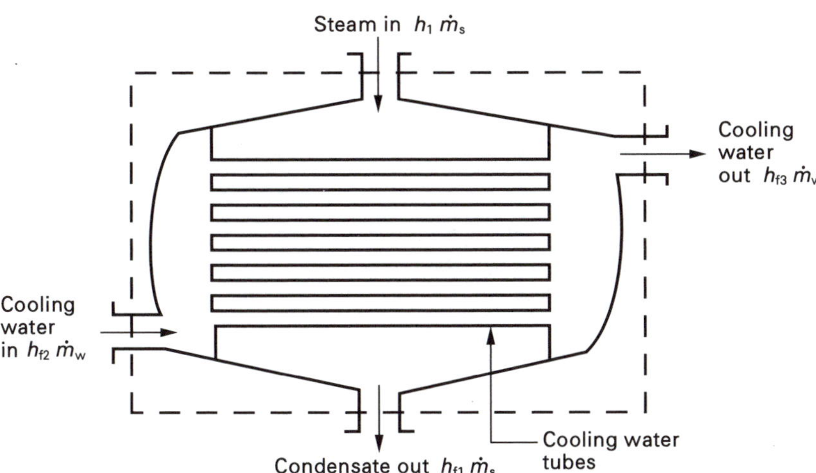

Fig. 10.12 Surface condenser flow diagram

The condenser is a constant pressure system and energy transfers will therefore appear as changes in enthalpy.

Thus, neglecting losses

Energy in = Energy out

or

$$\dot{m}_s h_1 + \dot{m}_w h_{f2} = \dot{m}_s h_{f1} + \dot{m}_w h_{f3} \quad [1]$$

from which

$$\dot{m}_s h_1 - \dot{m}_s h_{f1} = \dot{m}_w h_{f3} - \dot{m}_w h_{f2} \quad [2]$$

or

$$\dot{m}_s (h_1 - h_{f1}) = \dot{m}_w (h_{f3} - h_{f2}) \quad [3]$$

In the case of a jet condenser, let

$\dot{m}_s$ = mass of steam entering condenser, kg/unit time
h_2 = specific enthalpy of steam entering condenser, kJ/kg
$\dot{m}_w$ = mass of cooling water entering condenser, kJ/kg
h_{f2} = specific enthalpy of cooling water entering condenser, kJ/kg
h_{f3} = specific enthalpy of condensate and cooling water leaving condenser, kJ/kg

Then, neglecting losses

Energy in = Energy out

or

$$\dot{m}_s h_1 + \dot{m}_w h_{f2} = (\dot{m}_s + \dot{m}_w) h_{f3} \quad [4]$$

from which

$$\dot{m}_s h_1 - \dot{m}_s h_{f3} = \dot{m}_w h_{f3} - \dot{m}_w h_{f2} \quad [5]$$

or

$$\dot{m}_s (h_1 - h_{f3}) = \dot{m}_w (h_{f3} - h_{f2}) \quad [6]$$

Example 10.5 *A surface condenser operating at a pressure of 24 kN/m² condenses 1.8 tonne of steam per hour. The steam enters the condenser with a dryness fraction of 0.98 and is condensed, but not undercooled. Cooling water enters the condenser at a temperature of 21 °C and leaves at 57 °C. Determine the flow rate of the cooling water.*

SOLUTION
From equation [3]

$$\dot{m}_s (h_1 - h_{f1}) = \dot{m}_w (h_{f3} - h_{f2})$$

from which

$$\dot{m}_w = \frac{\dot{m}_s (h_1 - h_{f1})}{(h_{f3} - h_{f2})}$$

$$h_1 = 268.2 + 0.98 \times (2616.8 - 268.2)$$
$$= 268.2 + (0.98 \times 2348.6)$$
$$= 268.2 + 2301.6$$
$$= \mathbf{2569.8\ kJ/kg}$$

$$\therefore \quad \dot{m}_w = \frac{1.8 \times (2569.8 - 268.2)}{(238.6 - 88.1)}$$

$$= \frac{1.8 \times 2301.6}{150.5}$$

$$= \mathbf{27.5\ tonne/h}$$

10.13 The Carnot cycle and steam plant

In section 6.6 it was suggested that the Carnot cycle had the greatest efficiency possible between any two given limits of temperature. It is important, therefore, to see whether the Carnot cycle can be successfully applied to a steam plant. From this will be developed the **Rankine cycle** and the concepts of **reheat** and **feed heat**.

To understand the Carnot cycle in a steam plant, consider the P–V diagram in Fig. 10.13(a).

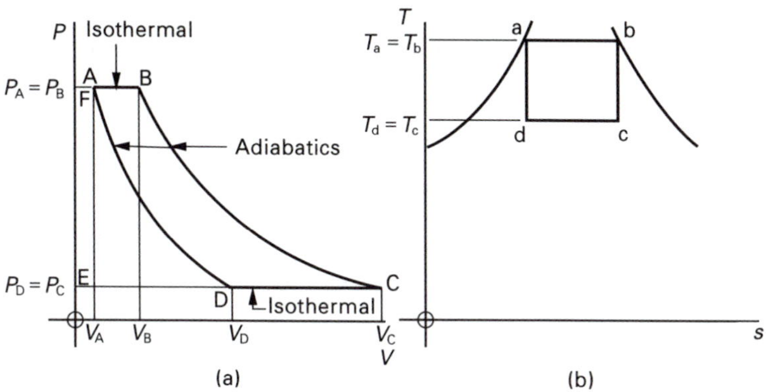

Fig. 10.13 Carnot cycle for steam plant: (a) P–V diagram; (b) T–s diagram

- **AB** Water at boiler pressure P_B and volume V_A is fed from the feed pump into the boiler. This is shown as process AF. In the boiler, the water is converted into steam at pressure P_B. The volume of the steam produced is V_B. This volume of steam V_B is then fed from the boiler into the engine or turbine. This is shown as process FB.

  The conversion of water into steam at constant pressure takes place at constant temperature, the saturation temperature T_B, as long as the steam does not enter the superheat phase. If the steam produced is either wet or dry saturated, the process is isothermal.

- **BC** The steam is expanded frictionlessly and adiabatically in the engine or turbine.

- **CD** The steam, after expansion, is passed from the engine or turbine into a condenser. This is shown as process CE. In the condenser the volume of the steam is reduced from V_C to V_D. This process takes place at constant condenser pressure P_C and at constant condenser saturation temperature T_C. This process is therefore isothermal.

- **DA** The partially condensed steam at pressure P_C and volume V_D is fed from the condenser into the feed pump. This is shown as process ED. In the feed pump the steam is compressed frictionlessly and adiabatically to boiler pressure P_B. This is shown as process DA. The compression converts the wet steam at condenser pressure into water at boiler pressure. This water is fed into the boiler, shown as process AF, and the cycle is repeated.

Now the *P–V* diagram is really composed of two diagrams. There is the engine or turbine diagram FBCE, whose area will give work output; there is also the feed pump diagram EDAF, whose area will give the required work input to run the feed pump.

The net work output from the plant, therefore, will be the net area of these two diagrams. This is the area ABCD.

Area ABCD is enclosed by two isothermal processes and two adiabatic processes, so this is a Carnot cycle. Its thermal efficiency will be given by $(T_B - T_C)/T_B$, the maximum efficiency possible between these temperature limits (see section 15.2).

The *T–s* diagram of the cycle is shown in Fig. 10.13(b).

- **ab** represents the constant temperature formation of the steam in the boiler.

- **bc** represents the frictionless adiabatic (isentropic) expansion of the steam in the engine or turbine.

- **cd** represents the condensation of the steam in the condenser.

- **da** represents the frictionless adiabatic (isentropic) compression of the steam in the feed pump back to water at boiler pressure at point a.

Now this cycle for operation in a steam plant is practical up to a point. The isothermal expansion of the steam in the boiler and the adiabatic expansion of the steam in the engine or turbine (especially in turbines) is reasonable.

The impractical part is in the handling of the steam in the condenser and feed pump. In the condenser, the steam is only partially condensed and condensation must be stopped at point d. Also the feed pump must be capable of handling both wet steam and water.

A slight modification to this cycle, however, will produce a cycle which is more practical, although it will have a reduced thermal efficiency. This cycle is the Rankine cycle and is usually accepted as the appropriate ideal cycle for steam plant.

10.14 The Rankine cycle

The modification made to the Carnot cycle to produce the Rankine cycle is that, instead of stopping the condensation in the condenser at some intermediate condition, the condensation is completed. This is shown in Fig. 10.14(a). On the *T–s* diagram, Fig. 10.14(b), the Carnot cycle would be abcg. For the Rankine cycle, however, condensation is continued until it is complete at point d, all water. This water can be successfully dealt with in a feed pump in which its pressure can be raised feeding it back into the boiler. The cycle thus becomes more practical.

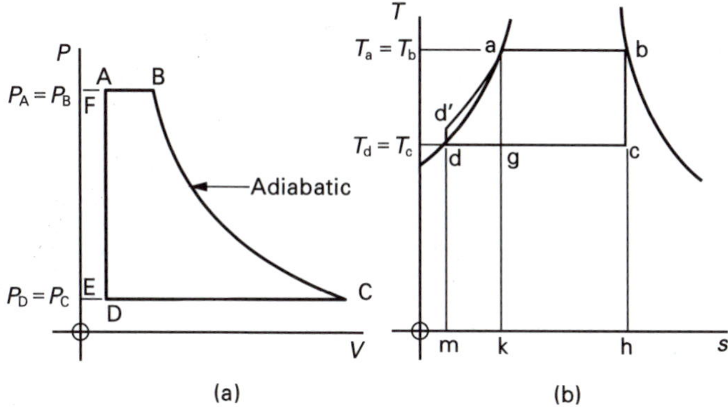

Fig. 10.14 Rankine cycle: (a) *P–V* diagram; (b) *T–s* diagram

This is shown exaggerated as process dd′ on the *T–s* diagram. In the boiler the temperature of the water is raised at boiler pressure, shown as process d′a, thus the cycle is completed. The complete Rankine cycle is therefore abcdd′a.

On the *P–V* diagram, Fig. 10.14(a), there are two cycles.

The work done in the engine or turbine is represented by the area FBCE. There is also the feed pump work, represented by the area EDAF. The feed pump work is negative because work must be put into the pump.

Hence

$$\text{Work done/cycle} = \text{Area ABCD} \qquad [1]$$

Using the steady-flow equation, neglecting changes in potential and kinetic energy, and putting $Q = 0$ for an adiabatic expansion, the energy equation becomes

$$h_1 = h_2 + W$$

or

$$\text{Specific work} = h_1 - h_2 \qquad [2]$$

Using symbols as in Fig. 10.14, then

$$\text{Specific work} = h_b - h_c$$
$$= \text{Area FBCE on } P\text{–}V \text{ diagram} \qquad [3]$$
$$\text{The feed pump work/unit mass} = \text{Area EDAF}$$
$$= (P_B - P_C)v_D \qquad [4]$$

Hence

$$\text{Net work done/cycle} = (h_b - h_c) - (P_B - P_C)v_D \qquad [5]$$

Now the heat transfer required in the boiler to convert the water at d′ into steam at b is

$$h_b - h_{d'} \qquad [6]$$

But the total energy of the water entering the boiler at d′ is

Liquid enthalpy at d + Feedpump work

or

$$h_{d'} = h_d + (P_B - P_C)v_D \qquad [7]$$

Substituting equation [7] in equation [6] gives the heat transfer required in the boiler as

$$h_b - [h_d + (P_B - P_C)v_D] = (h_b - h_d) - (P_B - P_C)v_D \qquad [8]$$

The thermal efficiency of the cycle is

$$\frac{\text{Work done/cycle}}{\text{Heat received/cycle}}$$

Hence, from equations [5] and [8], the thermal efficiency of the Rankine cycle is

$$\frac{(h_b - h_c) - (P_B - P_C)v_D}{(h_b - h_d) - (P_B - P_C)v_D} \qquad [9]$$

But the feed pump term $(P_B - P_C)v_D$, is small compared with the other energy quantities, so it can be neglected.

Thus, equation [9] gives the Rankine efficiency as

$$\frac{(h_b - h_c)}{(h_b - h_d)} \qquad [10]$$

The cycle is named after William John Rankine (1820–1872), a Glasgow university professor.

If the work done by the feed pump is neglected and if the steam expansion can be expressed in the form PV^n = constant, the P–V diagram for the Rankine cycle is as shown in Fig. 10.15. From this diagram

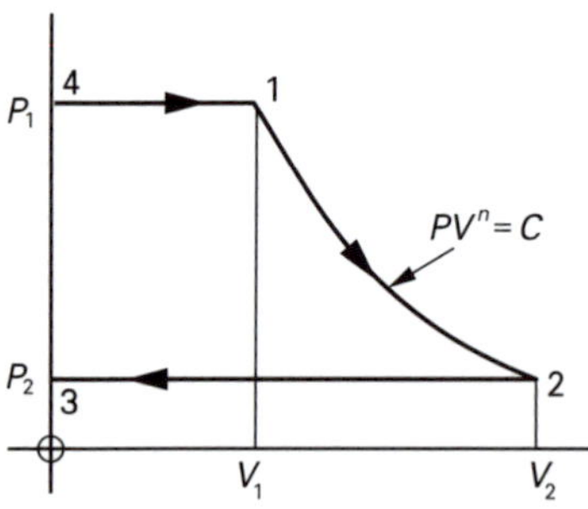

Fig. 10.15 Rankine cycle: P–V diagram

$$\text{Work done} = \text{Area under 4–1} + \text{Area under 1–2} - \text{Area under 2–3}$$

$$= P_1V_1 + \frac{(P_1V_1 - P_2V_2)}{n-1} - P_2V_2$$

$$= (P_1V_1 - P_2V_2) + \frac{(P_1V_1 - P_2V_2)}{n-1}$$

$$= (P_1V_1 - P_2V_2)\left(1 + \frac{1}{n-1}\right)$$

$$= (P_1V_1 - P_2V_2)\frac{(n-1)+1}{n-1}$$

$$\text{Work done} = \frac{n}{n-1}(P_1V_1 - P_2V_2) \qquad [11]$$

If superheated steam is used in the Rankine cycle then the appearance of the cycle on the T–s diagram is as shown in Fig. 10.16. The difference between this diagram and Fig. 10.14(b) is the inclusion of superheat line bc. The complete cycle is now abcdee′.

The thermal efficiency has the same form as before. Using the lettering of Fig. 10.16

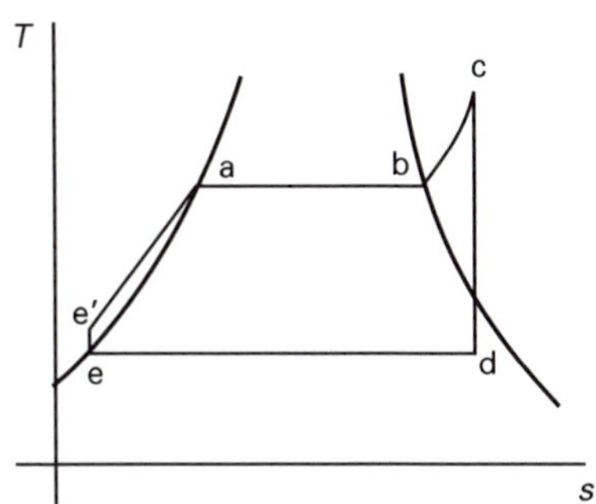

Fig. 10.16 Rankine cycle using superheated steam

$$\text{Rankine } \eta = \frac{h_c - h_d}{h_c - h_e} \qquad [12]$$

There is little gain in thermal efficiency as a result of using superheated steam instead of using saturated steam. The chief advantages of using superheated steam are as follows:

- There is little or no condensation loss in transmission.
- There is greater potential for enthalpy drop and hence for work done.

And by using superheated steam, there is a further departure from the Carnot cycle because the final temperature of the steam is above the constant saturation temperature of the boiler. The Rankine cycle will have a high work ratio ($\rightarrow 1$) because the net work done per cycle is very close to the positive work done per cycle; the feed pump work is very low by comparison (see section 15.1). The Rankine cycle will have a higher work ratio than the Carnot vapour cycle.

Example 10.6 *A steam turbine plant operates on the Rankine cycle. Steam is delivered from the boiler to the turbine at a pressure of 3.5 MN/m² and with a temperature of 350 °C. Steam from the turbine exhausts into a condenser at a pressure of 10 kN/m². Condensate from the condenser is returned to the boiler using a feed pump. Neglecting losses, determine*
(a) the energy supplied in the boiler per kilogram of steam generated
(b) the dryness fraction of the steam entering the condenser
(c) the Rankine efficiency

SOLUTION
First draw a diagram of the Rankine cycle (Fig. 10.17).

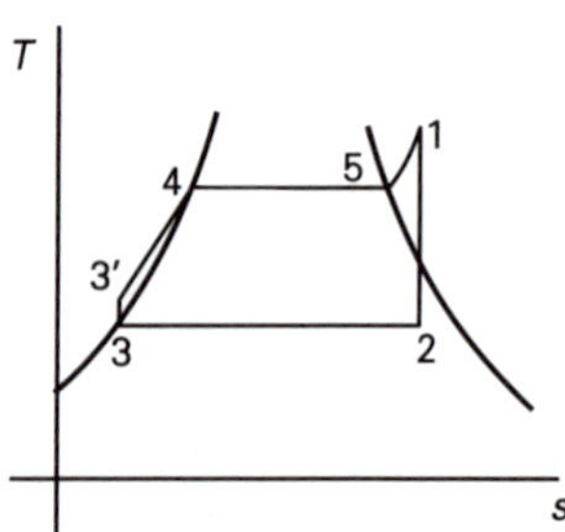

Fig. 10.17 Diagram for Example 10.6

(a)

$$\text{Energy supplied in boiler/kg steam} = (h_1 - h_3).$$

At 3.5 MN/m² and 350 °C

$$h_1 = 3139 - \frac{1.5}{2}(3139 - 3095)$$
$$= 3139 - \left(\frac{1.5}{2} \times 44\right)$$
$$= 3139 - 33$$
$$= \mathbf{3106\ kJ/kg}$$

At 10 kN/m², $h_3 = 191.8$ kJ/kg

$$\therefore \quad h_1 - h_3 = 3106 - 191.8$$
$$= \mathbf{2914.2\ kJ/kg}$$

This is the energy supplied in the boiler.

(b)
Expansion through the turbine is theoretically isentropic, so $s_1 = s_2$.

$$s_1 = 6.960 - \frac{1.5}{2}(6.960 - 6.587)$$
$$= 6.960 - \left(\frac{1.5}{2} \times 0.373\right)$$
$$= 6.960 - 0.28$$
$$= \mathbf{6.680\ kJ/kg\ K}$$

$$s_2 = s_{f2} + x_2(s_{g2} - s_{f2})$$
$$= 0.649 + x_2(8.151 - 0.649)$$
$$= 6.680$$

$$\therefore \quad x_2 = \frac{6.680 - 0.649}{8.151 - 0.649}$$

$$= \frac{6.031}{7.502}$$

$$= \mathbf{0.804}$$

This is the dryness fraction of the steam entering the condenser.

(c)

$$\text{Rankine } \eta = \frac{(h_1 - h_2)}{(h_1 - h_3)}$$

$$h_2 = h_{f2} + x_2 h_{fg2} = 191.8 + (0.803 \times 2392.9)$$
$$= 191.8 + 1921.5$$
$$= \mathbf{2113.3\ kJ/kg}$$

$$\therefore \quad \text{Rankine } \eta = \frac{3106 - 2113.3}{3106 - 191.8}$$

$$= \frac{992.7}{2914.2}$$

$$= 0.341$$
$$= \mathbf{34.1\%}$$

Example 10.7 *A steam plant operates on the Rankine cycle. Steam is supplied at a pressure of 1 MN/m² and with a dryness fraction of 0.97. The steam exhausts into a condenser at a pressure of 15 kN/m². Determine the Rankine efficiency. If the expansion of the steam is assumed to follow the law* $PV^{1.135} = C$, *estimate the specific work done and compare this with that obtained when determining the Rankine efficiency.*

SOLUTION

First draw a diagram, of the Rankine cycle (Fig. 10.18).

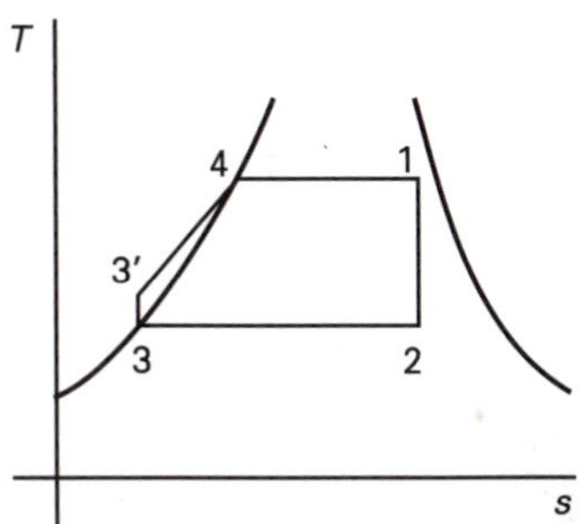

Fig. 10.18 Diagram for Example 10.7

$$\begin{aligned} h_1 &= 762.6 + (0.97 \times 2013.6) \\ &= 762.6 + 1953.2 \\ &= \mathbf{2715.8\ kJ/kg\ K} \end{aligned}$$

$$\begin{aligned} s_1 &= 2.138 + 0.97 \times (6.583 - 2.138) \\ &= 2.138 + (0.97 \times 4.445) = 2.138 + 4.332 \\ &= \mathbf{6.470\ kJ/kg\ K} \end{aligned}$$

Now $s_1 = s_2$

$$\therefore \quad 6.470 = 0.755 + x_2(8.009 - 0.755)$$

$$\begin{aligned} x_2 &= \frac{6.470 - 0.755}{8.009 - 0.755} = \frac{5.715}{7.254} \\ &= \mathbf{0.788} \end{aligned}$$

$$\begin{aligned} h_2 &= 226.0 + (0.788 \times 2373.2) \\ &= 226.0 + 1870.1 \\ &= \mathbf{2096.1\ kJ/kg\ K} \end{aligned}$$

(a)

$$\text{Rankine } \eta = \frac{h_1 - h_2}{h_1 - h_3} = \frac{2\,715.8 - 2\,096.1}{2\,715.8 - 2\,26.0} = \frac{619.7}{2\,489.8} = \mathbf{0.249}$$

$$= 0.249 \times 100$$

$$= \mathbf{24.9\ \%}$$

(b)

$$\text{Specific work done} = \frac{n}{n-1}(P_1V_1 - P_2V_2) \text{ from equation (11), page 305}$$

$$\begin{aligned} V_1 &= x_1V_{g1} = 0.97 \times 0.194\,3 = 0.188\ \text{m}^3/\text{kg} \\ V_2 &= x_2V_{g2} = 0.788 \times 10.02 = \mathbf{7.896\ m^3/kg} \end{aligned}$$

$$\begin{aligned} \therefore \quad \text{Specific work done} &= \frac{1.135}{1.135 - 1}[(1 \times 10^3 \times 0.188) - (15 \times 7.896)] \\ &= \frac{1.135}{0.135}(188 - 118.44) = \frac{1.135}{0.135} \times 69.56 \\ &= \mathbf{584.8\ kJ/kg} \end{aligned}$$

$$\begin{aligned} \text{Specific work done (from Rankine } \eta) = h_1 - h_2 &= 2715.8 - 2096.1 \\ &= \mathbf{619.7\ kJ/kg} \end{aligned}$$

10.15 The Rankine cycle for the steam engine

In steam plant, mainly using steam engines, the steam is not completely expanded down to condenser pressure. It is released at a higher pressure and then, theoretically, there is a pressure drop at constant volume down to condenser pressure. This is shown in Fig. 10.19 as AB on the *P–V* diagram and ab on the *T–s* diagram.

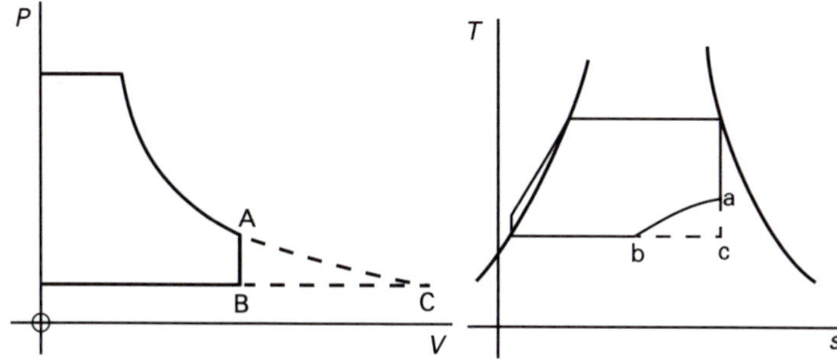

Fig. 10.19 Rankine cycle for a steam engine

This early release causes a drop in thermal efficiency because the energy received is the same but the work done is reduced. The reason for this early release is that at the lower pressures the specific volume of steam is high. The extra volume of the cylinder, which would be required to accommodate the rapidly expanding steam at the lower pressures, is not really justified when considering the small extra amount of work which is made available. The extra work is given by the area ABC on the *P–V* diagram.

The cycle with early release is called the **modified Rankine cycle**.

Example 10.8 *Steam is supplied from a boiler to a steam engine at a pressure of 1.1 MN/m² and at a temperature of 250 °C. It is expanded isentropically to a release pressure of 0.28 MN/m². Its pressure then falls at constant volume to 35 kN/m², when it is exhausted to a condenser. In the condenser the steam is condensed to water with no undercooling and this water is then pumped back into the boiler. Determine*

(a) the Rankine efficiency

(b) the specific steam consumption in kg/kWh

(c) the Carnot efficiency for the same temperature limits of the cycle

SOLUTION

First draw a diagram (Fig. 10.20).

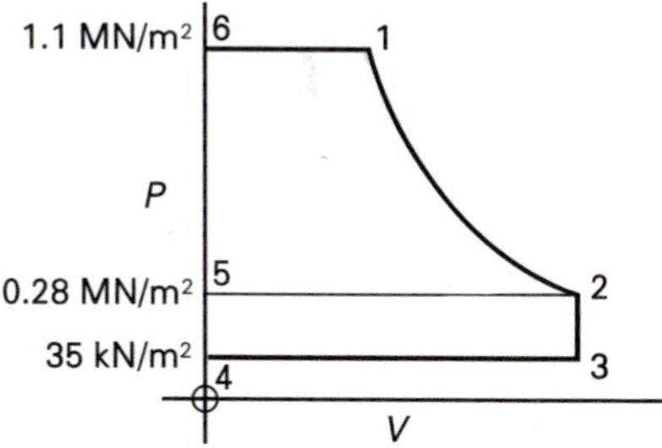

Fig. 10.20 Diagram for Example 10.8

(a)

At 1.1 MN/m² and 250 °C

$$\begin{aligned} h_1 &= 2943 - 0.1 \times (2943 - 2902) \\ &= 2943 - (0.1 \times 41) \\ &= 2943 - 4.1 \\ &= \mathbf{2938.9\ kJ/kg} \end{aligned}$$

For the isentropic 1–2, $s_1 = s_2$

$$\begin{aligned} s_1 &= 6.926 - 0.1 \times (6.926 - 6.545) \\ &= 6.926 - (0.1 \times 0.381) \\ &= 6.926 - 0.038 \\ &= \mathbf{6.888\ kJ/kg\ K} \end{aligned}$$

$$s_2 = s_{f2} + x_2(s_{g2} - s_{f2}) = 1.647 + x_2(7.014 - 1.647)$$

$$\therefore \quad 6.888 = 1.647 + x_2(7.014 - 1.647)$$

$$x_2 = \frac{6.888 - 1.647}{7.014 - 1.647} = \frac{5.241}{5.367} = \mathbf{0.977}$$

$$\begin{aligned} h_2 &= 551.4 + (0.977 \times 2170.1) \\ &= 551.4 + 2120.2 \\ &= \mathbf{2671.6\ kJ/kg} \end{aligned}$$

$$\begin{aligned} v_2 = x_2 v_{g2} &= 0.977 \times 0.646 \\ &= \mathbf{0.631\ m^3/kg} \end{aligned}$$

Work done = Area 61234 = Area 6125 + Area 5234

$$\begin{aligned} \text{Area } 6125 = h_1 - h_2 &= 2938.9 - 2671.6 \\ &= \mathbf{267.3\ kJ/kg} \end{aligned}$$

$$\begin{aligned} \text{Area } 5234 = v_2(P_2 - P_3) &= 0.631 \times (280 - 35) \\ &= 0.631 \times 245 \\ &= \mathbf{154.6\ kJ/kg} \end{aligned}$$

$$\begin{aligned} \therefore \quad \text{Work done} &= 267.3 + 154.6 \\ &= \mathbf{421.9\ kJ/kg} \end{aligned}$$

Energy received = Specific enthalpy of steam supplied – Specific enthalpy of water at condenser pressure

$$\begin{aligned} &= 2938.9 - 304.3 \\ &= \mathbf{2634.6\ kJ/kg} \end{aligned}$$

$$\therefore \quad \text{Rankine } \eta = \frac{421.9}{2634.6} = 0.16 = \mathbf{16\%}$$

(b)

1 kWh = $10^3 \times 3600$ kJ (see section 15.1)

$$\therefore \quad \text{Specific steam consumption} = \frac{3600}{421.9} = \mathbf{8.53\ kg/kWh}$$

(c)

Carnot $\eta = (T_1 - T_3)/T_1$ (see section 15.2)

$$\begin{aligned} T_1 &= 250 + 273 = 523\ \text{K} \\ t_3 &= 72.7\ ^\circ\text{C} \\ T_3 &= 72.7 + 273 = 345.7\ \text{K} \end{aligned}$$

$$\therefore \quad \text{Carnot } \eta = \frac{523 - 345.7}{523} = \frac{177.3}{523} = 0.339 = \mathbf{33.9\%}$$

10.16 Reheat

In modern steam turbine plant the pressure ratio through the turbine can be considerable, so any superheated steam supplied will soon become wet after partial expansion. Wet steam passing over turbine blades for long time periods will produce some corrosion and erosion of the blades. To avoid this, the superheated steam, after partial expansion in the turbine, is passed back to the boiler to be reheated at constant pressure to a higher temperature. It is then passed back to the turbine; the expansion which follows will be dry and superheated, thus largely eliminating the corrosion and erosion of the turbine blades.

Section 10.2 considered the concept of reheat while discussing the improvement of the thermal efficiency of a boiler. Whether or not there will be a change in thermal efficiency of the plant as a whole, as distinct from the boiler on its own, depends upon the reheat temperatures. The higher the reheat temperatures, the more likely there is to be a slight improvement in thermal efficiency of the plant. This will result from the higher potential for enthalpy drop, and hence work output, which can be obtained from the higher-temperature steam.

Example 10.9 *Steam is supplied to a turbine at a pressure of 6 MN/m² and at a temperature of 450 °C. It is expanded in the first stage to a pressure of 1 MN/m². The steam is then passed back to the boiler in which it is reheated at a pressure of 1 MN/m² to a temperature of 370 °C. It is then passed back to the turbine to be expanded in the second stage down to a pressure of 0.2 MN/m². The steam is then again passed back to the boiler in which it is reheated at a pressure of 0.2 MN/m² to a pressure of 320 °C. It is then passed back to the turbine to be expanded in the third stage down to a pressure of 0.02 MN/m². The steam is then passed to a condenser to be condensed, but not undercooled, at a pressure of 0.02 MN/m² and the condensate is then passed back to the boiler. Assuming isentropic expansions in the turbine and using Fig. 10.21 determine*

(a) the theoretical power per kilogram of steam per second passing through the turbine
(b) the thermal efficiency of the cycle
(c) the thermal efficiency of the cycle assuming there is no reheat

(a)
From Fig. 10.21 the following values of specific enthalpy are obtained

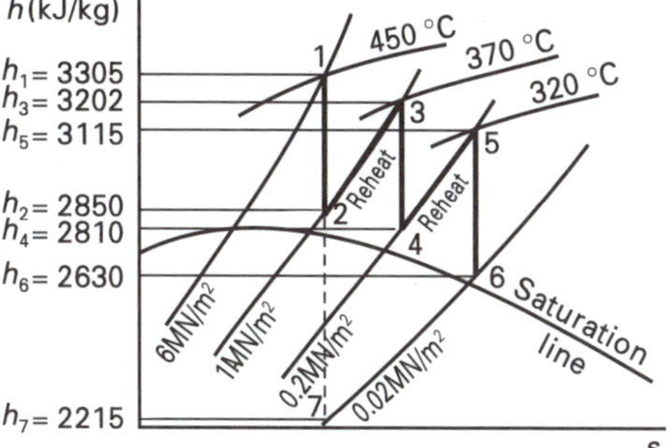

Fig. 10.21 Diagram for Example 10.9

$h_1 = 3305$ kJ/kg $\quad h_4 = 2810$ kJ/kg
$h_2 = 2850$ kJ/kg $\quad h_5 = 3115$ kJ/kg
$h_3 = 3202$ kJ/kg $\quad h_6 = 2630$ kJ/kg
$h_7 = 2215$ kJ/kg

The specific work per kilogram of steam passing through the turbine is

$$\begin{aligned}(h_1 - h_2) + (h_3 - h_4) + (h_5 - h_6) &= (3305 - 2850) + (3202 - 2810) + (3115 - 2630) \\ &= 455 + 392 + 485 \\ &= \mathbf{1332\ kJ}\end{aligned}$$

$$\begin{aligned}\therefore \quad \text{Theoretical power/kg steam/s} &= 1332 \text{ kJ/s} \\ &= \mathbf{1332\ kW}\end{aligned}$$

(b)

$$\begin{aligned}\text{Thermal efficiency} &= \frac{\text{Specific work}}{\text{Specific energy input}} \\ &= \frac{(h_1 - h_2) + (h_3 - h_4) + (h_5 - h_6)}{(h_1 - h_{f6}) + (h_3 - h_2) + (h_5 - h_4)} \\ &= \frac{1332}{(3305 - 251.5) + (3202 - 2850) + (3115 - 2810)} \\ &= \frac{1332}{3053.5 + 352 + 305} \\ &= \frac{1332}{3710.5} \\ &= 0.359 \\ &= \mathbf{35.9\%}\end{aligned}$$

Note that $h_{f6} = 251.5$ kJ/kg is the specific liquid enthalpy at 0.02 MN/m^2; it is obtained from steam tables.

(c)
If there is no reheat

$$\begin{aligned}\text{Thermal efficiency} &= \frac{h_1 - h_7}{h_1 - h_{f7}} \\ &= \frac{3305 - 2215}{3305 - 251.5} \\ &= \frac{1090}{3053.5} \\ &= 0.357 \\ &= \mathbf{35.7\%}\end{aligned}$$

There is little difference in the thermal efficiency, but the steam is at all times superheated in the case (a) when the steam is reheated.

10.17 Feed heat

In modern steam turbine plant, in which there is a large pressure ratio through the turbine, there is a considerable difference between the superheated steam temperature as supplied from the boiler to the turbine and the condensate temperature as it leaves the condenser on its way back to the boiler as feedwater. To increase the feedwater temperature on its way back to the boiler and, as a result, increase the thermal efficiency of the plant, the process of feed heating is introduced. Small quantities of steam are bled at various stages through the turbine; the bled steam passes through a feed heater in which it condenses in a heat transfer process with the feedwater. Thus the temperature of the feedwater is increased. The condensate from the bled steam is pumped into the feedwater main to be returned to the boiler.

In large steam turbine plants, several feed heaters are introduced. The process improves the thermal efficiency of the plant. During expansion of the steam through the stages of a turbine, the actual enthalpy drop in any one stage is less than theoretically obtainable during an isentropic expansion. This is mostly due to the effect of friction between the steam and the turbine blades; there is a similar effect in nozzles (Chapter 12) and gas turbines (Chapter 16). This effect is modelled by the **stage efficiency**, where

$$\text{Stage efficiency} = \frac{\text{Actual enthalpy drop in stage}}{\text{Isentropic enthalpy drop in stage}}$$

The use of stage efficiency is illustrated in Example 10.10.

Example 10.10 *Steam is supplied to a turbine at a pressure of 7 MN/m² and a temperature of 500 °C. Steam is bled for feed heating at pressures of 2 MN/m² and 0.5 MN/m². The condenser pressure is 0.05 MN/m². The stage efficiency of each section of the turbine can be taken as 82 per cent. In the feed heaters the feedwater has its liquid enthalpy raised to that of the corresponding bled steam. The bled steam is condensed but not undercooled and, in this state, on leaving the feed heater, is pumped into the feed main as it leaves the feed heater. The sequence is shown in Fig. 10.22. Determine*

(a) the mass of steam bled to each feed heater in kg/kg of supply steam

(b) the thermal efficiency of the arrangement

SOLUTION

From the enthalpy–entropy chart (Fig. 10.23) values of specific enthalpy are

$$h_1 = 3410 \text{ kJ/kg}$$
$$h_{2'} = 3045 \text{ kJ/kg}$$
$$h_1 - h_{2'} = 365 \text{ kJ/kg} \qquad \therefore \quad h_1 - h_2 = 0.82 \times 365 = 299 \text{ kJ/kg}$$

$$h_2 = 3410 - 299 = 3111 \text{ kJ/kg}$$
$$h_{3'} = 2790 \text{ kJ/kg}$$
$$h_2 - h_{3'} = 321 \text{ kJ/kg} \qquad \therefore \quad h_2 - h_3 = 0.82 \times 321 = 263 \text{ kJ/kg}$$

$$h_3 = 3111 - 263 = 2848 \text{ kJ/kg}$$
$$h_{4'} = 2450 \text{ kJ/kg}$$
$$h_3 - h_{4'} = 398 \text{ kJ/kg} \qquad \therefore \quad h_3 - h_4 = 0.82 \times 398 = 326 \text{ kJ/kg}$$

$$h_4 = 2848 - 326 = 2522 \text{ kJ/kg}$$

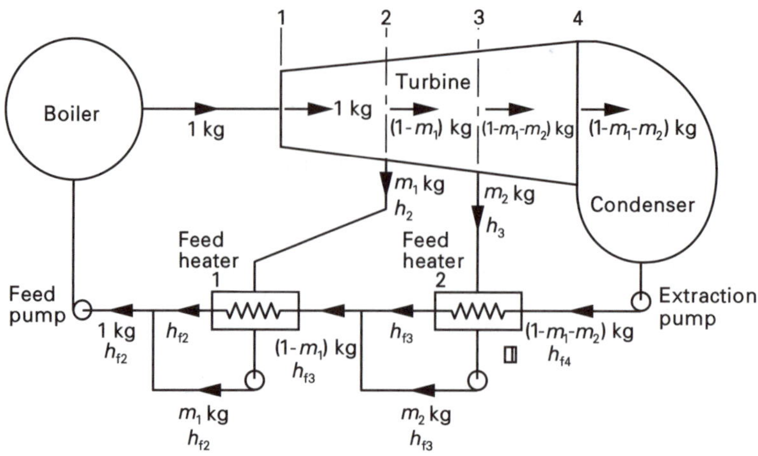

Fig. 10.22 Example 10.10: flow diagram

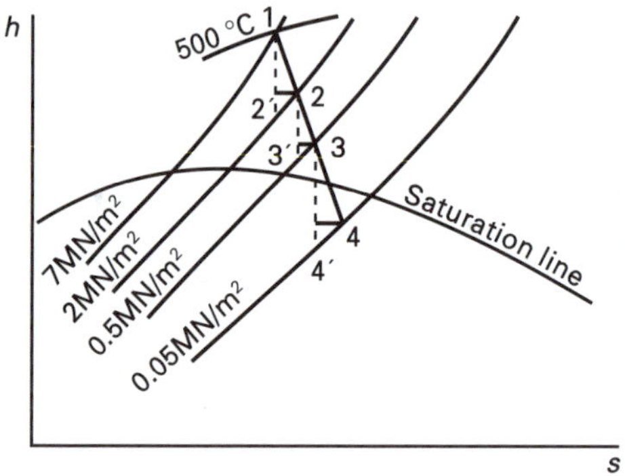

Fig. 10.23 Example 10.10: h–s chart

From steam tables

$$
\begin{aligned}
\text{at } 2\ \text{MN/m}^2,\ h_{f2} &= 908.6\ \text{kJ/kg}\\
0.5\ \text{MN/m}^2,\ h_{f3} &= 640.1\ \text{kJ/kg}\\
0.05\ \text{MN/m}^2,\ h_{f4} &= 340.6\ \text{kJ/kg}
\end{aligned}
$$

(a)

For feed heater 1

Let mass of bled steam = m_1 kg/kg supply steam

Enthalpy loss by bled steam = enthalpy gain by feedwater

$$
\begin{aligned}
m_1(h_2 - h_{f2}) &= (1 - m_1)(h_{f2} - h_{f3})\\
&= h_{f2} - h_{f3} - m_1 h_{f2} + m_1 h_{f3}\\
\therefore \quad m_1(h_2 - h_{f2} + h_{f2} - h_{f3}) &= h_{f2} - h_{f3}\\
m_1(h_2 - h_{f3}) &= h_{f2} - h_{f3}
\end{aligned}
$$

$$
\therefore \quad m_1 = \frac{h_{f2} - h_{f3}}{h_2 - h_{f3}}
$$

$$= \frac{908.6 - 640.1}{3111 - 640.1} = \frac{268.5}{2470.9}$$

$$= \mathbf{0.109\ kg/kg\ supply\ steam}$$

For feed heater 2

$$m_2(h_3 - h_{f3}) = (1 - m_1 - m_2)(h_{f3} - h_{f4})$$
$$= (1 - m_1)(h_{f3} - h_{f4}) - m_2(h_{f3} - h_{f4})$$

$$\therefore \quad m_2(h_3 - h_{f3} + h_{f3} - h_{f4}) = (1 - m_1)(h_{f3} - h_{f4})$$
$$m_2(h_3 - h_{f4}) = (1 - m_1)(h_{f3} - h_{f4})$$

$$\therefore \quad m_2 = (1 - m_1)\frac{(h_{f3} - h_{f4})}{(h_3 - h_{f4})}$$

$$= (1 - 0.109) \times \frac{640.1 - 340.6}{2848 - 340.6}$$

$$= 0.891 \times \frac{299.5}{2507.4}$$

$$= \mathbf{0.106\,4\ kg/kg\ supply\ steam}$$

(b)

The theoretical work done per kilogram of steam entering turbine, W, is given by

$$W = (h_1 - h_2) + (1 - m_1)(h_2 - h_3) + (1 - m_1 - m_2)(h_3 - h_4)$$
$$= (3410 - 3111) + (1 - 0.109) \times (3111 - 2848) + (1 - 0.109 - 0.106\,4) \times (2848 - 2522)$$
$$= 299 + (0.891 \times 263) + (0.784\,6 \times 326)$$
$$= 299 + 234 + 256$$
$$= \mathbf{789\ kJ/kg}$$

$$\text{Energy input to boiler} = h_1 - h_{f2}$$
$$= 3410 - 908.6$$
$$= \mathbf{2501.4\ kJ/kg}$$

$$\text{Thermal } \eta = \frac{\text{Theoretical work}}{\text{Energy input}}$$

$$= \frac{789}{2501.4}$$

$$= 0.315$$
$$= \mathbf{31.5\%}$$

Suppose there is no feed heating (see Fig. 10.24), then

$$h_1 = 3410\ \text{kJ/kg}$$
$$h_{5'} = 2370\ \text{kJ/kg}$$
$$h_1 - h_{5'} = 1040\ \text{kJ/kg}$$

Assuming an expansion efficiency of 82 per cent

$$h_1 - h_5 = 0.82 \times 1040 = \mathbf{853\ kJ/kg}$$

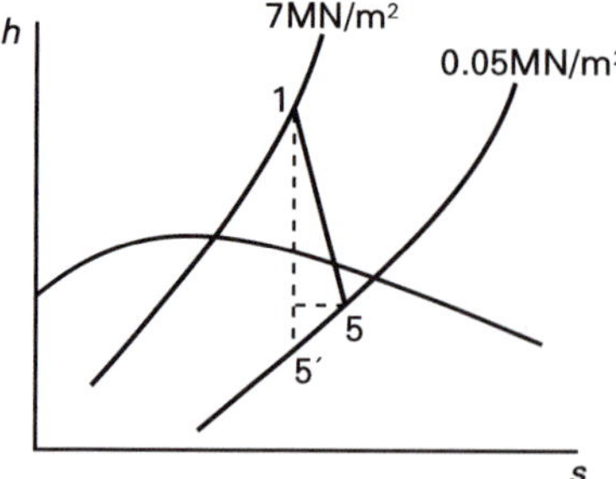

Fig. 10.24 Example 10.10: *h–s* chart with no feed heating

The theoretical work per kilogram of steam is 853 kJ/kg.

$$\text{Energy input} = h_1 - h_{f5}$$
$$= 3410 - 340.6$$
$$= \mathbf{3069.4\ kJ/kg}$$

$$\therefore \quad \text{Thermal } \eta = \frac{853}{3069.4}$$

$$= 0.278$$
$$= \mathbf{27.8\%}$$

This illustrates the thermal efficiency improvement which results from the use of a feed heating system.

Questions

1. A boiler with superheater generates 6000 kg/h of steam at a pressure of 1.5 MN/m^2, 0.98 dry at exit from the boiler and at a temperature of 300 °C on leaving the superheater. If the feedwater temperature is 80 °C and the overall efficiency of the combined boiler and superheater is 85 per cent, determine
   (a) the amount of coal or calorific value 30 000 kJ/kg used per hour
   (b) the equivalent evaporation from and at 100 °C for the combined unit
   (c) the heating surface required in the superheater if the rate of heat transmission may be taken as 450 000 kJ/m^2 of heating surface per hour

   [(a) 636 kg/h; (b) 7189 kg/h; (c) 3.85 m^2]

2. In a steam plant, the steam leaves the boiler at a pressure of 2.0 MN/m^2 and at a temperature of 250 °C. It is then expanded in a turbine and finally exhausted into a condenser. The pressure and condition of the steam at entry to the condenser are 14 kN/m^2 and 0.82 dry, respectively. Assuming the whole of the enthalpy drop is converted into useful work in the turbine, determine
   (a) the power developed for a steam flow of 15 000 kg/h
   (b) the thermal efficiency of the engine if the feed temperature is 50 °C.

   Estimate the mass of cooling water circulated in the condenser per kilogram of steam condensed, if the cooling water enters at 18 °C and leaves at 36 °C and the condensate leaves at 50 °C. How much energy is lost from the condensate per hour due to its being cooled below the saturation temperature corresponding to the condenser pressure?

   [(a) 3.058 MW; (b) 27.3%; 26.1 kg/kg steam; 160 500 kJ/h]

3. A boiler is to produce 6250 kg/h of steam superheated by 40 °C at a pressure of 2.1 MN/m^2. The temperature of the feedwater is 50 °C. If the thermal efficiency of the boiler is 70 per cent, how much fuel oil will be consumed in one hour? The calorific value of the fuel oil used is 45 000 kJ/kg, c_p of superheated steam is 2.093 kJ/kg K.

Extract from steam tables

Pressure (MN/m^2)	Sat. temp. t_f (°C)	Spec. enthalpy (kJ/kg)	
		h_f	h_{fg}
2.1	214.9	920.0	1.880.0

[531 kg]

4. The feedwater to a boiler enters an economiser at 32 °C and leaves at 120 °C, being fed into the boiler at this temperature. The steam leaves the boiler 0.95 dry at 2.0 MPa and passes through a superheater where its temperature is raised to 250 °C without change of pressure. The steam output is 8.2 kg/kg of coal burned and the calorific value of the coal is 28 000 kJ/kg. Determine the energy received per kilogram of water and steam in
(a) the economiser
(b) the boiler
(c) the superheater
Expressing the answers as percentages of the energy supplied by the coal.

Specific heat capacity of superheated steam = 2.093 kJ/kg K
Specific heat capacity of feedwater = 4.18 kJ/kg K

Extract from steam tables

Pressure (MPa)	Sat. temp. t_f (°C)	Spec. enthalpy (kJ/kg)		
		h_f	h_{fg}	h_g
2.0	212.4	909	1 890	2 799

[(a) 10.77%; (b) 64.2%; (c) 5.28%]

5. In a test on a gas-fired boiler, the following observations were made

Gas burned per hour	383 m^3
Water evaporated per hour	4375 kg
Boiler pressure	3.0 MN/m^2
Feedwater temperature	95 °C
Temperature of steam leaving boiler	260 °C
Calorific value of gas used	38.5 MJ/m^3

Determine
(a) the efficiency of the boiler
(b) the equivalent evaporation from and at 100 °C
Take c_p of superheated steam as 2.093 kJ/kg K.

Extract from steam tables

Pressure (MN/m^2)	Sat. temp. t_f (°C)	Spec. enthalpy (kJ/kg)	
		h_f	h_{fg}
3.0	233.8	1 008	1 795
0.103 25	100	419.1	2 256.7

[(a) 73%; (b) 4768 kg/h]

6. The equivalent evaporation of a boiler from and at 100 °C is 10.4 kg steam per kilogram of fuel. The calorific value of the fuel is 29 800 kJ/kg. Determine the efficiency of the boiler. If the boiler produces 15 000 kg steam per hour at 24 bar from feedwater at 40 °C, and the fuel consumption is 1650 kg/h, determine the condition of the steam produced.

[78.8%; 0.972]

7. A steam turbine is supplied with steam at a pressure of 2.5 MN/m^2 and at a temperature of 300 °C and exhausts into a condenser where the pressure is 20 kN/m^2. The steam consumption is 6000 kg/h and the thermal efficiency is 20 per cent. Determine the power developed by the turbine.

If the dryness fraction of the exhaust steam is 0.95, the temperature of the condensate is 60 °C and the mass of the cooling water is 90 000 kg/h, determine the rise in temperature of the cooling water in passing through the condenser.

[920 kW, 35.7 °C]

8. A steam boiler generates 5000 kg/h of superheated steam at 4.0 MN/m^2 and 350 °C from feedwater at 55 °C. If the boiler efficiency is 75 per cent, determine the fuel oil consumption in tonne/h if the calorific value of the fuel oil is 44 000 kJ/kg. If the mass of flue gas is 27.4 kg/kg oil burned and if the flue gas temperature is 310 °C, determine the percentage of the fuel energy which is carried away by the flue gas. Take air temperature = 18 °C, specific heat capacity of flue gas = 1.04 kJ/kg K.

[434 kg/h, 18.9%]

9. Steam at a pressure of 30 bar and temperature of 250 °C is fed to a steam turbine from a boiler. In the turbine the steam is expanded isentropically to a pressure of 1 bar. The steam is then exhausted into a condenser where it is condensed but not undercooled. The condensate is then pumped back into the boiler. Determine
   (a) the dryness fraction of the steam after expansion
   (b) the Rankine efficiency

[(a) 0.823, (b) 23.88%]

10. Steam at 1.7 MN/m^2 and with a temperature of 250 °C is fed into a steam engine in which it is expanded adiabatically to a release pressure of 0.35 MN/m^2. From this pressure it is released at constant volume to a condenser pressure of 50 kN/m^2. The steam is exhausted from the engine into the condenser where it is condensed but not undercooled. The condensate is pumped back into the boiler. The steam flow rate is 1500 kg/h. Determine
    (a) the power output of the engine
    (b) the Rankine efficiency
    (c) the Carnot efficiency for the same temperature limits of the cycle

[(a) 187 kW; (b) 17.4%; (c) 32.25%]

11. Exhaust steam from a turbine developing 3.0 MW enters a jet condenser with a dryness fraction 0.9. The condenser vacuum is 90 kN/m^2 when the reading of the barometer is 100 kN/m^2. The inlet temperature of the injection water is 18 °C and the mixture of condensate and injection water leaves at a temperature of 39 °C. The gaseous volume of mixture passing the air–vapour exit flange of the condenser is 0.28 m^3/s and its temperature is 36 °C. Given the engine uses 3.7 kg steam/MJ, determine
    (a) the quantity (kg/s) of cooling water required
    (b) the rate (kg/h) of air leakage into the condenser

    For air, take $R = 0.287$ kJ/kg K.

[(a) 274 kg/s; (b) 46.1 kg/h]

12. In a steam turbine plant, the initial pressure and temperature of the steam are 3 MN/m^2 and 320 °C, respectively and the exhaust pressure is 0.06 MN/m^2. After isentropic expansion to 1 MN/m^2 the steam is reheated isobarically to 250 °C and after further isentropic expansion in the turbine to 0.4 MN/m^2 it is again reheated isobarically to 200 °C. The steam is finally expanded isentropically in the turbine to exhaust pressure. Indicate on a sketch the expansion of the steam through the turbine and the reheat processes as they would appear on the enthalpy–entropy diagram. Determine the thermal efficiency of the arrangement and compare this with the Rankine efficiency for the same initial conditions and final pressure.

[26.5%; 26.8%]

13. Steam is delivered from a boiler to the high-pressure stage of a steam turbine at a pressure of 3.5 MN/m^2 and with a temperature of 370 °C. It is expanded in the high-pressure stage of the turbine down to a pressure of 0.6 MN/m^2 with a stage of efficiency of 85 per cent. The steam is then passed to a reheater in which it is reheated to a temperature of 320 °C at a constant pressure of 0.6 MN/m^2. At this temperature and pressure the steam then passes to the low-pressure stages of the turbine. In the low-pressure turbine the steam expands to a pressure of 0.2 MN/m^2 with stage efficiency 85 per cent at which pressure sufficient steam is bled to supply a feed heater. The remaining steam in the turbine expands to condenser pressure of 0.03 MN/m^2 with a stage efficiency of 85 per cent. In the condenser the steam is condensed but not undercooled. The condensate is fed through an extraction pump into the feed heater in which its liquid enthalpy is raised to that of the condensate of the bled steam. The bled steam from the turbine is condensed but not undercooled. Both condensate streams from the feed heater are joined into a common feed main which is fed through a feed pump back into the boiler. Sketch the circuit and, using the enthalpy–entropy chart, determine

(a) the work done per kilogram of steam entering the turbine

(b) the thermal efficiency of the arrangement

[(a) 826 kJ/kg; (b) 28.1%]

Chapter 11

The steam engine

11.1 General introduction

For the steam engine, an honoured place in history is assured. Early developments were made by pioneers such as French physicist Denis Papin (1647–1712) and English engineer Thomas Newcomen (1663–1729), and fundamental improvements came from Scottish engineer James Watt (1736–1819), to mention but a few. Out of these inventions developed the Industrial Revolution and the intensive studies of thermodynamics and other branches of organised science.

It has been said that the steam engine owes less to science than science owes to the steam engine.

In modern times, steam turbines and internal combustion engines have largely replaced the steam engine due to their higher power outputs, higher efficiencies and smaller bulk for a given power output. However, many steam engines are used in various countries of the world, particularly on railways. It is for this reason, and because the steam engine provides another exercise in the use of steam, that a chapter on the steam engine is included.

A steam plant has a low noise level, and atmospheric pollution can be low if a high-efficiency burner system is installed in the boiler unit. There is also a high degree of flexibility in the possible fuels used in the boiler unit.

Early steam engine plant produced steam at about atmospheric pressure, mostly in or introduced into a vertical engine cylinder, thus lifting the engine piston. The back of the piston was open to atmospheric pressure. By condensing the steam in the cylinder as a result of the application of cooling water, a vacuum was formed in the cylinder and the atmospheric pressure, acting on the back of the piston, produced a net force to drive the piston downward. Thus work was done by the engine when the piston ascended as the steam was formed, and again when the steam was condensed as the piston descended. This general technique was adopted for many years, and in various forms, until boilers were developed which were capable of producing steam at pressures above atmospheric and until engine valve gear was developed to introduce and exhaust the steam to and from the engine cylinder.

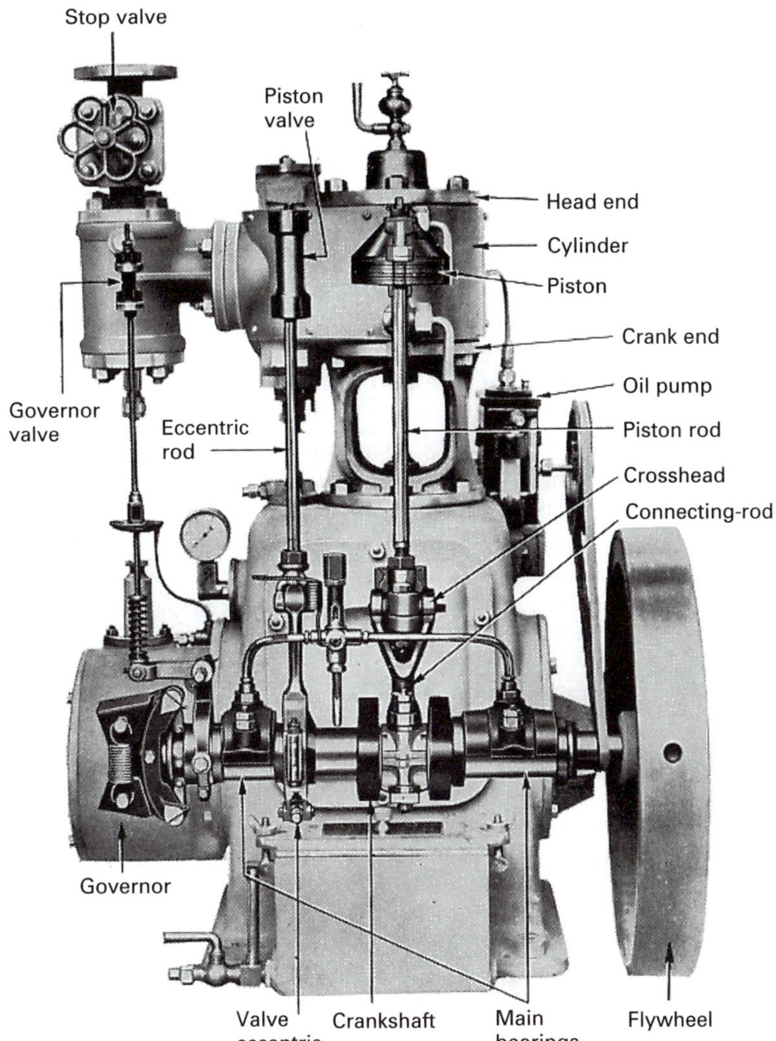

Fig. 11.1 Steam engine

11.2 General description of the steam engine

Figure 11.1 shows the basic elements of a present-day steam engine. It consists of a cylinder in which a **double-acting** piston operates. Double-acting means that both back and front faces of the piston are arranged to be working faces. Thus, when one piston face is working, the other piston face is exhausting. On the return stroke, what was the exhausting face now becomes the working face, and what was the working face now becomes the exhausting face. In this way there are two working strokes per revolution. How this is accomplished will be discussed in the section on valve gear. The end of the cylinder remote from the crank is called the **head end**. The end of the cylinder next to the crank is called the **crank end**. It is through the crank end that the piston rod passes. The crank end of the cylinder is a working end, so it is sealed and the piston rod passes out via a gland. The external end of the piston rod is connected to a **crosshead**. This crosshead reciprocates in guides and supports the piston rod end. The small end of the connecting-rod is connected to the crosshead. The reciprocating motion of the piston is thus transmitted, via the crosshead, to the connecting-rod. The big end of the connecting-rod is connected to the crank which turns on the crankshaft that runs in the main bearings. A flywheel is fitted to the crankshaft.

Control of the steam to and from the cylinder is by means of a valve which is assembled in the valve chest mounted on the side of the cylinder. A common way of operating this valve is to use an **eccentric** mounted on the crankshaft. The reciprocating motion of the eccentric is transmitted to the valve by a valve operating link called the **eccentric rod**.

In order to reduced condensation loss, the cylinder is often surrounded by a steam jacket. High-pressure steam from the main is fed to this steam jacket, which thus helps to maintain a high general cylinder temperature and cuts down working steam condensation loss. Condensate from the jacket is drained through a steam trap.

Note the **governor** fitted to the end of the crankshaft. The governor is connected via a linkage up to the governor valve. The governor serves to maintain nearly constant engine speed at all loads up to full load. The governor in Fig. 11.1 is a throttle governor. This type of governing of a steam engine is explained in section 11.7.

11.3 The steam engine valve

The steam engine valve is the means by which steam is controlled through the steam engine cycle. It also controls the admission and the exhaust of the steam from both the head end and crank end of the cylinder. The valve that is most commonly used in the steam engine is the piston valve. This is illustrated in Fig. 11.2.

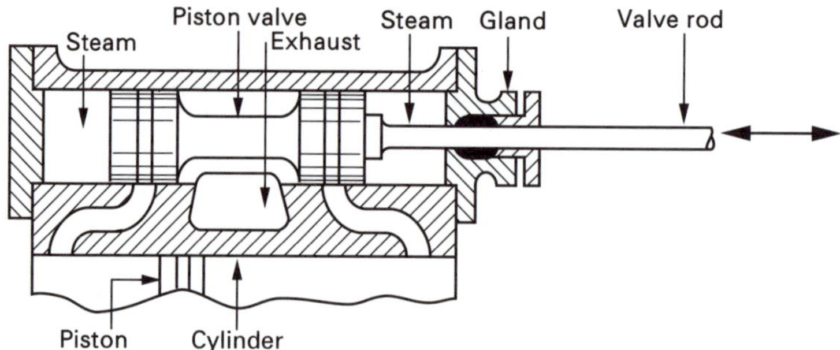

Fig. 11.2 Piston valve arrangement

The valve is essentially two pistons rigidly connected to each other. It is reciprocated in the valve cylinder by the valve rod, which passes out of the valve cylinder through a gland and is reciprocated by means of an appropriate mechanism. In the case of Fig. 11.1, the mechanism takes the form of an eccentric.

The valve in Fig. 11.2 is shown in the middle position. High-pressure steam is admitted to both outer ends of the valve. Steam from the engine cylinder ends exhausts through the space between the valve ends.

In Fig. 11.3(a) the valve is pulled over to the right. High-pressure steam is admitted through a port into the head end of the main engine cylinder, thus pushing the piston from left to right. At the same time, the position of the valve exposes the exhaust port at the crank end of the engine cylinder. As the main engine piston moves from left to right, it pushes steam to exhaust. In Fig. 11.3(b) the reverse is the case. The valve is now pushed over to the left and this allows high-pressure steam to be admitted to the crank end of the main cylinder while the head end is open to exhaust. Thus the arrangement allows for double-acting operation. The exhaust pressure in steam systems is sometimes called **back pressure**.

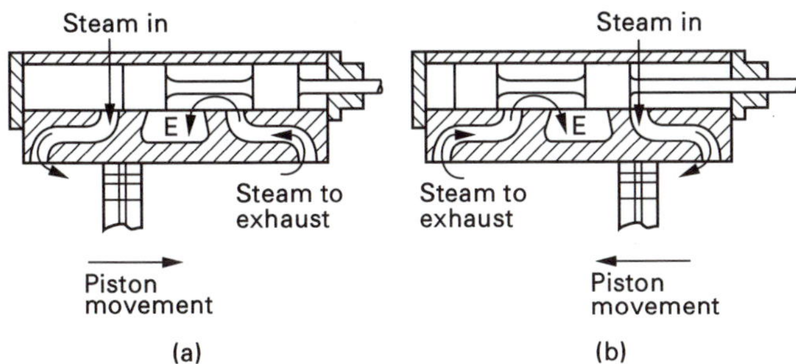

Fig. 11.3 Piston valve positioned (a) to the right and (b) to the left

11.4 The hypothetical steam engine indicator diagram

The hypothetical indicator diagram for a single cycle of a steam engine is shown in Fig. 11.4. Clearance has been neglected. The cycle of operations is as follows:

- **1–2** At point 1 some steam is admitted to the cylinder; this continues until point 2 is reached, where the steam is cut off.

  Point 1 is called the **point of admission** and point 2 is called the **point of cut-off**.

- **2–3** The steam is now locked in the cylinder, so it begins to expand.

  It is assumed to expand hyperbolically (i.e. according to the law $PV = C$) from point 2 to point 3. At point 3 the steam is released from the cylinder. Point 3 is called the **point of release**.

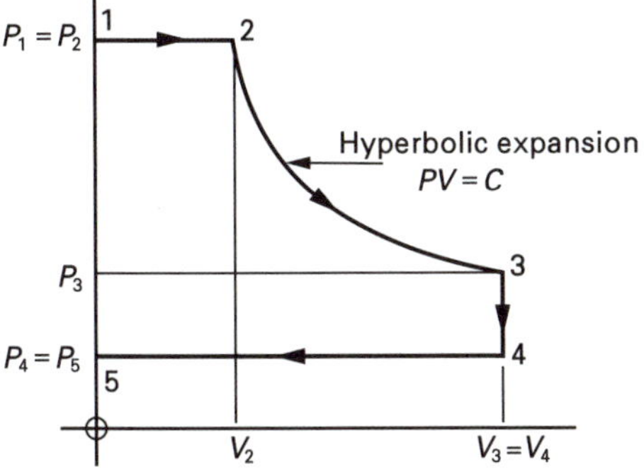

Fig. 11.4 Hypothetical steam engine indicator diagram

- **3–4** From point 3 to point 4 the steam pressure is reduced at constant volume $V_3 = V_4$.
- **4–1** From point 4 to point 5 the steam is exhausted at constant pressure, $P_4 = P_5$, from the cylinder.

The work done per cycle will be given by the area of the diagram.

$$\begin{aligned}\text{Work done/cycle} = \oint W &= \text{Area under 1–2} + \text{Area under 2–3} \\ &\quad - \text{Area under 4–5} \\ &= \text{Area 1234} \\ &= P_2V_2 + P_2V_2 \ln\frac{V_3}{V_2} - P_4V_4 \\ &= P_2V_2\left(1 + \ln\frac{V_3}{V_2}\right) - P_4V_4 \qquad [1]\end{aligned}$$

Now the ratio V_3/V_2 is often written r, called the expansion ratio or the number of expansions. Thus

$$\oint W = P_2V_2(1 + \ln r) - P_4V_4 \qquad [2]$$

The mean effective pressure of the cycle can now be obtained (see Chapter 15).

$$\begin{aligned}\text{Mean effective pressure} &= \frac{\text{Area of diagram}}{\text{Length of diagram}} \\ &= \frac{P_2V_2(1 + \ln r) - P_4V_4}{V_4} \\ &= \frac{P_2V_2}{V_4}(1 + \ln r) - \frac{P_4V_4}{V_4} \\ &= \frac{P_2V_2}{V_4}(1 + \ln r) - P_4 \qquad [3]\end{aligned}$$

and $V_4 = V_3$

$$\therefore \quad \frac{V_2}{V_4} = \frac{V_2}{V_3} = \frac{1}{r}$$

Hence

$$\text{Mean effective pressure} = \frac{P_2}{r}(1 + \ln r) - P_4 \qquad [4]$$

This is often written

$$P_m = \frac{P}{r}(1 + \ln r) - P_b \qquad [5]$$

where P_m = mean effective pressure
P = intake pressure
P_b = back or exhaust pressure
r = expansion ratio

11.5 Diagram factor

In any engine there will inevitably be losses which will result in a reduced output compared with the theoretical assessment. Figure 11.5 is a diagram of an actual steam engine cycle superimposed on a theoretical engine cycle.

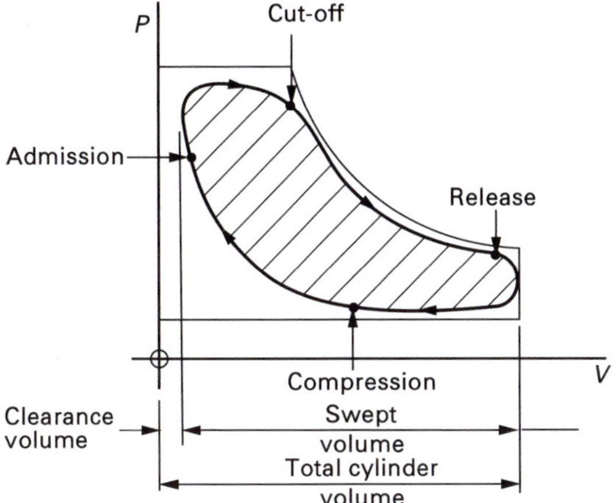

Fig. 11.5 Actual (shaded) and theoretical cycle for a steam engine

In any steam engine there must be a clearance volume, as illustrated. This clearance volume can be anywhere between 1/10 to 1/50 of the cylinder volume, to give some idea of its size.

Admission of the steam is arranged to commence slightly before the piston has reached the inner dead-centre position. This is done so that there will be a maximum pressure build-up for when the piston commences its working stroke, and so that the steam port is sufficiently wide open to give ample steam supply at the commencement of the working stroke.

Due to losses in the steam main and throttling in the valve, there is a pressure loss, so the admission line will be below the theoretical line.

Throttling in the valve is sometimes called wire drawing because damage can appear as grooves on the edges of the valve over long periods of operation. These grooves are as if a wire had been repeatedly drawn across the edge of the valve.

Now the valve, ports, cylinder and piston are exposed alternately to high-pressure, high-temperature working steam and low-pressure, low-temperature exhausting steam. Thus their mean working temperature will lie somewhere between. Upon admission, the working steam will meet working surfaces which are at a lower temperature, thus some steam will condense on these surfaces. This will result in an increased steam supply to make up for the loss. There is no corresponding increase in work output for this increased steam supply, so the thermal efficiency tends to be reduced.

Due to the lower admission pressure, the expansion line will be lower than the theoretical expansion line. In the earlier part of the expansion, this pressure reduction will be accentuated by further condensation on the cylinder walls. This will be counteracted somewhat in the latter part of the expansion; the steam temperature becomes lower than the mean cylinder temperature and there is re-evaporation of some of the condensed steam.

During the exhaust stroke, the pressure in the cylinder is slightly higher than the external exhaust pressure. This occurs because there must be a net positive pressure from inside to outside the cylinder in order to produce the force necessary to move the steam mass from the cylinder. And due to the higher mean cylinder temperature there is a tendency to evaporate some of the cylinder water film left behind as the result of condensation in the earlier part of the cycle. The exhaust line is therefore higher than the theoretical exhaust line.

In order to smooth the operation of the engine, the exhausting of the steam is stopped early at the point of compression. Here the exhaust valve is closed and a quantity of steam is locked up in the cylinder. This steam is called the **cushion steam**. The pressure of this cushion steam is raised and thus a smooth change from exhaust to admission is made. The cushion steam also tends to reduce steam consumption because, if it did not exist, more steam would be required to raise the pressure from exhaust to admission.

The net result of these phenomena is that the actual diagram is much more continuous and has a smaller area than the theoretical diagram.

The ratio

$$\frac{\text{Actual diagram area}}{\text{Theoretical diagram area}}$$

is called the **diagram factor**, k.

Thus

$$\text{Actual diagram area} = k \times \text{Theoretical diagram area}$$

From this, it follows that

$$\text{Actual work done} = k \times \text{Theoretical work done}$$

Also

$$\text{Actual mean effective pressure} = k \times \text{Theoretical mean effective pressure}$$

Example 11.1 *A single-cylinder, double-acting steam engine is required to develop 60 kW and the following assumptions are made*

Boiler pressure	1.25 MN/m^2
Back pressure	0.13 MN/m^2
Cut-off at	0.3 stroke
Diagram factor	0.82
Mechanical efficiency	78%
Mean piston speed	3 m/s
Stroke	1.25 times bore

Determine

(a) the bore of the cylinder

(b) the piston stroke

(c) the speed of the engine

(a)

$$P_m = \frac{P}{r}(1 + \ln r) - P_b$$

$r = 1/0.3 = \mathbf{3.33}$

$$\therefore \quad P_m = \frac{1.25}{3.33}(1 + \ln 3.33) - 0.13$$

$$= \frac{1.25}{3.33}(1 + 1.203) - 0.13$$

$$= \frac{1.25 \times 2.203}{3.33} - 0.13$$

$= 0.827 - 0.13$
$= 0.697\ \text{MN/m}^2$
$= \mathbf{697\ kN/m^2}$

$$\text{Actual indicated power developed} = \frac{60}{0.78}\ \text{kW}$$

$$\text{Theoretical indicated power developed} = \frac{60}{0.78 \times 0.82}$$

$$= \mathbf{93.8\ kW}$$

Now

Indicated power $= P_m LAN$ (see Chapter 17)

and

LN = mean piston speed = 3 m/s

$\therefore \quad 93.8 = 697 \times A \times 3$

where A = piston area, m^2

$$\therefore \quad A = \frac{93.8}{697.3} = 0.044\,9\ \text{m}^2 = (0.044\,9 \times 10^6)\ \text{mm}^2$$

Let d = diameter of piston = bore in mm

Then

$$\frac{\pi d^2}{4} = 0.044\,9 \times 10^6$$

or

$$d^2 = 0.044\,9 \times 10^6 \times \frac{4}{\pi} = 0.057 \times 10^6$$

$\therefore \quad d = \sqrt{(0.057 \times 10^6)} = 0.239 \times 10^3\ \text{mm} = \mathbf{239\ mm}$

(b)

Stroke $= 1.25d$
$= 1.25 \times 239$
$= \mathbf{299\ mm}$

(c)
Now

$$LN = \text{mean piston speed}$$

where $L = \text{stroke, m}$
$N = 2 \times \text{rev/s}$ (since engine is double-acting)

$$\therefore \quad 0.299 \times 2 \times \text{rev/s} = 3$$

$$\therefore \quad \text{rev/s} = \frac{3}{2 \times 0.299} = 5.02$$

$$\therefore \quad \text{rev/min} = 5.02 \times 60 = \mathbf{301.2}$$

The engine speed is 301.2 rev/min.

Example 11.2 *Dry saturated steam is supplied to a single-cylinder, double-acting steam engine. The inlet and exhaust pressures are 900 kN/m^2 and 140 kN/m^2, respectively. Cut-off occurs at 0.4 of the stroke and the diagram factor is 0.8. The stroke/bore ratio is 1.2:1, the engine speed is 4 rev/s and the expansion of the steam in the engine cylinder is assumed to be hyperbolic. The power output from the engine is 22.5 kW. Determine (a) the diameter of the cylinder and (b) the piston stroke.*

If the actual steam consumption is 1.5 times the theoretical requirement, determine (c) the actual steam consumption per hour and the indicated thermal efficiency.

(a)

$$P_m = \frac{P}{r}(1 + \ln r) - P_b$$

and

$$r = \frac{1}{0.4} = 2.5$$

$$\therefore \quad P_m = \frac{900}{2.5}(1 + \ln 2.5) - 140$$

$$= 360 \times (1 + 0.916) - 140$$
$$= (360 \times 1.916) - 140$$
$$= 689.8 - 140$$
$$= \mathbf{549.8\ kN/m^2}$$

$$\text{Actual } P_m = 0.8 \times 549.8$$
$$= \mathbf{439.8\ kN/m^2}$$

Let d = diameter of cylinder in mm, then stroke = $1.25d$ mm, and

$$\text{Indicated power} = P_m LAN$$

where L = stroke, m
A = area of piston, m^2
N = number of working strokes/s

$$\therefore \quad 22.5 = 439.8 \times \frac{1.2d}{10^3} \times \frac{\pi}{4}\left(\frac{d}{10^3}\right)^2 \times 2 \times 4$$

From this

$$d^3 = \frac{22.5 \times 4 \times 10^9}{439.8 \times 1.2 \times \pi \times 2 \times 4} = 0.006\,79 \times 10^9$$

$$d = \sqrt[3]{(0.006\,79 \times 10^9)} = 0.189 \times 10^3 = \mathbf{189\ mm}$$

(b)

$$\begin{aligned}\text{Piston stroke} &= 1.2d \\ &= 1.2 \times 189 \\ &= \mathbf{226.8\ mm}\end{aligned}$$

(c)
Now

$$\begin{aligned}\text{Stroke volume} &= \frac{\pi d^2}{4} \times 1.2d \\ &= \frac{\pi \times 0.189^2}{4} \times 0.226\,8 \\ &= \mathbf{0.006\,36\ m^3}\end{aligned}$$

Alternatively

$$\text{Indicated power} = P_m LAN$$

$$\begin{aligned}\therefore \quad LA = \text{Stroke volume} &= \frac{\text{Indicated power}}{P_m N} \\ &= \frac{22.5}{439.8 \times 8} \\ &= \mathbf{0.006\,36\ m^3}\end{aligned}$$

$$\therefore \quad \text{Volume of steam admitted/stroke} = 0.006\,36 \times 0.4 = \mathbf{0.002\,544\ m^3}$$

$$\text{Volume of steam consumed/h} = (0.002\,544 \times 240 \times 2 \times 60)\ \text{m}^2$$

$$\text{Specific volume of dry saturated steam at 900 kN/m}^2 = \mathbf{0.214\,8\ m^3/kg}$$

$$\begin{aligned}\text{Theoretical steam consumed/h} &= \frac{0.002\,544 \times 240 \times 2 \times 60}{0.2148} \\ &= \mathbf{341\ kg}\end{aligned}$$

$$\text{Actual steam consumed/h} = 341 \times 1.5 = \mathbf{511.5\ kg}$$

$$\text{Indicated thermal efficiency} = \frac{\text{Energy to indicated power/s}}{\text{Energy available from steam/s}}$$

$$\text{Energy available from steam/s} = \dot{m}(h_g - h_{fe})$$

where $\dot{m}$ = mass of steam used/s
h_g = specific enthalpy of inlet steam, kJ/kg
h_{fe} = specific liquid enthalpy at exhaust pressure, kJ/kg

So the indicated thermal efficiency is

$$\frac{22.5}{\frac{516.5}{3600}(2772.1 - 458.4)} = \frac{22.5 \times 3600}{511.5 \times 2313.7}$$

$$= 0.068$$

$$= \mathbf{6.8\%}$$

Example 11.3 *A single-cylinder, double-acting steam engine has a cylinder 250 mm diameter and the length of stroke is 375 mm. The engine is supplied with steam at 1.0 MPa, 0.96 dry, the exhaust pressure is 55 kPa and the expansion ratio is 6:1. The indicated power developed by the engine is 45 kW at 3.5 rev/s and the steam consumption is 460 kg/h. Determine*
(a) the diagram factor
(b) the indicated thermal efficiency of the engine

(a)

$$P_m = \frac{P}{r}(1 + \ln r) - P_b$$

$$= \frac{1000}{6}(1 + \ln 6) - 55$$

$$= \frac{1000}{6}(1 + 1.79) - 55$$

$$= \frac{1000 \times 2.79}{6} - 55$$

$$= 465 - 55$$

$$= \mathbf{410\ kPa}$$

Theoretical indicated power, $P_m LAN$, is

$$410 \times \frac{375}{10^3} \times \frac{\pi}{4} \times \left(\frac{250}{10^3}\right)^2 \times 2 \times 3.5 = \mathbf{52.8}$$

$$\therefore \quad \text{Diagram factor} = \frac{\text{Actual indicated power}}{\text{Theoretical indicated power}}$$

$$= \frac{45}{52.8}$$

$$= \mathbf{0.85}$$

(b)
Specific enthalpy of steam at 1.0 MPa, 0.96 dry is

$$h_f + xh_{fg} = 762.6 + (0.96 \times 2013.6)$$

$$= 762.6 + 1933$$

$$= \mathbf{2695.6\ kJ/kg}$$

Minimum possible specific enthalpy in engine is

$$h_f \text{ at } 55 \text{ kPa} = \mathbf{350.6\ kJ/kg}$$

So maximum energy available in steam in engine is

$$2695.6 - 350.6 = \mathbf{2345\ kJ/kg}$$

And mass of steam used per second is 460/3600 kg, thus

$$\text{Indicated thermal efficiency} = \frac{45}{\dfrac{460}{3600} \times 2345}$$

$$= 0.15$$
$$= \mathbf{15\%}$$

11.6 Steam engine governing

A steam engine must be capable of taking up or shedding load

- If load is to be taken up then more work must be done in the cylinder.
- If load is shed then less work must be done in the cylinder.

For a greater amount of work to be done in the cylinder, the indicator diagram area must be increased. Conversely, the indicator diagram area must be decreased for less work to be done in the cylinder.

To achieve these objectives, two possibilities arise. Either the intake steam pressure may be increased or decreased or the quantity of steam admitted at a constant intake pressure may be varied. Both these methods are used for the governing of a steam engine.

11.6.1 Throttle governing

Throttle governing is the method of controlling the engine output by varying the intake pressure to the engine. This is illustrated in Fig. 11.6 where the engine cut-off remains constant. By reducing the intake pressure, a smaller diagram is obtained and hence less output. The throttling of the steam is usually taken care of using a valve controlled by a governor on the engine.

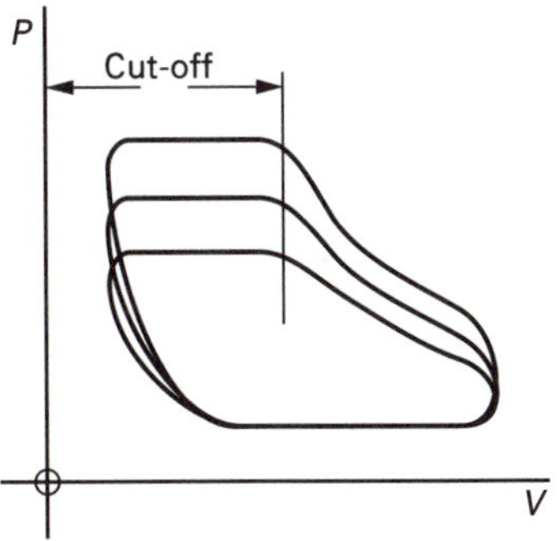

Fig. 11.6 Throttle governing

This is quite an effective way to govern an engine but is somewhat wasteful with regard to the steam. This is because the full steam pressure is not used at all loads below full-load. Also, due to the constant cut-off, large quantities of steam are used, as much at low loads as at high loads. Both effects will tend to lower the thermal efficiency of the engine.

11.6.2 Cut-off governing

Cut-off governing is the method of controlling the output of the engine by varying the cut-off. This is illustrated in Fig. 11.7. The intake steam pressure now remains constant, so full use is made of the steam pressure. But, by making the cut-off smaller, a smaller quantity of steam is admitted to the cylinder, so the diagram area is smaller. Hence a smaller amount of work is done in the cylinder. Cut-off governing produces a much more valuable use of the steam than throttle governing. It will, therefore, be generally more efficient.

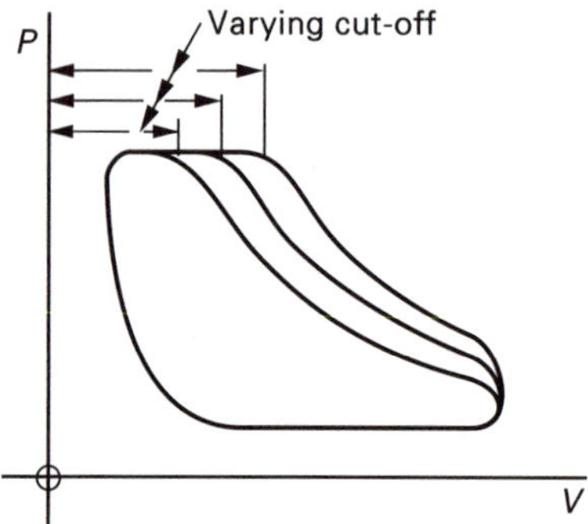

Fig. 11.7 Cut-off governing

11.7 Throttle governing and Willan's line

For a throttle-governed steam engine it is found that a linear relationship exists between the steam consumption and the indicated power output. A graph of this relationship will be a straight line and is illustrated in Fig. 11.8. The straight line is called **Willan's line**, after its discoverer.

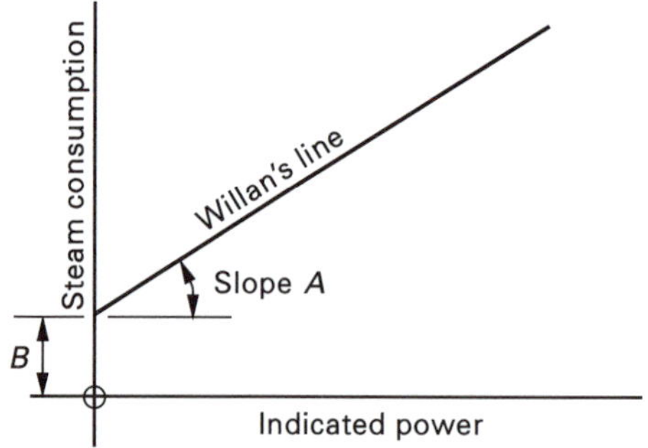

Fig. 11.8 Willan's line

The law of the line will be

$$W = (A \times \text{i.p.}) + B$$

where W = steam consumption
A = slope of the graph
i.p. = indicated power
B = intercept on the steam consumption axis

The indicated power is zero at B, so B represents loss due to condensation, leakage and so on.

Example 11.4 *A throttle-governed steam engine uses 276 kg of steam per hour when developing 11 kW. The steam used at no-load is 45 kg/h. Estimate the steam consumption when the engine output is 8 kW.*

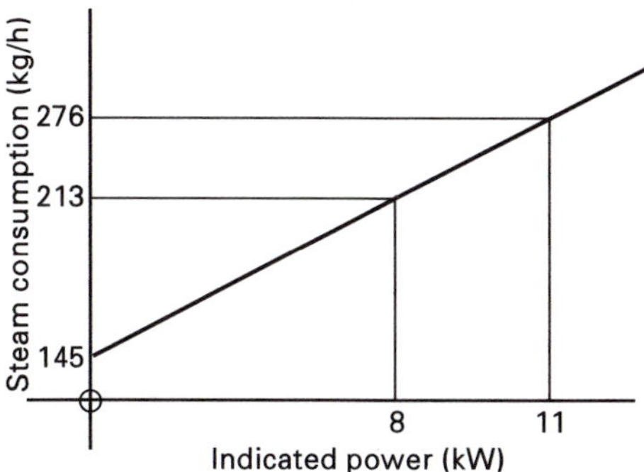

Fig. 11.9 Diagram for Example 11.4

SOLUTION

The law for Willan's line in Fig. 11.9 is given by

$$W = (A \times \text{i.p.}) + B$$

The slope of the line in this case is

$$A = \frac{276 - 45}{11} = \frac{231}{11} = \mathbf{21\ kg/kWh}$$

$$\therefore \quad W = (21 \times \text{i.p.}) + 45$$

Hence for an i.p. output of 8 kW

$$\begin{aligned} W &= (21 \times 8) + 45 \\ &= 168 + 45 \\ &= \mathbf{213\ kg/h} \end{aligned}$$

The steam consumption is 213 kg/h.

11.8 The effect of the condenser on the steam engine

The effect of exhausting a steam engine into a condenser which is working below atmospheric pressure is that more work is obtained from the engine using the same steam. This is illustrated by referring to Fig. 11.10.

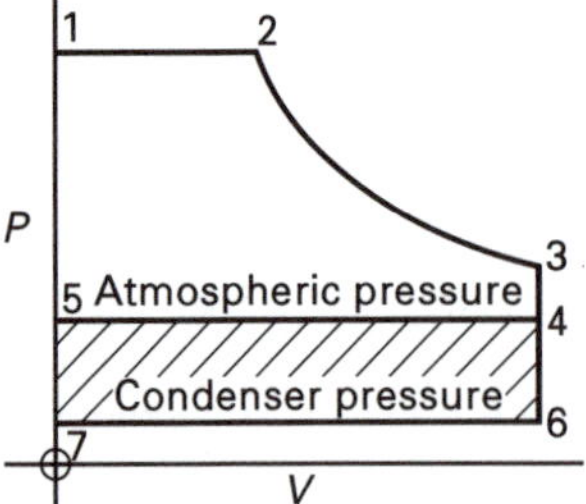

Fig. 11.10 Condenser: effect on steam engine

Diagram 12345 is the theoretical diagram for an engine exhausting to atmospheric pressure. But if the engine exhausts to a condenser, the diagram becomes 12367.

Now the same amount of steam has been used in both cases, but the condenser diagram is larger than the atmospheric diagram by the shaded area 5467 and hence there is a greater output using the condenser. And because there is a greater output using the same amount of steam, the thermal efficiency has been improved.

11.9 The indicator diagram for the double-acting steam engine

For the double-acting steam engine, there will be a separate indicator diagram for each side of the piston. The indicator diagrams already illustrated are for the head end of the engine.

Now when the head end is carrying out its working stroke, for the same piston direction, the crank end is exhausting. Also, when the head end is exhausting, the crank end is carrying out its working stroke.

The crank-end indicator diagram will therefore appear as illustrated in Fig. 11.11. It is shown superimposed on the head-end diagram. The crank-end diagram is really 180° out of phase with the head-end diagram.

Note that the total work done per revolution is the sum of the diagram areas taken separately.

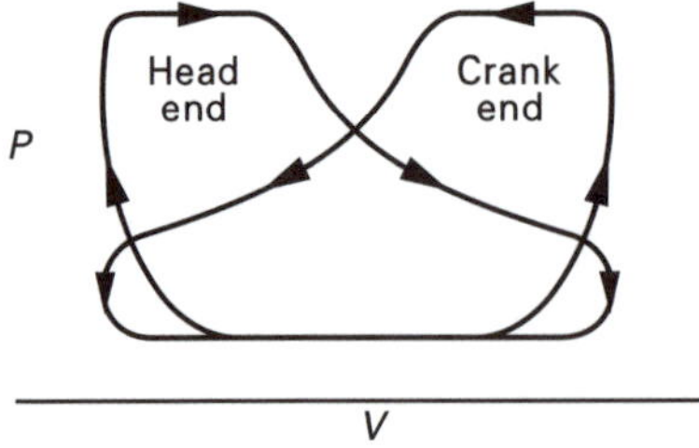

Fig. 11.11 Indicator diagram for a double-acting steam engine

The crank-end diagram will, in general, have a smaller area than the head-end diagram. This is because the effective piston area of the crank end is reduced by the area of the piston rod. Thus the crank-end work is smaller than the head-end work.

11.10 The compound steam engine

With the advent of the high-pressure boiler, difficulties arose in the use of the high-pressure steam in a single cylinder.

If a single cylinder is used with high-pressure steam, then not only has the cylinder to be designed to accommodate the high pressure but also to accommodate the large volume after expansion. This means that the cylinder will be of very heavy construction. Likewise the reciprocating parts must be large. The high-pressure range and the heavy reciprocating parts will create a considerable variation of torque. A heavy flywheel will be required to smooth this out.

There is also an increased balancing problem in a single-cylinder engine with heavy reciprocating masses. And due to the large pressure difference between the inlet and exhaust conditions, there will also be a large corresponding temperature difference. This large temperature difference will increase the condensation loss.

To overcome these difficulties, the compound steam engine was developed. In the compound steam engine, the steam is expanded through two or more cylinders. The steam is first admitted to the high-pressure cylinder in which it is only partially expanded. In this way the cylinder accommodating the high-pressure steam need not have such a large volume, so it can have a lighter construction. The exhaust steam from the high-pressure cylinder becomes the working steam for the following cylinder. The following cylinder must accommodate a larger steam volume, but the admission pressure is lower so, once again, lighter construction is possible. In each cylinder there will be a lower overall temperature range, so condensation loss is reduced.

The lighter cylinder construction means there are lighter reciprocating parts. And in each cylinder there is a lower pressure range. With two or more cylinders and a suitable crank arrangement, it is possible to obtain better balancing than with a single cylinder. Together, these result in a smoother torque output.

The smoother torque means that a smaller flywheel is required on a compound engine. If the cranks on a compound engine are not in phase or at 180° then the following cylinder will not be ready for the steam being exhausted from the cylinder before. In this case, a receiver is fitted between cylinders to hold the steam until the following cylinder is ready for it.

As the strokes of the cylinders of a compound engine are generally made equal and as the following cylinders accommodate the larger steam volumes, the diameters of the following cylinders are made larger.

A two-cylinder compound steam engine is called a **double-expansion engine**. This is illustrated in Fig. 11.12(b). A three-cylinder compound steam engine is called a **triple-expansion engine**. This is illustrated in Fig. 11.12(c). In a double-expansion engine the cylinders are called the high-pressure and low-pressure cylinders. In a triple-expansion engine the cylinders are called the high-pressure, intermediate-pressure and low-pressure cylinders.

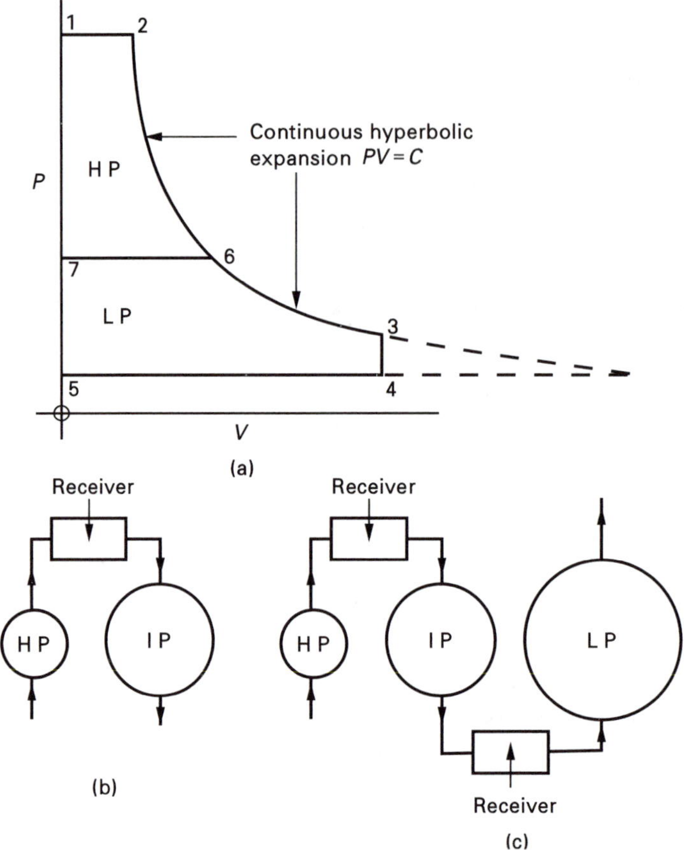

Fig. 11.12 (a) Hypothetical indicator diagram for a double-expansion engine; flow diagrams for (b) a double-expansion engine and (c) a triple-expansion engine

Figure 11.12(a) shows the hypothetical indicator diagram for a double-expansion engine:

- The high-pressure cylinder diagram is 1267.
- The low-pressure cylinder diagram is 76345.

The two diagrams are fitted together for convenience. In practice they are quite separate.

The expansion through the engine is assumed continuous and to be of hyperbolic form, $PV = C$. In the hypothetical diagram it is possible to assume that the expansion is complete in the high-pressure cylinder. This is shown as expansion 2–6. The reason for this is that the cylinder volume, which is 7–6, is not too large and the steam expansion can easily be accommodated. Complete expansion in the high-pressure cylinder also makes the high-pressure cylinder volume equal to the cut-off volume of the low-pressure cylinder.

But in the low-pressure cylinder there is usually an early release, as shown at 3. In this case the expansion in the low-pressure cylinder is not complete. Complete expansion would be as shown dotted. The reason for the early release in the low-pressure cylinder is that the extra large cylinder volume which is required for complete expansion does not compensate for the very small extra amount of work that is obtained. This small extra amount of work will be given by the area of the dotted section of the theoretical indicator diagram. The smaller low-pressure cylinder volume, 5–4, more than compensates for the small loss of work. In practice, the high-pressure cylinder also has an early release.

In choosing the intermediate pressure P_7, there are two generally accepted possibilities.

11.10.1 Equal initial piston loads

In a double-acting engine, the net pressure on the piston will be the difference between the pressures on either side of the piston.

For the high-pressure cylinder

$$\text{Net pressure} = P_1 - P_7$$

For the low-pressure cylinder

$$\text{Net pressure} = P_7 - P_5$$

Now

$$\text{Force on piston} = \text{net pressure on piston} \times \text{piston area}$$

Thus, for equal initial piston loads

$$(P_1 - P_7)A_{HP} = (P_7 - P_5)A_{LP} \qquad [1]$$

where A_{HP} = area of high-pressure piston
A_{LP} = area of low-pressure piston

Now, as already stated, the strokes of the pistons of a compound engine are usually made equal.

If the stroke is L, then from equation [1]

$$(P_1 - P_7)A_{HP}L = (P_7 - P_5)A_{LP}L \qquad [2]$$

But neglecting clearance

$$A_{HP}L = V_{HP} = \text{volume of high-pressure cylinder}$$

also

$$A_{LP}L = V_{LP} = \text{volume of low-pressure cylinder}$$

Hence, equation [2] becomes

$$(P_1 - P_7)V_{HP} = (P_7 - P_5)V_{LP} \qquad [3]$$

This is the necessary condition for equal initial piston loads.

11.10.2 Equal work per cylinder

For equal work per cylinder, the areas of the high-pressure and low-pressure diagrams must be equal.

For the high-pressure cylinder

$$\text{Work done} = \text{Area } 1267 = \text{Area under } 1\text{–}2 + \text{Area under } 2\text{–}6 - \text{Area under } 6\text{–}7$$

$$= P_2V_2 + P_2V_2 \ln\frac{V_6}{V_2} - P_6V_6$$

But $P_2V_2 = P_6V_6$ since $PV = C$.

$$\therefore \quad \text{Work done} = P_2V_2 + P_2V_2 \ln\frac{V_6}{V_2} - P_2V_2$$

$$= P_2V_2 \ln\frac{V_6}{V_2}$$

$$= P_2V_2 \ln r \qquad [4]$$

where $r = V_6/V_2$ = number of expansions in high-pressure cylinder

The low-pressure cylinder diagram is of the conventional shape, hence

$$\text{Work done in low-pressure cylinder} = P_6V_6(1 + \ln R) - P_4V_4 \qquad [5]$$

where $R = V_4/V_6$ = number of expansions in the low-pressure cylinder

Hence, for equal work per cylinder

$$P_2V_2 \ln r = P_6V_6(1 + \ln R) - P_4V_4 \qquad [6]$$

Note that the high-pressure cylinder volume is the same as the cut-off volume, 7–6, of the low-pressure cylinder.

Note also that the total work done is given by the total area 12345. This would also be the work done if the steam were admitted and expanded in the low-pressure cylinder only. This point is used in the solution of some steam engine problems.

Example 11.5 *A double-acting, compound, reciprocating steam engine has two cylinders and is supplied with steam at a pressure of 1.4 MN/m² and 0.9 dry. Exhaust from the engine is into a condenser working at a pressure of 35 kN/m². Each cylinder has equal initial piston loads. The diagram factor referred to the low-pressure cylinder is 0.8. The stroke in both cylinders is 350 mm. The diameters of the high- and low-pressure cylinders are 200 mm and 300 mm respectively. Expansion in the high-pressure cylinder is complete. The engine runs at 300 rev/min. Expansion in the engine is hyperbolic. Neglecting clearance, determine*

(a) the intermediate pressure

(b) the indicated power output

(c) the steam consumption of the engine in kg/h

SOLUTION

First draw a diagram (Fig. 11.13).

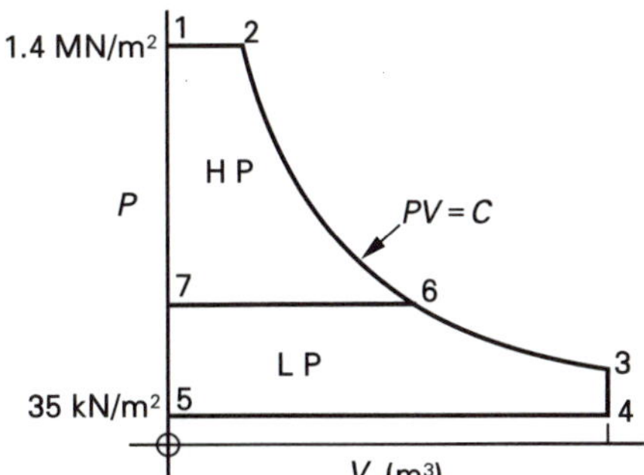

Fig. 11.13 Diagram for Example 11.5

(a)

For equal initial piston loads

$$(P_1 - P_7)\,A_{HP} = (P_7 - P_5)\,A_{LP}$$

$$(1400 - P_7)\frac{\pi \times (0.2)^2}{4} = (P_7 - 35)\frac{\pi \times (0.3)^2}{4}$$

$$\therefore \quad 1400 - P_7 = (P_7 - 35)\frac{(0.3)^2}{(0.2)^2}$$

$$= 2.25\,(P_7 - 35)$$
$$= 2.25\,P_7 - 78.75$$

$$\therefore \quad 3.25\,P_7 = 1478.75$$
$$P_7 = 1478.75/3.25$$
$$= \mathbf{455\ kN/m^2}$$

This is the immediate pressure.

(b)

Now $P_2V_2 = P_6V_6$

$$\therefore \quad V_2 = \frac{P_6V_6}{P_2} = \left(\frac{455}{1400} \times \frac{\pi \times (0.2)^2}{4} \times 0.35\right)\ \text{m}^3$$

This is the cut-off volume in the high-pressure cylinder.

$$\text{Volume of low-pressure cylinders} = \left(\frac{\pi \times (0.3)^2}{4} \times 0.35\right)\ \text{m}^3$$

Number of expansions through the engine is

$$R = \frac{\text{Volume of low-pressure cylinder}}{\text{Cut-off volume in the high-pressure cylinder}}$$

$$= \frac{\pi \times (0.3)^2 \times 0.34/4}{455 \times \pi \times (0.2)^2 \times 0.35/(1400 \times 4)}$$

$$= \frac{(0.3)^2 \times 1400}{(0.2)^2 \times 455}$$

$$= 2.25 \times \frac{1400}{455}$$

$$= \mathbf{6.92}$$

P_m referred to the low-pressure cylinder is

$$\frac{P}{R}(1 + \ln R) - P_b = \frac{1400}{6.92}(1 + \ln 6.92) - 35$$

$$= \frac{1400}{6.92}(1 + 1.934) - 35$$

$$= \frac{1400 \times 2.934}{6.92} - 35$$

$$= 594 - 35$$

$$= \mathbf{559\ kN/m^2}$$

Actual $P_m = 0.8 \times 559 = \mathbf{447.2\ kN/m^2}$

$$\text{Indicated power} = P_m LAN = 447.2 \times 0.35 \times \frac{\pi \times (0.3)^2}{4} \times 2 \times \frac{300}{60}$$

$$= \mathbf{110.6\ kW}$$

(c)
Volume of the low-pressure cylinder is

$$\pi \times \frac{(0.3)^2}{4} \times 0.35 = \mathbf{0.024\,7\ m^3}$$

So cut-off volume in high-pressure cylinder is

$$\frac{0.024\,7}{6.92} = \mathbf{0.003\,57\ m^3}$$

This is the volume of steam admitted per stroke. Hence

Volume of steam admitted/h $= (0.003\,57 \times 300 \times 2 \times 60)\ \text{m}^3$
Specific volume of admission steam $= xv_g = 0.9 \times 0.140\,7$
$= 0.126\,6\ \text{m}^3\text{kg}$

$$\therefore \quad \text{Steam consumption} = \frac{0.003\,57 \times 300 \times 2 \times 60}{0.126\,6}$$

$$= \mathbf{1015\ kg/h}$$

Example 11.6 *A double-acting, two-cylinder, compound reciprocating steam engine has a low-pressure cylinder 600 mm diameter with a stroke of 600 mm. Steam is supplied at a pressure of 1.1 MN/m² and it exhausts from the engine at a pressure of 28 kN/m². The engine runs at 4 rev/s and has a diagram factor referred to the low-pressure cylinder of 0.82. Expansion is hyperbolic throughout, clearance is neglected and expansion in the high-pressure cylinder is complete. The total expansion ratio through the engine is 8 and equal work is developed in each cylinder. Determine*

(a) the indicated power output of the engine
(b) the diameter of the high-pressure cylinder if the stroke is the same as that of the low-pressure cylinder
(c) the intermediate pressure

SOLUTION
First draw a diagram (Fig. 11.14).

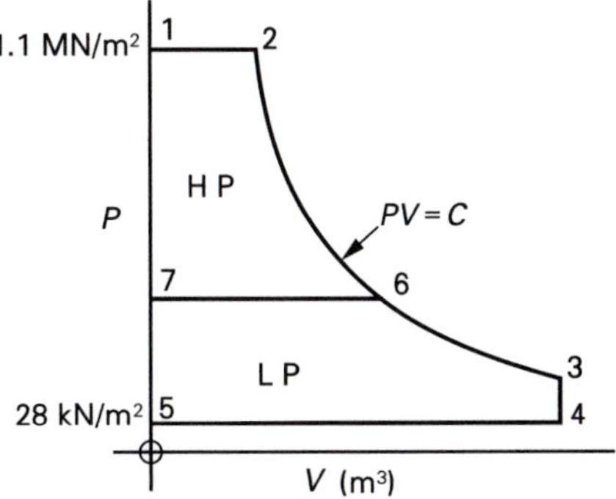

Fig. 11.14 Diagram for Example 11.6

(a)
P_m referred to low-pressure cylinder is

$$\frac{P}{R}(1+\ln R) - P_b = \frac{1100}{8}(1+\ln 8) - 28$$

$$= \frac{1100}{8}(1+2.08) - 28$$

$$= \frac{1100 \times 3.08}{8} - 28$$

$$= 424 - 28$$
$$= \mathbf{396\ kN/m^2}$$

$$\text{Actual } P_m = 0.82 \times 396 = \mathbf{325\ kN/m^2}$$

$$\therefore \quad \text{Indicated power} = P_m LAN = 325 \times 0.6 \times \frac{\pi \times (0.6)^2}{4} \times 2 \times 4$$

$$= \mathbf{441\ kW}$$

(b)
Work done by two cylinders per stroke is

$$\text{Total area of diagram} = \left(325 \times 0.6 \times \frac{\pi \times (0.6)^2}{2}\right) \text{kJ}$$

Since equal work is performed/cylinder, the work done in the high-pressure cylinder is

$$\frac{\text{Total area of diagram}}{2} = \frac{325 \times 0.6 \times \pi \times (0.6)^2}{2 \times 4}$$

$$= \mathbf{27.6\ kJ}$$

So the work done in the high-pressure cylinder is

$$\text{Area } 1267 = P_2 V_2 \ln \frac{V_6}{V_2} = \mathbf{27.6\ kJ}$$

and

$$V_2 = \frac{V_3}{8} = \frac{\pi \times (0.6)^2}{4} \times \frac{0.6}{8} = \mathbf{0.021\ 2\ m^3}$$

$$\therefore \quad 1100 \times 0.021\ 2 \ln \frac{V_2}{V_2} = 27.6$$

$$\ln \frac{V_6}{V_2} = \frac{27.6}{1100 \times 0.021\ 2}$$

$$= 1.18$$

$$\therefore \quad \frac{V_6}{V_2} = \mathbf{3.25}$$

From this

$$V_6 = 3.25\, V_2 = 3.25 \times 0.021\ 2 = \mathbf{0.069\ m^3}$$

The volume of the high-pressure cylinder is 0.069 m^3.

Let d = diameter of high-pressure cylinder in metres.

Stroke = 0.6 m

$$\text{Then } \frac{\pi d^2}{4} \times 0.6 = 0.069$$

$$d^2 = \frac{0.069 \times 4}{\pi \times 0.6} = 0.146\ 4$$

$$\therefore \quad d = \surd(0.146\ 4) = 0.383\text{ m} = \mathbf{383\ mm}$$

(c)

For expansion 2–6, $P_2V_2 = P_6V_6$ since expansion is hyperbolic.

$$\therefore \quad P_6 = \frac{P_2V_2}{V_6} = \frac{1100 \times 0.021\ 1}{0.069} = \mathbf{338\ kN/m^2}$$

The intermediate pressure is 338 kN/m^2.

Example 11.7 *A compound steam engine consists of one high-pressure cylinder and one low-pressure cylinder, both of which are double-acting. The engine develops 230 kW. The admission pressure is 1.4 MN/m² and the exhaust pressure is 35 kN/m². The total expansion ratio referred to the low-pressure cylinder is 12.5:1. The diameter of the low-pressure cylinder is 400 mm and the stroke in both cylinders is 500 mm. The diagram factor is 0.78, expansion is assumed to be hyperbolic and clearance is neglected. The expansion ratio in the high-pressure cylinder is 2.5:1. Determine*

(a) the speed of the engine in rev/s

(b) the diameter of the high-pressure cylinder

(a)
P_m referred to the low-pressure cylinder is

$$\frac{P}{R}(1+\ln R)-P_b=\frac{1400}{12.5}(1+\ln 12.5)-35$$

$$=\frac{1400}{12.5}(1+2.53)-35$$

$$=\frac{1400\times 3.53}{12.5}-35$$

$$=395-35$$

$$=\mathbf{360\ kN/m^2}$$

$$\text{Theoretical indicator power}=\frac{230}{0.78}=\mathbf{295\ kW}$$

$$\text{Indicated power}=P_mLAN$$

$$\therefore\quad 295=360\times 0.5\times\frac{\pi\times(0.4)^2}{4}\times 2\times \text{rev/s}$$

$$\therefore\quad \text{rev/s}=\frac{295\times 4}{360\times 0.5\times\pi\times(0.4)^2\times 2}=\mathbf{6.52}$$

The engine speed is 6.52 rev/s.

(b)

$$\text{Volume of low-pressure cylinder}=\frac{\pi\times(0.4)^2}{4}\times 0.5$$

$$=\mathbf{0.062\ 8\ m^2}$$

$$\therefore\quad \text{Cut-off volume in high-pressure cylinder}=\frac{0.0628}{12.5}$$

$$=\mathbf{0.005\ 02\ m^3}$$

$$\text{Total volume of high-pressure cylinder}=0.005\ 02\times 2.5$$
$$=\mathbf{0.012\ 55\ m^3}$$

If d = diameter of high-pressure cylinder in metres, then

$$\frac{\pi d^2}{4}=0.5\times 0.012\ 55$$

$$d^2=\frac{4\times 0.012\ 55}{\pi\times 0.5}=0.032$$

$$d=\sqrt{0.032}=0.179\ \text{m}=\mathbf{179\ mm}$$

Example 11.8 *A double-acting, compound steam engine has cylinders of 300 mm and 600 mm diameter and a common stroke of 400 mm. Steam is supplied at a pressure and a common stroke of 400 mm. Steam is supplied at a pressure of 1.1 MN/m^2 and cut-off in the high-pressure cylinder is at one-third stroke. The back pressure in the low-pressure cylinder is 32 kN/m^2. Average areas of indicator diagrams from the engine are HP 12.5 cm^2; indicator calibration 270 kN m^{-2}/cm; LP 11.4 cm^2; spring 80 kN m^{-2}/cm. The length of both diagrams is 75 cm. The engine speed is 2.7 rev/s. Determine*

(a) the actual and hypothetical mean effective pressures referred to the low-pressure cylinder
(b) the overall diagram factor
(c) the indicated power

(a)
For the high-pressure cylinder

$$P_{mH} = \frac{12.5}{7.5} \times 270 = \mathbf{450\ kN/m^2} \quad \text{(see Chapter 17)}$$

$$\text{Force on high-pressure piston} = P_{mH}\,A = \left(450 \times \frac{\pi \times (0.3)^2}{4}\right)\ \text{kN}$$

Since the stroke and rev/s are the same for both cylinders, then this force can be referred to the low-pressure cylinder by calculating from this force the pressure required over the low-pressure piston to produce this force.

$$\begin{aligned} P_{mH} \text{ referred to low-pressure cylinder} &= 450 \times \frac{\pi \times (0.3)^2/4}{\pi \times (0.6)^2/4} \\ &= 450 \times \frac{(0.3)^2}{(0.6)^2} \\ &= 450 \times \frac{0.09}{0.36} \\ &= \mathbf{112.5\ kN/m^2} \end{aligned}$$

For the low-pressure cylinder

$$P_{mL} = \frac{11.4}{7.5} \times 80 = \mathbf{121.6\ kN/m^2}$$

$$\text{Actual } P_{mA} \text{ referred to low-pressure cylinder} = 112.5 + 121.6 = \mathbf{234.1\ kN/m^2}$$

$$\text{Volume of high-pressure cylinder} = \left(\frac{\pi \times (0.3)^2}{4} \times 4\right)\ \text{m}^3$$

$$\text{Cut-off volume of high-pressure cylinder} = \left(\frac{\pi \times (0.3)^2}{4} \times \frac{0.4}{3}\right)\ \text{m}^3$$

$$\text{Volume of low-pressure cylinder} = \left(\frac{\pi \times (0.6)^2}{4} \times 4\right)\ \text{m}^3$$

Number of expansion through the engine is

$$R = \frac{\text{Volume of low-pressure cylinder}}{\text{Cut-off volume of high-pressure cylinder}}$$

$$= \frac{[\pi \times (0.6)^2/4] \times 0.4}{[\pi \times (0.3)^2/4] \times 0.4/3}$$

$$= \frac{(0.6)^2 \times 3}{(0.3)^2}$$

$$= \frac{0.36}{0.09} \times 3$$

$$= \mathbf{12}$$

$$\text{Hypothetical } P_m = \frac{P}{R}(1 + \ln R) - P_b$$

$$= \frac{1100}{12}(1 + \ln 12) - 32$$

$$= \frac{1100}{12}(1 + 2.48) - 32$$

$$= \left(\frac{1100}{12} \times 3.48\right) - 32$$

$$= 319 - 32$$
$$= \mathbf{287\ kN/m^2}$$

(b)

$$\text{Overall diagram factor} = \frac{P_{mA}}{P_m} = \frac{234.1}{287} = \mathbf{0.816}$$

(c)

$$\text{Indicated power} = P_{mA}LAN$$

$$= 234.1 \times 0.4 \times \frac{\pi \times (0.6)^2}{4} \times 2 \times 2.7$$

$$= \mathbf{143\ kW}$$

Example 11.9 *During a trial on a triple-expansion engine the following results were recorded.*

Cylinder	Diameter	Actual P_m (kN/m^2)
HP	300	590
IP	500	214
LP	900	88

Steam supply pressure	1.4 MN/m^2
Cut-off in HP cylinder	0.6 stroke
Back pressure in LP cylinder	20 kN/m^2

All cylinders have the same stroke. Determine
(a) the actual and hypothetical mean effective pressures referred to the low-pressure cylinder
(b) the overall diagram factor
(c) the percentage of the total indicated power developed in each cylinder

(a)
For the high-pressure cylinder

$$P_{mH} \text{ referred to LP cylinder} = 590 \times \frac{(0.3)^2}{(0.9)^2}$$

$$= 590 \times \frac{0.09}{0.81}$$

$$= \frac{590}{9}$$

$$= \mathbf{65.55\ kN/m^2}$$

For the intermediate-pressure cylinder

$$P_{mI} \text{ referred to LP cylinder} = 214 \times \frac{(0.5)^2}{(0.9)^2}$$

$$= 214 \times \frac{0.25}{0.81}$$

$$= \mathbf{66\ kN/m^2}$$

$$\therefore \quad \text{Actual } P_{mA} \text{ referred to the low-pressure cylinder} = 65.55 + 66 + 88$$
$$= \mathbf{219.55\ kN/m^2}$$

Total number of expansions through the engine is

$$\frac{900^2}{0.6 \times 300^2} = \frac{9}{0.6} = \mathbf{15}$$

Hypothetical P_m referred to the low-pressure cylinder is

$$\frac{P}{R}(1 + \ln R) - P_b = \frac{1400}{15}(1 + \ln 15) - 20$$

$$= \frac{1400}{15}(1 + 2.7) - 20$$

$$= \left(\frac{1400}{15} \times 3.7\right) - 20$$

$$= 345 - 20$$
$$= \mathbf{325\ kN/m^2}$$

(b)

$$\text{Overall diagram factor} = \frac{219.55}{325}$$

$$= \mathbf{0.675}$$

(c)

The percentage of the total indicated power developed in each cylinder is the same as the percentage analysis of the actual P_{mA} referred to the low-pressure cylinder. So the output from each cylinder is

$$\mathrm{HP} = \frac{65.55}{219.55} \times 100 = \mathbf{29.9\%}$$

$$\mathrm{IP} = \frac{66}{219.55} \times 100 = \mathbf{30.1\%}$$

$$\mathrm{LP} = \frac{88}{219.55} \times 100 = \mathbf{40\%}$$

Questions

1. A simple, double-acting, steam engine delivers a brake power of 37.5 kW at a speed of 2 rev/s. Steam is supplied at a pressure of 500 kN/m^2 and exhausts from the cylinder at a pressure of 34 kN/m^2. Cut-off occurs at one-third of the stroke. Assuming a diagram factor of 0.7 relative to the hypothetical diagram and a mechanical efficiency of 0.85, determine the swept volume of the cylinder in cubic metres.

   [0.05 m^3]

2. A double-acting steam engine is to develop an indicated power of 45.0 kW at 3 rev/s. The mean piston speed is to be 2.75 m/s, the steam supply 7 bar and the exhaust 1 bar. Taking a cut-off ratio of 0.5 and a diagram factor of 0.65, determine the bore and stroke of the engine. Neglect clearance.

   [Bore = 255 mm; stroke = 460 mm]

3. A double-acting, single-cylinder steam engine has a cylinder diameter of 230 mm and a piston stroke of 300 mm. Determine a suitable admission pressure if the engine is to develop an indicated power of 27 kW at a speed of 3.5 rev/s with a back pressure of 130 kN/m^2 and an expansion ratio of 2.5. Assume a diagram factor of 0.8.

   [674.5 kN/m^2]

4. Determine the cylinder diameter and the stroke of a single-cylinder, double-acting steam engine to develop an indicated power of 90 kW under the following conditions

Steam pressure	830 kN/m^2
Cut-off	$\frac{1}{3}$ stroke
Back pressure	100 kN/m^2
Diagram factor	0.7
Stroke	1.5 × cylinder diameter
Speed	2 rev/s

   [Diameter = 384 mm; stroke = 576 mm]

5. Determine the probable mean effective pressure of a steam engine taking steam at 14 bar and exhausting to a back pressure of 1 bar. Take a cut-off ratio of 0.4 and assume a diagram factor of 0.65.

   [6.32 bar]

6. A steam engine has a cylinder bore of 250 mm and a stroke of 300 mm. The clearance volume is 10 per cent of the swept volume. During the expansion of the steam in the engine cylinder, at 0.5 stroke the pressure is 380 kN/m^2 and at 0.9 stroke the pressure is 216 kN/m^2. The expansion is assumed to follow the law $PV^n = C$. Determine (a) the value of the index n and (b) the work done during the expansion between the two given pressures.

If the expansion had been hyperbolic from the pressure of 380 kN/m^2 at 0.5 stroke, determine (c) the work done to the position of 0.9 stroke.

[(a) 1.104; (b) 1.702 kJ; (c) 1.719 kJ]

7. A single-cylinder, double-acting steam engine is 152 mm bore by 203 mm stroke and the piston rod is 32 mm diameter. During a trial the engine develops a brake power of 13.5 kW and its mechanical efficiency is 81 per cent. The areas of the indicator diagrams are 1625 mm^2 at the head end and 1490 mm^2 at the crank end; the length is 75 mm and the indicator calibration is 16 MN m^{-2}/m in both cases. Determine the speed of the engine in rev/min. If the calorific value of the coal used is 25 100 kJ/kg and the overall efficiency from coal to brake power is 7 per cent, calculate the mass of coal used per hour.

[418 rev/min; 27.7 kg]

8. A single-cylinder, double-acting steam engine gives a brake power of 75 kW using steam at 1035 kN/m^2 with a cut-off at one-third stroke and a back pressure of 27.5 kN/m^2. The mechanical efficiency of the engine is 84 per cent and the diagram factor 0.7. If the mean piston speed is 4 m/s and the stroke is 1.2 times the bore, determine the bore and stroke of the engine.

[Bore = 388 mm; stroke = 466 mm]

9. A single-cylinder, double-acting steam engine is 250 mm bore by 300 mm stroke and runs at 3.5 rev/s. Steam is supplied at 1035 kN/m^2, the back pressure is 34 kN/m^2 and the diagram factor is 0.81. Determine the indicated power of the engine
   (a) if cut-off is at 0.25 stroke;
   (b) if cut-off is at 0.5 stroke.

[(a) 48.7 kW; (b) 70.3 kW]

10. A single-cylinder, double-acting steam engine gives an indicated power of 55 kW when running at 5 rev/s; the engine is 0.25 m bore and 0.30 m stroke. Steam is supplied at 860 kN/m^2 and the back pressure is 117 kN/m^2; cut-off is at 0.37 stroke. Determine the diagram factor.

    If a condenser is now fitted to the engine so that the back pressure is 34.5 kN/m^2, calculate the new indicated power if nothing else changes.

[0.72; 63.7 kW]

11. A throttle-governed steam engine developing an indicated power of 37.5 kW uses 1000 kg/h of steam. At no-load the steam consumption is 125 kg/h. Estimate the indicated power of the engine for a steam consumption of 750 kg/h.

[26.8 kW]

12. A two-cylinder, double-acting, compound steam engine develops an indicated power of 220 kW at a speed of 4.5 rev/s. The stroke in each cylinder is 0.5 m. Expansion is hyperbolic throughout and is complete in the high-pressure cylinder. The diagram factor referred to the low-pressure cylinder is 0.75. Steam is supplied at 12 bar and the engine exhausts at a pressure of 0.28 bar. Total expansion ratio through the engine is 10. Equal power is developed in the two cylinders. Neglecting clearance, determine the engine cylinder diameters.

[0.475 m; 0.373 m]

13. A two-cylinder, double-acting, compound steam engine is supplied with steam at a pressure of 1725 kN/m^2, the steam exhausts from the engine at a pressure of 41.5 kN/m^2. The low-pressure cylinder has a diameter of 0.45 m and a stroke of 0.4 m. The stroke in both cylinders is the same and the ratio of the cylinder volumes is 2.5:1. The diagram factor referred to the low-pressure cylinder is 0.78. Each cylinder has equal initial piston loads. The engine runs at 4.5 rev/s. Expansion in the engine is hyperbolic and the total expansion ratio through the engine is 9. Neglecting clearance, determine
    (a) the intermediate pressure

(b) the indicated power of the engine
(c) the diameter of the high-pressure cylinder

[(a) 523 kN/m^2; (b) 272 kW; (c) 0.285 m]

14. A double-acting, compound steam engine has cylinder diameters HP 300 mm, LP 600 mm and the stroke of both cylinders is 400 mm. When running at 160 rev/min the engine develops a brake power of 125 kW and its mechanical efficiency is 78 per cent. Steam is supplied at a pressure of 13.8 bar, cut-off in the HP cylinder is at one-third stroke and back pressure is 0.28 bar. Determine
    (a) the actual mean effective pressure
    (b) the hypothetical mean effective pressure
    (c) the overall diagram factor

[(a) 266 kN/m^2; (b) 373 kN/m^2; (c) 0.71)

15. A double-acting steam engine has a stroke of 0.45 m and the bore diameters are HP 0.33 m, LP 0.63 m. Steam is supplied at a pressure of 1200 kN/m^2, cut-off in the HP cylinder is at one-quarter stroke and back pressure is 21 kN/m^2. Assuming a diagram factor of 0.81 and mechanical efficiency of 82 per cent determine the brake power of the engine when running at 3 rev/s.

[157.6 kW]

16. A triple-expansion steam engine has cylinder diameters of 300 mm, 450 mm and 750 mm; the common stroke is 500 mm. Indicator diagrams taken from the cylinders give the tabulated results.

Cylinder	HP	IP	LP
Diagram area (mm^2)	1805	1875	2000
Diagram length (mm)	104	105	103
Indicator calibration (kg/mm^2/m)	40	16	5.35

The steam is supplied at a pressure of 1380 kN/m^2 and the back pressure is 28 kN/m^2; cut-off in the HP cylinder is at 0.6 stroke. Determine the actual and hypothetical mean effective pressure referred to the LP cylinder and hence find the overall diagram factor.

[317.8 kN/m^2; 415 kN/m^2; 0.765]

17. A double-acting, compound steam engine is to give an indicated power of 450 kW running at 2.5 rev/s. Both HP and LP cylinders have the same stroke and the mean piston speed is 4 m/s. Steam is supplied at 1240 kN/m^2, back pressure is 21 kN/m^2, the total number of expansions is 12 and the overall diagram factor is 0.8. Determine the common stroke and the bore of each cylinder if the bore of the LP cylinder is twice that of the HP cylinder. Determine also the cut-off point in the HP cylinder.

[0.8 m; 727 mm; 364 mm; $\frac{1}{3}$]

Chapter 12

Nozzles

12.1 General introduction

A nozzle is a device for accelerating a fluid substance to a high velocity by producing a drop in pressure of the substance.

Figure 12.1 shows two of the more general nozzle shapes. The entry area of the convergent–divergent nozzle (Fig. 12.1(a)) converges down to a minimum area called the **throat** then diverges to the exit area. The convergent nozzle (Fig. 21.1(b)) has a cross-section which converges down from the entry area to a minimum area at the exit.

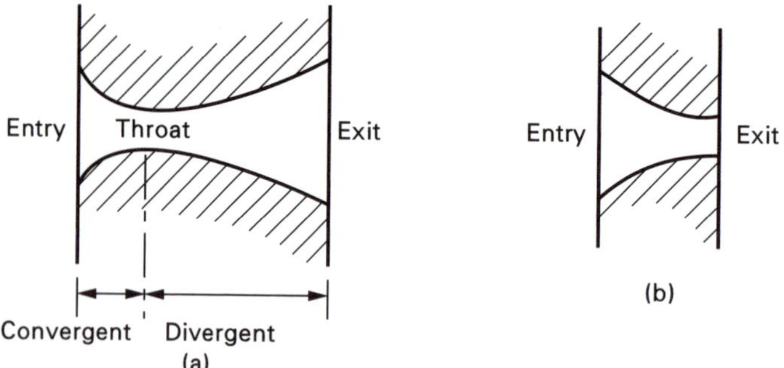

Fig. 12.1 Nozzle shapes: (a) convergent–divergent; (b) convergent

A diffuser is the reverse of a nozzle, in which the substance decelerates and the pressure is increased.

12.2 General flow analysis

Figure 12.2 shows a nozzle with entry conditions of area A_1, velocity C_1, pressure P_1, specific volume v_1, temperature T_1, specific enthalpy h_1. Exit conditions are A_2, C_2, P_2, v_2, T_2, h_2.

Neglecting change in potential energy and putting $W = 0$, as no work is done in a nozzle, the steady-flow energy equation becomes

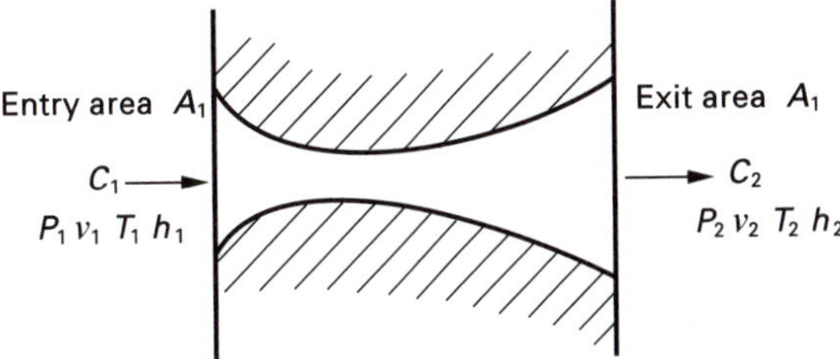

Fig. 12.2 Nozzle flow analysis

$$h_1 + \frac{C_1^2}{2} + Q = h_2 + \frac{C_2^2}{2} \qquad [1]$$

This is for unit mass flow.

Now the time taken for a substance to pass through a nozzle is very small, so there is little time for heat exchange between the substance and its surroundings. It is therefore reasonable to assume that $Q = 0$ and the flow is adiabatic. Equation [1] becomes

$$h_1 + \frac{C_1^2}{2} = h_2 + \frac{C_2^2}{2} \qquad [2]$$

from which

$$\frac{C_2^2 - C_1^2}{2} = h_1 - h_2 \qquad [3]$$

In many cases, the entry velocity to a nozzle is small compared with the exit velocity; C_1 can be sensibly neglected and equation [3] becomes

$$\frac{C^2}{2} = h_1 - h_2 \qquad [4]$$

where C = exit velocity, written C_2 above

From equation [4]

$$C = \sqrt{[2(h_1 - h_2)]} \qquad [5]$$

Now, for a gas

$$h_1 - h_2 = c_p(T_1 - T_2) = \frac{\gamma}{(\gamma - 1)} R(T_1 - T_2) \qquad [6]$$

since $c_p = \dfrac{\gamma R}{(\gamma - 1)}$

Substituting equation [6] in equation [5]

$$C = \sqrt{\left[2\frac{\gamma}{(\gamma - 1)} R(T_1 - T_2)\right]}$$

$$= \sqrt{\left[2\frac{\gamma}{(\gamma - 1)} (P_1 v_1 - P_2 v_2)\right]} \quad \text{since} \quad Pv = RT \qquad [7]$$

$$= \sqrt{\left[2\frac{\gamma}{(\gamma-1)}P_1v_1\left(1-\frac{P_2v_2}{P_1v_1}\right)\right]} \qquad [8]$$

For the adiabatic expansion of a gas

$$P_1v_1^\gamma = P_2v_2^\gamma$$

$$\therefore \quad \frac{v_2}{v_1} = \left(\frac{P_1}{P_2}\right)^{1/\gamma} \qquad [9]$$

Substituting equation [8] in equation [7]

$$C = \sqrt{\left\{2\frac{\gamma}{(\gamma-1)}P_1v_1\left[1-\frac{P_2}{P_1}\left(\frac{P_1}{P_2}\right)^{1/\gamma}\right]\right\}}$$

$$= \sqrt{\left\{2\frac{\gamma}{(\gamma-1)}P_1v_1\left[1-\frac{P_2}{P_1}\left(\frac{P_2}{P_1}\right)^{-1/\gamma}\right]\right\}}$$

$$= \sqrt{\left\{2\frac{\gamma}{(\gamma-1)}P_1v_1\left[1-\left(\frac{P_2}{P_1}\right)^{1-1/\gamma}\right]\right\}}$$

$$= \sqrt{\left\{2\frac{\gamma}{(\gamma-1)}P_1v_1\left[1-\left(\frac{P_2}{P_1}\right)^{(\gamma-1)/\gamma}\right]\right\}} \qquad [10]$$

Now, at any section of the nozzle

$$\dot{m}v = AC \quad \text{(see section 2.9)} \qquad [11]$$

where $\dot{m}$ = mass flow in kg/s
v = specific volume in m^3/kg
A = cross-sectional area in m^2
C = velocity in m/s

For continuity of flow $\dot{m}$ is constant at all sections of the nozzle.
From equation [11]

$$\frac{\dot{m}}{A} = \frac{C}{v} \qquad [12]$$

which is the mass flow per unit area at any section.

This equation shows that the mass flow $\dot{m}$ is constant at all sections of the nozzle and the velocity C and specific volume v vary through the nozzle, so the cross-sectional area A must also vary.

It also shows that if there is a maximum value of the mass flow per unit area, $\dot{m}/A$ the cross-sectional area must be a minimum because $\dot{m}$ is constant. This minimum value of A will be the throat of the nozzle.

Let,

P_t = throat nozzle
v_t = specific volume at the throat
c_t = velocity at the throat

then, from equation [10]

$$C_t = \sqrt{\left\{2\frac{\gamma}{(\gamma-1)}P_1 v_1\left[1-\left(\frac{P_t}{P_1}\right)^{(\gamma-1)/\gamma}\right]\right\}} \quad [13]$$

Also, from equation [12]

$$\frac{\dot{m}}{A_t} = \frac{C_t}{v_t} \quad [14]$$

where A_t = area at the throat

From equations [13] and [14], then

$$\frac{\dot{m}}{A_t} = \frac{1}{v_t}\sqrt{\left\{2\frac{\gamma}{(\gamma-1)}P_1 v_1\left[1-\left(\frac{P_t}{P_1}\right)^{(\gamma-1)/\gamma}\right]\right\}} \quad [15]$$

Now, for the adiabatic expansion of a gas $PV^\gamma = C$.

$$\therefore \quad P_1 v_1^\gamma = P_t v_t^\gamma$$

or

$$v_t = v_1\left(\frac{P_1}{P_t}\right)^{1/\gamma} \quad [16]$$

Substituting equation [16] in equation [15]

$$\frac{\dot{m}}{A_t} = \frac{1}{v_1\left(\frac{P_1}{P_t}\right)^{1/\gamma}}\sqrt{\left\{2\frac{\gamma}{(\gamma-1)}P_1 v_1\left[1-\left(\frac{P_t}{P_1}\right)^{(\gamma-1)/\gamma}\right]\right\}}$$

$$= \frac{1}{v_1}\left(\frac{P_t}{P_1}\right)^{1/\gamma}\sqrt{\left\{2\frac{\gamma}{(\gamma-1)}P_1 v_1\left[1-\left(\frac{P_t}{P_1}\right)^{(\gamma-1)/\gamma}\right]\right\}}$$

$$= \frac{1}{v_1}\sqrt{\left\{2\frac{\gamma}{(\gamma-1)}P_1 v_1\left(\frac{P_t}{P_1}\right)^{2/\gamma}\left[1-\left(\frac{P_t}{P_1}\right)^{(\gamma-1)/\gamma}\right]\right\}}$$

$$= \frac{1}{v_1}\sqrt{\left\{2\frac{\gamma}{(\gamma-1)}P_1 v_1\left[\left(\frac{P_t}{P_1}\right)^{2/\gamma}-\left(\frac{P_t}{P_1}\right)^{(\gamma+1)/\gamma}\right]\right\}}$$

For maximum $\dot{m}/A_t$, the value of

$$\left[\left(\frac{P_t}{P_1}\right)^{2/\gamma}-\left(\frac{P_t}{P_1}\right)^{(\gamma+1)/\gamma}\right]$$

must be a maximum. (γ, P_1 and v_1 are constants in this equation).

Let $P_t/P_1 = x$

Then for maximum $\dot{m}/A_t$

$$\frac{d}{dx}\left[x^{2/\gamma} - x^{(\gamma+1)/\gamma}\right] = 0$$

$$\frac{2}{\gamma}x^{(2/\gamma)-1} - \frac{(\gamma+1)}{\gamma}x^{[(\gamma+1)/\gamma]-1} = 0$$

$$\frac{2}{\gamma}x^{(2-\gamma)/\gamma} = \frac{(\gamma+1)}{\gamma}x^{1/\gamma}$$

$$\frac{x^{1/\gamma}}{x^{(2-\gamma)/\gamma}} = \frac{2}{(\gamma+1)}$$

$$x^{1/\gamma-[(2-\gamma)/\gamma]} = \frac{2}{(\gamma+1)}$$

$$x^{(\gamma-1)/\gamma} = \frac{2}{(\gamma+1)}$$

$$\therefore \quad x = \left(\frac{2}{\gamma+1}\right)^{\gamma/(\gamma-1)}$$

or

$$\frac{P_t}{P_1} = \left(\frac{2}{\gamma+1}\right)^{\gamma/(\gamma-1)} \qquad [17]$$

$\dfrac{P_t}{P_1}$ is called the **critical pressure ratio**

From equation [17]

$$P_t = \left(\frac{2}{\gamma+1}\right)^{\gamma/(\gamma-1)} P_1 \qquad P_t \text{ is the } \textbf{critical pressure} \qquad [18]$$

For air, the average value of $\gamma = 1.4$

$$\therefore \quad P_t = \left(\frac{2}{1.4+1}\right)^{1.4/(1.4-1)} P_1$$

$$= \left(\frac{2}{2.4}\right)^{1.4/0.4} P_1$$

$$= \frac{1}{1.2^{3.5}} P_1$$

$$= 0.528P_1 \qquad [19]$$

The critical temperature T_t may be obtained from the critical temperature ratio T_t/T_1.

$$\frac{T_t}{T_1} = \left(\frac{P_t}{P_1}\right)^{(\gamma-1)/\gamma} = \left[\left(\frac{2}{\gamma+1}\right)^{\gamma/(\gamma-1)}\right]^{(\gamma-1)/\gamma} = \frac{2}{\gamma+1}$$

$$\therefore \quad T_t = \left(\frac{2}{\gamma+1}\right)T_1 \qquad [20]$$

Note that from equation [5]

$$
\begin{aligned}
C_t &= \sqrt{[2(h_1 - h_t)]} \\
&= \sqrt{[2c_p(T_1 - T_t)]} \\
&= \sqrt{\left[2\frac{\gamma}{(\gamma-1)}R(T_1 - T_t)\right]} \\
&= \sqrt{\left[2\frac{\gamma}{(\gamma-1)}RT_t\left(\frac{T_1}{T_t} - 1\right)\right]} \\
&= \sqrt{\left[2\frac{\gamma}{(\gamma-1)}RT_t\left(\frac{\gamma-1}{2} - 1\right)\right]} \\
&= \sqrt{\left[2\frac{\gamma}{(\gamma-1)}RT_t\left(\frac{\gamma-1}{2}\right)\right]} \\
&= \sqrt{(\gamma R T_t)} \qquad [21]
\end{aligned}
$$

Now this can be shown to be the velocity of sound.

It was first established by Newton that the velocity of sound in a substance is given by

$$C = \sqrt{\left(\frac{\text{modulus of elasticity}}{\text{density}}\right)} \qquad [22]$$

Sound is transmitted through a substance by a succession of compressions and rarefactions of the substance through which it is passing. These changes in pressure will be very rapid, so they may be considered as being adiabatic.

For a gas, an adiabatic compression or expansion may be considered as following the law $PV^{\gamma} = \text{constant}$. Consider a gas at pressure P and volume V and let there be a small change in pressure δP which will be accompanied by a small change in volume $-\delta V$ (V varies inversely with P). For this is small change

$$PV^{\gamma} = (P + \delta P)(V - \delta V)^{\gamma} \qquad [23]$$

$$\text{Now } (V - \delta V)^{\gamma} = \frac{V^{\gamma}}{V^{\gamma}}(V - \delta V)^{\gamma}$$

$$= V^{\gamma}\left(\frac{V - \delta V}{V}\right)^{\gamma} = V^{\gamma}\left(1 - \frac{\delta V}{V}\right)^{\gamma} \qquad [24]$$

Substituting equation [24] in equation [23]

$$PV^{\gamma} = (P + \delta P)^{\gamma}\left(1 - \frac{\delta V}{V}\right)^{\gamma}$$

from which

$$P = (P + \delta P)\left(1 - \frac{\delta V}{V}\right)^{\gamma} \qquad [25]$$

Expanding $(1 - \delta V/V)^{\gamma}$ by the binomial theorem

$$\left(1 - \frac{\delta V}{V}\right)^{\gamma} = 1 - \gamma\frac{\delta V}{V} + \frac{\gamma(\gamma-1)}{1.2}\frac{\delta V^2}{V} - \ldots$$

and substituting into equation [25], neglecting higher powers of $\delta V/V$ as small

$$P = (P + \delta P)\left(1 - \gamma\frac{\delta V}{V}\right)$$
$$= (P + \delta P) - (P + \delta P)\gamma\frac{\delta V}{V}$$

from which

$$\gamma(P + \delta P) = [(P + \delta P) - P]\frac{V}{\delta V} = \frac{\delta P}{\delta V}V \quad [26]$$

As $\delta P \to 0$

$$\gamma P = \frac{dP}{dV}V = K \quad [27]$$

where K = the bulk modulus or modulus of elasticity

Substituting equation [27] in equation [22]

$$\text{Velocity of sound, } C = \sqrt{\left(\frac{\gamma P}{\rho}\right)} \quad [28]$$

where ρ = density = $1/v$
v = specific volume

$$\therefore \quad C = \sqrt{(\gamma P v)} = \sqrt{(\gamma R T)} \quad [29]$$

Comparing equation [29] with equation [21] shows that, at the throat of a nozzle, the velocity is the velocity of sound, or as it is sometimes called, **sonic velocity**. Sonic velocity is said to be Mach 1, sometimes written $Ma = 1$. If the velocity of a substance is 1.5 times the velocity of sound, it is said to be at Mach 1.5 or $Ma = 1.5$.

From the investigation into critical pressure, P_t, if $P_2 < P_t$, then the nozzle will be convergent–divergent, as shown in Fig. 12.1(a). From entry to the throat, the velocity will be subsonic ($Ma < 1$). At the throat, the velocity will be sonic ($Ma = 1$). From the throat to exit, the velocity will be supersonic ($Ma > 1$).

If $P_2 > P_t$, the nozzle will be convergent only, as shown in Fig. 12.1(b). Velocities will all be subsonic from inlet to exit ($Ma < 1$).

If sonic velocity is reached in a nozzle, the maximum mass flow per unit area has been reached for the nozzle (at the throat). This also means that the maximum mass flow rate through the nozzle has also been reached. The nozzle is said to be **choked**.

The temperature through the nozzle may be estimated using the equation

$$T_2 = T_1\left(\frac{P_2}{P_1}\right)^{\gamma - 1/\gamma} \quad [30]$$

Figure 12.3 plots pressure against distance along a convergent–divergent nozzle.

Curve ABC is obtained when the correct design exit pressure is applied to the nozzle. At B, the throat, the pressure is the critical pressure. But if the pressure is off-design, the exit pressures at E, G and J, which are above the design exit pressure, produce a condition of overexpansion. This is because pressures are produced in the nozzle which are below the exit pressure.

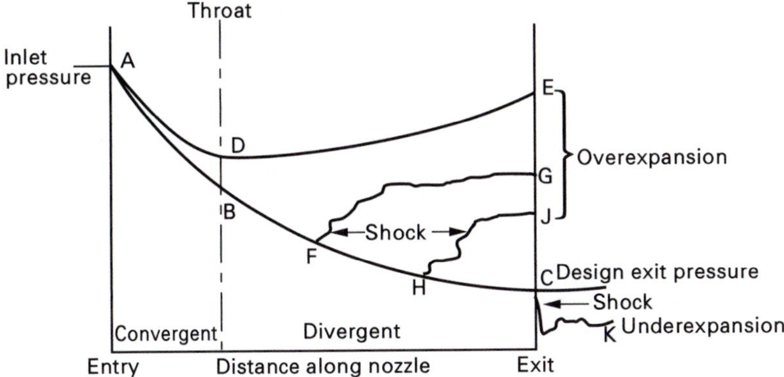

Fig. 12.3 Convergent–divergent nozzle: pressure plot

Curve ADE is a smooth curve obtained when exit pressure is somewhat above critical pressure; this would occur in a venturi tube. However, with curves ABFG and ABHJ, where the exit pressure is such that the expansion is through the critical pressure, the velocities become supersonic at F and H then decelerate from F to G and H to J. This produces an unstable shock condition as the pressure is increased or diffused. Note that this is in the divergent portion of the nozzle, which is where an acceleration should normally occur.

If an exit pressure is applied to the nozzle which is below design exit pressure, as at K, then a normal expansion ABC will occur through the nozzle followed by a shock reduction in pressure CK just outside the nozzle.

12.3 Steam flow through nozzles

The approach to the problem of steam flow through a nozzle will depend upon whether the steam flow can be considered as being **in equilibrium** or **supersaturated**. Figure 12.4(a) shows an enthalpy–entropy plot of an equilibrium expansion of steam through a nozzle from inlet pressure P_1 to exit pressure P_2. At point a the steam is superheated and, assuming frictionless adiabatic expansion in equilibrium through the nozzle, the steam will be just dry saturated at point b and will be wet at point c.

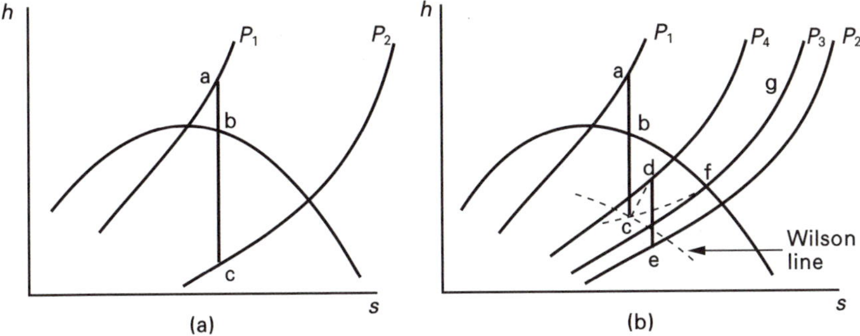

Fig. 12.4 *h–s* diagrams for steam flow through a nozzle: (a) equilibrium; (b) supersaturated

Now, the rate of expansion through a nozzle is very high and there is little time for phase change during the passage of the steam through the nozzle. So, considering Fig. 12.4(b), the steam is once again superheated at inlet pressure P_1 at point a. At point b, due to the rapid expansion, condensation does not commence and the steam continues to behave as a superheated vapour down to point c at some intermediate pressure P_3, less than the critical pressure. Point c is on the continuous curve gfc. Remember that a constant pressure line on the enthalpy–entropy chart is curved for superheated steam and straight for wet steam.

This non-equilibrium behaviour as a superheated vapour does not continue indefinitely; the restoration of equilibrium quickly occurs at point c, after the throat in the divergent portion of the nozzle. It is accompanied by a small increase of pressure to P_4. This is shown as cd, which also shows a small increase of entropy. From point d, equilibrium expansion occurs down to exit pressure P_2 at point e.

The steam during this type of expansion, in which the phase change is delayed, is said to be **supersaturated**. It is also said to be **supercooled** because at point c the steam temperature is lower than the corresponding saturation temperature at P_3. This type of expansion is not in equilibrium so it is also said to be **metastable**. An equilibrium expansion is **stable**.

Slightly higher flow rates are found during supersaturated expansion as against equilibrium expansion. This is because of the higher densities that occur due to supercooling.

It was first shown that there is a limit to supersaturated expansion by C.T. Wilson in 1897. This limit approximates to points where the dryness fraction is 0.94 at high pressures to 0.96 at low pressures. The locus of these points produces the **Wilson line**, as shown in Fig. 12.4(b).

12.4 Supersaturated expansion

During the supersaturated expansion of steam there is a continuity of phase up to the point where equilibrium conditions become established. Until the change to equilibrium conditions, the steam can be considered as a single phase, so it can be analysed rather like a gas.

The expansion will still be theoretically adiabatic but the index γ cannot be used because $\gamma = c_p/c_v$ and is for a gas only. For superheated steam, the adiabatic index has an average value, $n = 1.3$. For wet steam, the adiabatic index has an average value, $n = 1.135$.

Using the equations developed in section 12.2

$$C = \surd[2(h_1 - h_2)] \qquad [1]$$

$$= \sqrt{\left\{2\frac{n}{(n-1)}P_1 v_1\left[1 - \left(\frac{P_2}{P_1}\right)^{(n-1)/n}\right]\right\}} \qquad [2]$$

$$\dot{m}v = AC \qquad [3]$$

Critical pressure for superheated steam

$$P_t = \left(\frac{2}{n+1}\right)^{n/(n-1)} P_1$$

$$= \left(\frac{2}{1.3+1}\right)^{1.3/(1.3-1)} P_1$$

$$= \left(\frac{2}{2.3}\right)^{1.3/0.3} P_1$$

$$= \mathbf{0.546}\, \boldsymbol{P_1} \qquad [4]$$

Critical pressure for wet steam

$$P_t = \left(\frac{2}{1.135+1}\right)^{1.135/(1.135-1)} P_1$$

$$= \left(\frac{2}{2.135}\right)^{1.135/0.135} P_1$$

$$= \mathbf{0.577}\, \boldsymbol{P_1} \qquad [5]$$

The temperature through the nozzle during supersaturated expansion may be estimated using

$$T_2 = T_1 \left(\frac{P_2}{P_1}\right)^{(n-1)/n} \qquad [6]$$

12.5 Equilibrium expansion

Figure 12.5 is an enthalpy–entropy plot of equilibrium steam flow through a nozzle from inlet pressure P_1 to exit pressure P_2. Throat pressure is P_t.

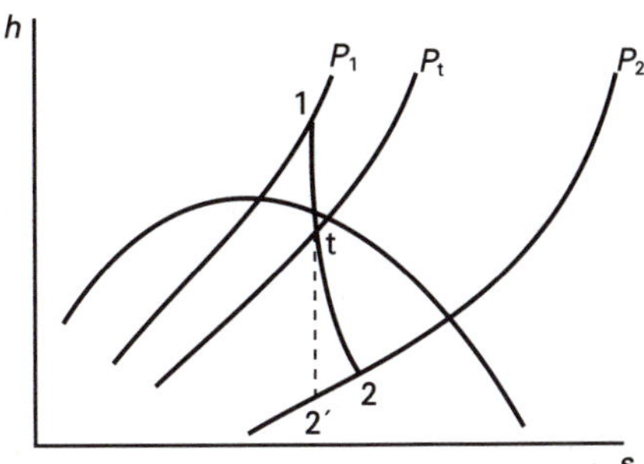

Fig. 12.5 Equilibrium steam flow through a nozzle: *h–s* diagram

From equation [5], section 12.2,

$$C = \sqrt{[2(h_1 - h_2)]} \qquad [1]$$

and if expansion is in equilibrium, this equation can be used to determine velocities through a nozzle. Values of specific enthalpy can be obtained either by calculation using steam tables or by readings taken from the enthalpy–entropy chart.

The throat pressure can be estimated by using equation [18], section 12.2, which is written as

$$P_t = \left(\frac{2}{n+1}\right)^{n/(n-1)} P_1 \qquad [2]$$

Since steam is being considered here, the adiabatic index is written n because γ has the value c_p/c_v and is for a gas only.

As illustrated in Fig. 12.5 the expansion from inlet to throat is mostly superheated. The index n for superheated steam has an average value of 1.3.

Equation [2] becomes

$$P_t = \left(\frac{2}{1.3+1}\right)^{1.3/(1.3-1)} P_1$$

$$= \left(\frac{2}{2.3}\right)^{1.3/0.3} P_1$$

$$= \mathbf{0.546}\, \boldsymbol{P_1} \qquad [3]$$

This is the critical pressure for superheated steam.

If the steam is wet from inlet to throat, the index n becomes 1.135 as an average value.

Equation [2] becomes

$$P_t = \left(\frac{2}{1.135+1}\right)^{1.135/(1.135-1)} P_1$$

$$= \left(\frac{2}{2.135}\right)^{1.135/0.135} P_1$$

$$= \mathbf{0.577}\, \boldsymbol{P_1} \qquad [4]$$

This is the critical pressure for wet steam.

Figure 12.1(a) illustrates a convergent–divergent nozzle, and the convergent portion of the nozzle is shown as being shorter than the divergent portion. In the divergent portion the velocity is higher than the convergent portion. The walls of the nozzle in the high-velocity portion diverge, so there is the possibility of breakaway of the steam (or gas) if the angle of the wall is too great. The divergent portion is therefore made long enough to prevent this. With the longer divergent portion, however, there is greater potential for friction than in the convergent portion. It is thus convenient to consider that expansion in the convergent portion is frictionless, as shown in Fig. 12.5 from inlet to throat. But from throat to exit, the effect of friction is internally to reheat the steam (or gas) as it were. Thus the steam (or gas) leaves with a higher specific enthalpy at 2 than if the steam expansion had been frictionless at 2′.

The ratio of the actual specific enthalpy drop to the frictionless specific enthalpy drop is called the **efficiency of expansion**. Referring to Fig. 12.5 gives

$$\text{Efficiency of expansion} = \frac{h_t - h_2}{h_t - h_{2'}} \qquad [5]$$

Also, for continuity of flow

$$\dot{m}V = AC \qquad [6]$$

Example 12.1 *Air enters a nozzle at a pressure of 3.5 MN/m² and at a temperature of 500 °C. It leaves at a pressure of 0.7 MN/m². The flow rate of air through the nozzle is 1.3 kg/s; expansion may be considered to be adiabatic and to follow the law* PV$^{\gamma}$ = *constant. Determine*
(a) the throat area
(b) the exit area
(c) the Mach number at exit
Take, $\gamma = 1.4$ *and* R = *0.287 kJ/kg K.*

(a)

$$v_1 = \frac{RT_1}{P_1} = \frac{0.287 \times 773}{3500}$$

$$= \mathbf{0.063\,4\ m^3/kg}$$

$$\text{Critical pressure, } P_t = \left(\frac{2}{\gamma + 1}\right)^{\gamma/(\gamma - 1)} P_1$$

$$= \left(\frac{2}{1.4 + 1}\right)^{1.4/0.4} P_1$$

$$= 0.528\, P_1$$
$$= 0.528 \times 3.5$$
$$= \mathbf{1.85\ MN/m^2}$$

The throat pressure is 1.86 MN/m².
Velocity at throat is

$$C_t = \sqrt{\left\{2 \frac{\gamma}{(\gamma - 1)} P_1 v_1 \left[1 - \left(\frac{P_2}{P_1}\right)^{(\gamma - 1)/\gamma}\right]\right\}}$$

$$= \sqrt{\left\{2 \times \frac{1.4}{0.4} \times 3.5 \times 10^6 \times 0.063\,4 \times \left[1 - \left(\frac{1.85}{3.5}\right)^{0.4/1.4}\right]\right\}}$$

$$= \sqrt{\left\{2 \times 3.5 \times 3.5 \times 10^6 \times 0.063\,4 \times \left[1 - \frac{1}{1.89^{1/3.5}}\right]\right\}}$$

$$= \sqrt{\left\{2 \times 3.5 \times 3.5 \times 10^6 \times 0.063\,4 \times \left[1 - \frac{1}{1.2}\right]\right\}}$$

$$= \sqrt{\{2 \times 3.5 \times 3.5 \times 10^6 \times 0.063\,4 \times (1 - 0.833)\}}$$
$$= \sqrt{\{2 \times 3.5 \times 3.5 \times 0.063\,4 \times 0.167 \times 10^6\}}$$
$$= \sqrt{\{0.259 \times 10^6\}}$$
$$= 0.508 \times 10^3$$
$$= \mathbf{508\ m/s}$$

$$\frac{\dot{m}}{A_t} = \frac{C_t}{v_t}$$

$$v_t = v_1\left(\frac{P_1}{P_t}\right)^{1/\gamma} = 0.063\,4 \times \left(\frac{3.5}{1.85}\right)^{1/1.4}$$

$$= 0.063\,4 \times 1.89^{1/1.4}$$
$$= 0.063\,4 \times 1.576$$
$$= 0.099\,9\ \mathrm{m^3/kg}$$

$$A_t = \frac{\dot{m}V_t}{C_t} = \frac{1.3 \times 0.999}{508}$$

$$= 0.000\,256 \text{ m}^2$$

$$= \mathbf{256\ mm^2}$$

The throat area is 256 mm^2.

(b)

At exit

$$C_2 = \sqrt{\left\{2 \times \frac{1.4}{0.4} \times 3.5 \times 10^6 \times 0.063\,4 \times \left[1 - \left(\frac{0.7}{3.5}\right)^{0.4/1.4}\right]\right\}}$$

$$= \sqrt{\left\{2 \times 3.5 \times 3.5 \times 10^6 \times 0.063\,4 \times \left[1 - \frac{1}{5^{1/3.5}}\right]\right\}}$$

$$= \sqrt{\left\{2 \times 3.5 \times 3.5 \times 10^6 \times 0.063\,4 \times \left[1 - \frac{1}{1.584}\right]\right\}}$$

$$= \sqrt{\{2 \times 3.5 \times 3.5 \times 10^6 \times 0.063\,4 \times (1 - 0.63)\}}$$

$$= \sqrt{\{2 \times 3.5 \times 3.5 \times 0.063\,4 \times 0.37 \times 10^6\}}$$

$$= \sqrt{\{0.574 \times 10^6\}}$$

$$= 0.76 \times 10^3$$

$$= \mathbf{760\ m/s}$$

$$v_2 = v_1 \left(\frac{P_1}{P_2}\right)^{1/\gamma} = 0.063\,4 \times \left(\frac{3.5}{0.7}\right)^{1/1.4}$$

$$= 0.063\,4 \times 5^{1/1.4}$$

$$= 0.063\,4 \times 3.16$$

$$= \mathbf{0.2\ m^3/kg}$$

$$A_2 = \frac{\dot{m}v_2}{C_2} = \frac{1.3 \times 0.2}{760}$$

$$= 0.000\,342 \text{ m}^2$$

$$= \mathbf{342\ mm^2}$$

The exit area is 342 mm^2.

(c)

$$\text{Mach number at exit} = \frac{C_2}{C_t} = \frac{760}{509}$$

$$= \mathbf{1.49}$$

Example 12.2 *Air flowing through a diverging tube, or diffuser, has at one section a temperature of 0 °C, pressure 140 kN/m^2 and velocity 900 m/s. A little further along the tube the velocity has fallen to 300 m/s. Assuming a frictionless adiabatic flow, what are the increases in pressure, temperature and internal energy per kilogram of air?*

Take $c_p = 1.006$ *kJ/kg K;* $C_v = 0.717$ *kJ/kg K.*

Solution

For frictionless adiabatic flow

$$\frac{C_2^2 - C_1^2}{2} = \frac{\gamma}{(\gamma - 1)} R(T_1 - T_2)$$

Now

$$\gamma = c_p - c_v = 1.006/0.717 = 1.4$$

Also

$$R = c_p - c_v = 1.006 - 0.717 = \mathbf{0.289\ kJ/kg\ K}$$

$$\therefore \quad \frac{300^2 - 900^2}{2} = \frac{1.4}{(1.4 - 1)} \times 0.289 \times (273 - T_2) \times 10^3$$

$$\left(\frac{9 - 81}{2}\right) \times 10^4 = 3.5 \times 0.289\,(273 - T_2) \times 10^3$$

$$(273 - T_2) = \frac{-36 \times 10^4}{3.5 \times 0.289 \times 10^3}$$

$$T_2 = 273 + \frac{36 \times 10^4}{3.5 \times 0.289 \times 10^3}$$

$$= 273 + \frac{360}{3.5 \times 0.289}$$

$$= 273 + 356$$
$$= \mathbf{629\ K}$$
$$t_2 = 629 - 273$$
$$= \mathbf{356\ ^\circ C}$$

Increase in temperature is $356 - 0 =$ **356 K**

For adiabatic flow

$$P_2 = P_1 \left(\frac{T_2}{T_1}\right)^{\gamma/(\gamma - 1)}$$

$$= 140 \times \left(\frac{629}{273}\right)^{1.4/0.4}$$

$$= 140 \times 2.3^{3.5}$$
$$= 140 \times 18.5$$
$$= 2590\ \text{kN/m}^2$$
$$= \mathbf{2.59\ MN/m^2}$$

$$\therefore \quad \text{Increase in pressure} = 2.59 - 0.14 = \mathbf{2.45\ MN/m^2}$$

$$\text{Increase in internal energy} = c_v(T_2 - T_1)$$
$$= 0.717 \times 356$$
$$= \mathbf{255\ kJ/kg}$$

Example 12.3 *A group of convergent–divergent nozzles are supplied with steam at a pressure of 2 MN/m² and a temperature of 325 °C. Supersaturated expansion according to the law* $PV^{1.3}$ *= constant occurs in the nozzle down to an exit pressure of 0.36 MN/m². Steam is supplied at the rate of 7.5 kg/s. Determine*

(a) the required throat and exit areas

(b) the degree of undercooling at exit

(a)

From tables, $v_1 = \mathbf{0.132\ m^3/kg}$

$$P_t = \left(\frac{2}{n+1}\right)^{n/(n-1)} P_1$$

$$= \left(\frac{2}{1.3+1}\right)^{1.3/(1.3-1)} P_1$$

$$= 0.546\, P_1$$

But $P_1 = 2\ \text{MN/m}^2$

$$\therefore \quad P_t = 0.546 \times 2 = \mathbf{1.092\ MN/m^2}$$

$$C_t = \sqrt{\left\{2\frac{n}{(n-1)}P_1 v_1\left[1-\left(\frac{P_t}{P_1}\right)^{(n-1)/n}\right]\right\}}$$

$$= \sqrt{\left\{2 \times \frac{1.3}{(1.3-1)} \times 2 \times 10^6 \times 0.132 \times \left[1-\left(\frac{1.092}{2}\right)^{(1.3-1)/1.3}\right]\right\}}$$

$$= \sqrt{\left\{2 \times 4.33 \times 2 \times 10^6 \times 0.132 \times \left[1-\frac{1}{1.83^{1/4.33}}\right]\right\}}$$

$$= \sqrt{\left\{2 \times 4.33 \times 2 \times 10^6 \times 0.132 \times \left[1-\frac{1}{1.15}\right]\right\}}$$

$$= \sqrt{\{2 \times 4.33 \times 2 \times 10^6 \times 0.132 \times (1-0.87)\}}$$
$$= \sqrt{\{2 \times 4.33 \times 2 \times 0.132 \times 0.13 \times 10^6\}}$$
$$= \sqrt{\{0.297 \times 10^6\}}$$
$$= 0.545 \times 10^3$$
$$= \mathbf{545\ m/s}$$

$$v_t = v_1\left(\frac{P_1}{P_t}\right)^{1/n} = 0.132 \times \left(\frac{2}{1.092}\right)^{1/1.3}$$

$$= 0.132 \times 1.83^{1/1.3}$$
$$= 0.132 \times 1.592$$
$$= \mathbf{0.21\ m^3/kg}$$

$$A_t = \frac{\dot{m}v_t}{C_t} = \frac{7.5 \times 0.21}{545}$$

$$= 0.002\,89\ \text{m}^2$$
$$= \mathbf{2890\ mm^2}$$

The throat area is 2890 mm².

At exit

$$C_2 = \sqrt{\left\{2 \times \frac{1.3}{(1.3-1)} \times 2 \times 10^6 \times 0.132 \times \left[1 - \left(\frac{0.36}{2}\right)^{(1.3-1)/1.3}\right]\right\}}$$

$$= \sqrt{\left\{2 \times 4.33 \times 2 \times 10^6 \times 0.132 \times \left[1 - \frac{1}{5.56^{1/4.33}}\right]\right\}}$$

$$= \sqrt{\left\{2 \times 4.33 \times 2 \times 10^6 \times 0.132 \times \left[1 - \frac{1}{1.486}\right]\right\}}$$

$$= \sqrt{\{2 \times 4.33 \times 2 \times 10^6 \times 0.132 \times (1 - 0.673)\}}$$
$$= \sqrt{\{2 \times 4.33 \times 2 \times 0.132 \times 0.327 \times 10^6\}}$$
$$= \sqrt{\{0.748 \times 10^6\}}$$
$$= 0.865 \times 10^3$$
$$= \mathbf{865\ m/s}$$

$$v_2 = v_1 \left(\frac{P_1}{P_t}\right)^{1/n} = 0.132 \times \left(\frac{2}{0.36}\right)^{1/1.3}$$

$$= 0.132 \times 5.56^{1/1.3}$$
$$= 0.132 \times 3.74$$
$$= \mathbf{0.494\ m^3/kg}$$

$$A_2 = \frac{\dot{m} v_2}{C_2} = \frac{7.5 \times 0.494}{865}$$

$$= 0.004\ 28\ \text{m}^2$$
$$= \mathbf{4280\ mm^2}$$

The exit area is 4280 mm^2.

(b)

$$T_2 = T_1 \left(\frac{P_2}{P_1}\right)^{(n-1)/n} = 598 \times \left(\frac{0.36}{2}\right)^{(1.3-1)/1.3}$$

$$= 598 \times 0.673$$
$$= \mathbf{402\ K}$$

$$t_2 = 402 - 273 = \mathbf{129\ °C}$$

At 0.36 MN/m^2, saturation temperature = **139.9 °C**

$\therefore$ Degree of undercooling at exit $= 139.9 - 129$
$= \mathbf{10.9\ K}$

Example 12.4 *Steam enters a group of convergent–divergent nozzles at a pressure of 2.2 MN/m² and with a temperature of 260 °C. Equilibrium expansion through the nozzles is to an exit pressure of 0.4 MN/m². Up to the throat of the nozzles the flow can be considered as frictionless. But from throat to exit there is an efficiency of expansion of 85 per cent. The rate of steam flow through the nozzles is 11 kg/s. Using the enthalpy–entropy chart for steam (Fig. 12.6), determine*

(a) the throat and exit velocities

(b) the throat and exit areas

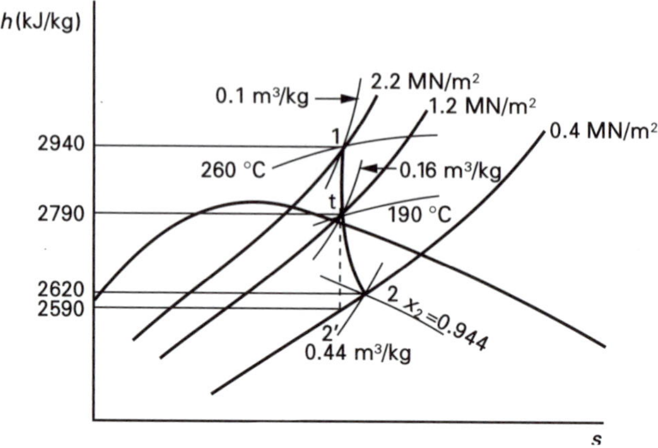

Fig. 12.6 Diagram for Example 12.4

(a)

$P_t = 0.546 \times 2.2 = \mathbf{1.2\ MN/m^2}$

From Fig. 12.6

$$h_1 = 2940 \text{ kJ/kg}$$
$$h_t = 2790 \text{ kJ/kg}$$

$$\begin{aligned} C_t &= \surd[2(h_1 - h_t)] \\ &= \surd\left[2 \times (2940 - 2790) \times 10^3\right] \\ &= \surd[2 \times 150 \times 10^3] \\ &= \surd[300 \times 10^3] \\ &= \surd[0.3 \times 10^6] \\ &= 0.548 \times 10^3 \\ &= \mathbf{548\ m^3} \end{aligned}$$

The throat velocity is 548 m/s.

From Fig. 12.6.

$$h_t = 2790 \text{ kJ/kg}$$
$$h_{2'} = 2590 \text{ kJ/kg}$$

$$0.85 = \frac{h_t - h_2}{h_t - h_{2'}}$$

$$\begin{aligned} \therefore \quad h_t - h_2 &= 0.85(h_t - h_{2'}) \\ &= 0.85 \times (2790 - 2590) \\ &= 0.85 \times 200 \\ &= \mathbf{170\ kJ/kg} \\ \therefore \quad h_2 &= 2790 - 170 \\ &= \mathbf{2620\ kJ/kg} \\ C_2 &= \surd[2(h_1 - h_2)] \\ &= \surd[2 \times (2940 - 2620) \times 10^3] \\ &= \surd[2 \times 320 \times 10^3] = \surd[640 \times 10^3] \\ &= \surd[0.640 \times 10^6] \\ &= 0.8 \times 10^3 \\ &= \mathbf{800\ m/s} \end{aligned}$$

The exit velocity is 800 m/s.

(b)
$\dot{m}v = AC$, therefore at the throat

$$A_t = \frac{\dot{m}v_t}{c_t}$$

From the chart, $v_t = 0.16\ \text{m}^3/\text{kg}$

$$\therefore \quad A_t = \frac{11 \times 0.16}{548} = 0.003\ 21\ \text{m}^2$$

$$= \mathbf{3210\ mm^2}$$

The throat area is 3210 mm^2. At exit

$$A_2 = \frac{\dot{m}v_2}{c_2}$$

From the chart, $v_2 = 0.44\ \text{m}^3/\text{kg}$

$$\therefore \quad A_2 = \frac{11 \times 0.44}{800} = 0.006\ 05\ \text{m}^2$$

$$= \mathbf{6050\ mm^2}$$

The exit area is 6050 mm^2.

Questions

1. Air enters a nozzle at a pressure of 3 MN/m^2 and with a temperature of 400 °C. It leaves at a pressure of 0.5 MN/m^2. The exit area is 5000 mm^2. Expansion through a nozzle is adiabatic according to the law PV^{γ} = constant. Determine
   (a) the mass flow through the nozzle
   (b) the throat area
   (c) the Mach number at exit
   Take $\gamma = 1.4$ and $R = 0.287$ kJ/kg K.
   [(a) 15.86 kg.s; (b) 3390 mm^2; (c) 1.55]
2. Air enters a nozzle with a pressure of 700 kN/m^2 and with a temperature of 180 °C. Exit pressure is 100 kN/m^2. The law connecting pressure and specific volume during the expansion in the nozzle is $PV^{1.3}$ = constant. Determine the velocity at exit from the nozzle. Take $c_p = 1.006$ kJ/kg K, $c_v = 0.717$ kJ/kg K.
   [576 m/s]
3. A convergent–divergent nozzle is supplied with steam at a pressure of 1.0 MN/m^2 and temperature 225 °C. Supersaturated expansion, according to the law $PV^{1.3}$ = constant, occurs in the nozzle down to an exit pressure of 0.32 MN/m^2. The exit diameter of the nozzle is 25 mm. If the flow through the nozzle is choked, determine
   (a) the exit velocity
   (b) the mass flow rate
   (c) the throat diameter
   [(a) 663 m/s; (b) 0.62 kg/s; (c) 23.6 mm]

4. Steam enters a group of convergent–divergent nozzles at a pressure of 3 MN/m^2 and a temperature of 300 °C. Equilibrium expansion occurs through the nozzles to an exit pressure of 0.5 MN/m^2. The exit velocity is 800 m/s. The steam flows at a rate of 14 kg/s. It is assumed that friction loss occurs in the divergent portion of the nozzles only. Using the enthalpy–entropy chart for steam, determine
   (a) the efficiency of expansion in the divergent portion of the nozzle
   (b) the total exit area
   (c) the throat velocity

   [(a) 0.8; (b) 6125 mm^2; (c) 529 m/s]

Chapter 13

Steam turbines

13.1 General introduction

If high-velocity steam is blown on to a curved blade, as shown in Fig. 13.1, the steam direction will be changed as it passes across the blade and it will leave as illustrated. As a result of its change of direction across the blade, the steam will impart a force to the blade. This force will be in the direction shown. Now if the blade were free, it would move off in the direction of the force. If a number of blades were fixed round the circumference of a disc and the disc were free to rotate on a shaft, steam blown across the blades, as illustrated, would cause the disc to rotate. This is the principle of the steam turbine.

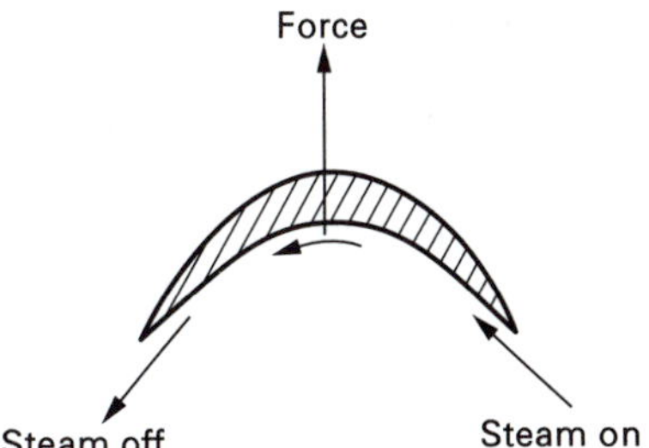

Fig. 13.1 Steam flow across a curved blade

Figure 13.2 illustrates the idea of the steam turbine. The blades are set round the circumference of the turbine disc. The tops of the blades are connected together, for rigidity, by means of the blade shroud ring. The turbine disc is free to rotate on a shaft. Set to the side of the blades, and at an angle to them, are steam nozzles. Using the nozzles, the high-pressure steam is made to give up some of its energy to produce a large increase in kinetic energy of the steam. The steam thus leaves the nozzles at a high velocity. It passes from the nozzles over the blades and thus the turbine disc rotates. Power can then be taken from the shaft. In practice, the turbine disc and the nozzles are fitted into a casing. The number of nozzles in use will be a function of the load on the turbine. The higher the load, the more steam must be used to sustain it. Thus more nozzles are put into service.

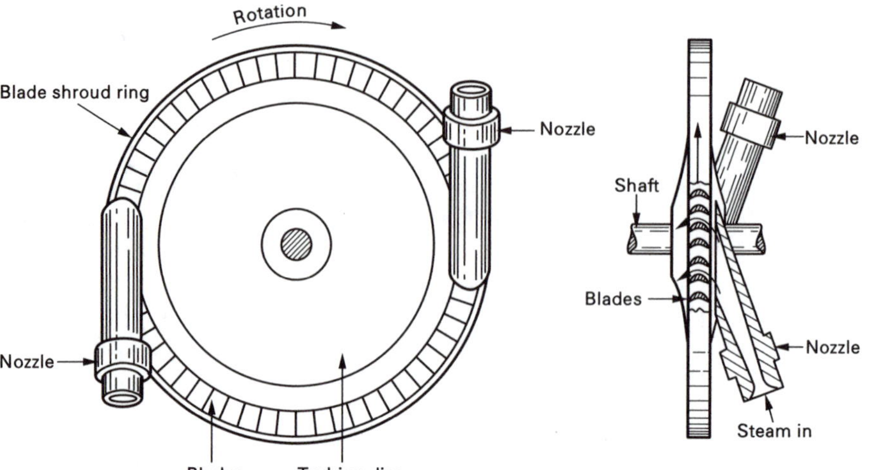

Fig. 13.2 Steam turbine: basic concept

The turbine just described is a simple turbine and was one of the first to be developed. It is called a **de Laval turbine** after its inventor.

This type of turbine usually rotates at a very high speed, some 300 to 400 rev/s. This high speed of rotation will restrict the size of the turbine disc for mechanical reasons such as centrifugal force. Thus, the de Laval turbine is of relatively small size, so it has a small power output. Also, due to the high speed of rotation, a direct drive between the turbine disc and external equipment is not generally possible. For this reason, a reduction gearbox is installed between the turbine disc and the external equipment.

A problem in steam turbine development has been to reduce the speed of rotation and at the same time to make full use of the energy in the steam, thus allowing the production of turbines of large size and high power output. Work in this direction has produced many turbine designs, but broadly they can be split into two basic types: **impulse turbines** and **reaction turbines**.

13.2 The impulse turbine

The simple de Laval turbine is an impulse turbine. The impulse turbine has two principal characteristics: it requires nozzles and the pressure drop of the steam takes place in the nozzles. The steam enters the turbine with a high velocity; the pressure in the turbine remains constant because the whole of the pressure drop has taken place in the nozzles. And the velocity of the steam is reduced as some of the kinetic energy in the steam is used up in producing work on the turbine shaft.

If the whole pressure drop from boiler to condenser pressure takes place in a single row of nozzles, as in the de Laval turbine, then the steam velocity entering the turbine is very high. If some of this velocity is used up in a single row of turbine blading, as in the de Laval turbine, then the speed of rotation is very high. In the impulse turbine this speed may be reduced by three techniques: velocity compounding, pressure compounding and pressure–velocity compounding.

13.3 Velocity compounding

In velocity compounding (Fig. 13.3(a)) the steam is expanded in a single row of nozzles, as before. The high-velocity steam leaving the nozzles passes on to the first row of moving blades where its velocity is only partially reduced. The steam leaving the first row of moving blades passes into a row of fixed blades which are mounted in the turbine casing. This row of fixed blades serves to redirect the steam back to the direction of motion such that it is correct for entry into a second row of moving blades which are mounted on the same turbine disc as the first row of moving blades. The steam velocity is again partially reduced in the second row of moving blades. These processes are shown in Fig. 13.3(b). Graphs of pressure and velocity through the turbine are included. Once again, all the pressure drop occurs in the nozzles; the pressure in the turbine remains constant. Only part of the velocity of the steam is used up in each row of blades, so a slower turbine results. But there is no loss of output because the rows of blades are connected to the same shaft. This turbine is sometimes called a **Curtis turbine**; it is quite common in the high-pressure stage of a large turbine. If necessary, further rows of fixed and moving blades may be added.

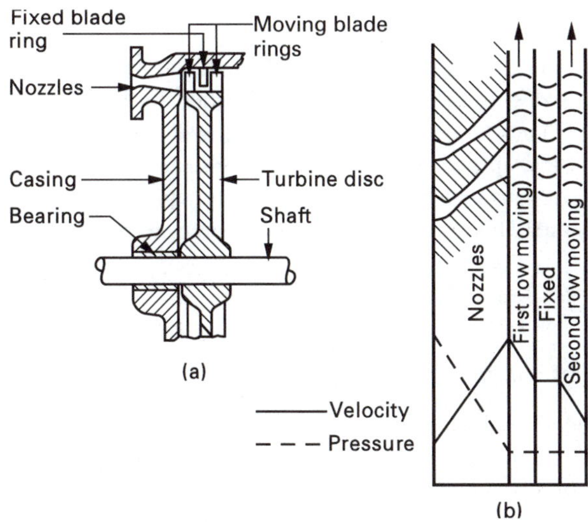

Fig. 13.3 Velocity compounding: (a) turbine and (b) graphs

13.4 Pressure compounding

In pressure compounding (Fig. 13.4(a)) the steam enters a row of nozzles where its pressure is only partially reduced and its velocity is increased. The high-velocity steam passes from the nozzles on to a row of moving blades where its velocity is reduced.

The steam then passes into a second row of nozzles where its pressure is again partially reduced and its velocity is again increased. This high-velocity steam passes from the nozzles on to a second row of moving blades where its velocity is again reduced. The steam then passes into a third row of nozzles and so on.

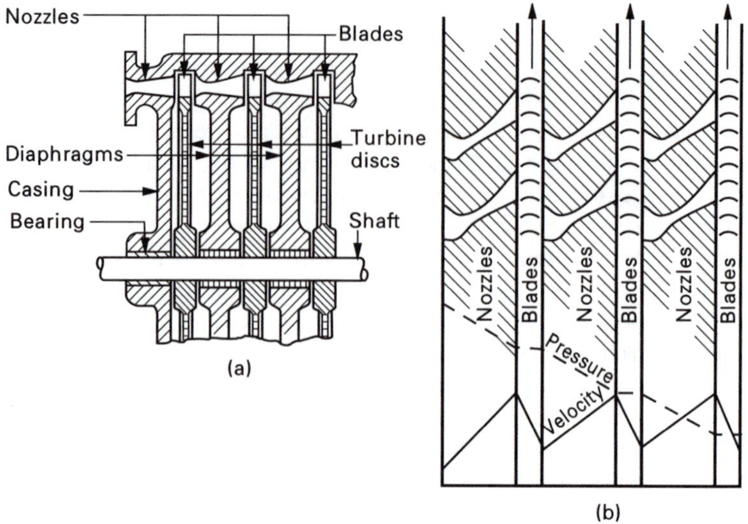

Fig. 13.4 Pressure compounding: (a) turbine and (b) graphs

Three stages of such a turbine are shown in Fig. 13.4(b). Once again, all the pressure drop occurs in the nozzles; the pressure remains constant in each turbine stage.

Since only part of the pressure drop occurs in each stage, the steam velocities will not be as high, so the turbine will run slower. But all stages are coupled to the same shaft, so there is no loss of output. This type of turbine is sometimes referred to as a **Rateau turbine**.

13.5 Pressure–velocity compounding

Pressure–velocity compounding (Fig. 13.5(a)) combines the techniques of pressure compounding and velocity compounding.

Steam is partially expanded in a row of nozzles where its velocity is increased. The steam then enters a few rows of velocity compounding (two rows are illustrated). From this stage the steam enters a second row of nozzles where its velocity is again increased. This is followed by another few rows of velocity compounding and so on. The processes are illustrated in Fig. 13.5(b). Once again, all the pressure drop takes place in the nozzles.

The turbine discs are shown with increasing diameters because all multi-stage turbines will generally increase in diameter from inlet to exhaust. The reason for this is as follows. As the pressure of steam falls, the specific volume increases. For continuity of mass flow, a greater area will be required to pass the steam. This can be accommodated either by increasing the diameter of the turbine discs or by increasing the height of the blades. Increasing the height of the turbine blades will ultimately be limited by their strength; eventually a disc diameter increase will be necessary.

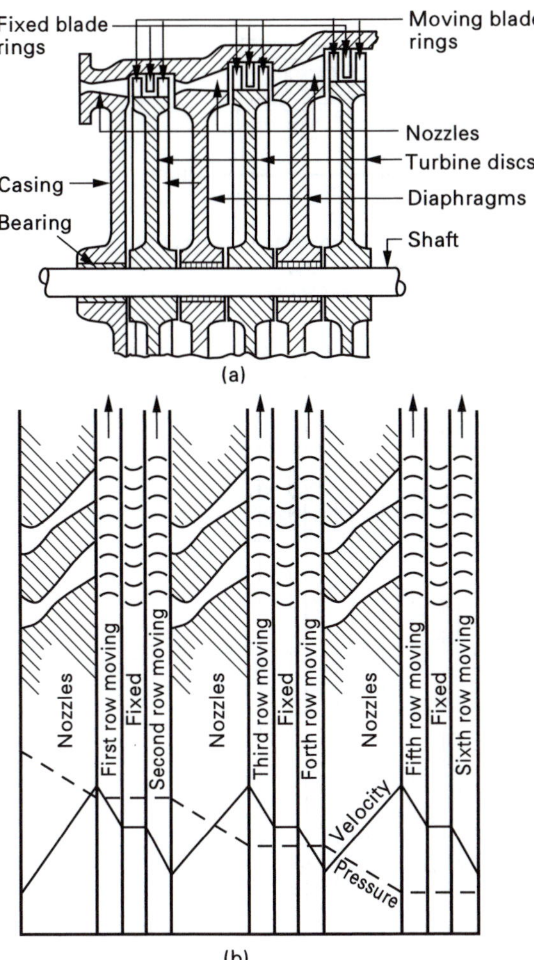

Fig. 13.5 Pressure–velocity compounding: (a) turbine and (b) graphs

A further point which will contribute to the increase in diameter is the velocity of the steam. If there is a general depreciation of velocity through the turbine then, once again, a greater area will be required to pass the steam in order to preserve the mass flow.

13.6 The reaction turbine

The construction of a reaction turbine (Fig. 13.6(a)) is somewhat different from that of the impulse turbine. Essentially the reaction turbine consists of rows of blades mounted on a drum. These drum blades are separated by rows of fixed blades mounted in the casing.

Unlike the impulse turbine, the reaction turbine has no nozzles as such. The fixed

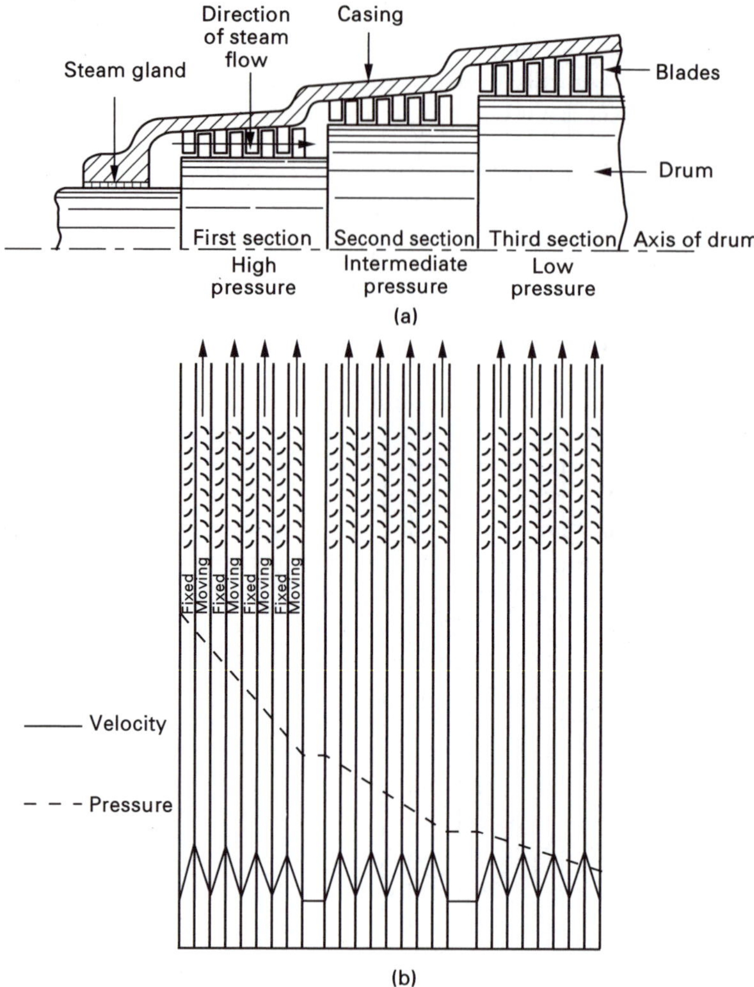

Fig. 13.6 Reaction turbine: (a) turbine and (b) graphs

blades act both as nozzles in which the velocity of the steam is increased and also as the means by which the steam is correctly directed onto the moving blades.

And unlike the impulse turbine, the steam in the reaction turbine enters the whole blade annulus, a condition called **full admission**.

The steam also expands in the moving blades of a reaction turbine with consequent pressure drop and velocity increase. This expansion in the moving blades of a reaction turbine gives an extra reaction to the moving blades, beyond that obtainable in an impulse turbine, other things being equal. It gives the turbine its name, the reaction turbine.

A characteristic of the reaction turbine is that the pressure drop occurs continuously through the turbine. This is unlike the impulse turbine, where the pressure drop takes place in the nozzles only, not in the turbine.

Changes in pressure and velocity through a reaction turbine are illustrated in Fig. 13.6(b). Three sections are shown. Each section increases in diameter as the pressure of the steam decreases, mainly due to increase in specific volume. The steam velocity in a reaction turbine is not very high, so the speed of the turbine is relatively low.

In a reaction turbine, a stage is made up of a row of fixed blades followed by a row of moving blades. Steam acceleration usually occurs in rows of fixed blades and rows of moving blades, so the steam passages between blades, both fixed and moving, are nozzle-shaped. Therefore there is an enthalpy drop in the steam during its passage through the blades; this produces the acceleration. The extent to which the enthalpy drop occurs in the moving blades is called the **degree of reaction**.

A common arrangement is to have 50 per cent of the enthalpy drop occurring in the moving blades, so the stage is said to have 50 per cent reaction. In the extreme cases, if no enthalpy drop occurs in the moving blades, it must all have occurred in the fixed blades, which is the necessary condition for an impulse turbine. Furthermore, if all the enthalpy drop occurred in the moving blades, the turbine would have 100 per cent reaction. Section 13.13 will illustrate the effect of the acceleration of the steam in the moving blade row.

A further point to note is that in the low-pressure sections of a reaction turbine the steam volume becomes very large. This greatly increased volume of steam becomes difficult to handle by increasing the blade height or drum diameter in a single turbine section. On large turbines, therefore, the low-pressure section is often made **double-flow**. In this case, the steam enters the centre of the section and divides to flow in opposite directions along the shaft axis. This is illustrated in Fig. 13.7.

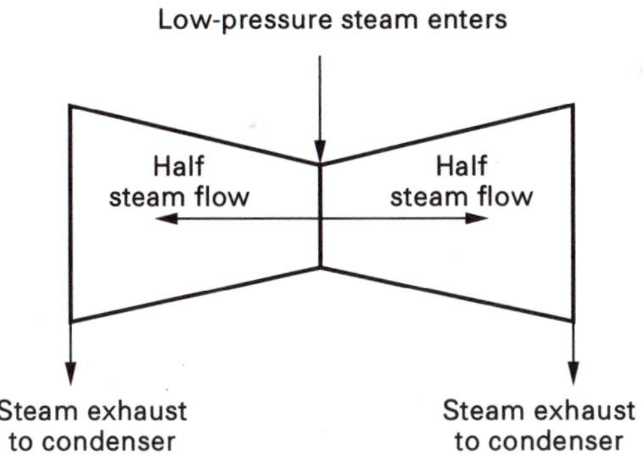

Fig. 13.7 Double-flow low-pressure turbine

In this way each half-section of the low-pressure turbine deals with only half the steam; its general dimensions are thereby reduced. This method also helps to balance end-thrusts which may appear along the turbine shaft. End-thrusts will occur along the turbine shaft due to pressure differences across the reaction turbine blading.

13.7 The practical steam turbine

Figure 13.8 shows two sections, or cylinders, of an actual steam turbine assembly. The top half of the casing is removed to show the blade assembly lying in the bottom casing. The two sections are the high-pressure section (smaller diameter) and the intermediate-pressure section (larger diameter). The low-pressure section (not illustrated) would attach to the large flange shown.

Fig. 13.8 Steam turbine assembly

The complete turbine, manufactured in this case by GEC Alsthom Power Generation, has an output of 135 MW with steam inlet conditions of 11.0 MN/m^2 and 540 °C. Steam turbines of smaller and larger outputs are also made. Rotational speed is usually of the order of 1500 rev/min or 3000 rev/min.

13.8 Velocity diagram for impulse turbine blade

Figure 13.9 shows a section of an impulse turbine blade. Inlet and exit velocity triangles are also included. The analysis of these velocity triangles is as follows. Consider the inlet triangle first. Recall that the impulse turbine uses a nozzle to produce a high velocity in the steam and also to direct this high velocity steam onto the turbine blades.

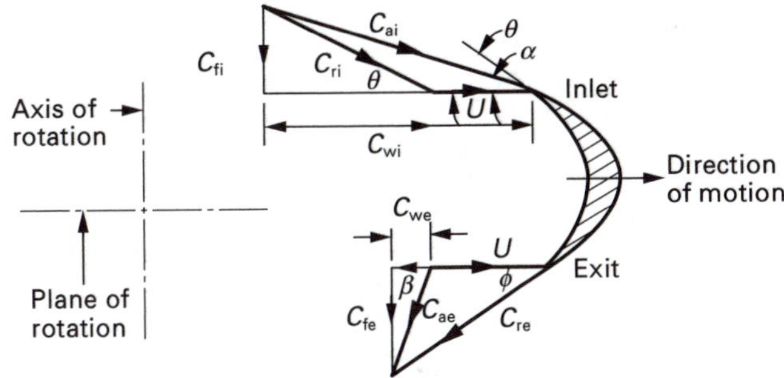

Fig. 13.9 Impulse turbine blade and velocity triangles

The velocity of the steam as it leaves the nozzle is called the **absolute velocity at inlet**, C_{ai}. The nozzle is inclined at an angle α to the plane of rotation of the turbine blades.

In operation, the turbine blades will be rotating. Let the mean blade speed be U. The mean blade speed will be the speed at the mean height of the blades.

Since the steam has velocity C_{ai} at angle α and the blade is moving with velocity U, the velocity of the steam relative to the blade, C_{ri}, will be obtained by compounding these two velocities as shown in the inlet velocity triangle. C_{ri} will be inclined to the plane of rotation of the blades at angle θ. If the steam is to enter the blades without shock, angle θ must be the inlet angle of the blades.

The component of C_{ai} in the plane of rotation of the blades, C_{wi}, is called the **velocity of whirl at inlet**. The component of C_{ai} which is along the axis of rotation, C_{fi}, is called the **velocity of flow at inlet**.

Now consider the exit conditions. The steam at inlet will have an ongoing velocity C_{ri} relative to the blade. As the steam passes over the blade, its direction will be changed and it will leave the blade with an exit relative velocity of C_{re} at angle ϕ to the plane of rotation, which is the exit angle of the blade. But the blade will be moving with mean blade speed U.

The steam at exit will therefore have two component velocities, C_{re} and U. The **absolute velocity at exit**, C_{ae}, will be obtained by compounding these two velocities as shown in the exit velocity triangle. C_{ae} will be at angle β to the plane of rotation. The component of C_{ae} in the plane of rotation, C_{we}, is called the **velocity of whirl at exit**. The component of C_{ae} which is along the axis of rotation, C_{fe}, is called the **velocity of flow at exit**.

13.9 The combination of inlet and exit velocity triangles

The mean blade speed, U, is common to both the inlet and the exit velocity triangles, so they can be combined into a single diagram as shown in Fig. 13.10. The exit velocity triangle remains as originally shown. The inlet triangle is drawn here as the other half of a parallelogram of velocities. The original triangle is shown dotted.

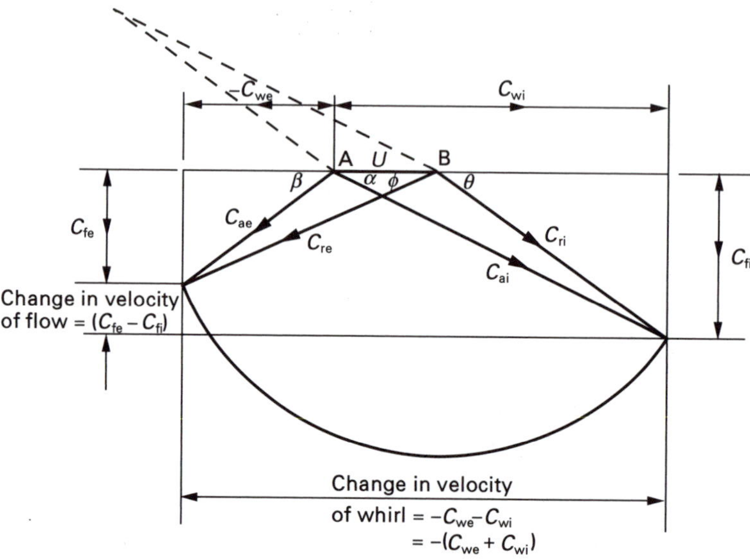

Fig. 13.10 Combination of inlet and exit velocity triangles

Note that both absolute velocities start from A. Both relative velocities start from B. If there is no friction in the blade then $C_{ri} = C_{re}$ and the extremities of both relative velocities lie on the arc of a circle, centre B, as shown.

13.10 Work done by the blades

The components of the absolute velocities in the direction of motion of the blades are the effective parts of the velocities in producing motion in the blades. The components concerned are the velocities of whirl.

Now, by Newton's laws

$$\begin{aligned} \text{Force} &= \text{range of change of momentum} \\ &= \text{mass} \times \text{change of velocity} \end{aligned} \qquad [1]$$

The change of velocity which produces the force on the blades in the plane of rotation is the change in velocity of whirl.

$$\begin{aligned} \text{Change in velocity of whirl} &= -C_{we} - C_{wi} \\ &= -(C_{we} + C_{wi}) \\ &= -(C_{wi} + C_{we}) \end{aligned} \qquad [2]$$

If $\dot{m}$ is the mass of steam flowing through the blades in kg/s, then

$$\text{Force to change the velocity of whirl from } C_{wi} \text{ to } -C_{we} = \dot{m} \times [-(C_{wi} + C_{we})] \quad [3]$$

The negative sign shows that this force acts in a direction opposite to the rotation.

By Newton's third law, to every action there is an equal and opposite reaction, the force in newtons imparted to the blades in the direction of rotation is given by

$$F = \dot{m}(C_{wi} + C_{we}) \quad [4]$$

when

$\dot{m}$ is measured in kg/s
C_{wi} and C_{we} are measured in m/s

so that

$$F = \frac{\text{kg}}{\text{s}}\frac{\text{m}}{\text{s}} = \frac{\text{kg m}}{\text{s}^2} = \text{N}$$

Now the mean velocity of the blades is U m/s and

Work done = force × distance

Hence

$$\text{Work done/s} = \text{power} = \dot{m}U(C_{wi} + C_{we})\frac{\text{N m}}{\text{s}} = \frac{\text{J}}{\text{s}} = \text{W} \quad [5]$$

In the expressions for work done there appears the term $(C_{wi} + C_{we})$. This shows the advantage of constructing a combined diagram as illustrated in Fig. 13.9. It shows that the overall length of the diagram is in fact $(C_{wi} + C_{we})$.

Now the kinetic energy of the steam supplied is

$$\frac{C_{ai}^2}{2} \text{ J/kg steam/s (W/kg steam)} \quad [6]$$

From this, the blade or diagram efficiency is

$$\frac{\text{Work done by blade/kg steam}}{\text{Energy supplied/kg steam}} = \frac{U(C_{wi} + C_{we})}{C_{ai}^2/2}$$

$$= \frac{2U(C_{wi} + C_{we})}{C_{ai}^2} \quad [7]$$

The velocity diagram includes the velocities of flow at inlet and exit. The velocity of flow is that velocity which passes the steam across the blades. If there is no change in velocity of flow across the blades, there is no end-thrust along the turbine shaft. But if the velocity of flow changes, there is an end-thrust along the shaft. The direction of this end-thrust will be a function of whether C_{fe} is greater than, or less than, C_{fi}. Generally, C_{fe} is less than C_{fi}.

Now, once again

$$\begin{aligned} \text{Force} &= \text{mass} \times \text{change of velocity} \\ &= \dot{m}(C_{fe} - C_{fi}) \\ &= \text{end-thrust} \end{aligned} \quad [8]$$

Note that $(C_{fe} + C_{fi})$ can be obtained directly, by measurement from the velocity diagram.

If $C_{fe} < C_{fi}$, the force is negative and, by Newton's third law, the end-thrust is along the turbine shaft in the direction of the velocity of flow. If $C_{fe} > C_{fi}$, the end-thrust is in the opposite direction to the velocity of flow.

13.11 The effect of friction

As the steam passes over the turbine blades, there will be friction between the steam and the blades. This friction will reduce the relative velocity with respect to the blades. Thus C_{re} will be less than C_{ri}. The loss is commonly expressed as a percentage loss of relative velocity.

To construct the velocity diagram, first construct the inlet triangle as before. The percentage loss due to friction is cut off C_{ri} to give a vector of length BC, shown in Fig. 13.11. With centre B swing round arc BD. From B draw a line at angle ϕ, the exit angle of the blade, to cut the arc in D. Then $BD = C_{re}$. Join AD which represents C_{ae}. ABD is the exit velocity triangle. The exit velocity triangle, neglecting friction, is shown dotted.

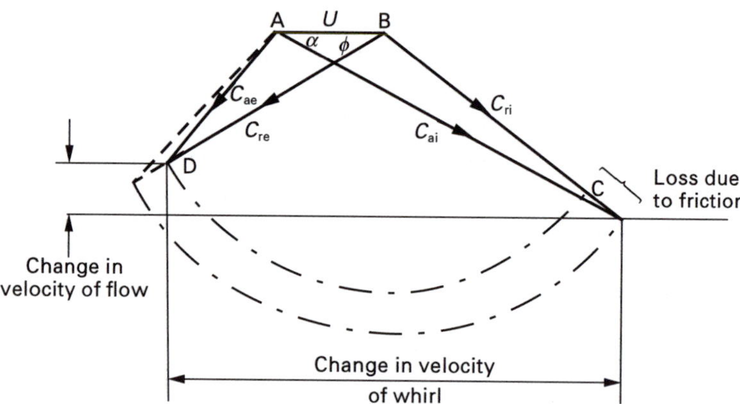

Fig. 13.11 Friction: effect on turbine blade

Note how friction has reduced the change in the velocity of whirl, and hence reduced the power output.

Example 13.1 *The nozzles of a simple impulse turbine are inclined at an angle of 20 ° to the direction of the path of the moving blades; the steam leaves the nozzles at 375 m/s and the blade speed is 165 m/s. Find suitable inlet and outlet angles for the blades in order that there shall be no axial thrust on the blades, allowing for the velocity of the steam in passing over the blades being reduced by 15 per cent.*

Determine the power developed for a steam flow of 1 kg/s at the blades and determine the kinetic energy of the steam finally leaving the wheel.

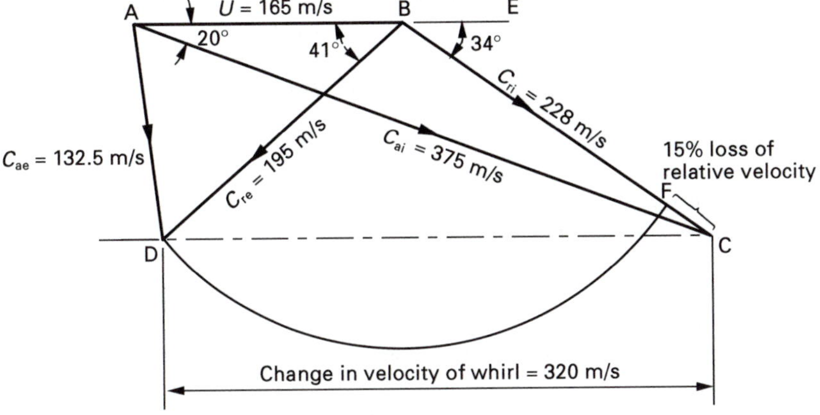

Fig. 13.12 Diagram for Example 13.1

SOLUTION

The construction of the velocity diagram (Fig. 13.12) is as follows. Set off AB to represent to some suitable scale the blade speed, $U = 165$ m/s. From A, and at 20° to AB, lay off a line AC, to correct scale, which represents the absolute velocity of the steam as it leaves the nozzles, $C_{ai} = 375$ m/s. Join BC to complete the inlet triangle. BC represents the relative velocity of the steam at inlet to the blade, C_{ri}. From the diagram $V_{ri} = 228$ m/s. In its passage over the blade, 15 per cent of this velocity is lost. Mark point F on BC such that the length FC represents 15 per cent of BC. With centre B, swing an arc FD having radius BF. In order that there shall be no axial thrust on the blade, there must be no change of velocity of flow. This means that the peaks of both the inlet and the outlet triangles must be at the same level. From C, draw horizontal line CD to cut the arc FD in D. Join BD and AD. Then triangle ABD will be the outlet triangle. BD represents the relative velocity at exit, C_{re}. AD represents the absolute velocity at exit, $C_{ae} = 132.5$ m/s.

From the velocity diagram

$$\text{Blade inlet angle} = \angle\,\text{EBC} = 34^\circ$$
$$\text{Blade outlet angle} = \angle\,\text{ABD} = 41^\circ$$

Power developed for a steam flow of 1 kg/s is

$$U \times (\text{change in velocity of whirl}) = 165 \times 320$$
$$= 52\,800 \text{ W}$$
$$= \mathbf{52.8\ kW}$$

Kinetic energy of the steam finally leaving the wheel is

$$\frac{C_{ae}^2}{2} = \frac{132.5^2}{2}$$
$$= 8778 \text{ J/kg s}$$
$$= \mathbf{8.778\ kW/kg}$$

Example 13.2 *Steam with a velocity of 600 m/s enters an impulse turbine row of blades at an angle of 25° to the plane of rotation of the blades. The mean blade speed is 255 m/s. The exit angle from the blades is 30°. There is a 10 per cent loss in relative velocity due to friction in the blades. Determine*

(a) the entry angle of the blades

(b) the work done per kilogram of steam per second

(c) the diagram efficiency

(d) the end-thrust per kilogram of steam per second

SOLUTION

Set off AB equal to $U = 255$ m/s using a suitable scale. From A, set off AC $= C_{ai} = 600$ m/s to the same scale and at angle 25° to AB. Join BC which is C_{ri}. Measure BC and cut off length CE which is 10 per cent of BC. With centre B and radius BE swing arc ED. Set off BD at 30° to AB to cut arc ED in D. Join AD. ABC is the inlet triangle. ABD is the exit triangle. That completes the velocity diagram (Fig. 13.13).

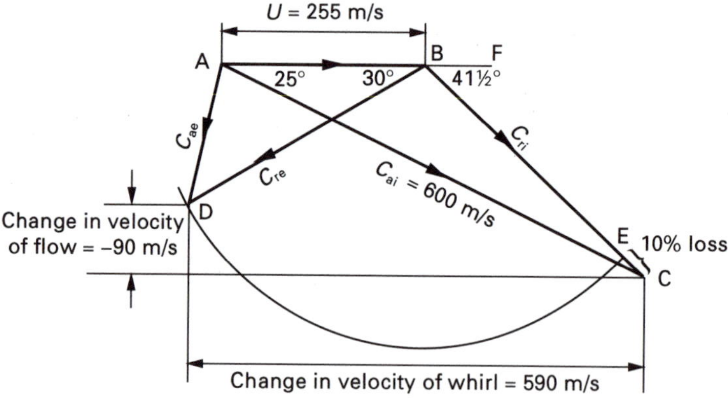

Fig. 13.13 Diagram for Example 13.2

(a)

$$\text{Entry angle of the blades} = \angle \text{FBC} = 41\tfrac{1}{2}^\circ$$

(b)

$$\text{Work done on the blades} = U(C_{wi} + C_{we}) \text{ W/kg}$$

From the diagram the change in velocity of whirl is $(C_{wi} + C_{we}) = 590$ m/s

$$\therefore \quad \text{Work done on the blade} = 255 \times 590 = 150\,450 \text{ W/kg} = \mathbf{150.45\ kW/g}$$

(c)

$$\text{Diagram efficiency} = \frac{2U(C_{wi} + C_{we})}{C_{ai}^{2}}$$

$$= \frac{2 \times 255 \times 590}{600^2}$$

$$= 0.836$$

$$= \mathbf{83.6\%}$$

(d)

$$\text{End-thrust} = (C_{fe} - C_{fi}) \text{ N/kg s}$$

From Fig. 13.13 the change in velocity of flow is

$$(C_{fe} - C_{fi}) = -90 \text{ m/s}$$

$$\therefore \quad \text{End-thrust} = \mathbf{-90 \text{ N/kg s}}$$

The negative sign shows that the end-thrust is along the shaft in the direction of steam flow.

13.12 Velocity diagram for two rows of blades

Figure 13.14 shows the velocity diagram for two rows of impulse blades having the same mean blade speed U. Subscript 1 indicates the first row of blades and subscript 2 indicates the second row of blades. Between these two rows of moving blades must be a fixed row of blades in order that the steam can be turned to the correct direction for entry into the second row of blades. The velocity diagram for the first row of blades is just as before. The absolute velocity at exit from the first row of blades is C_{ae1} at angle β_1, which is the entry angle of the fixed blade. With centre A and taking due account of loss due to friction, an arc is drawn to cut the absolute velocity at inlet to the second row of blades, C_{ai2}, which is at angle α_2, the exit angle of the fixed blade. From then on the construction for the second row of blades proceeds as before.

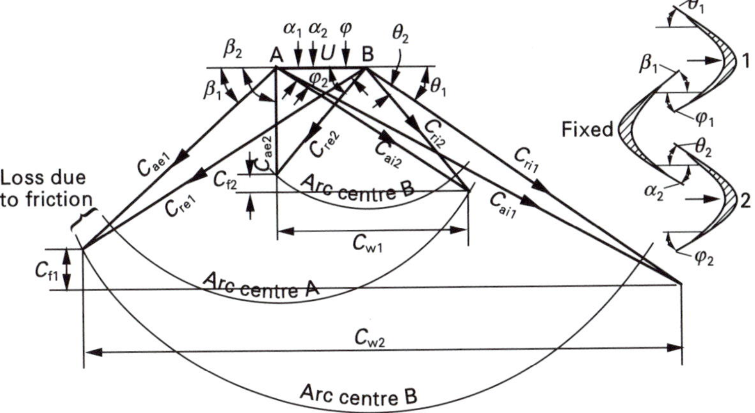

Fig. 13.14 Two rows of blades: velocity diagram

From equation [5] of section 13.10

$$\text{Power for the two rows of blades} = \dot{m}UC_{w1} + \dot{m}UC_{w2} = \dot{m}U(C_{w1} + C_{w2}) \qquad [1]$$

where C_{w1} = change in velocity of whirl of blade row 1
C_{w2} = change in velocity of whirl of blade row 2

From equation [7] of section 13.10

$$\text{Blade or diagram efficiency} = \frac{2U(C_{w1} + C_{w2})}{C_{ai1}^2} \qquad [2]$$

Note that the reference here is still to the kinetic energy of the input steam, i.e. $C_{ai1}^2/2$.

From equation [8] of section 13.10

$$\text{End-thrust} = \dot{m}(C_{f1} + C_{f2}) \qquad [3]$$

where C_{f1} = change in velocity of flow of blade row 1
C_{f2} = change in velocity of flow of blade row 2

Note that C_{f1} and C_{f2} can be positive or negative as explained in section 13.10.

Example 13.3 *Two rows of a velocity-compounded impulse turbine have a mean blade speed of 150 m/s and a nozzle velocity of 675 m/s. The nozzle angle is 20°. The exit angles of the first moving row, the fixed row and the second row of moving blades are 25°, 25° and 30°, respectively. There is a 10 per cent loss of velocity due to friction in all blades. The steam flow is 4.5 kg/s. Draw the velocity diagram to a suitable scale and determine*
(a) the power output of the turbine
(b) the diagram efficiency

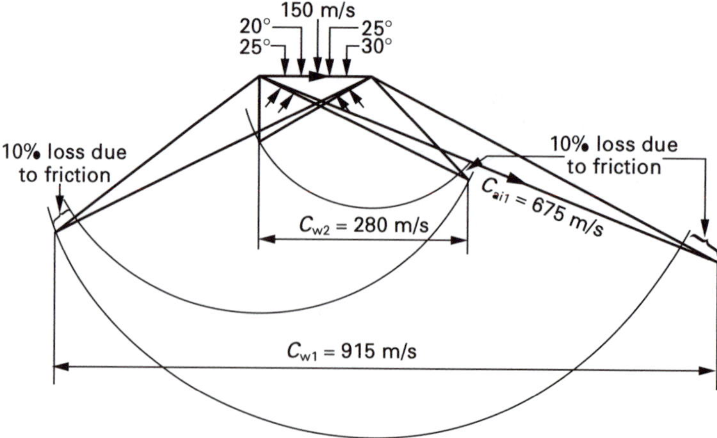

Fig. 13.15 Diagram for Example 13.3

SOLUTION

First draw the velocity diagram (Fig. 13.15). From the diagram

$$C_{w1} = 915 \text{ m/s}$$
$$C_{w2} = 280 \text{ m/s}$$

(a)

$$\begin{aligned}\text{Power of turbine} &= \dot{m}U(C_{w1} + C_{w2}) \\ &= 4.5 \times 150 \times (915 + 280) \\ &= 4.5 \times 150 \times 1195 \\ &= 806\,625 \text{ W} \\ &= \mathbf{806.625 \text{ kW}}\end{aligned}$$

(b)

$$\begin{aligned}\text{Diagram efficiency} &= \frac{2U(C_{w1} + C_{w2})}{C_{ai1}^2} \\ &= \frac{2 \times 150 \times 1195}{675^2} \\ &= 0.787 \\ &= \mathbf{78.7\%}\end{aligned}$$

13.13 Velocity diagram for reaction turbine stage

The velocity diagram for a reaction turbine stage is shown in Fig. 13.16. This diagram is symmetrical, showing equal accelerations in both fixed and moving blades, i.e. 50 per cent reaction. Note that due to acceleration in the moving blades $C_{re} > C_{ri}$ so, other things being equal, there is a greater change in velocity of whirl than for impulse blading. Velocities are, however, lower than would be found in impulse turbines. With the symmetrical diagram illustrated, there is no change in velocity of flow so there is no end-thrust due to this phenomenon. But in a reaction turbine there is a pressure drop across each stage. This pressure drop will combine with the presented area of the blade annulus to produce an end-thrust.

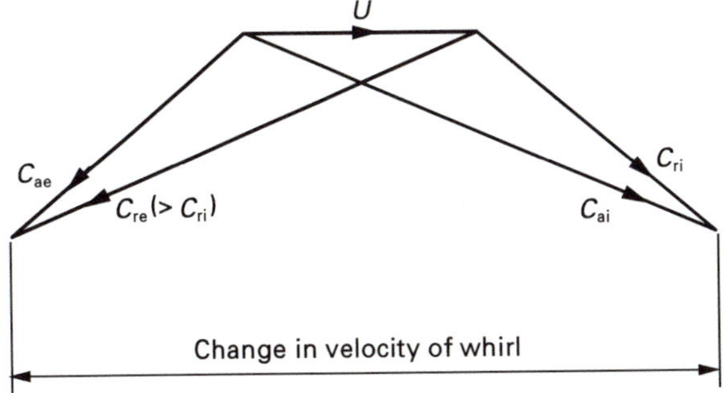

Fig. 13.16 Reaction turbine stage: velocity diagram

As with the impulse turbine

$$\text{Power} = \dot{m}U \text{ (change in velocity of whirl)} \qquad [1]$$

$$\text{End-thrust (due to velocity change)} = \dot{m} \text{ (change in velocity of flow)} \qquad [2]$$

However, the stage efficiency must be related to the energy available to the stage. In this case

$$\text{Energy available to the stage} = h \text{ J/kg}$$
$$= \text{specific enthalpy drop in stage}$$

$$\therefore \quad \text{Stage efficiency} = \frac{\text{Work done in stage}}{\text{Enthalpy drop in stage}}$$

$$= \frac{\dot{m}U(C_{wi} + C_{we})}{\dot{m}h}$$

from which

$$\text{Stage efficiency} = \frac{U(C_{wi} + C_{we})}{h} \qquad [3]$$

Example 13.4 *At a stage in a reaction turbine, the mean blade-ring diameter is 1 m and the turbine runs at a speed of 50 rev/s. The blades are designed for 50 per cent reaction with exit angles 30° and inlet angles 50°. The turbine is supplied with steam at the rate of 600 000 kg/h and the stage efficiency is 85 per cent. Determine*

(a) the power output of the stage

(b) the specific enthalpy drop in the stage in kJ/kg

(c) the percentage increase in relative velocity in the moving blades due to expansion in these blades

(a)

$$\text{Mean blade speed, } U = \pi dN = \pi \times 1 \times 50 = 157 \text{ m/s}$$

Laying off $U = 157$ m/s, exit angles $= 30°$ and inlet angles $= 50°$, the velocity diagram for the stage (Fig. 13.17) can be constructed. From the diagram

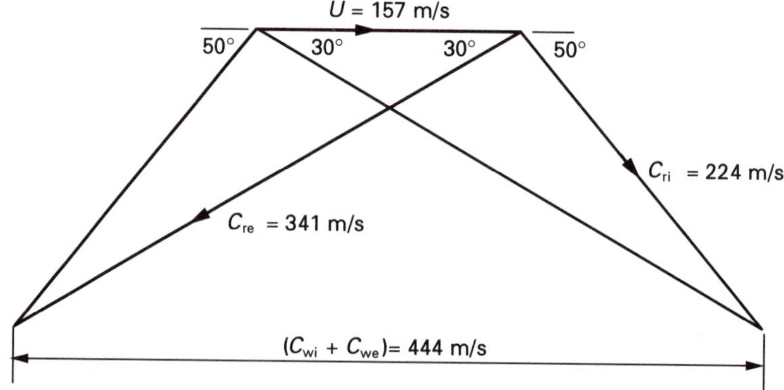

Fig. 13.17 Diagram for Example 13.4

Change in velocity of whirl $= (C_{wi} + C_{we}) = 444$ m/s

$$\therefore \quad \text{Power output of the stage} = \dot{m}U(C_{wi} + C_{we})$$

$$= \frac{600\,000}{3600} \times 157 \times 444$$

$$= (11.6 \times 10^6)\ \text{W}$$

$$= \mathbf{11.6\ MW}$$

(b)

$$\text{Stage } \eta = \frac{U(C_{wi} + C_{we})}{h}$$

$$\therefore \quad h = \frac{U(C_{wi} + C_{we})}{\text{Stage } \eta} = \frac{157 \times 444}{0.85 \times 10^3}$$

$$= \mathbf{82\ kJ/kg}$$

(c)

From the diagram, $C_{ri} = 224$ m/s and $C_{re} = 341$ m/s.

$$\therefore \quad \text{Increase in relative velocity} = \frac{341 - 224}{224} \times 100\%$$

$$= \frac{117}{224} \times 100\%$$

$$= \mathbf{52.2\%}$$

13.14 Blade height

The blade height at a particular section of a turbine will be a function of the following:

- The mass flow of steam through the section, $\dot{m}$ kg/s.
- The specific volume of the steam flowing through the section, v m^3/kg.
- The area through which the steam is passing, A m^2.
- The velocity of the steam as it passes the section, C m/s.

For the section

$$\dot{m}v = AC \qquad [1]$$

Actual calculation will be based on whether the turbine is impulse or reaction.

13.14.1 Blade height for impulse turbine

Figure 13.18(a) shows a plan view of two impulse turbine blades at pitch P apart. The inlet angle of the blades is θ. The projected pitch in the direction of the relative inlet velocity C_{ri} is $P \sin \theta$. Figure 13.18(b) shows a sketch of an impulse turbine velocity diagram for a single blade row. The velocity of flow at inlet is given by

$$C_{fi} = C_{ri} \sin \theta \qquad [2]$$

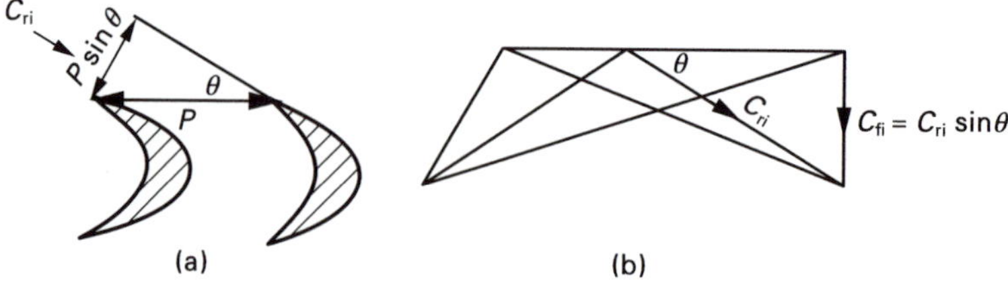

Fig. 13.18 Blade height in impulse turbine: (a) plan view of blades; (b) velocity diagram

For an impulse turbine blade row, let

$\dot{m}$ = mass flow of steam, kg/s
v = specific volume of the steam, m^3/kg
N = number of blades covered by nozzles
P = pitch of blades, m (taken at mean blade height)
H = blade height, m
θ = blade inlet angle
C_{ri} = relative velocity of steam at inlet, m/s
C_{fi} = velocity of flow at inlet, m/s

Then, from equation [1]

$$\begin{aligned} \dot{m}v &= AC \\ &= NP \sin\theta\, H \times C_{ri} \\ &= NPH \times C_{ri} \sin\theta \\ &= NPHC_{fi} \qquad \text{(from equation [2])} \end{aligned}$$

$$\therefore \quad H = \frac{\dot{m}v}{NPC_{fi}} \qquad [3]$$

Suppose the nozzle coverage of the blades is complete or, in later sections of the turbine, the steam has full admission, meaning that the steam flow is through the complete blade annulus. Then

$$\begin{aligned} NP &= \text{circumference at the mean blade diameter} \\ &= \pi d \end{aligned} \qquad [4]$$

where d = mean blade diameter, m

This is also the case in the reaction turbine

13.14.2 Reaction turbine

In this case, let

d = mean blade diameter, m

The reaction turbine has full admission, so from equations [3] and [4]

$$H = \frac{\dot{m}v}{\pi d C_{fi}} \qquad [5]$$

Example 13.5 *At a particular stage of a reaction steam turbine, the mean blade speed is 60 m/s and the steam is at a pressure of 350 kN/m² with a temperature of 175°C. Fixed and moving blades at this stage have inlet angles of 30° and exit angles of 20°. Determine*

(a) the blade height at this stage if the blade height is one-tenth the mean blade-ring diameter and the steam flow is 13.5 kg/s

(b) the power developed by a pair of fixed and moving blade rings at this stage

(c) the specific enthalpy drop in kJ/kg at the stage if the stage efficiency is 85 per cent

SOLUTION

First draw the velocity diagram for the stage (Fig. 13.19).

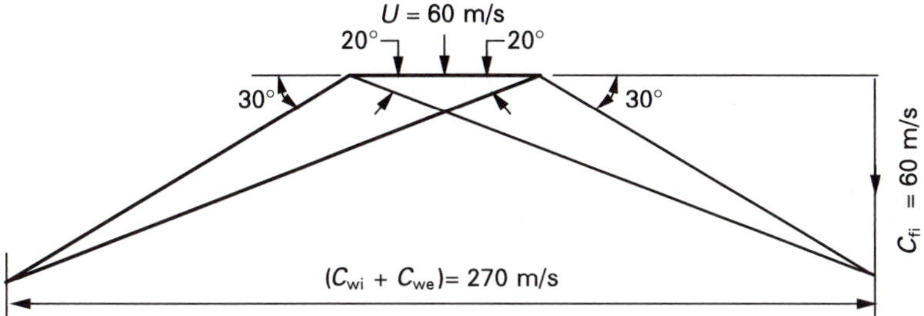

Fig. 13.19 Diagram for Example 13.5

(a)

At 350 kN/m² and with a temperature of 175 °C

$$v = \mathbf{0.589\ m^3/kg}$$

Since the blade height $H = d/10$, then $d = 10H$.

Now

$$H = \frac{\dot{m}v}{\pi d C_{fi}} = \frac{13.5 \times 0.589}{\pi \times 10H \times 60}$$

$$\therefore \quad H^2 = \frac{13.5 \times 0.589}{\pi \times 10 \times 60} = 0.004\ 218$$

$$\therefore \quad H = \sqrt{0.004\ 22} = 0.65\ \text{m} = \mathbf{65\ mm}$$

(b)

$$\text{Power} = \dot{m}U(C_{wi} + C_{we}) = 13.5 \times 60 \times 270$$
$$= 218\ 700\ \text{W}$$
$$= \mathbf{218.7\ kW}$$

(c)

$$\text{Stage } \eta = \frac{U(C_{wi} + C_{we})}{h}$$

$$\therefore \quad h = \frac{U(C_{wi} + C_{we})}{\text{Stage } \eta}$$

$$= \frac{60 \times 270}{0.85}$$

$$= 19\ 059\ \text{J/kg}$$
$$= \mathbf{19.059\ kJ/kg}$$

Questions

1. Steam leaves the nozzles of a single-stage impulse turbine with a velocity of 1000 m/s. The nozzles are inclined at an angle of 24° to the direction of motion of the turbine blades. The mean blade speed is 400 m/s and the blade inlet and exit angles are equal. The steam enters the blades without shock and the flow over the blades is considered to be frictionless. Determine
   (a) the inlet angle of the blades
   (b) the force exerted on the blades in the direction of their motion
   (c) the power developed when the steam flow rate is 4000 kg/h

   [(a) 39°; 9b) 1.133 kN; (c) 453.2 kW]

2. Steam issues from nozzles, inclined at an angle of 22° to the direction of motion, with a velocity of 680 m/s on to the blades of a single-stage impulse turbine. The mean diameter of the blade ring is 1.25 m. The blade angles at inlet and exit are the same and equal to 36°. Assuming that the steam enters the blades without shock and that the flow over the blades is frictionless, determine
   (a) the speed of the turbine rotor in rev/min
   (b) the absolute velocity of the steam leaving the blades
   (c) the torque on the turbine rotor when the flow of steam is 2500 kg/h

   [(a) 4584 rev/min; (b) 255 m/s; (c) 290.8 N m]

3. A single-row impulse turbine has a mean blade speed of 215 m/s. Nozzle entry angle is at 30° to the plane of rotation of the blades. The steam velocity from the nozzles is 550 m/s. There is a 15 per cent loss of relative velocity due to friction across the blades. The absolute velocity at exit is along the axis of the turbine. The steam flow through the turbine is at a rate of 700 kg/h. Determine
   (a) the inlet and exit angles of the blades
   (b) the absolute velocity of the steam at exit
   (c) the power output of the turbine

   [(a) 46°, 49°; (b) 240 m/s; (c) 19.9 kW]

4. A single-row impulse turbine has blades whose inlet angle is 40° and exit angle 37°. The mean blade speed is 230 m/s and the nozzles are inclined at an angle of 27° to the plane of rotation of the blades. There is a 10 per cent loss of relative velocity due to friction in the blades. The turbine uses 550 kg/h of steam. Determine
   (a) the nozzle velocity of the steam
   (b) the absolute velocity of the steam at exit
   (c) the power output of the turbine
   (d) the end-thrust on the turbine
   (e) the diagram efficiency

   [(a) 650 m/s; (b) 265 m/s; (c) 23.9 kW; (d) –7.26 N; (e) 74%]

5. At a particular stage of a reaction turbine the mean blade speed is 150 m/s. The exit angles of the fixed and moving blades are 20°. The inlet angles of the fixed and moving blades are 30°. The stage efficiency is 80 per cent. The pressure at entry to the stage is 15 bar and the temperature is 200°C. Determine
   (a) the specific enthalpy drop across the stage in kJ/kg
   (b) the drum diameter and blade height if the blade height is one-tenth of the drum diameter and the steam flow is 100 kg/s
   (c) the percentage increase in relative velocity across the blading as the result of the pressure drop across the blading

   [(a) 127 kJ/kg; (b) 521 mm, 52.1 mm; (c) 46%]

6. Two wheels of a velocity-compounded impulse turbine have a mean blade speed of 150 m/s. The nozzle angle is 23° and the steam leaves the nozzle with a velocity of 700 m/s. The exit angles of the first moving row, the fixed row and the second moving row of blades are 25°, 27° and 30°, respectively.
   There is a 10 per cent loss of velocity in all blades due to friction. Determine
   (a) the inlet angles of the first moving row, fixed, and second moving row of blades
   (b) the power output of the two wheels/kg of steam/s
   (c) the diagram efficiency

   [(a) 30°, 35°, 47°; (b) 190.5 kW; (c) 77.7%]

7. Two rows of a velocity-compounded impulse turbine have a mean blade speed of 170 m/s. The inlet and exit angles of the first moving row of blades are 30° and 25°, respectively. The inlet and exit angles of the second moving row of blades are 44° and 30°, respectively. The absolute discharge from the second row of moving blades is axial. There is a 15 per cent loss of velocity in all blades due to friction. Determine
   (a) the inlet and exit angles of the fixed blades
   (b) the inlet nozzle angle and velocity
   (c) the power output of the two rows of blades if the turbine uses 1.5 kg of steam per second

   [(a) 35°, 26°; (b) 24°, 835 m/s; (c) 366 kW]

Chapter 14

Air and gas compressors

14.1 General introduction

There are many processes used in a modern industrial society which have the need for a compressed gas. The majority of cases require the use of air as the compressed gas.

The compressors are of two general types, reciprocating or rotary.

14.2 The reciprocating air compressor

A single-stage reciprocating air compressor is illustrated in Fig. 14.1. It consists of a piston which reciprocates in a cylinder, driven through a connecting-rod and crank mounted in a crankcase. There are inlet and delivery valves mounted in the head of

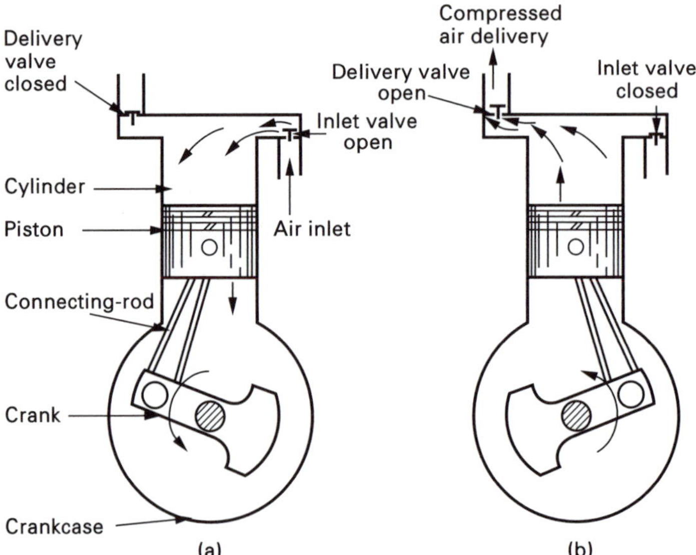

Fig. 14.1 Single-stage air compressor: (a) induction stroke; (b) compression stroke

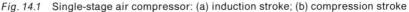

the cylinder. These valves are usually of the pressure differential type, meaning that they will operate as the result of the difference of pressure across the valve. The operation of this type of compressor is as follows.

In Fig. 14.1(a) the piston is moving down the cylinder. Any residual compressed air left in the cylinder after the previous compression will expand, eventually to reach a pressure slightly below intake pressure early on in the stroke. This means that the pressure outside the inlet valve is now higher than on the inside, so the inlet valve will lift off its seat. A stop is provided to limit its lift and to retain it within its valve seating. Thus a fresh charge of air will be aspirated into the cylinder for the remainder of the induction stroke, as it is called. During this stroke the delivery valve will remain closed because the compressed air on the outside of this valve is at a much higher pressure than the induction pressure. In Fig. 14.1(b) the piston is now moving upwards. At the beginning of this upward stroke, a slight increase in cylinder pressure will have closed the inlet valve. The inlet and delivery valves are now closed, so the pressure of the air will rapidly rise because it is now locked up in the cylinder. Eventually the pressure will become slightly greater than the pressure of the compressed air on the outside of the delivery valve, so the delivery valve will lift. The compressed air is now delivered from the cylinder for the remainder of the stroke. Once again, there is a stop on the delivery valve to limit its lift and to retain it within its seating. At the end of the compression stroke the piston again begins to move down the cylinder, the delivery valve closes, the inlet valve eventually opens and the cycle is repeated.

Air is locked up in the cylinder of a reciprocating compressor, so the pressure during compression can be very high. It is limited by the strength of the various parts of the compressor and the power of the driving motor. Note that with the reciprocating compressor there is intermittent flow of air.

Figure 14.2(a) shows a theoretical *P–V* diagram for a single-stage reciprocating air compressor, neglecting clearance. The processes are as follows:

- **4–1** Volume of air V_1 aspirated into compressor at pressure P_1 and temperature T_1.
- **1–2** Air compressed according to law $PV^n = C$ from pressure P_1 to pressure P_2. Volume decreases from V_1 to V_2. Temperature increases from T_1 to T_2.
- **2–3** Compressed air of volume V_2 and at pressure P_2 with temperature T_2 delivered from compressor.

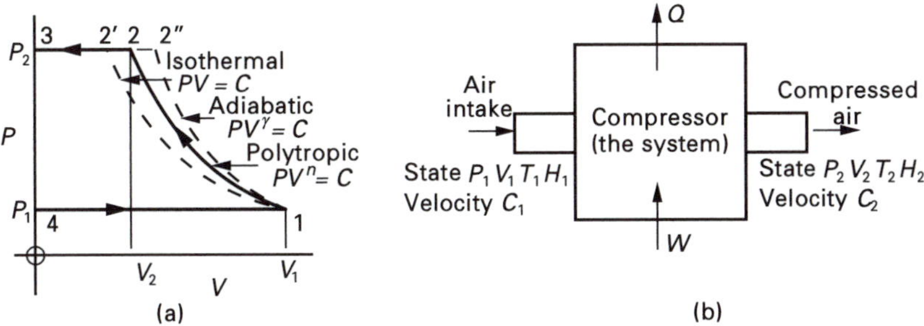

Fig. 14.2 Single-stage air compressor: (a) *P–V* diagram and (b) its theoretical circuit

During compression, due to its excess temperature above the compressor surroundings, the air will transfer some heat to the surroundings. The internal effect of friction is small in the reciprocating compressor, so neglecting friction, the index n is less than γ, the adiabatic index. Work must be input to an air compressor to keep it running, so every effort is made to reduce this input. Inspection of the P–V diagram (Fig. 14.2(a)) shows the frictionless adiabatic compression as 1–2″. If compression were along the isothermal 1–2′ instead of the polytropic 1–2, the work done, given by the area of the diagram, would be reduced and, in fact, would be a minimum. Isothermal compression cannot be achieved in practice, but an attempt is made to approach the isothermal case by cooling the compressor either by the addition of cooling fins or a water jacket to the compressor cylinder. For a reciprocating compressor, a comparison between the actual work done during compression and the ideal isothermal work done is made using the **isothermal efficiency**. This is defined as

$$\text{Isothermal efficiency} = \frac{\text{Isothermal work done during compression}}{\text{Actual work done during compression}} \qquad [1]$$

Thus, the higher the isothermal efficiency, the more nearly has the actual compression approached the ideal isothermal compression.

Neglecting the change of potential energy and writing $H =$ enthalpy of the actual mass passing through the compressor, and neglecting any small change of kinetic energy, the energy equation for the reciprocating compressor becomes

$$H_1 + Q = H_2 + W \quad \text{(see section 2.8)} \qquad [2]$$

or

$$W = (H_1 - H_2) + Q \qquad [3]$$

$$= mc_p(T_1 - T_2) + \frac{(\gamma - n)}{(\gamma - 1)}\frac{(P_1V_1 - P_2V_2)}{(n - 1)} \quad \text{(see Chapter 5)} \qquad [4]$$

Now

$$c_p - c_v = R \quad \text{and} \quad \frac{c_p}{c_v} = \gamma$$

$$\therefore \quad c_v = \frac{c_p}{\gamma}$$

Hence

$$c_p - \frac{c_p}{\gamma} = R$$

$$\therefore \quad c_p\left(1 - \frac{1}{\gamma}\right) = R$$

$$c_p\left(\frac{\gamma - 1}{\gamma}\right) = R$$

or

$$c_p = \frac{R\gamma}{(\gamma - 1)} \qquad [5]$$

Substituting equation [5] in [4]

$$W = \frac{\gamma}{(\gamma - 1)} mR(T_1 - T_2) + \frac{(\gamma - n)}{(\gamma - 1)} \frac{(P_1 V_1 - P_2 V_2)}{(n - 1)}$$

$$= \frac{\gamma}{(\gamma - 1)}(P_1 V_1 - P_2 V_2) + \frac{(\gamma - n)}{(\gamma - 1)} \frac{(P_1 V_1 - P_2 V_2)}{(n - 1)}$$

since $PV = mRT$

$$= \frac{(P_1 V_1 - P_2 V_2)}{(\gamma - 1)} \left[\gamma + \frac{(\gamma - n)}{(n - 1)}\right]$$

$$= \frac{(P_1 V_1 - P_2 V_2)}{(\gamma - 1)} \left[\frac{\gamma(n - 1) + (\gamma - n)}{(n - 1)}\right]$$

$$= \frac{(P_1 V_1 - P_2 V_2)}{(\gamma - 1)} \left[\frac{\gamma n - \gamma + \gamma - n}{(n - 1)}\right]$$

$$= \frac{(P_1 V_1 - P_2 V_2)}{(\gamma - 1)} \left[\frac{\gamma n - n}{(n - 1)}\right]$$

$$= \frac{(P_1 V_1 - P_2 V_2)}{(\gamma - 1)} \frac{n(\gamma - 1)}{(n - 1)}$$

$$= \frac{n}{(n - 1)}(P_1 V_1 - P_2 V_2) \qquad [6]$$

$$= \frac{n}{(n - 1)} mR(T_1 - T_2) \qquad [7]$$

since $PV = mRT$.

This result could have been arrived at by summing areas of the P–V diagram in Fig. 14.2(a).

$$\oint W = \text{Net area of diagram} \quad (\oint W \text{ means cycle work, see section 3.3})$$

$$= \text{Area 4123}$$

$$= \text{Area under 4–1} - \text{Area under 1–2} - \text{Area under 2–3}$$

$$= P_1 V_1 - \left[\frac{P_2 V_2 - P_1 V_1}{(n - 1)}\right] - P_2 V_2$$

$$= (P_1 V_1 - P_2 V_2) - \left[\frac{P_2 V_2 - P_1 V_1}{(n - 1)}\right]$$

$$= (P_1 V_1 - P_2 V_2) + \left[\frac{P_1 V_1 - P_2 V_2}{(n - 1)}\right]$$

$$= \left[1 + \frac{1}{(n - 1)}\right](P_1 V_1 - P_2 V_2)$$

$$= \left[\frac{n - 1 + 1}{(n - 1)}\right](P_1 V_1 - P_2 V_2)$$

$$\oint W = \frac{n}{(n-1)}(P_1V_1 - P_2V_2) \qquad [8]$$

Equation [8] can be modified to give

$$\oint W = \frac{n}{(n-1)}(P_1V_1 - P_2V_2)$$

$$= \frac{n}{(n-1)}P_1V_1\left(1 - \frac{P_2V_2}{P_1V_1}\right) \qquad [9]$$

Now $P_1V_1^n = P_2V_2^n$

$$\therefore \quad \frac{V_2}{V_1} = \left(\frac{P_1}{P_2}\right)^{1/n}$$

and substituting this into equation [9]

$$\oint W = \frac{n}{(n-1)}P_1V_1\left[1 - \frac{P_2}{P_2}\left(\frac{P_1}{P_2}\right)^{1/n}\right]$$

$$= \frac{n}{(n-1)}P_1V_1\left[1 - \frac{P_2}{P_1}\left(\frac{P_2}{P_1}\right)^{-1/n}\right]$$

$$= \frac{n}{(n-1)}P_1V_1\left[1 - \left(\frac{P_2}{P_1}\right)^{1-1/n}\right]$$

$$= \frac{n}{(n-1)}P_1V_1\left[1 - \left(\frac{P_2}{P_1}\right)^{(n-1)/n}\right] \qquad [10]$$

The solution to this equation will always come out negative, showing that work must be done on the compressor. Only the magnitude of the work done is required from the expression, so it is often written

$$\oint W = \frac{n}{(n-1)}P_1V_1\left[\left(\frac{P_2}{P_1}\right)^{(n-1)/n} - 1\right] \qquad [11]$$

$$= \frac{n}{(n-1)}mRT_1\left[\left(\frac{P_2}{P_1}\right)^{(n-1)/n} - 1\right] \qquad [12]$$

If the air delivery temperature T_2 is required, this can be obtained by using

$$\frac{T_2}{T_1} = \left(\frac{P_2}{P_1}\right)^{(n-1)/n}$$

or

$$T_2 = T_1\left(\frac{P_2}{P_1}\right)^{(n-1)/n} \qquad [13]$$

14.3 The effect of clearance volume

In practice all reciprocating compressors will have a clearance volume, which is the volume that remains in the cylinder after the piston has reached the end of its inward stroke. Figure 14.3 shows its effect.

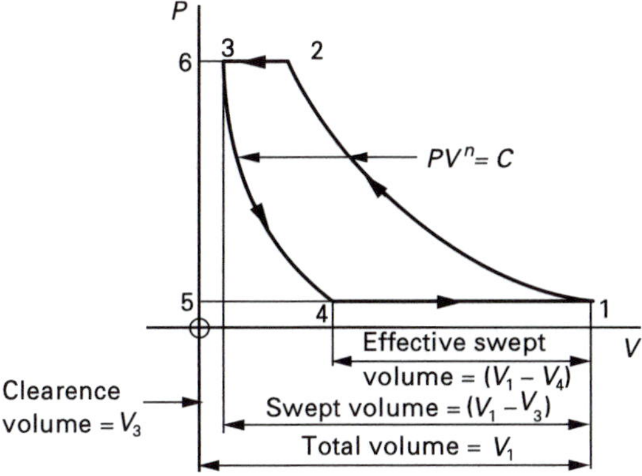

Fig. 14.3 Clearance volume: effect on reciprocating compressor

Commencing at 1 the cylinder is full of intake air, volume V_1, and the piston is about to commence its compression stroke. The air is compressed polytropically according to some law $PV^n = C$ to delivery pressure P_2 and volume V_2. At 2 the delivery valve theoretically opens and for the remainder of the stroke, 2 to 3, the compressed air is delivered from the cylinder. At 3 the piston has reached the end of its inward stroke, so delivery of compressed air ceases at 3. V_3 is the clearance volume and is filled at this stage with compressed air. As the piston begins the intake stroke, this residual compressed air will expand according to some polytropic law $PV^n = C$. It is not until the pressure has reduced to intake pressure at 4 that the inlet valve will begin to open, thus permitting the intake of a fresh charge of air. For the remainder of the intake stroke, a fresh charge is taken into the cylinder. This volume $(V_1 - V_4)$ is called the **effective swept volume**.

The ratio

$$\frac{\text{Effective swept volume}}{\text{Swept volume}} = \frac{(V_1 - V_4)}{(V_1 - V_3)} \qquad [1]$$

is called the **volumetric efficiency**; it is always less than unity because there has to be a clearance volume.

The volumetric efficiency will generally range from about 60 to 85 per cent.

The ratio

$$\frac{\text{Clearance volume}}{\text{Swept volume}} = \frac{V_3}{(V_1 - V_3)} \qquad [2]$$

is the **clearance ratio**. This will generally have a value of between 4 and 10 per cent. The greater the pressure ratio through a reciprocating compressor, the greater the effect of the clearance volume because the clearance air will now expand through a greater volume before intake conditions are reached. But the fixed cylinder size and stroke will mean that the effective swept volume, $(V_1 - V_4)$, will reduce as the pressure ratio increases, so the volumetric efficiency will also reduce.

This can also be shown as follows:

$$\text{Volumetric efficiency} = \frac{(V_1 - V_4)}{(V_1 - V_3)} = \frac{(V_1 - V_3) + (V_3 - V_4)}{(V_1 - V_3)}$$

$$= 1 + \frac{V_3}{(V_1 - V_3)} - \frac{V_4}{(V_1 - V_3)}$$

$$= 1 + \frac{V_3}{(V_1 - V_3)} - \left[\frac{V_4}{(V_1 - V_3)} \times \frac{V_3}{V_3}\right]$$

$$= 1 + \frac{V_3}{(V_1 - V_3)} - \left[\frac{V_3}{(V_1 - V_3)} \times \frac{V_4}{V_3}\right]$$

$$= 1 + \frac{V_3}{(V_1 - V_3)}\left[1 - \frac{V_4}{V_3}\right]$$

$$= 1 - \frac{V_3}{(V_1 - V_3)}\left[\frac{V_4}{V_3} - 1\right]$$

$$= 1 - \frac{V_3}{(V_1 - V_3)}\left[\left(\frac{P_2}{P_1}\right)^{1/n} - 1\right] \qquad [3]$$

since $V_4/V_3 = (P_2/P_1)^{1/n}$

This again shows that for fixed cylinder conditions, V_1 and V_3, the greater the pressure ratio P_2/P_1, the smaller the volumetric efficiency.

$$\text{Work done/cycle} = \text{Net area 1234}$$
$$= \text{Area 5126} - \text{Area 5436}$$

Assuming the polytropic index to be the same for both compression and clearance expansion, then

$$\text{Work done/cycle} = \oint W = \frac{n}{(n-1)} P_1 V_1 \left[\left(\frac{P_2}{P_1}\right)^{(n-1)/n} - 1\right]$$
$$- \frac{n}{(n-1)} P_4 V_4 \left[\left(\frac{P_3}{P_4}\right)^{(n-1)/n} - 1\right] \qquad [4]$$

But $P_4 = P_1$ and $P_3 = P_2$, therefore equation [4] becomes

$$\oint W = \frac{n}{(n-1)} P_1 V_1 \left[\left(\frac{P_2}{P_1}\right)^{(n-1)/n} - 1\right] - \frac{n}{(n-1)} P_1 V_4 \left[\left(\frac{P_2}{P_1}\right)^{(n-1)/n} - 1\right]$$

$$= \frac{n}{(n-1)} P_1 (V_1 - V_4) \left[\left(\frac{P_2}{P_1}\right)^{(n-1)/n} - 1\right] \qquad [5]$$

14.4 The actual reciprocating compressor diagram

Figure 14.4 shows an actual compressor diagram, 1234 is the theoretical *P–V* diagram already discussed.

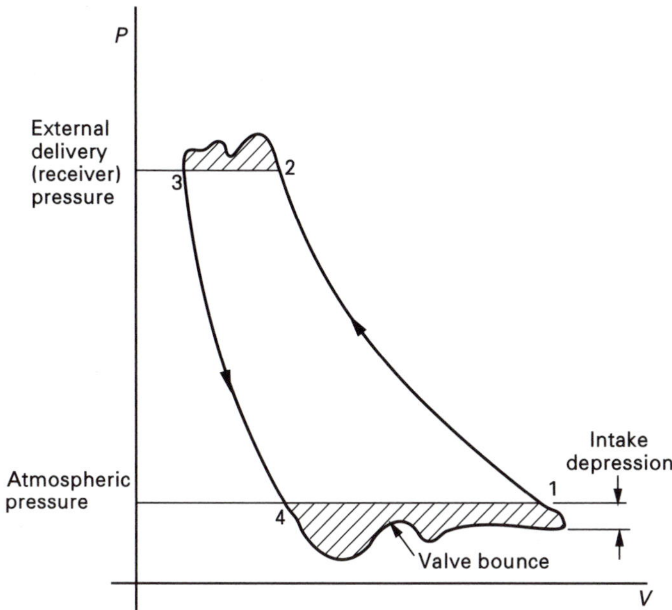

Fig. 14.4 Single-stage air compressor: actual *P–V* diagram

At 4, when the clearance air has reduced to atmospheric pressure, the inlet valve in practice will not open. There are two main reasons for this: there must be a pressure difference across the inlet valve in order to move it and there is the inertia of the inlet valve. Thus, the pressure drops away until the valve is forced off its seat. Some valve bounce will then set in, as shown by the wavy line, and eventually intake will become very nearly steady at some pressure below atmospheric pressure. This negative pressure difference, called the **intake depression**, settles naturally, showing that what is called suction is really the atmospheric air forcing its way into the cylinder against a reduced pressure.

A similar situation occurs at 2, at the beginning of compressed air delivery. There is a pressure rise followed by valve bounce; the pressure then settles at some pressure above external delivery pressure. Compressed air is usually delivered into a tank called the **receiver**, so external delivery pressure is sometimes called the **receiver pressure**. Other small effects at inlet and delivery would be gas inertia and turbulence.

The practical effects discussed are responsible for the addition of the two small shaded negative work areas shown in Fig. 14.4. These areas are in addition to the theoretical area 1234.

14.5 Free air delivery

If the volume of air delivered by an air compressor is reduced to atmospheric temperature and pressure, this volume of air is called the **free air delivery**.

Remember that due to mass flow continuity

$$\text{Delivered mass of air} = \text{Intake mass of air} \qquad [1]$$

Using the characteristic equation and assuming clearance

$$\frac{P_f V_f}{T_f} = \frac{P_1(V_1 - V_4)}{T_1} = \frac{P_2(V_2 - V_3)}{T_2} \qquad [2]$$

If clearance is neglected, equation [1] becomes

$$\frac{P_f V_f}{T_f} = \frac{P_1 V_1}{T_1} = \frac{P_2 V_2}{T_2} \qquad [3]$$

For convenience, P_f and T_f are often taken, as 0.101 325 MN/m^2 (101.325 kN/m^2 = 1.013 25 bar) and 288 K (15 °C).

Example 14.1 *A single-stage, single-acting, reciprocating air compressor has a bore of 200 mm and a stroke of 300 mm. It runs at a speed of 500 rev/min. The clearance volume is 5 per cent of the swept volume and the polytropic index is 1.3 throughout. Intake pressure and temperature are 97 kN/m^2 and 20 °C, respectively, and the compression pressure is 550 kN/m^2. With the aid of Fig. 14.5, determine*

(a) the free air delivered in m^3/min (free air conditions 101.325 kN/m^2 and 15 °C)
(b) the volumetric efficiency referred to the free air conditions
(c) the air delivery temperature
(d) the cycle power
(e) the isothermal efficiency, neglecting clearance

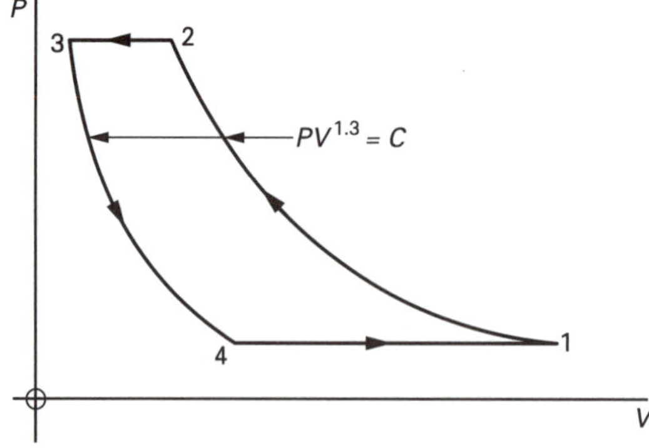

Fig. 14.5 Diagram for Example 14.1

(a)

$$\text{Swept volume, } (V_1 - V_3) = \left[\left(\frac{\pi \times 200^2}{4}\right) \times 300\right] \text{ mm}^3$$

$$= \left[\frac{(\pi \times 0.2^2)}{4} \times 0.3\right] \text{ m}^3$$

$$= \pi \times \frac{0.04}{4} \times 0.3$$

$$= \mathbf{0.009\ 425\ m^3}$$

$$V_3 = 0.05(V_1 - V_3) = 0.05 \times 0.009\ 425 = 0.000\ 471 \text{ m}^3$$

$$V_1 = (V_1 - V_3) + V_3 = 0.009\ 425 + 0.000\ 471 = 0.009\ 896 \text{ m}^3$$

$$V_4 = V_3\left(\frac{P_3}{P_4}\right)^{1/n} = 0.000\ 471 \times \left(\frac{550}{97}\right)^{1/1.3} = 0.000\ 471 \times 3.8 = 0.001\ 79 \text{ m}^3$$

$$\therefore \quad \text{Effective swept volume} = (V_1 - V_4) = 0.009\ 896 - 0.001\ 79$$

$$= \mathbf{0.008\ 106\ m^3}$$

$$\text{Effective swept volume/min} = 0.008\ 106 \times 500$$

$$= \mathbf{4.053\ m^3}$$

Now

$$\frac{P_1(V_1 - V_4)}{T_1} = \frac{P_f V_f}{T_f}$$

$$\therefore \quad V_f = \frac{P_1(V_1 - V_4)}{P_f T_1} T_f = \frac{97 \times 4.053 \times 288}{101.325 \times 293} = \mathbf{3.814\ m^3/min}$$

(b)

$$\text{Volumetric efficiency} = \frac{V_f}{500(V_1 - V_3)} = \frac{3.814}{500 \times 0.009\ 425}$$

$$= 0.809$$

$$= \mathbf{80.9\%}$$

(c)

$$T_2 = T_1\left(\frac{P_2}{P_1}\right)^{(n-1)/n} = 293 \times \left(\frac{550}{97}\right)^{(1.3-1)/1.3}$$

$$= 293 \times 5.67^{0.3/1.3}$$

$$= 293 \times 5.67^{1/4.33}$$

$$= 293 \times 1.493$$

$$= \mathbf{437.5\ K}$$

$$t_2 = 437.5 - 273 = \mathbf{164.5\ ^\circ C}$$

(d)

$$\text{Cycle power} = \frac{n}{(n-1)} P_1(V_1 - V_4)\left[\left(\frac{P_2}{P_1}\right)^{(n-1)/n} - 1\right] \times \frac{500}{60}$$

$$= \frac{1.3}{(1.3-1)} \times 97 \times 0.008\,106 \times \left[\left(\frac{550}{97}\right)^{(1.3-1)/1.3} - 1\right] \times \frac{500}{60}$$

$$= 4.33 \times 97 \times 0.008\,106 \times 0.493 \times \frac{500}{60}$$

$$= \mathbf{14\ kW}$$

(e)

Neglecting clearance

$$\oint W = \frac{n}{n-1} P_1 V_1 \left[\left(\frac{P_2}{P_1}\right)^{(n-1)/n} - 1\right]$$

$$\text{Isothermal } \oint W = P_1 V_1 \ln (P_2/P_1)$$

$$\therefore \quad \text{Isothermal efficiency} = \frac{P_1 V_1 \ln (P_2/P_1)}{\dfrac{n}{(n-1)} P_1 V_1 \left[\left(\dfrac{P_2}{P_1}\right)^{(n-1)/n} - 1\right]}$$

$$= \frac{\ln (P_2/P_1)}{\dfrac{n}{(n-1)} \left[\left(\dfrac{P_2}{P_1}\right)^{(n-1)/n} - 1\right]}$$

$$= \frac{\ln 5.67}{4.33 \times 0.943}$$

$$= \frac{1.735}{4.33 \times 0.493}$$

$$= 0.813$$
$$= \mathbf{81.3\%}$$

14.6 The multi-stage reciprocating compressor

If the delivery from a single-stage, reciprocating compressor is restricted, the delivery pressure will increase. But if the delivery pressure is increased too far, certain disadvantages will appear.

Referring to Fig. 14.6, assume that the single-stage compressor is compressing to pressure P_2, the complete cycle is 1234. Clearance air expansion will be 3–4 and the mass flow through the compressor will be controlled by the effective swept volume $(V_1 - V_4)$. Assume now that a restriction is placed on delivery. The delivery pressure becomes P_5, say, the cycle becomes 1567 and clearance air expansion becomes 6–7. The mass flow through the compressor is now controlled by effective swept volume $(V_1 - V_7)$, which is less than $(V_1 - V_4)$. In the limit, assuming the compressor to be strong enough, the compression 1–8 would take place, where V_8 is the clearance volume, in which case there would be no delivery. Consequently, as the delivery pressure for a single-stage, reciprocating compressor is increased, the mass flow through the compressor will increase.

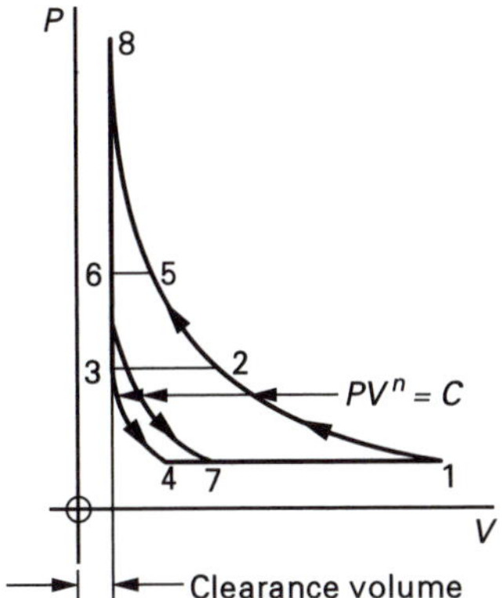

Fig. 14.6 Single-stage compressor: effect of increasing delivery pressure

And as the delivery pressure is increased, the delivery temperature will increase. Referring to Fig. 14.6, $T_8 > T_5 > T_2$. If high-temperature air is not a requirement of the compressed air delivered, any increase in temperature represents an energy loss.

If a single-stage machine is required to deliver high-pressure air, it will require heavy working parts in order to accommodate the high pressure ratio through the machine. This will increase the balancing problem and the high torque fluctuation will require a heavy flywheel installation.

Such disadvantages as described can largely be overcome by multi-stage compression. This is a series arrangement of cylinders in which the compressed air from the preceding cylinder becomes the intake air for the following cylinder. This is illustrated for a 3-stage compressor in Fig. 14.7.

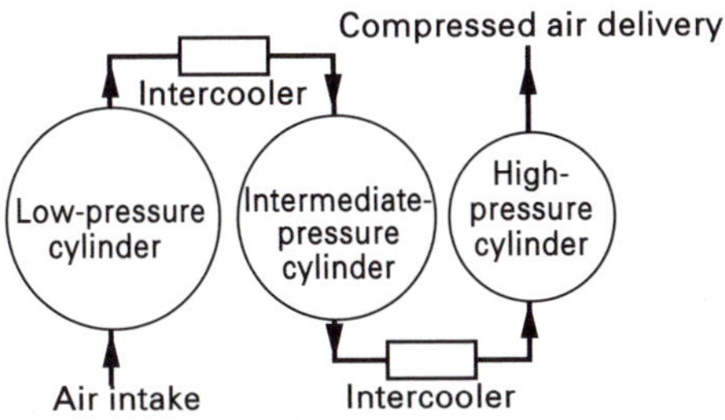

Fig. 14.7 Three-stage compressor

The low pressure ratio in the low-pressure cylinder means that the clearance air expansion is reduced and the effective swept volume of this cylinder is increased. This cylinder controls the mass flow through the machine because it introduces the air into the machine. This means there is greater mass flow through the multi-stage arrangement than through the single-stage machine.

By installing an intercooler between cylinders, in which the compressed air is cooled between cylinders, the final delivery temperature is reduced. This reduction in temperature means a reduction in internal energy of the delivered air (Joule's law). Since this energy must have come from the input energy required to drive the machine, it results in a decrease of input work requirement for a given mass of delivered air.

A multi-stage arrangement of cylinders can also be set up with better balancing and torque characteristics than a single-stage machine. It is common to find machines with either two or three stages of compression. The complexity of the machinery limits the number of stages.

It will be noted that Fig. 14.7 shows how the cylinder diameters decrease as the pressure increases. This is because, as the pressure increases, so the volume of a given mass of gas decreases. There is continuity of mass flow through a compressor, so each following cylinder will require a smaller volume due to its increased pressure range. This reduction in volume is usually accomplished by reducing the cylinder diameter.

Figure 14.8 illustrates cycle arrangements in the development of the ideal conditions required for multi-stage compression. For simplicity, clearance is neglected. The effect of clearance has been discussed in section 14.3. Referring to Fig. 14.8, the overall pressure range is P_1 to P_3. Cycle 8156 is that of the single-stage compressor. Cycles 8147 and 7456 are those of a two-stage compressor without intercooling between cylinders.

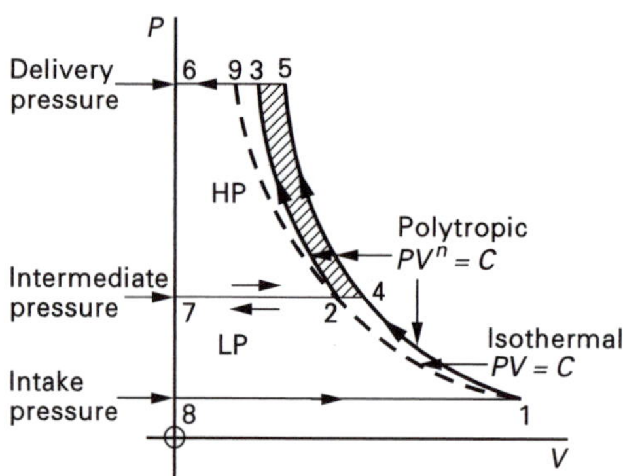

Fig. 14.8 Two-stage compressor: cycle diagram

Cycles 8147 and 7236 are those of a two-stage compressor with perfect intercooling between cylinders. Perfect intercooling means that, after the initial compression in the low-pressure (LP) cylinder, with its consequent temperature rise, the air is cooled in an intercooler back to its original temperature. Referring to Fig. 14.8, this means that $T_2 = T_1$, in which case point 2 lies on the isotherm through 1. This shows that multi-stage compression, with perfect intercooling, approaches more closely the ideal isothermal compression than does single-stage compression (see section 14.2).

14.7 Ideal conditions for multi-stage compressors

Consider Fig. 14.8: cycle 8156 is that of a single-stage compressor, neglecting clearance. For this cycle

$$\oint W = \frac{n}{(n-1)} P_1 V_1 \left[\left(\frac{P_5}{P_1} \right)^{(n-1)/n} - 1 \right] \qquad [1]$$

and

$$T_5 = T_1 \left(\frac{P_5}{P_1} \right)^{(n-1)/n} \qquad [2]$$

= delivery temperature

For a two-stage machine, without intercooling between cylinders, 8147 is the low-pressure cycle and 7456 is the high-pressure cycle. For this arrangement

$$\oint W = \frac{n}{(n-1)} P_1 V_1 \left[\left(\frac{P_4}{P_2} \right)^{(n-1)/n} - 1 \right] + \frac{n}{(n-1)} P_4 V_4 \left[\left(\frac{P_5}{P_4} \right)^{(n-1)/n} - 1 \right] \qquad [3]$$

This will give the same result as equation [1]. The final delivery temperature will also be as given by equation [2] because there is no intercooling. For the two-stage machine with perfect intercooling, 8147 is the low-pressure cycle and 7236 is the high-pressure cycle.

In this case

$$\oint W = \frac{n}{(n-1)} P_1 V_1 \left[\left(\frac{P_4}{P_1} \right)^{(n-1)/n} - 1 \right] + \frac{n}{(n-1)} P_2 V_2 \left[\left(\frac{P_3}{P_2} \right)^{(n-1)/n} - 1 \right] \qquad [4]$$

Delivery temperature is given by

$$T_3 = T_2 \left(\frac{P_3}{P_2} \right)^{(n-1)/n} = T_1 \left(\frac{P_3}{P_2} \right)^{(n-1)/n}, \quad \text{since } T_2 = T_1 \qquad [5]$$

And since $T_2 = T_1$

$$P_2 V_2 = P_1 V_1 \qquad [6]$$

Also

$$P_4 = P_2 \qquad [7]$$

Substituting equations [6] and [7] in equation [4]

$$\oint W = \frac{n}{(n-1)} P_1 V_1 \left[\left(\frac{P_2}{P_1} \right)^{(n-1)/n} + \left(\frac{P_3}{P_2} \right)^{(n-1)/n} - 2 \right] \qquad [8]$$

Now Fig. 14.8 shows the shaded area 2453, the work saving which occurs as the result of using an intercooler. As intermediate pressure $P_2 \to P_1$, area 2453 $\to 0$. And as $P_2 \to P_3$, area 2453 $\to 0$. This means there exists an intermediate pressure P_2 which makes area 2453 a maximum; this is the condition when $\oint W$ is a minimum.

Inspection of equation [8] shows that for minimum $\oint W$

$$\left(\frac{P_2}{P_1}\right)^{(n-1)/n} + \left(\frac{P_3}{P_2}\right)^{(n-1)/n}$$

must be a minimum because all other parts of the equation are constant in this consideration; P_2 is the variable.

Hence, for minimum $\oint W$

$$\frac{\mathrm{d}}{\mathrm{d}P_2}\left[\left(\frac{P_2}{P_1}\right)^{(n-1)/n} + \left(\frac{P_3}{P_2}\right)^{(n-1)/n}\right] = 0$$

Differentiating with respect to P_2

$$\frac{1}{P_1^{n-1/n}} \times \left(\frac{n-1}{n}\right) P_2^{(n-1/n)-1} + P_3^{n-1/n} \times -\left(\frac{n-1}{n}\right) P_2^{-[(n-1)/n]-1} = 0$$

$$\frac{1}{P_1^{(n-1)/n}} \times \left(\frac{n-1}{n}\right) P_2^{-1/n} = P_3^{(n-1)/n} \times \left(\frac{n-1}{n}\right) P_2^{(-2n+1)/n}$$

$$\frac{P_2^{-1/n}}{P_2^{(-2n+1)/n}} = (P_1P_3)^{(n-1)/n}$$

$$P_2^{-1/n} P_2^{-[(-2n+1)/n]} = (P_1P_3)^{(n-1)/n}$$

$$P_2^{-1/n} P_2^{(2n-1)/n} = (P_1P_3)^{(n-1)/n}$$

$$P_2^{(2n-2)/n} = (P_1P_3)^{(n-1)/n}$$

$$P_2^{2(n-1)/n} = (P_1P_3)^{(n-1)/n}$$

$$\therefore \quad P_2^2 = P_1P_3 \qquad [9]$$

From which

$$P_2 = (P_1P_3)^{1/2} = \surd(P_1P_3) \qquad [10]$$

and

$$\frac{P_2}{P_1} = \frac{P_3}{P_2} \qquad [11]$$

or, pressure ratio/stage is equal.

P_2 obtained from equation [10] will give the ideal intermediate pressure which, with perfect intercooling, will give the minimum $\oint W$.

With these ideal conditions, substituting equations [6], [7] and [11] into equation [4] shows that there is equal work per cylinder.

Hence

$$\oint W = \frac{2n}{(n-1)} P_1V_1\left[\left(\frac{P_2}{P_1}\right)^{(n-1)/n} - 1\right] \qquad [12]$$

Substituting equation [10] in equation [12]

$$\oint W = \frac{2n}{(n-1)} P_1 V_1 \left\{ \left[\frac{(P_1 P_3)^{\frac{1}{2}}}{P_1} \right]^{(n-1)/n} - 1 \right\}$$

$$= \frac{2n}{(n-1)} P_1 V_1 \left\{ \left[\left(\frac{P_3}{P_1} \right)^{1/2} \right]^{(n-1)/n} - 1 \right\}$$

$$= \frac{2n}{(n-1)} P_1 V_1 \left[\left(\frac{P_3}{P_1} \right)^{(n-1)/2n} - 1 \right]$$

Note that P_3/P_1 is the pressure ratio through the compressor. Now, from the analysis of compressors so far:

for a single-stage machine

$$\oint W = \frac{n}{(n-1)} P_1 V_1 \left[\left(\frac{P_2}{P_1} \right)^{(n-1)/n} - 1 \right]$$

for a two-stage machine

$$\oint W = \frac{2n}{(n-1)} P_1 V_1 \left[\left(\frac{P_3}{P_2} \right)^{(n-1)/2n} - 1 \right]$$

So it seems reasonable to assume that for a three-stage machine

$$\oint W = \frac{3n}{(n-1)} P_1 V_1 \left[\left(\frac{P_4}{P_1} \right)^{(n-1)/3n} - 1 \right]$$

and for an x-stage machine

$$\oint W = \frac{xn}{(n-1)} P_1 V_1 \left[\left(\frac{P_{x+1}}{P_2} \right)^{(n-1)/xn} - 1 \right] \quad [14]$$

Note that in each case P_{x+1}/P_1 is the pressure ratio through the compressor. For an ideal compressor, there is equal work per cylinder, so for an x-stage machine

$$\oint W = \frac{xn}{n-1} P_1 V_1 \left[\left(\frac{P_2}{P_1} \right)^{(n-1)/n} - 1 \right] \quad [15]$$

Equation [11] is used to determine the intermediate pressures for an x-stage machine running under ideal conditions. It shows that the pressure ratio per stage is equal.

Hence, for an x-stage machine

$$\frac{P_2}{P_1} = \frac{P_3}{P_2} = \cdots = \frac{P_{x+1}}{P_x} = k, \quad \text{say} \quad [16]$$

From this

$$\begin{aligned} P_2 &= kP_1 \\ P_3 &= kP_2 = k^2 P_1 \\ P_4 &= kP_3 = k^3 P_1 \\ &\vdots \\ P_{x+1} &= kP_x = k^x P_1 \end{aligned}$$

$$\therefore \quad k^x = \left(\frac{P_{x+1}}{P_1}\right)$$

or

$$k = \sqrt[x]{\left(\frac{P_{x+1}}{P_1}\right)} = \sqrt[x]{(\text{Pressure ratio through compressor})} \qquad [17]$$

Substituting the value of k in equation [16] will determine the intermediate pressures.

Example 14.2 *A two-stage, single-acting, reciprocating compressor takes in air at the rate of 0.2 m³/s. Intake pressure and temperature are 0.1 MN/m² and 16 °C, respectively. The air is compressed to a final pressure of 0.7 MN/m². The intermediate pressure is ideal and intercooling is perfect. The compression index is 1.25 and the compressor runs at 10 rev/s.*

Neglecting clearance, determine

(a) the intermediate pressure
(b) the total volume of each cylinder
(c) the cycle power

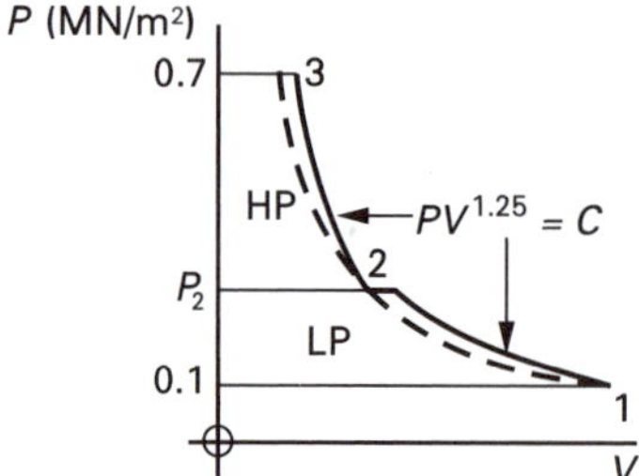

Fig. 14.9 Diagram for Example 14.2

SOLUTION

First draw a diagram (Fig. 14.9).

(a)

$$\begin{aligned} P_2 &= \sqrt{P_1 P_2} = \sqrt{0.1 \times 0.7} \\ &= \sqrt{0.07} \\ &= \mathbf{0.265\ MN/m^2} \end{aligned}$$

(b)

$$\text{Total volume of LP cylinder} = \frac{0.2}{10} = 0.02\ \text{m}^3 = \mathbf{20\ litres}$$

Intercooling is perfect, so 2 lies on the isothermal through 1.

$$\therefore \quad P_1V_1 = P_2V_2$$

or

$$V_2 = \frac{P_1V_1}{P_2} = \frac{0.1 \times 0.02}{0.265} = 0.007\,5\ \text{m}^3 = \mathbf{7.5\ litres}$$

The total volume of the HP cylinder is 7.5 litres.

(c)

$$\text{Cycle power} = \frac{2n}{(n-1)} P_1V_1 \left[\left(\frac{P_2}{P_1} \right)^{(n-1)/n} - 1 \right] \quad V_1 = \text{vol/s}$$

$$= 2 \times \frac{1.25}{(1.25-1)} \times 0.1 \times 0.2 \times \left[\left(\frac{0.265}{0.1} \right)^{(1.25-1)/1.25} - 1 \right]$$

$$= 2 \times \frac{1.25}{0.25} \times 0.02 \times (2.65^{1/5} - 1)$$

$$= 2 \times 5 \times 0.02 \times (1.215 - 1)$$
$$= 10 \times 0.02 \times 0.215$$
$$= 0.043\ \text{MW}$$
$$= \mathbf{42\ kW}$$

Example 14.3 *A three-stage, single-acting, reciprocating air compressor has a low-pressure cylinder of 450 mm bore and 300 mm stroke. The clearance volume of the low-pressure cylinder is 5 per cent of the swept volume. Intake pressure and temperature are 1 bar and 18 °C, respectively; the final delivery pressure is 15 bar. Intermediate pressures are ideal and intercooling is perfect. The compression and expansion index can be taken as 1.3 throughout. Determine*

(a) the intermediate pressures

(b) the effective swept volume of the low-pressure cylinder

(c) the temperature and the volume of air delivered per stroke at 15 bar

(d) the work done per kilogram of air

Take R = *0.29 kJ/kg K.*

SOLUTION

First draw a diagram (Fig. 14.10).

(a)

$$\frac{P_2}{P_1} = \frac{P_3}{P_2} = \frac{P_4}{P_3} = k = \sqrt[3]{P_4/P_1}$$

$$k = \sqrt[3]{15/1} = 2.466$$

$$\therefore \quad P_2 = kP_1 = 2.466 \times 1 = \mathbf{2.466\ bar}$$
$$P_3 = kP_2 = 2.466 \times 2.466 = \mathbf{6.081\ bar}$$

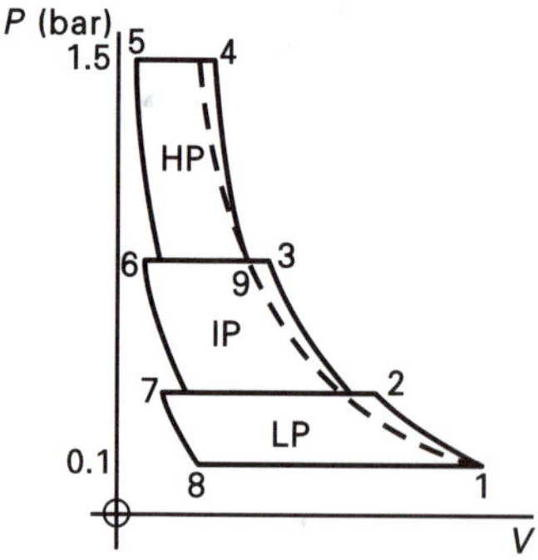

Fig. 14.10 Diagram for Example 14.3

(b)

$$\text{Swept volume of LP cylinder} = V_1 - V_7$$
$$= \pi \times \frac{0.45^2}{4} \times 0.3$$
$$= \pi \times 0.050\,6 \times 0.3$$
$$= \mathbf{0.047\,7\ m^3}$$

$$V_7 = 0.05 \times 0.047\,7 = 0.002\,39\ \text{m}^3$$

$$\therefore \quad V_1 = (V_1 - V_7) + V_7 = 0.0477 + 0.002\,39 = 0.050\,09\ \text{m}^3$$

$$P_7 V_7^{1.3} = P_8 V_8^{1.3}$$

$$\therefore \quad V_8 = V_7 \left(\frac{P_7}{P_8}\right)^{1/1.3} = 0.002\,39 \times 2.466^{1/1.3} = 0.004\,78\ \text{m}^3$$

The effective swept volume of the low-pressure cylinder is

$$V_1 - V_8 = 0.050\,09 - 0.004\,78$$
$$= 0.045\,31\ \text{m}^3$$
$$= \mathbf{45.31\ litres}$$

(c)

$$\frac{T_4}{T_9} = \left(\frac{P_4}{P_9}\right)^{(n-1)/n}$$

$$\therefore \quad T_4 = T_9 \left(\frac{P_4}{T_9}\right)^{(n-1)/n}$$

Intercooling is perfect, so $T_9 = T_1$.

$$\therefore \quad T_4 = 291 \times 2.466^{0.3/1.3}$$
$$= 291 \times 1.232$$
$$= 358.5\ \text{K}$$

$$t_4 = 358.5 - 273 = \mathbf{85.5\ ^\circ C}$$

The delivery temperature is 85.5 °C.

$$\frac{P_4(V_4 - V_5)}{T_4} = \frac{P_1(V_1 - V_8)}{T_1}$$

$$\therefore \quad V_4 - V_5 = \frac{P_1 T_4}{P_4 T_1}(V_1 - V_8)$$

$$= \frac{1}{15} \times \frac{358.5}{291} \times 0.045\,31$$

$$= 0.003\,72 \text{ m}^3$$

$$= \mathbf{3.72\ litres}$$

The delivery volume per stroke is 3.72 litres.

(d)

$$\text{Work/kg of air} = \frac{3n}{(n-1)} RT_1 \left[\left(\frac{P_2}{P_1}\right)^{(n-1)/n} - 1\right]$$

$$= \frac{3 \times 1.3}{(1.3 - 1)} \times 0.29 \times 291 \times (1.232 - 1)$$

$$= 3 \times 4.33 \times 0.29 \times 291 \times 0.232$$

$$= \mathbf{254.3\ kJ}$$

14.8 Rotary air compressors

There are three basic types of rotary air compressor: the radial or centrifugal compressor, the axial-flow compressor and the positive-displacement compressor or blower.

A general arrangement of a radial compressor is shown in Fig. 14.11. It consists of an impeller rotating within a casing, usually at high speed (something like 20 000–30 000 rev/min in some cases). The impeller consists of a disc onto which radial blades are attached. The blades break up the air into cells. The impeller is shrouded by the casing. If the impeller is rotated, the cells of air will also be rotated with the

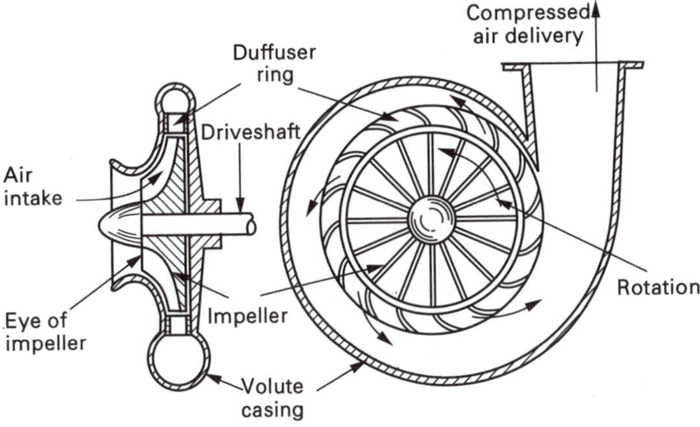

Fig. 14.11 Radial compressor

impeller. Centrifugal force means that the air in the cells will move out from the outside edge of the impeller and more air will move into the centre of the impeller to take its place. The centre is called the eye of the impeller. As it moves away from the outside edge of the impeller, the air passes into a diffuser ring which helps to direct it into the volute. The air is also decelerated in the diffuser ring, producing a pressure rise in the air because, theoretically, there is no energy loss to the airstream.

The volute is the collecting device for this compressor. Its section increases round the compressor so that, as the air is collected round the volute, a greater section will be required to pass the increasing quantity of air. A duct leads away from the volute to take the compressed air out of the compressor. This type of compressor is a continuous-flow device and will deal with large quantities of air through a moderate pressure range. Pressure compression ratios of some 4 or 6:1 are common.

The general arrangement of the axial-flow compressor is shown in Fig. 14.12. In this type of compressor there are alternate rows of fixed and moving blades. The fixed blades are fixed in an outer casing, whereas the moving blades are fixed to a central drum which can be rotated by a driveshaft. The moving blades can be looked at in a simple way as a set of fans in series. These blades progress the air through the compressor; the preceding fan boosts the following fan, as it were. The fixed blades act as guide vanes and diffusers. The angles of all blade rows are set such that there is a smooth progression of air from blade row to blade row. The air passes axially along the compressor, hence its name. Air is removed by suitable ducting at the end of the compressor. Once again, this type of compressor runs at high speed (10 000–30 000 rev/min) and generally deals with large quantities of air. Pressure compression ratios of 10:1 or more can be obtained. This compressor design is commonly used in aircraft gas turbines.

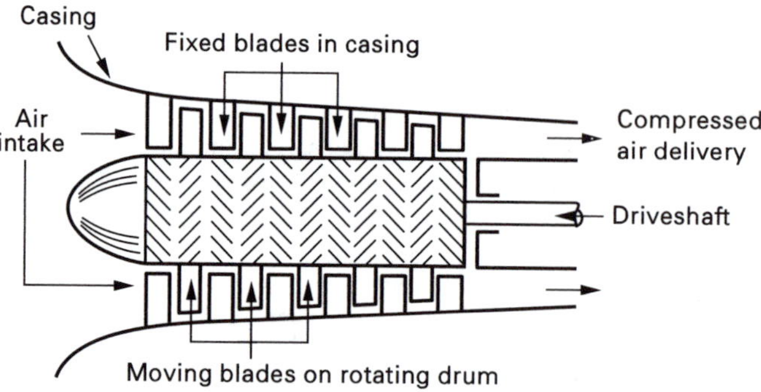

Fig. 14.12 Axial-flow compressor

The compression occurs so rapidly in rotary and axial-flow compressors that there is little time for heat exchange between the gas and its surroundings, so the compression will be very nearly adiabatic. But due to its high velocity through the compressor, the air will encounter considerable friction, internally and with the compressor walls; there will be turbulence and shock due to changes in direction. Friction and turbulence will generate internal energy within the air and produce a temperature higher than the theoretical adiabatic temperature. The index n for a rotary compressor will therefore be greater than the adiabatic index γ. It should be noted that although n is greater than γ in the rotary compressor the compression is nevertheless adiabatic. The adiabatic compression is defined as a compression carried out such that no heat is received or rejected from or to the surroundings during the progress of the compression. Neglecting any small loss of heat to the surroundings, during compression in the rotary compressor, the compression is adiabatic. The temperature increase over the theoretical adiabatic compression temperature is an internal effect, so it does not modify the adiabatic nature of the compression.

Figure 14.13(a) is a P–V diagram for a rotary compressor; Fig. 14.13(b) is the corresponding T–s diagram. The pressure range is from lower pressure P_1 to higher pressure P_2.

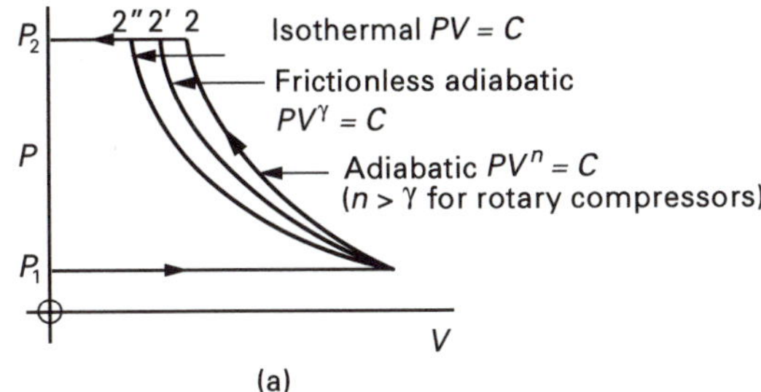

(a)

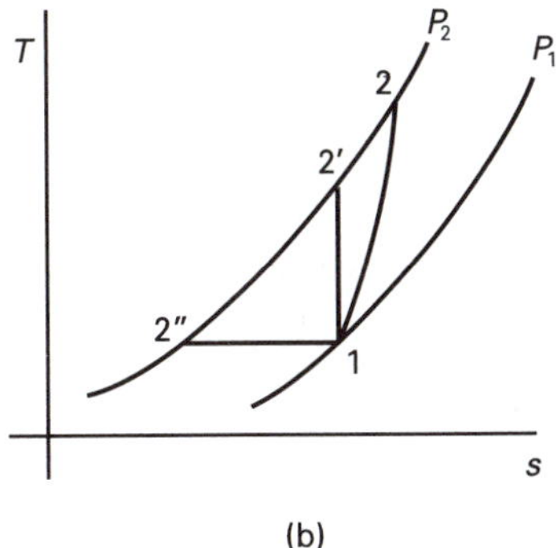

(b)

Fig. 14.13 Rotary compressor: (a) P–V diagram; (b) T–s diagram

The actual adiabatic compression is shown as 1–2. The frictionless adiabatic compression (isentropic compression see section 7.8) is shown as 1–2′. The isothermal compression is shown as 1–2″. Note that $T_2 > T_{2'} > T_{2''}$ and that $W_{1-2} > W_{1-2'} > W_{1-2''}$.

Consider the steady-flow energy equation applied to the rotary compressor

$$u_1 + P_1 v_1 + \frac{C_1^2}{2} + Q = u_2 + P_2 v_2 + \frac{C_2^2}{2} + W \qquad [1]$$

or

$$h_1 + \frac{C_1^2}{2} + Q = h_2 + \frac{C_2^2}{2} + W \quad (\text{since } u + Pv = h) \qquad [2]$$

The compression is adiabatic, so $Q = 0$ by definition. And across the compressor rotor there is little or no change of velocity, so the kinetic energy terms can be neglected. Thus, the energy equation becomes

$$h_1 = h_2 + W \qquad [3]$$

from which

$$\text{Specific work} = W = (h_1 - h_2)$$
$$= c_p(T_1 - T_2) \quad \text{(see section 5.8)} \qquad [5]$$

For a mass flow $\dot{m}$

$$W = \dot{m}c_p(T_1 - T_2) \qquad [6]$$

This equation will be negative, showing that work must be done on the compressor.

For a rotary compressor, the ratio

$$\frac{\text{Frictionless adiabatic work}}{\text{Actual adiabatic work}} = \frac{W_{1-2'}}{W_{1-2}} \qquad [7]$$

is called the **isentropic efficiency**.

Hence, from equation [6]

$$\text{Isentropic } \eta = \frac{\dot{m}c_p(T_1 - T_{2'}}{\dot{m}c_p(T_1 - T_2)}$$
$$= \frac{(T_1 - T_{2'})}{(T_1 - T_2)} \qquad [8]$$

There are several designs of positive-displacement compressor or blower. A well-known design is the Roots blower, illustrated in Fig. 14.14. There are two rotors which mesh together in the same way as gearwheels. Two rotors each with two lobes are shown. The two rotors are driven together through external gearing. Rotors with more than two lobes are sometimes used when an increase in pressure ratio is required. The rotors rotate in a casing.

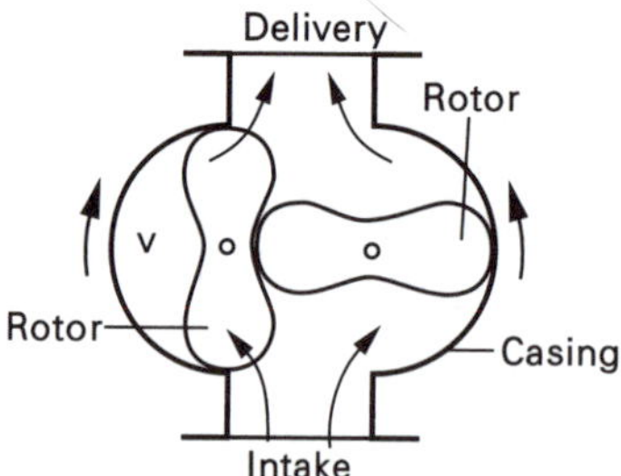

Fig. 14.14 Roots blower: positive-displacement compressor

The operation is as follows. Gas is taken in at the intake and, as the rotors are rotated, a volume of gas V becomes trapped between the rotor and the casing. This volume V is transported to the delivery side of the machine. As the gas is delivered, the compressed gas already on the delivery side compresses it to delivery pressure.

The volume of gas transported is $4V$ per revolution. The pressure ratio through the machine is usually low, say 2:1.

An approximate P–V diagram for the Roots blower is given in Fig. 14.15. From the diagram

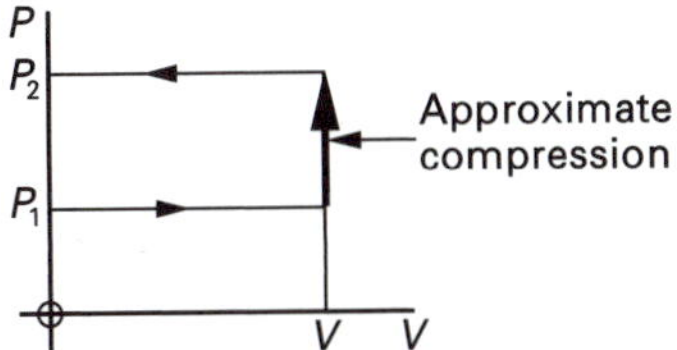

Fig. 14.15 Roots blower: approximate *P–V* diagram

$$\text{Work done to compress volume } V = V(P_2 - P_1) \qquad [9]$$

$$\text{Work done/revolution} = 4V(P_2 - P_1) \qquad [10]$$
(for two-lobe rotors)

Figure 14.16 is a diagram of a vane pump. It consists of a circular casing in which a drum rotates about a centre eccentric to the centre of the casing. Slots are cut in the drum into which vanes are fitted. During rotation the vanes remain in contact with the casing.

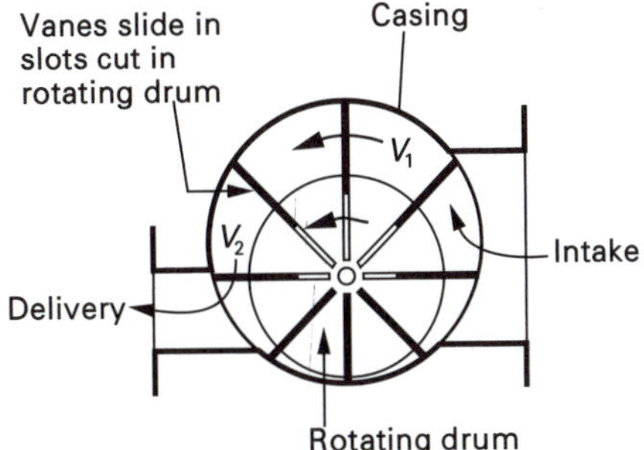

Fig. 14.16 Vane pump

In operation, as the drum rotates, a volume of gas V_1 is trapped between the vanes, the drum and the casing. The space between the drum and casing reduces as delivery is reached and the gas has a reduced volume V_2. The gas has therefore been partially compressed. The remainder of compression is obtained by back pressure from the already compressed gas, as in the Roots blower. An approximate P–V diagram for the vane pump is shown in Fig. 14.17.

Compression 1–2, in which the volume is reduced from V_1 to V_2 and pressure is increased from P_1 to P_2, is assumed to be according to the law $PV^{\gamma} = C$.

Compression 2–3, due to back pressure from the already compressed gas, is assumed to be at constant volume.

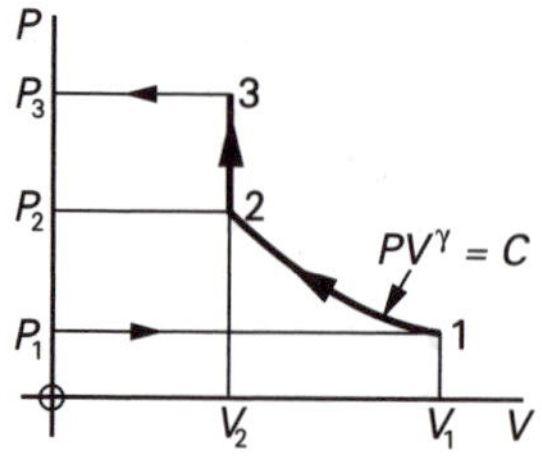

Fig. 14.17 Vane pump: approximate *P–V* diagram

For compression 1–2

$$P_2 V_2^\gamma = P_1 V_1^\gamma \quad \therefore \quad P_2 = P_1 (V_1/V_2)^\gamma \qquad [11]$$

$$\text{Work}_{1-02} = \frac{\gamma}{(\gamma-1)} P_1 V_1 \left[\left(\frac{P_2}{P_1} \right)^{(\gamma-1)/\gamma} - 1 \right] \qquad [12]$$

For the compression 2–3

$$\text{Work}_{2-3} = V_2 (P_3 - P_2) \qquad [13]$$

$$\text{Total work}_{1-3} = \frac{\gamma}{(\gamma-1)} P_1 V_1 \left[\left(\frac{P_2}{P_1} \right)^{(\gamma-1)/\gamma} - 1 \right] + V_2 (P_3 - P_2) \qquad [14]$$

If there are N vanes, then

$$\text{Work/revolution} = N \left\{ l \frac{\gamma}{(\gamma-1)} P_1 V_1 \left[\left(\frac{P_2}{P_1} \right)^{(\gamma-1)/\gamma} - 1 \right] + V_2 (P_3 - P_2) \right\} \qquad [15]$$

Example 14.4 *A rotary compressor has a pressure compression ratio of 5:1. It compresses air at the rate of 10 kg/s. The initial pressure and temperature are 100 kN/m² and 20 °C, respectively. The isentropic efficiency of the compressor is 0.85.*
Determine
(a) the final pressure and temperature
(b) the energy, in kilowatts, required to drive the compressor
Take $\gamma = 1.4$ and $c_p = 1.005$ kJ/kg K.

(a)
Let

$T_{2'}$ = frictionless absolute temperature after compression

then

$$T_{2'} = T_1 \left(\frac{P_2}{P_1} \right)^{(\gamma-1)/\gamma} = 293 \times 5^{0.4/1.4} = 293 \times 5^{1/3.5}$$

$$= 293 \times 1.584$$

$$= \mathbf{464\ K}$$

$$\text{Isentropic efficiency} = \frac{\text{Frictionless adiabatic work}}{\text{Actual adiabatic work}}$$

$$= \frac{\dot{m}c_p(T_{2'} - T_1)}{\dot{m}c_p(T_2 - T_1)}$$

$$= \frac{(T_{2'} - T_1)}{(T_2 - T_1)}$$

$$\therefore \quad 0.85 = \frac{464 - 293}{T_2 - 293} = \frac{171}{T_2 - 293}$$

$$\therefore \quad T_2 = 293 + \frac{171}{0.85} = 293 + 201 = \mathbf{494\ K}$$

$$t_2 = 494 - 273 = \mathbf{221\ ^\circ C}$$

The final temperature is 221 °C.

$$\text{Final pressure} = 100 \times 5 = \mathbf{500\ kN/m^2}$$

(b)

$$\begin{aligned}\text{Energy to drive} &= \dot{m}c_p(T_1 - T_2) \\ &= 10 \times 1.005 \times (293 - 494) \\ &= -10 \times 1.005 \times 201 \\ &= \mathbf{-2020\ kW}\end{aligned}$$

The negative sign indicates energy input.

Example 14.5 *A supercharger on a petrol engine deals with an air–fuel mixture of ratio 14:1. It compresses the mixture from a pressure of 93 kN/m² to a pressure of 200 kN/m²; the initial temperature is 15 °C. The density of the mixture at the initial conditions is 1.3 kg/m³. The engine uses 0.68 kg fuel/min. The isentropic efficiency of the compressor is 82 per cent. Determine the power absorbed in driving the compressor. Take γ for the mixture = 1.38.*

SOLUTION

$$\text{For the mixture, } R = \frac{P_1 V_1}{m T_1}$$

and at the initial conditions 1 m^3 has a mass of 1.3 kg.

$$\therefore \quad R = \frac{93 \times 1}{1.3 \times 288} = \mathbf{0.248\ kJ/kg\ K}$$

For the mixture

$$c_p = \frac{\gamma R}{(\gamma - 1)} = \frac{1.38 \times 0.248}{(1.38 - 1)}$$

$$= \frac{1.38 \times 0.248}{0.38}$$

$$= \mathbf{0.901\ kJ/kg\ K}$$

$$T_{2'} = T_1\left(\frac{P_2}{P_1}\right)^{(\gamma-1)/\gamma} = 288 \times \left(\frac{200}{93}\right)^{0.38/1.38} = 288 \times 2.15^{1/3.63}$$

$$= 288 \times 1.235$$

$$= \mathbf{355.7\ K}$$

$$0.82 = \frac{T_{2'} - T_1}{T_2 - T_1}$$

$$\therefore \quad T_2 - T_1 = \frac{T_{2'} - T_1}{0.82} = \frac{355.7 - 288}{0.82} = \frac{67.7}{0.82} = \mathbf{82.6\ K}$$

For every 1 kg of fuel there will be 15 kg of mixture because the air–fuel mixture has a ratio of 14:1. Hence

$$\text{Mass flow, } \dot{m} = \frac{0.68}{60} \times 15 = \mathbf{0.17\ kg/s}$$

$$\therefore \quad \text{Power absorbed by compressor} = \dot{m}c_p(T_2 - T_1)$$

$$= 0.17 \times 0.902 \times 82.6$$

$$= \mathbf{12.67\ kW}$$

Example 14.6 *A Roots blower has an air capacity of 1 kg/s. The pressure ratio through the blower is 2:1 with an intake pressure and temperature of 1 bar and 70 °C, respectively. Determine the power required to drive the blower in kilowatts. Take* R = *0.29 kJ/kg K.*

SOLUTION

$$P_1\dot{V}_1 = \dot{m}RT_1$$

$$\therefore \quad \dot{V}_1 = \frac{\dot{m}RT_1}{P_1} = \frac{1 \times 0.29 \times 10^3 \times 343}{1 \times 10^5} = \mathbf{0.995\ m^3/s}$$

$$\therefore \quad \text{Power required} = \dot{V}_1(P_2 - P_1)$$

$$= 0.995 \times 10^5$$

$$= 99\,500\ \text{W}$$

$$= \mathbf{99.5\ kW}$$

Example 14.7 *A vane pump has the same air capacity, pressure ratio and intake conditions as the Roots blower of Example 14.6. In the vane pump the volume is reduced to 0.7 of the intake volume and the air is then delivered. Determine the power required to drive the vane pump in kilowatts. Take $\gamma = 1.4$.*

SOLUTION

$$P_1V_1^{\gamma} = P_2V_2^{\gamma} \quad \therefore \quad P_2 = P_1\left(\frac{V_1}{V_2}\right)^{\gamma} = 1 \times \left(\frac{1}{0.7}\right)^{1.4}$$

$$= 1 \times 1.43^{1.4}$$

$$= 1 \times 1.65$$

$$= \mathbf{1.65\ bar}$$

$\dot{V}_2 = 0.7\,\dot{V}_1 = 0.7 \times 0.995 = \mathbf{0.696\ m^3/s}$

$$\text{Power} = \frac{\gamma}{(\gamma - 1)} P_1 \dot{V}_1 \left[\left(\frac{P_2}{P_1}\right)^{(\gamma-1)/\gamma} - 1\right] + \dot{V}_2 (P_3 - P_2)$$

$$= \frac{1.4}{(1.4 - 1)} \times 1 \times 10^5 \times 0.995 \times \left[\left(\frac{1.65}{1}\right)^{0.4/1.4} - 1\right] + 0.696(2 - 1.65) \times 10^5$$

$$= [(3.5 \times 0.995 \times 0.154) + (0.696 \times 0.35)] \times 10^5$$
$$= (0.536 + 0.244) \times 10^5$$
$$= 0.780 \times 10^5$$
$$= 78\,000\ \text{W}$$
$$= \mathbf{78\ kW}$$

14.9 *Total or stagnation temperature and pressure*

If a gas stream is brought to rest, its total energy remains constant, so the loss of kinetic energy of the stream will appear as an increase in its temperature and pressure. This happens during the process of diffusion and in the vicinity of a temperature-sensing bulb suspended in a gas stream. The stream is brought to rest at the bulb, so the temperature recorded is higher than it would be if the device were moving with the stream. The temperature recorded with a fixed bulb is called the **total or stagnation** temperature. If the bulb were moving with the stream it would record the **static** temperature.

Stagnation temperatures can be used in rotary compressors if it is required to include inlet and exit velocities.

Equation [2] of section 14.8 shows the energy balance for the rotary compressor

$$h_1 + \frac{C_1^2}{2} + Q = h_2 + \frac{C_2^2}{2} + W \qquad [1]$$

For adiabatic compression $Q = 0$.

Also

$$(h_1 - h_2) = c_p(T_1 - T_2) \quad \text{(see section 5.8)}$$

Thus, equation [1] can be rewritten

$$c_p T_1 + \frac{C_1^2}{2} = c_p T_2 + \frac{C_2^2}{2} + W \qquad [2]$$

Let

$$c_p T + \frac{C^2}{2} = c_p T_t \qquad [3]$$

where T_t = total temperature

From this

$$T_t = T + \frac{C^2}{2c_p} \qquad [4]$$

Substituting equation [3] in equation [2]

$$c_p T_{t1} = c_p T_{t2} + W$$

from which

$$W = c_p (T_{t1} - T_{t2}) \qquad [5]$$

There is assumed to be no energy loss, so it can be assumed that the stream is brought to rest adiabatically. The total or stagnation pressure, P_t, can therefore be determined from

$$\frac{P_1}{P} = \left(\frac{T_t}{T}\right)^{\gamma/(\gamma - 1)} \qquad [6]$$

from which

$$P_t = P\left(\frac{T_t}{T}\right)^{\gamma/(\gamma - 1)} \qquad [7]$$

where P = static pressure

Example 14.8 *A rotary compressor is used to supercharge a petrol engine. The static pressure ratio across the rotor is 2.5:1. Static inlet pressure and temperature are 60 kN/m² and 5 °C, respectively. The air–fuel ratio is 13:1 and the engine consumes 0.04 kg fuel/s. For the air–fuel mixture take $\gamma = 1.39$ and $c_p = 1.005$ kJ/kg K. The compressor has an isentropic efficiency of 84 per cent. Using the static properties given, estimate the power required to drive the compressor. Taking the exit velocity from the compressor as 120 m/s and assuming that the air–fuel mixture is adiabatically brought to rest in the engine cylinders, estimate the total temperature and pressure of the mixture at the beginning of the compression stroke in the engine cylinder.*

SOLUTION

For the compressor

$$\frac{T_1}{T_{2'}} = \left(\frac{P_1}{P_2}\right)^{(\gamma - 1)/\gamma}$$

$$\therefore \quad T_{2'} = T_1\left(\frac{P_2}{P_1}\right)^{(\gamma - 1)/\gamma} = 278 \times 2.5^{0.39/1.39} = 278 \times 2.5^{1/3.56}$$

$$= 278 \times 1.294$$

$$= \mathbf{359.7\ K}$$

$$(T_2 - T_1) = \frac{(T_{2'} - T_1)}{0.84} = \frac{(359.7 - 278)}{0.84} = \frac{81.7}{0.84} = 97.3\ \text{K}$$

$$\therefore \quad T_2 = 278 + 97.3 = 375.3\ \text{K}$$

$$t_2 = 375.3 - 273 = \mathbf{102.3\ °C}$$

Mass of air–fuel mixture/s $= 0.04 \times (13 + 1)$

$= 0.04 \times 14$

$= \mathbf{0.56\ kg}$

$\therefore$ Power to drive compressor $= \dot{m}c_p(T_2 - T_1)$
$= 0.56 \times 1.005 \times 97.3$
$= \mathbf{54.8\ kW}$

$$T_{t2} = T_2 + \frac{C_2^2}{2c_p}$$

$$= 375.3 + \frac{120^2}{2 \times 1.005 \times 10^3}$$

$$= 375.3 + 7.16$$
$$= \mathbf{382.46\ K}$$

$$t_{t2} = 382.46 - 273 = \mathbf{109.46\ °C}$$

The temperature in the engine cylinder is 109.46 °C.

$$P_{t2} = P_2 \times \left(\frac{T_{t2}}{T_2}\right)^{\gamma(\gamma-1)} = 150 \times \left(\frac{382.46}{375.3}\right)^{3.56}$$

$$= 150 \times 1.019^{3.56}$$
$$= 150 \times 1.07$$
$$= \mathbf{160.5\ kN/m^2}$$

The pressure in the engine cylinder is 160.5 kN/m^2.

Questions

1. A single-stage, single-acting air compressor compresses 7 litres of air per second from a pressure of 0.101 3 MN/m^2 to a pressure of 1.4 MN/m^2. Compression follows the law $PV^{1.3} = C$. The mechanical efficiency of the compressor is 82 per cent. The effect of clearance can be neglected.

   Determine the power required to drive the compressor.

   [3.12 kW]

2. A single-stage, single-acting, reciprocating air compressor has a bore and stroke of 150 mm. The clearance volume is 6 per cent of the swept volume and the speed is 8 rev/s. Intake pressure is 100 kN/m^2 and the delivery pressure is 550 kN/m^2. The polytropic index is 1.32 throughout. Determine
   (a) the theoretical volumetric efficiency referred to intake conditions
   (b) the volume of air delivered per second at 550 kN/m^2
   (c) the air power of the compressor.

   [(a) 84.2%; (b) 4.9 litres; (c) 3.75 kW]

3. A three-stage, single-acting, reciprocating air compressor takes in 0.5 m^3 of air per second at a pressure of 100 kN/m^2 and at a temperature of 20 °C. The air is compressed to a delivery pressure of 2 MN/m^2. The intermediate pressures are ideal and intercooling between stages is perfect. The compression index can be taken as 1.25 in all stages. The compressor runs at 8.5 rev/s. Neglecting clearance, determine
   (a) the intermediate pressures
   (b) the total volume of each cylinder
   (c) the air power of the compressor

   [(a) 271.4 kN/m^2, 736.6 kN/m^2; (b) 0.058 8 m^3, 0.021 7 m^3, 0.008 m^3; (c) 165.8 kW]

4. A two-stage, single-acting, reciprocating air compressor has a low-pressure cylinder 250 mm diameter with a 250 mm stroke. The clearance volume of the low-pressure cylinder is 5 per cent of the stroke volume of the cylinder. The intake pressure and temperature are 100 kN/m^2 and 18 °C, respectively. Delivery pressure is 700 kN/m^2 and the compressor runs at 5 rev/s. The polytropic index is 1.3 throughout. The intermediate pressure is ideal and intercooling is complete. The overall efficiency of the plant, including the electric driving motor is 70 per cent. Take $R = 0.29$ kJ/kg K. Determine
   (a) the air mass flow rate through the compressor
   (b) the energy input to drive the motor

   [(a) 0.069 kg/s; (b) 18.1 kW]

5. A two-stage, single-acting, reciprocating air compressor delivers 0.07 m^3 of free air per second (free air conditions 101.325 kN/m^2 and 15 °C). Intake conditions are 95 kN/m^2 and 22 °C. Delivery pressure from the compressor is 1300 kN/m^2. The intermediate pressure is ideal and there is perfect intercooling. The compression index is 1.25 in both cylinders. The overall mechanical and electrical efficiency is 75 per cent. Neglecting clearance, determine
   (a) the energy input to the driving motor
   (b) the heat transfer per second in the intercooler
   (c) the percentage saving in work by using a two-stage intercooled compressor instead of a single-stage compressor.

   Take $c_p = 1.006$ kJ/kg K, $R = 0.287$ kJ/kg K.

   [(a) 29.1 kW; (b) 6.99 kJ/s; (c) 13%]

6. A rotary compressor is used as a supercharger to an aero-engine. When flying at a particular altitude the engine uses 4.6 kg of fuel per minute. The air–fuel mixture is compressed from a pressure of 55 kN/m^2 to a pressure of 100 kN/m^2 with an isentropic efficiency of 87 per cent. The volume of the mixture produced by 1 kg of fuel occupies 11.7 m^3 at 0 °C and 101.3 kN/m^2. The air–fuel ratio is 145:1 and the initial temperature of the mixture is −2 °C. If γ for the mixture is 1.37, determine the power required to drive the supercharger.

   [67.6 kW]

7. A rotary air compressor has an inlet static pressure and temperature of 100 kN/m^2 and 20 °C, respectively. The compressor has an air mass flow rate of 2 kg/s through a pressure ratio of 5:1. The isentropic efficiency of compression is 85 per cent. Exit velocity from the compressor is 150 m/s. Neglecting change of velocity through the compressor, determine the power required to drive the compressor. Estimate the total temperature and pressure at exit from the compressor.

   Take $\gamma = 1.4$ and $c_p = 1.005$ kJ/kg K.

   [404 kW, 232.2 °C, 541.5 kN/m^2]

8. A vane pump aspirates 1.25 m^3 of air and compresses it through an overall pressure ratio of 3:1. The pressure ratio due to volume reduction in the pump is 2:1. Intake pressure and temperature are 95 kN/m^2 and 18 °C, respectively. Determine the power required to drive the pump and compare this power with that required to drive a Roots blower which has the same mass flow rate and overall pressure ratio. Take $\gamma = 1.4$, $R = 0.287$ kJ/kg K.

   [163.4 kW, 237.5 kW]

Chapter 15

Ideal gas power cycles

15.1 General introduction

If a substance passes through a series of processes such that it is eventually returned to its original state, the substance is said to have been taken through a cycle (see section 1.13). During a cycle there will be some heat transfer and some work transfer to and from the substance. After performing a cycle, the substance is returned to its original state, so by the first law of thermodynamics

$$\oint Q = \oint W \qquad \text{(see section 3.3)} \tag{1}$$

Thus, for a cycle, the net work transfer can be determined by an analysis of the net heat transfer, or

$$\text{Net work done} = \text{Net heat received} - \text{Net heat rejected} \tag{2}$$

$$\text{The ratio } \frac{\text{Net work done}}{\text{Net heat received}}$$

is called the **thermal efficiency** (see section 1.30), or

$$\text{Thermal } \eta = \frac{\oint W}{\oint Q} \tag{3}$$

where $\oint W$ = net work done
$\oint Q$ = net heat received

And because the area under a process on a pressure–volume graph is equal to the work done, the net area of a pressure–volume diagram of a cycle is equal to the net work of the cycle. This, therefore, gives another method by which the net work of a cycle can be determined.

The equation

$$\text{Thermal } \eta = \frac{\oint W}{\oint Q}$$

gives the **theoretical** or **ideal** thermal efficiency. The **actual** thermal efficiency of a practical cycle is given by

$$\text{Actual thermal } \eta = \frac{\text{Actual work done}}{\text{Thermal energy from fuel}} \qquad [4]$$

This is always less than the theoretical thermal efficiency.

A practical cycle is carried out in an engine or turbine and will incur many losses, including heat transfer loss, fuel combustion loss, non-uniform energy distribution in the working substance, friction, leakage, the need to keep temperatures within practical working limits, the running of auxiliary equipment such as pumps, alternators, valve gear and cooling equipment.

The ratio of the actual thermal efficiency and the ideal thermal efficiency is called the **relative efficiency** or **efficiency ratio**, thus

$$\text{Relative efficiency} = \frac{\text{Actual thermal efficiency}}{\text{Ideal thermal efficiency}} \qquad [5]$$

Another useful concept is the **work ratio**. This is defined as

$$\text{Work ratio} = \frac{\text{Net work done}}{\text{Positive work done}} = \frac{\oint W}{\text{Positive work done}} \qquad [6]$$

where

$$\text{Net work done} = \text{Positive work done} - \text{Negative work done} = \oint W.$$

Note that, from equation [6], if the negative work is reduced then the work ratio tends to 1.

A cycle with good ideal thermal efficiency together with a good work ratio suggests good overall efficiency potential in a practical power plant using the cycle.

Work ratio can give comparative indication of plant size. A plant which has a low work ratio suggests that the work components of the plant are larger when compared with a plant which has a higher work ratio and similar power output.

Work ratio is most commonly applied to such cycles as arranged in steam plant and gas turbines. Such plants are composed of separate units each performing a particular function. In a steam plant there is the boiler, the engine or turbine, the condenser and the feed pump (see Chapter 10). In a gas turbine there is the compressor, the combustion chamber (or chambers) and the turbine (see Chapter 16).

Another comparison between cycles is the **specific steam** or **fuel consumption**.

In the case of steam plant

$$\text{Specific steam consumption} = \frac{\text{Mass flow of steam in kg/h}}{\text{Power output in kW}} \qquad [7]$$

This gives the mass of steam used per unit power output in kg/kWh.

In the case of internal combustion engines, such as the gas turbine and the petrol or diesel engine, the specific fuel consumption is used, where

$$\text{Specific fuel consumption} = \frac{\text{Mass of fuel used in kg/h}}{\text{Power output in kW}} \qquad [8]$$

This gives the mass of fuel used per unit power output in kg/kWh.

Thus a cycle which has a lower specific steam or fuel consumption indicates that it has better energy conversion performance than a cycle with a higher specific steam or fuel consumption.

The specific steam or fuel consumption can also be determined from a knowledge of the specific work output.

If

$$\text{Specific work output} = w \frac{\text{kJ}}{\text{kJ (steam or fuel)}}$$

then

$$\frac{1}{w}\frac{\text{kg}}{\text{kJ}} = \frac{1}{w}\frac{\text{kg}}{\text{kWs}} = \frac{3600}{w}\frac{\text{kg}}{\text{kWh}}$$

$$= \text{specific steam or fuel consumption} \qquad [9]$$

(1 kW = 1 kJ/s $\therefore$ 1 kJ = 1 kWs.)

In the case of reciprocating engines, such as steam, petrol and diesel engines, which use a piston–crank mechanism, comparison between cycles can be made using the **mean effective pressure**. This is the theoretical pressure which, if it were maintained constant throughout the volume change of the cycle (engine stroke of a practical cycle), would give the same work output as that obtained from the cycle.

Figure 15.1 illustrates a cycle plotted on a P–V diagram and operating between the volume limits V_1 and V_2. The area of the cycle diagram will determine the work done $= \oint W$. This is called the **indicated** work done. The stroke volume of the diagram is $V_1 - V_2$.

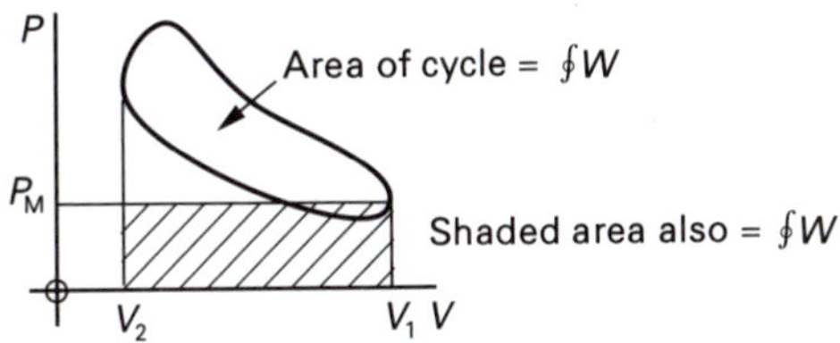

Fig. 15.1 Mean effective pressure

The mean effective pressure is determined by the equation

$$P_M = \frac{\oint W}{V_1 - V_2}, \qquad [10]$$

where P_M = mean effective pressure

Note that $\oint W/(V_1 - V_2)$ is also the work done per unit swept volume.

Thus a cycle with a higher mean effective pressure will indicate that it has better work characteristics than a cycle with a lower mean effective pressure.

If the effects of the fuel used are neglected, the gas may be considered as a close approximation of air alone. Thus, the theoretical cycles such as the constant volume, the constant pressure and the Diesel cycles are sometimes called the **air standard cycles** and the related efficiencies are called **air standard efficiencies**.

15.2 Survey of important gas power cycles

The following survey lists and illustrates some important gas power cycles together with their date of origin and the names of the people most commonly associated with their development. All the cycles were originally developed in the nineteenth century.

The illustrations of the cycles are P–V and T–s diagrams. They are listed in date order and each is accompanied by a brief note on past and present applications.

15.2.1 The Cayley cycle

The Cayley cycle (Fig. 15.2) consists of two constant pressure processes, 1–2 and 3–4, together with two polytropic processes, 2–3 and 4–1.

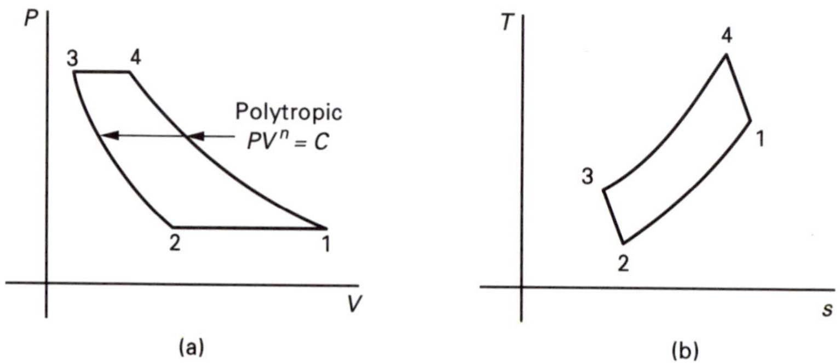

Fig. 15.2 Cayley cycle: (a) P–V diagram; (b) T–s diagram

It was probably the first cycle to use air, heated using an external furnace, for the development of mechanical power. It was originally developed in 1807 by Sir George Cayley, a British engineer. Further use of the cycle was subsequently made in 1883 by another British engineer named Buckett.

15.2.2 The Stirling cycle

The Stirling cycle (Fig. 15.3) consists of two isothermal processes, 1–2 and 3–4, together with two constant volume processes, 2–3 and 4–1 (see section 15.9).

The cycle was developed for use in air engines from 1815 by two Scottish brothers, James Stirling, an engineer, and Dr Robert Stirling, a church minister. A regenerator

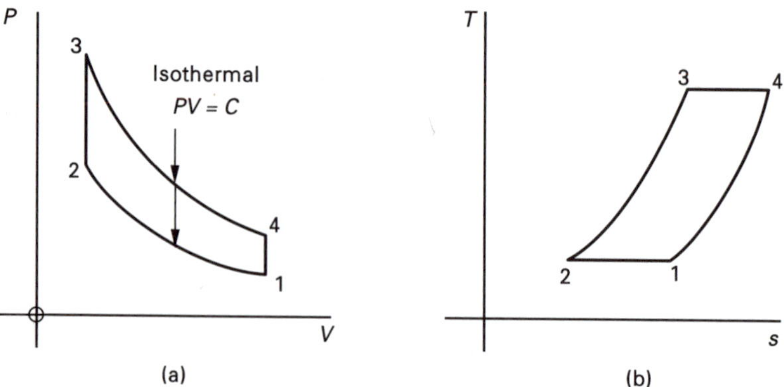

Fig. 15.3 Stirling cycle: (a) P–V diagram; (b) T–s diagram

was developed and used (see section 15.9) which improved upon the theoretical efficiency of the Carnot cycle (see section 15.3). Further work on this cycle was carried out by Wilhelm Lehmann in 1866 and the general theory of the cycle was developed by Professor Gustav Schmidt in 1871.

There is still some current interest in this cycle because it has the potential for high thermal efficiency.

15.2.3 *The Carnot cycle*

The Carnot cycle (Fig. 15.4) was conceived by Sadi Carnot in 1824 (see section 6.6). The cycle consists entirely of reversible processes (see Chapter 6).

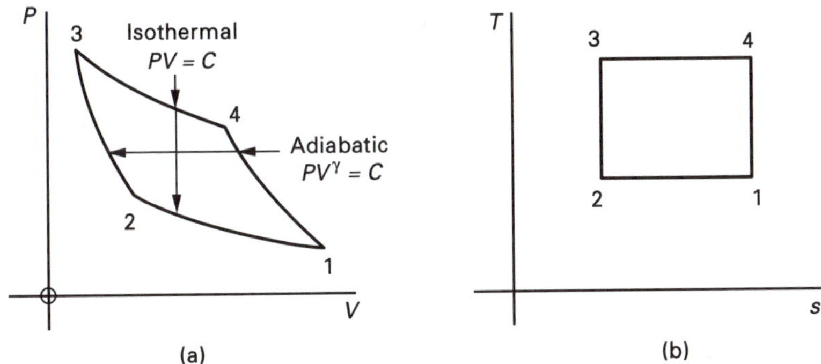

Fig. 15.4 Carnot cycle: (a) *P–V* diagram; (b) *T–s* diagram

The cycle consists of two isothermal processes, 1–2 and 3–4, together with two adiabatic processes, 2–3 and 4–1 (see section 13.3). No practical engine was ever built to run on this cycle.

It is important because it is composed of reversible processes, which are the most thermodynamically efficient processes, so its thermal efficiency establishes the maximum thermal efficiency possible within the temperature limits of the cycle (see section 15.3).

15.2.4 *The constant pressure cycle*

The constant pressure cycle (Fig. 15.5) consists of two constant pressure processes, 1–2 and 3–4, together with two adiabatic processes, 2–3 and 4–1 (see section 13.4).

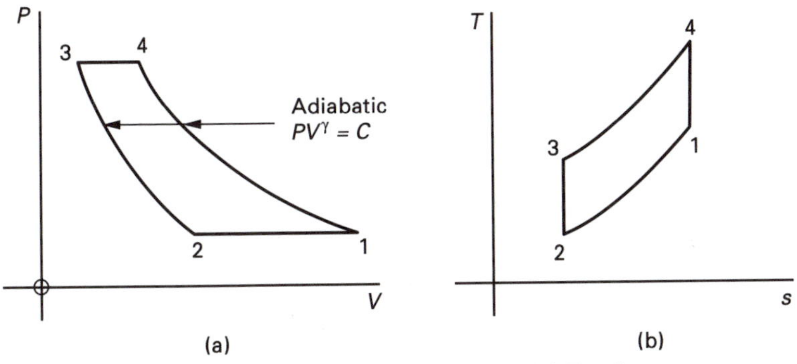

Fig. 15.5 Constant pressure cycle: (a) *P–V* diagram; (b) *T–s* diagram

The cycle was originated for use in a hot-air engine by John Ericsson in 1833 (see section 15.9).

James Prescott Joule, a British physicist immortalised in the SI unit of energy, gave some thought to the use of this cycle. There is no evidence that Joule ever constructed an engine to use the cycle. George Brayton, an American engineer, used the cycle in an internal combustion engine in 1876.

The cycle is of considerable present-day importance in that it is the basic cycle for the modern gas turbine (see Chapter 6). The reversed constant pressure cycle is also of importance because it was used in the early development of the process of refrigeration. It was used by Windhausen in 1870, Paul Giffard in 1877 and in the Bell–Coleman refrigerator of 1881 (see Chapter 18).

15.2.5 The Ericsson cycle

The Ericsson cycle (Fig. 15.6) consists of two isothermal processes, 2–3 and 4–1, together with two constant pressure processes, 1–2 and 3–4 (see section 15.10).

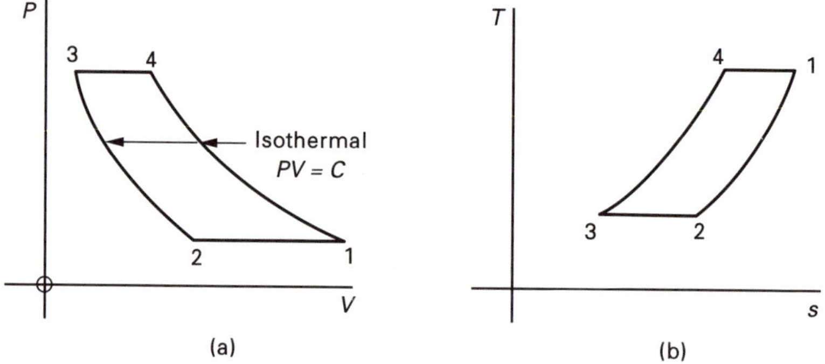

Fig. 15.6 Ericsson cycle: (a) *P–V* diagram; (b) *T–s* diagram

The cycle was originated by John Ericsson in 1853 for use in a hot-air engine. John Ericsson developed and constructed many engines for use with hot air and also with steam.

It is important because the two constant pressure processes are bounded by the same temperature limits which make regeneration possible, thus improving upon the theoretical thermal efficiency of the Carnot cycle (see section 15.3).

15.2.6 The Lenoir cycle

The Lenoir cycle (Fig. 15.7) consists of two constant pressure flow processes, 4–1 and 1–2, one constant volume non-flow process, 2–3, and one adiabatic non-flow process, 3–4.

The cycle was developed by Étienne Lenoir, a French engineer, for use in a gas engine. An air–gas mixture was drawn into a cylinder of constant pressure, 1–2. It was ignited at about midstroke using a spark. Combustion of the gas produced a rapid pressure rise, theoretically at constant volume, 2–3. The combustion products then expanded, theoretically adiabatically, 3–4, to be exhausted from the cylinder at constant pressure, 4–1.

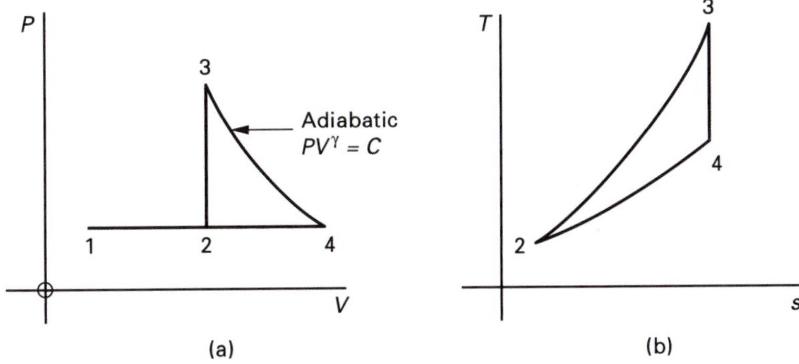

Fig. 15.7 Lenoir cycle: (a) *P–V* diagram; (b) *T–s* diagram

The engine was made to be double-acting, and slide valves similar to those in steam engines were used.

15.2.7 The constant volume cycle

The constant volume cycle (Fig. 15.8) consists of two adiabatic processes, 1–2 and 3–4, together with two constant volume processes, 2–3 and 4–1 (see section 15.5).

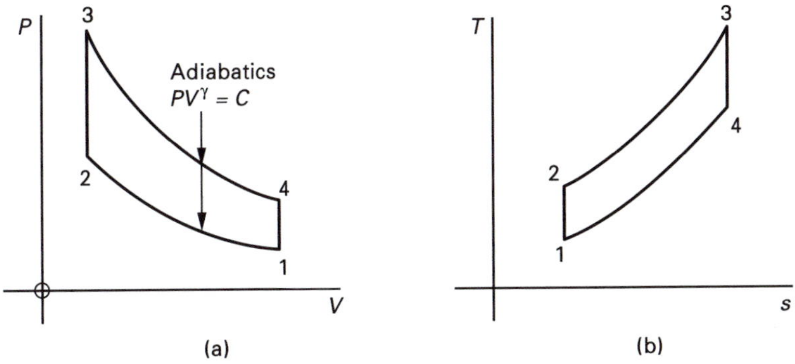

Fig. 15.8 Constant volume cycle: (a) *P–V* diagram; (b) *T–s* diagram

The cycle was originally conceived by a Frenchman named Beau de Rochas in 1862. And in 1867 it was successfully applied to an internal combustion engine by Dr Nikolaus August Otto, a German engineer, so it is often called the Otto cycle.

The cycle is of importance because it is the theoretical cycle on which the modern petrol engine is based.

15.2.8 The Reitlinger cycle

The Reitlinger cycle (Fig. 15.9) consists of two isothermal processes, 1–2 and 3–4, together with two polytropic processes, 2–3 and 4–1.

It was conceived by Edmund Reitlinger, an Austrian professor, in 1873. He observed that cycles containing isothermal processes had maximum work potential and that, with perfect regeneration, the Carnot thermal efficiency could be realised.

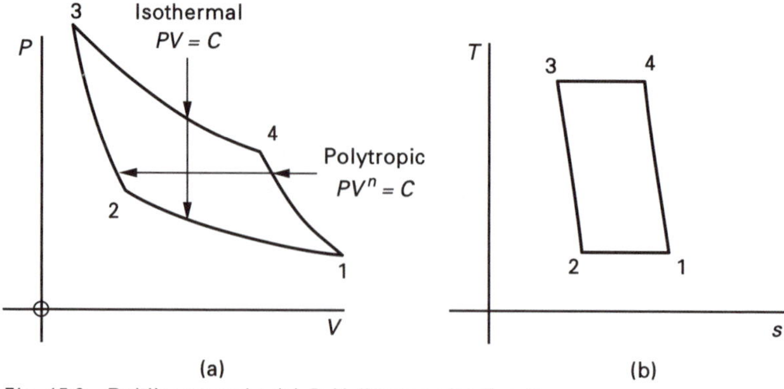

Fig. 15.9 Reitlinger cycle: (a) P–V diagram; (b) T–s diagram

The Reitlinger cycle contains two polytropic processes in place of the two adiabatic processes in the Carnot cycle. It is said, therefore, that Reitlinger generalised the Carnot cycle.

15.2.9 The Atkinson cycle

The Atkinson cycle (Fig. 15.10) consists of two adiabatic processes, 2–3 and 4–1, together with one constant volume process, 3–4, and one constant pressure process, 1–2 (see section 15.8).

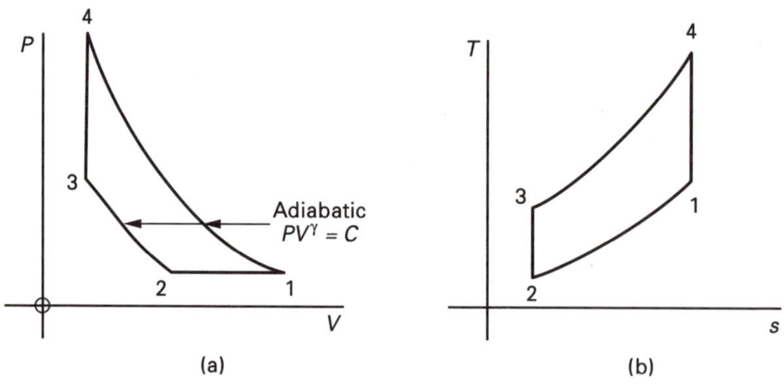

Fig. 15.10 Atkinson cycle: (a) P–V diagram; (b) T–s diagram

The cycle was conceived in 1886 by J. Atkinson, a British engineer, for use in a gas engine. The cycle was also used in the Holzwarth gas turbine in 1908.

15.2.10 The Diesel cycle

The Diesel cycle (Fig. 15.11) consists of two adiabatic processes, 1–2 and 3–4, one constant pressure process, 2–3, and one constant volume process, 4–1 (see section 15.6).

The cycle was conceived of by Rudolf Diesel, a German engineer, born in Paris, who from about 1893 onward pioneered the development of fuel injection oil engines. The diesel engine is named after him. Rudolf Diesel mysteriously disappeared while crossing the English Channel in 1913.

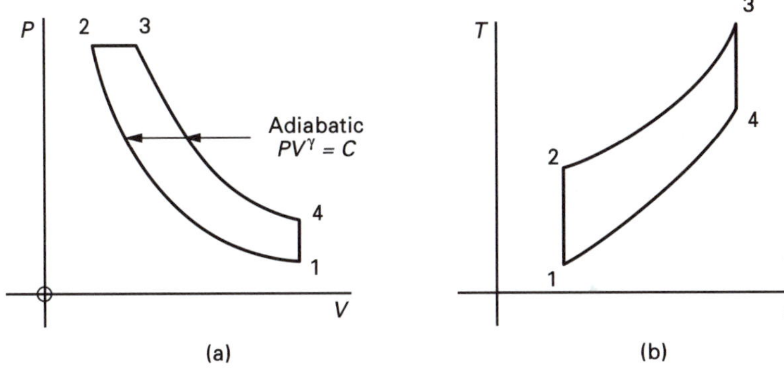

Fig. 15.11 Diesel cycle: (a) *P–V* diagram; (b) *T–s* diagram

15.2.11 The Crossley cycle

The Crossley cycle (Fig. 15.12) consists of two polytropic processes, 1–2 and 3–4, and two constant volume processes, 2–3 and 4–1.

The cycle is named after a British manufacturing company, Crossley, who made internal combustion engines in 1896. It is similar to the Otto cycle except that the two adiabatic processes of the Otto cycle are replaced by two polytropic processes. The polytropic processes were considered nearer to those which occurred in practice.

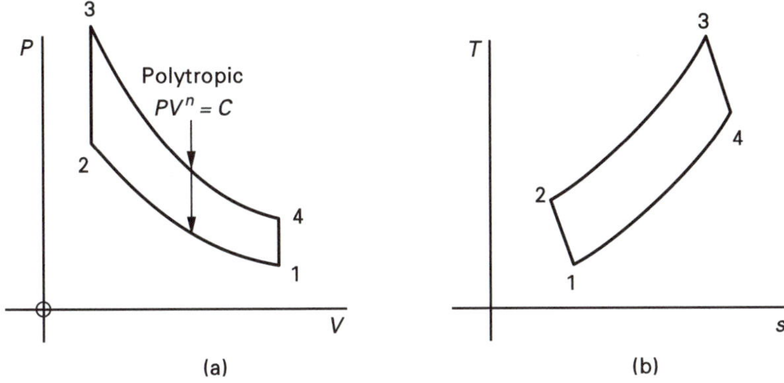

Fig. 15.12 Crossley cycle: (a) *P–V* diagram; (b) *T–s* diagram

15.3 The Carnot cycle for a gas

Carnot conceived a cycle made up of thermodynamically reversible processes. By calculating the thermal efficiency of this cycle it is possible to establish the maximum possible efficiency between the temperature limits taken (see sections 6.6 and 15.2).

Figure 15.13(a) illustrates the Carnot cycle on a *P–V* diagram and Fig. 15.13(b) on a *T–s* diagram. It consists of four reversible processes:

- **1–2 Isothermal expansion**
  Pressure falls from P_1 to P_2.
  Volume increases from V_1 to V_2.
  Temperature remains constant at $T_1 = T_2$.

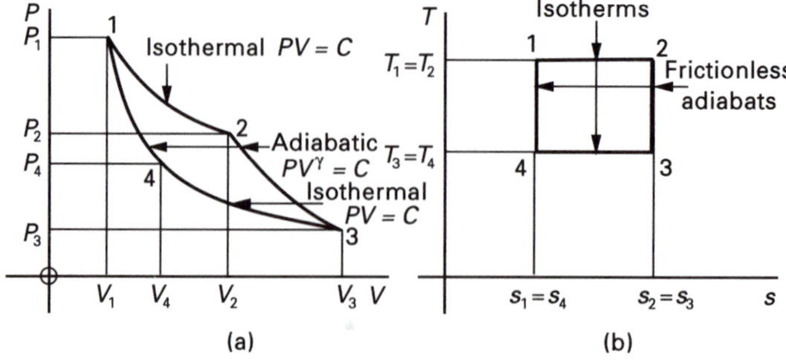

Fig. 15.13 Carnot cycle: (a) *P–V* diagram; (b) *T–s* diagram

$$\text{Work done} = P_1V_1 \ln \frac{V_2}{V_1} = mRT_1 \ln \frac{V_2}{V_1}$$

For an isothermal process, $Q = W$

$$\therefore \quad \text{Heat received} = mRT_1 \ln \frac{V_2}{V_1}$$

- **2–3 Adiabatic expansion**
  Pressure falls from P_2 to P_3.
  Volume increases from V_2 to V_3.
  Temperature falls from T_2 to T_3.

$$\text{Work done} = \frac{P_2V_2 - P_3V_3}{(\gamma - 1)} = \frac{mR(T_2 - T_3)}{(\gamma - 1)}$$

For an adiabatic process, $Q = 0$

$\therefore$ No heat transfer during this process

- **3–4 Isothermal compression**
  Pressure increases from P_3 to P_4.
  Volume reduced from V_3 to V_4.
  Temperature remains constant at $T_3 = T_4$.

$$\text{Work done} = P_3V_3 \ln \frac{V_4}{V_3} = -P_3V_3 \ln \frac{V_3}{V_4}$$
$$= -mRT_3 \ln \frac{V_3}{V_4}$$

For an isothermal process, $Q = W$

$$\therefore \quad \text{Heat rejected} = mRT_3 \ln \frac{V_3}{V_4}$$

- **4–1 Adiabatic compression**
  Pressure increases from P_4 to P_1.
  Volume reduced from V_4 to V_1.
  Temperature increases from T_4 to T_1.

$$\text{Work done} = \frac{P_4V_4 - P_1V_1}{(\gamma - 1)} = -\frac{(P_1V_1 - P_4V_4)}{(\gamma - 1)}$$

$$= -\frac{mR(T_1 - T_4)}{(\gamma - 1)}$$

For the adiabatic process, $Q = 0$

$\therefore$ No heat transfer during this process

Note that this process returns the gas to its original state at 1.

The net work done during this cycle may be determined by summing the areas beneath the various processes, taking the expansions as positive areas and the compressions as negative areas. Thus

$$\text{Net work done/cycle} = \oint W$$

$$= \text{Area under 1–2} + \text{Area under 2–3} - \text{Area under 3–4} - \text{Area under 4–1}$$

$$= \text{Area 1234}$$

$$= \text{Area enclosed by cycle}$$

or

$$\oint W = mRT_1 \ln \frac{V_2}{V_1} + \frac{mRT(T_2 - T_3)}{(\gamma - 1)} - mRT_3 \ln \frac{V_3}{V_4} - \frac{mR(T_1 - T_4)}{(\gamma - 1)} \quad [4]$$

Now $T_1 = T_2$ and $T_3 = T_4$, from the isotherms.

$$\therefore \quad \frac{mR(T_2 - T_3)}{(\gamma - 1)} = \frac{mR(T_1 - T_4)}{(\gamma - 1)}$$

Hence, from equation [1]

$$\oint W = mRT_1 \ln \frac{V_2}{V_1} - mRT_3 \ln \frac{V_3}{V_4} \quad [2]$$

Now for adiabatic 1–4,

$$\frac{T_1}{T_4} = \left(\frac{V_4}{V_1}\right)^{(\gamma - 1)} \quad [3]$$

for adiabatic 2–3

$$\frac{T_2}{T_3} = \left(\frac{V_3}{V_2}\right)^{(\gamma - 1)} \quad [4]$$

But $T_1 = T_2$ and $T_3 = T_4$

$$\therefore \quad \frac{T_1}{T_4} = \frac{T_2}{T_3} \quad [5]$$

Hence, from equations [3] and [4]

$$\frac{V_4}{V_1} = \frac{V_3}{V_2} \quad \text{or} \quad \frac{V_2}{V_1} = \frac{V_3}{V_4} \quad [6]$$

Substituting equation [6] in equation [2]

$$\oint W = mR \ln \frac{V_2}{V_1}(T_1 - T_3) \qquad [7]$$

This is positive work done and this is always the case if the processes of a cycle proceed in a clockwise direction. Net external work can thus be obtained from such cycles.

If the processes proceed in an anticlockwise direction then the work done is negative, in which case equation [7] now becomes

$$\oint W = -mR \ln \frac{V_2}{V_1}(T_1 - T_3) \qquad [8]$$

Negative work means that net external work must be put in to carry out such cycles.

Now

$$\text{Thermal } \eta = \frac{\text{Heat received} - \text{Heat rejected}}{\text{Heat received}}$$

So from the analysis given above

$$\text{Thermal } \eta = \frac{mRT_1 \ln (V_2/V_1) - mRT_3 \ln (V_3/V_4)}{mRT_1 \ln (V_2/V_1)}$$

$$= \frac{mR \ln (V_2/V_1)(T_1 - T_3)}{mR \ln (V_2 - V_1)T_1}$$

And since $V_2/V_1 = V_3/V_4$ from equation [6]

$$\text{Thermal } \eta = \frac{T_1 - T_3}{T_1} \qquad [9]$$

$$= \frac{\text{Max. abs. temp.} - \text{Min. abs. temp.}}{\text{Max. abs. temp.}} \qquad [10]$$

Now from equation [9]

$$\text{Thermal } \eta = 1 - \frac{T_3}{T_1} \qquad [11]$$

and from equations [3], [4] and [5]

$$\frac{T_1}{T_3} = \left(\frac{V_4}{V_1}\right)^{(\gamma - 1)} = \left(\frac{V_3}{V_2}\right)^{(\gamma - 1)} = r_v^{(\gamma - 1)}$$

where r_v = adiabatic compression and expansion volume ratio

So from equation [11]

$$\text{Thermal } \eta = 1 - \frac{1}{r_v^{(\gamma - 1)}} \qquad [12]$$

This thermal efficiency gives the maximum possible thermal efficiency obtainable between any two given temperature limits.

From the T–s diagram (Fig. 15.13(b))

$$\text{Heat received from 1 to 2} = T_1(s_1 - s_2) \qquad [13]$$
$$= \text{Area under 1–2}$$

$$\text{Heat rejected from 3 to 4} = T_3(s_3 - s_4) \qquad [14]$$
$$= \text{Area under 3–4}$$

$$\text{Thermal efficiency} = \frac{\text{Heat received} - \text{Heat rejected}}{\text{Heat received}}$$

$$= \frac{T_1(s_2 - s_1) - T_3(s_3 - s_4)}{T_1(s_2 - s_1)}$$

$$= \frac{(T_1 - T_3)(s_2 - s_1)}{T_1(s_2 - s_1)} \quad \text{since } (s_2 - s_1) = (s_3 - s_4)$$

$$= \frac{T_1 - T_3}{T_1} \qquad [15]$$

Note that this solution for the thermal efficiency of the Carnot cycle is somewhat simpler than that given in the previous work. Also

$$\oint W = \text{Heat received} - \text{Heat rejected}$$
$$= (T_1 - T_2)(s_2 - s_1) \qquad [16]$$

Now, the Carnot cycle has the maximum thermal efficiency obtainable within given temperature limits, so it is possible to suggest that if any engine working between the same temperature limits has a thermal efficiency lower than that of the Carnot cycle, then thermal efficiency improvement for the engine is theoretically possible. All practical engines have a thermal efficiency much lower than the Carnot thermal efficiency.

The ultimate aim, in practice, should be an attempt to reach an efficiency as near 100 per cent as possible. How could this be achieved?

$$\text{Carnot thermal } \eta = \frac{\text{Max. abs. temp.} - \text{Min. abs. temp.}}{\text{Max. abs. temp.}}$$

For this equation to be a maximum it must equal unity, in which case the thermal efficiency would be 100 per cent.

Consider the minimum absolute temperature as being fixed. If the maximum absolute temperature is increased, the magnitude of the right-hand quotient gets larger. Eventually, as the maximum absolute temperature $\rightarrow \infty$, the quotient $\rightarrow 1$. It is quite impossible to have an infinitely high maximum absolute temperature. In any case, contemporary engine materials will not stand up to continuous exposure at the high temperatures which are obtainable.

Now consider the maximum absolute temperature as being fixed. If the minimum absolute temperature is reduced to zero (to the absolute zero of temperature) then, once again, the quotient $\rightarrow 1$ and the Carnot efficiency $\rightarrow 100$ per cent. This too is impossible because all working substances will have liquefied and solidified before reaching this low temperature. In any case, the absolute zero of temperature is such a difficult temperature to achieve that it is out of the question as the sink temperature of an engine.

It would appear from this that a thermal efficiency of 100 per cent is impossible to achieve in practice. Hence all practical engines are inefficient. However, a guide has been given as to how the thermal efficiency of engines delivering net work may be improved. That is, to spread the maximum and minimum temperatures as far apart as possible, consistent with the satisfactory safe operation of the engine.

In section 15.1, the concept of work ratio for a cycle was discussed. This was defined as

$$\text{Work ratio} = \frac{\text{Net work done}}{\text{Positive work done}}$$

From equation [7] for the net positive work Carnot cycle

$$\text{Net work done} = \oint W = mR \ln \frac{V_2}{V_1}(T_1 - T_3)$$

The positive work done during the Carnot cycle occurs during process 1–2 and 2–3 (see Fig. 15.13(a)). The area under these processes shown on the P–V diagram gives the positive work done, thus

$$\text{Positive work done} = mRT_1 \ln \frac{V_2}{V_1} + \frac{mR(T_1 - T_3)}{(\gamma - 1)} \qquad [17]$$

Hence, for the net positive work Carnot cycle

$$\text{Work ratio} = \frac{mR \ln \frac{V_2}{V_1}(T_1 - T_3)}{mRT_1 \ln \frac{V_2}{V_1} + \frac{mR(T_1 - T_3)}{(\gamma - 1)}}$$

$$= \frac{\ln \frac{V_2}{V_1}(T_1 - T_3)}{T_1 \ln \frac{V_2}{V_1} + \frac{(T_1 - T_3)}{(\gamma - 1)}} \qquad [18]$$

In section 13.1 it was suggested that good ideal thermal efficiency together with good work ratio is required of a cycle, if possible. It is unfortunate that the Carnot cycle has a low work ratio even though it has the highest ideal thermal efficiency. The Carnot cycle is sometimes called the **constant temperature cycle** because heat is transferred during the isothermal processes only.

Example 15.1 *A Carnot cycle using a gas has temperature limits 400 °C and 70 °C. Determine the thermal efficiency of the cycle.*

SOLUTION

From equation [15], section 15.3

$$\text{Thermal } \eta = \frac{T_1 - T_3}{T_1}$$

Now $T_1 = 400 + 273 = \mathbf{673\ K}$
$T_3 = 70 + 273 = \mathbf{343\ K}$

$$\therefore \quad \text{Thermal } \eta = \frac{673 - 343}{673}$$

$$= \frac{330}{673}$$

$$= 0.49$$
$$= \mathbf{49\%}$$

Example 15.2 *The overall volume expansion ratio of a Carnot cycle is 15. The temperature limits of the cycle are 260 °C and 21 °C. Determine*
(a) the volume ratios of the isothermal and adiabatic processes
(b) the thermal efficiency of the cycle
Take $\gamma = 1.4$.

SOLUTION
First draw a diagram (Fig. 15.14).

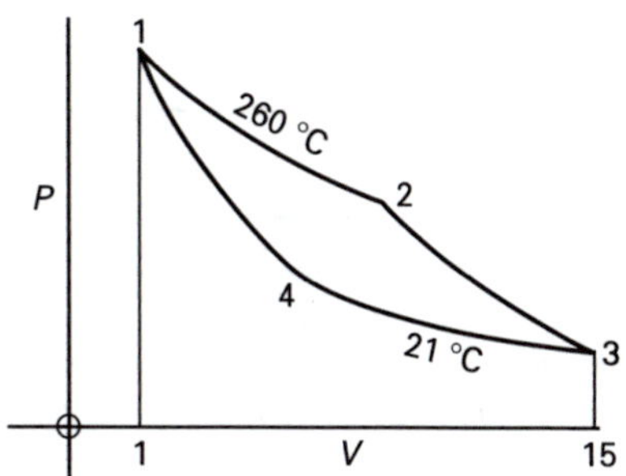

Fig. 15.14 Diagram for Example 15.2

(a)
For the adiabatic processes

$$\frac{T_1}{T_4} = \frac{T_2}{T_3} = \left(\frac{V_4}{V_1}\right)^{(\gamma - 1)} = \left(\frac{V_3}{V_2}\right)^{(\gamma - 1)}$$

$$\therefore \quad \frac{V_4}{V_1} = \frac{V_3}{V_2} = \left(\frac{T_1}{T_4}\right)^{1/(\gamma - 1)} = \left(\frac{T_2}{T_3}\right)^{1/(\gamma - 1)}$$

and $T_1 = T_2 = 260 + 273 = \mathbf{533\ K}$
$T_3 = T_4 = 21 + 273 = \mathbf{294\ K}$

$$= \left(\frac{533}{294}\right)^{1/(1.4 - 1)}$$

$$= 1.812^{1/0.4}$$
$$= \mathbf{4.426}$$

The volume ratio of adiabatics r_v is 4.426

$$\text{Volume ratio of isothermals} = \frac{V_3}{V_4} = \frac{V_3}{V_1}\frac{V_1}{V_4}$$

$$= \frac{15}{4.426}$$

$$= \mathbf{3.39}$$

(b)

$$\text{Thermal efficiency} = \frac{T_1 - T_4}{T_1}$$

$$= \frac{533 - 294}{533}$$

$$= \frac{239}{533}$$

$$= 0.448$$
$$= \mathbf{44.8\%}$$

Alternatively

$$\text{Thermal efficiency} = 1 - \frac{1}{r_v^{(\gamma-1)}} = 1 - \frac{1}{4.426^{(1.4-1)}} = 1 - \frac{1}{4.426^{0.4}}$$

$$= 1 - \frac{1}{1.812}$$

$$= 1 - 0.552$$
$$= 0.448$$
$$= \mathbf{44.8\%}$$

Example 15.3 *One kilogram of air is taken through a Carnot cycle. The initial pressure and temperature of the air are 1.73 MN/m^2 and 300 °C, respectively. From the initial conditions, the air is expanded isothermally to three times its initial volume and then further expanded adiabatically to six times its initial volume. Isothermal compression, followed by adiabatic compression, completes the cycle. Determine*

(a) the pressure, volume and temperature at each corner of the cycle
(b) the thermal efficiency of the cycle
(c) the work done per cycle
(d) the work ratio

Take R = *0.29 kJ/kg K,* $\gamma = 1.4$.

SOLUTION

First draw a diagram (Fig. 15.15).

(a)

For the isothermal process 1–2

Pressure $P_1 = 1.73$ MN/m^2; Temperature $t_1 = 300$ °C

Now $P_1V_1 = mRT_1$ and $T_1 = 300 + 273 = \mathbf{573\ K}$

$$\therefore \quad V_1 = \frac{mRT_1}{P_1} = \frac{1 \times 0.29 \times 573}{1730}$$

$$= \mathbf{0.096\ m^3}$$

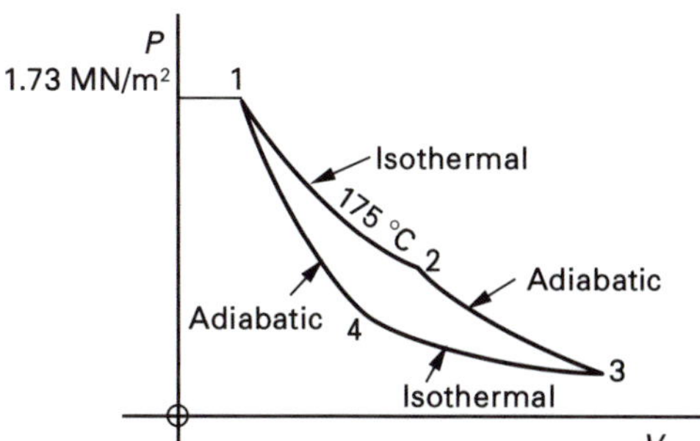

Fig. 15.15 Diagram for Example 15.3

Temperature remains constant at $t_2 = 300$ °C. The volume becomes three times the initial volume.

$$\therefore \quad V_2 = 3 \times V_1 = 3 \times 0.096 = \mathbf{0.288\ m^3}$$

For an isothermal process, $P_1 V_1 = P_2 V_2$

$$\therefore \quad P_2 + \frac{P_1 V_1}{V_2} = \frac{P_1}{3} = \frac{1730}{3} = \mathbf{576.7\ kN/m^2}$$

For the adiabatic process 2–3

$$\text{Volume } V_3 = 6 \times V_1 = 6 \times 0.096 = \mathbf{0.576\ m^3}$$

For the adiabatic process, $T_1/T_3 = (V_3/V_2)^{(\gamma-1)}$

$$\therefore \quad T_3 = T_2\left(\frac{V_2}{V_3}\right)^{(\gamma-1)} = 573 \times \left(\frac{0.288}{0.576}\right)^{(1.4-1)}$$

$$= 573 \times \left(\frac{1}{2}\right)^{0.4} = \frac{573}{2^{0.4}}$$

$$= \frac{573}{1.32}$$

$$= \mathbf{434\ K}$$

$$t_3 = 434 - 273 = \mathbf{161\ °C}$$

Also $P_2 V_2^{\gamma} = P_3 V_3^{\gamma}$

$$\therefore \quad P_3 = P_2\left(\frac{V_2}{V_3}\right)^{\gamma} = \frac{576.7}{2^{1.4}}$$

$$= \frac{576}{2.639}$$

$$= \mathbf{218.53\ kN/m^2}$$

For the isothermal process 3–4

$$t_3 = t_4 = \mathbf{161\ °C}$$

Now for both adiabatic processes, the temperature ratio is the same because they both have the same end temperatures.

$$\therefore \quad \frac{T_1}{T_4} = \frac{T_2}{T_3} = \left(\frac{V_4}{V_1}\right)^{(\gamma-1)} = \left(\frac{V_3}{V_2}\right)^{(\gamma-1)}$$

$$\therefore \quad \frac{V_4}{V_1} = \frac{V_3}{V_2} = 2$$

$$\therefore \quad V_4 = 2V_1 = 2 \times 0.096 = \mathbf{0.192\ m^3}$$

For the isothermal process, $P_3V_3 = P_4V_4$

$$\therefore \quad P_4 = P_3\frac{V_3}{V_4} = 218.53 \times \frac{0.576}{0.192} = 218.3 \times 3 = \mathbf{655.6\ kN/m^2}$$

Alternatively, $P_1V_1^{\gamma} = P_4V_4^{\gamma}$

$$\therefore \quad P_4 = P_1\left(\frac{V_1}{V_4}\right)^{\gamma} = 1730 \times \left(\frac{1}{2}\right)^{1.4}$$

$$= \frac{1730}{2^{1.4}} = \frac{1730}{2.639}$$

$$= \mathbf{655.6\ kN/m^2}$$

(b)

$$\text{Thermal efficiency} = \frac{T_1 - T_3}{T_1}$$

$$= \frac{573 - 464}{573}$$

$$= \frac{109}{573}$$

$$= 0.19$$
$$= \mathbf{19\%}$$

(c)

$$\oint W = mR \ln \frac{V_2}{V_1}(T_1 - T_3)$$

$$= 1 \times 0.29 \times \ln 3 \times (573 - 464)$$
$$= 1 \times 0.29 \times 1.099 \times 109$$
$$= \mathbf{34.7\ kJ}$$

Alternatively

$$\oint W = \text{Heat received} \times \text{Thermal efficiency}$$

$$= mRT_1 \ln \frac{V_2}{V_1} \times 0.19$$

$$= 1 \times 0.29 \times 573 \times 1.099 \times 0.19$$
$$= \mathbf{34.7\ kJ}$$

(d)

$$\text{Work ratio} = \frac{\ln \dfrac{V_2}{V_1}(T_1 - T_3)}{T_1 \ln \dfrac{V_2}{V_1} + \dfrac{(T_1 - T_3)}{(\gamma - 1)}}$$

$$= \frac{1.099 \times 109}{(573 \times 1.099) + \dfrac{109}{0.4}}$$

$$= \frac{119.79}{629.73 + 272.5}$$

$$= \frac{119.79}{902.23}$$

$$= \mathbf{0.133}$$

Note that this is a very low work ratio.

15.4 The constant pressure cycle

A general discussion of this cycle is given in section 15.3. Figure 15.16(a) is the *P–V* diagram of the constant pressure cycle; Fig. 15.16(b) shows the corresponding *T–s* diagram. The cycle is arranged as follows:

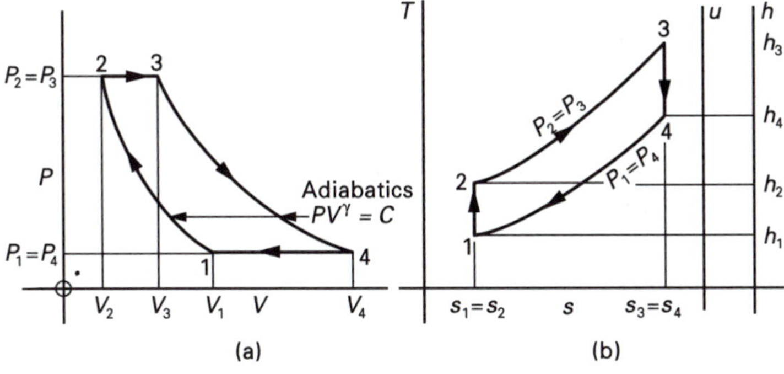

Fig. 15.16 Constant pressure cycle: (a) *P–V* diagram; (b) *T–s* diagram

- **1–2 Adiabatic compression according to the law $PV^\gamma = C$**
  Pressure increases from P_1 to P_2.
  Temperature increases from T_1 to T_2.
  Volume decreases from V_1 to V_2.
  Entropy remains constant at $s_1 = s_2$.

- **2–3 Constant pressure heat addition**
  Pressure remains constant at $P_2 = P_3$.
  Temperature increases from T_2 to T_3.
  Volume increases from V_2 to V_3.
  Entropy increases from s_2 to s_3.

- **3–4 Adiabatic expansion according to the law $PV^{\gamma} = C$**
  Pressure decreases from P_3 to P_4.
  Temperature decreases from T_3 to T_4.
  Volume increases from V_3 to V_4.
  Entropy remains constant at $s_3 = s_4$.

- **4–1 Constant pressure heat rejection**
  Pressure remains constant at $P_4 = P_1$.
  Temperature decreases from T_4 to T_1.
  Volume decreases from V_4 to V_1.
  Entropy decreases from s_4 to s_1.
  This process completes the cycle and returns the gas to its original state.

In practice this cycle has been mostly considered for use in engines which do not carry out all processes in a single unit, e.g. a cylinder. Compression and expansion have been arranged in separate units.

Figure 15.17(a) illustrates the general arrangement of such an engine. A reciprocating air compressor is connected to a reciprocating expander using a coupling rod. In the past, the connection was often made using an oscillating beam. The net work output appears at the piston rod.

Air from the compressor is fed through a heater in which it is expanded, theoretically at constant pressure. From the heater the air is fed to the expander.

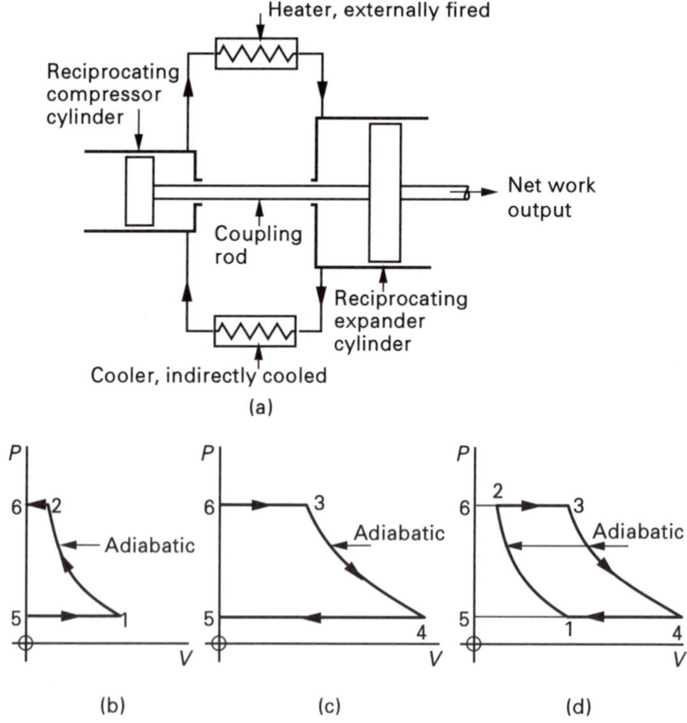

Fig. 15.17 Practical constant pressure cycle: (a) circuit diagram; (b) *P–V* diagram for compressor; (c) *P–V* diagram, for expander; (d) combined *P–V* diagram

From the expander the air enters a cooler, theoretically at constant pressure. The air is then fed to the compressor for recirculation. The system is closed; the same air is circulated through the engine, neglecting losses. Figure 15.17(b) and (c) show the P–V diagrams for the compressor and expander. The combined diagram is shown in Fig. 15.17(d). The net diagram for the engine is given by 1234, which is the same as Fig. 15.16. The net diagram 1234 is called the **constant pressure cycle** and its area will give the theoretical net work output.

Note that the expander must provide the internal work for the compressor as well as the net work output. By replacing the reciprocating compressor and expander with a rotary compressor and a turbine, this same arrangement becomes the basic design for a constant pressure gas turbine. The gas turbine is discussed in Chapter 16. And by reversing the airflow and driving the machine, this arrangement can be used as a refrigerator (see Chapter 18).

Referring to Fig. 15.16, an analysis of the properties at state points, 1, 2, 3 and 4 will now be made. It will be assumed that P_1, V_1 and T_1 are known.

- P_1, V_1, T_1
- Assume that the volume ratio V_1/V_2 is known.

$$T_1/T_2 = (V_2/V_1)^{(\gamma-1)}$$

$$\therefore \quad T_2 = T_1\left(\frac{V_1}{V_2}\right)^{(\gamma-1)} = T_1 r_v^{(\gamma-1)}$$

where $r_v = V_1/V_2 =$ adiabatic compression volume ratio

Also $P_1 = V_1^{\gamma} = P_2 V_2^{\gamma}$

$$\therefore \quad P_2 = P_1\left(\frac{V_1}{V_2}\right)^{\gamma} = P_1 r_v^{\gamma}$$

- $P_3 = P_2$ because the pressure remains constant

$$V_3/T_3 = V_2/T_2$$

$$\therefore \quad T_3 = T_2\frac{V_3}{V_2} = \frac{V_3}{V_2}T_1 r_v^{(\gamma-1)}$$

- $T_4/T_3 = (V_3/V_4)^{(\gamma-1)}$

$$\therefore \quad T_4 = T_3\left(\frac{V_3}{V_4}\right)^{(\gamma-1)}$$

Now consider adiabatic processes 1–2 and 3–4. Both have the same pressure ratios, $P_2/P_1 = P_3/P_4$.

For adiabatic 1–2

$$\frac{P_2}{P_1} = \left(\frac{V_1}{V_2}\right)^{\gamma}$$

For adiabatic 3–4

$$\frac{P_3}{P_4} = \left(\frac{V_4}{V_3}\right)^{\gamma}$$

But $P_2/P_1 = P_3/P_4$, so it follows that

$$\frac{V_1}{V_2} = \frac{V_4}{V_3} = r^{v}$$

= adiabatic compression and expansion volume ratio

Hence

$$T_4 = \frac{T_3}{r_v^{(\gamma-1)}} = \frac{V_3}{V_2} T_1 \frac{r_v^{(\gamma-1)}}{r_v^{(\gamma-1)}} = \frac{V_3}{V_2} T_1$$

Note also that for constant pressure process 4–1, $V_4/T_4 = V_1/T_1$

$$\therefore \quad T_4 = \frac{V_4}{V_1} T_1$$

So it follows that

$$\frac{V_4}{V_1} = \frac{V_3}{V_2} = \text{constant pressure process volume ratios}$$

This could also have been obtained from the fact that $V_1/V_2 = V_4 V_3 = r_v$, so $V_4/V_1 = V_3/V_2$.

Also $P_4 V_4^{\gamma} = P_3 V_3^{\gamma}$

$$\therefore \quad P_4 = P_3 \left(\frac{V_3}{V_4}\right)^{\gamma} = \frac{P_3}{r_v^{\gamma}}$$

The work done during the cycle may be obtained as follows:

- Processes 2–3 and 3–4 are expansions and give positive work done.
- Processes 4–1 and 1–2 are compressions and give negative work done.
- The net work done during the cycle will be the sum of the work done during these processes.

Hence

$$\oint W = \text{Area under 2–3} + \text{Area under 3–4} - \text{Area under 4–1} - \text{Area under 1–2}$$

$$= P_2(V_3 - V_2) + \frac{(P_3V_3 - P_4V_4)}{(\gamma - 1)} - P_1(V_4 - V_1) - \frac{(P_2V_2 - P_1V_1)}{(\gamma - 1)}$$

$$= P_2(V_3 - V_2) - P_1(V_4 - V_1) + \frac{(P_3V_3 - P_4V_4) - (P_2V_2 - P_1V_1)}{(\gamma - 1)} \qquad [1]$$

$$= mR(T_3 - T_2) - mR(T_4 - T_1) + \frac{mR}{(\gamma - 1)}[(T_3 - T_4) - (T_2 - T_1)]$$

$$= mR\left[(T_3 - T_2) - (T_4 - T_1) + \frac{(T_3 - T_4) - (T_2 - T_1)}{(\gamma - 1)}\right]$$

$$= mR\left[(T_3 - T_4) - (T_2 - T_1) + \frac{(T_3 - T_4) - (T_2 - T_1)}{(\gamma - 1)}\right]$$

$$= mR[(T_3 - T_4) - (T_2 - T_1)]\left(1 + \frac{1}{\gamma - 1}\right)$$

$$= mR[(T_3 - T_4) - (T_2 - T_1)]\left(\frac{\gamma - 1 + 1}{\gamma - 1}\right)$$

$$= mR\frac{\gamma}{(\gamma - 1)}[(T_3 - T_4) - (T_2 - T_1)] \qquad [2]$$

Alternatively, the work done may be obtained from

$$\oint W = \text{Heat received} - \text{Heat rejected}$$

In this cycle, heat is received during constant pressure process 2–3 and rejected during constant pressure process 4–1.

No heat is received or rejected during the adiabatic processes. Hence

$$\oint W = mc_p(T_3 - T_2) - mc_p(T_4 - T_1)$$

$$= mc_p[(T_3 - T_2) - (T_4 - T_1)] \qquad [3]$$

Again, the work done may be obtained from

$$\oint W = \text{Heat received} \times \text{Thermal } \eta$$

$$\therefore \quad \oint W = mc_p(T_3 - T_2) \times \text{Thermal } \eta \qquad [4]$$

The thermal efficiency may be obtained from

$$\text{Thermal } \eta = 1 - \frac{\text{Heat rejected}}{\text{Heat received}}$$

$$= 1 - \frac{mc_p(T_4 - T_1)}{mc_p(T_3 - T_2)}$$

and assuming that c_p remains constant

$$\text{Thermal } \eta = 1 - \frac{(T_4 - T_1)}{(T_3 - T_2)} \qquad [5]$$

Equation [5] gives the thermal efficiency in terms of temperature.

Also, substituting temperatures in terms of T_1

$$\text{Thermal } \eta = 1 - \frac{(V_3/V_2)T_1 - T_1}{(V_3/V_2)T_1 r_v^{(\gamma-1)} - T_1 r_v^{(\gamma-1)}}$$

$$= 1 - \frac{T_1(V_3/V_2 - 1)}{T_1 r_v^{(\gamma-1)}(V_3/V_2 - 1)}$$

$$= 1 - \frac{1}{r_v^{(\gamma-1)}} \qquad [6]$$

Now consider the adiabatic processes 1–2 and 3–4. Each has the same pressure ratio.

$$\frac{T_2}{T_1} = \left(\frac{P_2}{P_1}\right)^{(\gamma-1)/\gamma} = \left(\frac{V_1}{V_2}\right)^{(\gamma-1)} = r_v^{(\gamma-1)}$$

also

$$\frac{T_3}{T_4} = \left(\frac{P_3}{P_4}\right)^{(\gamma-1)/\gamma} = \left(\frac{V_4}{V_3}\right)^{(\gamma-1)} = r_v^{(\gamma-1)}$$

and since $P_2/P_1 = P_3/P_4$, then

$$\frac{T_2}{T_2} = \frac{T_3}{T_4} = r_v^{(\gamma-1)} \quad \text{or} \quad \frac{T_1}{T_2} = \frac{T_4}{T_3} = \frac{1}{r_v^{(\gamma-1)}}$$

Hence, from equation [6]

$$\text{Thermal } \eta = 1 - \frac{T_1}{T_2} \qquad [7]$$

$$= 1 - \frac{T_4}{T_3} \qquad [8]$$

Now for a constant pressure process

Heat received or rejected = Change of enthalpy

So using a T–s chart

Heat received = $h_3 - h_2$
Heat rejected = $h_4 - h_1$

$$\text{Thermal } \eta = 1 - \frac{\text{Heat rejected}}{\text{Heat received}}$$

$$= 1 - \frac{(h_4 - h_1)}{(h_3 - h_2)} \qquad [9]$$

Also

$$\oint W = \text{Heat received} - \text{Heat rejected}$$

$$= (h_3 - h_2) - (h_4 - h_1) \qquad [10]$$

This will give the work done per unit mass per cycle because the chart will be made out for unit mass of gas.

Example 15.4 *In an ideal constant pressure cycle, using air, the overall volume ratio of the cycle is 8:1. The volume ratio of the adiabatic compression is 6:1. The pressure, volume and temperature of the air at the beginning of the adiabatic compression are 100 kN/m², 0.084 m³ and 28 °C, respectively.*

Take, $\gamma = 1.4$, $c_p = 1.006$ *kJ/kg K. Determine for the cycle*

(a) the pressure, volume and temperature at cycle state points
(b) the heat received
(c) the work done
(d) the thermal efficiency
(e) the Carnot efficiency within the same temperature limits as the cycle
(f) the work ratio
(g) the mean effective pressure

SOLUTION

First draw a diagram (Fig. 15.18)

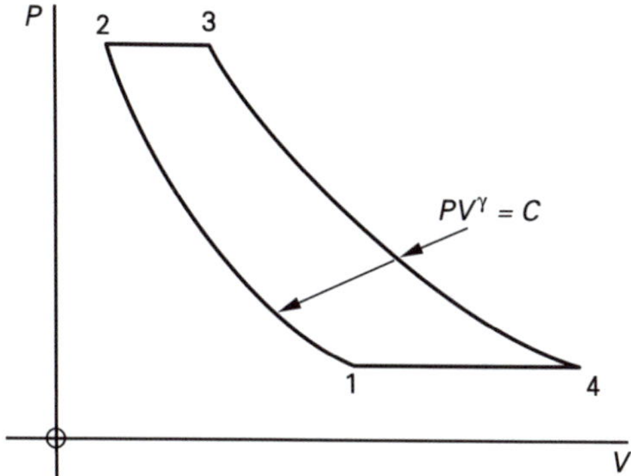

Fig. 15.18 Diagram for Example 15.4

(a)

$$V_1 = \mathbf{0.084\ m^3}$$

$$\therefore \quad V_2 = \frac{0.084}{6} = \mathbf{0.014\ m^3}$$

$$V_4 = 8 \times V_2 = 8 \times 0.014 = \mathbf{0.112\ m^3}$$

Now

$$P_1 V_1^{\gamma} = P_2 V_2^{\gamma}$$

$$\therefore \quad P_2 = P_1 \left(\frac{V_1}{V_2}\right)^{\gamma} = 100 \times 6^{1.4} = 100 \times 12.29 = \mathbf{1229\ kN/m^2}$$

$$\frac{T_1}{T_2} = \left(\frac{V_2}{V_1}\right)^{(\gamma - 1)} \quad \text{and} \quad T_1 = 28 + 273 = \mathbf{301\ K}$$

$$\therefore \quad T_2 = T_1 \left(\frac{V_1}{V_2}\right)^{(\gamma - 1)} = 301 \times 6^{0.4}$$

$$= 301 \times 2.048$$

$$= \mathbf{616\ K}$$

$$t_2 = 616 - 273 = \mathbf{343\ ^\circ C}$$

$$P_3 = P_2 = \mathbf{1229\ kN/m^2}$$

$$r_v = V_1/V_2 = V_4/V_3 = \mathbf{6}$$

$$\therefore \quad V_3 = \frac{V_4}{6} = \frac{0.112}{6} = \mathbf{0.018\ 7\ m^3}$$

$$\frac{V_3}{T_3} = \frac{V_2}{T_2}$$

$$\therefore \quad T_3 = T_2 \frac{V_3}{V_2} = 616 \times \frac{0.018\ 7}{0.014} = \mathbf{823\ K}$$

$$t_3 = 823 - 273 = \mathbf{550\ ^\circ C}$$

$$P_4 = \mathbf{100\ kN/m^2}, \quad V_4 = \mathbf{0.112\ m^3}$$

$$\frac{T_3}{T_4} = \left(\frac{V_4}{V_3}\right)^{(\gamma-1)}$$

$$\therefore \quad T_4 = T_3\left(\frac{V_3}{V_4}\right)^{(\gamma-1)} = \frac{823}{2.048} = \mathbf{401\ K}$$

$$t_4 = 402 - 273 = \mathbf{129\ ^\circ C}$$

Alternatively

$$\frac{V_4}{T_4} = \frac{V_1}{T_1}$$

$$\therefore \quad T_4 = T_1\frac{V_4}{V_1} = 301 \times \frac{0.112}{0.084} = \mathbf{402\ K}$$

Hence

$$\begin{array}{lll} P_1 = 100\ \text{kN/m}^2 & V_1 = 0.084\ \text{m}^3 & t_1 = 28\ ^\circ\text{C} \\ P_2 = 1229\ \text{kN/m}^2 & V_2 = 0.014\ \text{m}^3 & t_2 = 343\ ^\circ\text{C} \\ P_3 = 1229\ \text{kN/m}^2 & V_3 = 0.018\ 7\ \text{m}^3 & t_3 = 550\ ^\circ\text{C} \\ P_4 = 100\ \text{kN/m}^2 & V_4 = 0.112\ \text{m}^3 & t_4 = 129\ ^\circ\text{C} \end{array}$$

(b)

$$R = c_p\left(\frac{\gamma - 1}{\gamma}\right) = 1.006 \times \frac{0.4}{1.4} = \mathbf{0.287\ kJ/kg\ K}$$

$$m = \frac{P_1V_1}{RT_1} = \frac{100 \times 0.084}{0.287 \times 301} = \mathbf{0.097\ 2\ kg}$$

$$\begin{aligned} \therefore \quad \text{Heat received} &= mc_p(T_3 - T_2) \\ &= 0.097\ 2 \times 1.006 \times (823 - 616) \\ &= 0.097\ 2 \times 1.006 \times 207 \\ &= \mathbf{20.24\ kJ} \end{aligned}$$

(c)

$$\oint W = P_2(V_3 - V_2) - P_1(V_4 - V_1) + \frac{(P_3V_3 - P_4V_4) - (P_2V_2 - P_1V_1)}{(\gamma - 1)}$$

$$= 1229 \times (0.018\ 7 - 0.014) - 100 \times (0.112 - 0.084)$$

$$+ \frac{[(1229 \times 0.018\ 7) - (100 \times 0.112)] - [(1229 \times 0.014) - (100 \times 0.084)]}{(1.4 - 1)}$$

$$= (1229 \times 0.004\ 7) - (100 \times 0.028) + \frac{(23 - 11.2) - (17.2 - 8.4)}{0.4}$$

$$= 5.78 - 2.8 + \frac{11.8 - 8.8}{0.4}$$

$$= 2.98 + 7.5$$

$$= \mathbf{10.48\ kJ}$$

Alternatively

$$\oint W = mR\frac{\gamma}{(\gamma - 1)}[(T_3 - T_4) - (T_2 - T_1)]$$

$$= 0.097\,2 \times 0.287 \times \frac{1.4}{(1.4 - 1)}[(823 - 402) - (616 - 301)]$$

$$= 0.097\,2 \times 0.287 \times 3.5 \times (421 - 315)$$
$$= 0.097\,2 \times 0.287 \times 3.5 \times 106$$
$$= \mathbf{10.4\ kJ}$$

(d)

$$\text{Thermal } \eta = 1 - \frac{1}{r_v^{(\gamma-1)}} = 1 - \frac{1}{6^{0.4}}$$
$$= 1 - \frac{1}{2.048}$$
$$= 1 - 0.488$$
$$= 0.512$$
$$= \mathbf{51.2\%}$$

Alternatively

$$\text{Thermal } \eta = 1 - \frac{T_1}{T_2} = 1 - \frac{301}{616}$$
$$= 0.512$$
$$= \mathbf{51.2\%}$$

Alternatively

$$\text{Thermal } \eta = 1 - \frac{T_4}{T_3} = 1 - \frac{402}{823}$$
$$= 0.512$$
$$= \mathbf{51.2\%}$$

Note also that

$$\oint W = \text{Heat received} \times \text{Thermal } \eta$$
$$= 20.24 \times 0.512$$
$$= \mathbf{10.4\ kJ}$$

The small difference in $\oint W$ shown by different calculation methods is due to slight cumulative error.

(e)

$$\text{Carnot } \eta = \frac{T_3 - T_1}{T_3}$$
$$= \frac{823 - 301}{823}$$
$$= \frac{522}{823}$$
$$= 0.634$$
$$= \mathbf{63.4\%}$$

(f)

$$\text{Work ratio} = \frac{\text{Net work done/cycle}}{\text{Positive work done/cycle}}$$

The positive work done per cycle is

$$\begin{aligned}
&\text{Area under 2–3} \\
&+\text{Area under 3–4} = P_2(V_3 - V_2) + \frac{(P_3V_3 - P_4V_4)}{(\gamma - 1)} \\
&= 1229 \times (0.018\,7 - 0.014) + \frac{(1229 \times 0.018\,7) - (100 \times 0.112)}{(1.4 - 1)} \\
&= (1229 \times 0.004\,7) + \frac{(23 - 11.2)}{0.4} \\
&= 5.78 + \frac{11.8}{0.4} = 5.78 + 29.5 \\
&= \mathbf{35.28\ kJ}
\end{aligned}$$

$$\therefore \quad \text{Work ratio} = \frac{10.48}{35.28} = \mathbf{0.297}$$

(g)

$$\begin{aligned}
\text{Mean effective pressure} &= \frac{\oint W}{(V_4 - V_2)} \\
&= \frac{10.48}{0.112 - 0.014} \\
&= \frac{10.48}{0.098} \\
&= \mathbf{106.94\ kN/m^2}
\end{aligned}$$

Example 15.5 *A gas turbine operating on a simple constant pressure cycle has a pressure compression ratio of 8:1. The turbine thermal efficiency is 60 per cent of ideal. The fuel used has a calorific value of 43 MJ/kg. If $\gamma - 1.4$, determine*

(a) the actual thermal efficiency of the turbine

(b) the specific fuel consumption of the turbine in kg/kWh

(a)

$$\text{Ideal thermal efficiency} = 1 - \frac{1}{r_v^{(\gamma - 1)}}$$

$$\text{where } r_v = \frac{V_1}{V_2}$$

But the pressure ratio only is given

$$\frac{P_2}{P_1} = \mathbf{8}$$

Now

$$\frac{T_1}{T_2} = \left(\frac{P_1}{P_2}\right)^{(\gamma-1)/\gamma} = \left(\frac{V_2}{V_1}\right)^{(\gamma-1)}$$

$$\therefore \quad \left(\frac{1}{r_v}\right)^{(\gamma-1)} = \left(\frac{V_2}{V_1}\right)^{(-1)} = \left(\frac{P_1}{P_2}\right)^{(\gamma-1)/\gamma}$$

So ideal thermal efficiency is

$$\begin{aligned} 1 - \left(\frac{P_1}{P_2}\right)^{(\gamma-1)/\gamma} &= 1 - \frac{1}{8^{(1.4-1)/1.4}} \\ &= 1 - \frac{1}{8^{1/3.5}} \\ &= 1 - \frac{1}{1.81} \\ &= 1 - 0.552 \\ &= 0.448 \\ &= \mathbf{44.8\%} \end{aligned}$$

$\therefore$ Actual thermal efficiency $= 44.8 \times 0.6 = \mathbf{26.88\%}$

(b)

$$\begin{aligned} \text{Energy to work/kg of fuel} &= 43\,000 \times 0.268\,8 \\ &= \mathbf{11\,558\ kJ} \end{aligned}$$

Energy equivalent of 1 kWh $= 1$ kJ/s $\times$ 3600 s $=$ **3600 kJ**

$$\begin{aligned} \therefore \quad \text{Specific fuel consumption} &= \frac{3600}{11\,558} \\ &= \mathbf{0.311\ kg/kWh} \end{aligned}$$

15.5 The constant volume cycle

A general discussion of this cycle is given in section 15.2. Figure 15.19(a) is a P–V diagram of the constant volume cycle; Fig. 15.19(b) is the corresponding T–s diagram. The cycle is arranged as follows:

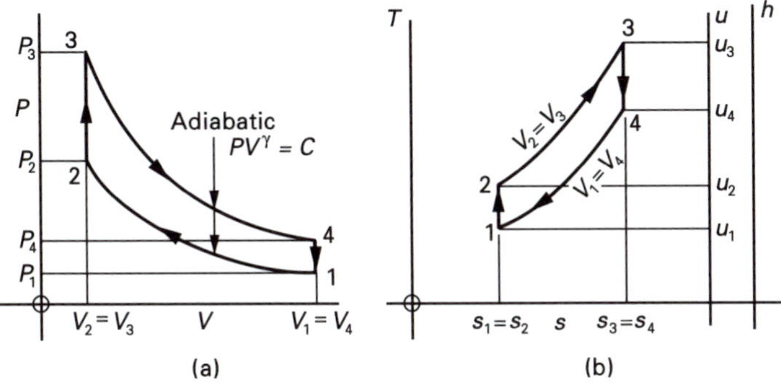

Fig. 15.19 Constant volume cycle: (a) P–V diagram; (b) T–s diagram

- **1–2 Adiabatic compression of the gas according to the law $PV^{\gamma} = C$**
  Pressure increases from P_1 to P_2.
  Volume decreases from V_1 to V_2.
  Temperature increases from T_1 to T_2.
  Entropy remains constant at $s_1 = s_2$.
- **2–3 Constant volume heat addition**
  Volume remains constant at $V_2 = V_3$.
  Pressure increases from P_2 to P_3.
  Temperature increases from T_2 to T_3.
  Entropy increases from s_2 to s_3.
- **3–4 Adiabatic expansion of the gas according to the law $PV^{\gamma} = C$**
  Pressure decreases from P_3 to P_4.
  Volume increases from V_3 to V_4.
  Temperature decreases from T_3 to T_4.
  Entropy remains constant at $s_3 = s_4$.
- **4–1 Constant volume heat rejection**
  Volume remains constant at $V_4 = V_1$.
  Pressure decreases from P_4 to P_1.
  Temperature decreases from T_4 to T_1.
  Entropy decreases from s_4 to s_1.
  This process completes the cycle and returns the gas to its original state at 1.

An analysis of the properties P, V and T at state points, 1, 2, 3 and 4 will now be made and it will be assumed that P_1, V_1 and T_1 are known.

- P_1, V_1, T_1.
- Assume that the volume ratio V_1/V_2 is known. Recall that the constant volume cycle is the theoretical cycle for a petrol or gas engine. The ratio V_1/V_2 for a petrol or gas engine is often called the compression ratio of the engine. In fact, the ratio V_1/V_2 is a volume ratio. It is sometimes called the geometric compression ratio.

$$T_1/T_2 = (V_2/V_1)^{(\gamma - 1)}$$

$$\therefore \quad T_2 = T_1\left(\frac{V_1}{V_2}\right)^{(\gamma-1)} = T_1 r_v^{(\gamma-1)}$$

$$\text{where } r_v = \begin{cases} V_1/V_2 = \text{adiabatic compression volume ratio} \\ V_4/V_3 = \text{adiabatic expansion volume ratio} \end{cases}$$

Also $P_1 V_1^{\gamma} = P_2 V_2^{\gamma}$

$$\therefore \quad P_2 = P_1\left(\frac{V_1}{V_2}\right)^{\gamma} = P_1 r_v^{\gamma}$$

- $V_3 = V_2$ because the volume remains constant.

$$P_3/T_3 = P_2/T_2$$

$$\therefore \quad T_3 = T_2\left(\frac{P_3}{P_2}\right) = \frac{P_3}{P_2} T_1 r_v^{(\gamma-1)}$$

- $T_3/T_4 = (V_4/V_3)^{(\gamma-1)} = r_v^{(\gamma-1)}$

$$\therefore \quad T_4 = \frac{T_3}{r_v^{(\gamma-1)}} = \frac{P_3}{P_2} T_1 \frac{r_v^{(\gamma-1)}}{r_v^{(\gamma-1)}} = \frac{P_3}{P_2} T_1$$

Also $P_4 V_4^{\gamma} = P_3 V_3^{\gamma}$

$$\therefore \quad P_4 = P_3 \left(\frac{V_3}{V_4}\right)^{\gamma} = \frac{P_3}{r_v^{\gamma}}$$

Also, from the constant volume process 4–1, $P_4/T_4 = P_1/T_1$

$$\therefore \quad T_4 = \frac{P_4}{P_1} T_1 = \frac{P_3}{P_2} T_1$$

From this it follows that $P_4/P_1 = P_3/P_2$.

The work done during the cycle may be obtained as follows:

- Process 3–4 is an expansion which gives positive work done.
- Process 1–2 is a compression which gives negative work done.
- The net work done is the sum of the work done by these two processes.

$$\therefore \quad \text{Net work done} = \oint W = \text{Area under 3–4} - \text{Area under 1–2}$$

$$= \text{Area 1234}$$

$$= \text{Area of diagram}$$

$$\therefore \quad \oint W = \frac{(P_3V_3 - P_4V_4)}{(\gamma - 1)} - \frac{(P_2V_2 - P_1V_1)}{(\gamma - 1)}$$

$$= \frac{(P_3V_3 - P_4V_4) - (P_2V_2 - P_1V_1)}{(\gamma - 1)} \qquad [1]$$

$$= \frac{mR}{(\gamma - 1)}[(T_3 - T_4) - (T_2 - T_1)] \qquad [2]$$

(since $PV = mRT$)

Alternatively, the cycle work done can be determined using

$$\oint W = \oint Q$$

or

$$\text{Cycle work done} = \oint W = \text{Heat received} - \text{Heat rejected} \qquad [3]$$

Equation [3] will hold good for any cycle.

Now in this cycle heat is received and rejected only during constant volume processes. Hence the name, constant volume cycle.

There are also two adiabatic processes, but during an adiabatic process no heat is received or rejected. Hence the adiabatic processes do not appear when the discussion is on heat received or rejected. This will be true in any other cycle where adiabatic processes appear.

In this cycle

Heat is received from 2–3 $= mc_v(T_3 - T_2)$ [4]

Heat is rejected from 4–1 $= mc_v(T_4 - T_1)$ [5]

$$\therefore \quad \oint W = mc_v(T_3 - T_2) - mc_v(T_4 - T_1)$$

$$= mc_v[(T_3 - T_2) - (T_4 - T_1)] \quad [6]$$

The thermal efficiency of the cycle may be determined from

$$\text{Thermal efficiency} = \frac{\text{Heat received} - \text{Heat rejected}}{\text{Heat received}} \quad [7]$$

$$= 1 - \frac{\text{Heat rejected}}{\text{Heat received}} \quad [8]$$

$$= 1 - \frac{mc_v(T_4 - T_1)}{mc_v(T_3 - T_2)}$$

and assuming that c_v remains constant

$$\text{Thermal } \eta = 1 - \frac{(T_4 - T_1)}{(T_3 - T_2)} \quad [9]$$

Equation [9] gives the thermal efficiency in terms of temperatures. Substituting temperatures in terms of T_1 in equation [9]

$$\text{Thermal } \eta = 1 - \frac{(P_3/P_2)T_1 - T_1}{(P_3/P_2)T_1 r_v^{(\gamma-1)} - T_1 r_v^{(\gamma-1)}}$$

$$= 1 - \frac{T_1(P_3/P_2 - 1)}{T_1 r_v^{(\gamma-1)}(P_3/P_2 - 1)}$$

$$= 1 - \frac{1}{r_v^{(\gamma-1)}} \quad [10]$$

Now consider the adiabatic processes 1–2 and 3–4.

$$\frac{T_2}{T_1} = \left(\frac{V_1}{V_2}\right)^{(\gamma-1)} \quad \text{and} \quad \frac{T_3}{T_4} = \left(\frac{V_4}{V_3}\right)^{(\gamma-1)}$$

But $V_1/V_2 = V_4/V_3 = r_v$

$$\therefore \quad \frac{T_2}{T_1} = \frac{T_3}{T_4} = r_v^{(\gamma-1)} \quad \text{or} \quad \frac{T_1}{T_2} = \frac{T_4}{T_3} = \left(\frac{1}{r_v}\right)^{(\gamma-1)}$$

So from equation [10]

$$\text{Thermal } \eta = 1 - \frac{T_1}{T_2} \quad [11]$$

$$= 1 - \frac{T_3}{T_4} \quad [12]$$

Note that from equation [7]

$$\text{Heat received} - \text{Heat rejected} = \text{Heat received} \times \text{Thermal } \eta \qquad [13]$$

Substituting equation [3] into equation [13]

$$\oint W = \text{Heat received} \times \text{Thermal } \eta \qquad [14]$$

This is another way in which the work done may be determined.

In the case of this cycle,

$$\oint W = mc_v(T_3 - T_2) \times \text{Thermal } \eta \qquad [15]$$

The cycle can also be analysed by using the temperature–entropy chart.

By tracing the points of the cycle round the chart, the various properties P, V, T, u, h at state points 1, 2, 3 and 4 can be found.

Now for the constant volume process, it has been shown that

$$\text{Heat transferred} = \text{Change of internal energy} \quad \text{(see section 5.7)}$$

Hence

$$\text{Heat received from 2–3} = (u_3 - u_2) \qquad [16]$$

$$\text{Heat rejected from 4–1} = (u_4 - u_1) \qquad [17]$$

From equations [3], [16] and [17]

$$\oint W = (u_3 - u_2) - (u_4 - u_1) \qquad [18]$$

From equations [8], [16] and [17]

$$\text{Thermal } \eta = 1 \frac{(u_4 - u_1)}{(u_3 - u_2)} \qquad [19]$$

Remember that the chart is made out for unit mass of gas, so equations [16], [17] and [18] will give values per unit mass per cycle.

On a practical note, $r_v = V_1/V_2$ must increase for the thermal efficiency to increase (i.e. the so-called compression ratio must increase). In a practical engine, the ability to increase r_v is limited because of high material loading, high temperatures and fuel combustion problems.

In the case of the petrol engine, the addition of tetraethyl lead to the fuel has helped to prevent knocking or pinking, thus enabling the use of higher compression ratios. The use of tetraethyl lead is now in disfavour because of atmospheric pollution and possible damage to health. Research and development into fuel combustion techniques are making it possible to use lead-free petrol as well as to increase compression ratios.

Example 15.6 *An engine works on the constant volume cycle. It has a bore of 80 mm and a stroke of 85 mm. The clearance volume of the engine is 0.06 litre. The actual thermal efficiency of the engine is 22 per cent. Determine the relative efficiency of the engine. Take $\gamma = 1.4$.*

SOLUTION

Suppose the engine has bore d and the stroke l, then

$$\text{Stroke volume} = V_1 - V_2$$

$$= \frac{\pi d^2}{4} \times l$$

$$= \frac{\pi \times 80^2}{4} \times 85$$

$$= \mathbf{427.257\ mm^3}$$

$$\text{Clearance volume} = V_1 = 0.06 \times 10^6$$

$$= \mathbf{60\ 000\ mm^3}$$

$$\text{Total volume} = V_1 = (V_1 - V_2) + V_2$$

$$= 427\ 257 + 60\ 000$$

$$= \mathbf{487\ 257\ mm^3}$$

$$\text{Ideal thermal } \eta = 1 - \frac{1}{r_v^{(\gamma-1)}}$$

and

$$r_v = \frac{V_1}{V_2} = \frac{487\ 257}{60\ 000}$$

$$= \mathbf{8.12}$$

$$\therefore \quad \text{Ideal thermal } \eta = 1 - \frac{1}{8.12^{(1.4-1)}} = 1 - \frac{1}{8.12^{0.4}}$$

$$= 1 - \frac{1}{2.311}$$

$$= 1 - 0.433$$

$$= 0.567$$

$$= \mathbf{56.7\%}$$

$$\text{Relative efficiency} = \frac{\text{Actual thermal efficiency}}{\text{Ideal thermal efficiency}}$$

$$= \frac{0.22}{0.567}$$

$$= 0.388$$

$$= \mathbf{38.8\%}$$

Example 15.7 *One kilogram of air is taken through a constant volume cycle. At the commencement of the adiabatic compression, the pressure and temperature are 103 kN/m² and 100 °C respectively. The adiabatic compression has a volume ratio of 6:1. The maximum pressure of the cycle is 3.45 MN/m². Determine for the cycle*

(a) the pressure, volume and temperature at each of the cycle process change points
(b) the heat transferred to the air
(c) the heat rejected by the air
(d) the ideal thermal efficiency
(e) the work done
(f) the mean effective pressure

For the air, take R = *0.287 kJ/kg K,* $\gamma = 1.4$.

SOLUTION

First draw a diagram (Fig. 15.20).

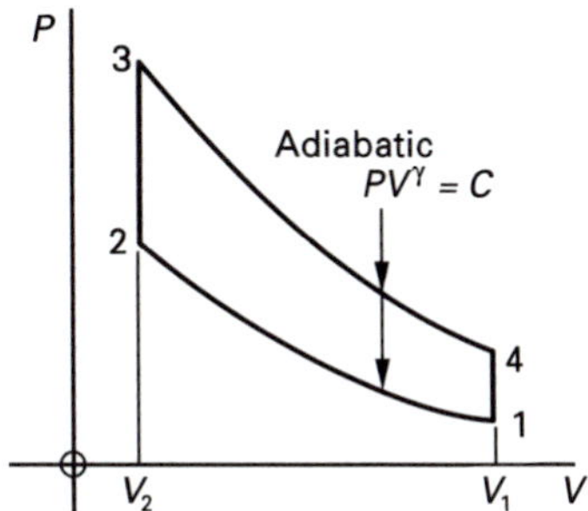

Fig. 15.20 Diagram for Example 15.7

(a)

For point 1

Pressure $= P_1 = \mathbf{103\ kN/m^2}$; temperature $= t_1 = \mathbf{100\ °C}$

$$P_1V_1 = mRT_1 \quad \text{and} \quad T_1 = 100 + 273 = \mathbf{373\ K}$$

$$\therefore \quad V_1 = \frac{mRT_1}{P_1} = \frac{1 \times 0.287 \times 373}{103} = \mathbf{1.039\ m^3}$$

For point 2

$$P_1V_1^\gamma = P_2V_2^\gamma$$

$$\therefore \quad P_2 = P_1\left(\frac{V_1}{V_2}\right)^\gamma = 103 \times 6^{1.4}$$

$$= 103 \times 12.27$$

$$= \mathbf{1263.8\ kN/m^2}$$

$$V_2 = \frac{V_1}{6} = \frac{1.039}{6} = \mathbf{0.173\ m^3}$$

$$P_1V_1/T_1 = P_2V_2/T_2$$

$$\therefore \quad T_2 = \frac{P_2 V_2}{P_1 V_1} T_1 = \frac{1263.8}{103} \times \frac{0.173}{1.039} \times 373$$

$$= \mathbf{762\ K}$$

$$t_2 = 762 - 273 = \mathbf{489\ ^\circ C}$$

This result could also have been achieved by using

$$\frac{T_1}{T_2} = \left(\frac{P_1}{P_2}\right)^{(\gamma-1)/\gamma} = \left(\frac{V_2}{V_1}\right)^{(\gamma-1)}$$

For point 3

$$V_3 = V_2 = \mathbf{0.173\ m^3},\ P_3 = \mathbf{3450\ kN/m^2}$$

$$P_3/T_3 = P_2/T_2$$

$$\therefore \quad T_3 = \frac{P_3}{P_2} T_2 = \frac{3450}{1263.8} \times 762 = \mathbf{2080\ K}$$

$$t_3 = 2080 - 273 = \mathbf{1807\ ^\circ C}$$

For point 4

$$P_3 V_3^\gamma = P_4 V_4^\gamma$$

$$\therefore \quad P_4 = P_3 \left(\frac{V_3}{V_4}\right)^\gamma = 3450 \times \left(\frac{1}{6}\right)^{1.4}$$

$$= \frac{3450}{12.27}$$

$$= \mathbf{281.2\ kN/m^2}$$

$$V_4 = V_1 = \mathbf{1.039\ m^3}$$

$$P_4/T_4 = P_1/T_1$$

$$\therefore \quad T_4 = \frac{P_4}{P_1} T_1 = \frac{281.2}{103} \times 373 = \mathbf{1018\ K}$$

$$t_4 = 1018 - 273 = \mathbf{745\ ^\circ C}$$

This result could also have been achieved by using

$$\frac{T_3}{T_4} = \left(\frac{P_3}{P_4}\right)^{(\gamma-1)/\gamma}$$

Tabulated results

Change point	Pressure (kN/m^2)	Volume (m^3)	Temperature (°C)
1	103	1.039	100
2	1 263.8	0.173	489
3	3 450	0.173	1 807
4	281.2	1.039	745

(b)

$$\text{Now } c_v = \frac{R}{(\gamma - 1)} = \frac{0.287}{(1.4 - 1)} = \frac{0.287}{0.4} = \mathbf{0.717\ kJ/kg\ K}$$

Heat transferred to the air between state points 2 and 3 is

$$\begin{aligned} mc_v(T_3 - T_2) &= 1 \times 0.717 \times (2080 - 762) \\ &= 1 \times 0.717 \times 1318 \\ &= \mathbf{945\ kJ} \end{aligned}$$

(c)
Heat rejected from the air between state points 3 and 4 is

$$\begin{aligned} mc_v(T_4 - T_1) &= 1 \times 0.717 \times (1080 - 373) \\ &= 1 \times 0.717 \times 645 \\ &= \mathbf{462.5\ kJ} \end{aligned}$$

(d)

$$\begin{aligned} \text{Ideal thermal } \eta &= \frac{\text{Heat received} - \text{Heat rejected}}{\text{Heat received}} \\ &= 1 - \frac{\text{Heat rejected}}{\text{Heat received}} \\ &= 1 - \frac{462.5}{945} \\ &= 1 - 0.489 \\ &= 0.511 \\ &= \mathbf{51.1\%} \end{aligned}$$

Alternatively

$$\begin{aligned} \text{Ideal thermal } \eta &= 1 - \frac{1}{r_v^{(\gamma-1)}} \\ &= 1 - \frac{1}{6^{0.4}} \\ &= 1 - \frac{1}{2.048} \\ &= 1 - 0.489 \\ &= 0.511 \\ &= \mathbf{51.1\%} \end{aligned}$$

(e)

$$\begin{aligned} \oint W &= \text{Heat received} - \text{Heat rejected} \\ &= 945 - 462.5 \\ &= \mathbf{482.5\ kJ} \end{aligned}$$

Alternatively

$$\oint W = \frac{(P_3V_3 - P_2V_2)}{(\gamma - 1)} - \frac{(P_2V_2 - P_1V_1)}{(\gamma - 1)}$$

$$= \frac{(P_3V_3 - P_2V_2) - (P_2V_2 - P_1V_1)}{(\gamma - 1)}$$

$$= \frac{[(3450 \times 0.173) - (281.2 \times 1.039)] - [(1263.8 \times 0.173) - (103 \times 1.039)]}{(1.4 - 1)}$$

$$= \frac{[596.8 - 292.2] - [218.6 - 107]}{0.4}$$

$$= \frac{[304.6 - 111.6]}{0.4} = \frac{193}{0.4}$$

$$= \mathbf{482.5\ kJ}$$

(f)

$$\text{Mean effective pressure} = \frac{\oint W}{(V_1 - V_2)}$$

$$= \frac{482.5}{(1.039 - 0.173)}$$

$$= \frac{482.5}{0.866}$$

$$= \mathbf{557.16\ kN/m^2}$$

Example 15.8 *The pressure, volume and temperature at the beginning of the compression of a constant volume cycle are 101 kN/m², 0.003 m³ and 18 °C, respectively. The maximum pressure of the cycle is 4.5 MN/m². The volume ratio of the cycle is 19:1. The cycle is repeated 3000 times/min. Determine for the cycle*

(a) the pressure, volume and temperature at each of the cycle process change points
(b) the thermal efficiency
(c) the theoretical output in kilowatts
(d) the mean effective pressure
(e) the Carnot efficiency within the cycle temperature limits

Take $c_p = 1.006$ *kJ/kg K,* $c_v = 0.716$ *kJ/kg K. The diagram is as Fig. 15.20.*

(a)

$P_1 = 101\ \text{kN/m}^2$, $V_1 = 0.003\ \text{m}^3$, $t_1 = 18\ °\text{C}$, all given.

For process 1–2

$$\frac{V_1}{V_2} = r_v = \mathbf{9}$$

$$P_1V_1^\gamma = P_2V_2^\gamma \quad \text{and} \quad \gamma = c_p/c_v = 1.006/0.716 = \mathbf{1.405}$$

$$\therefore \quad P_2 = P_1\left(\frac{V_1}{V_2}\right)^\gamma = 101 \times 9^{1.405}$$

$$= 101 \times 21.91$$
$$= 2212.9\ \text{kN/m}^2$$
$$= \mathbf{2.212\,9\ MN/m^2}$$

$$\frac{T_2}{T_1} = \left(\frac{V_1}{V_2}\right)^{(\gamma-1)} \text{ and } T_1 = 18 + 273 = \mathbf{291\ K}$$

$$\therefore \quad T_2 = T_1\left(\frac{V_1}{V_2}\right)^{(\gamma-1)} = 291 \times 9^{(1.405-1)}$$

$$= 291 \times 9^{0.405}$$

$$= 291 \times 2.435$$

$$= \mathbf{708.6\ K}$$

$$t_2 = 708.6 - 273$$

$$= \mathbf{435.6\ ^\circ C}$$

$$V_2 = V_1/9 = 0.003/9$$

$$= \mathbf{0.000\ 33\ m^3}$$

For process 2–3

$$V_3 = V_2 = \mathbf{0.000\ 33\ m^3}$$

$$\frac{P_3}{T_3} = \frac{P_2}{T_2} \quad \text{and} \quad P_3 = \mathbf{4.5\ MN/m^2}$$

$$\therefore \quad T_3 = T_2\frac{P_3}{P_2} = 708.6 \times \frac{4.3}{2.212\ 9}$$

$$= \mathbf{1441\ K}$$

$$t_3 = 1441 - 273$$

$$= \mathbf{1168\ ^\circ C}$$

For process 3–4

$$V_4 = V_1 = \mathbf{0.003\ m^3}$$

$$P_4 = P_3\left(\frac{V_3}{V_4}\right)^{\gamma} = \frac{500}{9^{1.405}}$$

$$= \frac{4500}{21.91}$$

$$= \mathbf{205.4\ kN/m^2}$$

$$T_4 = T_3\left(\frac{V_3}{V_4}\right)^{(\gamma-1)} = \frac{1441}{9^{0.405}}$$

$$= \frac{1441}{2.435}$$

$$= \mathbf{591.8\ K}$$

$$t_4 = 591.8 - 273$$

$$= \mathbf{318.8\ ^\circ C}$$

Tabulated results

Change point	Pressure (kN/m^2)	Volume (m^3)	Temperature (°C)
1	101	0.003	18
2	2 212.9	0.000 33	435.6
3	4 500	0.000 33	1 168
4	205.4	0.003	318.8

(b)

$$\text{Thermal } \eta = 1 - \frac{1}{r_v^{(\gamma - 1)}} \quad \text{and} \quad r_v = \mathbf{9} \qquad \text{(equation [10])}$$

$$\text{Now } \gamma = \frac{c_p}{c_v} = \frac{1.006}{0.716} = \mathbf{1.405}$$

$$\therefore \quad \text{Thermal } \eta = 1 - \frac{1}{9^{(1.405 - 1)}}$$

$$= 1 - \frac{1}{9^{0.405}}$$

$$= 1 - \frac{1}{2.439}$$

$$= 1 - 0.41$$
$$= 0.59$$
$$= \mathbf{59\%}$$

Alternatively

$$\text{Thermal } \eta = 1 - \frac{(T_4 - T_1)}{(T_3 - T_2)} \quad \text{(equation [9])}$$

$$= 1 - \frac{(591.8 - 291)}{(1441 - 708.6)}$$

$$= 1 - \frac{300.8}{732.4}$$

$$= 1 - 0.41$$
$$= \mathbf{0.59}$$

(c)

$$\oint W = \text{Heat received} - \text{Heat rejected} \qquad \text{(equation [3])}$$
$$= mc_v[(T_3 - T_2) - (T_4 - T_1)] \qquad \text{(equation [6])}$$

and

$$m = \frac{P_1 V_1}{RT_1} = \frac{101 \times 0.003}{(1.006 - 0.716) \times 291}$$

$$= \frac{101 \times 0.003}{0.29 \times 291}$$

$$= \mathbf{0.003\ 6\ kg}$$

$$\therefore \quad \oint W = 0.003\ 6 \times 0.716 \times [(1441 - 708.6) - (591.8 - 291)]$$
$$= 0.003\ 6 \times 0.716 \times (732.4 - 300.8)$$
$$= 0.003\ 6 \times 0.716 \times 413.6$$
$$= \mathbf{1.112\ kJ}$$

Work done/min = (1.112 × 3000) kJ

$$\therefore \quad \text{Theoretical output} = \frac{1.112 \times 3000}{60}$$

$$= \mathbf{55.6\ kW} \qquad (1\ \text{W} = 1\ \text{J/s})$$

(d)

$$\text{Mean effective pressure} = \frac{\oint W}{(V_1 - V_2)} \qquad \text{(equation [10], section 15.1)}$$

$$= \frac{1.112}{(0.003 - 0.000\,33)}$$

$$= \frac{1.112}{0.002\,67}$$

$$= \mathbf{416.5\ kN/m^2}$$

(e)

$$\text{Carnot } \eta = \frac{T_3 - T_1}{T_3}$$

$$= \frac{1441 - 291}{1441}$$

$$= \frac{1150}{1441}$$

$$= 0.798$$
$$= \mathbf{79.8\%}$$

15.6 *The Diesel cycle*

This cycle is named after Rudolf Diesel (1858–1913). A general discussion of this cycle is given in section 15.2. Figure 15.21(a) is the *P–V* diagram of the Diesel cycle; Fig. 15.21(b) is the corresponding *T–s* diagram. The cycle is arranged as follows:

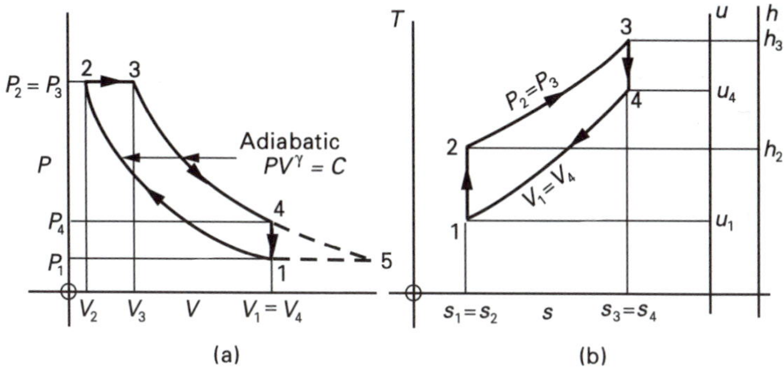

Fig. 15.21 Diesel cycle: (a) *P–V* diagram; (b) *T–s* diagram

- **1–2 Adiabatic compression according to the law $PV^\gamma = C$**
  Pressure increases from P_1 to P_2.
  Temperature increases from T_1 to T_2.
  Volume decreases from V_1 to V_2.
  Entropy remains constant at $s_1 = s_2$.

- **2–3 Constant pressure heat addition**
  Pressure remains constant at $P_2 = P_3$.
  Temperature increases from T_2 to T_3.
  Volume increases from V_2 to V_3.
  Entropy increases from s_2 to s_3.
- **3–4 Adiabatic expansion according to the law $PV^{\gamma} = C$**
  Pressure decreases from P_3 to P_4.
  Temperature decreases from T_3 to T_4.
  Volume increases from V_3 to V_4.
  Entropy remains constant at $s_3 = s_4$.
- **4–1 Constant volume heat rejection**
  Volume remains constant at $V_4 = V_1$.
  Temperature decreases from T_4 to T_1.
  Pressure decreases from P_4 to P_1.
  Entropy decreases from s_4 to s_1.
  This process completes the cycle and returns the gas to its original state.

This cycle is sometimes called the **modified constant pressure cycle**. Inspection of Fig. 15.17(a) will show the reason. If, instead of cutting off the expansion at 4, the gas were allowed to expand completely to 5, then in order to return the gas to its original state at 1, constant pressure heat rejection would have to take place from 5 to 1.

This is shown dotted. The diagram 1235 is the constant pressure cycle and, by cutting off the part 145, it is modified into the Diesel cycle. In practice, by cutting off part 145 of the cycle, a considerable saving in cylinder volume would be obtained. The area 145 represents a small amount of work which does not really justify the increase of the cylinder volume from V_1 to V_5.

An analysis of the properties at state points 1, 2, 3 and 4 can be made. Again it is assumed that P_1, V_1 and T_1 are known.

- P_1, V_1, T_1.
- Assume that the volume ratio V_1/V_2 is known.

$$T_1/T_2 = (V_2/V_1)^{\gamma - 1}$$

$$\therefore \quad T_2 = T_1\left(\frac{V_1}{V_2}\right)^{(\gamma - 1)} = T_1 r_v^{(\gamma - 1)}$$

where $r_v = V_1/V_2$ = adiabatic compression volume ratio

Also $P_1 V_1^{\gamma} = P_2 V_2^{\gamma}$

$$\therefore \quad P_2 = P_1\left(\frac{V_1}{V_2}\right)^{\gamma} = P_1 r_v^{\gamma}$$

- $P_3 = P_2$ because the pressure remains constant

$$V_3/T_3 = V_2/T_2$$

$$\therefore \quad T_3 = T_2\frac{V_3}{V_2} = \frac{V_3}{V_2} T_1 r_v^{(\gamma - 1)} = \beta T_1 r_v^{(\gamma - 1)}$$

where $\beta = V_3/V_2$ = cut-off ratio

- $T_4/T_3 = (V_3/V_4)^{(\gamma-1)}$

$$\therefore \quad T_4 = T_3\left(\frac{V_3}{V_4}\right)^{(\gamma-1)}$$

Now $V_3/V_4 = V_3/V_1$, since $V_4 = V_1$, and

$$\frac{V_3}{V_1} = \frac{V_3}{V_2}\frac{V_2}{V_1} = \frac{\beta}{r_v}$$

$$\therefore \quad T_4 = T_3\left(\frac{\beta}{r}\right)^{(\gamma-1)} = \beta T_1 r_v^{(\gamma-1)}\frac{\beta^{(\gamma-1)}}{r_v^{(\gamma-1)}}$$

or

$$T_4 = \beta^{\gamma} T_1$$

Also $P_4 V_4^{\gamma} = P_3 V_3^{\gamma}$

$$\therefore \quad P_4 = P_3\left(\frac{V_3}{V_4}\right)^{\gamma} = P_3\left(\frac{\beta}{r_v}\right)^{\gamma}$$

The work done during the cycle may be determined as follows:

- Processes 2–3 and 3–4 are expansions and hence give positive work done.
- Process 1–2 is a compression and hence will give negative work done.
- The net work done during the cycle will be the sum of the work done during these processes.

Hence

$$\oint W = \text{Area under } 2\text{–}3 + \text{Area under } 3\text{–}4 - \text{Area under } 1\text{–}2$$

$$= P_2(V_3 - V_2) + \frac{(P_3V_3 - P_4V_4)}{(\gamma-1)} - \frac{(P_2V_2 - P_1V_1)}{(\gamma-1)}$$

$$= P_2(V_3 - V_2) + \frac{(P_3V_3 - P_4V_4) - (P_2V_2 - P_1V_1)}{(\gamma-1)} \qquad [1]$$

$$= mR(T_3 - T_3) + \frac{mR}{(\gamma-1)}[(T_3 - T_4) - (T_2 - T_1)]$$

$$= mR\left[(T_3 - T_2) + \frac{(T_3 - T_4) - (T_2 - T_1)}{(\gamma-1)}\right] \qquad [2]$$

Alternatively, the work done may be obtained from

$$\oint W = \text{Heat received} - \text{Heat rejected}$$

In this cycle, heat is received during constant pressure process 2–3 and rejected during constant volume process 4–1.

No heat is received or rejected during the adiabatic processes.

Hence

$$\oint W = mc_p(T_3 - T_2) - mc_v(T_4 - T_1) \qquad [3]$$

Alternatively

$$\oint W = \text{Heat received} \times \text{Thermal } \eta$$

or

$$\oint W = mc_p(T_3 - T_2) \times \text{Thermal } \eta \qquad [4]$$

The thermal efficiency may be determined from

$$\text{Thermal } \eta = 1 - \frac{\text{Heat rejected}}{\text{Heat received}}$$

$$= 1 - \frac{mc_v(T_4 - T_1)}{mc_p(T_3 - T_2)}$$

$$= 1 - \frac{1}{\gamma}\frac{(T_4 - T_1)}{(T_3 - T_2)} \qquad [5]$$

This gives the thermal efficiency in terms of temperatures.
Also, substituting temperatures in terms of T_1 into equation [5]

$$\text{Thermal } \eta = 1 - \frac{1}{\gamma}\frac{(\beta^{\gamma} T_1 - T_1)}{[\beta T_1 r_v^{(\gamma-1)} - T_1 r^{(\gamma-1)}]}$$

$$= 1 - \frac{1}{\gamma}\frac{T_1(\beta^{\gamma} - 1)}{T_1 r_v^{(\gamma-1)}(\beta - 1)}$$

$$= 1 - \frac{1}{r_v^{(\gamma-1)}}\frac{(\beta^{\gamma} - 1)}{\gamma(\beta - 1)]} \qquad [6]$$

Now, for a constant pressure process

Heat received or rejected = Change of enthalpy

For a constant volume process

Heat received or rejected = Change of internal energy

So using the *T–s* chart

Heat received = $h_3 - h_2$

Heat rejected = $u_2 - u_1$

$$\text{Thermal } \eta = 1 - \frac{\text{Heat rejected}}{\text{Heat received}}$$

$$= 1 - \frac{(u_4 - u_1)}{(h_3 - h_2)} \qquad [7]$$

Also

$$\oint W = \text{Heat received} - \text{Heat rejected}$$

$$= (h_3 - h_2) - (u_4 - u_1) \qquad [8]$$

Once again this gives the work done per unit mass per cycle because the chart will be made out for unit mass of gas.

It should be noted here that the Diesel cycle does not in fact follow the cycle of a modern diesel engine. The actual cycle is more closely approximated by the dual combustion cycle (section 15.7).

Example 15.9 *At the beginning of compression of an ideal Diesel cycle the gas has a temperature and pressure of 40 °C and 90 kN/m², respectively. The volume ratio of compression is 16:1. The maximum temperature of the cycle is 1400 °C. Determine for the cycle*

(a) the pressure and temperature at each of the cycle process change points
(b) the work done per kilogram of gas
(c) the thermal efficiency
(d) the work ratio
(e) the mean effective pressure
(f) the Carnot efficiency within the cycle temperature limits

Take $\gamma = 1.4$, $c_p = 1.004$ kJ/kg K.

SOLUTION

First draw a diagram (Fig. 15.22)

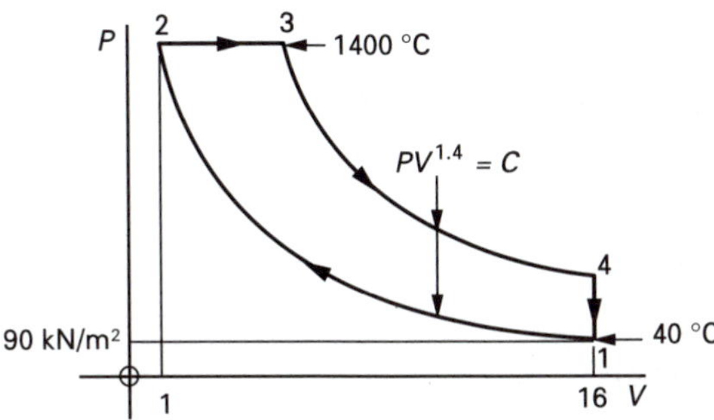

Fig. 15.22 Diagram for Example 15.9

(a)

$$P_1 V_1^{\gamma} = P_2 V_2^{\gamma}$$

$$\therefore \quad P_2 = P_1 \left(\frac{V_1}{V_2}\right)^{\gamma} = 90 \times 16^{1.4} = 90 \times 48.5 = \mathbf{4365\ kN/m^2}$$

$$\frac{T_1}{T_2} = \left(\frac{V_2}{V_1}\right)^{(\gamma - 1)} \quad \text{and} \quad T_1 = 40 + 273 = \mathbf{313\ K}$$

$$\therefore \quad T_2 = T_1 \left(\frac{V_1}{V_2}\right)^{(\gamma - 1)} = 313 \times 6^{0.4} = 313 \times 3.031 = \mathbf{948.7\ K}$$

$$t_2 = 675.7 - 273 = \mathbf{402.7\ ^\circ C}$$

$$P_3 = P_2 = \mathbf{4365\ kN/m^2}$$

$$t_3 = \mathbf{1400^\circ C}$$

For the constant pressure process

$$\frac{V_2}{T_2} = \frac{V_3}{T_3} \quad \text{and} \quad T_3 = 1400 + 273 = \mathbf{1673\ K}$$

$$\therefore \quad V_3 = V_2 \frac{T_3}{T_2} = 1 \times \frac{1673}{948.7} = \mathbf{1.763\ volumes}$$

For the adiabatic expansion

$$P_3 V_3^{\gamma} = P_4 V_4^{\gamma}$$

$$\therefore \quad P_4 = P_3 \left(\frac{V_3}{V_4}\right)^{\gamma} = 4365 \times \left(\frac{1.763}{16}\right)^{1.4}$$

$$= \frac{4365}{9.075^{1.4}}$$

$$= \frac{4365}{21.93}$$

$$= \mathbf{199.04\ kN/m^2}$$

For the constant volume process

$$\frac{P_1}{T_1} = \frac{P_4}{T_4}$$

$$\therefore \quad T_4 = T_1 \frac{P_4}{P_1} = \frac{199.04}{90} = \mathbf{692.2\ K}$$

$$t_4 = 692.2 - 273 = \mathbf{419.2\ °C}$$

Hence

$P_1 = 90$ kN/m^2 $\qquad t_1 = 40$ °C
$P_2 = 4365$ kN/m^2 $\qquad t_2 = 402.7$ °C
$P_3 = 4365$ kN/m^2 $\qquad t_3 = 1400$ °C
$P_4 = 199.04$ kN/m^2 $\qquad t_4 = 419.2$ °C

(b)

$$\oint W/\text{kg} = c_p(T_3 - T_2) - c_v(T_4 - T_1) \qquad \text{(equation [3])}$$

$$\text{Now} \quad \frac{c_p}{c_v} = \gamma$$

$$\therefore \quad c_v = \frac{c_p}{\gamma} = \frac{1.004}{1.4} = \mathbf{0.717\ kJ/kg\ K}$$

$$\therefore \quad \oint W/\text{kg} = 1.004 \times (1673 - 948.7) - 0.717 \times (692.2 - 313)$$

$$= (1.004 \times 724.3) - (0.717 \times 379.2)$$

$$= 727.2 \times 271.9$$

$$= \mathbf{455.3\ kJ}$$

(c)

$$\text{Thermal } \eta = 1 - \frac{1}{\gamma}\frac{(T_4 - T_1)}{(T_3 - T_2)}$$

$$= 1 - \frac{1}{1.4}\frac{(692.2 - 313)}{(1673 - 948.7)}$$

$$= 1 - \frac{379.2}{1.4 \times 727.2}$$

$$= 1 - 0.372$$
$$= 0.628$$
$$= \mathbf{62.8\%}$$

(d)

$$\text{Work ratio} = \frac{\oint W}{\text{Positive work done}} \qquad \text{(see equation [6], section 15.1)}$$

$$\text{Positive work done} = c_p(T_3 - T_2) + \frac{R(T_3 - T_4)}{(\gamma - 1)}$$

Also $(c_p - c_v) = R = (1.004 - 0.717) = \mathbf{0.287\ kJ/kg\ K}$

$$\therefore \quad \text{Positive work done} = 1.004 \times (1673 - 948.7) + \frac{0.287 \times (1673 - 692.2)}{(1.4 - 1)}$$

$$= (1.004 \times 724.3) + \frac{(0.287 \times 980.8)}{0.4}$$

$$= 727.2 + 703.7$$
$$= \mathbf{1430.9\ kJ/kg}$$

$$\therefore \quad \text{Work ratio} = \frac{455.3}{1430.9}$$

$$= \mathbf{0.318}$$

(e)

$$\text{Mean effective pressure} = \frac{\oint W}{(V_1 - V_2)} \qquad \text{(see equation [10], section 15.1)}$$

Now, for 1 kg of gas

$$P_1 V_1 = R T_1$$

$$\therefore \quad V_1 = \frac{RT_1}{P_1} = 0.287 \times \frac{313}{90} = \mathbf{0.998\ m^3}$$

$$\text{Hence } V_2 = \frac{V_1}{16} = \mathbf{0.062\ m^3}$$

$$\therefore \quad \text{Mean effective pressure} = \frac{455.3}{(0.998 - 0.062)}$$

$$= \frac{455.3}{0.936}$$

$$= \mathbf{486.4\ kN/m^2}$$

(f)

$$\text{Carnot } \eta = \frac{(T_3 - T_1)}{T_3}$$

$$= \frac{(1673 - 313)}{1673}$$

$$= \frac{1360}{1673}$$

$$= 0.813$$
$$= \mathbf{81.3\%}$$

Example 15.10 *In an ideal Diesel cycle the volume ratios of the adiabatic expansion and compression are 7.5:1 and 15:1, respectively. The pressure and temperature at the beginning of compression are 98 kN/m^2 and 44 °C, respectively. The pressure at the end of the adiabatic expansion is 258 kN/m^2. Determine*
(a) the maximum temperature attained during the cycle
(b) the thermal efficiency of the cycle
Take $\gamma = 1.4$.

SOLUTION
First draw a diagram (Fig. 15.23).

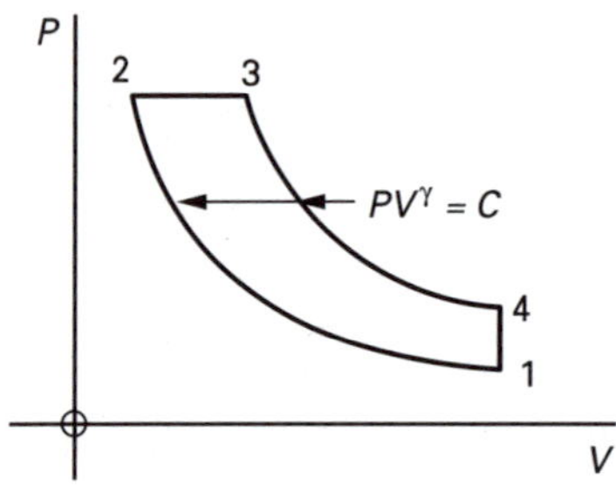

Fig. 15.23 Diagram for Example 15.10

(a)

$$\frac{P_4}{T_4} = \frac{P_1}{T_1} \quad \text{and} \quad T_1 = 44 + 273 = \mathbf{317\ K}$$

$$\therefore \quad T_4 = T_1 \frac{P_4}{P_1} = 317 \times \frac{258}{98} = \mathbf{834.6\ K}$$

For the expansion

$$\therefore \quad \frac{T_3}{T_4} = \left(\frac{V_4}{V_3}\right)^{(\gamma - 1)}$$

$$\therefore \quad T_3 = T_4\left(\frac{V_4}{V_3}\right)^{(\gamma-1)} = 834.6 \times 7.5^{0.4}$$

$$= 834.6 \times 2.239$$

$$= \mathbf{1\,868.7\ K}$$

$$t_3 = 1868.7 - 273 = \mathbf{1595.7\ °C}$$

This is the maximum temperature attained during the cycle.

(b)

$$\frac{T_2}{T_1} = \left(\frac{V_1}{V_2}\right)^{(\gamma-1)}$$

$$\therefore \quad T_2 = T_1\left(\frac{V_1}{V_2}\right)^{(\gamma-1)} = 317 \times 15^{0.4} = 317 \times 2.95 = \mathbf{935\ K}$$

$$\text{Thermal } \eta = 1 - \frac{(T_4 - T_1)}{\gamma(T_3 - T_2)}$$

$$= 1 - \frac{(834.6 - 317)}{1.4(1868.7 - 935)}$$

$$= 1 - \frac{517.6}{1.4 \times 933.7}$$

$$= 1 - 0.396$$

$$= 0.604$$

$$= \mathbf{60.4\%}$$

Example 15.11 *An oil engine works on the ideal Diesel cycle. The overall volume ratio of compression is 11:1 and constant pressure energy addition ceases at 10 per cent of the stroke. The pressure and temperature at the commencement of compression are 96 kN/m² and 18 °C, respectively. The engine uses 0.05 m³ of air per second. Determine*

(a) the thermal efficiency of the cycle

(b) the indicated power of the cycle

Take $\gamma = 1.4$.

SOLUTION

First draw a diagram (Fig. 15.24).

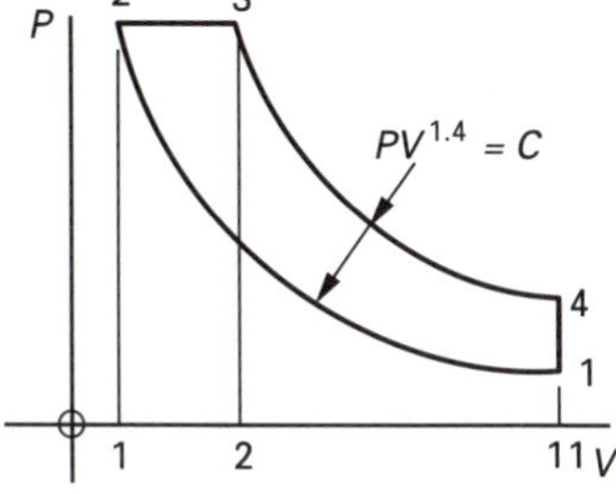

Fig. 15.24 Diagram for Example 15.11

(a)

$$\text{Swept volume} = 11 - 1 = \textbf{10 volumes}$$

10% of swept volume $= 10 \times 0.1 =$ **1 volume**

$\therefore$ Constant pressure energy addition ceases at $1 + 1 =$ **2 volumes**

$$\text{Thermal } \eta = 1 - \frac{1}{r_v^{(\gamma-1)}} \frac{1}{\gamma} \frac{(\beta^{\gamma} - 1)}{(\beta - 1)}$$

and $r_v = V_1/V_2 = 11, \beta = V_3/V_2 = 2.$

$$\therefore \text{ Thermal } \eta = 1 - \frac{1}{11^{(1.4-1)}} \times \frac{1}{1.4} \times \frac{(2^{1.4} - 1)}{(2 - 1)}$$

$$= 1 - \frac{1}{11^{0.4}} \times \frac{(2.64 - 1)}{1.4}$$

$$= 1 - \frac{1.64}{2.61 \times 11.4}$$

$$= 1 - 0.449$$
$$= 0.551$$
$$= \textbf{55.1\%}$$

(b)

Let $V_1 - V_2 = 0.05\ \text{m}^3$
then $V_2 = 0.05 \times 0.1 = \mathbf{0.005\ m^3}$
From this

$V_1 = 11 \times 0.005 = \mathbf{0.055\ m^3}$
$V_3 = 2V_2 = 2 \times 0.005 = \mathbf{0.01\ m^3}$
$V_4 = V_1 = \mathbf{0.055\ m^3}$
$P_1 V_1^{\gamma} = P_2 V_2^{\gamma}$

$$\therefore \quad P_2 = P_1 \left(\frac{V_1}{V_2}\right)^{\gamma} = 96 \times 11^{1.4} = 96 \times 28.7 = \mathbf{2755\ kN/m^2}$$

$P_3 V_3^{\gamma} = P_4 V_4^{\gamma}$ and $P_3 = P_2$

$$\therefore \quad P_4 = P_3 \left(\frac{V_3}{V_4}\right)^{\gamma} = 2760 \times \left(\frac{2}{11}\right)^{1.4} = \frac{2760}{5.5^{1.4}} = \frac{2760}{10.88} = \mathbf{253.7\ kN/m^2}$$

$\oint W/\text{s} =$ indicated power

$$= P_2(V_3 - V_2) + \frac{(P_3V_3 - P_4V_4) - (P_2V_2 - P_1V_1)}{(\gamma - 1)}$$

$$= 2755 \times (0.01 - 0.005) + \frac{[(2755 \times 0.01) - (253.7 \times 0.005)] - [(2755 \times 0.005) - (96 \times 0.005)]}{(1.4 - 1)}$$

$$= (2755 \times 0.005) + \frac{(27.55 - 13.95) - (13.78 - 5.28)}{0.4}$$

$$= 13.78 + \frac{(13.6 - 8.5)}{0.4} = 13.78 + \frac{5.1}{0.4}$$

$$= 13.78 + 12.75$$
$$= \mathbf{26.53\ kW}$$

15.7 The dual combustion cycle

This cycle is sometimes called the **composite cycle**. Heat is received partly at constant volume and partly at constant pressure, hence the name, dual combustion cycle. Heat is rejected at constant volume.

It fits the actual cycle of an oil engine rather more closely than the Diesel cycle. This is because an actual oil engine burns part of the fuel very nearly at constant volume and burns the rest very nearly at constant pressure. Another name for the cycle is the **high-speed Diesel cycle**.

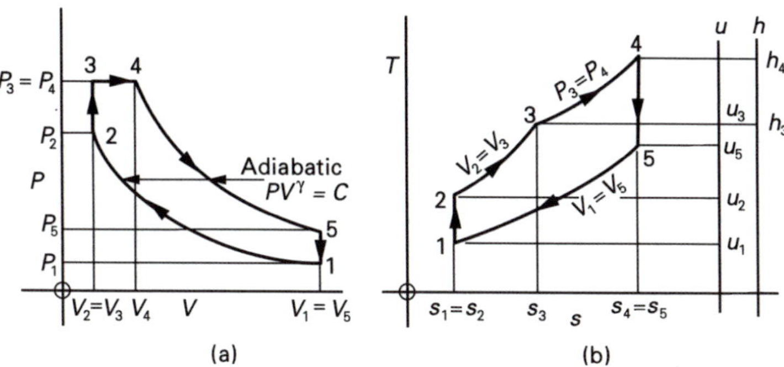

Fig. 15.25 Dual combustion cycle: (a) *P–V* diagram; (b) *T–s* diagram

Figure 15.25(a) is a *P–V* diagram of the dual combustion cycle; Fig. 15.25(b) is the corresponding *T–s* diagram. The cycle is arranged as follows:

- **1–2 Adiabatic compression according to the law $PV^{\gamma} = C$**
  Pressure increases from P_1 to P_2.
  Temperature increases from T_1 to T_2.
  Volume decreases from V_1 to V_2.
  Entropy remains constant at $s_1 = s_2$.

- **2–3 Constant volume heat addition**
  Volume remains constant at $V_2 = V_3$.
  Pressure increases from P_2 to P_3.
  Temperature increases from T_2 to T_3.
  Entropy increases from s_2 to s_3.

- **3–4 Constant pressure heat addition**
  Pressure remains constant at $P_3 = P_4$.
  Volume increases from V_3 to V_4.
  Temperature increases from T_3 to T_4.
  Entropy increases from s_3 to s_4.

- **4–5 Adiabatic expansion according to the law $PV^{\gamma} = C$**
  Pressure decreases from P_4 to P_5.
  Temperature decreases from T_4 to T_5.
  Volume increases from V_4 to V_5.
  Entropy remains constant at $s_4 = s_5$.

- **5–1 Constant volume heat rejection**
  Volume remains constant at $V_5 = V_1$.
  Pressure decreases from P_5 to P_1.
  Temperature decreases from T_5 to T_1.
  Entropy decreases from s_5 to s_1.
  This process completes the cycle and returns the gas to its original state.

An analysis of the properties at state points 1, 2, 3, 4 and 5 can be made. Assume P_1, V_1 and T_1 are known.

- P_1, V_1, T_1.
- Assume that the volume ratio V_1/V_2 is known.

$$T_1/T_2 = (V_2/V_1)^{(\gamma-1)}$$

$$\therefore \quad T_2 = T_1\left(\frac{V_1}{V_2}\right)^{\gamma} = T_1 r_v^{(\gamma-1)}$$

where r_v = adiabatic compression volume ratio

Also $P_1 V_1^{\gamma} = P_2 V_2^{\gamma}$

$$\therefore \quad P_2 = P_1\left(\frac{V_1}{V_2}\right)^{\gamma} = P_1 r_v^{\gamma}$$

- $P_3/T_3 = P_2/T_2$ because the volume remains constant at $V_2 = V_3$.

$$\therefore \quad T_3 = \frac{P_3}{P_2} T_2 = \alpha T_1 r_v^{(\gamma-1)}$$

where $\alpha = P_3/P_2$ = constant volume heat addition pressure ratio

- $V_3/T_3 = V_4/T_4$ because the pressure remains constant at $P_3 = P_4$.

$$\therefore \quad T_4 = \frac{V_4}{V_3} T_3 = \alpha\beta T_1 r_v^{(\gamma-1)}$$

where $\beta = V_4/V_3$ = constant pressure heat addition cut-off ratio

- $T_5/T_4 = (V_4/V_5)^{(\gamma-1)}$

$$\therefore \quad T_5 = T_4\left(\frac{V_4}{V_5}\right)^{(\gamma-1)}$$

Now $V_5 = V_1$

$$\therefore \quad \frac{V_4}{V_5} = \frac{V_4}{V_1} = \frac{V_4}{V_3}\frac{V_3}{V_1}$$

Also $V_3 = V_2$

$$\therefore \quad \frac{V_4}{V_5} = \frac{V_4}{V_3}\frac{V_2}{V_1} = \frac{\beta}{r_v}$$

since $V_4/V_3 = \beta$ and $V_1/V_2 = r_v$

$$T_5 = \alpha\beta T_1 r_v^{(\gamma-1)} \frac{\beta^{(\gamma-1)}}{r_v^{(\gamma-1)}} = \alpha\beta^{\gamma} T_1 \qquad \text{(from above)}$$

Note also that for the constant volume process 5–1, $P_5/T_5 = P_1/T_1$.

$$\therefore \quad T_5 = \frac{P_5}{P_1} T_1$$

So it appears that

$$\frac{P_5}{P_1} = \alpha\beta^{\gamma}$$

Now this can be shown as follows:

$$\alpha\beta^{\gamma} = \frac{P_3}{P_2}\left(\frac{V_4}{V_3}\right)^{\gamma}$$

$$P_3 V_4^{\gamma} = P_4 V_4^{\gamma} \quad \text{since} \quad P_3 = P_4$$

and

$$P_4 V_4^{\gamma} = P_5 V_5^{\gamma} \quad \text{since } PV^{\gamma} = C \text{ for process 4–5}$$

Also

$$P_2 V_3^{\gamma} = P_2 V_2^{\gamma} \quad \text{since} \quad V_2 = V_3$$

and

$$P_2 V_2^{\gamma} = P_1 V_1^{\gamma} \quad \text{since } PV^{\gamma} = C \text{ for process 1–2}$$

$$\therefore \quad \alpha\beta^{\gamma} = \frac{P_5 V_5^{\gamma}}{P_1 V_1^{\gamma}} = \frac{P_5}{P_1} \quad \text{since} \quad V_1 = V_5$$

Note also that $P_5 V_5^{\gamma} = P_4 V_4^{\gamma}$

$$\therefore \quad P_5 = P_4\left(\frac{V_4}{V_5}\right)^{\gamma} = P_4\left(\frac{\beta}{r_v}\right)^{\gamma}$$

The work done during the cycle may be obtained as follows:

- Processes 3–4 and 4–5 are expansions and hence give positive work done.
- Process 1–2 is a compression and hence gives negative work done.
- The net work done during the cycle will be the sum of the work done during these processes.

Hence

$$\oint W = \text{Area under 3–4} + \text{Area under 4–5} - \text{Area under 1–2}$$

$$= P_3(V_4 - V_3) + \frac{(P_4V_4 - P_5V_5)}{(\gamma - 1)} - \frac{(P_2V_2 - P_1V_1)}{(\gamma - 1)}$$

$$= P_3(V_4 - V_3) + \frac{(P_4V_4 - P_5V_5) - (P_2V_2 - P_1V_1)}{(\gamma - 1)} \qquad [1]$$

$$= mR(T_4 - T_3) + \frac{mR}{(\gamma - 1)}[(T_4 - T_5) - (T_2 - T_1)]$$

$$= mR\left[(T_4 - T_3) + \frac{(T_4 - T_5) - (T_2 - T_1)}{(\gamma - 1)}\right] \qquad [2]$$

Alternatively, the net work done may be obtained from

$$\oint W = \text{Heat received} - \text{Heat rejected}$$

In this cycle, heat is received during constant volume process 2–3 and constant pressure process 3–4 and heat is rejected during constant volume process 5–1.

No heat is received or rejected during the adiabatic processes.

$$\therefore \quad \oint W = mc_v(T_3 - T_2) + mc_p(T_4 - T_3) - mc_v(T_5 - T_1) \qquad [3]$$

Alternatively

$$\oint W = \text{Heat received} \times \text{Thermal } \eta$$

$$= [mc_v(T_3 - T_2) + mc_p(T_4 - T_3)] \times \text{Thermal } \eta \qquad [4]$$

The thermal efficiency may be determined from

$$\text{Thermal } \eta = 1 - \frac{\text{Heat rejected}}{\text{Heat received}}$$

$$= 1 - \frac{mc_v(T_5 - T_1)}{mc_v(T_3 - T_2) + mc_p(T_4 - T_3)}$$

$$= 1 - \frac{(T_5 - T_1)}{(T_3 - T_2) + \gamma(T_4 - T_3)} \qquad [5]$$

This gives the thermal efficiency in terms of temperatures.

Substituting temperatures in terms of T_1 into equation [5]

$$\text{Thermal } \eta = 1 - \frac{(\alpha\beta^{\gamma} T_1 - T_1)}{[\alpha T_1 r_v^{(\gamma-1)} - T_1 r_v^{(\gamma-1)}] + \gamma[\alpha\beta T_1 r_v^{(\gamma-1)} - \alpha T_1 r_v^{(\gamma-1)}]}$$

$$= 1 - \frac{T_1(\alpha\beta^{\gamma} - 1)}{T_1 r_v^{(\gamma-1)}[(\alpha - 1) + \alpha\gamma(\beta - 1)]}$$

$$= 1 - \frac{1}{r_v^{(\gamma-1)}}\left[\frac{(\alpha\beta^{\gamma} - 1)}{(\alpha - 1) + \alpha\gamma(\beta - 1)}\right] \qquad [6]$$

Note that this expression contains the expressions for thermal efficiency of the constant volume cycle and the Diesel cycle.

If $\beta = 1$ there is no constant pressure heat addition because $V_4 = V_3$; substituting this into equation [6]

$$\text{Thermal } \eta = 1 - \frac{1}{r_v^{(\gamma-1)}}$$

i.e. the constant volume cycle.

If $\alpha = 1$ there is no constant volume heat addition because $P_3 = P_2$; substituting this into equation [6]

$$\text{Thermal } \eta = 1 - \frac{1}{r_v^{(\gamma-1)}} \frac{(\beta^{\gamma} - 1)}{\gamma(\beta - 1)}$$

i.e. the Diesel cycle.

For the dual combustion cycle, using the *T–s* chart

Heat received at constant volume $= u_3 - u_2$
Heat received at constant pressure $= h_4 - h_3$
Heat rejected at constant volume $= u_5 - u_1$

$$\text{Thermal } \eta = 1 - \frac{\text{Heat rejected}}{\text{Heat received}}$$

$$= 1 - \frac{(u_5 - u_1)}{(u_3 - u_2) + (h_4 - h_3)} \qquad [7]$$

$$\oint W = (u_3 - u_2) + (h_4 - h_3) - (u_5 - u_1) \qquad [8]$$

Again, this gives the work done per unit mass per cycle because the chart will be made out for unit mass of gas.

Example 15.12 *At the beginning of the compression of an ideal dual combustion cycle the pressure and temperature of the gas are 103 kN/m² and 22°C, respectively. The volume ratio of the compression is 16:1. The heat added during the constant volume process is 244 kJ/kg of gas. Heat is added during the constant pressure process for 3 per cent of the expansion stroke. Determine*

(a) the pressure and temperature at the end of compression
(b) the pressure and temperature at the end of the constant volume process
(c) the temperature at the end of the constant pressure process
Take $\gamma = 1.4$, $c_v = 0.707$ *kJ/kg K.*

SOLUTION

First draw a diagram (Fig. 15.26).

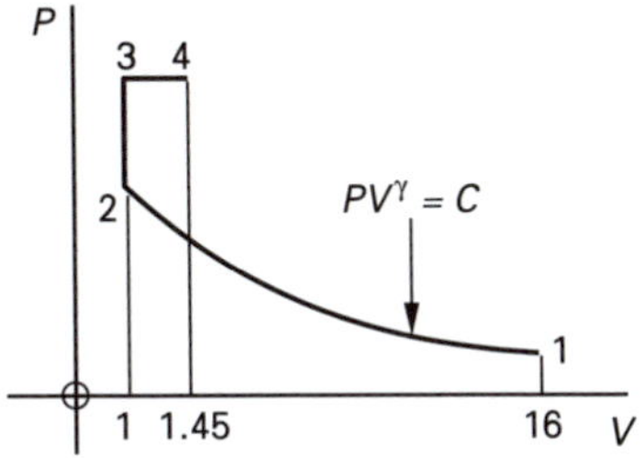

Fig. 15.26 Diagram for Example 15.12

(a)

For the compression

$$P_1 V_1^{\gamma} = P_2 V_2^{\gamma}$$

$$\therefore \quad P_2 = P_1 \left(\frac{V_1}{V_2}\right)^{\gamma} = 103 \times 16^{1.4}$$

$$= 103 \times 48.5$$
$$= \mathbf{5000\ kN/m^2}$$
$$= \mathbf{5\ MN/m^2}$$

$$\frac{T_1}{T_2} = \left(\frac{V_2}{V_1}\right)^{(\gamma-1)} \quad \text{and} \quad T_1 = 22 + 273 = \mathbf{295\ K}$$

$$\therefore \quad T_2 = T_1\left(\frac{V_1}{V_2}\right)^{(\gamma-1)} = 295 \times 16^{0.4} = 295 \times 3.03 = \mathbf{894\ K}$$

$$t_2 = 894 - 273 = \mathbf{621\ ^\circ C}$$

(b)
For the constant volume process

$$\text{Heat added/kg gas} = c_v(T_3 - T_2)$$

$$\therefore \quad 244 = 0.717(T_3 - 894)$$

from which

$$T_3 = \frac{244}{0.717} + 894$$

$$= 340 + 894 = \mathbf{1234\ K}$$

$$t_3 = 1234 - 273 = \mathbf{961\ ^\circ C}$$

Also

$$\frac{P_3}{T_3} = \frac{P_2}{T_3}$$

$$\therefore \quad P_3 = P_2\frac{T_3}{T_2} = 5 \times \frac{1234}{894} = \mathbf{6.9\ MN/m^2}$$

(c)
For the constant pressure process

$$\text{Stroke} = 16 - 1 = \mathbf{15\ volumes}$$

$$\therefore \quad 3\%\ \text{of stroke} = 15 \times 0.03 = \mathbf{0.45\ volumes}$$

So the volume at the end of the constant pressure heat addition is

$$V_4 = 1 + 0.45 = \mathbf{1.45\ volumes}$$

And

$$\frac{V_4}{T_4} = \frac{V_3}{T_3}$$

$$\therefore \quad T_4 = T_3\frac{V_4}{V_3} = 1234 \times \frac{1.45}{1} = \mathbf{1789\ K}$$

$$t_4 = 1789 - 273 = \mathbf{1516\ ^\circ C}$$

Example 15.13 *An ideal dual combustion cycle has a volume ratio for the adiabatic compression of 15:1. At the beginning of the adiabatic compression the pressure, volume and temperature of the gas are 97 kN/m², 0.084 m³ and 28 °C, respectively. The maximum pressure and temperature of the cycle are 6.2 MN/m² and 1320 °C, respectively. For the cycle, determine*

(a) the pressure, volume and temperature at the cycle process change points
(b) the net work done
(c) the thermal efficiency
(d) the heat received
(e) the work ratio
(f) the mean effective pressure
(g) the Carnot efficiency within the cycle temperature limits

Take $c_p = 1.005$ *kJ/kg K,* $c_v = 0.717$ *kJ/kg K.*

SOLUTION

First draw a diagram (Fig. 15.27).

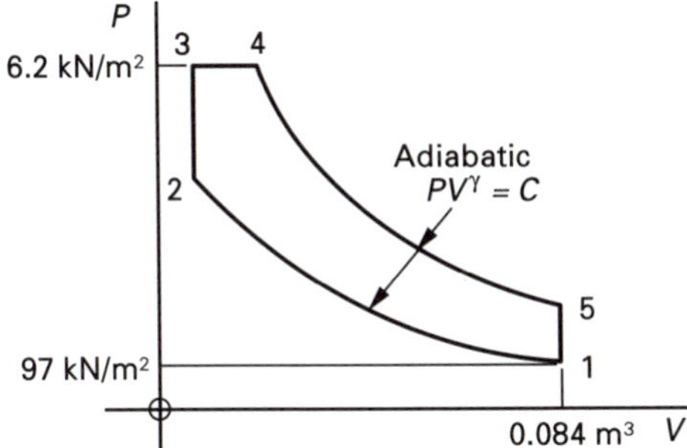

Fig. 15.27 Diagram for Example 15.13

(a)

$$P_1 = 97 \text{ kN/m}^2,\ V_1 = 0.084 \text{ m}^3,\ t_1 = 28\ ^\circ\text{C}$$

$$V_2 = \frac{0.084}{15} = \mathbf{0.005\,6\ m^3}$$

$$c_p/c_v = \gamma = \frac{1.005}{0.717} = \mathbf{1.4}$$

$$P_1 V_1^\gamma = P_2 V_2^\gamma$$

$$\therefore \quad P_2 = P_1 \left(\frac{V_1}{V_2}\right)^\gamma = 97 \times 15^{1.4}$$
$$= 97 \times 44.3$$
$$= 4297 \text{ kN/m}^2$$
$$= \mathbf{4.297\ MN/m^2}$$

$$\frac{T_1}{T_2} = \left(\frac{V_2}{V_1}\right)^{(\gamma-1)} \quad \text{and} \quad T_1 = 28 + 273 = \mathbf{301\ K}$$

$$\therefore \quad T_2 = T_1\left(\frac{V_1}{V_2}\right)^{(\gamma-1)} = 301 \times 15^{0.4}$$
$$= 301 \times 2.954$$
$$= \mathbf{889\ K}$$

$$t_2 = 889 - 273 = \mathbf{616\ {}^\circ C}$$

$$\frac{P_3}{T_3} = \frac{P_2}{T_2}$$

$$\therefore \quad T_3 = T_2\frac{P_3}{P_2} = 889 \times \frac{6.2}{4.297} = \mathbf{1283\ K}$$

$$t_3 = 1283 - 273 = \mathbf{1010\ {}^\circ C}$$

$$V_3 = V_2 = \mathbf{0.005\,6\ m^3}$$

$$\frac{V_4}{T_4} = \frac{V_3}{T_3} \quad \text{and} \quad T_4 = 1320 + 273 = \mathbf{1593\ K}$$

$$\therefore \quad V_4 = V_3\frac{T_4}{T_3} = 0.005\,6 \times \frac{1593}{1283} = \mathbf{0.006\,95\ m^3}$$

$$P_4 = P_3 = \mathbf{6.2\ MN/m^2} \quad V_5 = V_1 = \mathbf{0.084\ m^3}$$

$$P_4V_4^\gamma = P_5V_5^\gamma$$

$$\therefore \quad P_5 = P_4\left(\frac{V_4}{V_5}\right)^\gamma = 6.2 \times \left(\frac{0.006\,95}{0.084}\right)^{1.4}$$
$$= \frac{6.2}{12.09^{1.4}}$$
$$= \frac{6.2}{32.8}$$
$$= 0.189\ \text{MN/m}^2$$
$$= \mathbf{189\ kN/m^2}$$

$$\frac{T_5}{T_4} = \left(\frac{V_4}{V_5}\right)^{(\gamma-1)}$$

$$\therefore \quad T_5 = T_4\left(\frac{V_4}{V_5}\right)^{(\gamma-1)} = 1593 \times \left(\frac{0.006\,95}{0.084}\right)^{0.4}$$
$$= \frac{1593}{12.09^{0.4}}$$
$$= \frac{1593}{2.71}$$
$$= \mathbf{588\ K}$$

$$t_5 = 588 - 273 = \mathbf{315\ {}^\circ C}$$

Hence

$P_1 = 97$ kN/m^2	$V_1 = 0.084$ m^3	$t_1 = 28$ °C
$P_2 = 4.297$ MN/m^2	$V_2 = 0.005\,6$ m^3	$t_2 = 616$ °C
$P_3 = 6.2$ MN/m^2	$V_3 = 0.005\,6$ m^3	$t_3 = 1010$ °C
$P_4 = 6.2$ MN/m^2	$V_4 = 0.006\,95$ m^3	$t_4 = 1320$ °C
$P_5 = 189$ kN/m^2	$V_5 = 0.084$ m^3	$t_5 = 315$ °C

(b)

$$\oint W = P_3(V_4 - V_3) + \frac{(P_4V_4 - P_5V_5) - (P_2V_2 - P_1V_1)}{(\gamma - 1)}$$

$$= 6200 \times (0.006\,95 - 0.005\,6)$$

$$+ \frac{[(6200 \times 0.006\,95) - (189 \times 0.084)] - [(4297 \times 0.005\,6) - (97 \times 0.084)]}{(1.4 - 1)}$$

$$= (6200 \times 0.001\,35) + \frac{(43.1 - 15.9) - (24.1 - 8.15)}{0.4}$$

$$= 8.37 + \frac{(27.2 - 15.95)}{0.4}$$

$$= 8.37 + 28.13$$

$$= \mathbf{36.5\ kJ}$$

(c)

$$\text{Thermal } \eta = 1 - \frac{(T_5 - T_1)}{(T_3 - T_2) + \gamma(T_4 - T_3)}$$

$$= 1 - \frac{(588 - 301)}{(1283 - 889) + 1.4(1593 - 1283)}$$

$$= 1 - \frac{287}{394 + (1.4 \times 310)} = 1 - \frac{287}{394 + 434}$$

$$= 1 - \frac{287}{828}$$

$$= 1 - 0.347$$

$$= 0.653$$

$$= \mathbf{65.3\%}$$

Alternatively

$$\text{Thermal } \eta = 1 - \frac{1}{r_v^{(\gamma-1)}} \cdot \frac{(\alpha\beta^{\gamma} - 1)}{(\alpha + 1) + \alpha\gamma(\beta - 1)}$$

$$r_v = 15$$

$$\alpha = \frac{P_3}{P_2} = \frac{6.2}{4.297} = \mathbf{1.443}$$

$$\beta = \frac{V_4}{V_3} = \frac{0.006\,95}{0.005\,6} = \mathbf{1.241}$$

$$\therefore \quad \text{Thermal } \eta = 1 - \frac{1}{15^{0.4}} \times \frac{(1.443 \times 1.241^{1.4}) - 1}{(1.443 - 1) + (1.443 \times 1.4) \times (1.241 - 1)}$$

$$= 1 - \frac{1}{2.95} \times \frac{(1.443 \times 1.353) - 1}{0.443 + (2.02 \times 0.241)}$$

$$= 1 - \frac{1}{2.95} \times \frac{1.952 - 1}{0.448 + 0.487}$$

$$= 1 - \frac{1}{2.95} \times \frac{0.952}{0.93}$$

$$= 1 - 0.347$$

$$= \mathbf{0.653}$$

(d)

$$\oint W = \text{Heat received} \times \text{Thermal } \eta$$

$$\therefore \quad \text{Heat received} = \frac{\oint W}{\text{Thermal } \eta}$$

$$= \frac{36.5}{0.653}$$

$$= \mathbf{55.9\ kJ}$$

Alternatively

$$R = c_p - c_v = 1.005 - 0.717 = \mathbf{0.288\ kJ/kg\ K}$$

$$m = \frac{P_1 V_1}{RT_1} = \frac{97 \times 0.084}{0.288 \times 301} = \mathbf{0.095\ kg}$$

$$\begin{aligned}\text{Heat received} &= mc_v(T_3 - T_2) + mc_p(T_4 - T_3)\\ &= [0.094 \times 0.717 \times (1283 - 889)] + [0.094 \times 1.006 \times (1593 - 1283)]\\ &= (0.094 \times 0.717 \times 394) + (0.094 \times 1.006 \times 310)\\ &= 26.6 + 29.3\\ &= \mathbf{55.9\ kJ}\end{aligned}$$

(e)

$$\text{Work ratio} = \frac{\oint W}{\text{Positive work done}}$$

$$= \frac{36.5}{P_3(V_4 - V_3) + \dfrac{(P_4 V_4 - P_5 V_5)}{(\gamma - 1)}}$$

$$= \frac{36.5}{837 + \dfrac{27.2}{0.4}}$$

$$= \frac{36.5}{837 + 68}$$

$$= \frac{36.5}{76.37}$$

$$= \mathbf{0.478}$$

(f)

$$\text{Mean effective pressure} = \frac{\oint W}{(V_1 - V_2)}$$

$$= \frac{36.5}{(0.084 - 0.005\ 6)}$$

$$= \frac{36.5}{0.078\ 4}$$

$$= \mathbf{465.6\ kN/m^2}$$

(g)

$$\text{Carnot } \eta = \frac{T_4 - T_1}{T_4}$$

$$= \frac{1593 - 301}{1593}$$

$$= \frac{1292}{1593}$$

$$= 0.81$$

$$= \mathbf{81\%}$$

15.8 The Atkinson cycle

This cycle is named after J. Atkinson, a British Engineer who carried out work on the gas engine during the mid 1880s. A peculiarity of the Atkinson gas engine was its alternate short and long strokes which were obtained from a special link mechanism. The short stroke was the compression stroke and the long stroke was the expansion stroke. In this way, Atkinson hoped to obtain more work than in the constant volume cycle. In fact, the amount of extra work is so small that it does not really justify the consequent increase in cylinder volume and the more complex link mechanism required to give the alternate short and long stroke.

This cycle has also been used as a gas turbine cycle; a noteworthy example is the **Holtzwarth constant volume** or **explosion** gas turbine of about 1908. In this type of gas turbine, air was compressed in a compressor and passed to a constant volume combustion chamber. Fuel was injected into the combustion chamber in which it was burnt, or **exploded**, and the pressure was increased at a constant volume. The high-pressure gas was released to expand through a gas turbine then exhausted into the atmosphere. This type of gas turbine has not found great favour, although more work was done on its development in Switzerland in about 1930.

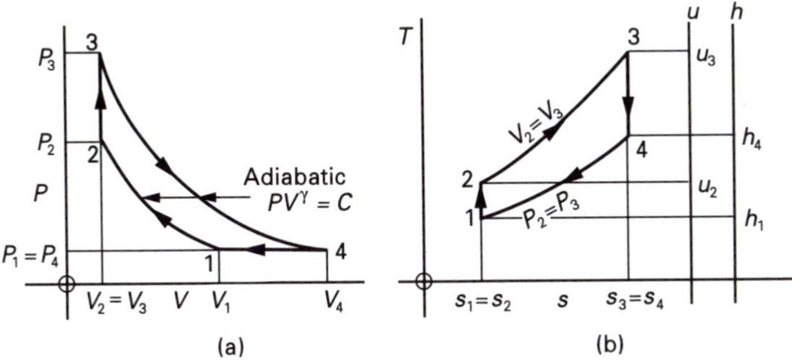

Fig. 15.28 Atkinson cycle: (a) *P–V* diagram; (b) *T–s* diagram

Figure 15.28(a) is a *P–V* diagram of the Atkinson cycle; Fig. 15.28(b) is the corresponding *T–s* diagram. The cycle is arranged as follows:

- **1–2 Adiabatic compression according to the law $PV^\gamma = C$**
  Pressure increases from P_1 to P_2.
  Temperature increases from T_1 to T_2.
  Volume decreases from V_1 to V_2.
  Entropy remains constant at $s_1 = s_2$.

- **2–3 Constant volume heat addition**
  Pressure increases from P_2 to P_3.
  Temperature increases from T_2 to T_3.
  Volume remains constant at $V_2 = V_3$.
  Entropy increases from s_2 to s_3.

- **3–4 Adiabatic expansion according to the law $PV^\gamma = C$**
  Pressure decreases from P_3 to P_4.
  Temperature decreases from T_3 to T_4.
  Volume increases from V_3 to V_4.
  Entropy remains constant at $s_3 = s_4$.

- **4–1 Constant pressure heat rejection**
  Pressure remains constant at $P_4 = P_1$.
  Temperature decreases from T_4 to T_1.
  Volume decreases from V_4 to V_1.
  Entropy decreases from s_4 to s_1.
  This process completes the cycle and returns the gas to its original state.

An analysis of the properties at state points 1, 2, 3 and 4 can be made. It is assumed that P_1, V_1 and T_1 are known.

- P_1, V_1, T_1.

- Assume that the volume ratio V_1/V_2 is known.

$$T_1/T_2 = (V_2/V_1)^{(\gamma-1)}$$

$$\therefore \quad T_2 = T_1\left(\frac{V_1}{V_2}\right)^{(\gamma-1)} = T_1 r_v^{(\gamma-1)}$$

where $r_v = V_1/V_2$ = adiabatic compression volume ratio

$$P_1/V_1^\gamma = P_2/V_2^\gamma$$

$$\therefore \quad P_2 = P_1\left(\frac{V_1}{V_2}\right)^\gamma = P_1 r_v^\gamma$$

- $P_3/T_3 = P_2/T_2$ because the volume remains constant at $V_3 = V_2$.

$$\therefore \quad T_3 = \frac{P_3}{P_2}T_2 = \frac{P_3}{P_2}T_1 r_v^{(\gamma-1)}$$

- $T_4/T_3 = (V_3/V_4)^{(\gamma-1)}$

$$\therefore \quad T_4 = T_3\left(\frac{V_3}{V_4}\right)^{\gamma-1} = \frac{T_3}{R_v^{(\gamma-1)}}$$

where $R_v = V_4/V_3$ = adiabatic expansion volume ratio

$$\therefore \quad T_4 = \frac{P_3}{P_2} T_1 \left(\frac{r_v}{R_v}\right)^{(\gamma-1)}$$

Now

$$\frac{P_2}{P_1} = \left(\frac{V_1}{V_2}\right)^{\gamma} = r_v^{\gamma} \quad \text{also} \quad \frac{P_3}{P_4} = \left(\frac{V_4}{V_3}\right)^{\gamma} = R_v^{\gamma}$$

and dividing

$$\frac{P_3}{P_4}\frac{P_1}{P_2} = \left(\frac{R_v}{r_v}\right)^{\gamma}$$

But $P_1 = P_4$

$$\therefore \quad \frac{P_3}{P_4} = \left(\frac{R_v}{r_v}\right)^{\gamma}$$

From this

$$T_4 = \left(\frac{R_v}{r_v}\right)^{\gamma} T_1 \left(\frac{r_v}{R_v}\right)^{(\gamma-1)} = T_1 \frac{R_v}{r_v}$$

Note also that from the constant pressure process, $V_4/T_4 = V_1/T_1$.

$$\therefore \quad T_4 = T_1 \frac{V_4}{V_1}$$

From this it appears that $V_4/V_1 = R_v/r_v$.

Now

$$\frac{R_v}{r_v} = \frac{V_4}{V_3}\frac{V_2}{V_1} \quad \text{and} \quad V_2 = V_3$$

$$\therefore \quad \frac{R_v}{r_v} = \frac{V_4}{V_1}$$

The work done during the cycle may be obtained as follows:

- Process 3–4 is an expansion and hence gives positive work done.
- Processes 4–1 and 1–2 are compressions and hence give negative work done.
- The net work done will be the sum of the work done during these processes.

Hence

$$\oint W = \text{Area under } 3-4 - \text{Area under } 4-1 - \text{Area under } 1-2$$

$$= \frac{P_3V_3 - P_4V_4}{(\gamma-1)} - P_1(V_4 - V_1) - \frac{(P_2V_2 - P_1V_2)}{(\gamma-1)}$$

$$= \frac{(P_3V_3 - P_4V_4) - (P_2V_2 - P_1V_1)}{(\gamma-1)} - P_1(V_4 - V_1) \qquad [1]$$

$$= mR\left[\frac{(T_3 - T_4) - (T_2 - T_1)}{(\gamma-1)} - (T_4 - T_1)\right] \qquad [2]$$

R in this case is the characteristic gas constant.

Alternatively, the work done may be obtained from

$$\oint W = \text{Heat received} - \text{Heat rejected}$$

In this cycle, heat is received during constant volume process 2–3. Heat is rejected during the constant pressure process 4–1.

No heat is received or rejected during the adiabatic processes.

$$\therefore \quad \oint W = mc_v(T_3 - T_2) - mc_p(T_4 - T_1) \qquad [3]$$

Alternatively

$$\oint W = \text{Heat received} \times \text{Thermal } \eta$$
$$= mc_v(T_3 - T_2) \times \text{Thermal } \eta \qquad [4]$$

The thermal efficiency may be determined from

$$\text{Thermal } \eta = 1 - \frac{\text{Heat rejected}}{\text{Heat received}}$$
$$= 1 - \frac{mc_p(T_4 - T_1)}{mc_v(T_3 - T_2)}$$
$$= 1 - \frac{\gamma(T_4 - T_1)}{(T_3 - T_2)} \qquad [5]$$

This gives the thermal efficiency in terms of temperature.

Substituting temperature in terms of T_1 into equation [5]

$$\text{Thermal } \eta = 1 - \frac{\gamma(T_1 R_v/r_v - T_1)}{(R_v/r_v)^\gamma T_1 r_v^{(\gamma-1)} - T_1 r_v^{(\gamma-1)}}$$
$$= 1 - \frac{\gamma T_1(R_v/r_v - 1)}{T_1 r_v^{(\gamma-1)}[(R_v/r_v)^\gamma - 1]}$$
$$= 1 - \frac{1}{r_v^{(\gamma-1)}} \frac{\gamma(R_v/r_v - 1)}{(R_v/r_v)^\gamma - 1} \qquad [6]$$

For this cycle, using the T–s chart

Heat received at constant volume $= u_3 - u_2$
Heat rejected at constant pressure $= h_4 - h_1$

$$\text{Thermal } \eta = 1 - \frac{\text{Heat rejected}}{\text{Heat received}}$$
$$= 1 - \frac{(h_4 - h_1)}{(u_3 - u_2)} \qquad [7]$$

Also

$$\oint W = \text{Heat received} - \text{Heat rejected}$$
$$= (u_3 - u_2) - (h_4 - h_1) \qquad [8]$$

Again this gives the work done per unit mass per cycle because the chart will be made out for unit mass of gas.

Example 15.14 *At the beginning of the adiabatic compression of an Atkinson cycle the pressure, volume and temperature are 101 kN/m², 14 litres and 15 °C, respectively. At the end of the adiabatic compression the volume is 2.8 litres. The maximum pressure of the cycle is 1.85 MN/m². Determine*

(a) the thermal efficiency
(b) the heat received
(c) the heat rejected
(d) the net work
(e) the work ratio
(f) the mean effective pressure
(g) the Carnot efficiency within the cycle temperature limits

Take $\gamma = 1.4$, $R = 0.29$ *kJ/kg K.*

SOLUTION

First draw a diagram (Fig. 15.29).

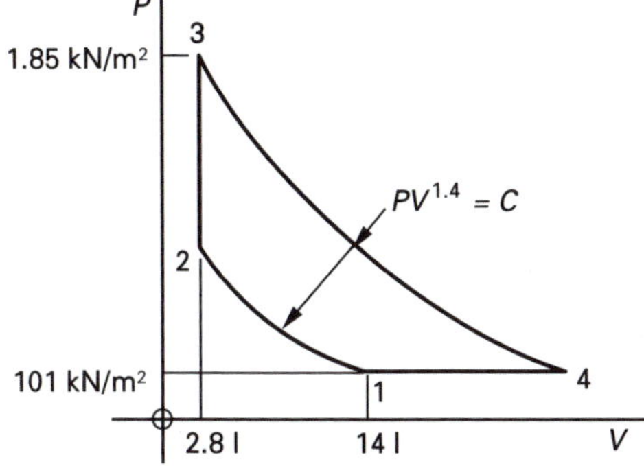

Fig. 15.29 Diagram for Example 15.14

(a)

$$T_1 = 273 + 15 = \mathbf{288\ K}$$

$$\frac{T_1}{T_2} = \left(\frac{V_2}{V_1}\right)^{(\gamma-1)}$$

$$\therefore \quad T_2 = T_1\left(\frac{V_1}{V_2}\right)^{(\gamma-1)} = 288 \times \left(\frac{14}{2.8}\right)^{(1.4-1)}$$

$$= 288 \times 5^{0.4}$$
$$= 288 \times 1.904$$
$$= \mathbf{548\ K}$$

$$P_1V_1^{\gamma} = P_2V_2^{\gamma}$$

$$\therefore \quad P_2 = P_1\left(\frac{V_1}{V_2}\right)^{\gamma} = 101 \times 5^{1.4} = 101 \times 9.52 = \mathbf{961.5\ kN/m^2}$$

$$\frac{P_2}{T_3} = \frac{P_2}{T_2}$$

$$\therefore \quad T_3 = T_2 \frac{P_3}{P_2} = 548 \times \frac{1850}{961.5} = \mathbf{1054\ K}$$

$$\frac{T_3}{T_4} = \left(\frac{P_3}{P_4}\right)^{(\gamma-1)/\gamma}$$

$$\therefore \quad T_4 = T_3 \left(\frac{P_4}{P_3}\right)^{(\gamma-1)/\gamma} = 1054 \times \left(\frac{101}{1850}\right)^{(1.4-1)/0.4}$$

$$= \frac{1054}{18.3^{1/3.5}}$$

$$= \frac{1054}{2.29}$$

$$= \mathbf{460\ K}$$

$$\text{Thermal } \eta = 1 - \frac{\gamma(T_4 - T_1)}{(T_3 - T_2)}$$

$$= 1 - \frac{1.4 \times (460 - 288)}{(1054 - 548)}$$

$$= 1 - \frac{1.4 \times 172}{506}$$

$$= 1 - 0.476$$
$$= 0.524$$
$$= \mathbf{52.4\%}$$

(b)

$$m = \frac{P_1 V_1}{R T_1} = \frac{101 \times 0.014}{0.29 \times 288} = \mathbf{0.016\ 9\ kg}$$

$$\text{Heat received} = m c_v (T_3 - T_2)$$

$$c_v = \frac{R}{(\gamma - 1)} = \frac{0.29}{0.4} = \mathbf{0.725\ kJ/kg\ K}$$

$$\therefore \quad \text{Heat received} = 0.016\ 9 \times 0.725 \times (1054 - 548)$$
$$= 0.016\ 9 \times 0.725 \times 506$$
$$= \mathbf{6.19\ kJ/cycle}$$

(c)

$$\text{Heat rejected} = m c_p (T_4 - T_1)$$

$$\frac{c_p}{c_v} = \gamma$$

$$c_p = \gamma c_v = 1.4 \times 0.725 = \mathbf{1.015\ kJ/kg\ K}$$

$$\therefore \quad \text{Heat rejected} = 0.016\ 9 \times 1.015 \times (460 - 288)$$
$$= 0.016\ 9 \times 1.015 \times 172$$
$$= \mathbf{2.95\ kJ/cycle}$$

(d)

$$\oint W = \text{Heat received} - \text{Heat rejected}$$
$$= 6.19 - 2.95$$
$$= \mathbf{3.24\ kJ/cycle}$$

Alternatively

$$\oint W = \text{Heat received} \times \text{Thermal efficiency}$$
$$= 6.19 - 0.524$$
$$= \mathbf{3.24\ kJ/cycle}$$

(e)

$$\text{Work ratio} = \frac{\oint W}{\text{Positive work done}}$$

$$= \frac{3.24}{\dfrac{(P_3V_3 - P_4V_4)}{(\gamma - 1)}}$$

$$= \frac{3.24 \times (1.4 - 1)}{(1850 \times 0.002\,8) - (101 \times 0.014)}$$

$$= \frac{3.24 \times 0.4}{(5.18 - 1.414)}$$

$$= \frac{1.296}{3.766}$$

$$= \mathbf{0.344}$$

(f)

$$\text{Mean effective pressure} = \frac{\oint W}{(V_4 - V_2)}$$

Now $\dfrac{V_4}{T_4} = \dfrac{V_1}{T_1}$ since $P_4 = P_1$

$$\therefore \quad V_4 = V_1\frac{T_4}{T_1} = 0.014 \times \frac{460}{288} = \mathbf{0.022\ m^3}$$

$$\therefore \quad \text{Mean effective pressure} = \frac{3.24}{(0.022 - 0.002\,8)}$$

$$= \frac{3.24}{0.019\,2}$$

$$= \mathbf{168.75\ kN/m^2}$$

(g)

$$\text{Carnot } \eta = \frac{T_3 - T_1}{T_3}$$

$$= \frac{1054 - 288}{1054}$$

$$= \frac{766}{1054}$$

$$= 0.727$$

$$= \mathbf{72.7\%}$$

15.9 The Stirling and Ericsson cycles

The Stirling cycle is named after Dr Robert Stirling and his brother James Stirling who, as early as 1815, obtained a British patent on an air engine (see section 15.2). In 1845 one such engine was used in a Dundee foundry. The air was used in a closed system and heat was supplied by a furnace through a heating surface. In the system used, the heat transfer into the air was slow and eventually the heater surface burned out, the engine was abandoned.

Figure 15.30 shows the *P–V* and *T–s* diagrams; the Stirling cycle is composed of two isothermal processes and two constant volume processes. Its importance here lies in the process of regeneration which the Stirling brothers used on their engines. The two constant volume processes, 2–3 and 4–1, are bounded by the isotherms and therefore have the same temperature limits. Thus, neglecting change in specific heat capacity, the heat transfer required for constant volume process 2–3 is equal to the heat transfer rejection for constant volume process 4–1, or $mc_v(T_3 - T_2) = mc_v(T_4 - T_1)$.

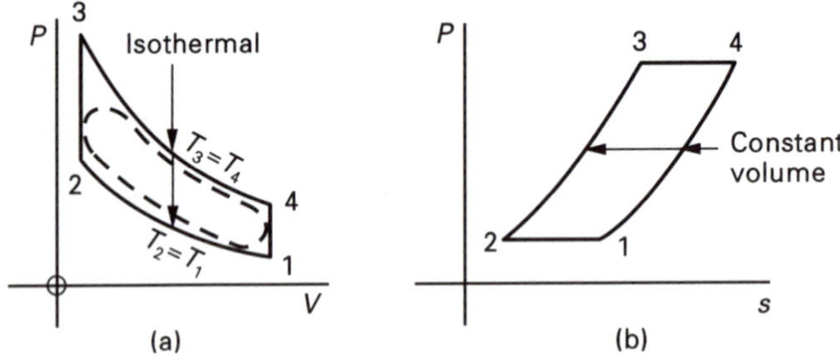

Fig. 15.30 Stirling cycle: (a) *P–V* diagram; (b) *T–s* diagram

Now, during the process 4–1, this amount of energy is rejected and during process 2–3, the same amount of energy is required. Suppose a device could be developed within the engine, to store the energy rejected during process 4–1 then give it up during process 2–3. Once the necessary warm-up of the engine has taken place and temperature limits have been established, these two processes will be self-perpetuating; no subsequent heat energy transfer will be required from an external source through the system boundary. There is such a device; it is called a **regenerator** and the process is called **regeneration**. In the case of the Stirling engine, the regenerator consisted of a matrix of sheet-iron plates maintained at the high temperature at one end by the furnace and the low temperature at the other end by a water-cooler. Thus the necessary temperature gradient was maintained through the matrix, which was sufficiently bulky that the necessary heat energy transfer during the successive processes did not substantially modify the temperatures. With the regenerator installed, external heat transfer is not required to carry out the constant volume processes, so the thermal efficiency relies only on the two isothermal processes, which is the same as for the Carnot cycle. Hence, with the process of regeneration included, the theoretical Stirling cycle thermal efficiency has the highest thermal efficiency possible, $(T_3 - T_1)/T_3$. By including the regeneration process, a cycle which would otherwise have been irreversible has been made reversible and now has the highest ideal thermal efficiency.

From time to time there is a revival of interest in the Stirling cycle with its process of regeneration. Modern research has shown the possibility of engines with thermal efficiencies as high as 40 per cent and with good power outputs; some have, in fact, been constructed. They are achieved using improved manufacturing techniques and by using gases other than air, e.g. hydrogen, at general pressures much higher than atmospheric to increase density and hence heat transfer properties.

Note that in Fig. 15.30(a) the practical cycle is shown dotted and has an oval shape. It does not attain the high ideal efficiency. In a practical Stirling engine, air or some other gas is used in a closed cycle and is heated and cooled indirectly. Thus the cycle is independent of the type of fuel or energy source. Hence, nuclear fuel and solar radiation have been considered as possible energy sources together with the more conventional fuels such as coal, oil, gas and wood. The reversed Stirling cycle has been successfully applied in the liquefaction of air.

The Ericsson cycle is named after John Ericsson (1803–1889), a Swedish engineer who spent some time in England and constructed a steam locomotive at the same time as George Stephenson. Ericsson eventually left England to live in America. In America his interest centred on hot-air engines (see section 15.2).

Figure 15.31 shows the P–V and T–s diagrams of the Ericsson cycle; it is composed of two isothermal processes and two constant pressure processes.

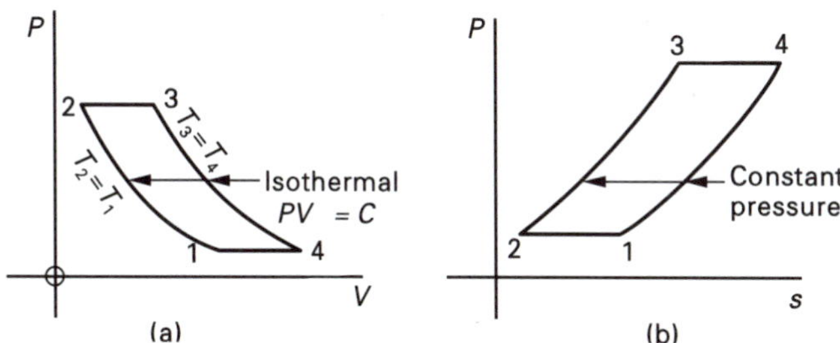

Fig. 15.31 Ericsson cycle: (a) P–V diagram; (b) T–s diagram

Once again, because the two constant pressure processes are bounded by the same temperature limits

$$mc_p(T_3 - T_2) = mc_p(T_4 - T_1)$$

and the process of regeneration is again possible.

By including the process of regeneration, the Ericsson cycle becomes a reversible cycle and has the highest thermal efficiency possible, $(T_3 - T_1)/T_3$.

Example 15.15 *An ideal Stirling cycle, using air and including regeneration, has at the beginning of isothermal compression a pressure, volume and temperature of 110 kN/m^2, 0.05 m^3 and 30 °C, respectively. The minimum volume of the cycle is 0.005 m^3. The maximum temperature of the cycle is 700 °C. Determine for the cycle*

(a) the net work done

(b) the ideal thermal efficiency

(c) the thermal efficiency if the process of regeneration is not included

Take R $= 0.289$ *kJ/kg K*, $c_v = 0.718$ *kJ/kg K.*

SOLUTION

First draw a diagram (Fig. 15.32).

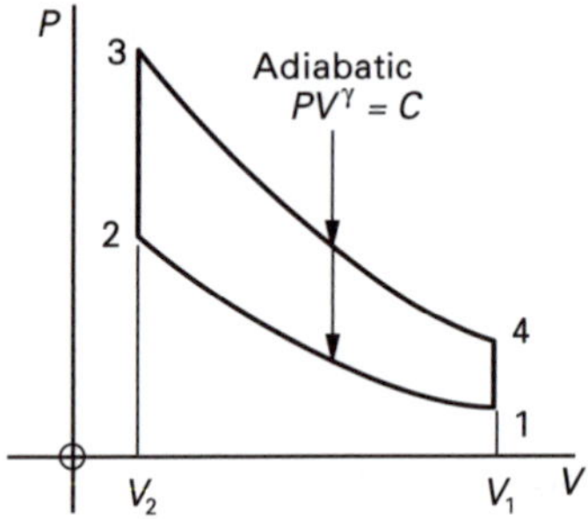

Fig. 15.32 Diagram for Example 15.15

(a)

$$\oint W = mR \ln \frac{V_1}{V_2}(T_3 - T_1) \quad \text{and} \quad T_3 = 700 + 273 = \mathbf{973\ K}$$
$$T_1 = 30 + 273 = \mathbf{303\ K}$$

$$m = \frac{P_1 V_1}{RT_1} = \frac{110 \times 0.05}{0.289 \times 303} = \mathbf{0.063\ kg}$$

$$\therefore \quad \oint W = 0.063 \times 0.289 \ln \frac{0.05}{0.005}(973 - 303)$$
$$= 0.063 \times 0.289 \ln 10 \times 670$$
$$= 0.063 \times 0.289 \times 2.3 \times 670$$
$$= \mathbf{28.1\ kJ}$$

(b)

$$\text{Ideal thermal } \eta = \frac{T_3 - T_1}{T_3}$$
$$= \frac{973 - 303}{973}$$
$$= \frac{670}{973}$$
$$= 0.689$$
$$= \mathbf{68.9\%}$$

(c)

$$\text{Thermal } \eta = 1 - \frac{\text{Heat rejected}}{\text{Heat received}}$$

$$= 1 - \left[\frac{mc_v(T_4 - T_1) + mRT_1 \ln \dfrac{V_1}{V_2}}{mc_v(T_3 - T_2) + mRT_3 \ln \dfrac{V_4}{V_3}}\right]$$

$$= 1 - \left[\frac{c_v(T_4 - T_1) + RT_1 \ln \dfrac{V_1}{V_2}}{c_v(T_3 - T_2) + RT_3 \ln \dfrac{V_4}{V_3}}\right]$$

$$= 1 - \frac{(0.718 \times 670) + (0.289 \times 303 \ln 10)}{(0.718 \times 670) + (0.289 \times 973 \ln 10)}$$

$$= 1 - \frac{48.1 + (0.289 \times 303 \times 2.3)}{481.1 + (0.289 \times 973 \times 2.3)}$$

$$= 1 - \frac{(481.1 + 201.4)}{(481.1 + 646.8)}$$

$$= 1 - \frac{685.5}{1127.9}$$

$= 1 - 0.605$
$= 0.395$
$=$ **39.5%**

Example 15.16 *At the beginning of the isothermal compression of an ideal Ericsson cycle, using air and including regeneration, the pressure, volume and temperature are 100 kN/m², 0.08 m³ and 20 °C, respectively. The volume ratio of the isothermal compression is 5:1. After the isothermal compression, the volume of the air is doubled during the constant pressure process. Determine for the cycle*
(a) the maximum temperature
(b) the net work done
(c) the ideal thermal efficiency
(d) the thermal efficiency if the process of regeneration is not included
Take R $= 0.287$ *kJ/kg K,* $c_p = 1.006$ *kJ/kg K.*

SOLUTION
First draw a diagram (Fig. 15.33).

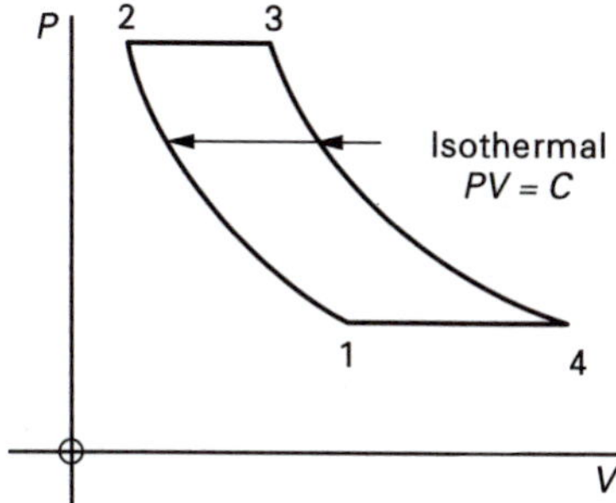

Fig. 15.33 Diagram for Example 15.16

(a)

$T_1 = 20 + 273 = \mathbf{293\ K}$

$$\frac{V_3}{T_3} = \frac{V_2}{T_2} \quad \text{and} \quad T_2 = T_1$$

$$\therefore \quad T_3 = T_2\frac{V_3}{V_2} = 293 \times 2$$

$$= \mathbf{586\ K}$$

$$t_3 = 586 - 273$$

$$= \mathbf{313\ ^{\circ}C}$$

The maximum temperature is 313 °C.

(b)

$$\oint W = mR(T_3 - T_2) + mRT_3 \ln \frac{V_4}{V_3} - mR(T_4 - T_1) - mRT_1 \ln \frac{V_1}{V_2}$$

$$= mR \ln \frac{V_1}{V_2}(T_3 - T_1) \quad \text{since} \quad T_3 = T_4$$

$$T_1 = T_2$$

$$\frac{V_4}{V_3} = \frac{V_1}{V_2}$$

$$\text{Now} \quad m = \frac{P_1V_1}{RT_1} = \frac{100 \times 0.08}{0.287 \times 293} = \mathbf{0.095\ kg}$$

$$\therefore \quad \oint W = 0.095 \times 0.287 \times \ln 5 \times (586 - 293)$$

$$= 0.095 \times 0.287 \times 1.61 \times 293$$

$$= \mathbf{12.86\ kJ}$$

(c)

$$\text{Ideal thermal } \eta = \frac{T_3 - T_1}{T_3}$$

$$= \frac{(586 - 293)}{586}$$

$$= \frac{293}{586}$$

$$= 0.5$$

$$= \mathbf{50\%}$$

(d)

$$\text{Thermal } \eta = 1 - \frac{\text{Heat rejected}}{\text{Heat received}}$$

$$= 1 - \left[\frac{mc_p(T_4 - T_1) + mRT_1 \ln \dfrac{V_1}{V_2}}{mc_p(T_3 - T_2) + mRT_3 \ln \dfrac{V_4}{V_3}}\right]$$

$$= 1 - \left[\frac{c_p(T_4 - T_1) + RT_1 \ln \dfrac{V_1}{V_2}}{c_p(T_3 - T_2) + RT_3 \ln \dfrac{V_4}{V_3}}\right]$$

$$= 1 - \frac{(1.006 \times 293) + (0.287 \times 293 \ln 5)}{(1.006 \times 293) + (0.287 \times 586 \ln 5)}$$

$$= 1 - \frac{294.8 + 135.4}{294.8 + 270.8}$$

$$= 1 - \frac{430.2}{565.6}$$

$$= 1 - 0.76$$
$$= 0.24$$
$$= \mathbf{24\%}$$

15.10 Miscellaneous cycles

Example 15.17 *One kilogram of air at a temperature of 230 °C is expanded isothermally from a pressure of 3.45 MN/m² to a pressure of 2 MN/m². It is then expanded adiabatically to a pressure of 140 kN/m². The air is then cooled at constant pressure and finally restored to its initial state by adiabatic compression. Determine for the cycle*

(a) the net work done

(b) the thermal efficiency

Take $\gamma = 1.4$, $c_p = 1.006$ kJ/kg K.

SOLUTION

First draw a diagram (Fig. 15.34)

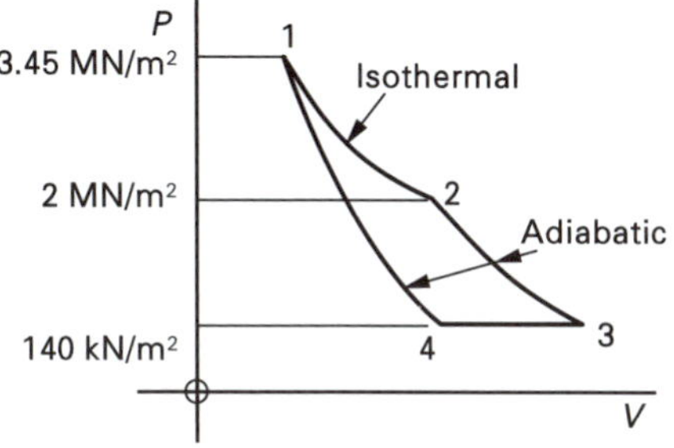

Fig. 15.34 Diagram for Example 15.17

(a)

Since 1–2 is isothermal, $t_2 = 230\ °\text{C}$

$$\frac{T_2}{T_3} = \left(\frac{P_2}{P_3}\right)^{(\gamma-1)/\gamma} \quad \text{and} \quad T_2 = 230 + 273 = \mathbf{503\ K}$$

$$\therefore \quad T_3 = T_2\left(\frac{P_3}{P_2}\right)^{(\gamma-1)/\gamma} = 503 \times \left(\frac{140}{2000}\right)^{(1.4-1)/1.4}$$

$$= \frac{503}{14.3^{1/3.5}}$$

$$= \frac{503}{2.14}$$

$$= \mathbf{235\ K}$$

$$\frac{T_1}{T_4} = \left(\frac{P_1}{P_4}\right)^{(\gamma-1)/\gamma}$$

$$\therefore \quad T_4 = T_1\left(\frac{P_4}{P_1}\right)^{(\gamma-1)/\gamma} = 503 \times \left(\frac{140}{3450}\right)^{(1.4-1)/\gamma}$$

$$= \frac{503}{24.64^{1/3.5}}$$

$$= \frac{503}{2.5}$$

$$= \mathbf{201\ K}$$

Heat received during isothermal process/kg is $RT_1 \ln(P_1/P_2)$ and

$$R = c_p\left(\frac{\gamma-1}{\gamma}\right) = 1.006\left(\frac{1.4-1}{1.4}\right) = 1.006 \times \frac{0.4}{1.4} = \mathbf{0.287\ kJ/kg\ K}$$

So heat received during isothermal process per kilogram is

$$0.287 \times 503 \times \ln 1.725 = 0.287 \times 503 \times 0.545$$
$$= \mathbf{78.7\ kJ}$$

Heat rejected during constant pressure process per kilogram is

$$c_p(T_3 - T_4) = 1.006 \times (235 - 201)$$
$$= 1.006 \times 34$$
$$= \mathbf{34.2\ kJ}$$

There is no heat received or rejected during the adiabatic processes.

$$\oint W = \text{Heat received} - \text{Heat rejected}$$

$$\therefore \quad \oint W = 78.7 - 34.2 = \mathbf{44.5\ kJ}$$

(b)

$$\text{Thermal } \eta = \frac{\text{Heat received} - \text{Heat rejected}}{\text{Heat received}}$$

$$= \frac{44.3}{78.7}$$

$$= 0.563$$
$$= \mathbf{56.3\%}$$

Example 15.18 *A gas cycle consists of the following four processes:*

- **1–2** Adiabatic (isentropic) compression from initial temperature T_1 through volume ratio r_v.
- **2–3** Heat transfer at constant pressure until the volume at state point 3 is equal to twice the volume at 2.
- **3–4** Isothermal expansion to the original volume.
- **4–1** Constant volume heat transfer to original state point 1.

For this cycle, sketch the pressure–volume and temperature–entropy diagrams and show that

$$\text{Thermal } \eta = 1 - \frac{1/\gamma[2 - 1/r_v^{(\gamma-1)}]}{1 + 2[(\gamma-1)/r_v]\ln r_v/2}$$

If $r_v = 5$, *determine the thermal efficiency of the cycle and the Carnot efficiency between the same temperature limits. Take* $\gamma = 1.4$.

SOLUTION

The sketch diagrams are shown in Fig. 15.35.

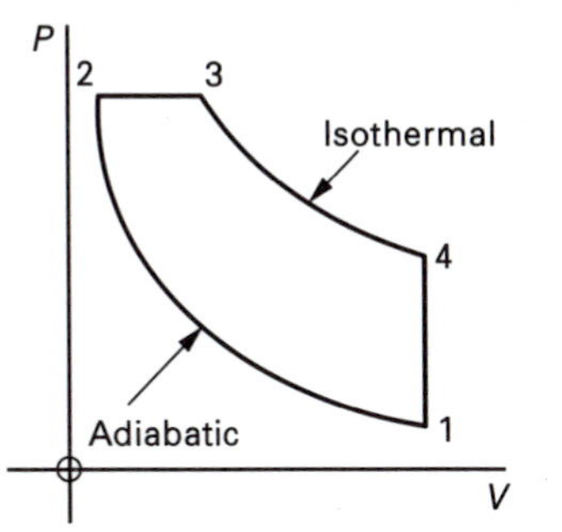

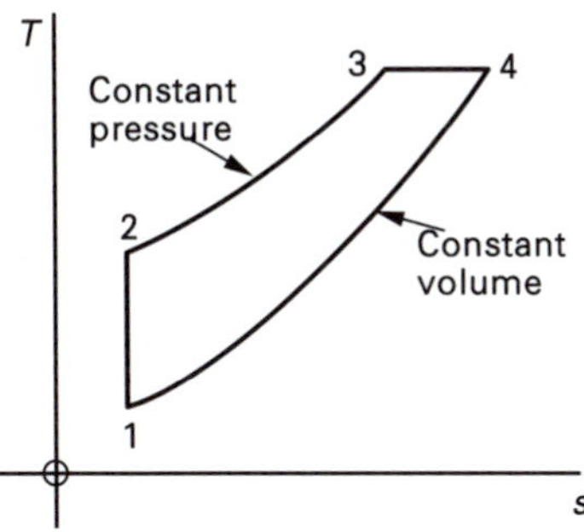

Fig. 15.35 Diagrams for Example 15.18

To solve the temperatures at the state points

- T_1
- $T_1/T_2 = (V_1/V_2)^{(\gamma-1)} = r_v^{(\gamma-1)}$

$\therefore \quad T_2 = T_1 r_v^{(\gamma-1)}$

since $r_v = V_1/V_2$

- $V_3/T_3 = V_2/T_2$

  $\therefore \quad T_3 = T_2 \dfrac{V_3}{V_2} = 2T_1 r_v^{(\gamma-1)}$

  since $V_3 = 2V_2$ and $T_2 = T_1 r_v^{(\gamma-1)}$

- $T_4 = T_3$ because process 3–4 is isothermal

  $\therefore \quad T_4 = 2T_1 r_v^{(\gamma-1)}$

To analyse the heat transfer through the cycle.

- **1–2** No heat transfer since the process is adiabatic
- **2–3** Heat received $= mc_p(T_3 - T_2)$
- **3–4** Heat received $= mRT_3 \ln V_4/V_3$
- **4–1** Heat rejected $= mc_v(T_4 - T_1)$

$$\text{Thermal } \eta = 1 - \frac{\text{Heat rejected}}{\text{Heat received}}$$

$$= 1 - \frac{mc_v(T_4 - T_1)}{mc_p(T_3 - T_2) + mRT_3 \ln (V_4/V_3)}$$

Now $V_4 = V_1$, $V_3 = 2V_2$ and $V_1/V_2 = r_v$, so

$$\frac{V_4}{V_3} = \frac{V_1}{2V_2} = \frac{r_v}{2}$$

$$\therefore \quad \text{Thermal } \eta = 1 - \frac{c_v[2T_1 r_v^{(\gamma-1)} - T_1]}{c_p[2T_1 r_v^{(\gamma-1)} - T_1 r_v^{(\gamma-1)}] + c_p[(\gamma-1)/\gamma]2T_1 r_v^{(\gamma-1)} \ln (r_v/2)}$$

Note that $R = c_p(\gamma - 1)/\gamma$ from $c_p/c_v = \gamma$ and $(c_p - c_v) = R$.
Simplifying

$$\text{Thermal } \eta = 1 - \frac{c_v T_1 r_v^{(\gamma-1)}[2 - 1/r_v^{(\gamma-1)}]}{c_p T_1 r_v^{(\gamma-1)}\{(2-1) + 2[(\gamma-1)/\gamma] \ln (r_v/2)]\}}$$

or

$$\text{Thermal } \eta = 1 - \frac{1/\gamma[2 - 1/r_v^{(\gamma-1)}]}{1 + 2[(\gamma-1)/\gamma] \ln (r_v/2)}$$

If $r_v = 5$

$$\text{Thermal } \eta = 1 - \frac{\dfrac{1}{1.4}\left(2 - \dfrac{1}{5^{0.4}}\right)}{1 + 2 \times \left(\dfrac{0.4}{1.4} \times \ln \dfrac{5}{2}\right)}$$

$$= 1 - \frac{\dfrac{1}{1.4}\left(2 - \dfrac{1}{1.4}\right)}{1 + 2 \times (0.286 \times \ln 2.5)}$$

$$= 1 - \frac{(2 - 0.526)/1.4}{1 + 2 \times (0.286 \times 0.916)}$$

$$= 1 - \frac{1.474/1.4}{1 + 0.524}$$

$$= 1 - \frac{1.053}{1.524}$$

$$= 1 - 0.69$$
$$= 0.31$$
$$= \mathbf{31\%}$$

$$\text{Carnot } \eta = \frac{\text{Max. abs. temp.} - \text{Max. abs. temp.}}{\text{Max. abs. temp.}}$$

$$= 1 - \frac{\text{Min. abs. temp.}}{\text{Max. abs. temp.}}$$

$$= 1 - \frac{T_1}{T_3}$$

$$= 1 - \frac{T_1}{2T_1 r_v^{(\gamma-1)}}$$

$$= 1 - \frac{1}{2r_v^{(\gamma-1)}}$$

$$= 1 - \frac{1}{2 \times 5^{(1.4-1)}} = 1 - \frac{1}{2 \times 5^{0.4}} = 1 - \frac{1}{2 \times 1.9} = 1 - \frac{1}{3.8}$$

$$= 1 - 0.263$$
$$= 0.737$$
$$= \mathbf{73.7\%}$$

Questions

1. The temperatures of the two isothermal processes of a Carnot cycle are 350 °C and 15 °C, respectively. Determine the thermal efficiency of the cycle.

   [53.7%]

2. 0.23 kg of gas is taken through a Carnot cycle whose temperature limits are 300 °C and 50 °C. If the volume ratio of the isothermal processes is 2.5:1, determine for the cycle
   (a) the thermal efficiency
   (b) the net work done
   (c) the work ratio
   Take $R = 0.28$ kJ/kg K, $\gamma = 1.4$.

   [(a) 43.6%; (b) 14.75 kJ; (c) 0.2]

3. The high temperature of a Carnot cycle is 400 °C and the cycle has a thermal efficiency of 55 per cent. The volume ratio of the isothermal processes is 2.8:1. Determine for the cycle
   (a) the low temperature
   (b) the volume ratio of the adiabatic processes
   (c) the overall volume ratio
   Take $\gamma = 1.4$.

   [(a) 30 °C; (b) 7.34:1; (c) 20.6:1]

4. The pressures at the beginning and end of the adiabatic compression of a constant volume cycle are 103.5 kN/m^2 and 1.85 MN/m^2, respectively. Determine for the cycle
   (a) the overall volume ratio
   (b) the thermal efficiency
   Take $\gamma = 1.4$.

   [(a) 7.87:1; (b) 56.1%]

5. An internal combustion engine working on the constant volume cycle has a cylinder bore of 100 mm and a stroke of 95 mm. The clearance volume is 0.001 m^3. Determine for the cycle
   (a) the ideal thermal efficiency
   (b) the actual thermal efficiency if the relative efficiency is 45%
   Take $\gamma = 1.4$.
   [(a) 57.4%; (b) 25.8%]

6. The pressure, volume and temperature at the beginning of the compression of a constant volume (Otto) cycle are 105 kN/m^2, 0.002 m^3 and 25 °C, respectively. The maximum temperature of the cycle is 1250 °C. The volume ratio of the cycle is 8:1. The cycle is repeated 4000 times/min. Determine for the cycle
   (a) the theoretical output in kilowatts
   (b) the thermal efficiency
   (c) the mean effective pressure
   (d) the Carnot efficiency within the same temperature limits
   Take $c_p = 1.007$ kJ/kg K, $c_v = 0.717$ kJ/kg K.
   [(a) 54.3 kW; (b) 57%; (c) 465.1 kN/m^2; (d) 80.4%]

7. 0.5 kg of air is taken through a constant pressure cycle. The pressure and temperature of the air at the beginning of the adiabatic compression are 96.5 kN/m^2 and 15 °C, respectively. The pressure ratio of compression is 6:1. Constant pressure heat addition occurs after the adiabatic compression until the volume is doubled. Determine for the cycle
   (a) the thermal efficiency
   (b) the heat received
   (c) the net work done
   (d) the work ratio
   (e) the mean effective pressure
   [(a) 40%; (b) 241 kJ; (c) 96.4 kJ; (d) 0.466; (e) 134.3 kN/m^3]

8. In an ideal Diesel cycle the pressure and temperature at the beginning of the adiabatic compression are 98.5 kN/m^2 and 60 °C, respectively. The maximum pressure attained during the cycle is 4.5 MN/m^2 and the heat received during the cycle is 580 kJ/kg gas. Determine for the cycle
   (a) the volume ratio of compression
   (b) the temperature at the end of compression
   (c) the temperature at the end of heat reception
   (d) the temperature at the end of the adiabatic expansion
   (e) the net work done per kilogram of gas
   (f) the thermal efficiency
   (g) the work ratio
   (h) the mean effective pressure
   (i) the Carnot efficiency within the cycle temperature limits
   [(a) 15.3:1; (b) 719 °C; (c) 1297 °C; (d) 360 °C; (e) 365 kJ/kg; (f) 63%; (g) 0.435; (h) 402 kN/m^2; (i) 79%]

9. At the commencement of the compression stroke, the cylinder of a Diesel engine is charged with air at a pressure of 96.5 kN/m^2 and at a temperature of 65 °C. Compression takes place to 1/14 the original volume according to the law $PV^{1.35}$ = constant. Fuel is then injected; the mass of fuel injected is 1.40 that of the air in the cylinder. Combustion takes place at constant pressure. The calorific value of the fuel oil is 44 000 kJ/kg. Determine
   (a) the theoretical pressure and temperature after compression

(b) the theoretical temperature after combustion
(c) the fraction of the stroke at which combustion is theoretically complete

For the mixture, take $c_p = 1.003$ kJ/kg K.

[(a) 3403 kN/m^2, 579 °C; (b) 1649 °C; (c) 0.096 6]

10. An engine working on an ideal Diesel cycle has a clearance volume of 0.000 25 m^3. It has a bore and stroke of 152.5 mm and 200 mm, respectively. At the beginning of the adiabatic compression the air in the cylinder has a pressure of 100 kN/m^2 and a temperature of 20 °C, respectively. The maximum temperature of the cycle is 1090 °C. Determine
(a) the temperature and pressure at the end of the adiabatic compression
(b) the temperature and pressure at the end of the adiabatic expansion
(c) the thermal efficiency of the cycle

Take $\gamma = 1.4$.

[(a) 606 °C, 4680 kN/m^2; (b) 269 °C, 185 kN/m^2; (c) 63.3%]

11. In an ideal dual combustion cycle the pressure, volume and temperature at the beginning of the adiabatic compression are 93 kN/m^2, 0.05 m^3 and 24 °C, respectively. The volume ratio of the adiabatic compression is 9:1. The constant volume heat addition pressure ratio is 1.5:1 and the constant pressure heat addition volume ratio is 2:1. Determine for the cycle
(a) the pressure, volume and temperature at the cycle process change points
(b) the thermal efficiency
(c) the net work done
(d) the work ratio
(e) the mean effective pressure
(f) the Carnot cycle within the cycle temperature limits

Take $c_p = 1.05$ kJ/kg K, $c_v = 0.775$ kJ/kg K.

[(a) $P_1 = 93$ kN/m^2, $V_1 = 0.05$ m^3, $t_1 = 24$ °C
$P_2 = 1823$ kN/m^2, $V_2 = 0.005\,6$ m^3, $t_2 = 379$ °C
$P_3 = 2735$ kN/m^2, $V_3 = 0.005\,6$ m^3, $t_3 = 705$ °C
$P_4 = 2735$ kN/m^2, $V_4 = 0.011\,2$ m^3, $t_4 = 1683$ °C
$P_5 = 360$ kN/m^2, $V_5 = 0.05$ m^3, $t_5 = 877$ °C
(b) 48.3%; (c) 35 kJ; (d) 0.685; (e) 788 kN/m^2; (f) 84.8%]

12. In an ideal Atkinson cycle the volume ratios of the adiabatic expansion and the adiabatic compression are 9:1 and 5:1, respectively. At the beginning of the adiabatic compression the pressure and temperature of the gas are 96.5 kN/m^2 and 27 °C, respectively. Determine for the cycle
(a) the pressure and temperature at the cycle process change points
(b) the thermal efficiency
(c) the net work done per kilogram of gas
(d) the work ratio
(e) the mean effective pressure
(f) the Carnot efficiency within the cycle temperature limits

Take $\gamma = 1.38$, $c_v = 0.716$ kJ/kg K.

[(a) $P_1 = 96.5$ kN/m^2, $t_1 = 27$ °C
$P_2 = 888$ kN/m^2, $t_2 = 279$ °C
$P_3 = 1998$ kN/m^2, $t_3 = 969$ °C
$P_4 = 96.5$ kN/m^2, $t_4 = 267$ °C
(b) 52%; (c) 260 kJ; (d) 0.487; (e) 73.4 kN/m^2; (f) 75.8%]

13. At the beginning of a cycle using air, the pressure, volume and temperature are 550 kN/m^2, 0.025 m^3 and 15 °C, respectively. The air is heated at constant pressure until the temperature is 204 °C. It is then expanded adiabatically until the temperature

becomes 120 °C. The air is cooled at constant volume to a temperature of 15 °C then restored to its original volume by isothermal compression. Determine for the cycle

(a) the heat received

(b) the work done

(c) the thermal efficiency

Take $c_p = 1.00$ kJ/kg K, $\gamma = 1.4$.

[(a) 31.6 kJ; (b) 5.54 kJ; (c) 17.5%]

14. A gas is taken through a four-process cycle as follows:

1–2 Isothermal compression through a volume ratio r_v

2–3 Constant pressure (isobaric) expansion through a volume ratio β

3–4 Adiabatic (isentropic) expansion to the initial volume

4–1 Constant volume (isochoric) pressure reduction to the initial state at 1

Sketch the cycle on *P–V* and *T–s* diagrams and show that the thermal efficiency is given by

$$\text{Thermal } \eta = 1 - \frac{(\gamma - 1)\ln r_v + \left[\dfrac{\beta^{\gamma}}{r_v^{(\gamma-1)}} - 1\right]}{\gamma(\beta - 1)}$$

If $r = 8$ and $\beta = 2$, determine the thermal efficiency of the cycle and also the Carnot efficiency within the same temperature limits.

[30.5%, 50%]

Chapter 16

Internal combustion engines

16.1 General introduction

In the internal combustion engine, combustion takes place within the engine itself, hence its name. This is unlike a steam turbine, where steam is introduced to the turbine after having been raised externally in a boiler. The combustion is external; it takes place in the boiler, not the turbine.

All internal combustion engines aspirate air into which is introduced a measured quantity of fuel. This fuel burns within the engine and in such a way that it produces a gas containing a high energy. This gas can then be made to expand within the engine and, using a suitable mechanism, work may be obtained. Because of the absence of external auxiliary equipment, such as a boiler, which is generally large, the internal combustion engine is of a much more compact size for a given output.

Internal combustion engines run on gaseous fuels and liquid fuels; they may be found as reciprocating piston engines, turbines and rotary engines.

16.2 Reciprocating engine details

Figure 16.1 shows the mechanical elements of the reciprocating engine. The reciprocating element is a piston in a cylinder. The piston and the cylinder have a very close fit, and to ensure the assembly is gas-tight, the piston is fitted with piston rings. They spring out and wear to the cylinder bore, so they greatly assist in making the piston gas-tight. The cylinder is bored in the cylinder block which may be cast integral with the crankcase (as shown) or bolted on top. The top of the cylinder is sealed by bolting on to it the cylinder head. The seal between the cylinder head and the cylinder is usually made by inserting a gasket between them. The recess in the cylinder head, above the cylinder, is known as the combustion space; it is here that combustion is initiated. Sometimes, however, the cylinder head is left flat and the combustion space is cast as a bowl in the top of the piston. The engine is then called a bowl-in-piston engine. Due to combustion, the cylinder head, piston and cylinder block will become very hot. It is necessary to cool the engine to preserve the engine materials and lubrication. In Fig. 16.1 coolant passages are shown cast in the walls of the cylinder block, round the cylinder and in the cylinder head round the combustion space. Water is the usual coolant which circulates through these coolant passages and this maintains the engine at a workable temperature.

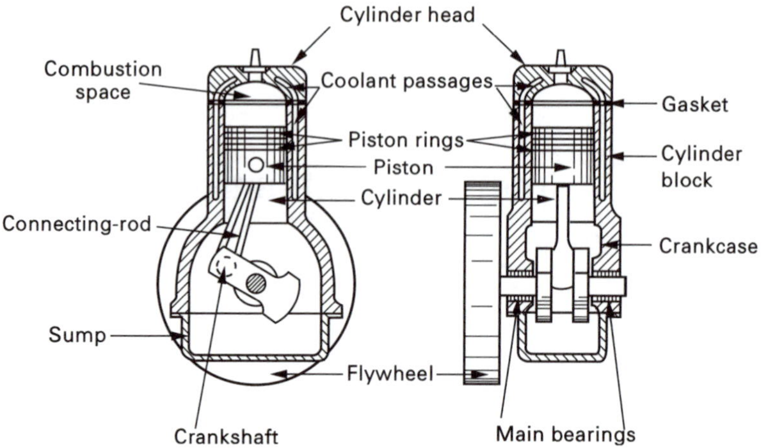

Fig. 16.1 Reciprocating internal combustion engine

Another method of cooling is that used in the air-cooled engine. This method is illustrated in Fig. 16.2. It consists simply of casting fins on to the cylinder and the cylinder head. These fins greatly increase the surface area of the cylinder and the cylinder head; if air is passed over the fins, effective engine cooling is obtained.

Now the reciprocating motion of the piston in the cylinder is to be converted into a rotary motion. This is accomplished by using a connecting-rod to couple the piston to a crankshaft. The connecting rod is free to oscillate on a pin in the piston. This pin is called the gudgeon pin (or wrist pin) and the end of the connecting-rod which fits over this gudgeon pin is called the small end of the connecting rod. The other end of the connecting rod is fitted over the crankpin at the end of the crank. This end of the connecting-rod, which oscillates on the crankpin, is called the big end. Now the crank is fitted, or cast, with two shafts; the whole crank assembly is called the crankshaft. This crankshaft rotates in the main bearings which are set in the crankcase. On the outside of the crankcase (sometimes inside) is fitted a flywheel. The flywheel is fitted in order to help smooth out the uneven torque which is generated in the reciprocating engine. If the piston reciprocates in the cylinder, it will impart a rotary motion to the crankshaft.

The bottom of the engine is closed in by the sump. This often contains the oil which is pumped round the engine for lubrication; the engine is then wet-sump lubricated. In other engines the lubricating oil is kept in a separate tank from which

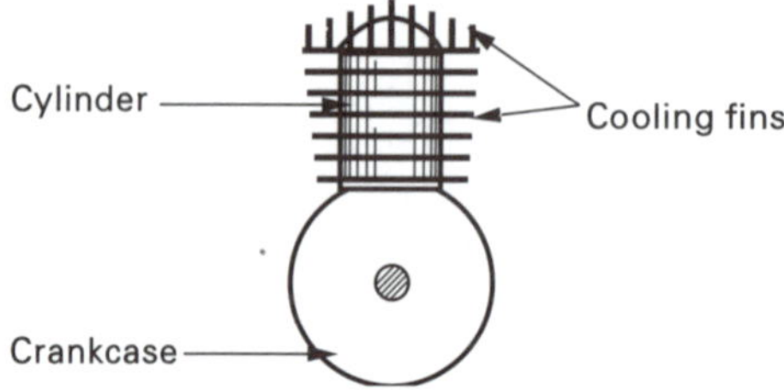

Fig. 16.2 Air-cooled internal combustion engine

it circulates at pressure round the engine and is then pumped back to the oil tank. Such engines are said to be dry-sump engines because oil is not allowed to collect in the sump.

The engine illustrated in Fig. 16.1 is single-cylinder. The crankshaft has a single crank and is called a single-throw crankshaft. There are many other basic engine designs; some of the more conventional designs are illustrated in Fig. 16.3. Figure 16.3(a) shows the single-cylinder arrangement already discussed. Figure 16.3(b) shows the multicylinder, in-line arrangement. Here, four cylinders are shown in a single block. There is a common crankcase for these cylinders. There is also a common crankshaft which has a crank situated beneath each cylinder and hence is called a multithrow crankshaft. The crankshaft in the figure would be four-throw. In-line

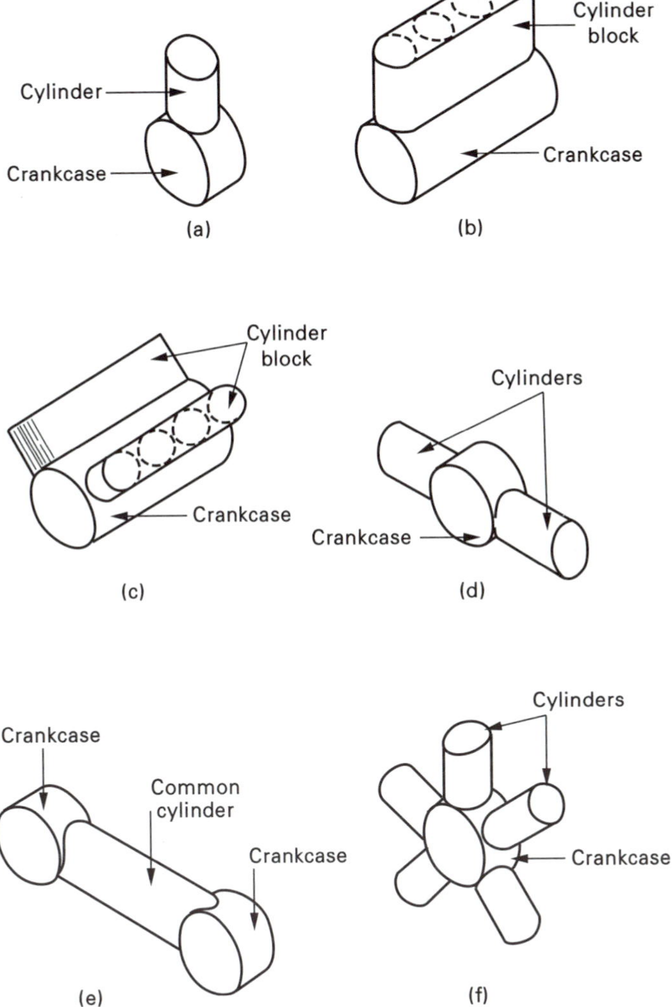

Fig. 16.3 Engine configurations: (a) single-cylinder; (b) multicylinder in-line; (c) multicylinder V8; (d) horizontally opposed; (e) opposed piston; (f) radial

engines of up to eight cylinders are made. Most commonly they have two, four, six or eight cylinders.

Figure 16.3(c) shows the V-class engine. This is an extension of the in-line engine. There are two in-line cylinder blocks, set at an angle to each other on a common crankcase. These cylinder blocks form a V with each other, hence the name. There is a common crankshaft to both cylinder blocks. The engine illustrated is a V8, four cylinders to each block.

Figure 16.3(d) shows the horizontally opposed engine. Cylinders are fitted diametrically opposite to each other on a common crankcase and with a common crankshaft.

Figure 16.3(e) shows the opposed-piston engine. This is basically two separate engines. They are fitted together, head-on, so there is a common cylinder with a common combustion space. This common combustion space will occur between the opposed pistons. There are two crankcases and two crankshafts. The output of these engines is usually taken from a shaft geared to the two crankshafts.

Figure 16.3(f) shows the radial engine. Here there are several cylinders which fit radially on to a common crankcase. There is a single-throw crankshaft on to which fits one connecting-rod, called the master rod. The other connecting-rods fit on to the master rod. The radial engine was common in aircraft before the takeover of gas turbines.

16.3 The two-stroke cycle

As the name implies, all the events in the two-stroke cycle are completed in two strokes. In two strokes, the crankshaft makes one revolution, so the two-stroke cycle is complete in one revolution. The engine and its cycle are shown in Fig. 16.4.

Control of admission and exhaust in this engine is by ports let into the side of the cylinder and also by the piston. The piston in this type of engine is also the engine valve.

The crankcase is made gas-tight, since the incoming air–fuel mixture passes through the crankcase on its way into the cylinder. There is an inlet port in the bottom of the cylinder through which the air–fuel mixture passes into the crankcase. A transfer port is led from the crankcase into the cylinder through which the air–fuel mixture is transferred from crankcase to cylinder. There is a further port, the exhaust port, from cylinder to atmosphere, through which the combustion products are exhausted to atmosphere.

Referring to Fig. 16.4, the cycle of events is as follows. In Fig. 16.4(a) the piston is moving upwards and the piston is also sealing the transfer and exhaust ports. There is a fresh air–fuel charge in the cylinder above the piston and this is being compressed. The crankcase is hermetically sealed, so there will be a reduction in pressure produced in the crankcase as the piston rises. This reduction in pressure means that an air–fuel charge will be sucked into the crankcase as soon as the piston uncovers the inlet port in the bottom of the cylinder. Just before the top of the stroke, the compressed charge above the piston is ignited. The reason for timing the ignition just before the end of the compression stroke will be discussed in section 16.16.

In Fig. 16.4(b) the piston has been pushed down by the rapidly expanding products of combustion. This is the power stroke. Eventually, as the piston descends, the exhaust

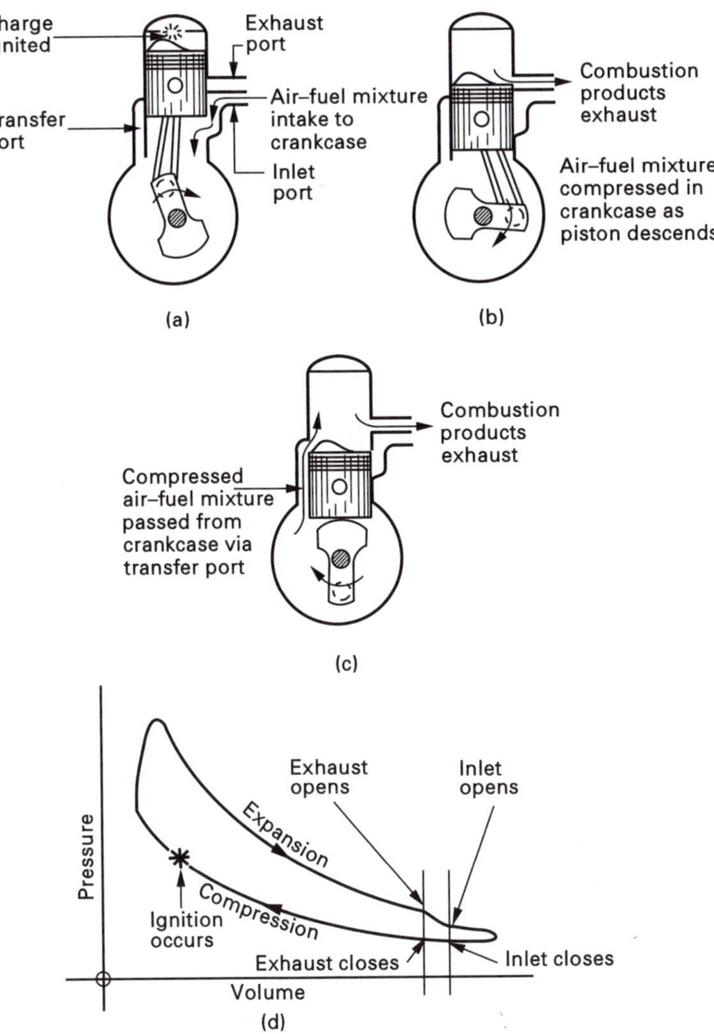

Fig. 16.4 Two-stroke engine: (a) compression stroke; (b) power stroke; (c) scavenging; (d) indicator diagram

port will be uncovered (as shown); the combustion products are still above atmospheric pressure, so they will rapidly expand through the exhaust port into the atmosphere. Sometimes the top of the piston in the two-stroke engine is shaped to assist in deflecting the gases to and from the ports (as shown); the piston is called a **deflector piston**. Also, during the power stroke, the descending piston will eventually cover the inlet port. This will lock up an air–fuel charge in the hermetically sealed crankcase. As the piston descends further, this air–fuel charge will be compressed in the crankcase.

In Fig. 16.4(c) the piston has descended to the bottom of its stroke. The transfer port has been opened and the compressed air–fuel charge in the crankcase is transferred to the cylinder via the transfer port. The deflector piston deflects this charge to the top of the cylinder; this displaces further combustion products out through the exhaust port. Displacing the combustion products out of the cylinder is called **scavenging**.

By deflecting the incoming charge to the top of the cylinder, it ensures that, as nearly as possible, there is not a direct blow through to exhaust with the consequent loss of fuel. There is always a slight loss in this way in the two-stroke engine.

Scavenging in some large diesel engines, e.g. in ships, is assisted by blowing compressed air through the cylinder. Rotary compressors are commonly used, run by exhaust-driven gas turbines.

From the bottom of its stroke, the piston will now ascend. In doing so, it will again close off the transfer and exhaust ports and once more a charge will be compressed in the cylinder; eventually the inlet port will open and a fresh air–fuel charge will be sucked into the crankcase. The cycle is now complete.

Figure 16.4(d) is a sketch of an indicator diagram taken from a two-stroke cycle engine cylinder. The various events, as they occur, are indicated. Note the reduction in pressure in the expansion line as the exhaust port opens. And note the similarity of shape between the indicator diagram and the theoretical constant volume cycle.

16.4 The four-stroke cycle

In an engine running on the four-stroke cycle there are mechanically operated valves which control admission and exhaust to and from the engine cylinder.

The inlet and exhaust ports run into the combustion space at the top of the cylinder and are circular in section. The valves, called poppet valves, are circular discs and are operated through a central spindle. The edges of the valves are chamfered to mate with similarly chamfered faces cut at the combustion space orifices of the inlet and exhaust ports. The valves are held central by, and operate through, valve guides. Valve springs hold them shut in their closed position, with the chamfered faces mating. The opening and closing of the valves is controlled by cams which are fixed to a camshaft or shafts. The camshaft is operated by means of a gear, belt or chain drive from the crankshaft.

If the valves open downward into the combustion space, as shown in Fig. 16.5, the engine is said to be an **overhead valve engine**. If the valves open upward into the sides of the combustion space, the engine is said to be a **side-valve engine**. Sometimes the camshaft is set up over the top of the valves in an overhead valve engine. In this case the engine is said to be an **overhead camshaft engine**.

Referring to Fig. 16.5, the operation of the four-stroke cycle is as follows. In Fig. 16.5(a) the inlet valve is fully open and the exhaust valve is closed. The piston is descending so it is sucking a fresh air–fuel charge into the cylinder through the open inlet valve. Toward the bottom of the suction stroke the inlet valve begins to close.

In Fig. 16.5(b) both the inlet and exhaust valves are closed. The piston is ascending and is compressing the fresh air–fuel charge into the combustion space. Ignition of the charge occurs toward the top of this stroke; once the stroke is completed, the piston again begins to descend on the power stroke.

In Fig. 16.5(c) both the inlet and exhaust valves are closed. The air–fuel charge has been ignited and the combustion products are rapidly expanding, pushing the piston down on its power stroke. Toward the bottom of the power stroke the exhaust cam has rotated such that it begins to lift the exhaust valve while the inlet valve remains closed.

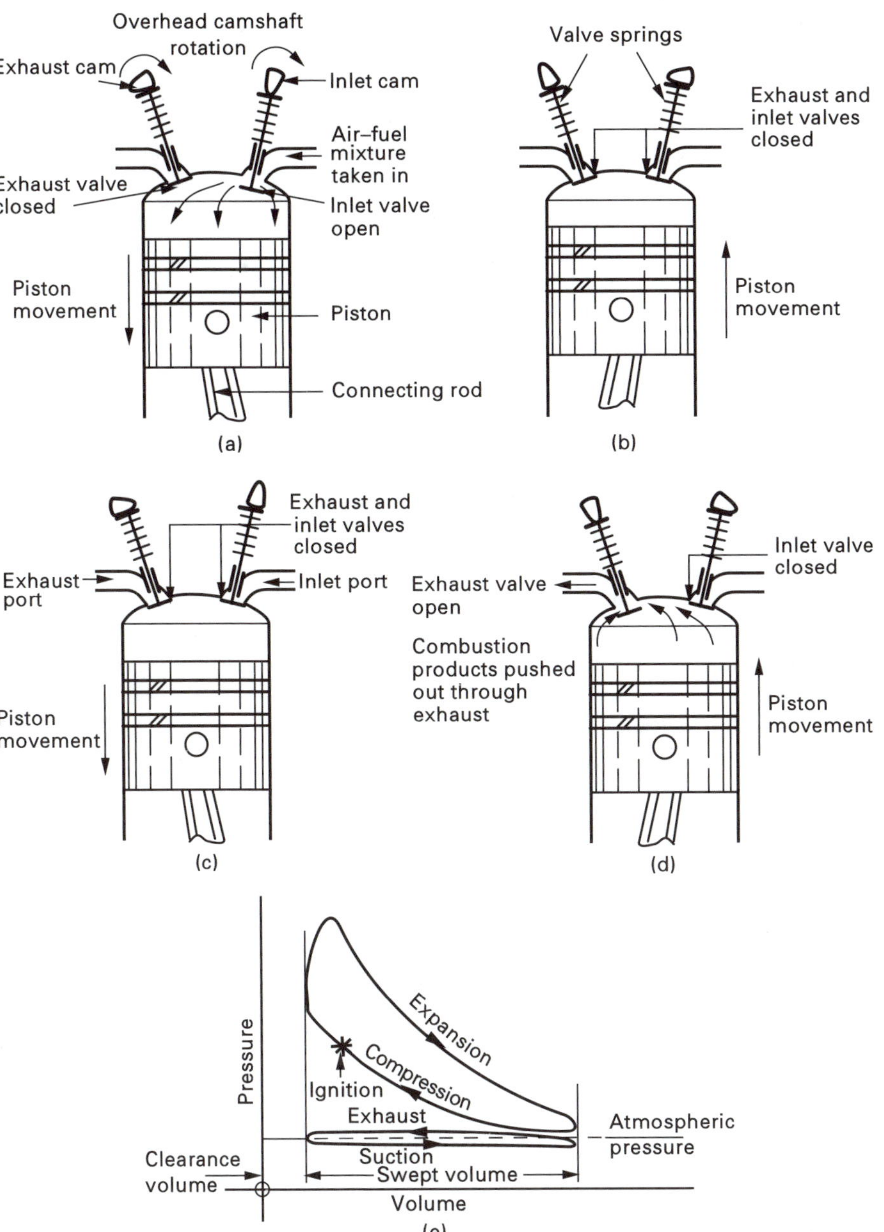

Fig. 16.5 Four-stroke engine: (a) suction stroke, piston descends; (b) compression stroke, piston ascends; (c) working stroke, piston descends; (d) exhaust stroke, piston ascends; (e) *P*–*V* diagram

In Fig. 16.5(d) the exhaust valve is fully open and the piston is ascending. As the piston ascends, it pushes the combustion products out through the open exhaust port. Toward the top of the exhaust stroke the exhaust valve begins to close and the inlet valve begins to open as a result of cam action.

In order to complete the cycle, the piston has made four strokes, two up and two down; hence its name, the **four-stroke cycle**. At the same time the crankshaft has made two revolutions. For this cycle, then, there is one power stroke in two revolutions. The valves open once only during the cycle, so the cams must rotate once only in two revolutions of the crankshaft. Thus the camshaft is geared down 1:2 to the crankshaft.

Figure 16.5(e) shows the *P–V* diagram taken from a four-stroke engine cylinder. Compression, ignition and expansion appear in a similar manner to the two-stroke cycle. But after expansion in this cycle, there is a definite exhaust stroke which will appear as a line on the *P–V* diagram. The atmospheric pressure line (drawn on this diagram) is slightly below the exhaust pressure line. This is always the case because the piston builds up a pressure inside the cylinder above atmospheric pressure in order that there will be a positive net pressure from inside the cylinder to outside. The combustion products will then move from inside the cylinder to outside and will thus be exhausted. During the suction stroke, which follows the exhaust stroke, the suction pressure is slightly below atmospheric pressure. This occurs as the piston descends and there will therefore be a net pressure from outside to inside the cylinder. Thus a fresh charge will move into the cylinder. The difference between atmospheric and suction pressures is called the **intake depression**. After the suction stroke, the compression stroke follows, and so on. Note the four distinct lines: compression, expansion, exhaust and suction which clearly show the four-stroke nature of this cycle.

This is the cycle which Otto worked on in the early days of internal combustion engines. For this reason it is sometimes called the **Otto cycle** (see section 15.2). The compression and expansion lines enclose an area very similar to that of the theoretical constant volume cycle. Furthermore, two distinct areas are enclosed by the lines of this diagram. The first, and by far the largest, is the area enclosed by the compression and expansion lines. Now the area under lines drawn on a *P–V* diagram give work done. The area under the expansion line is greater than the area under the compression line; the difference between them is the area enclosed by the two lines. Hence there is a positive work output. Note that the arrows show a clockwise movement round this area; clockwise movement on any pressure–volume diagram indicates a positive area and a positive work output.

The reverse is the case on the second enclosed area. This is bounded by the exhaust and suction lines. The exhaust area is larger than the suction area. During the exhaust stroke, the piston does the work in pushing out the combustion products. During the suction stroke, the atmosphere pushes in the fresh charge. The exhaust work is greater than the suction work, so the enclosed area is negative. Note that the progress round this area is anticlockwise; anticlockwise movement on any *P–V* diagram indicates a negative area requiring a work input. The exhaust–suction area is sometimes called the **pumping loop**. It has been opened up a little on the diagram to make it clear. Usually it appears rather narrower than illustrated.

16.5 Piston position

16.5.1 Vertical engine

When the piston of a vertical reciprocating engine is at the top of its stroke it is said to be at its **top dead centre** position; this is often written TDC. When the piston is at the bottom of its stroke it is said to be at its **bottom dead centre** position; this is often written BDC. The connecting-rod and the crank are in line at TDC and BDC.

16.5.2 Horizontal engine

When the piston of a horizontal, reciprocating engine is as far in the cylinder as it will go at the end of its stroke, it is said to be at its **inner dead centre** position; this is often written IDC. When the piston is as far out of the cylinder as it will go at the end of its stroke, it is said to be at its **outer dead centre** position; this is often written ODC. The connecting-rod and the crank are in line at IDC and ODC.

16.6 Valve timing diagrams for reciprocating internal combustion engines

Valve timing diagrams give the phasing of the valve operations with respect to the angular position of the crank.

16.6.1 Two-stroke cycle timing diagram

Figure 16.6 shows the timing diagram for a two-stroke cycle engine. It consists of a circle upon which are marked the angular positions of the various cycle events. The diagram is for a vertical engine; for a horizontal engine the diagram would appear on its side. With the two-stroke cycle the inlet and exhaust ports open and close at equal angles on either side of the BDC position. This is because the piston in this type of engine is also the inlet and exhaust valve, so port opening and closing will occur at equal angles on either side of the dead centre position. Angles shown are representative only.

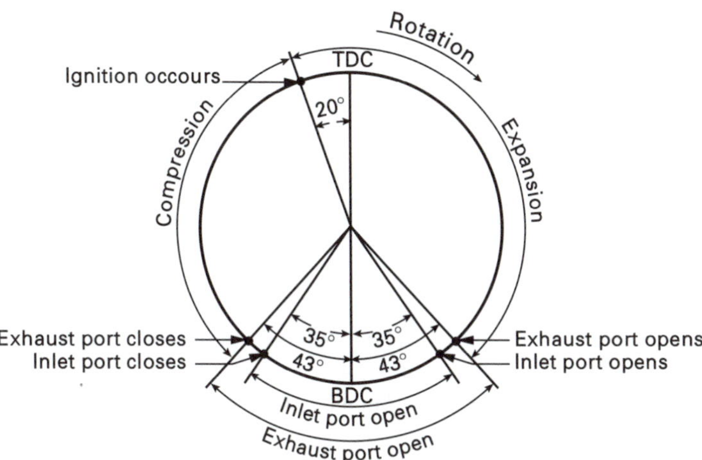

Fig. 16.6 Two-stroke cycle: timing diagram

16.6.2 Four-stroke cycle timing diagram

Figure 16.7 shows the timing diagram for a four-stroke cycle engine. It really consists of two circles, one superimposed on the other because the four-stroke cycle is completed in two revolutions. The inlet and exhaust valves do not begin to operate at the dead centre positions but at angles on either side of them.

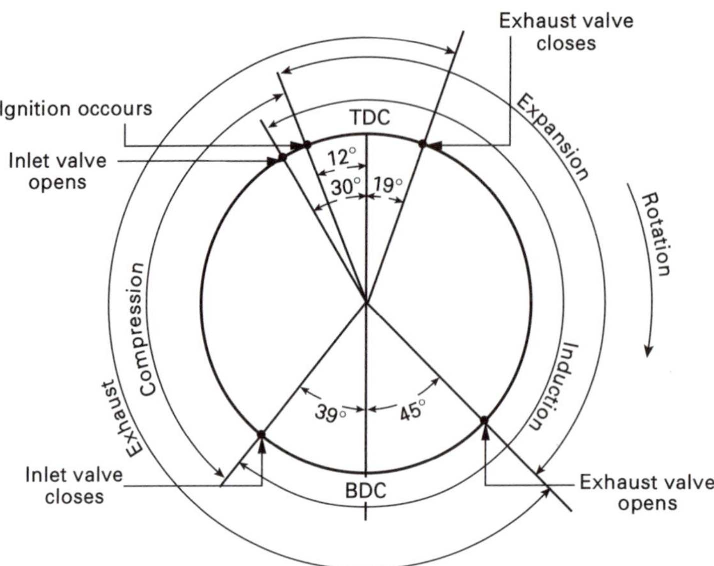

Fig. 16.7 Four-stroke cycle: timing diagram

The inlet valve opens before TDC and closes after BDC. This is arranged in an attempt to get as much air–fuel mixture into the cylinder as possible. When the inlet valve opens the air–fuel mixture outside the valve has to be accelerated up to inlet velocity; this takes time. In order that maximum inlet velocity will occur at the earliest possible moment in the induction stroke, the inlet valve is opened early. The moving air–fuel mixture possesses kinetic energy; this energy is used at the end of the induction stroke to produce a ramming effect by closing the inlet valve some degrees after BDC. The ramming effect pushes more air–fuel mixture into the cylinder, increasing the potential output of the engine.

In order to exhaust as much of the combustion products as possible, the exhaust valve opens early, some degrees before BDC. Thus some of the exhaust gas leaves by virtue of its excess pressure above atmospheric, so it is flowing freely from the cylinder by the time the piston commences the exhaust stroke. By closing the exhaust valve late, some degrees after TDC, the kinetic energy of the exhaust gas can be utilised to assist in maximum exhausting of the cylinder before the exhaust valve closes. The inlet valve begins to open before the exhaust valve closes. This is called **valve overlap**. Angles shown are representative only. For a horizontal engine the diagram would appear on its side.

16.7 *The petrol engine*

As the name implies, the petrol engine uses petrol (gasoline) for fuel. Mechanically, it can be made either as a two-stroke or a four-stroke cycle engine. The petrol must be correctly metered into the intake air and, when the air–fuel mixture is in the cylinder, it must be ignited at the correct instant. This means there must be a fuel metering device and an ignition system for the petrol engine. The fuel metering, in a petrol engine, is carried out using a **carburettor** or a **fuel injection system**. The **ignition system** can be a **coil** or **magneto**, and is generally called a **high-tension ignition system**. It can be controlled either mechanically or electronically.

16.8 The carburettor and fuel injection

A simple carburettor is illustrated in Fig. 16.8. It consists of a tube through which air is aspirated into the engine by the suction stroke of the piston, or pistons in the case of a multicylinder engine. Into this tube is introduced a venturi. The venturi is another tube whose cross-section reduces to a minimum, called the throat, then increases to its original size.

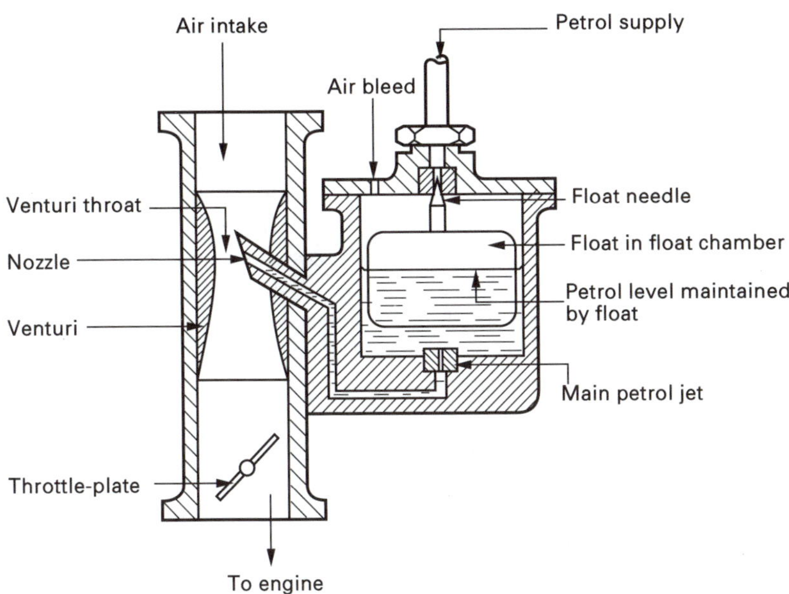

Fig. 16.8 Simple carburettor

Air passing through a venturi must satisfy the continuity of mass flow because the same mass is passing all sections per unit time. It follows that there must be an increase in velocity at the throat in order to pass the same mass as at the larger sections. An increase in velocity means an increase in kinetic energy. Now the total energy content of a fluid mass in motion remains constant (by conservation of energy). But there has been an increase in kinetic energy at the throat, so this must be accompanied by a corresponding reduction in pressure at the throat. It is this reduction in pressure which is used to suck petrol into the intake air flowing through the venturi (see Chapter 12).

Leading into the throat of the venturi is a nozzle. Drillings connect the nozzle to the float chamber, a petrol well fixed to the side of the tube containing the venturi. The level of petrol in the float chamber is controlled by a float. To the top of the float is fixed a needle which acts as a valve for closing off the petrol supply. This needle valve will control the flow of petrol into the float chamber; as petrol is used, the petrol level will fall and so will the float. This will open the needle valve and petrol will flow in from the supply. The float will then rise, once more restricting the petrol supply. When the engine is stopped, the float controls the petrol level such that the petrol supply is completely cut off when the level is just below the nozzle height in the venturi. This stops the leakage of petrol through the nozzle under stationary conditions. At the bottom of the float chamber is the main petrol jet. The size of this jet is such that it meters the correct amount of fuel to the nozzle in the venturi.

The action of the carburettor is as follows. With air flowing into the engine through the venturi, there is a reduction in pressure in the throat of the venturi. There is an air bleed hole in the cover plate of the float chamber. This allows atmospheric pressure to operate on the surface of the petrol in the float chamber. Now the petrol is in communication with the venturi throat, so there is a pressure difference between the petrol surface in the float chamber and the petrol surface at the venturi nozzle. This pressure difference is called the **intake depression**. The intake depression causes petrol to be pushed into the venturi throat, where it mixes with the air, the correct amount of petrol having been metered by the main petrol jet. The air–fuel mixture is then aspirated into the engine.

Engine control is facilitated by the throttle-plate situated in the intake tube. By rotating this throttle-plate the intake cross-section can be increased or decreased at will; this will control the amount of air–fuel mixture which is aspirated into the engine.

In the simple carburettor shown in Fig. 16.8, the air–fuel ratio would vary with engine speed if the carburettor were fitted to a variable speed engine, such as used in a motor car. Carburettors used on variable speed engines have many compensating devices to help maintain an air–fuel ratio as near to ideal as possible throughout the engine speed range.

Some petrol engines have a fuel injection system fitted in which the petrol is injected directly into the intake airstream using a pump and nozzle assembly. Fuel pump and nozzle arrangements are discussed later in connection with the oil or diesel engine. The fuel in an oil engine is injected directly into the engine cylinder, but the fuel in a petrol engine is injected into the intake airstream. Thus the fuel injection system on a petrol engine takes the place of a carburettor (see also section 16.11).

16.9 Early ignition systems

Various ignition systems were used in early engines to ignite the charge. Notable among these were the following.

16.9.1 Flame ignition

In flame ignition the cylinder charge was directly ignited by means of a flame applied to an opening operated by valve gear. The engines were very slow and were of very low compression. Flame ignition was mostly used on gas engines.

16.9.2 Hot-tube ignition

In hot-tube ignition a tube of refractory material, sealed at the end where it entered the cylinder, was heated to red heat by applying a flame down the inside of the tube. Ignition occurred when the charge in the cylinder came into contact with the tube. Hot-tube ignition was mostly used on gas engines.

16.9.3 Hot-bulb ignition

Hot-bulb ignition was similar to hot-tube ignition, but the tube was made of metal and, after starting, was maintained at high temperature by the burning cylinder charge. It was used over a limited speed and loading on some types of oil engine.

16.9.4 Breaker points

Breaker points were set inside the combustion chamber and connected to an external electrical circuit. When the breaker points opened, a low-tension arc flashed across them, igniting the charge. They were used on gas engines.

16.10 Modern ignition systems

The development of the high-speed, high-compression, internal combustion engine required a reliable high-speed ignition system. This demand was met by the development of the high-tension ignition system which uses the fixed-gap spark-plug as the ignition source. The electrical energy to the spark-plug is supplied by external systems which can be divided as follows:

- coil ignition
- magneto ignition
- electronic ignition

Before examining the external systems, consider the spark-plug itself. Figure 16.9 shows a section of one type of spark-plug. It consists of a porcelain insulator with an electrode running along its axis and an external contact at the top. The central electrode protrudes for a short length through the bottom of the insulator. Surrounding the bottom part of the insulator, and making a gas-tight seal, is a metal screw which usually has spanner flats so it can be tightened. On to the bottom of the metal screw is welded a metal tongue which bends over to lie across the end of the protruding central electrode but with a small gap between it and the electrode. It is across this gap that the high-tension electric spark jumps to ignite the charge in the engine cylinder; the gap is called the spark-gap.

The spark-plug is screwed into the cylinder head such that the spark-gap protrudes slightly into the combustion space. A sealing washer makes a gas-tight seal between the spark-plug and cylinder head. The plug is pulled down hard on to this washer.

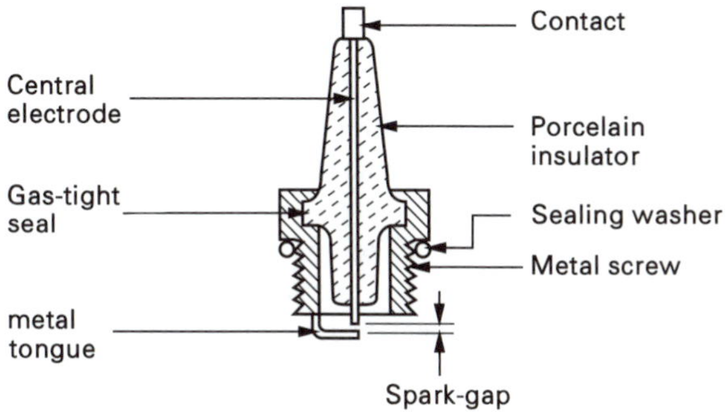

Fig. 16.9 Spark-plug

Electrical contact from the external circuit is made to the contact on top of the plug insulator. The other side of the electrical circuit is connected to the engine. The flow of a high-tension current will then produce a high-intensity spark across the gap, which is now in the combustion space in the engine cylinder head. The charge in the cylinder will thus be ignited.

A schematic diagram of a typical coil ignition system is shown in Fig. 16.10. The power for such a circuit is often supplied from a 6 V or 12 V battery. There are two circuits: the primary and the secondary. The primary circuit consists of a battery, one side of which is connected to earth. The earth is usually the engine itself. The other side of the battery is connected, via an ammeter and ignition switch in series, to the primary winding of a coil. This coil is, in effect, an autotransformer. The other end of the primary winding is connected to one side of the contact-breaker. The contact-breaker is, in effect, a spring-closed switch operated by a cam. The other side of the contact-breaker is connected to earth. A capacitor is connected across the points of the contact-breaker to earth. The points are the contact surfaces of the contact-breaker. This completes the primary circuit.

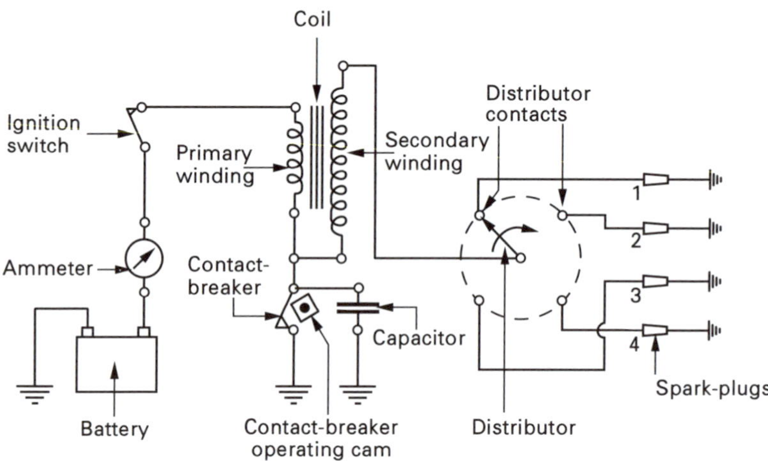

Fig. 16.10 Coil ignition system

The secondary circuit is as follows. The bottom end of the secondary winding is connected to the bottom end of the primary winding. The top end of the secondary winding is connected to the centre of the distributor rotor. The distributor rotor is a rotating contact which sweeps past, and makes contact with, fixed contacts which in turn are connected to the spark-plugs in the engine. The other side of each spark-gap is connected to the engine earth. This completes the secondary circuit.

With the ignition switch closed and the engine running, a current will flow in the primary circuit as soon as the contact breaker closes. The build-up of current in the primary winding of the coil is relatively slow because it must overcome the counter e.m.f. of its own magnetic field. The contact-breaker cam, which is continuously revolving, now opens the contact-breaker. Immediately, the magnetic field in the coil begins to collapse. This collapse of the magnetic field tends to keep the current flowing in the primary winding, so the capacitor is rapidly charged. As soon as it is fully charged, the capacitor will begin to discharge through the primary winding, increase the rate of collapse of the magnetic field. This rapidly collapsing magnetic field will induce a current flow in the secondary winding of the coil which, because of its high turn ratio, will produce a very high voltage in the secondary circuit. At the same time that the contact-breaker in the primary circuit opens, the distributor rotor connects with a contact to a spark-plug so a high voltage will be set up across the spark-gap. This will cause a high-intensity arc to spread across the gap, igniting the cylinder charge.

The capacitor across the contact-breaker not only helps to collapse the magnetic field, it also prevents excessive arcing across the contact-breaker. This is because the major proportion of the current flow in the primary circuit is discharged into the capacitor instead of producing an arc across the points as they open.

Figure 16.10 shows the circuit for a four-cylinder engine. It will be noted that the contact-breaker cam is square; its four lobes effect four openings of the contact-breaker per revolution of the cam. The distributor has four contacts. Usually the number of cam lobes and the number of distributor contacts is the same as the number of engine cylinders. Both the cam and the distributor rotor are motored in-phase and are usually mounted on the same shaft.

- In the two-stroke cycle engine they are motored at engine speed.
- In the four-stroke cycle engine they are motored at half engine speed.

A schematic diagram of a typical magneto ignition system is shown in Fig. 16.11. In this type of ignition system a battery is not required because the magneto acts as

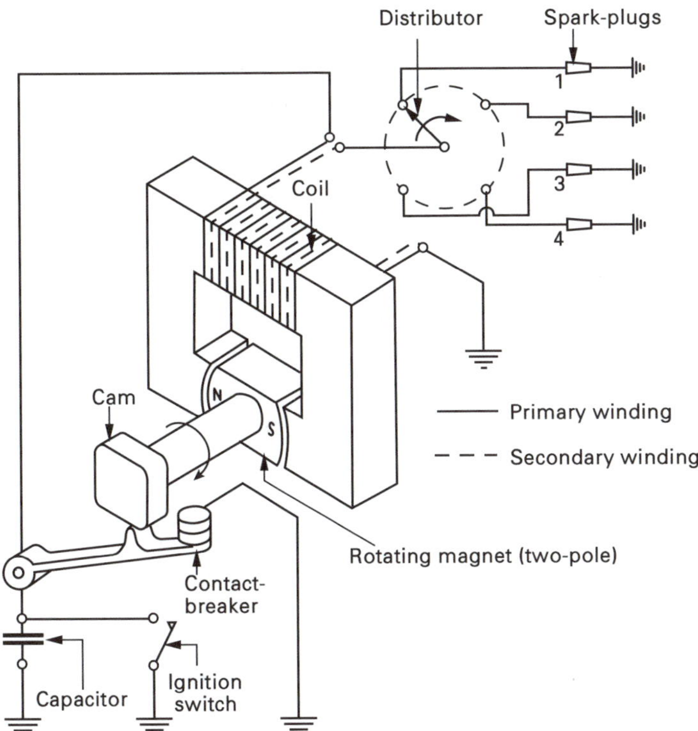

Fig. 16.11 Magneto ignition system

its own generator. It may consist either of rotating magnets in fixed coils or rotating coils in fixed magnets; the former is illustrated in Fig. 16.11. The rotating magnet is a two-pole magnet; as the poles pass between the shoes of the coil, a rapid change of magnetic flux in the coil takes place. This induces a current in the primary winding of the coil. When the current is at its maximum, the contact-breaker is opened by a cam on the magneto rotor shaft; the circuit then operates in the same way as the coil system.

There will be two current reversals per revolution of the rotating magnet shown, so two spark-plugs can be fed per revolution of the magneto. The direction of the current through the spark-plug is not important.

The principle and position of the ignition switch is changed in this circuit. The switch is placed across the contact-breaker; the engine 'on' position is actually open circuit. If the switch is closed, any current build-up in the primary circuit is shorted to earth, so the circuit is out of operation.

In the magneto, the speed of the rotor must be governed to cover the number of magnetic poles used and the number of engine cylinders. When feeding a four-cylinder, four-stroke cycle engine, the two-pole arrangement of Fig. 16.11 would cause the rotor to rotate at engine speed, giving four current reversals in two revolutions of the engine. If the same magneto fed an eight-cylinder, four-stroke cycle engine, the rotor would have to run at twice the engine speed.

The magneto must be turned over at a sufficient speed to supply the necessary current. The lowest speed at which the necessary current is supplied is called the coming-in speed. It is usually about 100 rev/min. If this speed is unattainable, usually at the start, a boost from a battery or a hand-cranked magneto is required.

Both the coil and magneto arrangements described so far have a mechanically operated contact-breaker. This mechanical system suffers from several disadvantages, including wear, burnt points due to electric arcing, operational speed limitation and the need for periodic readjustment or renewal.

The electronic ignition system works in almost the same way as described but, instead of the mechanically operated contact-breaker, a breakerless system is substituted. Such a system is triggered (turned on and off) using a magnetic sensor or a photoelectric device which takes the place of the mechanical contact breaker. Associated with the trigger is an electronic control unit. Usually transistor operated, it provides the electrical impulses required by the rest of the ignition system, which then functions in the manner already described. Once initially set, an electronic ignition system should have no requirement for further adjustment.

16.11 Petrol engine management

Modern petrol engines are now largely electronically controlled by an electronic control module. The electronic control is exercised over the fuel injection and the ignition systems.

Figure 16.12 illustrates a type of electronic control system. A similar design is also offered by Lucas Engine Management Systems. Fuel is pressurised and pumped into a control rail, from where it is distributed to the cylinder injectors. The injectors are controlled from an electronic module.

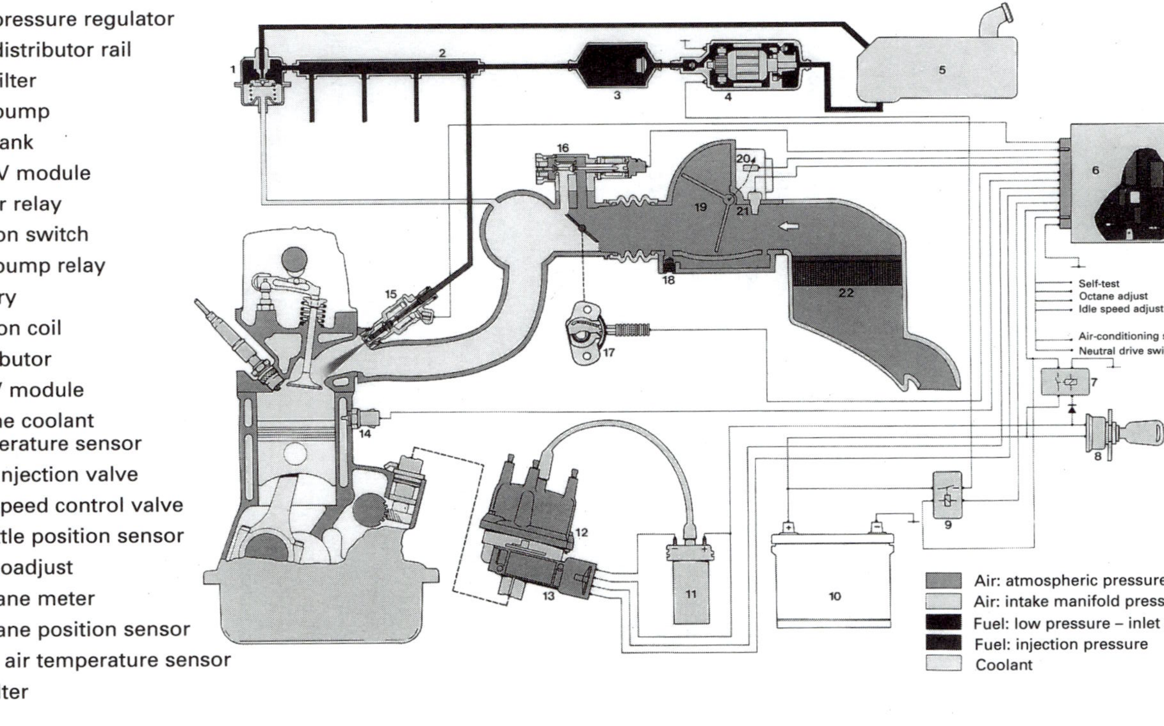

1 Fuel pressure regulator
2 Fuel distributor rail
3 Fuel filter
4 Fuel pump
5 Fuel tank
6 EEC IV module
7 Power relay
8 Ignition switch
9 Fuel pump relay
10 Battery
11 Ignition coil
12 Distributor
13 TFI IV module
14 Engine coolant temperature sensor
15 Fuel injection valve
16 Idle speed control valve
17 Throttle position sensor
18 Idle coadjust
19 Air vane meter
20 Air vane position sensor
21 Vane air temperature sensor
22 Air filter

Fig. 16.12 Petrol engine: electronic control system

A solenoid on the injectors controls the time (timing) and the amount of opening of the nozzle (metering). The fuel is sprayed into the intake air just behind the open inlet valve. In the arrangement shown, electrical power to the spark-plugs is supplied through a coil and breakerless distributor.

A further modern development now employs a distributorless ignition system. The correct timing is achieved using a crankshaft position sensor, an inductive pulse generator mounted adjacent to a specially prepared flywheel. Signals from the pulse generator are processed by the control module. Many engine adjustments in the control system are possible in order to set and improve the engine performance and emission.

16.12 The complete petrol engine

Figure 16.13 shows a cutaway view of a high-speed, four-cylinder, four-stroke cycle petrol engine made by the Ford Motor Company. The various elements already discussed are clearly shown. This is an overhead valve and camshaft engine, so the valves are mounted in the cylinder head and open downwards into the combustion space of the engine cylinder. The valves are operated directly by an overhead camshaft which, in turn, is rotated by a belt drive on the front of the engine. The belt has teeth which engage in gears initially driven from a gear on the end of the crankshaft. The belt also has a drive to the distributor. All components are coupled to give the correct speed and operational sequence for the engine cycle. Note the piston, connecting-rod and crankshaft assembly.

Observe the lubricating oil pump mounted at the bottom of the engine in the lubricating oil sump. The lubricating oil pump delivers lubricating oil, at pressure, to the various engine bearings and cylinders. An oil filter is mounted on the remote side of the engine in order to maintain the lubricating oil in a clean condition.

Note the carburettor mounted on the top of the engine. Fuel (petrol) to the carburettor is fed from the fuel pump mounted on the side of the engine. This fuel pump maintains a supply of fuel from the supply tank.

On the top of the carburettor is mounted an air filter assembly, which helps ensure that only dust- and grit-free air is aspirated into the carburettor. In the air intake of the air filter assembly is mounted an air temperature sensor valve. If the air temperature is low, especially when starting the engine, the sensor valve allows some warm air, ducted from the outside of the hot exhaust system, to mix with the cooler intake air. Thus the actual air aspirated into the carburettor is warmed.

The cooling water pump and cooling water outlet are also illustrated. A cooling air fan is mounted on the front of the engine. Cooling air fans are mostly thermostatically controlled, either mechanically, as illustrated, or electrically through a drive motor. Under cool conditions, the fan will free-wheel, thus absorbing little energy. When the engine becomes quite hot, the fan will cut-in to assist engine cooling.

Note the alternator mounted on the side of the engine, run by a V-belt driven from a pulley on the end of the engine crankshaft. The alternator has a rectification and control unit installed and will provide direct current (d.c.) power to the engine electrical system and for battery charging. Spark-plugs (not illustrated in Fig. 16.13) will be mounted in the cylinder head; they protrude into each combustion space in order to ignite the intake air–fuel mixture at the appropriate moment in the engine cycle.

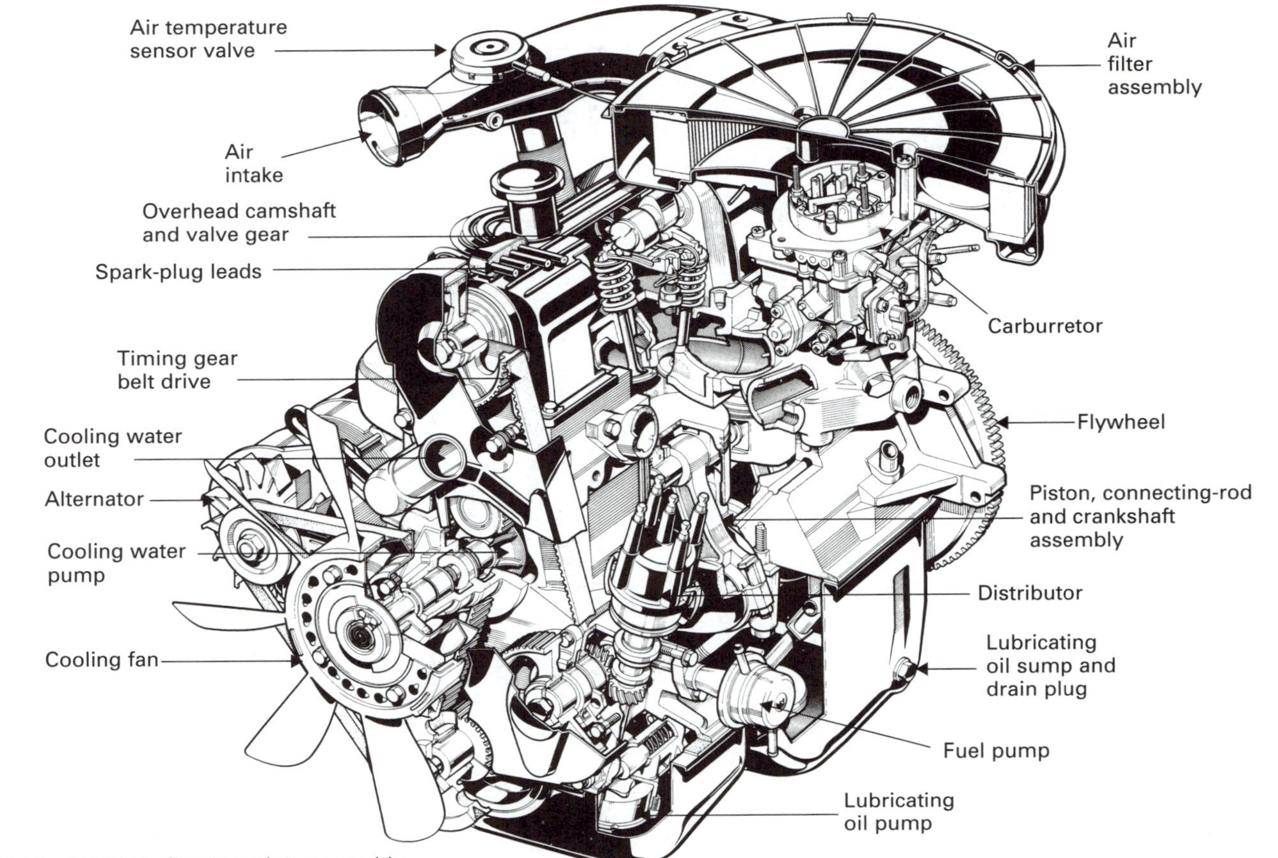

Fig. 16.13 Petrol engine: complete assembly

Figure 16.14 illustrates another petrol engine, made by the Ford Motor Company, which contains components similar to those already described. But this engine is fitted with a turbocharger. This, in effect, is a small gas turbine (see section 16.19). In this unit, exhaust gas from the engine passes through a small centrifugal gas turbine which, in turn, runs a small centrifugal air compressor at high rotational speed. Atmospheric air enters the compressor and is compressed through a pressure ratio of up to 1.5:1. If air is compressed quickly (almost adiabatically), its temperature is increased, so at its final pressure, its density (mass per unit volume) is reduced. To compensate for this, the compressed air is passed through an air cooler unit which reduces the compressed air temperature and thus increases the compressed air density. The cooled compressed air passes to the engine intake control unit.

The turbocharger thus increases the pressure and the air mass delivered at intake to the engine cylinders. With correct fuel control, it enables a greater power output, fuel economy and cleaner exhaust emission to be available.

The engine shown in Fig. 16.14 has an electronically controlled fuel injection system which will, at any time, be matched to the engine air intake, speed, and external prevailing conditions.

Note that both the engines illustrated in Figs. 16.13 and 16.14, indeed all reciprocating engines, have a flywheel attached to the external drive from the engine crankshaft. The flywheel evens out the fluctuations of torque that are generated; it also provides a surface upon which the plates of a drive clutch can operate e.g. in a motor vehicle.

Most four-stroke engines have two valves per cylinder, one inlet and one exhaust. But some modern engines incorporate four valves, two inlet and two exhaust. Thus a four-cylinder engine has 16 valves. This is illustrated in Fig. 16.15. The multivalve cylinder arrangement helps engine breathing and assists in improving engine output, combustion efficiency and emission control. The tappets are self-adjusting.

Figure 16.16 shows a typical cooling arrangement for internal combustion engines used in motor vehicles. The cylinders are each surrounded by a cast-in water jacket and the cooling water is passed into and out of the engine via a radiator. The radiator consists of a large nest of vertical thin-walled metal tubes which have metal fins attached across them to increase the cooling area. The bottom and top ends of the radiator tubes fit into headers which are coupled to the engine by rubber hoses. The large number of radiator tubes break up the water into a large number of streams which can be rapidly cooled by the passage of air over the tubes. Cooled water sinks to the bottom of the radiator while hot water from the engine rises into the top of the radiator. The circulation is therefore set up round the engine and radiator; it is often assisted by a pump impeller. Water circulation of this kind is called thermosyphon impeller-assisted cooling.

A fan is mounted on the engine to assist the free passage of air through the radiator. This is very important if the engine is stationary. Note also the pressure cap on the radiator top. Modern practice is to get the engine working as hot as possible. The pressure cap, which is also a safety relief valve, will allow a pressure slightly above atmospheric to build up in the radiator, so the boiling point will rise above 100 °C. This will enable the water to operate successfully as a cooling medium at higher temperatures than normal. The pressure cap also cuts down loss of water due to evaporation.

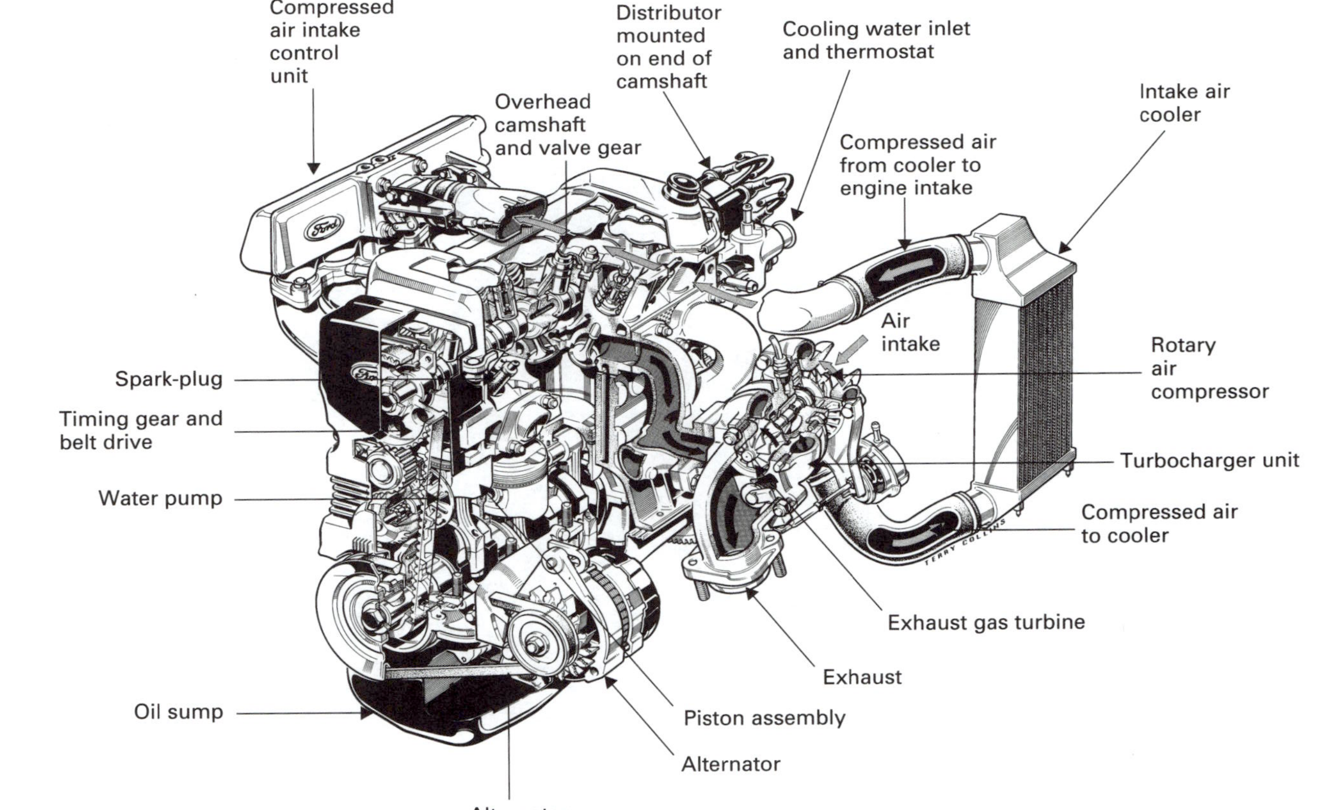
Compressed air intake control unit
Distributor mounted on end of camshaft
Cooling water inlet and thermostat
Overhead camshaft and valve gear
Intake air cooler
Compressed air from cooler to engine intake
Air intake
Rotary air compressor
Spark-plug
Timing gear and belt drive
Turbocharger unit
Water pump
Compressed air to cooler
Exhaust gas turbine
Exhaust
Oil sump
Piston assembly
Alternator
Alternator belt drive

Fig. 16.14 Petrol engine with turbocharger

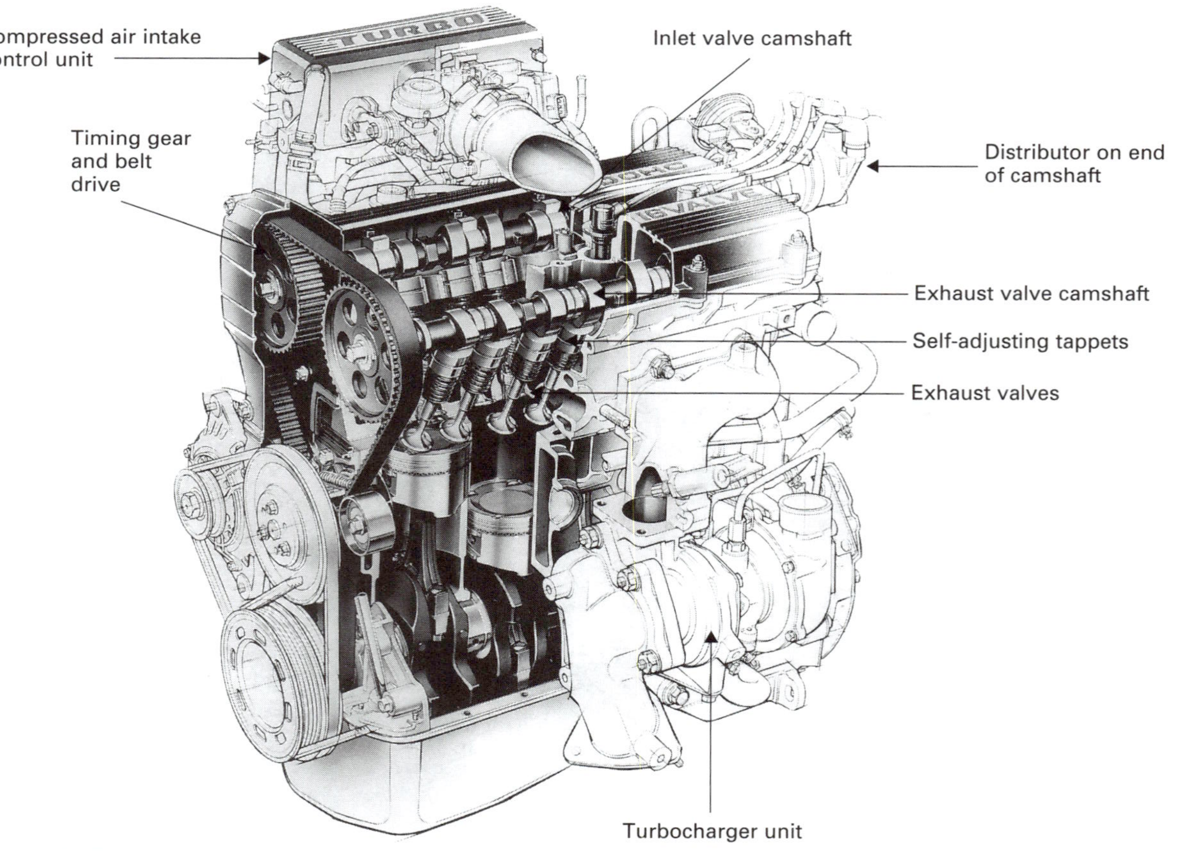

Fig. 16.15 Four cylinder engine with 16 valves

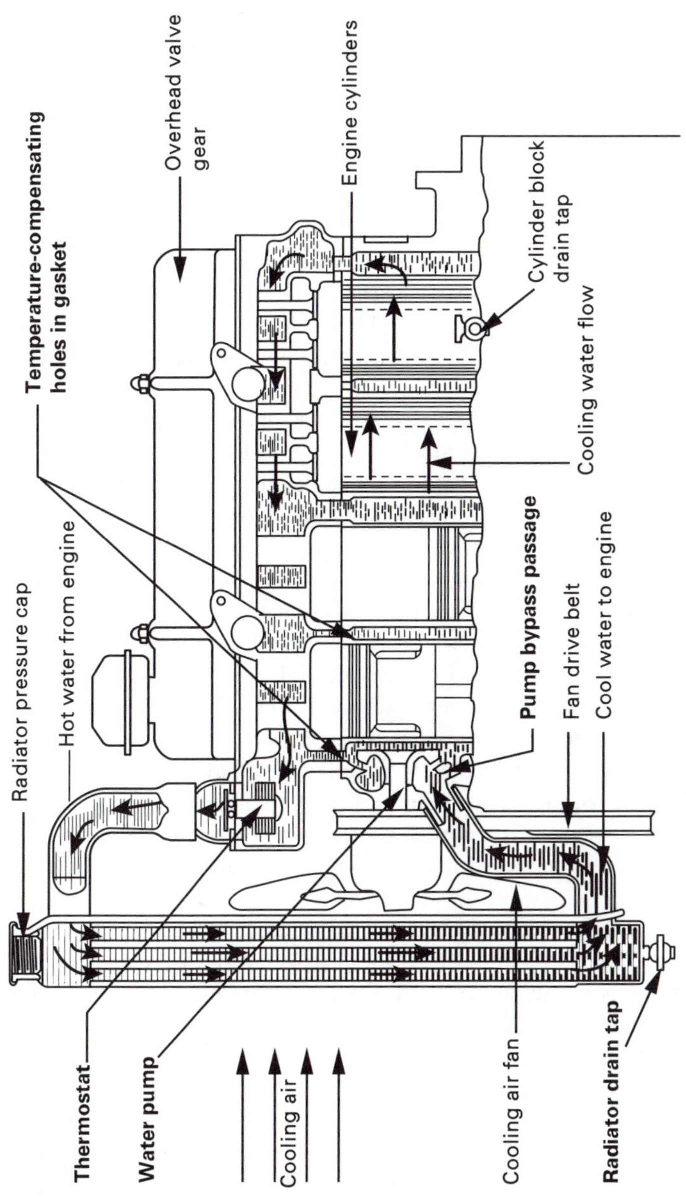

Fig. 16.16 IC engine: cooling system

A thermostat is set in the hot-water outlet from the engine cylinders. It will allow a fast engine warm-up from engine start. The thermostat will usually open at about 98 °C and will assist in maintaining a steady operating temperature for the engine.

Note that in some modern engines the pressure cap is fitted to a separate coolant reservoir. The reservoir takes care of coolant expansion and contraction. Also, in some vehicles the radiator has transverse flow. Flow connections are also made to the vehicle heating system. In cool climates an antifreeze solution (commonly ethylene glycol) is added to the cooling water to lower the freezing point of the coolant in subzero temperatures.

16.13 The oil engine

The oil engine is very often called the diesel engine, after Rudolf Diesel who pioneered the Diesel cycle. The basic mechanical elements of the oil engine are the same as those already described. The method of introduction of the fuel and its ignition are different from the petrol engine, but the oil engine is similar in that it can be either two-stroke or four-stroke.

Two important temperatures are relevant for the ignition of liquid fuels. The **flashpoint** is the lowest temperature at which enough fuel vapour is given off to mix with air and will ignite if a flame or spark is applied to the mixture. The air–fuel mixture must be of the correct proportions. The **ignition temperature** is the temperature at which the fuel of an air–fuel mixture of correct proportions will spontaneously ignite.

Petrol has a low flashpoint and a higher ignition temperature and is used in the petrol engine with its associated electrical ignition system. The oil fuel used in oil engines is variously called diesel oil, fuel oil or gas oil. It is sometimes called DERV, which stands for diesel engine road vehicle. This oil will have a relatively low ignition temperature and does not need a low flashpoint. The oil engine requires no ignition system comparable to that used in the petrol engine.

The operation of the oil engine is as follows. Firstly, apart from all normal features, the engine requires a fuel pump which will pump the fuel to a high pressure. Secondly, an injector nozzle is required, protruding into the engine combustion space. Through this nozzle, at the correct moment, fuel from the pump will be sprayed into the engine cylinder.

The engine aspirates air only during the suction stroke. It also has a high compression ratio, so during the compression stroke, the air is highly compressed and attains a high temperature, above the ignition temperature of the fuel. Toward the end of the compression stroke a measured quantity of high-pressure fuel is sprayed into the compressed air in the cylinder. The fuel spontaneously ignites then the cycle proceeds as before.

The oil pressure at admission must be very high in order to achieve successful atomisation and penetration of the compressed air in the cylinder. Nozzle opening pressures are of the order of 14.5–26 MN/m^2. The fuel line pressure from pump to nozzle will probably be more than double, some 35–70 MN/m^2.

Oil engine control is by means of the quantity of fuel injected. The fuel pump is therefore designed such that the fuel quantity can be varied to meet the load demand at any instant. The quantity of air aspirated is much the same at all engine loads, so the air–fuel mixture strength varies considerably from no-load to full-load conditions.

16.14 The fuel pump

Typical fuel pump operation is illustrated in Fig. 16.17. The pumping element is a plunger which reciprocates in a barrel. A hole is bored through the centre of the plunger for a short length. An angular groove is cut on the outside of the plunger and extends for only part of the way round the plunger. This groove and the central hole are connected by a hole drilled between them. Two ports are drilled in the barrel. The left-hand port is slightly higher than the right-hand port. The left-hand port is the main fuel supply and the right-hand port is the spillway.

In Fig. 16.17(a) the pump plunger has just descended and both ports are uncovered. Fuel has thus been sucked into the barrel and fills the space above the plunger, the central hole and the angular groove. In Fig. 16.17(b) the plunger is ascending. Both ports are now closed and the fuel oil above the plunger has now been compressed into the delivery to the injector.

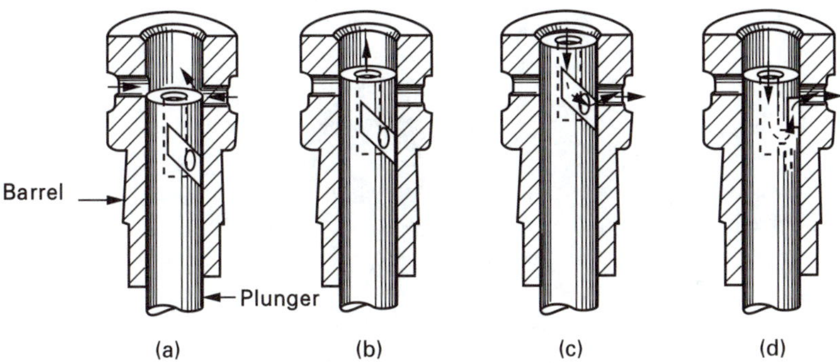

Fig. 16.17 Diesel fuel pump: (a) fuel sucked in; (b) fuel compressed into injector; (c) delivery ends and plunger completes stroke; (d) the engine is stopped when the plunger is rotated so the angular groove meets the spillway port just before it covers the fuel port

In Fig. 16.17(c) high-pressure fuel has been delivered to the injector and the plunger, still ascending, reaches a point where the top edge of the angular groove meets the spillway port. This port is then in direct communication, via the angular groove and the central hole, with the oil above the plunger. The pressure of this oil immediately drops, so delivery to the engine is terminated. The plunger continues to rise, completing its stroke, but it will not deliver any further fuel to the engine.

The distance from the top of the angular groove to the top of the plunger varies. This varying distance is used to control the engine. By rotating the plunger, any one of these distances can be selected to match up with the spillway port. If a short distance is selected, a small quantity of fuel is delivered. If a long distance is selected, a large quantity of fuel is delivered. The rotation of the plunger is controlled either by an arm or by a rack-and-pinion at the bottom of the plunger. A control rod actuates either the arm or the rack. The plunger reciprocation is constant in stroke and is usually cam operated. The plunger is usually force-closed on to the cam surface using a spring.

The engine is stopped by rotating the plunger as shown in Fig. 16.17(d). The plunger is now in such a position that the angular groove meets the spillway port just before it covers the fuel port on the left. No pressure build-up is possible for the remainder of the stroke because the oil above the plunger is at all times in contact with the spillway port, via the central hole and angular groove. Under these conditions no high-pressure oil is delivered to the engine, so the engine stops.

A complete pump unit to feed a four-cylinder oil engine is illustrated in Fig. 16.18. It shows that each cylinder has its own pump and all pumps are enclosed in one pump unit. The operating cams are located on one camshaft at the bottom of the pump unit. A supply fuel pump is mounted on the side to pump fuel from the main tank to the pump unit fuel gallery at the top. The control rod in this pump rotates levers on the bottom of the pump plungers. This controls the fuel quantity pumped to the engine cylinders. The engine is governed by a pneumatic governor fitted on to the front of the pump unit. The pneumatic governor is connected to the engine intake and is controlled by intake depression.

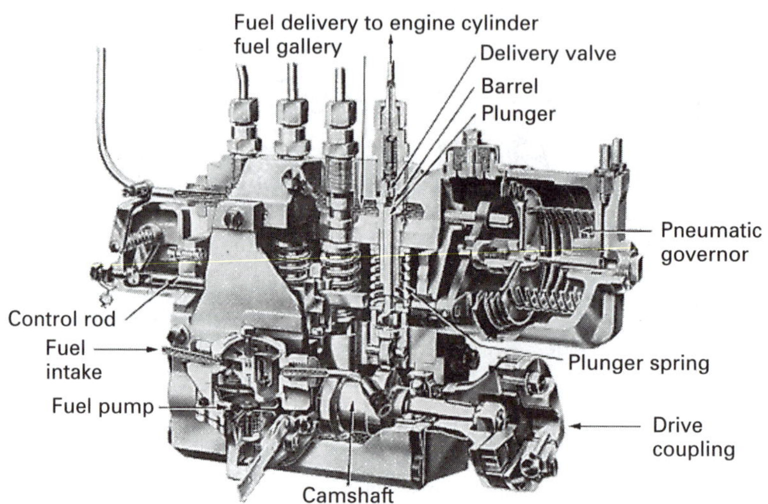

Fig. 16.18 Four-cylinder diesel engine: fuel pump

In the case of a four-stroke cycle engine, the camshaft of the pump is motored at half engine speed. In the two-stroke cycle engine, the camshaft of the pump is motored at engine speed.

The pump illustrated in Fig. 16.18 is not the only possible multicylinder pump arrangement. Another arrangement is to have a pump with a single plunger and a delivery distribution system whereby the engine cylinders are separately fed with fuel at the correct time. The arrangement is called a **distributor pump**. Distributor pumps are commonly fitted on the smaller, high-speed, motor vehicle diesel engines.

Figure 16.19 shows a fuel pump for a large ship's diesel engine, manufactured by MAN B&W Diesel. Apart from the pump's larger size, many of the features are the same as those already described. But this pump also has a timing rack-and-pinion by which the pump cylinder can be partially rotated, thus changing the engine cycle delivery time to the engine cylinder. This gives flexibility in engine control. At the top of the pump is a puncture valve; if the valve is opened, it stops the fuel flow to the engine cylinder and thus stops the engine. Each engine cylinder is fed by its own individual pump.

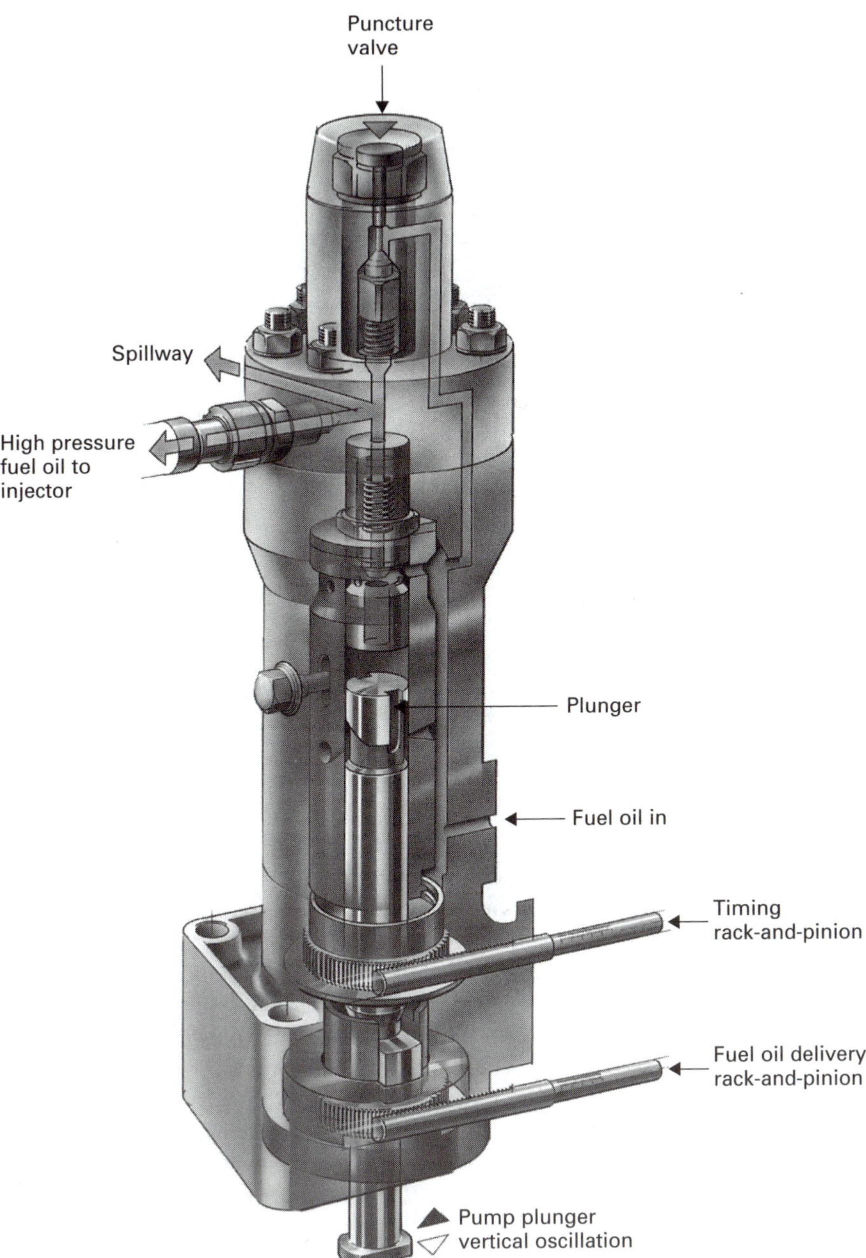

Fig. 16.19 Large ship's diesel engine: fuel pump

16.15 The injector nozzle

Each cylinder of an oil engine has an injector nozzle which protrudes into the combustion space of the cylinders. Like the spark-plug, the injector nozzle is made gas-tight with the cylinder. A typical injector nozzle is shown in Fig. 16.20.

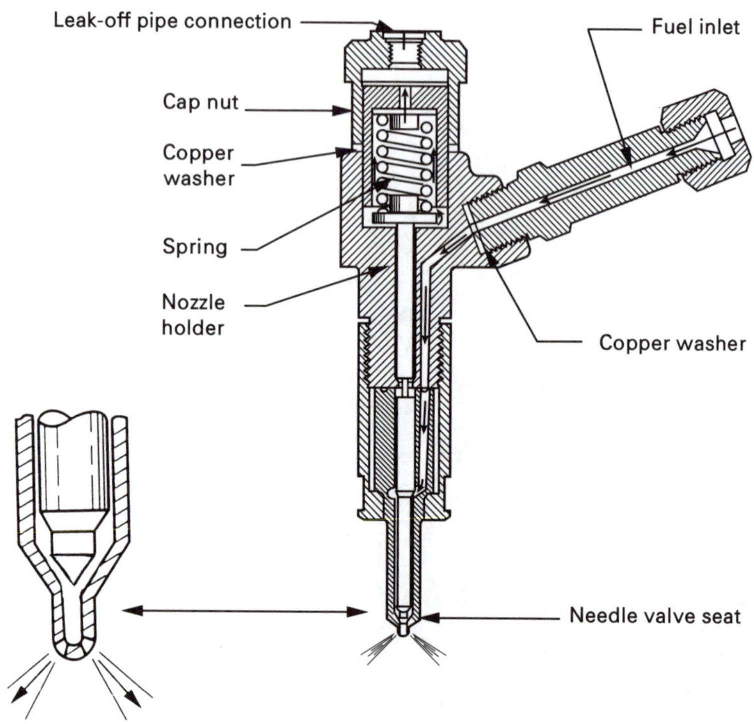

Fig. 16.20 Injector nozzle

High-pressure fuel passes into the injector at the fuel inlet. It passes down the injector body and reaches the needle-valve seat. The pressure operates against the area of the spindle above this seat, so the valve needle is lifted against the action of the spring at the top of the injector. High-pressure oil is then forced through very small holes into the engine cylinder. Because of the high pressure and the small hole size, the fuel is very finely atomised as it enters the engine cylinder. This greatly assists the rapid and successful burning of the fuel. The pressure is set by using the spring at the top of the injector nozzle.

A small amount of fuel leaks back between the body of the injector and the needle-valve stem. This provides the necessary lubrication. Eventually the fuel will find its way out of the top of the injector through a leak-off pipe; it will return to the main supply.

16.16 The complete oil (diesel) engine

Figure 16.21 shows a complete oil engine made by the Ford Motor Company. The essential features are much the same as those of the petrol engine already described. An oil engine is, however, generally of heavier construction due to the high compression ratio and hence the higher pressures associated with its operation.

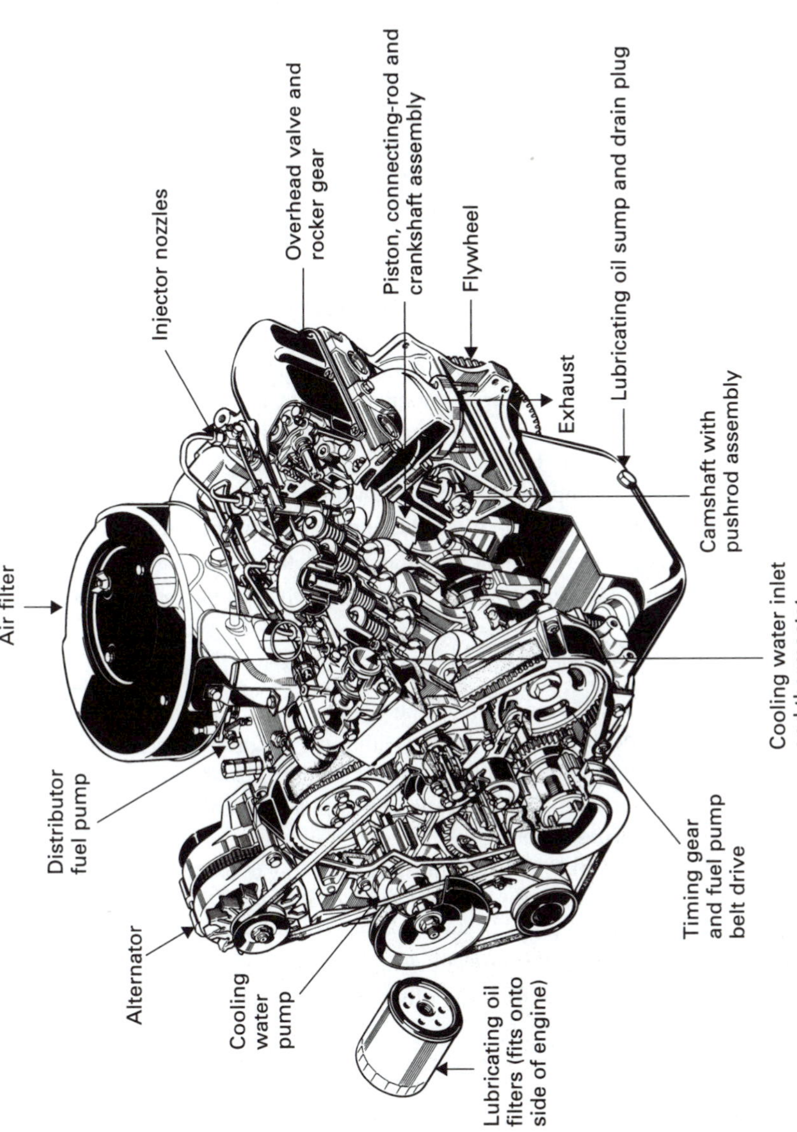
Air filter
Distributor fuel pump
Alternator
Cooling water pump
Lubricating oil filters (fits onto side of engine)
Injector nozzles
Overhead valve and rocker gear
Piston, connecting-rod and crankshaft assembly
Flywheel
Exhaust
Lubricating oil sump and drain plug
Camshaft with pushrod assembly
Cooling water inlet and thermostat
Timing gear and fuel pump belt drive

Fig. 16.21 Complete oil (diesel) engine

Commonly, oil engine cylinders are longer, so the cylinder block is taller than for a petrol engine because of the higher compression.

Note the distributor fuel pump which feeds the nozzles protruding into the combustion space of the cylinders. The overhead valves illustrated in Fig. 16.21 are operated through rocker gear and pushrods which engage with a camshaft lower down in the engine.

Large oil engines, such as those used on ships, have similar general operational features. But because of their size, the rotational speed is low, say of the order of 100 rev/min.

The cross-section of a large ship's diesel engine is shown in Fig. 16.22. This is a two-stroke cycle engine manufactured by MAN B&W Diesel A/S. The air intake is pumped into the engine through gills at the bottom of a long cylinder. The intake air helps to purge the products of combustion from the previous working stroke. The exhaust is from the top of a long cylinder and is controlled by a hydraulically operated exhaust valve. The bottom end of the piston rod connects to a crosshead, connecting-rod, crankshaft assembly.

The diesel engines on large ships can be very massive and usually have 4–12 cylinders. In their larger sizes they can have an approximate height of 14 m, length 20 m, and width 5 m, with a dry mass of up to 2000 tonnes. Outputs can be up to 50 MW with a specific fuel oil consumption of 760 litres/kWh, (see chapter 15). The fuel used is usually a low-grade, residual oil of approximate calorific value 43 000 kJ/kg. To improve the flow characteristics, the fuel oil is often heated before entering the engine system.

A small quantity of a lower viscosity diesel fuel is sometimes added to the residual oil to help improve easy flow, and subsequent atomisation, from the injectors into the engine cylinders.

16.17 Reciprocating engine firing order

Engine firing order means the order in which ignition occurs in the various engine cylinders. The firing order is chosen to give a uniform torque and hence a uniform distribution of firing per revolution of the engine. This will naturally depend upon the design of the engine, its cylinder arrangement and its crankshaft design. If, as sometimes occurs, two firing orders give the same result then the one with the least tendency to wind up the crankshaft and camshaft is usually chosen.

Typical firing orders for some engines are given in the table below. In numbering the engine cylinders for in-line engines, No. 1 cylinder is usually the cylinder in front. For multibank engines such as V8 and horizontally opposed engines, cylinders are usually numbered down one side then the other. For radial engines the cylinders are usually numbered consecutively.

Table of engine firing orders

Cylinder arrangement	Number of cylinders	Firing order	
In-line	4	1,2,4,3	1,3,4,2
In-line	6	1,3,5,6,4,2	1,4,2,6,3,5
		1,2,4,6,5,3	1,5,3,6,2,4
V8	8	1,5,4,8,6,3,7,2	
Horizontally opposed	4	1,3,2,4	1,4,2,3
Radial	7	1,3,5,7,2,4,6	
Radial	9	1,3,5,7,9,2,4,6,8	

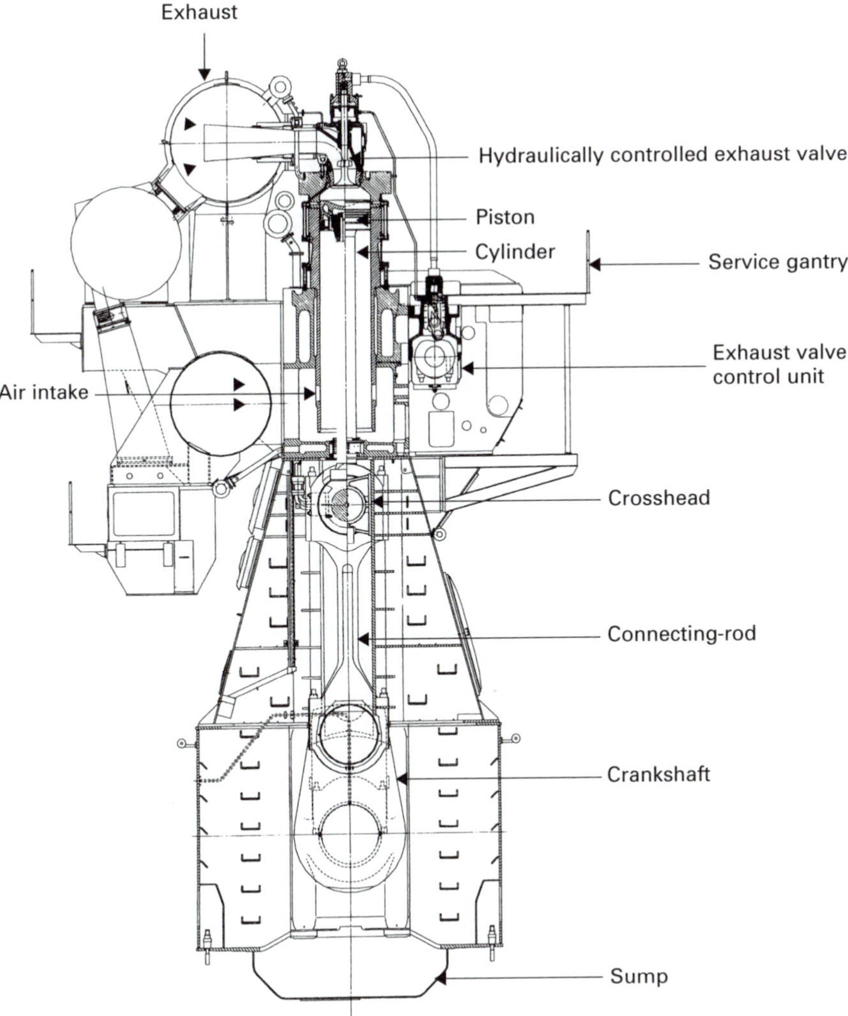

Fig. 16.22 Cross-section of large ship's diesel engine

16.18 Combustion in the reciprocating internal combustion engine

The combustion period in the reciprocating, internal combustion engine occurs a few degrees before and after the piston reaches the end of the compression stroke, or top dead centre (TDC).

The normal pressure–volume diagram does not illustrate this region clearly enough for combustion investigation. For this reason, the open-cycle or pressure–crank angle diagram has been developed.

Figure 16.23 shows the constant volume cycle illustrated on the pressure–crank angle diagram:

- ab represents the compression from 0° to 180°.

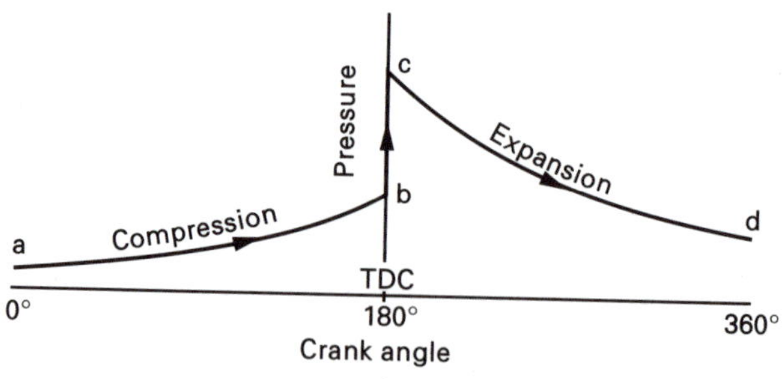

Fig. 16.23 Theoretical constant volume cycle: pressure–crank angle diagram

- bc represents the constant volume heat reception at TDC.
- cd represents the expansion from 180° to 360°.

In this diagram combustion is assumed to be instantaneous. But in actual engines, combustion passes through definite phases which occupy definite times. These phases may be illustrated rather more clearly on the pressure–crank angle diagram because any region under investigation may be opened out by extending the base scale.

In an actual diagram (Fig. 16.24) the compression and expansion lines will remain fundamentally the same. But the combustion line, bc, will show quite clearly that the combustion does not take place at constant volume; it occurs partly before TDC and partly after. This region is very important; the smoothness of combustion will determine whether or not the engine will run smoothly.

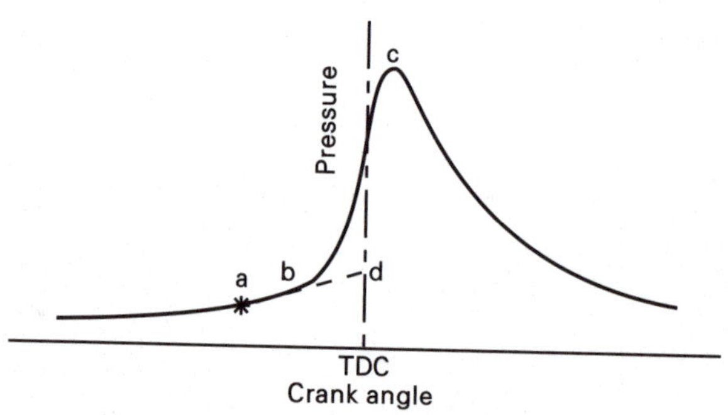

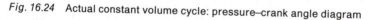

Fig. 16.24 Actual constant volume cycle: pressure–crank angle diagram

The open-cycle diagram is best displayed in practice on an electronic indicator. The display, as in Fig. 16.24, appears on the screen of a visual display unit, (VDU). Variation in cylinder pressure is detected using a transducer inserted in the engine cylinder head. The pressure variation is converted into variation of electric impulse by the transducer. These impulses are fed to the oscilloscope, in which they are amplified, and connected to the vertical shift, or y axis, of the display unit.

The horizontal shift, or x axis, is obtained either by a transducer set on the crankshaft of the engine or by a timebase circuit built into the oscilloscope. The display can be calibrated for pressure, time and crank angle.

The electronic indicator is very accurate, is immediate in response and does not have the disadvantage of inertia, friction and spring oscillation which occur in mechanical indicators. It also has the advantage that any part of the diagram can be amplified and brought into prominence on the display unit so that closer investigation is possible. Photographs can provide a permanent record.

When a fuel is ignited in a cylinder, it does not begin to burn at any appreciable rate until after a measurable period of time. Referring to Fig. 16.24, assume that during the compression stroke the fuel is ignited at a. The curve does not immediately begin to rise, signifying a high rate of pressure rise due to rapid combustion. Instead, it continues on the same compression curve as if ignition had not taken place. Assuming no ignition, the curve would continue along ad. But after period ab, the curve begins to rise sharply upwards, showing that combustion at a high rate has now begun. Combustion will eventually cease at c and the expansion will then begin. Period ab is called the **delay period of combustion** or, sometimes, the **period of incubation**. For a given air–fuel ratio, other things being equal, the delay time is very nearly constant for a given fuel.

Because of the delay period of ignition, it becomes necessary to ignite the air–fuel charge in the engine cylinder some degrees before the crankshaft reaches the TDC position. The number of degrees before TDC at which ignition occurs is called the **angle of advance**. If the angle of advance is increased, the engine ignition is said to have been overadvanced. If the angle of advance is decreased, the engine ignition is said to have been retarded.

The optimum angle of advance allows combustion to cease just after TDC, so that the maximum possible pressure is built up at a point just at the beginning of the expansion stroke. This is shown as the normal curve in Fig. 16.25; it is a smooth curve indicating smooth engine running.

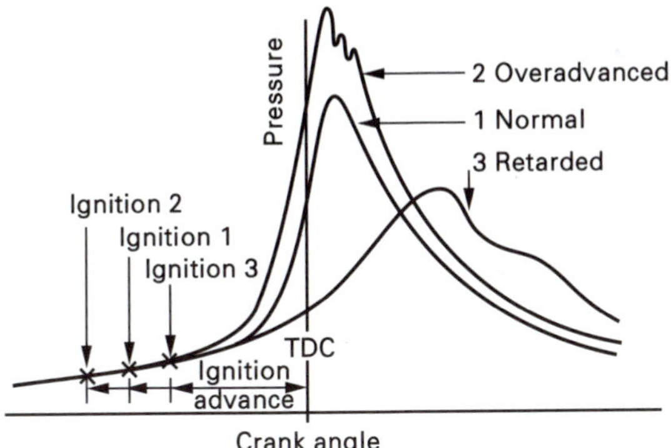

Fig. 16.25 Constant volume cycle: effect of ignition

If, however, the engine ignition is overadvanced, combustion is initiated too early and the cylinder pressure begins to rise rapidly while the piston is still trying to complete its compression stroke. This will create excessive cylinder pressures and may even produce shock waves in the cylinder, as illustrated by the ragged top on curve 2 of Fig. 16.25. If these shock waves are formed, an audible metallic sound will be heard, commonly called **pinking**. An overadvanced engine will run rough; it will tend to overheat and there will be a loss of power output.

If, on the other hand, the engine ignition is retarded (curve 3), combustion will be initiated late. In fact, combustion will still be occurring while the piston is sweeping out its power stroke. Maximum pressure will occur late and will not be as high as in the normal case. A retarded engine produces a low power output, and due to the late burning, the engine will run hot; in extreme cases, exhaust valve and port damage may result.

If the speed of an engine is increased but the ignition setting remains the same, the engine becomes retarded. This is because the delay time is very nearly constant. Thus, rapid combustion occurs later, in terms of crank angle, because the crankshaft will sweep out a greater number of degrees in a given time when it is speeded up. To compensate, the engine ignition should be advanced as the engine speed is increased. On many petrol engines there is an automatic device fitted which takes care of this matter.

Another ignition advance adjustment is required at part-load conditions. In this case, less air, and hence less fuel, enters the cylinders. At lower pressures, fuel burns more slowly, so the ignition is advanced to improve combustion efficiency at part-load. This is normally accomplished by a diaphragm control unit connected to sense the intake manifold depression.

Much emphasis is now given to the control of exhaust emission gases. In the petrol engine, for example, an average air–fuel ratio is of the order of 14:1 by mass. In the past, to improve engine power output, rich mixtures – meaning rich in fuel – were employed. Thus, the air–fuel ratio was reduced. This produced unwanted atmospheric emission pollutants in the form of unused hydrocarbons, carbon monoxide and oxides of nitrogen. Diesel engine fuel may contain some sulphur, consequently, some oxides of sulphur may be in the exhaust emission. Many devices are now used to clean up the exhaust of the internal combustion engine. These include exhaust gas recirculation, air injection to oxidise the raw exhaust, and a catalyst in the exhaust system to chemically treat the exhaust. There is also continuous monitoring of the exhaust tailpipe.

It is now useful to introduce the excess air factor, defined as

$$\text{Excess air factor} = \lambda \text{ (lambda)} = \frac{\text{Air supplied}}{\text{Theoretical air required}}$$

The theoretical amount of air required for the complete combustion of a fuel is called the **stoichiometric air** (see Chapter 8).

- If the air supplied = the stoichiometric air, $\lambda = 1$.
- If the air supplied < the stoichiometric air, $\lambda < 1$; this is a rich mixture, a fuel-rich mixture.
- If the air supplied > the stoichiometric air, $\lambda > 1$; this is a weak or lean mixture, a fuel-lean mixture.

Generally, in the case of the petrol engine, if $\lambda > 1.2$, the fuel will not ignite. This is sometimes called the **lean misfire limit** (LML).

Much combustion chamber design effort takes place to produce good combustion with lean mixtures – the **lean-burn engine**. This attempts to produce a good combustion mixture at the ignition point while preserving a lean overall air–fuel mixture. The improved combustion is accomplished by good combustion chamber design, high intake swirl and sculptured piston crowns and multivalve cylinders, to give some examples. Further control is achieved by the introduction of electronic subroutines to monitor engine performance.

16.19 Octane and cetane numbers

In the petrol engine, the ability of the fuel to perform well during combustion in the engine cylinder is indicated by the **octane number** of the fuel.

Under certain operational conditions, the air–fuel mixture can be prone to pinking. The higher the volume compression ratio of the engine, the more prone to pinking the engine becomes. However, the higher the volume compression ratio, the higher the theoretical thermal efficiency. Certain additives mixed with petrol enable it to operate at the higher compressions. In the main these additives have been small quantities of tetraethyl lead or tetramethyl lead (approximately 0.15 to 0.4 g/l). However, lead is known to have toxic properties, and it leaves an engine in vapour form to become an atmospheric contaminant and pollutant. Thus, petrols are now made for the lead to be reduced or removed, and some petrol is now marketed as being unleaded or green. Efforts are made to improve the combustion properties of unleaded fuel by the introduction of other additives such as benzene, but they can also be toxic.

The probable combustion performance of petrol is usually given by the **octane number**, obtained from an arbitrary scale of 0–100. The scale is derived from a mix of 2,2,4-trimethylpentane (*iso*-octane), which is given a value of 100, and *n*-heptane, which is given the value 0. 2,2,4-trimethylpentane has a low pinking characteristic, whereas heptane has a high pinking characteristic. The octane number of a fuel is the percentage of 2,2,4-trimethylpentane in a blend with *n*-heptane that appears to give the same engine performance as the given fuel. The octane value, or rating, of a fuel with a number greater than 100 is obtained by extrapolation from known data.

There are two accepted octane number scales, according to the test method used. They are the **research octane number** (RON) and the **motor octane number** (MON). Usually it is the RON that is quoted. However, the MON is probably from the more severe test since this test attempts to simulate vehicle operational conditions. RON and MON numbers of standard petrols are given in the following table.

Number of stars	RON	MON
2	90	80
3	94	82
4	97	86
5	100	86

Unleaded petrol is probably in the range 93–96 RON. The performance of diesel fuel oil is indicated by the **cetane number**. Under certain severe operational conditions, a diesel engine can produce diesel knock. This has a certain similarity to pinking in a petrol engine. A high cetane number suggest the ability of the diesel fuel to resist knock. The cetane number is the percentage of cetane in a blend of cetane and 1-methylnaphthalene having the same ignition and performance quality as a given fuel.

16.20 The gas turbine

The concept of the gas turbine began to be developed in the latter half of the nineteenth century. Ideas on gas turbines had accumulated earlier than this and, in fact, a type of gas turbine was patented in 1791 by John Barber of Nuneaton. The first gas turbine to run under its own power, and simultaneously to deliver external power, appears to have been in 1905 in Paris. But only in recent years have there been great strides in turbine design and use.

Gas turbines are very widely used in aircraft. The are increasingly used for the generation of electrical power and there are many installations in ships as propulsion units. Attempts are also being made to develop gas turbines as engines for automobiles.

There have been many designs for gas turbines, but the arrangement that has proved most successful in the continuous combustion, constant pressure, gas turbine. Its three basic elements are illustrated in Fig. 16.26(a): an air compressor, a turbine and a combustion chamber (sometimes there are a number of combustion chambers instead of a single large chamber).

The method of operation is as follows. Air enters the air compressor in which it is compressed through a pressure compression ratio of some 6:1 or 10:1. There are some installations in which the pressure compression ratio is as high as 20:1 or even 40:1. The air compressors are usually of the rotary type and are either radial or axial flow. An axial-flow compressor is illustrated.

The compressed air is passed from the air compressor into the combustion chamber through a duct. If there are several combustion chambers, the take-off volute from the air compressor will have ducts feeding the combustion chambers equispaced around it.

In the combustion chamber, fuel (either a fuel oil, such as gas oil or kerosine, or a gas) is passed into a burner and burnt continuously. Thus the air passing through the combustion chamber has its temperature and volume increased while its pressure remains constant. The combustion products are then passed from the combustion chamber into a turbine in which they are expanded. From the turbine, the combustion products are passed out to exhaust.

Figure 16.26(a) shows that the turbine is coupled back to the air compressor by a coupling shaft. On the other side of the turbine there is a coupling by which the turbine can be coupled to drive some external equipment. Part of the turbine output is used to drive the air compressor, so the net output appears for driving external equipment. The air compressor, shaft and turbine assembly is called a **spool**.

Due to the continuous combustion which occurs in the combustion chamber, steps are taken to ensure that temperatures do not become too high. This is usually dealt with by supplying considerable excess air above that required for complete combustion. A special shroud is usually built round the burner in order to meter the air to the combustion space. This ensures there is good burning of the fuel and that further air is mixed with the very hot combustion products further down the combustion chamber. This brings the final combustion product temperature down to something workable before entry to the turbine. Much research and development in metallurgy was necessary to develop metals which could withstand the high temperatures and stresses in gas turbines.

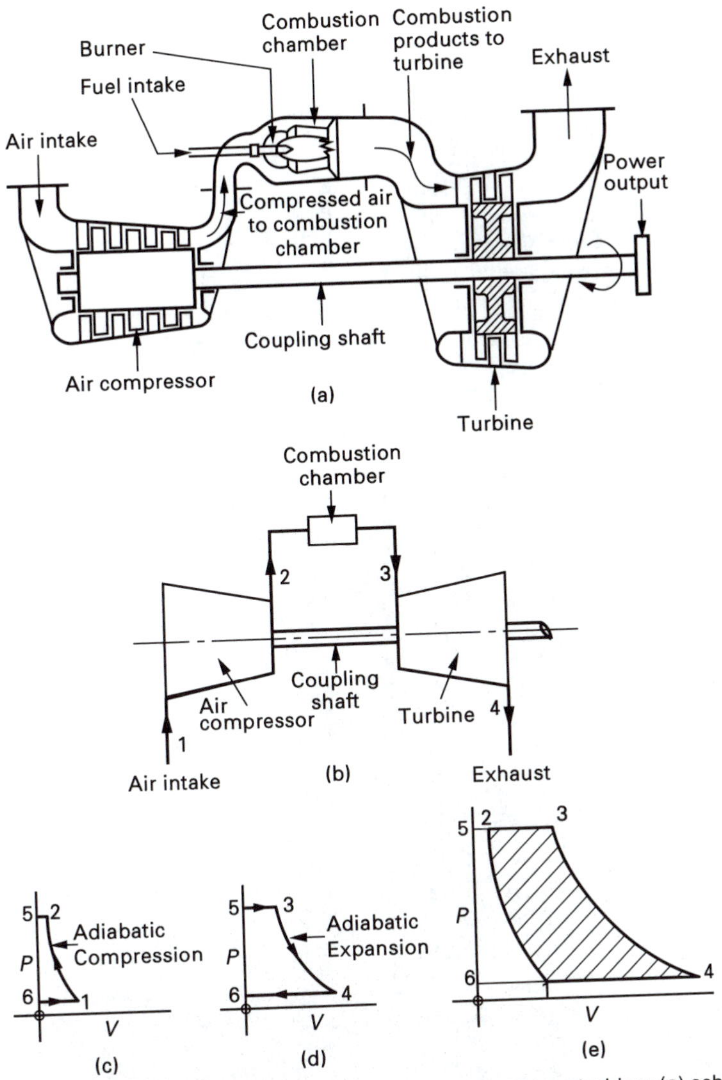

Fig. 16.26 Continuous combustion constant pressure gas turbine: (a) schematic; (b) flow diagram; (c) *P–V* diagram for compressor; (d) *P–V* diagram for expander; (e) combined *P–V* diagram

The turbine illustrated in Fig. 16.26(a) is arranged to develop shaft power, so the turbine would be designed to extract as much energy from the combustion products as possible before they are passed to exhaust.

As a practical example, a large gas turbine with a high power output is illustrated in Fig. 16.27. This is a model V84.3A manufactured by Siemens AG for their advanced combined-cycle (GUD) power plants. (GUD is a registered trade mark for Siemens combined-cycle power plants.) The combustion chamber arrangement consists of a hybrid burner-ring combustor (HBR) which contains 24 burners. The combustor is capable of processing gaseous and oil fuels. Some parameters of the turbine are given in the following table.

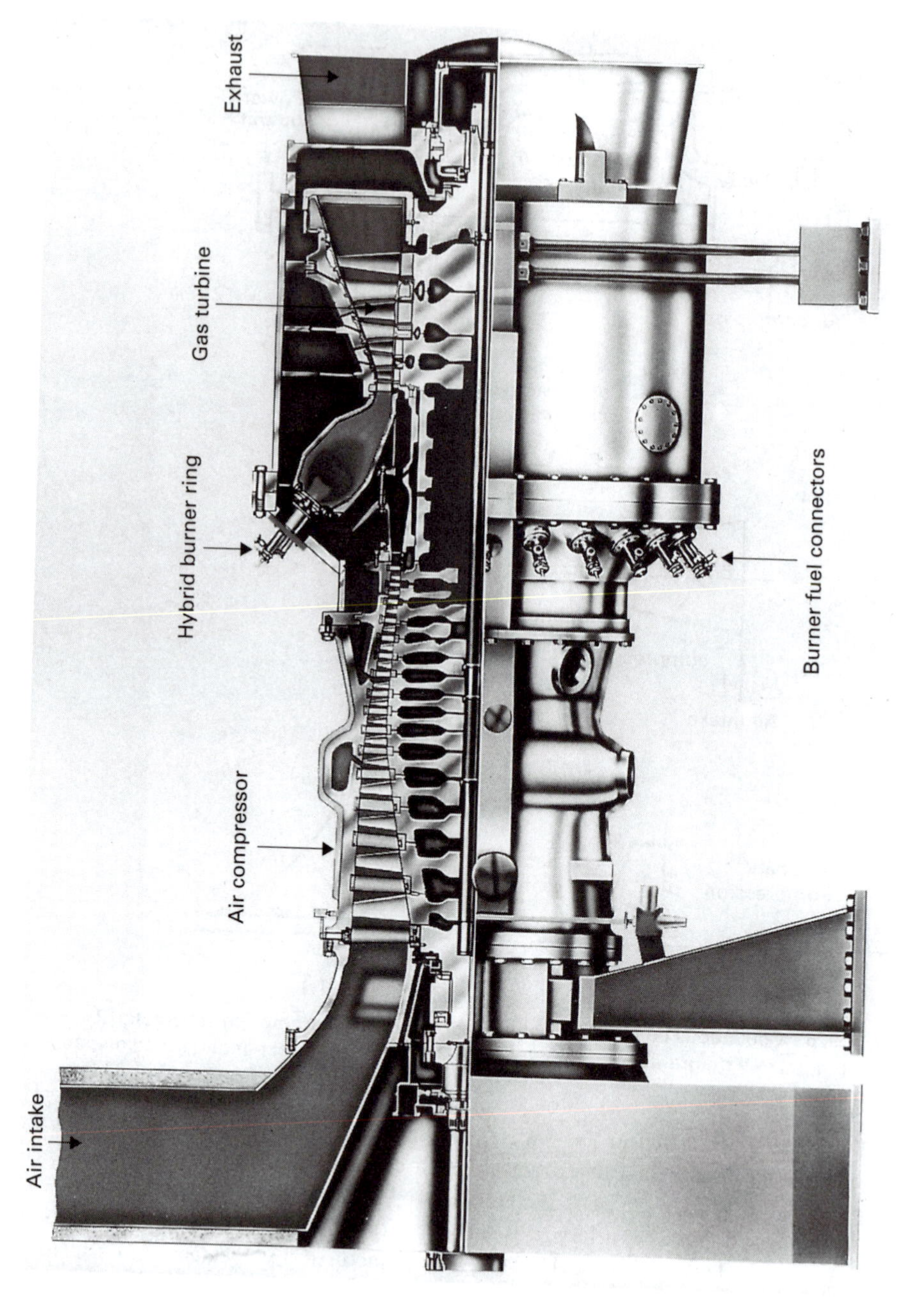

Fig. 16.27 Large gas turbine with high power output

Output power (MW)	170
Thermal efficiency (%)	38
Heating rate (kJ/kWh)	9 474
Fuel oil consumption (tonne/h)	35
Fuel gas consumption (m^3/h)	42 400
Exhaust gas rate (kg/s)	454
Exhaust gas temperature (°C)	562

The hybrid burner-ring is shown in Fig. 16.28. The word *hybrid* refers to the fact that the burners can accommodate both oil and gas. Figure 16.29 shows a schematic diagram of one type of combined-cycle power plant (part of the GUD range by Siemens). It shows the arrangement of a gas turbine in tandem with a three-stage steam turbine. The electric generator lies between the two turbine arrangements. Gas or oil can be used as fuel and the exhaust gas from the gas turbine feeds a three-stage boiler system. Steam from the boiler system feeds the three-stage steam turbine. One system, using the gas turbine described, has an overall thermal efficiency of 57.9 per cent with an electrical output of 254 MW. Figure 16.30 illustrates a block elevation diagram of the practical arrangement of the power plant.

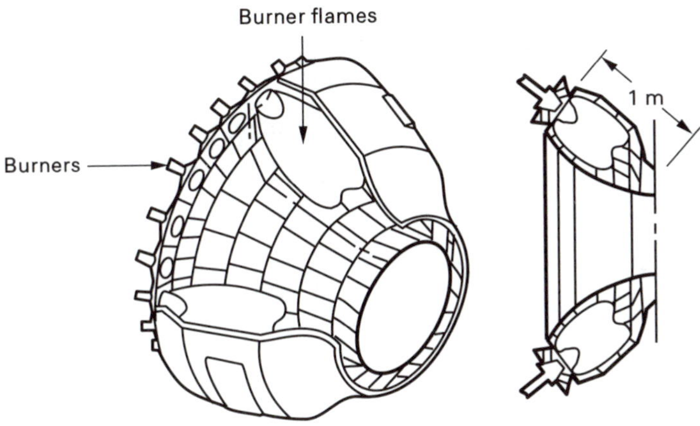

Fig. 16.28 Hybrid burner ring

Figure 16.31 shows basic arrangements for some modern power plants fired by fossil fuel. Many arrangements are possible and plants can be tailor-made for specific needs. These plants allow considerable control to be exercised over noxious emissions.

The gas turbine has a very wide use as an aircraft propulsion unit. Gas turbines are variously designated as turbojet, turbofan, turboprop and turboshaft. The turbojet is the simplest form, consisting of a compressor, combustion chamber(s) and turbine. It is used to produce a high-velocity jet for aircraft propulsion.

The turbofan is the most commonly used arrangement in aircraft. Part of the air intake is bypassed around the outside of the combustion chamber arrangement. The remaining air passes through the combustion chamber system. The bypass flow either rejoins the hot flow downstream of the combustion chamber system or is exhausted

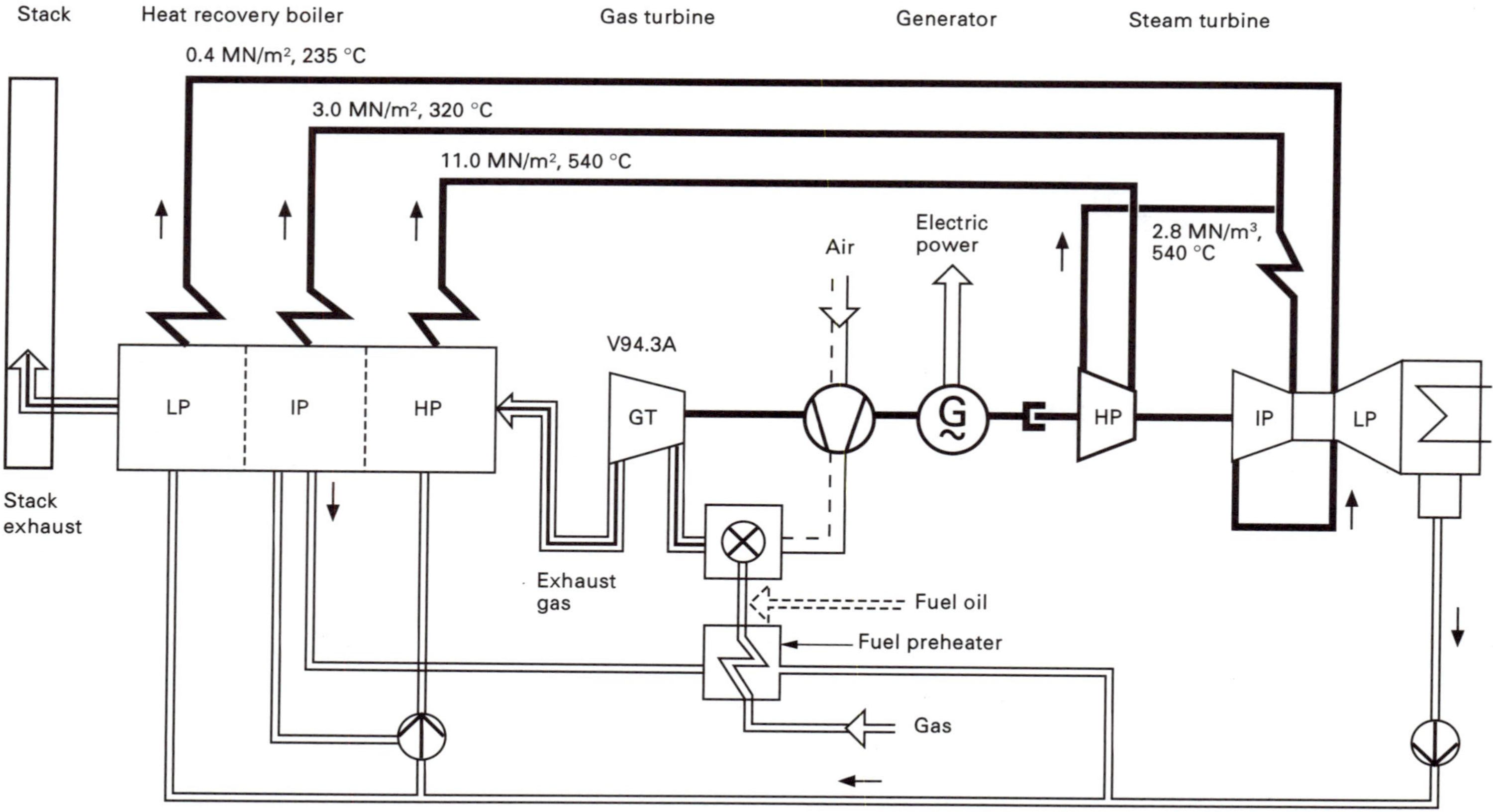

Fig. 16.29 Combined-cycle power plant: schematic

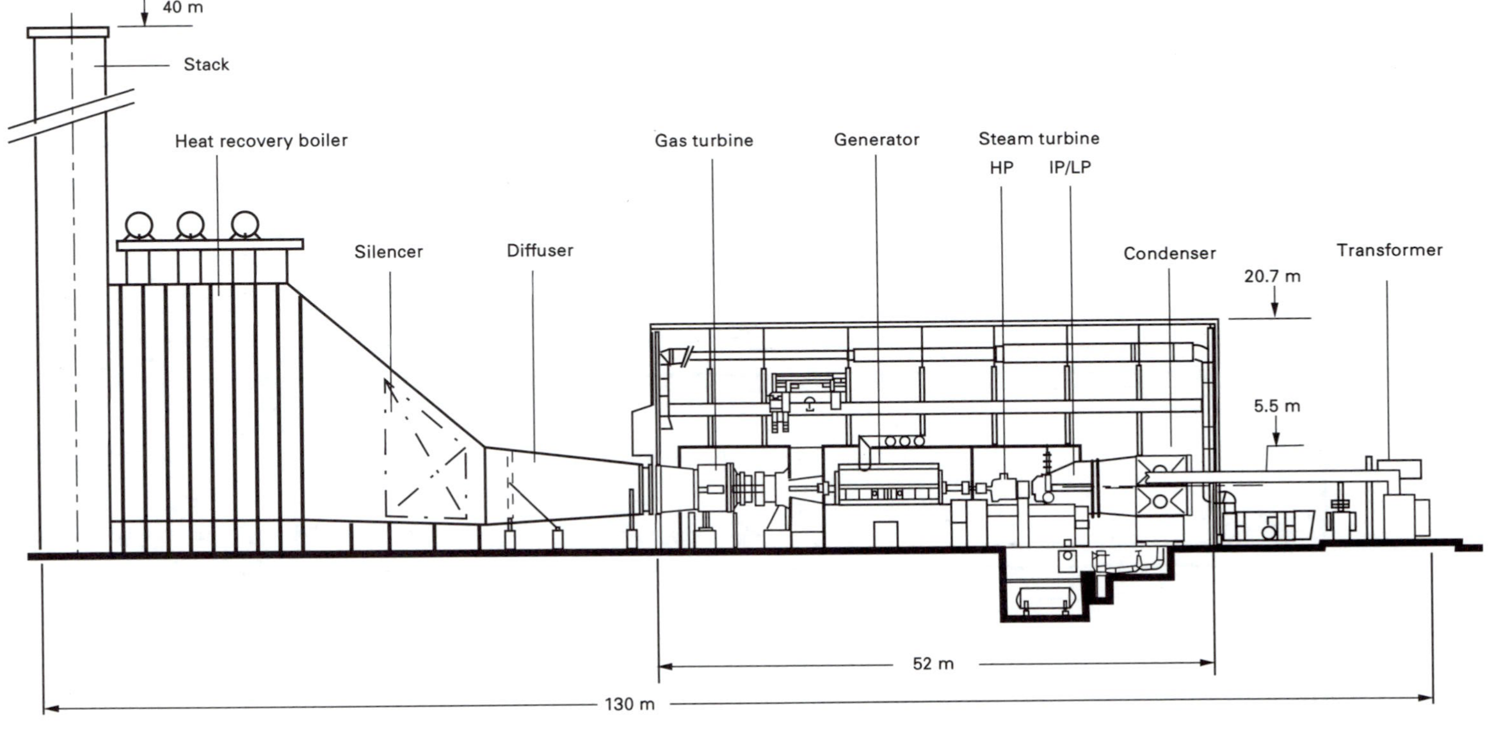

Fig. 16.30 Combined-cycle power plant: block arrangement

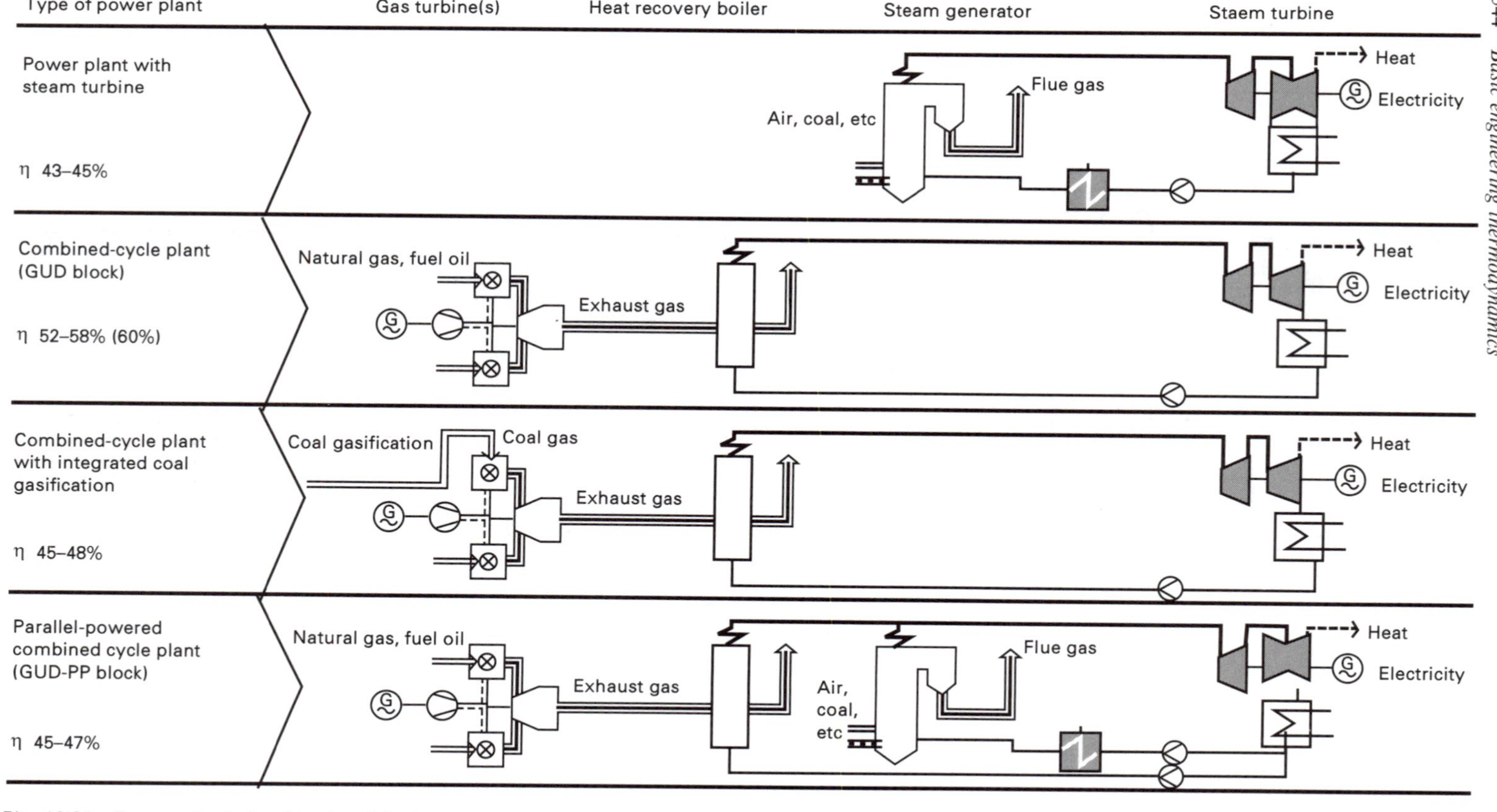

Fig. 16.31 Power plants fired by fossil fuel

through an annulus surrounding the hot exhaust. The arrangement with turbofan gives a lower jet velocity, improved lower-speed propulsive efficiency and specific fuel consumption plus a lower noise level. Sometimes the fan section has variable pitch, which helps to reduce fan noise and improves the engine control and flexibility.

Figure 16.32 is a cutaway illustration of a modern turbofan engine used for aircraft. The engine illustrated is the Trent 700 manufactured by Rolls-Royce. Air passes into the engine through the turbofan and then divides. Some air passes into the air compressor of the engine but most of it becomes the bypass air. It travels round the outside of the engine, through cowling, ultimately to join the jet exhaust at the rear of the engine. The compressed air from the air compressor in the engine passes into the combustion chamber ring then through the gas turbine to become the jet exhaust and to combine with the bypass air. Some general statistics of the Trent 700 are given in the table.

Table of Trent 700 parameters

Rotational speed (rev/min)	3300
Engine mass (tonne)	6
Overall diameter (m)	3.66
Length (m)	7.9
Take-off thrust (kg)	34 000–40 800
Fuel flow rate (kg/h)	14 290
Cruise fuel consumption (kg/h per kg thrust)	0.545
Air mass flow rate (kg/s)	900–1200
Overall pressure ratio	39:1
Air bypass ratio	5.3:1

The turboprop and turboshaft engines are similar in arrangement. They usually have an additional turbine which produces power for external use. In the case of the turboprop unit used on aircraft, the engine will provide power to drive a propeller, and there will be some residual thrust obtained from the exhaust. The combined effect of propeller and jet thrusts is sometimes called the **effective power** or the **total equivalent power**. The turboshaft engine is used to power external equipment by tapping the shaft power. It is used in ships, power and pumping stations, hovercraft and helicopters.

Some jet engines in aircraft are used with **vectored thrust**. This means that the jet nozzle assembly can be changed in direction. Such arrangements are used on vertical and short take-off and landing (V/STOL) aircraft. Other jet engines on aircraft will employ **reverse thrust**. This is a device in the jet exhaust for altering the jet thrust to a more reversed direction, thus assisting in slowing down an aircraft after landing.

Gas turbines are not self-starting machines. In the reciprocating internal combustion engine it is necessary only to turn the engine over one compression; the engine will fire and will pick up speed on its own. The gas turbine will not start simply by turning the burner on. It must first be motored up to some minimum speed, called the coming-in speed, before the fuel is turned on. When this speed has been reached, the fuel is turned on, ignited, and the turbine will then pick up speed on its own.

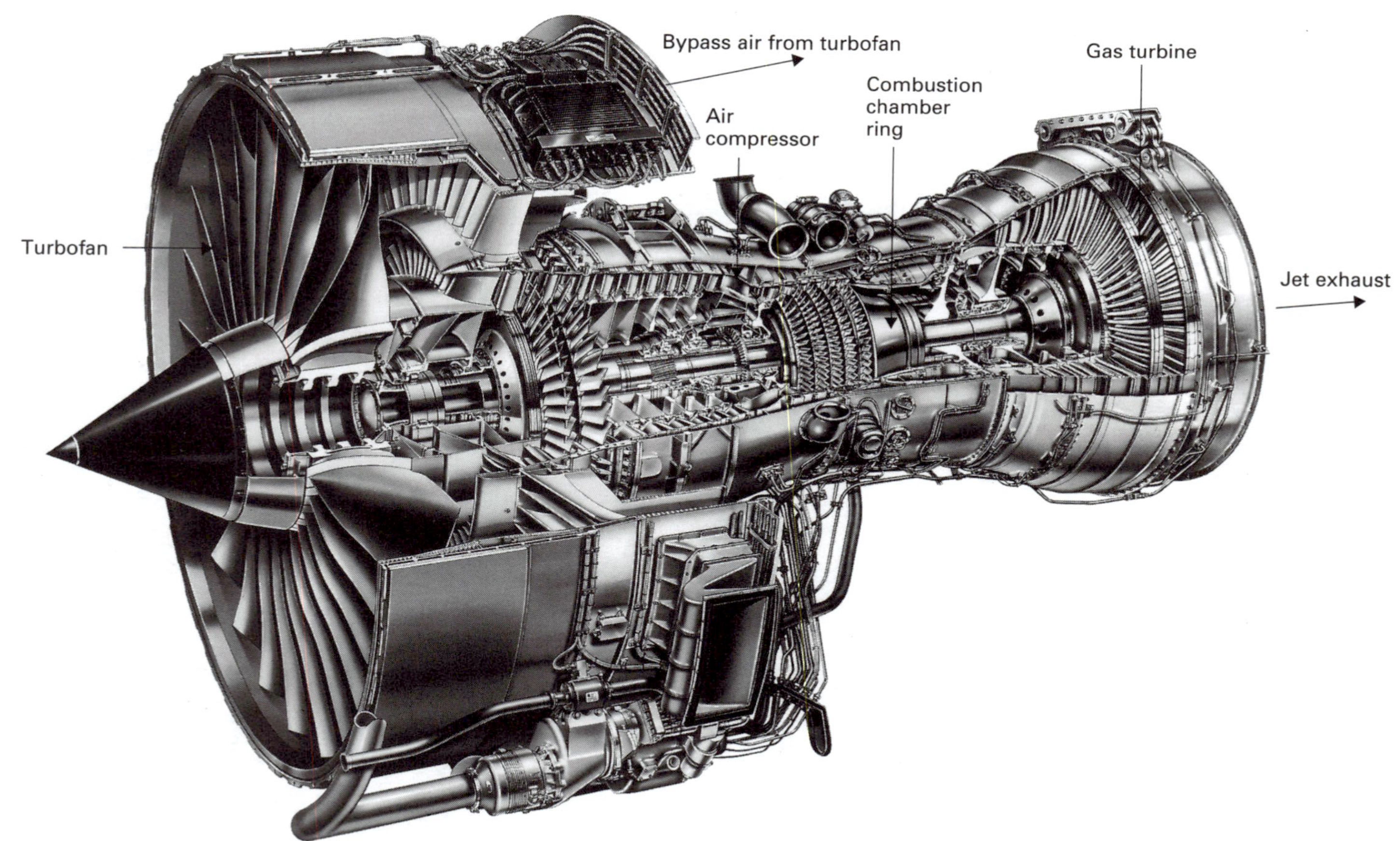

Fig. 16.32 Modern turbofan engine

The turbine rotor is usually motored up to coming-in speed by a starter motor. Usually an electric motor, it may be a small turbine. One design is driven for a very short period by firing a special cartridge. The gas from the cartridge drives the small starter turbine which is temporarily engaged with the main turbine rotor by gearing. It is disengaged when the main turbine starts.

If there is more than one combustion chamber fitted to the turbine, they are usually cross-connected in some way. This is a safety measure in case one of the burners should become extinguished, called a flameout. This is dangerous because fuel will continuously be pumped into the extinguished combustion chamber and, if not immediately ignited, will accumulate, eventually to cause an explosion when the combustion mixture reaches the exhaust from the other combustion chambers. An explosion could wreck the engine. This is prevented by cross-coupling. There have also been other devices developed to perform the same function in the event of flameout.

Gas turbine speeds vary considerably, from as low as 3000 rev/min to as high as 35 000 rev/min. Reduction gearboxes are fitted to the high-speed turbines to reduce this speed for coupling to external equipment.

Some aircraft gas turbines are fitted with an afterburner in the jet exhaust nozzle. Burning fuel in the afterburner produces additional thrust on aircraft take-off or for military engagement. The fuel used for aircraft gas turbines is usually paraffin (kerosine). In all gas turbine arrangements there is some internal power requirement necessary to run such auxiliaries as the fuel pump, the lubricating oil pump and the electrical generation.

16.21 Gas turbine calculations

Figure 16.26(b) is the schematic diagram of the turbine shown in Fig. 16.26(a). This type of diagram is very commonly used when discussing gas turbines. The compressor is coupled to the turbine, so

$$\text{Net work output} = \text{Turbine output} - \text{Compressor work} \qquad [1]$$

When discussing the rotary air compressor in Chapter 14, it was shown that

$$\text{Compressor work} = \dot{m}_a c_{pa} (T_2 - T_1) \qquad [2]$$

where $\dot{m}_a$ = mass of airflow per second
c_{pa} = specific heat capacity of air at constant pressure
T_2 = final compression absolute temperature
T_1 = intake absolute temperature

It will be remembered that the final compression temperature is above the normal adiabatic compression temperature due to turbulence, friction, etc., which occurs in the compressor.

The frictionless adiabatic temperature is calculated using gas laws and is obtained from

$$T'_2 = T_1 \left(\frac{P_2}{P_1}\right)^{(\gamma_a - 1)/\gamma_a} \qquad [3]$$

where γ_a = adiabatic index for air

The connection between the frictionless adiabatic compression temperature, T'_2, and the final compression temperature, T_2, is made through the adiabatic or isentropic efficiency equation

$$\text{Isentropic } \eta_{\text{comp}} = \frac{T'_2 - T_1}{T_2 - T_1} \qquad [4]$$

T_2 may be calculated from the isentropic efficiency of the compressor. Figure 16.26(c) shows the air compressor P–V diagram. In the air compressor the airflow is continuous, so by considering the flow of a particular mass of air, the air compressor diagram will be as follows:

- **6–1** Air taken into compressor.
- **1–2** Air compressed in compressor.
- **2–5** Compressed air delivered from compressor.

Now to consider the turbine. By a similar analogy to that used for the air compressor, the turbine output is obtained from

$$\text{Turbine output} = \dot{m}_t c_{pt}(T_3 - T_4) \qquad [5]$$

where $\dot{m}_t$ = mass of combustion products through turbine per second
c_{pt} = specific heat capacity of combustion products at constant pressure
T_3 = inlet absolute temperature to turbine
T_4 = exhaust absolute temperature from turbine

The final exhaust temperature from the turbine will be above the frictionless adiabatic exhaust temperature as a result of the turbulence, friction, etc., which occurs in the turbine.

Similar to the air compressor, these two temperatures are connected by

$$\text{Isentropic } \eta_{\text{turb}} = \frac{T_3 - T_4}{T_3 - T'_4} \qquad [6]$$

where Isentropic η_{turb} = isentropic efficiency of the turbine
T'_4 = frictionless adiabatic absolute exhaust temperature

T'_4 may be calculated from the gas law equation

$$T'_4 = T_3\left(\frac{P_4}{P_3}\right)^{(\gamma_t - 1)/\gamma_t} \qquad [7]$$

where γ_t = adiabatic index for the combustion products through the turbine

Figure 16.26(d) shows the turbine P–V diagram. For a particular mass flow, taken to be the same as through the air compressor, the turbine diagram will be as follows:

- **5–3** Combustion products enter turbine.
- **3–4** Combustion products expand in turbine.
- **4–6** Combustion products exhaust from turbine.

The pressure range in the compressor and the turbine is the same, so both P–V

diagrams can be combined as shown in Fig. 16.26(e). This diagram clearly shows the net output as the shaded area 1234. Note that 1234 is the same as the constant pressure cycle.

The T–s diagram is shown in Fig. 16.33(a). Strictly, however, since most gas turbines are **open circuit**, meaning that fresh air is continuously entering while exhaust is continuously leaving, the T–s diagram should be like Fig. 16.33(b). Here, the constant pressure process, 4–1, is shown dotted because it does not exist within the gas turbine arrangement (see section 16.23). However, it is conventional to use the full diagram (Fig. 16.33(a)) for calculation purposes.

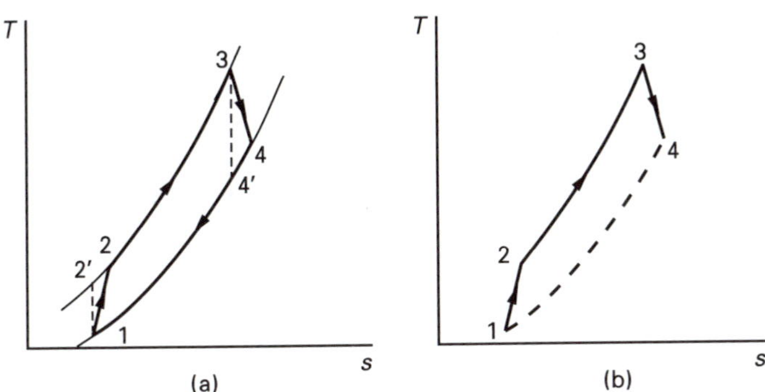

Fig. 16.33 Simple gas turbine T–s diagrams (a) used for making calculations and (b) with a broken line for process 4–1 because there is a continuous airflow

From equations [2] and [5]

$$\text{Net turbine output} = \dot{m}_t c_{pt}(T_3 - T_4) - \dot{m}_a c_{pa}(T_2 - T_1) \quad [8]$$

Now the mass of fuel used is usually small compared with the mass of air, so the mass of fuel is often neglected. If this is the case, then

$$\dot{m}_t = \dot{m}_a = \dot{m} \quad \text{(say)} \quad [9]$$

If the fuel is neglected, it can be considered that

$$c_{pt} = c_{pa} = c_p \quad \text{(say)} \quad [10]$$

Substituting equations [9] and [10] in equation [8], then

$$\begin{aligned}\text{Net turbine output} &= \dot{m}c_p(T_3 - T_4) - \dot{m}c_p(T_2 - T_1)\\ &= \dot{m}c_p[(T_3 - T_4) - (T_2 - T_1)]\end{aligned} \quad [11]$$

If $\dot{m}$ = mass of air in kg/s, then from equation [11]

$$\text{Net power output of turbine} = \dot{m}c_p[(T_3 - T_4) - (T_2 - T_1)]\ \text{kW} \quad [12]$$

The energy received in the gas turbine is in the combustion chamber at constant pressure. In the combustion chamber the temperature is raised from T_2 to T_3. If the fuel mass is neglected as before, then the energy received at constant pressure in the combustion chamber is

$$\dot{m}c_p(T_3 - T_2) = \text{change in enthalpy} \quad [13]$$

Now

$$\text{Thermal efficiency} = \frac{\text{Net work output}}{\text{Energy input}}$$

which from equations [11] and [13] becomes

$$\text{Thermal } \eta = \frac{\dot{m}c_p[(T_3 - T_4) - (T_2 - T_1)]}{\dot{m}c_p(T_3 - T_2)} \qquad [14]$$

$$= \frac{(T_3 - T_4) - (T_2 - T_1)}{(T_3 - T_2)} \qquad [15]$$

Another point that arises with the gas turbine cycle is that of the maximum possible work output within given temperature limits.

Consider the cycles A, B and C in Fig. 16.34. The temperature limits of each cycle are low temperature T_1 and high temperature T_3. The maximum pressure ratio possible between these temperature limits is given by

$$\frac{P_3}{P_1} = \left(\frac{T_3}{T_1}\right)^{\gamma/(\gamma - 1)} = r_{p\max} \qquad [16]$$

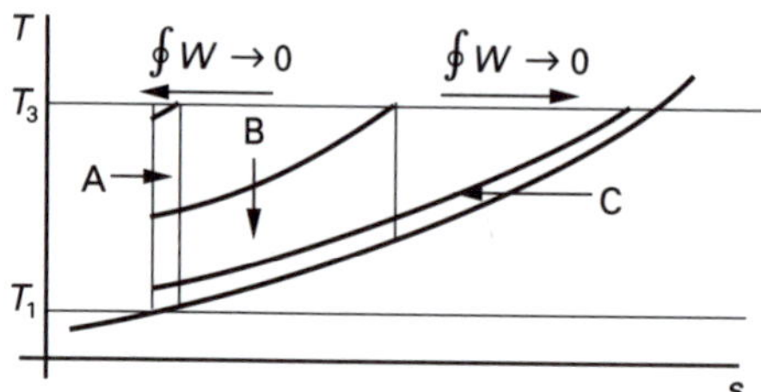

Fig. 16.34 Simple gas turbine: maximum work output

Consider cycle A shown in Fig. 16.34. Here the pressure ratio $r_p \to r_{p\max}$ but the net cycle work $\oint W \to 0$.

Consider cycle C. Here the pressure ratio $r_p \to 0$ and, once again, the net cycle work $\oint W \to 0$.

Somewhere in between, therefore, a cycle must exist in which $\oint W =$ a maximum. This is illustrated as cycle B. The problem now is to determine the pressure ratio, r_p, which will give the maximum net cycle work.

Equation [11] gives the net cycle work as

$$\oint W = \dot{m}c_p[(T_3 - T_4) - (T_2 - T_1)]$$

The pressure ratio of the cycle is given by

$$r_p = \frac{P_2}{P_1} = \frac{P_3}{P_4}$$

and

$$\frac{T_2}{T_1} = \left(\frac{P_2}{P_1}\right)^{(\gamma-1)/\gamma} = r_p^{(\gamma-1)/\gamma}$$

$$\therefore \quad T_2 = T_1 r_p^{(\gamma-1)/\gamma} \qquad [17]$$

also

$$\frac{T_3}{T_4} = \left(\frac{P_3}{P_4}\right)^{(\gamma-1)/\gamma} = r_p^{(\gamma-1)/\gamma}$$

$$\therefore \quad T_4 = \frac{T_3}{r_p^{(\gamma-1)/\gamma}} \qquad [18]$$

Substituting equations [17] and [18] into equation [11],

$$\oint W = \dot{m}c_p\left[\left(T_3 - \frac{T_3}{r_p^{(\gamma-1)/\gamma}}\right) - (T_1 r_p^{(\gamma-1)/\gamma} - T_1)\right] \qquad [19]$$

Assuming m, c_p, T_1 and T_3 to be constant, differentiating equation [19] with respect to r_p and equating to zero, it is possible to determine the value of r_p to give maximum $\oint W$. Thus

$$\frac{\mathrm{d}}{\mathrm{d}r_p}\left[\left(T_3 - \frac{T_3}{r_p^{(\gamma-1)/\gamma}}\right) - (T_1 r_p^{(\gamma-1)/\gamma} - T_1)\right] = 0 \quad \text{for max.} \quad \oint W \qquad [20]$$

or

$$-\left[-\left(\frac{\gamma-1}{\gamma}\right)\right] T_3 r_p^{-(\gamma-1/\gamma)-1} - \left(\frac{\gamma-1}{\gamma}\right) T_1 r^{(\gamma-1/\gamma)-1} = 0$$

from which

$$\left(\frac{\gamma-1}{\gamma}\right) T_3 r_p^{(-2\gamma+1)/\gamma} - \left(\frac{\gamma-1}{\gamma}\right) T_1 r_p^{-1/\gamma} = 0$$

$$\left(\frac{\gamma-1}{\gamma}\right) T_3 r_p^{(-2\gamma+1)/\gamma} = \left(\frac{\gamma-1}{\gamma}\right) T_1 r_p^{-1/\gamma}$$

or

$$T_3 r_p^{(-2\gamma+1)/\gamma} = T_1 r_p^{-1/\gamma}$$

hence

$$\frac{r_p^{-1/\gamma}}{r_p^{(-2\gamma+1)/\gamma}} = \frac{T_3}{T_1}$$

from which

$$r_p^{-(1/\gamma)-[-(2\gamma+1)/\gamma]} = \frac{T_3}{T_1}$$

or

$$r_p^{(2\gamma-2)/\gamma} = \frac{T_3}{T_1}$$

$$r_p^{2[(\gamma-1)/\gamma} = \frac{T_3}{T_1}$$

hence

$$r_p = \left(\frac{T_3}{T_1}\right)^{\gamma/[2(\gamma-1)]}$$

or

$$r_p = \sqrt{\left(\frac{T_3}{T_1}\right)^{\gamma(\gamma-1)}} \qquad [21]$$

Substituting equation [11] into equation [21] shows that within the given temperature limits of T_1 and T_3, the maximum net cycle work is obtained when the pressure ratio of the cycle is given by the expression

$$r_p = \sqrt{r_{\text{max}}} \qquad [22]$$

Note that from equations [15] and [17] and [22]

$$T_2 = T_1 r_p^{(\gamma-1)/\gamma} = \frac{T_3}{r_{p\text{max}}^{(\gamma-1)/\gamma}} r_p^{(\gamma-1)/\gamma} = \frac{T_3}{r_p^{[2(\gamma-1)/\gamma]}} r_p^{(\gamma-1)/\gamma} = \frac{T_3}{r_p^{(\gamma-1)/\gamma}}$$

But from equation [18]

$$T_4 = \frac{T_3}{r_p^{(\gamma-1)/\gamma}}$$

Hence it follows that when the maximum net cycle work is obtained within the given temperature limits

$$T_2 = T_4 \qquad [23]$$

Furthermore, for this cycle

$$\text{Thermal } \eta = \frac{(T_3 - T_4) - (T_3 - T_1)}{(T_3 - T_2)} \qquad [24]$$

and

$$\text{Work ratio} = \frac{(T_3 - T_4) - (T_2 - T_1)}{(T_3 - T_4)} \qquad [25]$$

And $T_2 = T_4$ for the pressure ratio which gives maximum net work output, so for that case

$$\text{Thermal } \eta = \text{Work ratio} \qquad [26]$$

Example 16.1 *In a continuous combustion, constant pressure, gas turbine, air is taken into a rotary compressor at a pressure of 100 kN/m² and temperature 18°C. It is compressed through a pressure ratio of 8:1 with an isentropic efficiency of 85 per cent. From the compressor, the compressed air is passed to a combustion chamber where its temperature is raised to 1000°C. From the combustion chamber, the high-temperature air is passed to a gas turbine in which it is expanded down to 100 kN/m² with an isentropic efficiency of 88 per cent. From the turbine, the air is passed to exhaust. If the air used is 4.5 kg/s and neglecting the mass of fuel as small, determine*

(a) the net power output of the turbine plant if the turbine is coupled to the compressor

(b) the thermal efficiency of the plant

(c) the work ratio

Take cp = *1.006 kJ/kg K,* $\gamma = 1.4$.

(a)

For the compressor

$$T_2' = T_1\left(\frac{P_2}{P_1}\right)^{(\gamma-1)/\gamma} = 291 \times 8^{(1.4-1)/1.4}$$
$$= 291 \times 8^{1/3.5}$$
$$= 291 \times 1.811$$
$$= \mathbf{527\ K}$$

$$\text{Isentropic } \eta_{\text{comp}} = \frac{T_2' - T_1}{T_2 - T_1}$$

$$\therefore \quad T_2 - T_1 = \frac{T_2' - T_1}{\text{Isentropic } \eta_{\text{comp}}} = \frac{527 - 291}{0.85} = \frac{236}{0.85} = \mathbf{278\ K}$$

$$\therefore \quad T_2 = 291 + 278 = \mathbf{569\ K}$$
$$t_2 = 569 - 273 = \mathbf{296\ ^\circ C}$$

For the turbine

$$T_4' = T_3\left(\frac{P_4}{P_3}\right)^{(\gamma-1)/\gamma} = \frac{1273}{8^{(1.4-1)/1.4}} = \frac{1273}{1.811} = \mathbf{703\ K}$$

$$\text{Isentropic } \eta_{\text{turb}} = \frac{T_3 - T_4}{T_3 - T_4'}$$

$$\therefore \quad T_3 - T_4 = (T_3 - T_4') \times \text{Isentropic } \eta_{\text{turb}}$$
$$= (1273 - 703) \times 0.88$$
$$= 570 \times 0.88$$
$$= \mathbf{502\ K}$$

$$\therefore \quad T_4 = 1273 - 502 = \mathbf{771\ K}$$
$$t_4 = 771 - 273 = \mathbf{498\ ^\circ C}$$

$$\text{Net power output} = \dot{m}c_p\left[(T_3 - T_4) - (T_2 - T_1)\right]$$
$$= 4.5 \times 1.006(502 - 278)$$
$$= 4.5 \times 1.006 \times 224$$
$$= \mathbf{1014\ kW}$$

(b)

$$\text{Thermal } \eta = \frac{(T_3 - T_4) - (T_2 - T_1)}{(T_3 - T_2)}$$

$$= \frac{502 - 278}{1273 - 569}$$

$$= 0.32$$
$$= \mathbf{32\%}$$

(c)

$$\text{Work ratio} = \frac{\text{Net cycle work}}{\text{Positive cycle work}}$$

$$= \frac{1014}{\dot{m}c_p(T_3 - T_4)}$$

$$= \frac{1014}{4.5 \times 1.006 \times 502}$$

$$= \frac{1014}{2273}$$

$$= \mathbf{0.446}$$

Example 16.2 *A gas turbine plant has temperature limits 1080 °C and 10 °C. Compression in the compressor and expansion in the turbine are isentropic. Determine*
(a) the pressure ratio which will give the maximum net work output
(b) the maximum net specific work output
(c) the thermal efficiency at maximum work output
(d) the work ratio at maximum work output
(e) the Carnot efficiency within the cycle temperature limits
Take $\gamma = 1.41$ *and* cp $= 1.007$ *kJ/kg K.*

SOLUTION
First draw a diagram (Fig. 16.35).

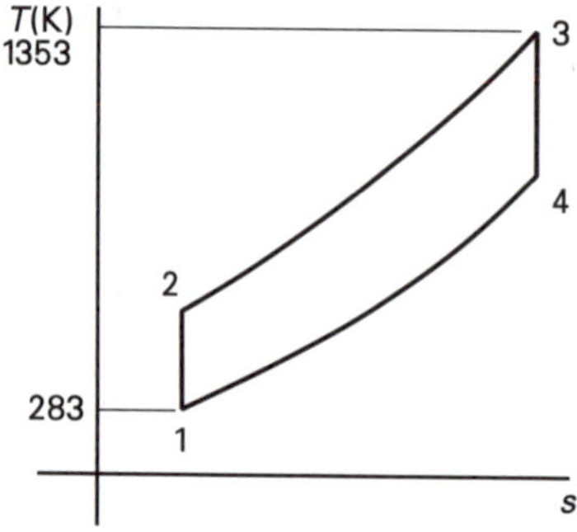

Fig. 16.35 Diagram for Example 16.2

(a)

$T_3 = 1080 + 273 = \mathbf{1353\ K}$
$T_1 = 10 + 273 = \mathbf{283\ K}$

$$\text{Maximum pressure ratio} = r_{p\max} = \left(\frac{T_3}{T_1}\right)^{\gamma/(\gamma-1)} \quad \text{(see equation [16])}$$

$$= \left(\frac{1353}{283}\right)^{1.41/(1.41-1)}$$

$$= 4.78^{1.41/0.41}$$
$$= 4.78^{3.44}$$
$$= \mathbf{217.4}$$

For maximum net work output

$$\text{Pressure ratio} = r_p = \sqrt{r_{p\max}} \quad \text{(see equation [22])}$$
$$= \sqrt{217.4}$$
$$= \mathbf{14.74}$$

(b)

$$\frac{T_2}{T_1} = r_p^{(\gamma-1)/\gamma}$$

$$\therefore \quad T_2 = T_1 r_p^{(\gamma-1)/\gamma} = 283 \times 14.74^{(1.41-1)/1.41}$$
$$= 283 \times 14.74^{0.41/1.41}$$
$$= 283 \times 14.74^{1/3.44}$$
$$= 283 \times 2.19$$
$$= \mathbf{620\ K}$$

And from equation [23] $T_4 = T_2 = 620$ K.

Maximum net specific work output is

$$c_p[(T_3 - T4) - (T_2 - T_1) = 1.007\,[(1353 - 620) - (620 - 283)]$$
$$= 1.007\,(733 - 337)$$
$$= 1.007 \times 396$$
$$= \mathbf{399\ kJ}$$

(c)

$$\text{Thermal } \eta = \frac{399}{c_p(T_3 - T_2)}$$

$$= \frac{399}{1.007(1353 - 620)}$$

$$= \frac{399}{1.007 \times 733}$$

$$= \frac{399}{738}$$

$$= 0.54$$
$$= \mathbf{54\%}$$

(d)
From equation [26]

$$\text{Work ratio} = \text{Thermal } \eta = \mathbf{0.54}$$

(e)

$$\text{Carnot } \eta = \frac{T_3 - T_1}{T_3} = \frac{1353 - 283}{1353} = \frac{1070}{1353} = 0.79 = \mathbf{79\%}$$

16.22 *Thermal efficiency improvement in gas turbines*

The exhaust leaves the gas turbine with quite a high temperature and high residual energy content. When the power is required in the shaft, the energy in the exhaust will go to waste unless some effort is made to reclaim it. It is impossible to reclaim all the energy in the exhaust. A common method is shown in Fig. 16.36.

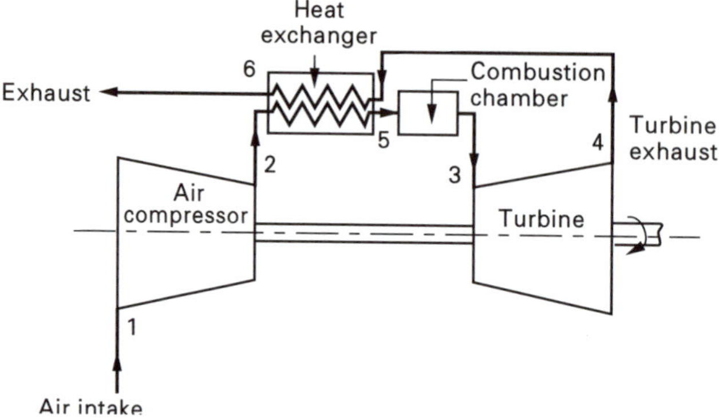

Fig. 16.36 Gas turbine with exhaust energy recovery

The basic units of the gas turbine circuit are as already discussed. In this arrangement, however, a heat exchanger is fitted in the compressed air line from the air compressor to the combustion chamber. The compressed air passes through this heat exchanger on its way to the combustion chamber. And the exhaust from the turbine passes through the heat exchanger on its way out to atmosphere.

Before it reaches the combustion chamber, the compressed air receives some heat and has its temperature elevated by heat exchange from the exhaust gases. This means that the quantity of fuel used in the combustion chamber can now be reduced because the temperature rise now required in the combustion chamber is reduced.

The output from the turbine is hardly affected by the introduction of the heat exchanger. On the other hand, the fuel used has been reduced. This means there has been an improvement in the thermal efficiency because there is much the same output with reduced fuel input.

The energy recovered from the exhaust in actual gas turbines varies from 50–90 per cent. They operate most commonly at 70–80 per cent recovery. The percentage recovery of the heat exchanger is called its **effectiveness**. The thermal efficiency of gas turbines without heat exchange is usually in the range 10–25 per cent, mostly from 15–20 per cent. With a heat exchanger fitted, the thermal efficiency is pushed up to 20–30 per cent and even as high as 35 per cent in a few cases.

The *T–s* diagram for the turbine arrangement with heat exchanger is illustrated in Fig. 16.37. The maximum exhaust temperature drop available in the exchanger is $(T_4 - T_2)$ because T_2 is the lowest temperature in the exchanger.

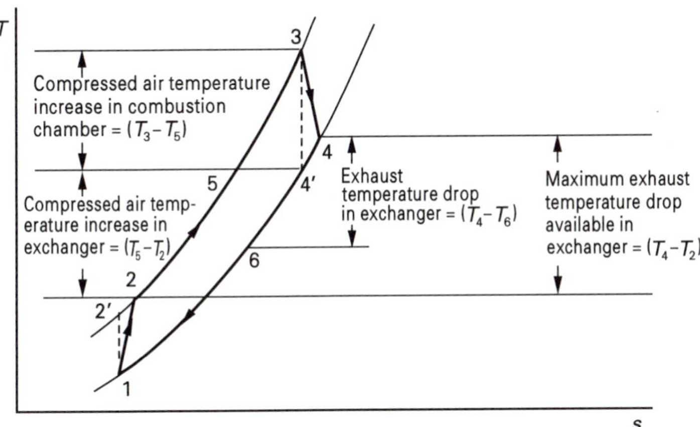

Fig. 16.37 Gas turbine with heat exchanger: *T–s* diagram

Let the effectiveness of the exchanger be k.

Then

$$k\dot{m}c_p(T_4 - T_2) = \dot{m}c_p(T_4 - T_6) = \dot{m}c_p(T_5 - T_2) \qquad [1]$$

and assuming $\dot{m}$ and c_p to be constant throughout, then

$$k(T_4 - T_2) = (T_4 - T_6) = (T_5 - T_2) \qquad [2]$$

Equation [2] will enable the exhaust temperature from the exchanger, T_6, and the compressed air temperature at entry to the exchanger, T_5, to be determined.

As before

$$\text{Net turbine output} = \dot{m}c_p[(T_3 - T_4) - (T_2 - T_1)] \qquad [3]$$

However, the energy required from the fuel is that required to increase the temperature from T_5 to T_3.

$$\text{Required energy} = \dot{m}c_p(T_3 - T_5) \qquad [4]$$

This is evidently less than would be required if no heat exchanger were fitted, when the required temperature increase required would be from T_2 to T_3.

Thus, with a heat exchanger, the thermal efficiency of the plant is increased.

$$\text{Thermal } \eta = \frac{\dot{m}c_p[(T_3 - T_4) - (T_2 - T_1)]}{\dot{m}c_p(T_3 - T_5)} \qquad [5]$$

and assuming $\dot{m}$ and c_p constant throughout

$$\text{Thermal } \eta = \frac{(T_3 - T_4) - (T_2 - T_1)}{T_3 - T_5} \qquad [6]$$

Example 16.3 *An open-circuit, continuous combustion, constant pressure, gas turbine takes in air at a pressure of 101 kN/m^2 and at a temperature of 15 °C. The air is compressed in a rotary compressor through a pressure ratio of 6:1. The air then passes at constant pressure through a heat exchanger of effectiveness 65 per cent. From the heat exchanger the air passes at constant pressure through a combustion chamber in which its temperature is raised to 870 °C. From the combustion chamber the air passes through a gas turbine in which it is expanded to a pressure of 101 kN/m^2 and it then passes through the heat exchanger to exhaust. The isentropic efficiency of the compressor is 85 per cent while that of the turbine is 80 per cent. Neglect the mass of the fuel and take the air mass flow rate as 4 kg/s. Determine*

(a) the net power output of the plant
(b) the exhaust temperature from the heat exchanger
(c) the thermal efficiency of the plant
(d) the thermal efficiency of the plant if there were no heat exchanger
(e) the work ratio

Take $\gamma = 1.4$ and $c_p = 1.005$ kJ/kg K.

SOLUTION

First draw a diagram (Fig. 16.38).

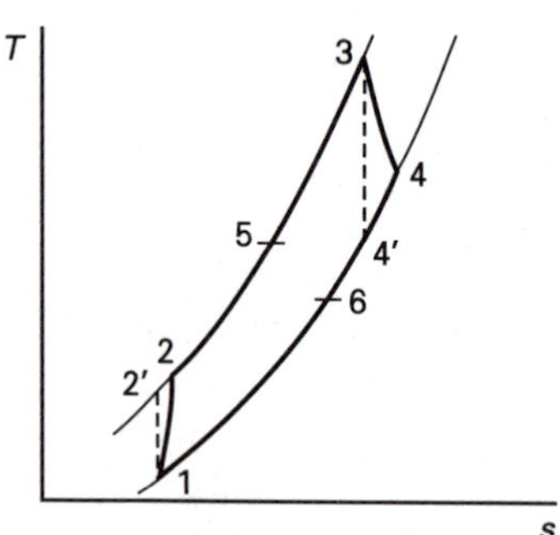

Fig. 16.38 Diagram for Example 16.3

(a)

For the compressor

$$T_2' = T_1\left(\frac{P_2}{P_1}\right)^{(\gamma-1)/\gamma} = 288 \times 6^{(1.4-1)/1.4}$$
$$= 288 \times 6^{0.4/1.4}$$
$$= 288 \times 6^{1/3.5}$$
$$= 288 \times 1.67$$
$$= \mathbf{481\ K}$$

$$0.85 = \frac{T_2' - T_1}{T_2 - T_1}$$

$$\therefore \quad T_2 - T_1 = \frac{T_2' - T_1}{0.85} = \frac{481 - 288}{0.85} = \frac{193}{0.85} = \mathbf{227\ K}$$

$$\therefore \quad T_2 = 227 + 288 = \mathbf{515\ K}$$
$$t_2 = 515 - 273 = \mathbf{242\ ^\circ C}$$

For the turbine

$$T_4' = T_3\left(\frac{P_4}{P_3}\right)^{(\gamma-1)/\gamma} = 1143 \times \left(\frac{1}{6}\right)^{(1.4-1)/1.4} = \frac{1143}{1.67} = \mathbf{684\ K}$$

$$0.80 = \frac{T_3 - T_4}{T_3 - T_4'}$$

$$\begin{aligned}\therefore \quad T_3 - T_4 = 0.80(T_3 - T_4') &= 0.80(1143 - 684)\\ &= 0.80 \times 459\\ &= \mathbf{367\ K}\end{aligned}$$

$$\therefore \quad T_4 = 1143 - 367 = \mathbf{776\ K}$$
$$t_4 = 776 - 273 = \mathbf{503\ ^\circ C}$$

$$\begin{aligned}\text{Net power output} &= \dot{m}c_p\,[(T_3 - T_4) - (T_2 - T_1)]\\ &= 4 \times 1.005 \times (367 - 227)\\ &= 4 \times 1.005 \times 140\\ &= \mathbf{563\ kW}\end{aligned}$$

(b)

Maximum temperature drop available for heat transfer $= (T_4 - T_2)$

$$\begin{aligned}\text{Actual temperature drop} &= 0.65(T_4 - T_2)\\ &= 0.65(776 - 515)\\ &= 0.65 \times 261\\ &= \mathbf{170\ K}\end{aligned}$$

$$\begin{aligned}\therefore \quad \text{Exhaust temperature from heat exchanger} &= T_4 - 170\\ &= 776 - 170\\ &= \mathbf{606\ K}\end{aligned}$$

$$t_6 = 606 - 273 = \mathbf{333\ ^\circ C}$$

(c)

$$\text{Thermal } \eta = \frac{(T_3 - T_4) - (T_2 - T_1)}{(T_3 - T_5)}$$

$$T_5 = T_2 + 170 = 515 + 170 = \mathbf{685\ K}$$

$$\begin{aligned}\therefore \quad \text{Thermal } \eta &= \frac{367 - 227}{1143 - 685}\\ &= \frac{140}{458}\\ &= 0.306\\ &= \mathbf{30.6\%}\end{aligned}$$

(d)
With no heat exchanger

$$\text{Thermal } \eta = \frac{(T_3 - T_4) - (T_2 - T_1)}{(T_3 - T_2)}$$

$$= \frac{367 - 227}{1143 - 515}$$

$$= \frac{140}{628}$$

$$= 0.223$$
$$= \mathbf{22.3\%}$$

(e)

$$\text{Work ratio} = \frac{\text{Net cycle work}}{\text{Positive cycle work}}$$

$$= \frac{563}{\dot{m}c_p(T_3 - T_4)}$$

$$= \frac{563}{4 \times 1.005 \times 367}$$

$$= \frac{563}{1475}$$

$$= \mathbf{0.38}$$

16.23 Gas turbine cycles in general

Gas turbines are capable of many other arrangements different from those already discussed. Each arrangement is made to fit the turbine to the particular installation required.

Figure 16.39 shows the turbine split into two components. First there is a turbine which is coupled back to run the compressor. The combination of the air compressor, combustion chamber and compressor turbine (enclosed by the dotted line in Fig. 16.39) is called the **gas generator unit**.

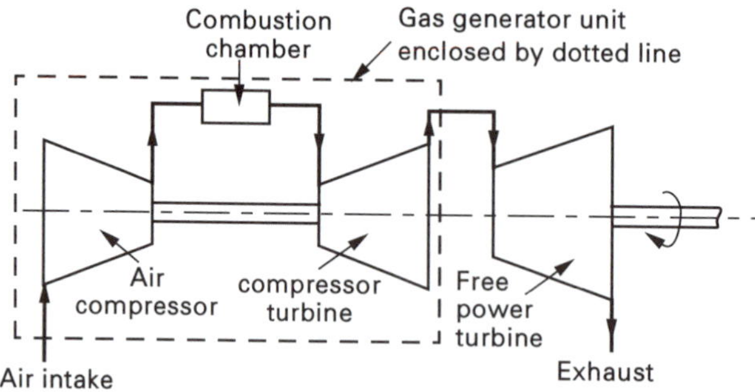

Fig. 16.39 Gas turbine with free power turbine

The exhaust from the compressor turbine passes into a second turbine which is completely separate. This is the free power turbine which is coupled to drive external equipment. The advantage of this arrangement is that the free power turbine is completely independent of the air compressor. In this way all units of the plant can be designed to run at their most efficient speed. This is especially so if the free power turbine speed is to be some fixed value such as 3000 rev/min. This is a common speed required in the generation of electrical power.

The turbines discussed so far may be put under the general heading of **open-cycle** or **open-circuit** gas turbines. This name is given to them because air is taken from the atmosphere and the exhaust products are passed out to atmosphere. Work has also been done on the **closed-cycle**, or **closed-circuit**, gas turbines. The basic principle of this type of turbine is shown in Fig. 16.40. It shows the usual turbine–compressor arrangement, but now the air operates in a closed circuit. It is the same air being circulated all the time, with the exception of any make-up necessary due to leakage.

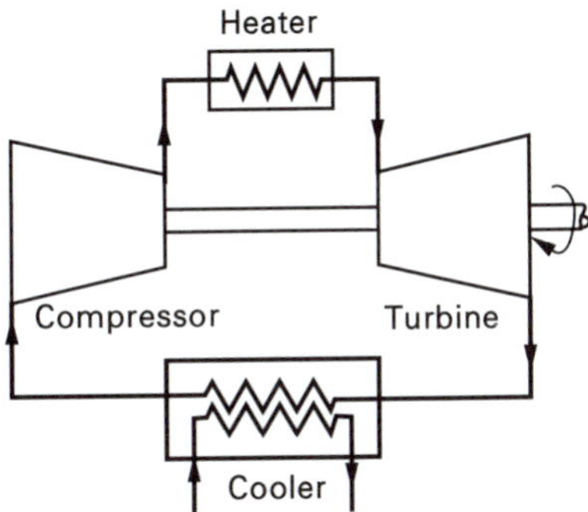

Fig. 16.40 Closed-circuit gas turbine

The compressed air passes from the compressor through a heater in which the air is indirectly heated. The heated air passes into the turbine and after expansion into a cooler, where it is cooled then recirculated back to the compressor. The air in the circuit, even on the low-pressure side of the circuit, is usually compressed above atmospheric conditions. In this condition it has a higher density and has better heat transfer properties.

This circuit has the advantage that the heater can use any type of fuel, solid, liquid or gas, because there is no direct mixing with the working air. It is also possible to use a nuclear reactor to provide the necessary energy. Then the circulating gas would probably be changed to some gas such as helium, which is relatively immune to the effects of radioactivity.

The disadvantage of this circuit is that it needs a large supply of cooling water for the cooler and the use of indirect heating leads to a reduced thermal efficiency. On small plants the closed circuit has a lower thermal efficiency than the corresponding open circuit.

It is probably more advantageous to use the closed cycle in large turbine installations in which many efficiency improvement devices can be installed.

Example 16.4 *Air enters a gas turbine plant at a pressure of 100 kN/m^2 and a temperature of 19 °C. It is compressed in a rotary compressor through a pressure ratio of 8:1 with an isentropic efficiency of 85%. The compressed air then passes into a combustion chamber in which its temperature is raised at constant pressure to 980 °C. From the combustion chamber the air enters the compressor turbine whose isentropic efficiency is 88 per cent. In the compressor turbine a pressure–temperature drop occurs such that the power output from the compressor turbine is just sufficient to drive the compressor. From the compressor turbine the air enters a free power turbine in which it is expanded to a pressure of 100 kN/m^2 with an isentropic efficiency of 86 per cent. The air then passes to exhaust. Air is used at the rate of 7 kg/s. Neglecting the mass of fuel and taking $\gamma = 1.4$ and $c_p = 1.006$ kJ/kg K, determine*

(a) the pressure and temperature as the air leaves the compressor turbine
(b) the power output from the free power turbine
(c) the thermal efficiency of the plant
(d) the work ratio
(e) the Carnot efficiency within the cycle temperature limits

SOLUTION

First draw a diagram (Fig. 16.41).

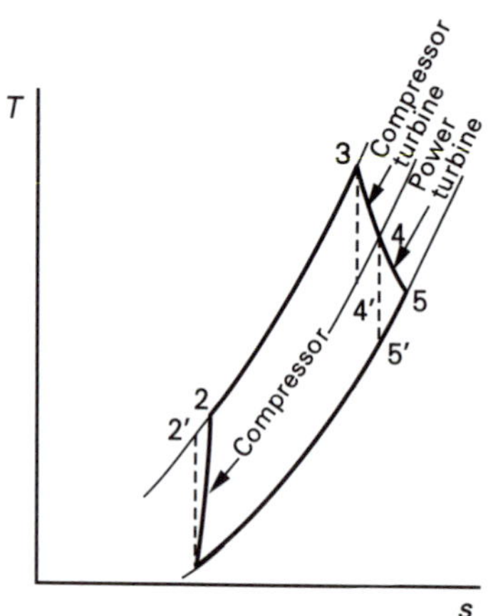

Fig. 16.41 Diagram for Example 16.4

(a)

For the compressor

$$T_2' = T_1\left(\frac{P_2}{P_1}\right)^{(\gamma-1)/\gamma} = 292 \times 8^{(1.4-1)/1.4} = 292 \times 8^{0.4/1.4}$$
$$= 292 \times 8^{1/3.5}$$
$$= 292 \times 1.81$$
$$= \mathbf{529\ K}$$

$$0.85 = \frac{T_2' - T_1}{T_2 - T_1}$$

$$\therefore \quad T_2 - T_1 = \frac{T_2' - T_1}{0.85} = \frac{529 - 292}{0.85} = \frac{237}{0.85} = \mathbf{279\ K}$$

$$T_2 = 279 + T_1 = 279 + 292 = \mathbf{571\ K}$$
$$t_2 = 571 - 273 = \mathbf{298\ °C}$$

For the compressor turbine

$$T_3 - T_4 = T_2 - T_1$$

This is because

Compressor turbine power = Compressor power

$$\therefore \quad T_3 - T_4 = \mathbf{279\ K}$$

$$T_4 = T_3 - 279 = 1253 - 279 = \mathbf{974\ K}$$

$$t_4 = 974 - 273 = \mathbf{701\ ^\circ C}$$

The temperature as the air leaves the compressor turbine is 701 °C.

$$0.88 = \frac{T_3 - T_4}{T_3 - T'_4}$$

$$\therefore \quad T_3 - T'_4 = \frac{T_3 - T_4}{0.88} = \frac{279}{0.88} = \mathbf{317\ K}$$

$$T'_4 = T_3 - 317 = 1253 - 317 = \mathbf{936\ K}$$

$$\frac{T_3}{T'_4} = \left(\frac{P_3}{P'_4}\right)^{(\gamma - 1)/\gamma}$$

$$\therefore \quad P_4 = P_3\left(\frac{T'_4}{T_3}\right)^{\gamma/(\gamma - 1)} \quad \text{and} \quad P_3 = 8 \times 100 = \mathbf{800\ kN/m^2}$$

$$\therefore \quad P_4 = 800 \times \left(\frac{936}{1253}\right)^{1.4/(1.4 - 1)}$$

$$= \frac{800}{1.34^{3.5}} = \frac{800}{2.79}$$

$$= \mathbf{287\ kN/m^2}$$

The pressure as the air leaves the compressor turbine is 287 kN/m^2.

(b)

For the free power turbine

$$T'_5 = T_4\left(\frac{P_5}{P_4}\right)^{(\gamma - 1)/\gamma} = 974 \times \left(\frac{100}{287}\right)^{(1.4 - 1)/1.4}$$

$$= \frac{974}{2.87^{1/3.5}} = \frac{974}{1.35}$$

$$= \mathbf{721\ K}$$

$$0.86 = \frac{T_4 - T_5}{T_4 - T'_5}$$

$$\begin{aligned} \therefore \quad T_4 - T_5 &= 0.86(T_4 - T'_5) \\ &= 0.86 \times (974 - 721) \\ &= 0.86 \times 253 \\ &= \mathbf{218\ K} \end{aligned}$$

$$\therefore \quad T_5 = T_4 - 218 = 974 - 218 = \mathbf{756\ K}$$
$$t_5 = 756 - 273 = \mathbf{483\ ^\circ C}$$

Power output from the free power turbine is

$$\dot{m}c_p(T_4 - T_5) = 7 \times 1.006 \times 218$$
$$= \mathbf{1535\ kW}$$

(c)

$$\text{Thermal } \eta = \frac{\dot{m}c_p(T_4 - T_5)}{\dot{m}c_p(T_3 - T_2)}$$
$$= \frac{(T_4 - T_5)}{(T_3 - T_2)}$$
$$= \frac{218}{1253 - 571} = \frac{218}{682}$$
$$= 0.32$$
$$= \mathbf{32\%}$$

(d)

$$\text{Work ratio} = \frac{\text{Net cycle work}}{\text{Positive cycle work}}$$
$$= \frac{\dot{m}c_p(T_4 - T_5)}{\dot{m}c_p(T_3 - T_5)}$$
$$= \frac{(T_4 - T_5)}{(T_3 - T_5)}$$
$$= \frac{218}{1253 - 756} = \frac{218}{497}$$
$$= \mathbf{0.44}$$

(e)

$$\text{Carnot } \eta = \frac{T_3 - T_1}{T_3} = \frac{1253 - 242}{1253}$$
$$= \frac{961}{1253}$$
$$= 0.77$$
$$= \mathbf{77\%}$$

Example 16.5 *An aircraft is flying at an altitude of 6100 m. At this altitude the atmospheric pressure and temperature are 470 bar and −22.4 °C, respectively. The aircraft is propelled by a gas turbine. The pressure ratio through its air compressor is 30:1. The air mass flow rate is 80 kg/s. The gas turbine produces power for the air compressor and engine auxiliaries. The*

ratio of turbine power to compressor power is 1.25:1. In the combustion chamber the compressed air temperature is raised to 960 °C. Both the compressor and the gas turbine have an isentropic efficiency of 86 per cent. Neglecting the mass of fuel, determine
(a) the pressure and temperature of the air after compression
(b) the power developed by the gas turbine
(c) the temperature and pressure of the air entering the exhaust jet as it leaves the gas turbine
Take $c_p = 1.05$ *kJ/kg K,* $\gamma = 1.41$.

(a)

$$P_2 = 30\ P_1$$
$$= 30 \times 470 = 14\ 100 \text{ mbar}$$
$$= \mathbf{14.1\ bar}$$

For the compressor

$$T_{2'} = T_1\left(\frac{P_2}{P_1}\right)^{(\gamma-1)/\gamma} \quad \text{and} \quad T_1 = 273 - 22.4 = \mathbf{250.6\ K}$$

$$\therefore \quad T_{2'} = 250.6 \times 30^{(1.41-1)/1.41}$$
$$= 250.6 \times 30^{1/3.44}$$
$$= 250.6 \times 2.688$$
$$= \mathbf{673.6\ K}$$

$$\text{Isentropic } \eta = \frac{T_{2'} - T_1}{T_2 - T_1}$$

$$\therefore \quad T_2 - T_1 = \frac{T_{2'} - T_1}{\text{Isentropic } \eta} = \frac{673.6 - 250.6}{0.86} = \frac{423}{0.86} = \mathbf{491.9\ K}$$

$$\therefore \quad T_2 = 250.6 + 491.9 = \mathbf{742.5\ K}$$
$$t_2 = 742.5 - 273 = \mathbf{469.5\ ^\circ C}$$

The temperature after compression is 469.5 °C.

(b)

$$\text{Temperature drop in turbine} = 1.25\ (T_2 - T_1)$$
$$= 1.25 \times 491.9$$
$$= \mathbf{614.88\ K}$$

$$\therefore \quad \text{Power developed by turbine} = 80 \times 1.05 \times 614.88$$
$$= 51\ 650 \text{ kW}$$
$$= \mathbf{51.65\ MW}$$

(c)

$$\text{Temperature of air leaving turbine, } t_4 = 960 - 614.88$$
$$= \mathbf{345.12\ ^\circ C}$$

$$\text{Isentropic temperature drop} = \frac{614.88}{0.86} = \mathbf{714.98\ K}$$

$$t_{4'} = 960 - 714.98 = \mathbf{245.02\ ^\circ C}$$
$$T_{4'} = 245.02 + 273 = \mathbf{518.02\ K}$$
$$T_3 = 960 + 273 = \mathbf{1233\ K}$$

$$P_4 = P_3\left(\frac{T_{4'}}{T_3}\right)^{\frac{\gamma}{\gamma-1}} = 14.1 \times \left(\frac{518.02}{1233}\right)^{1.41/0.41}$$

$$= 14.1 \times \frac{1}{2.38^{3.44}}$$

$$= \frac{14.1}{19.74}$$

$$= \mathbf{0.714\ bar}$$

The air pressure as it leaves the gas turbine is 0.714 bar.

Example 16.6 *The gas turbine in a CHP generating plant has an output of 150 MW with a thermal efficiency of 35 per cent. The fuel oil used has a calorific value of 43 MJ/kg. The exhaust gas flow rate from the gas turbine is 400 kg/s and its temperature is 550 °C. The exhaust gas from the gas turbine passes through a steam generator boiler plant and leaves at a temperature of 90 °C. The steam generated is at a pressure of 10 MN/m² and a temperature of 450 °C. The feedwater temperature to the boiler is 140 °C. The generated steam passes through a steam turbine system with an isentropic efficiency of 0.86 to exhaust at a pressure of 0.5 MN/m². The boiler system has a thermal efficiency of 92 per cent. Determine*

(a) the mass of fuel oil used by the gas turbine in tonne/h

(b) the mass flow of steam from the boiler in tonne/h

(c) the theoretical output from the steam turbine in MW

(d) the overall theoretical thermal efficiency of the plant

For the turbine exhaust gas, take $c_p = 1.1$ kJ/kg K.

(a)

$$\text{Energy requirement from fuel in gas turbine} = \frac{(150 \times 3600)}{0.35}\ \text{MJ/h}$$

$$\therefore \quad \text{Fuel required} = \frac{150 \times 3600}{0.35 \times 43 \times 10^3} = \mathbf{35.9\ tonne/h}$$

(b)

$$\text{Energy available from gas turbine exhaust} = 400 \times 3600 \times 1.1\ (550 - 90)$$
$$= (400 \times 3600 \times 1.1 \times 460)\ \text{kJ/h}$$

$$\text{Energy transferred to steam} = (400 \times 3600 \times 1.1 \times 4.60 \times 0.92)\ \text{kJ/h}$$

$$\text{Energy required to raise steam} = (3244 - 588.5)$$
$$= \mathbf{2655.5\ kJ/kg}$$

$$\therefore \quad \text{Mass flow of steam} = \frac{400 \times 3600 \times 1.1 \times 460 \times 0.92}{2655.5 \times 10^3}$$

$$= \mathbf{252.4\ tonne/h}$$

(c)

For the steam turbine

$$h_1 = 3244\ \text{kJ/kg}$$

and

$$s_1 = s_2 = 6.424\ \text{kJ/kg K}$$
$$s_2 = s_{f2} + x_2(s_{g2} - s_{f2})$$

$$\therefore \quad x_2 = \frac{(s_2 - s_{f2})}{(s_{g2} - s_{f2})} = \frac{(6.424 - 1.86)}{(6.819 - 1.86)}$$

$$= \frac{4.564}{4.959} = \mathbf{0.92}$$

$$\therefore \quad h_{2'} = 640.1 + 0.92(2747.5 - 640.1)$$
$$= 640.1 + (0.92 \times 2107.4)$$
$$= 640.1 + 1930.8$$
$$= \mathbf{2570.9\ kJ/kg}$$

$$\text{Theoretical steam turbine output} = 0.86(3244 - 2570.9)$$
$$= (0.86 \times 673.1)$$
$$= \mathbf{578.9\ kJ/kg}$$

$$\therefore \quad \text{Total theoretical steam turbine output} = \frac{578.9 \times 252 \times 10^3}{3600 \times 10^3}$$

$$= \mathbf{40.58\ MW}$$

(d)

$$\text{Overall theoretical thermal } \eta = \frac{(150 + 40.58)0.35}{150}$$

$$= \frac{190.58 \times 0.35}{150}$$

$$= 0.445$$
$$= \mathbf{44.5\%}$$

16.24 Rotary engines

Many attempts have been made, and are being made, to develop rotary engines which will produce direct rotary motion, thus dispensing with the reciprocating piston, connecting-rod and crankshaft of the conventional reciprocating engine. Most attempts have met with little success. Some success has, however, been obtained from the Wankel engine developed by NSU of Germany. Its elements are illustrated in Fig. 16.42. The engine consists of a triangular rotor which rotates eccentrically in a casing. The apexes of the rotor have gas seals which mate with, and slide freely across, the inside surface of the casing. The operation is as follows.

In Fig. 16.42(a) a fresh air–fuel charge is being taken into space A. In space B there is a compressed charge ignited by the spark-plug. In space C there is an expanding gas which is just beginning to exhaust. A slightly later event is shown in Fig. 16.42(b). In space A a fresh charge is being compressed. In space B the burnt charge is just at the beginning of expansion. In space C the exhaust is about to end and a fresh charge is about to enter.

The arrangement has no mechanically operated valves, so it produces events similar to those of the two-stroke cycle engine. A high mass flow rate is possible through such an engine which produces a fairly high power to weight ratio.

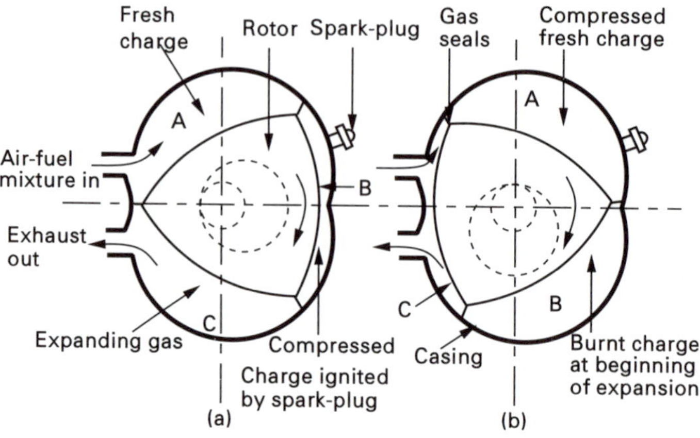

Fig. 16.42 Wankel engine: (a) fuel being taken in; (b) a short time later

Disadvantages appear in obtaining good gas seal between the rotor and the casing. Temperature distribution round the casing can be uneven because the fresh charge, burning charge and expanding gas always occur at their same respective positions. Some combustion problems can occur in the narrow combustion space.

Questions

1. A petrol engine has a volume ratio of compression of 8:1. When 0.1 and 0.8 of the compression stroke have been completed, the cylinder pressures are 103.5 kN/m^2 and 451 kN/m^2, respectively. Determine the value of the index n for the compression curve.

   [1.28]

2. A continuous combustion, constant pressure, gas turbine takes in air at a pressure of 93 kN/m^2 with a temperature of 20 °C. A rotary air compressor compresses the air to a pressure of 552 kN/m^2 with an isentropic efficiency of 83 per cent. The compressed air is passed to a combustion chamber in which its temperature is increased to 870 °C. From the combustion chamber the high temperature air passes into a gas turbine in which it is expanded to 93 kN/m^2 with an isentropic efficiency of 80 per cent. For an airflow of 10 kg/s and neglecting the fuel mass as small, determine
   (a) the net power output of the plant if the turbine is coupled to the compressor
   (b) the thermal efficiency of the plant
   (c) the work ratio
   Take, $\gamma = 1.4$ and $c_p = 1.00$ kJ/kg K.

   [(a) 1310 kW; (b) 21%; (c) 0.36]

3. An open-circuit, continuous combustion, constant pressure gas turbine received air from the atmosphere at 15 °C and compresses it to 9.5 times the intake pressure in a rotary compressor whose isentropic efficiency is 80 per cent. The compressed air then passes through a heat exchanger, whose effectiveness is 60 per cent, in which there is heat transfer from the turbine exhaust to the compressed air. From the heat exchanger the air passes through a combustion chamber at constant pressure in which its temperature is raised to 1015 °C. From the combustion chamber the air expands through a gas turbine to atmospheric pressure and at this pressure passes through the

heat exchanger to exhaust. The isentropic efficiency of the turbine is 82 per cent. The flow of air through the plant is at the rate of 2.5 kg/s. Neglect the mass of fuel and take $\gamma = 1.4$ and $c_p = 1.006$ kJ/kg K, throughout. Determine

(a) the net power output of the turbine if the turbine is coupled to the compressor
(b) the temperature of the air leaving the heat exchanger to enter the combustion chamber
(c) the thermal efficiency of the plant
(d) the work ratio
(e) the Carnot efficiency within the cycle temperature limits

[(a) 412 kW; (b) 452 °C; (c) 29%; (d) 0.34; (e) 78%]

4. An open-circuit, continuous combustion, constant pressure gas turbine takes in air at 20 °C and compresses it to 11 times the intake pressure in a rotary compressor whose isentropic efficiency is 82 per cent. A heat exchanger then transfers 70 per cent of the heat available for this purpose from the turbine exhaust to the compressed air. The compressed air then passes through a combustion chamber at constant pressure in which its temperature is raised to 1100 °C. The air is then expanded through a gas turbine, whose isentropic efficiency is 85 per cent. The pressure ratio through the turbine is 6:1. From the turbine the air passes through the heat exchanger to exhaust. The turbine is coupled to the compressor. Neglect the mass of fuel and take:

$\gamma = 1.39$, $c_p = 1.006$ kJ/kg K up to entry to the turbine
$\gamma = 1.34$, $c_p = 1.01$ kJ/kg K through the turbine and heat exchanger to exhaust

Determine

(a) the thermal efficiency of the arrangement
(b) the percentage increase in fuel required if the heat exchanger were removed and all other conditions remained the same.

[(a) 32%; (b) 24%]

5. A gas turbine plant has a maximum work pressure ratio of 13:1. The inlet temperature is 8 °C. Compression in the compressor and expansion in the turbine are isentropic. For the plant, determine

(a) the maximum temperature
(b) the net specific work output
(c) the thermal efficiency
(d) the work ratio
(e) the Carnot efficiency within the cycle temperature limits

Take $\gamma = 1.405$ and $c_p = 1.005$ kJ/kg K.

[(a) 957 °C; (b) 345 kJ; (c) 53%; (d) 0.53; (e) 77%]

6. A CHP plant consists of a gas turbine and a steam turbine working in tandem to produce a combined output. The exhaust from the gas turbine feeds a boiler which raises steam for use in the steam turbine. The theoretical overall thermal efficiency of the plant is 45 per cent. Steam flow from the boiler is 220 tonne/h. The steam feed to the steam turbine is at a pressure of 8 MN/m^2 and at a temperature of 400 °F. Feedwater temperature to the boiler is at a temperature of 143.6 °C. Expansion of the steam in the steam turbine has as isentropic efficiency of 0.88 and the steam exhaust pressure is 0.3 MN/m^2. The output from the gas turbine is 125 MW. The fuel used is natural gas with a calorific value of 38 MJ/m^3. Determine

(a) the theoretical power output of the steam turbine in MW
(b) the theoretical fuel gas requirement of the gas turbine in m^3/h

[(a) 36.06 MW; (b) 33 907 m^3/h]

7. An aircraft is flying at an altitude of 3000 m. Atmospheric temperature and pressure are −4.8 °C and 697 mbar, respectively. The aircraft is powered by a turboprop gas turbine. The engine compressor has a pressure ratio of 12:1. After compression the air temperature is raised to 1100 °C in the combustion chamber. The gas turbine arrangement is in two parts: the first follows the combustion chamber and runs the air compressor, the second runs the propeller unit. The pressure drop in the second part of the gas turbine is 85 per cent of that potentially available. The isentropic efficiency of the compressor and turbine units can each be taken as 0.84. Determine
   (a) the temperature of the air after compression
   (b) the temperature and pressure of the air after expansion in the first part of the turbine
   (c) the temperature and pressure of the air after expansion in the second part of the turbine
   (d) the power output of the second part of the turbine for an air mass flow rate of 20 kg/s

   Take $c_p = 1.02$ kJ/kg K, $\gamma = 1.405$; neglect the fuel mass.

   [(a) 329.5 °C; (b) 765.7 °C, 2.55 bar; (c) 503.2 °C, 1.575 bar; (d) 4041 kW]

Chapter 17

Engine and plant trials

17.1 General introduction

When choosing an engine, or any engineering plant for that matter, it is necessary to refer to the relevant performance characteristics. Engine performance characteristics will have been determined during a series of trials then tabulated and, where possible, graphed. Some of the general performance characteristics and their method of determination are now discussed.

17.2 Torque

An engine is required to drive external equipment, so it is important to know how much torque the engine will deliver at the various running conditions. The torque, which is usually determined in newton-metres (N m), is measured by coupling a measuring device, called a **brake** or **dynamometer**, to the engine output shaft. Four of the most common types of brake are described in the following sections. The rope and Prony brakes are now rarely used; they are included only to illustrate basic principles.

17.3 The rope brake

A typical rope brake arrangement is illustrated in Fig. 17.1. It consists of a rope (sometimes a leather or webbing band) which is wrapped around the flywheel. To one end of the rope is attached a spring balance which is attached to the ground. To the other end of the rope is attached a mass carrier. This mass carrier is allowed to hang freely, although its bottom end is usually connected to the ground by a loosely hanging chain. This is a safety chain in case the rope at any time tends to get caught up on the flywheel, in which case the chain will hold back the mass. Note that the spring balance is always on the side of the flywheel such that, when rotating, the flywheel turns toward the balance. The mass carrier is always on the other side of the flywheel such that the flywheel tends to lift the carrier.

For a heavy load on the engine, a large mass is placed on the mass carrier. For light loads, some mass can be removed from the mass carrier. For no load, the rope brake is usually removed from the flywheel.

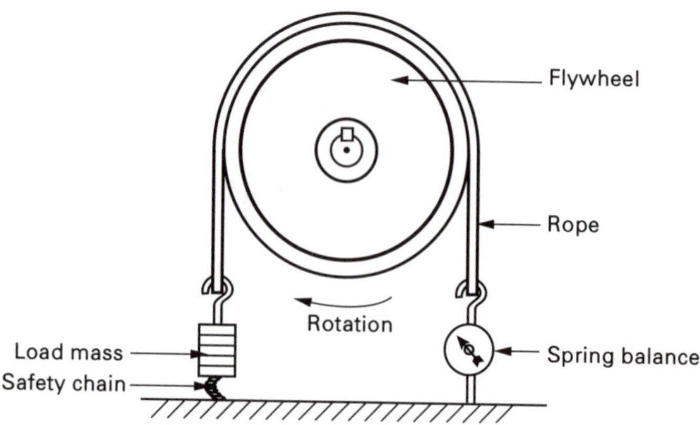

Fig. 17.1 Rope brake

Due to the friction between the rope and flywheel rim, and especially at the higher loads, the flywheel rim will get hot. In small installations this may not be serious, but in larger installations a water trough is built on the flywheel. Water is poured into the trough and centrifugal action causes it to form a thin layer on the inside of the flywheel rim. The friction heat transfer evaporates the water; this has a cooling effect which keeps down the temperature of the flywheel. This brake is normally used for relatively slow speed engines.

Now, let D = diameter of flywheel, m

d = diameter of rope, m

M = load mass, kg

m_s = spring balance reading, kg

then the net tangential force (in N) that the engine is working against at the flywheel rim is given by

$$g(M - m_s) = 9.81(M - m_s)$$

This is assuming the standard gravitational acceleration is 9.81 m/s^2.

This force occurs at a radius $(D + d)/2$, hence the torque (in N m) developed by the engine, T, is given by

$$T = \text{Tangential force} \times \text{Radius at which tangential force acts}$$

$$= 9.81(M - m_s)\left(\frac{D + d}{2}\right)$$

Sometimes the diameter of the rope is very small compared with the diameter of the flywheel, then the diameter of the rope can be neglected and the torque (in N m) is given by

$$T = 9.81(M - m_s)\frac{D}{2}$$

17.4 *The Prony brake*

The Prony brake is illustrated in Fig. 17.2. It consists of brake shoes which are clamped on to the flywheel rim by means of tie-bolts. Rim pressure can be adjusted using nuts, operating through compression springs, on the tie-bars. A load-bar extends from the top of the brake and to its end is attached a mass hanger. A mass is hung on to this mass hanger and the load arm is kept horizontal.

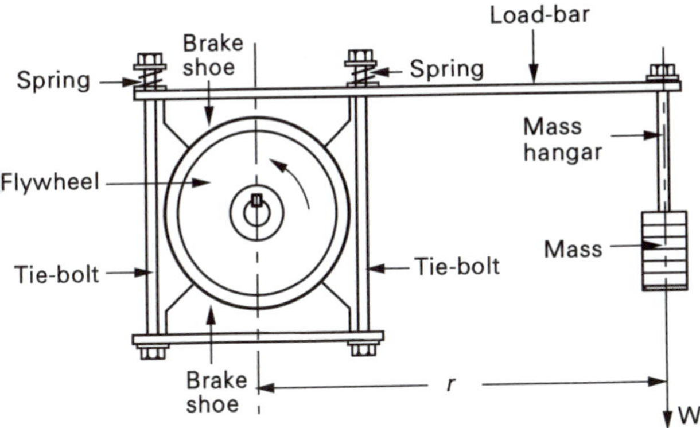

Fig. 17.2 Prony brake

As in the case of the rope brake, the flywheel is sometimes cooled by fitting a water trough to the flywheel rim.

If M = mass on hanger, kg
r = distance from centre of flywheel to hanger, m
T = torque developed by engine, N m

then

$$T = 9.81\ Mr$$

The Prony brake is usually used for relatively low-speed engines.

17.5 *The hydraulic dynamometer*

There are many designs of hydraulic dynamometers. The basic elements are illustrated in Fig. 17.3. They work on what is known as the torque reaction principle.

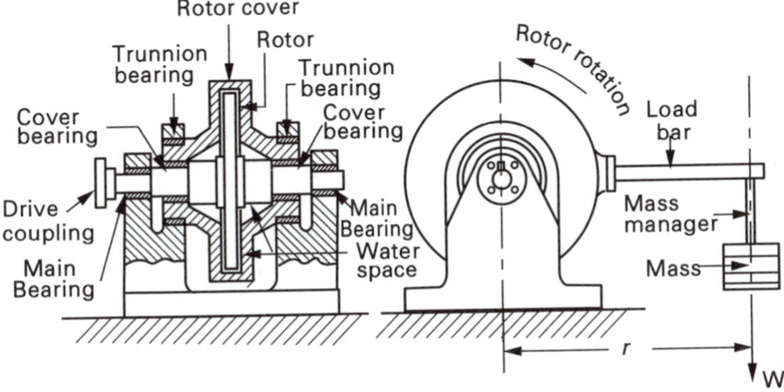

Fig. 17.3 Hydraulic dynamometer

A rotor is mounted in a rotor cover which is suitably hydraulically sealed by bearings and glands. The shaft of the rotor passes out from the cover bearings and is mounted in the main bearings which are fixed on the dynamometer bed. The rotor cover is not fixed, but is free to rotate in trunnion bearings which are again mounted on the dynamometer bed.

From the side of the rotor cover extends a load-bar on to the end of which is attached a mass hanger. The principle of operation is as follows. The rotor sometimes has holes projecting blades and sometimes cups. The rotor cover may have matching blades or cups. The space between the rotor and the cover is either filled or part-filled with water, according to the design.

If the rotor is now rotated, a reaction will be set up with the water and the cover; the cover will tend to rotate in the trunnion bearings with the rotor. The cover rotation is prevented by hanging suitable masses on the mass hanger. Thus a torque reaction is set up and this torque, which is the torque generated by the engine, can be measured.

If M = mass on hanger, kg
r = distance from centre of dynamometer to hanger, m
T = torque developed by engine, N m

then

$$T = 9.81\,Mr$$

The amount of torque absorbed by the dynamometer can be varied by control of the water. In some dynamometers, the water depth is varied: the greater the depth, the greater the load. In others, plates are introduced between the rotor and the cover. The plates are controlled from the outside. If more plate is introduced, there is less interference between the rotor and the cover, so less load is absorbed by the dynamometer. If the plates are withdrawn, more interference will occur, so more load will be absorbed.

In some hydraulic dynamometers there is a fixed quantity of water in the cover, but in others the cover is usually filled and means are provided for the free flow of water through the cover. This dissipates friction energy generated between the rotor and the water.

17.6 The electrical dynamometer

This dynamometer is, in many ways, similar to the hydraulic dynamometer. It is of the torque reaction type, but the reaction is not hydraulic; it is magnetic, between the rotating armature of a generator and the magnetic field set up by the field coils mounted in the outer casing of the generator.

The outer casing of the generator is not fixed as is usually the case; like the cover of the hydraulic dynamometer, it is mounted in trunnion bearings. As the armature of the generator is rotated by the engine, its reaction with the magnetic field tends to pull the field coils and casing round with it. This rotation is prevented in the same way as in the hydraulic dynamometer – a load-bar with mass hanger fitted to the outside of the casing. Torque is measured as the product of the force required to keep the load-bar horizontal and the distance from the hanger to the centre of the generator, so

$$T = 9.81\ Mr$$

where M = mass on hanger, kg
r = distance from centre of dynamometer to hanger, m
T = torque developed by engine, N m

The electrical power generated by this type of dynamometer is usually dissipated through banks of electrical resistors. The load can be increased or decreased on the dynamometer, and hence the engine under test, by switching the resistors in or out.

Due to the fact that the field coils are free to rotate with the casing in the trunnion bearings, the dynamometer is often called a **swinging field dynamometer**. In both the hydraulic and electrical dynamometers the load hanger could be replaced by a calibrated spring-loaded device.

17.7 *Brake power (b.p.)*

Using a brake or dynamometer it is possible to determine the useful work output of an engine

Now

$$\text{Work done} = \text{force} \times \text{distance}$$

If

F = tangential force on the brake in newtons, N
r = radius at which the force acts in metres, m

then during one revolution of the brake

$$\begin{aligned}\text{Work done} &= F \times 2\pi r \\ &= 2\pi Fr \\ &= 2\pi T\ \text{N m} = 2\pi T\ \text{J} \qquad (1\ \text{J} = 1\ \text{N m})\end{aligned}$$

If the engine is running at N rev/s, then

$$\begin{aligned}\text{Work done/s on the brake} &= 2\pi NT\ \text{N m/s} \\ &= 2\pi NT\ \text{watts (W)} \\ &= \text{brake power (b.p.)}\end{aligned}$$

Any one of the brakes or dynamometers described can be used for determining the torque T, so the brake power (b.p.) can be determined.

It is possible to estimate the brake power from an engine coupled to an electrical dynamometer from a knowledge of the electrical output of the dynamometer. Remember that the electrical dynamometer is really a generator.

Now, if

Generated voltage = V volts
Current output = I amps

then

$$\text{Power output} = VI \text{ watts (W)}$$

Neglecting electrical losses, this will generally be very close to the brake power of the engine.

17.8 Indicated power (i.p.)

The indicated power (i.p.) of an engine is the power actually developed in the cylinders. The brake power will, in fact, be less than the indicated power because losses occur from cylinders to shaft, such as friction and running auxiliary equipment (e.g. fuel and oil pumps).

In order to determine the indicated power, it is necessary to know the work conditions in the cylinders. The area of a pressure–volume diagram represents work. If the engine cycle can be obtained in the form of a pressure–volume diagram, direct from the cylinders, it would therefore be possible to obtain the work done by the engine as indicated by the cylinder operations. And knowing the number of cycles per second, it will be possible to determine the indicated power output from the cylinders.

A scaled-down pressure–volume diagram of an engine cycle can be obtained while the engine is running by using an engine indicator. The engine indicator can appear in various forms: mechanical, part mechanical and part electrical, and electronic. Most modern engine indicators are electronic.

17.9 The engine indicator

Most modern engine indicators are electronic. Suitable transducers and sensors, which monitor such parameters as cylinder pressure, temperature and crank angle, are strategically placed on or in the engine. Signals from the transducers and sensors are transmitted to an electronic processing unit which displays performance information on a visual display unit (VDU). Information can be verbal or by calibrated diagram; it can also be in colour. Verbal information can be cylinder pressure and engine rotational speed, calibrated diagrams can be pressure–volume (Figs 16.4(d) and 16.5(e)) or pressure–crank angle (Fig. 16.24). The electronic engine indicator can be used for continuous operational performance monitoring.

17.10 Indicated mean effective pressure (IMEP)

The indicated mean effective pressure (IMEP) is that constant pressure which, if it acted over the full length of the stroke, would produce the same amount of work done on the piston as is actually obtained during a complete engine cycle (see also section 15.1).

Figure 17.4 is a P–V diagram of a four-stroke cycle internal combustion engine. Two separate areas are shown. The positive diagram area is the area which produces work output. The negative diagram area is the area which requires work input. The net area gives the theoretical positive cycle work output. In general the negative area is small compared with the positive area, so it is frequently neglected. This is the case with the four-stroke cycle engine.

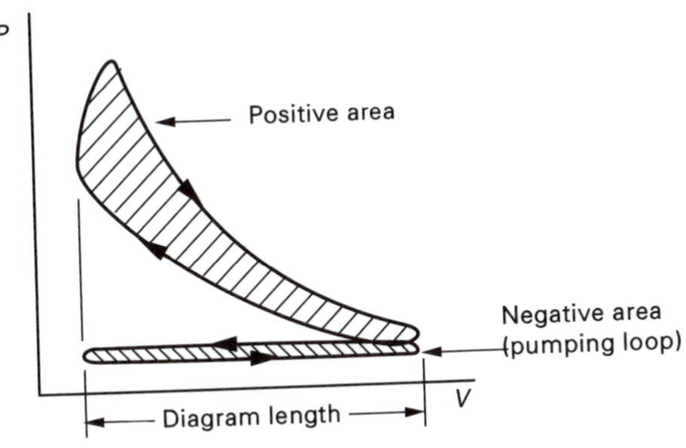

Fig. 17.4 Determination of IMEP on a *P–V* diagram

In the case of the two-stroke cycle engine, only the positive diagram area is present.

Now, let

Diagram area = A, mm^2
Diagram length = l, mm
Indicator calibration = C, N/m^2 per millimetre mean diagram height

then IMEP (in N/m^2) is given by

$$\text{IMEP} = \frac{AC}{l} \qquad [1]$$

17.11 *Calculation of indicated power (i.p.)*

From a knowledge of the IMEP of an engine, the i.p. can now be calculated.

Let P_M = IMEP, N/m^2
A = area of piston, m^2
L = length of stroke, m
N = number of working strokes per second

Then

Force on piston = ($P_M \times A$) newtons (N)

Work done/working stroke = force × length of stroke
= ($P_M \times A \times L$) N m
= ($P_M \times A \times L$) joules (J)

Work done/s = work done/stroke × number of working strokes/s
= ($P_M \times A \times L \times N$) J/s
= ($P_M LAN$) J/s
= ($P_M LAN$) watts (W) = i.p.

Consider the number of working strokes in three different cases:

- For the four-stroke cycle engine

$$N = \frac{\text{rev/s}}{2}$$

because there is only one working stroke in two revolutions.

- For the two-stroke cycle engine

$$N = \text{rev/s}$$

because there is one working stroke in each revolution.

- For the double-acting steam engine or double-acting two-stroke cycle oil engine

$$N = 2 \times \text{rev/s}$$

because there are two working strokes in each revolution (assumes both ends have same IMEP; if they have different IMEP, each end must be treated separately).

17.12 Friction power (f.p.)

The i.p. is greater than the b.p. This is because there are losses from cylinder to shaft, mainly as a result of friction in the various reciprocating and rotating parts. In the four-stroke cycle, internal combustion engine there are also some pumping losses during exhaust and induction. Also, there will be some loss to brake output as a result of driving auxiliaries such as the oil pump and fuel pump. But a large part of the loss is due to friction.

The difference between the i.p. and the b.p. is sometimes cumulatively called the **friction power** (f.p.).

Hence

$$\text{i.p.} - \text{b.p.} = \text{f.p.} \qquad [1]$$

from which

$$\text{i.p.} - \text{f.p.} = \text{b.p.} \qquad [2]$$

17.13 Indicated power by Morse test

In the absence of an engine indicator, a close estimate of the i.p. of a multicylinder internal combustion engine can be determined from the Morse test. The method can only be used on engines with more than one cylinder; this will become clear during its description.

The engine under test is coupled to a suitable brake and, at the test condition, the b.p. output, b, is determined. The first cylinder is now cut out. This can be achieved in a petrol engine by shorting out the spark-plug of the first cylinder; for an oil engine the fuel supply to the first cylinder is interrupted.

As a result of cutting out the first cylinder, the engine speed will drop. Load is now removed from the brake in order to restore the original speed. The b.p. under this new condition is now determined from the new brake load. The first cylinder operation is now restored and the second cylinder is now cut out. The engine speed is corrected, if necessary, to its original value and the b.p. with this second cylinder cut out is now determined. This procedure is adopted for each cylinder in turn.

Take the case of a four-cylinder engine.

Let i_1, i_2, i_3, i_4 = i.p. of each individual cylinder
f_1, f_2, f_3, f_4 = f.p. of each individual cylinder

Now, in the case when all cylinders are working, i.e. normal operation

$$\text{Total b.p.} = b = (i_1 - f_1) + (i_2 - f_2) + (i_3 - f_3) + (i_4 - f_4)$$
$$= (i_1 + i_2 + i_3 + i_4) - (f_1 + f_2 + f_3 + f_4) \qquad [1]$$

When the first cylinder is cut out, i_1 is cut out but the friction and other losses of this first cylinder, f_1, remain.

Thus the b.p. with the first cylinder cut out is

$$b_1 = (i_2 + i_3 + i_4) - (f_1 + f_2 + f_3 + f_4) \qquad [2]$$

Subtracting equation [2] from equation [1]

$$b - b_1 = i_1 \qquad [3]$$

Thus the i.p. of the first cylinder is determined.

In a similar way, when the second cylinder is cut out

$$b - b_2 = i_2$$

This is repeated for each cylinder in turn.

Finally, the total i.p. is given by

$$\text{Total i.p.} = i_1 + i_2 + i_3 + i_4 \qquad [4]$$

The i.p. so determined is not quite accurate because the friction and pumping losses when a cylinder is working are not the same as when it is not working. But the method still gives a close approximation.

17.14 Mechanical efficiency

The mechanical efficiency of an engine is defined as the ratio of the b.p. to the i.p. or

$$\text{Mechanical efficiency} = \frac{\text{b.p.}}{\text{i.p.}}$$

The closer this value approaches unity the better, for this implies smaller losses.

It is common to express the mechanical efficiency as a percentage by multiplying the ratio b.p./i.p. by 100. Normal values are in the region 70–80 per cent, but sometimes values lie outside this range.

Mechanical efficiency is often abbreviated to mech η or η_{mech}.

17.15 Brake mean effective pressure (BMEP)

Sometimes the brake mean effective pressure (BMEP) is quoted for an engine. This is obtained by multiplying the IMEP by the mech η, or

$$\text{BMEP} = P_{\text{Mb}} = \text{IMEP} \times \text{mech } \eta$$
$$= P_{\text{M}} \times \text{mech } \eta \qquad [1]$$

The BMEP gives that amount of the IMEP which has been effective in producing output at the brake.

If the BMEP is known, then the b.p. can be determined by using an equation of the same form as that used in the case of i.p.

$$\text{b.p.} = (P_{\text{Mb}} LAN) \text{ watts (W)} \qquad [2]$$

17.16 Fuel consumption

A knowledge of the fuel consumed by an engine and the time it takes to consume this fuel is essential when assessing the qualities of the engine. In the case of a steam engine or turbine, it will be the steam consumed. Consumption is usually measured in kg/h (or kg/s).

In the case of an engine using gas, the fuel used will be measured in m^3/h (or m^3/s). For petrol and oil engines, the fuel is run through a special measuring device. This may be a reservoir of fuel of known quantity, and the time for the engine to consume this measured quantity of fuel is taken. Alternatively, the fuel may flow through a special flowmeter which is calibrated to give the fuel consumption by direct reading, usually kg/h (or kg/s).

For an engine using gas, the gas first passes through a gas meter before entering the engine. A measured quantity of gas is timed in this case. For the steam engine or turbine the exhaust steam is condensed in a condenser and is measured in tanks at the outlet of the condenser. Alternatively, the steam as condensate is passed through a direct reading flowmeter.

An important derivative from the fuel consumption or steam consumption is called the indicated or brake specific fuel or steam consumption. This is obtained by dividing the consumption by the i.p. or b.p. as appropriate (see also section 15.1).

Thus, in the case of the petrol or oil engine

$$\text{Indicated specific fuel consumption} = \frac{\text{Fuel used in kg/h}}{\text{i.p.}} \qquad [1]$$

This will give the fuel (in kg) required to deliver one watt for one hour at the load considered; its units are kg/Wh (or kg/kWh).

$$\text{Brake specific fuel consumption} = \frac{\text{Fuel used in kg/h}}{\text{b.p.}} \qquad [2]$$

This will give the fuel (in kg) required to deliver one watt for one hour at the load considered; its units are kg/Wh (or kg/kWh).

For an engine using gas

$$\text{Indicated specific fuel consumption} = \frac{\text{Gas used in m}^3\text{/h}}{\text{i.p.}}\ \text{m}^3\text{/Wh (or m}^3\text{/kWh)} \qquad [3]$$

Similarly

$$\text{Brake specific fuel consumption} = \frac{\text{Gas used in m}^3\text{/h}}{\text{b.p.}}\ \text{m}^3\text{/Wh (or m}^3\text{/kWh)} \qquad [4]$$

These give the fuel required (in m^3) to deliver one watt (or one kilowatt) for one hour, at the load considered.

In the case of the steam engine or turbine

$$\text{Indicated specific steam consumption} = \frac{\text{Steam used in kg/h}}{\text{i.p.}}\ \text{kg/Wh (or kg/kWh)} \qquad [5]$$

$$\text{Brake specific steam consumption} = \frac{\text{Steam used in kg/h}}{\text{b.p.}}\ \text{kg/Wh (or kg/kWh)} \qquad [6]$$

These give the steam required (in kg) to deliver one watt (or one kilowatt) for one hour at the load considered.

The importance of the specific fuel consumption is that it is possible to compare the amount of fuel or steam required per hour, to deliver one watt (or one kilowatt) at a particular load with the amount of fuel required per hour to deliver one watt (or one kilowatt) at another load. In each case it is one watt (or one kilowatt), so if at one load less fuel is required than at the other load, the engine is running more efficiently at the lower specific fuel condition.

If the time interval is measured in seconds, then

$$1\ \text{kg/Ws} = 1\ \text{kg} \Big/ \frac{\text{J}}{\text{s}} \cdot \text{s} = 1\ \text{kg/J} \qquad [7]$$

17.17 Thermal efficiency

The thermal efficiency of an engine is important because it determines how efficiently the fuel is being used in the engine. There are two cases to consider: the **indicated thermal efficiency** and the **brake thermal efficiency.** Consider first the internal combustion engine; its thermal efficiencies are

$$\text{Indicated thermal efficiency} = \frac{\text{Energy equivalent of the i.p./s}}{\text{Energy supplied by fuel/s}} \qquad [1]$$

$$= \frac{\text{i.p.}}{\dot{m} \times CV} \qquad [2]$$

where $\dot{m}$ = mass of fuel per second (volume for gases)

CV = calorific value of fuel

$$\text{Brake thermal efficiency} = \frac{\text{Energy equivalent of the b.p./s}}{\text{Energy supplied by fuel/s}} \qquad [3]$$

$$= \frac{\text{b.p.}}{\dot{m} \times CV} \qquad [4]$$

In the case of the steam engine or turbine the efficiencies are

$$\text{Indicated thermal efficiency} = \frac{\text{Energy equivalent of the i.p./s}}{\text{Useful energy received/s}} \qquad [5]$$

Now the engine or turbine receives steam with specific enthalpy h. From a theoretical standpoint the steam in the engine could be effectively utilised until it finally became water at the exhaust pressure. In this condition, it would be of no further use to the engine or turbine and would now have liquid enthalpy h_f. Hence the useful energy which could theoretically be extracted per kilogram from the steam in the engine would be $(h - h_f)$. If the steam engine uses $\dot{m}$ kg steam per second, then

$$\text{Useful energy received/s} = \dot{m}(h - h_f) \qquad [6]$$

Hence

$$\text{Indicated thermal efficiency} = \frac{\text{i.p.}}{\dot{m}(h - h_f)} \qquad [7]$$

In a similar way

$$\text{Brake thermal efficiency} = \frac{\text{b.p.}}{\dot{m}(h - h_f)} \qquad [8]$$

17.18 Relative efficiency

The **relative efficiency**, or **efficiency ratio** as it is sometimes called, is the ratio of the actual efficiency obtained from an engine or plant to the theoretical efficiency of the engine or plant.

Hence

$$\text{Relative efficiency} = \frac{\text{Actual efficiency}}{\text{Theoretical efficiency}} \qquad [1]$$

In the case of the steam engine, the theoretical efficiency is obtained from the theoretical indicator diagram. In the case of steam plant the theoretical efficiency is taken as the Rankine efficiency. In the case of internal combustion engines the theoretical efficiency is taken from the theoretical cycle. But sometimes the theoretical efficiency is taken from the constant volume cycle $= 1 - 1/r_v^{(\gamma - 1)}$. This cycle has the highest efficiency of the practical internal combustion cycles, so that is why it is proposed as the basis of comparison for internal combustion engine cycles. Air is the gas theoretically used, consequently, the cycle is sometimes callėd the **air standard cycle**. An average value of γ for air is 1.4, therefore

$$\text{Air standard efficiency} = 1 - \frac{1}{r_v^{(1.4-1)}} \qquad [2]$$

$$= 1 - \frac{1}{r_v^{0.4}} \qquad [3]$$

17.19 The energy balance or energy audit

In any thermal engine or plant, a quantity of energy is supplied from the fuel in a given time. It is the purposes of the energy balance, or energy audit, to trace the distribution of this energy through the engine or plant. The energy balance is often taken over a period of time and the various energy quantities are often expressed as percentages of the energy supplied from the fuel. The energy balance is often plotted on a graph or displayed on a Sankey diagram.

Consider the piston internal combustion engine.

The energy supplied from the fuel (in J/s) is given by

$$\text{Energy supplied} = \dot{m} \times CV \qquad [1]$$

where, $\dot{m}$ = mass of fuel, kg/s

CV = calorific value of fuel, J/kg

The general distribution of this energy through the engine will be

Energy to b.p. = b.p. (W) = b.p. (J/s) [2]

Energy to coolant = [mass of cooling water (kg/s) × specific heat capacity of water (J/kg K) × temp. rise of cooling water (K)] J/s [3]

Energy to exhaust = [mass of exhaust (kg/s) × specific heat capacity of exhaust (J/kg K) × temperature difference between exhaust and atmosphere (K)] J/s [4]

Energy lost to surrounding air

This is not directly measured but is obtained by difference, in other words

Energy lost to surrounding air = energy from fuel – (energy to b.p. + energy to coolant + energy to exhaust)

The friction losses should not be treated separately in the balance (obtained from the equation f.p. = i.p. – b.p.). Any energy to friction will be dissipated in the coolant, exhaust, radiation, etc. Since each of these quantities has been separately measured and friction energy is a hidden part of these measurements, its separate treatment would cause it to be included twice. It should therefore be omitted from the energy balance.

Also i.p. = b.p. + f.p.: so b.p. and the energy to friction (f.p.) have been included in the balance, so energy to i.p. has been included in the balance. The energy to i.p. should not be treated as a separate item.

There will be an energy balance for each load condition of the engine. Results are usually tabulated appropriately or graphed as shown in Fig. 17.5. Here it is assumed that an energy balance has been drawn up for no load and for six other load conditions. The energies at each load condition are plotted, one on top of the other, each expressed as a percentage of the energy supplied. At each condition, therefore, they will total 100 per cent. Smooth curves are drawn through the plotted points. The graph will then clearly illustrate the distribution of energy through the engine at all load conditions. The energy to b.p. varies from zero at no load to a maximum at full load. But as a percentage of the energy supplied, it is very small; its maximum is somewhere in the region of 20–40 per cent. The energy to b.p. expressed as a

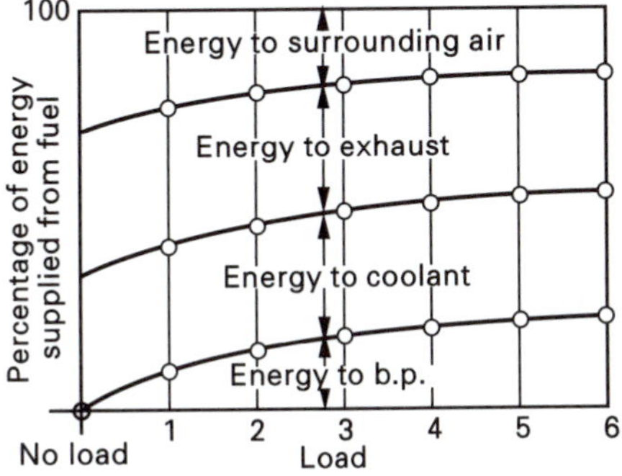

Fig. 17.5 Energy balance graph

percentage of the energy supplied is, of course, the brake thermal efficiency; it is a very low figure because there is considerable energy loss. It is unfortunately true that most of the energy supplied to engine plant is lost. Much effort is made to develop equipment to reclaim some of this waste energy. More details are given in the sections on boilers, the gas turbine and the Stirling and Ericsson cycles.

A further method of representing an energy balance is by means of the Sankey diagram. This is a stream or flow diagram in which the width of the stream represents the energy quantity being considered, usually as a percentage of the energy supplied. A Sankey diagram for the piston internal combustion engine is shown in Fig. 17.6. It starts at the bottom with a stream whose width represents the energy input from the fuel, marked 100 per cent. Moving up the diagram, the coolant loss stream is led off to the left. The width of this stream represents the percentage loss to the coolant. Next the exhaust loss stream is led off to the left and finally the surroundings loss is led off to the left. The loss streams finally meet as a single loss stream, as shown. Of the original vertical stream, only the b.p. output stream remains at the top of the diagram. The figures on the diagram are percentages of the original energy supplied in the fuel.

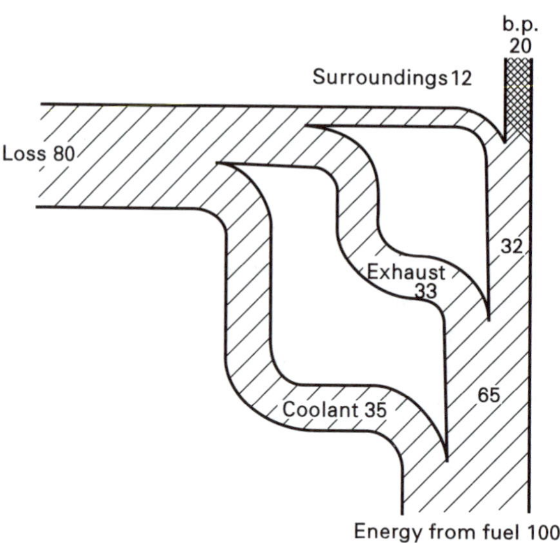

Fig. 17.6 IC engine: Sankey diagram

The energy to exhaust is sometimes measured direct by the installation of an exhaust gas calorimeter. This consists of a heat exchanger unit in which the exhaust gas is cooled by cooling water passing through it. In this case

Energy (enthalpy) lost by exhaust gas = Heat transferred to cooling water

Hence, if $\dot{m}$ = mass of cooling water, kg/s
t_1 = inlet temperature of cooling water, °C
t_2 = outlet temperature of cooling water, °C
c_p = specific heat capacity of water, J/kg

then

$$\text{Energy to exhaust} = \dot{m}c_p(t_2 - t_1) \text{ J/s} \quad [5]$$

If the exhaust temperature is dropped to atmospheric temperature through the calorimeter, this will give the total energy to exhaust. If this is not the case and only part of the temperature drop is obtained, the total energy to exhaust is estimated by proportion.

Thus, if t_c = exhaust temperature, °C
t_a = atmospheric temperature, °C
t_3 = exhaust temperature leaving the calorimeter, °C

then

$$\text{Total energy to exhaust} = \dot{m}c_p(t_2 - t_1) \times \frac{(t_c - t_a)}{(t_a - t_3)} \text{ J/s} \quad [6]$$

The steam engine and the steam turbine can also be analysed for energy balance. The energy supplied to the engine or turbine is obtained from the useful enthalpy content of the steam being admitted (see section 17.17). The energy will be divided up into energy to exhaust, surroundings and b.p. The main bulk of the energy will appear in the exhaust steam. A Sankey diagram for the steam engine or turbine is shown in Fig. 17.7. The percentages shown are generally representative.

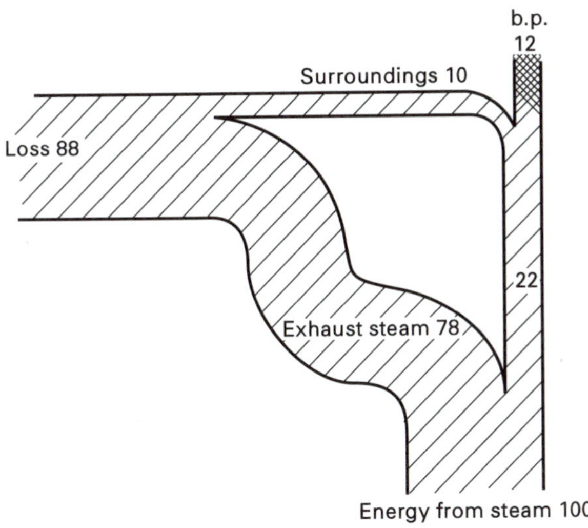

Fig. 17.7 Steam engine or turbine: Sankey diagram

Figure 17.8 is a Sankey diagram for the energy balance for a steam plant. Both boiler and turbine are included in the diagram. Losses are again fed off as streams to the left, and the b.p. stream appears at the top. Representative percentage figures appear on the diagram.

Figure 17.9 is a Sankey diagram for a basic CHP plant. Energy from a gas turbine exhaust supplies a steam turbine plant. Outputs from the gas and steam turbines are shown combined as are the combined losses of the plant as a whole. The percentages are representative.

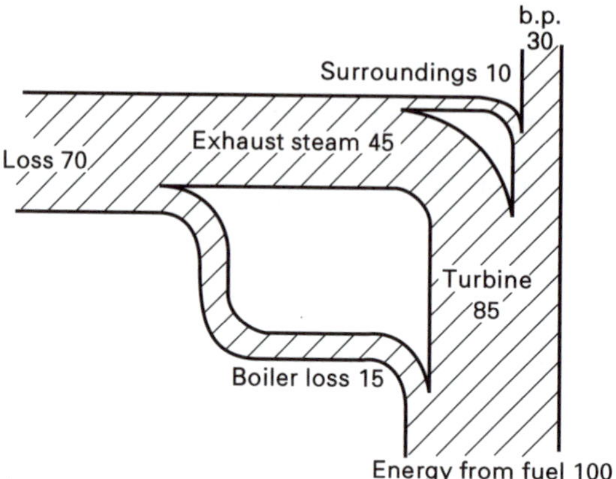

Fig. 17.8 Steam plant: Sankey diagram

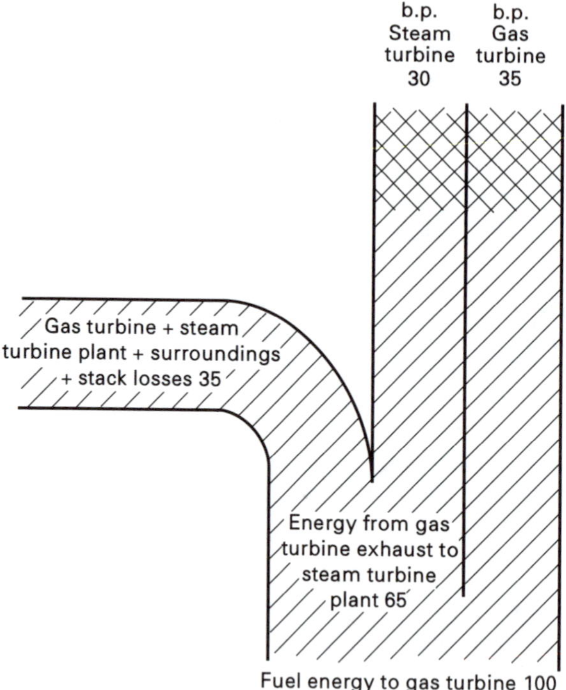

Fig. 17.9 Basic CHP plant: Sankey diagram

17.20 Typical graph shapes

17.20.1 Torque

For a variable speed engine, the torque increases to a maximum then decreases as the speed increases. The general shape of the graph is shown in Fig. 17.10(a). It is generally a fairly shallow curve.

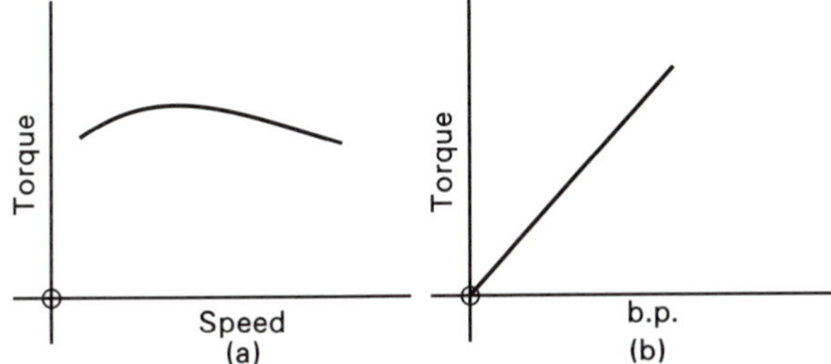

Fig. 17.10 Torque graphs (a) for a variable speed engine and (b) for a constant speed engine

For a constant speed engine the torque will increase in proportion to the b.p. and a torque–b.p. graph results in a straight line, as shown in Fig. 17.10(b).

17.20.2 b.p. and i.p.

For a variable speed engine, both the b.p. and the i.p. increase as the speed increases. The b.p. is lower than the i.p. and both graphs generally curve slightly toward the top, as shown in Fig. 17.11(a). The difference between any two corresponding speed values will give the f.p. loss.

For a constant speed engine, if the i.p. is plotted against b.p., as shown in Fig. 17.11(b), a straight or nearly straight line of increasing value will occur. The graph passes through the i.p. axis with a positive value when the b.p. is zero. This gives the no-load i.p.

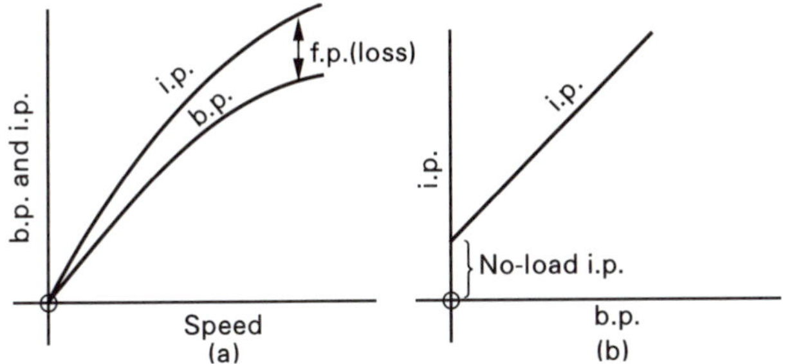

Fig. 17.11 Graphs of i.p. (a) for a variable speed engine and (b) for a constant speed engine

17.20.3 Mechanical efficiency

The mechanical efficiency graph appears with a similar shape for both the variable and constant speed engines. It increases from zero to a maximum then generally begins to fall off in a smooth curve, as shown in Fig. 17.12. There is a maximum mechanical efficiency condition for most engines. Beyond this condition, friction is beginning to increase rapidly and the engine is becoming overloaded.

For a variable speed engine, mechanical efficiency is plotted against speed. For a constant speed engine, mechanical efficiency is plotted against b.p. The efficiency curve for a steam engine, may not show a decline up to full load.

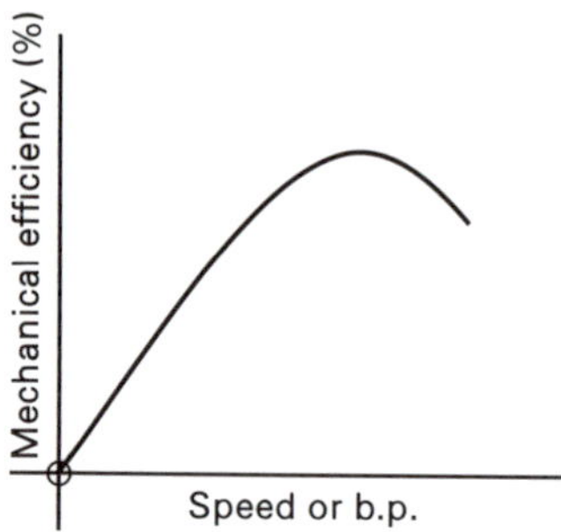

Fig. 17.12 Mechanical efficiency for variable speed and constant speed engines

17.20.4 Fuel and steam consumption

For a variable speed engine – usually an internal combustion engine – the fuel consumption is usually plotted in kg/h against speed. It results in a graph of increasing value, as shown in Fig. 17.13(a). It is usually a fairly straight line.

For a constant speed engine – an internal combustion engine, steam engine, or steam turbine – fuel or steam consumption in kg/h is plotted against b.p., as shown in Fig. 17.13(b). Once again, it results in a fairly straight graph of increasing value. It crosses the fuel or steam axis with a positive value at zero b.p. This is the no-load fuel or steam required to keep the engine motoring.

From the consumption graph, a cost graph can also be determined if the cost of the fuel is known. It will have a similar shape to the consumption graph.

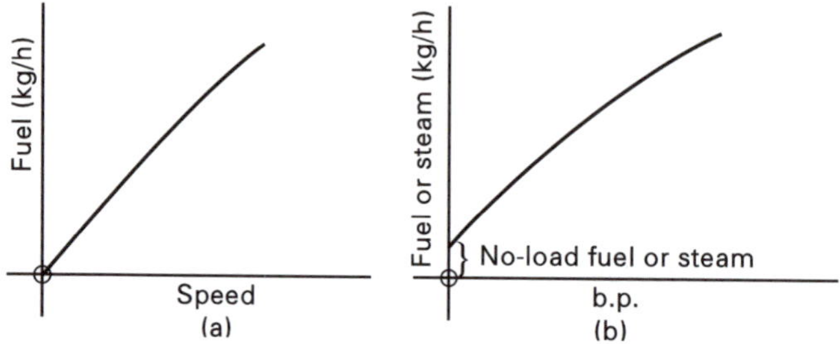

Fig. 17.13 Fuel consumption (a) for a variable speed engine and (b) for a constant speed engine

17.20.5 Specific fuel or steam consumption

For both variable and constant speed engines, the specific fuel or steam consumption graph, either for i.p. or b.p., results in a curve like Fig. 17.14. For a variable speed engine – usually an internal combustion engine – the specific fuel consumption (in kg/Wh) is plotted against speed. For the constant speed engine – an internal combustion engine or steam engine – the specific fuel consumption (in kg/Wh) is plotted against b.p.

The curve has a drooping characteristic, displaying a minimum value. This minimum value will show the operating speed or b.p. at which the engine is operating at its best and most economical condition. It will correspond with the maximum thermal efficiency condition.

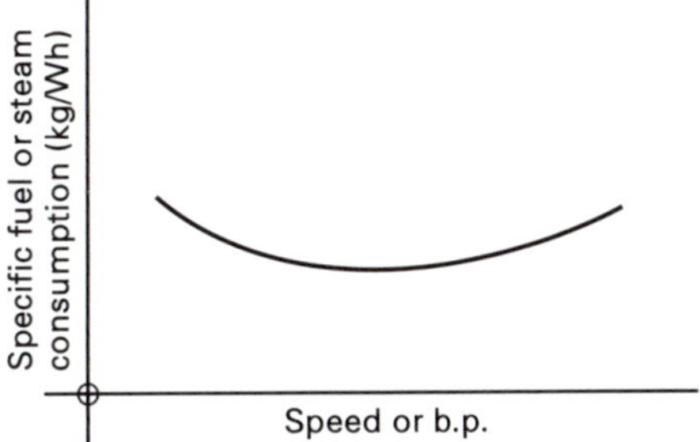

Fig. 17.14 Specific fuel or steam consumption

17.20.6 Thermal efficiency

Thermal efficiency, usually on b.p. basis, is plotted against speed for the variable speed engine or b.p. for the constant speed engine. The curve increases to a maximum value then decreases as shown in Fig. 17.15. It displays a maximum value of the thermal efficiency, the condition at which the engine is running at its best and with the maximum economy.

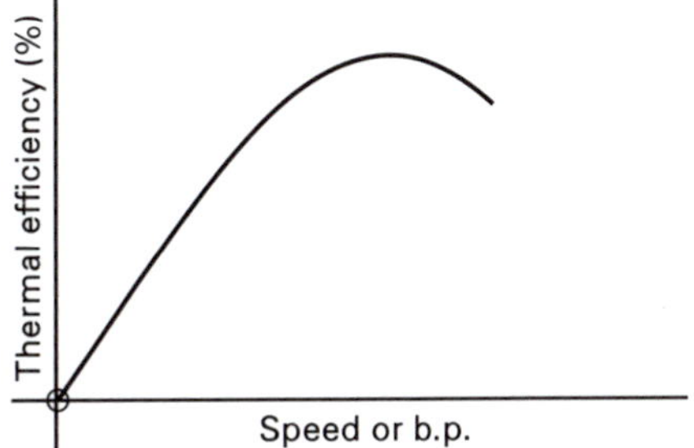

Fig. 17.15 Thermal efficiency

Example 17.1 *During a test on a four-stroke cycle oil engine the following data and results were obtained:*

Mean height of indicator diagram	21 mm
Indicator calibration	27 kN/m^2 per mm
Swept volume of cylinder	14 litres
Speed of engine	6.6 rev/s
Effective brake load	77 kg
Effective brake radius	0.7 m
Fuel consumption	0.002 kg/s
Calorific value of fuel	44 000 kJ/kg
Cooling water circulation	0.15 kg/s
Cooling water inlet temperature	38 °C
Cooling water outlet temperature	71 °C
Specific heat capacity of water	4.18 kJ/kg K
Energy to exhaust gases	33.6 kJ/s

Determine the indicated and brake outputs and the mechanical efficiency. Draw up an overall energy balance in kJ/s and as a percentage.

SOLUTION

Indicated mean effective pressure is

$$P_M = 27 \times 21 = \mathbf{567\ kN/m^2}$$

$$\text{Indicated power} = P_M LA \frac{N}{2}$$

$N/2$ in this case because the engine is four-stroke.

$$\begin{aligned}\text{Now } LA &= \text{swept volume of cylinder}\\ &= 14 \text{ litres}\\ &= 14 \times 10^{-3}\ \text{m}^3\end{aligned}$$

$$\therefore \quad \text{Indicated power} = 567 \times \frac{14}{10^3} \times \frac{6.6}{2} = \mathbf{26.2\ kW}$$

$$\begin{aligned}\text{Brake power} &= 2\pi NT\\ &= (2\pi \times 6.6 \times 9.81 \times 77 \times 0.7)\ \text{W}\\ &= (2\pi \times 6.6 \times 9.81 \times 77 \times 0.7 \times 10^{-3})\ \text{kW}\\ &= \mathbf{22\ kW}\end{aligned}$$

$$\therefore \quad \text{Mech } \eta = \frac{22}{26.2}$$

$$= 0.84$$

$$= \mathbf{84\%}$$

$$\text{Energy from fuel} = 44\,000 \times 0.002 = \mathbf{88\ kJ/s}$$

$$\text{Energy to brake power} = 22\ \text{kW} = \mathbf{22\ kJ/s}$$

$$\begin{aligned}\text{Energy to coolant} &= 0.15 \times 4.18 \times (71 - 38)\\ &= 0.15 \times 4.18 \times 33\\ &= \mathbf{20.7\ kJ/s}\end{aligned}$$

$$\text{Energy to exhaust (given)} = \mathbf{33.6\ kJ/s}$$

$$\begin{aligned}\text{Energy to surroundings, etc.} &= 88 - (22 + 20.7 + 33.6)\\ &= 88 - 76.3\\ &= \mathbf{11.7\ kJ/s}\end{aligned}$$

The energy balance can now be tabulated.

	kJ/s	Percentage
Energy from fuel	88	100
Energy to brake power	22	25
Energy to coolant	20.7	23.5
Energy to exhaust	33.6	38.2
Energy to surroundings, etc.	11.7	13.3

Example 17.2 *The following data were obtained from a test on a single-cylinder, double-acting steam engine fitted with a rope brake:*

Cylinder diameter	200 mm
Stroke	250 mm
Speed	5 rev/s
Effective diameter of brake wheel	0.75 m
Stop valve pressure	800 kN/m^2
Dryness fraction of steam supplied	0.97
Brake load	136 kg
Spring balance reading	90 N
Mean effective pressure	232 kN/m^2
Condenser pressure	10 kN/m^2
Steam consumption	3.36 kg/min
Condenser cooling water	113 kg/min
Temperature rise of condenser cooling water	11 K
Condensate temperature	38 °C

Determine

(a) b.p.

(b) i.p.

(c) mechanical efficiency

(d) indicated thermal efficiency

(e) brake specific steam consumption

(f) draw up a complete energy account for the test on a one-minute basis taking 0 °C as datum

Extract from steam tables

Pressure (kN/m^2)	Sat. temp. t_f (°C)	Specific enthalpy (kJ/kg)			Spec. vol. v_g (m^3/kg)
		h_f	h_{fg}	h_g	
800	170.4	720.9	2 046.5	2 767.5	0.240 3
10	45.8	191.8	2 392.9	2 584.8	14.67

Take the specific heat capacity of water 4.18 kJ/kg K and use the extract from steam tables.

(a)

$$\text{b.p.} = 2\pi NT$$

$$= 2\pi \times 5 \times [(9.81 \times 136) - 90] \times \frac{0.75}{2}$$

$$= 2\pi \times 5 \times (1334 - 90) \times \frac{0.75}{2}$$

$$= 2\pi \times 5 \times 1224 \times \frac{0.75}{2}$$

$$= 14\ 656\ W$$

$$= \mathbf{14.656\ kW}$$

(b)

$$\text{i.p.} = P_M LAN$$

$$= 232 \times \frac{250}{10^3} \times \frac{\pi}{4} \times \left(\frac{200}{10^3}\right)^2 \times 5 \times 2 \qquad \text{(double-acting so } \times 2\text{)}$$

$$= 232 \times 0.25 \times \frac{\pi}{4} \times 0.2^2 \times 10$$

$$= \mathbf{18.2\ kW}$$

(c)

$$\text{Mechanical efficiency} = \frac{\text{b.p.}}{\text{i.p.}} = \frac{15.656}{18.2}$$

$$= 0.805$$

$$= \mathbf{80.5\%}$$

(d)

$$\text{Indicated thermal efficiency} = \frac{18.2}{\frac{3.36}{60}[720.9 + (0.97 \times 2046.5) - 191.8]}$$

$$= \frac{18.2}{\frac{3.36}{60}(720.9 + 1985 - 191.8)}$$

$$= \frac{18.2}{\frac{3.36}{60}(2705.9 - 191.8)}$$

$$= \frac{18.2}{\frac{3.36}{60} \times 2514.1}$$

$$= 0.129\ 3$$

$$= \mathbf{12.93\%}$$

(e)

$$\text{Brake specific steam consumption} = \frac{3.36 \times 60}{14.656}$$

$$= \mathbf{13.76\ kg/kWh}$$

(f)

Energy balance reckoned from 0 °C

$$\text{Energy supplied/min} = 3.36 \times 2705.9$$

$$= \mathbf{9902\ kJ}$$

$$\text{Energy to bp/mm} = 14.656 \times 60 = \mathbf{879\ kJ}$$

$$\text{Energy to condenser cooling water/min} = 113 \times 4.18 \times 11$$

$$= \mathbf{5196\ kJ}$$

$$\text{Energy to condensate/min} = 3.36 \times 4.18 \times 38$$

$$= \mathbf{534\ kJ}$$

$$\text{Energy to surroundings, etc./min} = 9092 - (879 + 5196 + 534)$$

$$= 9092 - 6609$$

$$= \mathbf{2483\ kJ}$$

Example 17.3 *In a trial of an oil engine the following data were obtained:*

Duration of trial	30 min
Speed	1750 rev/min
Brake torque	330 N m
Fuel consumption	9.35 kg
Fuel calorific value	42 300 kJ/kg
Jacket cooling water circulation	483 kg
Inlet temperature	17 °C
Outlet temperature	77 °C
Air consumption	182 kg
Exhaust temperature	486 °C
Atmospheric temperature	17 °C
Mechanical efficiency	83%
Mean specific heat capacity of the exhaust gas	1.25 kJ/kg K

Determine
(a) the brake power
(b) the brake specific fuel consumption
(c) the indicated thermal efficiency
(d) the energy balance, expressing the various items in kJ/min.
Take the specific heat capacity of water as 4.18 kJ/kg K.

(a)

$$\text{b.p.} = 2\pi NT$$

$$= \left(2\pi \times \frac{1750}{60} \times 330\right)\text{W}$$

$$= \left(2\pi \times \frac{1750}{60} \times \frac{330}{10^3}\right)\text{kW}$$

$$= \mathbf{60.5\ kW}$$

(b)

$$\text{Brake specific fuel consumption} = \frac{9.35 \times 2}{60.5}$$

$$= \mathbf{0.309\ kg/kWh}$$

(c)

$$\text{i.p.} = \frac{60.5}{0.83} = \mathbf{72.8\ kW}$$

$$\text{Indicated thermal efficiency} = \frac{72.8}{\dfrac{9.35 \times 2}{3600} \times 42\,300}$$

$$= 0.332$$

$$= \mathbf{33.2\%}$$

(d)

Energy balance for one minute

$$\text{Energy from fuel} = \frac{9.35}{30} \times 42\,300 = \mathbf{13\,200\ kJ}$$

$$\text{Energy to b.p.} = 60.5 \times 60 = \mathbf{3630\ kJ}$$

$$\begin{aligned}\text{Energy to cooling water} &= \frac{483}{30} \times 4.18 \times (77 - 17)\\ &= \frac{483}{30} \times 4.18 \times 60\\ &= \mathbf{4038\ kJ}\end{aligned}$$

$$\begin{aligned}\text{Energy to exhaust} &= \left(\frac{182 + 9.35}{30}\right) \times 1.25 \times (486 - 17)\\ &= \frac{191.35}{30} \times 1.25 \times 469\\ &= \mathbf{3739\ kJ}\end{aligned}$$

$$\begin{aligned}\text{Energy to surroundings, etc.} &= 13\,184 - (3630 + 4038 + 3739)\\ &= 13\,184 - 11\,407\\ &= \mathbf{1777\ kJ}\end{aligned}$$

Example 17.4 *During a trial on a six-cylinder petrol engine, a Morse test was carried out as the means of estimating the indicated power of the engine. The brake power output was 52 kW, running at full load, all cylinders in. The table shows the measured brake power outputs when each cylinder was cut out in turn and the load reduced to bring the engine back to its original speed.*

Cylinder cut	1	2	3	4	5	6
Brake power (kW)	40.5	40.2	40.1	40.6	40.7	40.0

From this data, estimate
(a) the indicated power of the engine
(b) the mechanical efficiency of the engine

(a)

$$\begin{aligned}\text{i.p. cylinder } 1 &= 52 - 40.5 = 11.5\\ \text{i.p. cylinder } 2 &= 52 - 40.2 = 11.8\\ \text{i.p. cylinder } 3 &= 52 - 40.1 = 11.9\\ \text{i.p. cylinder } 4 &= 52 - 40.6 = 11.4\\ \text{i.p. cylinder } 5 &= 52 - 40.7 = 11.3\\ \text{i.p. cylinder } 6 &= 52 - 40.0 = \underline{12.0}\\ \text{Total i.p.} &= \mathbf{69.9}\end{aligned}$$

(b)

$$\text{Mechanical efficiency} = \frac{\text{b.p.}}{\text{i.p.}}$$

$$= \frac{52}{69.9}$$

$$= 0.744$$
$$= \mathbf{74.4\%}$$

Example 17.5 *During a trial on a four-cylinder petrol engine running at 50 rev/s the brake load was 267 N when all cylinders were working. When each cylinder was cut out in turn and the speed returned to 50 rev/s the brake readings were 178 N, 187 N, 182 N, and 182 N. Using these readings, determine the brake power of the engine and estimate its indicated power and mechanical efficiency. For the brake, b.p.* = FN/455 *where* F = *brake load in newtons and* N = *rev/s. The following results were also obtained during the trial:*

Fuel consumption	0.568 litre in 130 s
Specific gravity of fuel	0.72
Calorific value of fuel	43 000 kJ/kg
Air : fuel ratio	14 : 1
Exhaust temperature	760 °C
Specific heat capacity of exhaust gas	1.015 kJ/kg K
Cooling water inlet temperature	18 °C
Cooling water outlet temperature	56 °C
Cooling water flow rate	0.28 kg/s
Ambient temperature	21 °C

From these results, draw up an energy balance in kJ/s and as a percentage of the energy supplied.

SOLUTION

$$\text{b.p. engine} = \frac{267 \times 50}{455} = \mathbf{29.3\ kW}$$

$$\text{b.p. cylinder 1 out} = \frac{178 \times 50}{455} = \mathbf{19.6\ kw}$$

$$\therefore\quad i_1 = 29.3 - 19.6 = \mathbf{9.7\ kW}$$

$$\text{b.p. cylinder 2 out} = \frac{187 \times 50}{455} = \mathbf{20.6\ kW}$$

$$\therefore\quad i_2 = 29.3 - 20.6 = \mathbf{8.7\ kW}$$

$$\text{b.p. cylinder 3 out} = \frac{182 \times 50}{455} = \mathbf{20\ kW}$$

$$\therefore\quad i_3 = 29.3 - 20 = \mathbf{9.3\ kW}$$

$$\text{b.p. cylinder 4 out} = \frac{182 \times 50}{455} = \mathbf{20\ kW}$$

$$\therefore\quad i_4 = 29.3 - 20 = \mathbf{9.3\ kW}$$

Total i.p. $= 9.7 + 8.7 + 9.3 + 9.3 = \mathbf{37.0\ kW}$

$$\text{Mech } \eta = \frac{\text{b.p.}}{\text{i.p.}} = \frac{29.3}{37.0}$$
$$= 0.792$$
$$= \mathbf{79.2\%}$$

$$\text{Mass of fuel/s} = \left(\frac{0.568}{130} \times 0.72\right)\text{kg}$$

$$\therefore \quad \text{Energy from fuel/s} = \frac{0.568}{130} \times 0.72 \times 43\,000 = 135.3\ \text{kJ}$$

$$\text{Mass of exhaust/s} = 15 \times \frac{0.568}{130} \times 0.72$$
$$= \mathbf{0.047\ kg}$$

$$\text{Energy to exhaust/s} = 0.047 \times 1.015 \times (760 - 21)$$
$$= 0.047 \times 1.015 \times 739$$
$$= \mathbf{35.3\ kJ}$$

$$\text{Energy to cooling water/s} = 0.28 \times 4.18 \times (56 - 18)$$
$$= 0.28 \times 4.18 \times 38$$
$$= \mathbf{44.5\ kJ}$$
$$\text{Energy to b.p./s} = \mathbf{29.3\ kJ}$$

$$\text{Energy to surroundings, etc./s} = 135.3 - (35.3 + 44.5 + 29.3)$$
$$= 135.3 - 109.1$$
$$= \mathbf{26.2\ kJ}$$

The energy balance can now be tabulated.

	kJ/s	Percentage
Energy from fuel	135.3	100
Energy to brake power	29.3	21.7
Energy to exhaust	35.3	26.0
Energy to coolant	44.5	32.9
Energy to surroundings, etc.	26.2	19.4

Example 17.6 *A two-stroke cycle diesel engine of a large ship has eight cylinders, each of 850 mm bore and with a stroke of 2200 mm. The BMEP of each cylinder is 15 bar and the engine runs at 95 rev/min with a specific fuel oil consumption of 0.2 kg/kWh. The calorific value of the fuel oil is 43 000 kJ/kg. Determine*

(a) the brake power of the engine in MW
(b) the fuel consumption of the engine in tonne/h
(c) the brake thermal efficiency of the engine

(a)

b.p. $= P_{Mb}LAN$ (see section 17.15)

$$\therefore \quad \text{b.p.} = 1500 \times 2.2 \times \frac{\pi}{4} \times (0.85)^2 \times \frac{95}{60} \times 8$$
$$= 23\,719\ \text{kW}$$
$$= \mathbf{23.719\ MW}$$

(b)

$$\text{Fuel consumption} = \frac{0.2 \times 23\,719}{10^3}$$

$$= \mathbf{4.74\ tonne/h}$$

(c)

$$\text{Fuel used} = \frac{(4.74 \times 10^3)}{3600}\ \text{kg/s}$$

$$\text{Brake thermal } \eta = \frac{\text{Energy to b.p./s}}{\text{Energy from fuel/s}} \qquad \text{(see section 17.17)}$$

$$= \frac{23\,719 \times 3600}{4.74 \times 10^3 \times 43\,000}$$

$$= 0.42$$

$$= \mathbf{42\%}$$

Questions

1. The following results were obtained during a test on a two-cylinder, four-stroke cycle, oil engine over a period of one hour:

Cylinder diameter	108 mm
Piston stroke	135 mm
Speed	16.5 rev/s
Brake torque	90 N m
Fuel consumption	2.5 kg
Calorific value of fuel	45 500 kJ/kg
Mechanical efficiency	82%

On leaving the cylinders, the exhaust gases were passed through an exhaust gas calorimeter and raised the temperature of 215 kg of water from 15 °C to 61 °C. Determine

(a) the brake thermal efficiency
(b) the percentage of the energy supplied by the fuel which is carried away by the cooling water, friction and radiation
(c) the indicated mean effective pressure
(d) the indicated specific fuel consumption

[(a) 29.5%; (b) 34%; (c) 558 kN/m^2; (d) 0.22 kg/kWh]

2. A six-cylinder, four-stroke cycle, marine oil engine has cylinders of diameter 610 mm and a piston stroke of 1250 mm. When the engine speed is 2 rev/s it uses 340 kg of fuel oil of calorific value 44 200 kJ/kg in one hour. The cooling water amounts to 19 200 kg/h, entering at 15 °C and leaving at 63 °C. The torque transmitted at the engine coupling is 108 kN m and the indicated mean effective pressure is 775 kN/m^2. Determine

(a) the indicated power
(b) the brake power
(c) the percentage of the energy supplied per kilogram of fuel lost to the cooling water
(d) the brake thermal efficiency
(e) the brake mean effective pressure
(f) the mechanical efficiency
(g) the brake specific fuel consumption

[(a) 1699 kW; (b) 1357 kW; (c) 25.6%; (d) 32.5%; (e) 619 kN/m^2; (f) 80%; (g) 0.25 kg/kWh]

3. A throttle-governed steam turbine uses 1820 kg of steam per hour when developing an output power of 150 kW and 4000 kg/h when developing 375 kW. Estimate the thermal efficiency of the turbine when developing an output power of 225 kW assuming the steam supply is dry and saturated at 1.0 MN/m^2 and the exhaust is at 0.028 MN/m^2.

[12.7%]

Extract from steam tables

Pressure (bar)	Sat. temp. t_f (°C)	Specific enthalpy (kJ/kg)	
		h_f	h_g
2.8	67.5	283	2 622
10.0	179.9	763	2 778

4. The following data were obtained from a test on a single-cylinder, double-acting steam engine fitted with a rope brake.

Cylinder diameter	203 mm
Stroke	254 mm
Speed	5 rev/s
Effective diameter of brake wheel	0.75 m
Stop valve pressure	800 kN/m^2
Dryness fraction of steam supply	0.97
Brake load	1.425 kN
Spring balance reading	93.5 N
Mean effective pressure from indicator card	235 kN/m^2
Condenser pressure	10 kN/m^2
Steam consumption	3.5 kg/min
Condenser cooling water	90 kg/min
Temperature rise of condenser cooling water	14 °C
Temperature of condensate	38 °C

Determine

(a) the brake power
(b) the indicated power
(c) the mechanical efficiency
(d) the indicated thermal efficiency
(e) the steam consumption in kg/kWh
(f) taking 0 °C as the datum, draw up an energy account for the test on a 1 s basis

Extract from steam tables

Pressure (kN/m^2)	Sat. temp. t_f (°C)	Specific enthalpy (kJ/kg)			Spec. vol. v_g (m^3/kg)
		h_f	h_{fg}	h_g	
10	45.8	192	2 392	2 584	14.67
800	170.4	721	2 048	2 769	0.240 3

[(a) 15.7 kW; (b) 19.3 kW; (c) 81%; (d) 13.2%; (e) 10.9 kg/kWh; (f) energy supplied, 146.8 kJ; brake, 15.7 kJ; cooling water, 87.8 kJ; condensate, 9.3 kJ; surroundings, 34 kJ]

5. An internal combustion engine working on the constant volume cycle has a bore of 150 mm and a stroke of 165 mm. The clearance volume is 0.5 litres. The fuel used has a calorific value of 45 250 kJ/kg and the brake specific fuel consumption is 0.334 kg/kWh. The mechanical efficiency of the engine is 81 per cent. Determine
   (a) the volume ratio of compression
   (b) the ideal thermal efficiency
   (c) the indicated thermal efficiency
   (d) the efficiency ratio

   Take $\gamma = 1.4$.

   [(a) 6.84:1; (b) 53.7%; (c) 29.4%; (d) 0.547]

6. A diesel engine has a geometric compression ratio of 14:1 and an efficiency ratio of 0.7 when referred to the air standard efficiency. The fuel consumption is 5.65 kg/h. Take γ as 1.4 and the calorific value of the fuel oil as 44 500 kJ/kg. Determine the indicated power developed.

   [31.85 kW]

7. In a test on a two-stroke, heavy-oil, marine engine, the following observations were made:

Oil consumption	4.05 kg/h
Calorific value of oil	43 000 kJ/kg
Net brake load	579 N
Mean brake diameter	1 m
Mean effective pressure	275 kN/m^2
Cylinder diameter	0.20 m
Stroke	0.250 m
Speed	6 rev/s

   Calculate
   (a) the mechanical efficiency
   (b) the indicated thermal efficiency
   (c) the brake thermal efficiency
   (d) the quantity of jacket water required per minute if 30 per cent of the energy supplied by the fuel is absorbed by this water. Permissible rise in temperature is 25 °C.

   [(a) 84.2%; (b) 26.8%; (c) 22.6%; (d) 8.33 kg/min]

8. An engine working on the constant volume cycle has a geometric compression ratio of 8. It uses petrol having a calorific value of 44 000 kJ/kg. If the brake thermal efficiency of the engine is 60 per cent of the air standard efficiency, determine the specific fuel consumption in kg/kWh. Take $\gamma = 1.4$.

   [0.241 kg/kWh]

9. A four-cylinder, four-stroke, petrol engine has a bore of 75 mm and a stroke of 90 mm. It operates on the constant volume cycle and has a volume ratio of compression of 8.5:1. The efficiency ratio is 55 per cent. When running at 50 rev/s, the engine develops a brake mean effective pressure of 725 kN/m^2 and used 9.2 kg of fuel per hour of calorific value 44 000 kJ/kg. Determine
   (a) the indicated thermal efficiency
   (b) the brake thermal efficiency
   (c) the mechanical efficiency
   (d) the indicated specific fuel consumption

   [(a) 31.6%; (b) 25.6%; (c) 81%; (d) 0.259 kg/kWh]

10. In a trial on a steam plant the following results were obtained:
    Steam/h, 1 150 kg at 12 bar having a dryness fraction 0.98
    Feedwater temperature, 20 °C
    Coal/h, 183 kg of calorific value, 26 500 kJ/kg

    The steam was supplied to a turbo-alternator unit of output 180 kW. Determine
    (a) the thermal efficiency of the boiler
    (b) the overall plant efficiency
    (c) the steam used per kilowatt-hour
    (d) the coal used per kilowatt-hour

    Extract from steam tables

Pressure (MN/m^2)	Sat. temp. t_f (°C)	Specific enthalpy (kJ/kg)		
		h_f	h_{fg}	h_g
1.2	188.0	798	1986	2784

[(a) 63%; (b) 13.4%; (c) 6.39 kg; (d) 1.017 kg]

11. During a trial on a four-cylinder, compression ignition oil engine, a Morse test was carried out in order to estimate the indicated power of the engine. At full load, with all cylinders working, the engine developed a brake power of 45 kW. The measured brake power outputs (kW), when each cylinder was cut out in turn and the load reduced to bring the engine back to the original speed, were as follows:

1	2	3	4
31	32	31.8	31.2

    From this data, estimate
    (a) the indicated power of the engine
    (b) the mechanical efficiency of the engine

[(a) 54 kW; (b) 83.3%]

12. A four-cylinder, four-stroke cycle petrol engine has a stroke of 95.0 mm and cylinders of diameter 64.5 mm. The volume ratio of compression is 7.5:1. When running at 42 rev/s, the engine consumes 6.5 kg of petrol per hour. The petrol has a calorific value of 44 200 kJ/kg. The brake power developed is 16.5 kW. Determine
    (a) the brake thermal efficiency
    (b) the cylinder clearance volume
    (c) the brake mean effective pressure

[(a) 20.7%; (b) 0.477×10^{-4} m^3; (c) 633 kN/m^2]

13. In a trial of a six-cylinder petrol engine a Morse test was carried out. When running at full load, all cylinders working, the brake power was 56 kW. The measured brake powers (kW) when each cylinder was cut out in turn and the load reduced to bring the engine back to its original speed were as follows:

1	2	3	4	5	6
44.2	44.0	43.9	44.3	44.1	43.7

    The following are further recordings taken during the trial:

Fuel consumption (i.p. basis)	0.342 litres/kWh
Calorific value of fuel	42 000 kJ/kg
Specific gravity of fuel	0.76

Air-fuel ratio by mass	15.2:1
Exhaust temperature	400 °C
Specific heat capacity of exhaust gas	1.1 kJ/kg K
Ambient temperature	18 °C
Cooling water inlet temperature	18 °C
Cooling water outlet temperature	56 °C
Cooling water flow rate	34 kg/min

From the above data, estimate the mechanical efficiency of the engine and draw up an energy balance as a percentage of the energy supplied by the fuel.

[Mech η, 78%; brake, 25.7%; exhaust, 16.2%; coolant, 41.3%; radiation etc., 16.8%]

14. A two stroke cycle diesel engine has 10 cylinders. The bore and stroke of each cylinder are 800 mm and 2000 mm, respectively. The BMEP of each cylinder is 16 bar and the engine rotates at 100 rev/min. The brake thermal efficiency of the engine is 35 per cent and the calorific value of the fuel oil is 42 800 kJ/kg. Determine

    (a) the brake power of the engine in kW
    (b) the fuel consumption of the engine in tonne/h
    (c) the specific fuel oil consumption of the engine in kg/kWh

[(a) 26 808 kW; (b) 6.44 tonne/h; (c) 0.24 kg/kWh]

Chapter 18

Refrigeration

18.1 General introduction

If a body is to be maintained at a temperature lower than its surrounding or ambient temperature, any heat transfer which will naturally occur down the temperature gradient from the surroundings to the body (second law of thermodynamics) must be transferred back to the surroundings. Unless this is done, the temperature of the body will increase compared to that of its surroundings.

Now the transfer of heat from a colder to a hotter body is contrary to the second law of thermodynamics; this implies that external energy is required to effect such a transfer. The external energy can be supplied by a heating device or a compressor (pump); either will produce the necessary increase in temperature,

The cyclic process by which natural heat transfer down a temperature gradient is returned up the temperature gradient, using a supply of external energy, is the process of **refrigeration**. The production of very low temperatures is usually known as **cryogenics**. In any **refrigerator**, as the plant is called, an amount of energy will be removed from the cold body by the refrigeration process. This is called the **refrigeration effect**. The ratio

$$\frac{\text{Refrigerating effect}}{\text{External energy supplied}}$$

is called the **coefficient of performance** (COP). This definition is similar to that used for efficiency. The term *efficiency* is not used here because very often COP $>$ 1; the term *coefficient* is preferred for such cases.

The various heat transfers associated with the refrigeration process are illustrated in Fig. 18.1. Note that the high temperature is higher than the ambient temperature so that heat transfer can take place.

The heat transfer from the high temperature to the low refrigeration temperature takes place in two stages. There is a natural heat transfer to the surroundings from the high temperature to ambient temperature. This is followed by a natural heat transfer from ambient temperature to the low refrigeration temperature. The heat transfer from the low temperature to the high temperature requires external energy and takes place directly.

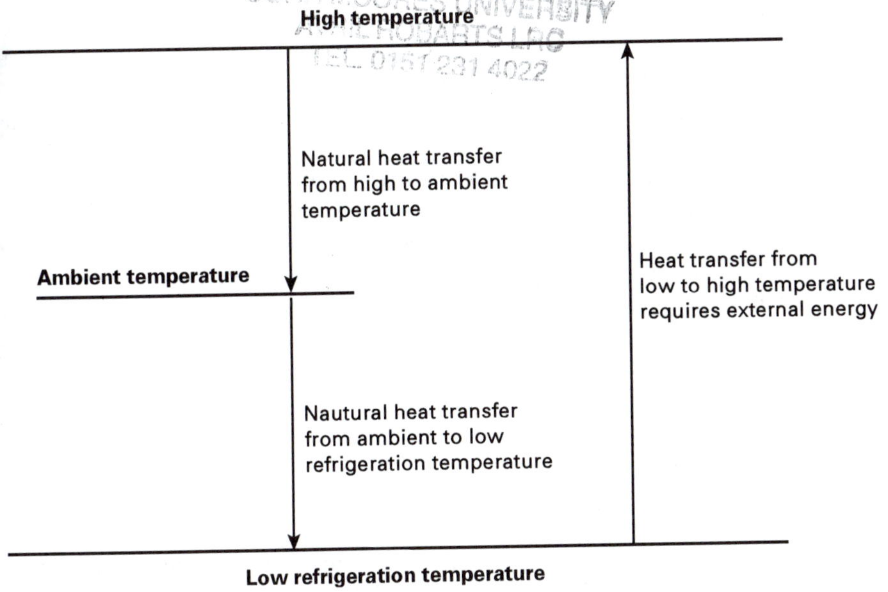

Fig. 18.1 Heat transfers during refrigeration

The refrigeration cycle is the reverse of the heat engine cycle. In the heat engine cycle, energy is received at high temperature and rejected at low temperature; work is obtained from the cycle. In the refrigeration cycle, energy is received at low temperature and rejected at high temperature; work (or heat) is required to perform the cycle. Due to the transfer of energy from low to high temperature, the refrigerator is sometimes called a **heat pump**.

18.2 Refrigerants

The working substances which flow through refrigerators are called **refrigerants**. Refrigerants remain in the liquid phase at suitable pressures and subzero temperatures (< 0 °C); this is a crucial property. It is usual that heat transfer into the liquid refrigerant at low pressure and subzero temperature evaporates the refrigerant. This is called the **refrigerating effect**. Heat transfer from the refrigerant, at high pressure and temperature, condenses the refrigerant.

Since about 1992 the refrigerants industry has experienced something of an upheaval. Before then, probably the most commonly used refrigerants were the chlorofluorocarbons (CFCs). Examples are freon 12 (dichlorodifluoromethane, CCl_2F_2) and methyl chloride (CH_3Cl). However, the world is becoming more ecologically conscious, and it has been discovered that the release of CFCs into the atmosphere produces significant ozone depletion in the upper stratosphere. This is mainly caused by increased upper atmospheric loading of the chlorine released by CFCs.

The depletion of ozone seems to have produced major effects in the earth's polar regions. Holes have appeared in the upper ozone layer, particularly in the southern hemisphere during spring and summer.

The presence of ozone in the upper stratosphere is very important because ozone attenuates the incoming ultraviolet (UV) light from the sun. Ozone depletion may lead to greater UV exposure at the earth's surface, and too much exposure can have a damaging effect on living organisms. Prolonged exposure can seriously affect human skin, causing sunburn, especially to light-coloured skin.

A further consequence of the discharge of CFCs into the atmosphere is their contribution to the so-called greenhouse effect or the global warming of the earth's atmosphere. The CFCs tend to absorb infrared radiation from the earth's surface into the atmosphere. Weight-for-weight, the greenhouse effect of CFCs appear to be greater than for carbon dioxide (CO_2), another greenhouse gas.

In 1987 the atmospheric damage caused by CFCs was eventually recognised at an international conference in Montreal, Canada. The conference issued the Montreal Protocol to restrict the production and consumption of CFCs, and has reconvened several times to update it.

The protocol is upheld by monitoring the atmospheric condition and reporting the findings. It provides important points of reference concerning the use and non-use of particular refrigerants.

Refrigerants are commonly classified under R numbers. The hydrofluorocarbons (HFCs) CH_2F_2 and CF_3CH_2F are classified as R32 and R134a, respectively. Ammonia (NH_4) is R717. New and existing refrigerants should have the requirements of non-toxicity, non-flammability, stability and low impact on the environment. Many refrigerants are marketed under trade names and some are patented.

Property tables and charts are produced for the various refrigerants similar to those produced for water and steam. Such tables and charts can be obtained from the manufacturers. All new refrigerants should have zero ozone depletion potential (ODP). Some substitute refrigerants to replace CFCs are as follows:

- The HFC R134a (CF_3CH_2F) to replace the CFC R12 (CCl_2F_2) and the CFC/HFC blend R500 (CCl_2F_2/CH_3CHF_2)
- R22/R124/R152a ($CHCl_2F_2/CHClFCF_3/CH_3CHF_2$) blends to replace R12 and R500
- R123 ($CHCl_2CF_3$) to replace R11 (CCl_3F)
- R717 (ammonia-NH_3), HFC blends, propane ($CH_3CH_2CH_3$) and R134a as possible replacements for R22 ($CHClF_2$)

A refrigerant which is made from a blend of various chemicals is called **azeotropic** if its thermal behaviour is as though it were a single chemical substance. Also, if an old CFC refrigerant can be directly replaced by a new environmentally friendly refrigerant, the new refrigerant is called a **drop-in replacement**.

The following manufacturers are among those producing new ranges of refrigerants:

- Du Pont de Nemours International SA, also associated with British Oxygen Co. (BOC), produce the SUVA range.
- ICI Chemicals and Polymers Ltd produce the KLEA range.
- Rhône-Poulenc Chemicals produce the ISCEON range

18.3 The reversed Carnot cycle

Figure 18.2 shows a *T–s* diagram of a reversed Carnot cycle.

The processes of the cycle are as follows:

- **4–1** Isothermal expansion at low temperature $T_1 = T_4$

For an isothermal process, $Q = W$

$\therefore \quad T_1(s_1 - s_4) = Q_{4\text{-}1} = W_{4\text{-}1} =$ Area 5416

- **1–2** Isentropic compression from T_1 to T_2
  The compression is also adiabatic
  $\therefore \quad Q_{1-2} = 0 \quad \text{and} \quad W_{1-2} = -U_{1-2}$
- **2–3** Isothermal compression at high temperature $T_2 = T_3$, also
  $-(T_2(s_2 - s_3) = -T_2(s_1 - s_4) = -Q_{2-3} = -W_{2-3} =$ Area 5623
  (negative sign because heat transfer is negative, i.e. heat is lost)
- **3–4** Isentropic expansion from T_3 to T_4 (same as T_2 to T_1)
  The expansion is also adiabatic
  $\therefore \quad Q_{3-4} = 0 \quad \text{and} \quad W_{3-4} = -U_{3-4}$

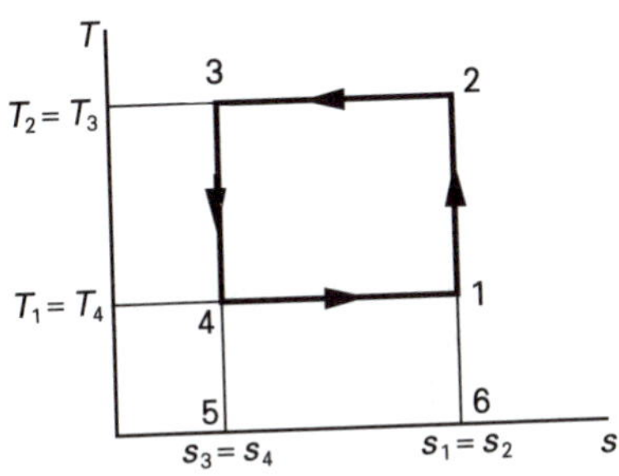

Fig. 18.2 Reversed Carnot cycle

For this cycle

Heat received at low temperature = Refrigerating effect = $T_1(s_1 - s_4)$

Now, for a cycle

$$\oint W = \oint Q$$

or

Net work = Heat received – Heat rejected

In this case

$$\text{Net work} = \oint W = T_1(s_1 - s_4) - T_2(s_1 - s_4)$$

$$= -(T_2 - T_1)(s_1 - s_4)$$

The negative sign shows that work must be supplied in order to perform the cycle. Thus, the external energy supplied to perform the cycle is

$$(T_2 - T_1)(s_1 - s_4)$$

For a refrigeration cycle

$$\text{COP} = \frac{\text{Refrigerating effect}}{\text{External energy supplied}}$$

So in this case

$$\text{COP} = \frac{T_1(s_1 - s_4)}{(T_2 - T_1)(s_1 - s_4)}$$

$$= \frac{T_1}{T_2 - T_1}$$

Now, the Carnot cycle is composed of reversible processes which are the most efficient thermodynamic processes possible (see Chapter 6). Hence, the reversed Carnot cycle will have the highest COP possible between any given limits of temperature.

Note that the equation

$$\text{COP} = \frac{T_1}{T_2 - T_1}$$

can be rewritten

$$\text{COP} = \frac{1}{(T_2/T_1) - 1}$$

This shows that as $T_1 \to T_2$, so $\text{COP} \to \infty$. Thus, to improve the COP of a refrigerator, the limits of temperature must be as close as possible or, in other words, do not refrigerate at a lower temperature than is necessary. Furthermore, as $T_1 \to T_2$ so $\text{COP} \to \infty$, which shows it is possible to have COP values > 1.

18.4 Reversed air engines

Some early attempts to carry out the refrigeration process were made using reversed air engines. Successful refrigerators resulted from the use of the reversed constant pressure cycle. These were the Bell–Coleman, together with other refrigerators, of about 1880 (see section 15.2).

The P–V and T–s diagrams for the reversed constant pressure cycle are illustrated in Fig. 18.3. For the adiabatic process, $Q = 0$. The constant pressure process, 4–1, produces the refrigerating effect.

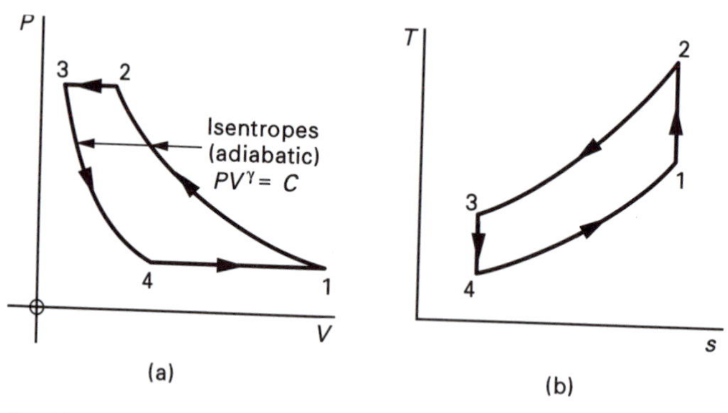

Fig. 18.3 Reversed constant pressure cycle: (a) P–V diagram; (b) T–s diagram

$$\text{Refrigerating effect} = \dot{m}c_p(T_1 - T_4) \quad [1]$$

Energy is rejected as heat transfer during process 2–3, given by

$$\text{Energy rejected} = \dot{m}c_p(T_2 - T_3) \quad [2]$$

$$\begin{aligned}\text{External energy supplied} &= \text{Energy rejected} - \text{Refrigerating effect}\\ &= \dot{m}c_p(T_2 - T_3) - \dot{m}c_p(T_1 - T_4)\\ &= \dot{m}c_p[(T_2 - T_3) - (T_1 - T_4)] \quad [3]\\ &\quad \text{(assuming } c_p \text{ constant)}\end{aligned}$$

From this

$$\begin{aligned}\text{COP} &= \frac{\dot{m}c_p(T_1 - T_4)}{\dot{m}c_p[(T_2 - T_3) - (T_1 - T_4)]}\\ &= \frac{T_1 - T_4}{(T_2 - T_3) - (T_1 - T_4)} \quad [4]\end{aligned}$$

Now for the adiabatic processes

$$\frac{T_3}{T_4} = \left(\frac{P_3}{P_4}\right)^{(\gamma-1)/\gamma} \quad \text{and} \quad \frac{T_2}{T_1} = \left(\frac{P_2}{P_1}\right)^{(\gamma-1)/\gamma}$$

But $P_3 = P_2$ and $P_4 = P_1$

$$\therefore \quad \frac{T_3}{T_4} = \frac{T_2}{T_1} \quad [5]$$

from equation [4]

$$\text{COP} = \frac{(T_1 - T_4)}{(T_2 - T_3)} - 1 \quad [6]$$

from equation [5]

$$T_4 = \frac{T_3 T_1}{T_2} \quad \text{and} \quad T_3 = \frac{T_4 T_2}{T_1} \quad [7]$$

Substituting equation [7] into equation [6]

$$\begin{aligned}\text{COP} &= \frac{T_1 - (T_3 T_1/T_2)}{T_2 - (T_4 T_2/T_1)} - 1\\ &= \frac{T_1[1 - (T_3/T_2)]}{T_2[1 - (T_4/T_1)]} - 1 \quad [8]\end{aligned}$$

But from equations [5]

$$\frac{T_3}{T_2} = \frac{T_4}{T_1}$$

so equation [8] becomes

$$\begin{aligned}\text{COP} &= \frac{T_1}{T_2} - 1\\ &= \frac{T_1}{T_2 - T_1} \quad [9]\end{aligned}$$

This COP is less than for the reversed Carnot cycle within the same temperature limits.

In this case

$$\text{COP (Carnot)} = \frac{T_4}{T_2 - T_4}$$

Air is rarely used as a refrigerant. It has the disadvantage of a moisture content. This moisture will freeze at 0 °C and could eventually block the refrigerator pipework and valves.

Air-drying equipment can be installed, but it is doubtful whether the moisture can be totally removed. Also, air has poor heat transfer properties and a refrigeration plant requires good heat transfer. Nevertheless, air-conditioning plant, in which cooled air is required, can use a circuit similar to Fig. 18.3. Turbomachinery would probably be used instead of reciprocating machinery in many cases. Cooled air in this case is passed directly into the air-conditioned chamber.

18.5 The vapour compression refrigerator

In the vapour compression refrigerator, as the name implies, liquid refrigerants are used which are alternately evaporated and condensed. Using a liquid refrigerant, the reversed Carnot cycle could be closely approximated. This is illustrated in Fig. 18.4(a). It was shown in Chapter 4 (on two-phase systems) that the temperature remains constant during the evaporation of a liquid at constant pressure. Referring to Fig. 18.4, a wet low-pressure, low-temperature refrigerant enters the evaporator at 4 and is evaporated to a nearly dry state at 1. This evaporation produces the refrigerating effect.

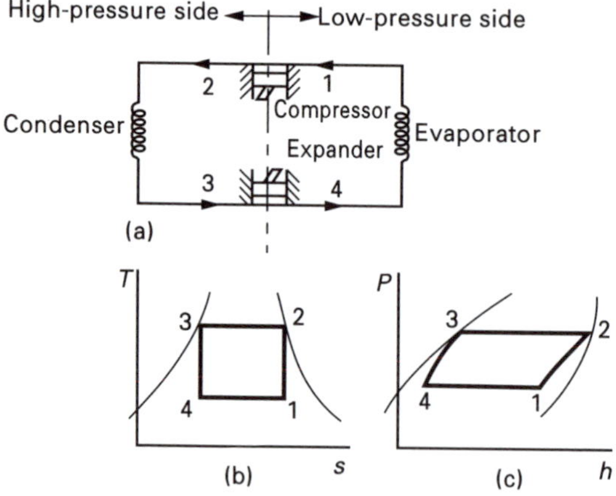

Fig. 18.4 Approximation to the reversed Carnot cycle (a) circuit diagram; (b) *T–s* diagram; (c) *P–h* diagram

The refrigerant then enters a compressor in which it is compressed, theoretically isentropically, to 2. As illustrated, the refrigerant would then be dry saturated at a higher pressure and temperature. The refrigerant then passes through a condenser at constant pressure and temperature and is condensed to liquid at 3. The refrigerant then passes through an expander in which it is expanded, theoretically isentropically, back to its original low-pressure, low-temperature, wet state at 4.

Temperature–entropy (T–s) and pressure–enthalpy (P–h) diagrams of the cycle are shown in Fig. 18.4(b) and (c). It is common practice, however, to use a throttle valve or regulator in place of the expander, as illustrated in Fig. 18.5. Most vapour compression refrigerators have this basic arrangement. The throttling process 3–4 moves the cycle away from the reversed Carnot cycle but the refrigerator has now become a more simple and practical arrangement.

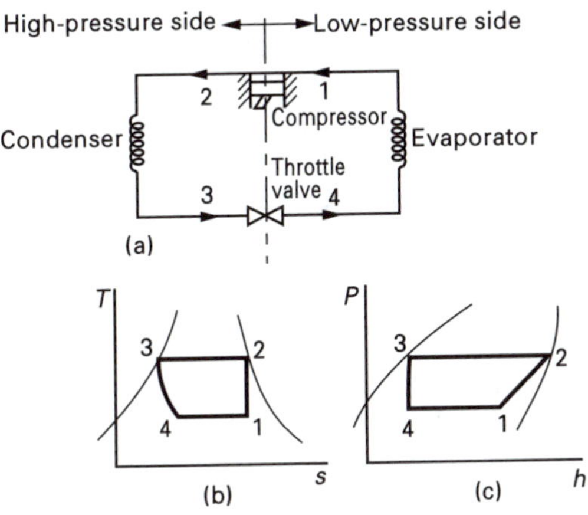

Fig. 18.5 Refrigerator using throttle valve: (a) circuit diagram; (b) T–s diagram; (c) P–h diagram

In large refrigeration plant the evaporator may be suspended in a secondary refrigerant such as brine. The heat exchange then takes place in two stages: between the cold chamber and the secondary refrigerant, which is pumped round the cold chamber, then between the secondary refrigerant and the primary refrigerant in the evaporator of the refrigerator. Again, in large refrigeration plant, the condenser may be water-cooled or have forced-draught air cooling using fans.

In small refrigeration plant, such as the domestic refrigerator, the evaporator is suspended directly in the cold chamber and the condenser is suspended in the surrounding atmospheric air. Also, in small refrigeration plant, the throttling process may be accomplished by using a short length of capillary tubing. This produces a fixed low temperature in the evaporator. The control of the cold chamber temperature is obtained by using a thermostat in the cold chamber. When the required temperature is reached in the cold chamber, controls connected to the thermostat, switch off the motor driving the refrigerator. The temperature in the cold chamber then slowly rises and the thermostat switches on the motor; the process is then repeated. If a throttle valve is fitted, there is a control on the evaporator temperature.

Figure 18.6 shows the T–s and P–h diagrams of the type of cycle more commonly used in the vapour compression refrigerator. The modifications made to the cycle already illustrated in Fig. 18.5 produce a more effective operation of the plant. Entry to the compressor is at 1, where the refrigerant is shown as being dry saturated. Sometimes there is a slight degree of superheat, which increases the refrigerating

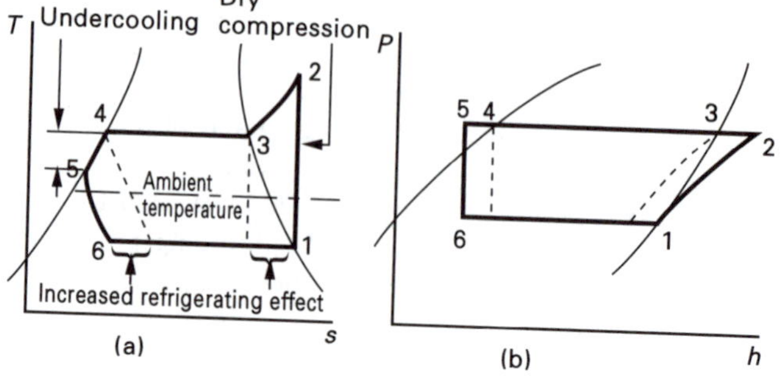

Fig. 18.6 More effective refrigeration: (a) *T–s* diagram, (b) *P–h* diagram

effect and produces dry compression in the refrigerator, shown as process 1–2. This means there is no loss of mass flow due to evaporation of the liquid refrigerant in the compressor during the induction stroke. If liquid refrigerant washes lubricant from the cylinder walls and carries it into the other sections of the plant, there may be a reduction of heat transfer.

A further improvement can be obtained by undercooling (or subcooling) the refrigerant after condensation, shown as process 4–5. The refrigerant is cooled toward the ambient temperature, producing a wetter vapour at 6, after the throttling process, therefore an improved refrigerating effect. The refrigerating effect per unit time is called the **duty** of the refrigerator. It depends upon the end states of the refrigerant in the evaporator and also the mass flow rate of the refrigerant.

Tables of properties for various refrigerants are similar to tables for steam (or water substance, as it is sometimes called). The refrigerant tables have their own reference state: commonly the specific enthalpy and specific entropy are considered to be zero at −40 °C.

Some refrigeration plants have a more complex circuit arrangement than shown in Fig. 18.6.

18.6 Calculations for the vapour compression refrigerator

The cycle illustrated in Fig. 18.7 is representative of a typical vapour compression cycle. Tables of properties are available for refrigerants, so the properties of state points 7, 8, 4 and 3 may be looked up in the relevant tables.

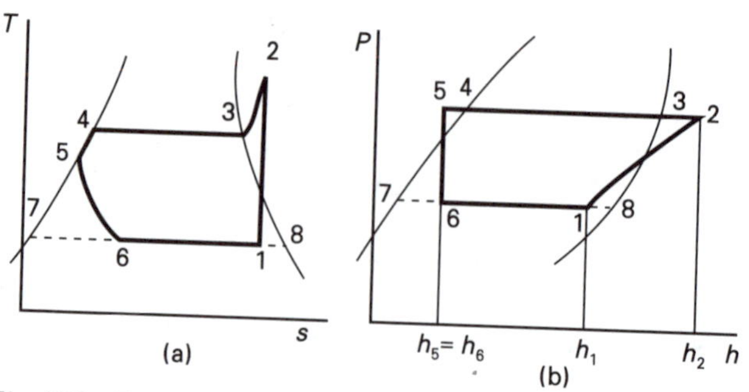

Fig. 18.7 Typical vapour compression refrigerator cycle: (a) *T–s* diagram; (b) *P–h* diagram

- **At state point 1**
This is at exit from the evaporator and entry to the compressor.

$$h_1 = h_{f7} + x_1 (h_{g8} - h_{f7})$$
$$s_1 = s_{f7} + x_1 (s_{g8} - s_{f7})$$

Compression 1–2 is considered as being theoretically isentropic, so

$$s_1 = s_2$$

The specific volume of the refrigerant at entry to the compressor at 1, together with the compressor characteristics, will control the mass flow of refrigerant through the refrigerator.

$$v_1 = x_1 v_{g8}$$

- **At state point 2**
This is at delivery from the compressor and entry to the condenser; h_2 may be determined from superheat tables.
Alternatively

If c_{pV} = specific heat capacity of the superheated vapour

$$h_2 = h_{g3} + c_{pV} (T_2 - T_3)$$

Since the compression is isentropic

$$s_2 = s_1$$

and s_2 may be determined from superheat tables.
Alternatively

$$s_2 = s_{g3} + c_{pV} \ln \frac{T_2}{T_3}$$

- **At state point 5**
This is at exit from the condenser and entry to the throttle valve.

If c_{pL} = specific heat capacity of the liquid

$$h_5 = h_{f4} - c_{pL} (T_4 - T_5)$$

Alternatively, h_5 may be looked up in the tables as the specific enthalpy of the liquid refrigerant at saturation temperature T_5.
Now 5–6 is a throttling process, so

$$h_5 = h_6$$

- **At state point 6**
This is at exit from the throttle valve and entry to the evaporator.
Because of the throttling process

$$h_6 = h_5$$

Alternatively

$$h_6 = h_{f7} + x_6 (h_{g8} - h_{f7})$$

Alternatively

$$h_6 = h_1 - (h_1 - h_6)$$
$$= h_1 - \text{specific refrigerating effect}$$

from the information obtained

$$\text{Theoretical COP} = \frac{h_1 - h_6}{h_2 - h_1}$$

In a refrigerator trial

$$\text{Actual COP} = \frac{\text{Actual refrigerating effect}}{\text{Actual energy input}}$$

Example 18.1 *A vapour compression refrigerator uses the refrigerant ISCEON 69-S (Rhône-Poulenc) and operates between the pressure limits of 462.47 kN/m² and 1785.90 kN/m². At entry to the compressor the refrigerant is dry saturated and after compression it has a temperature of 59 °C. The compressor has a bore and stroke of 75 mm and runs at 8 rev/s with a volumetric efficiency of 80 per cent. The temperature of the liquid refrigerant as it leaves the condenser is 32 °C and its specific heat capacity is 1.32 kJ/kg K. The specific heat capacity of the superheated vapour may be assumed constant. Determine*

(a) the coefficient of performance of the refrigerator

(b) the mass flow of the refrigerant in kg/h

(c) the cooling water required by the condenser in kg/h if the cooling water temperature rise is limited to 12 °C

Take the specific heat capacity of water as 4.187 kJ/kg K. The relevant properties of the refrigerant 69-S are given in the table.

Pressure (kN/m²)	Sat temp. t_f (°C)	Spec. enthalpy (kJ/kg)		Spec. vol (m³/kg)		Spec. entropy (kJ/kg K)	
		h_f	h_g	v_f	v_g	s_f	s_g
462.47	−10	35.732	231.40	0.000 8079	0.045 73	0.141 8	0.861 4
1 785.90	40	99.270	246.40	0.000 9487	0.011 05	0.353 7	0.809 3

SOLUTION

First draw a diagram (Fig. 18.8).

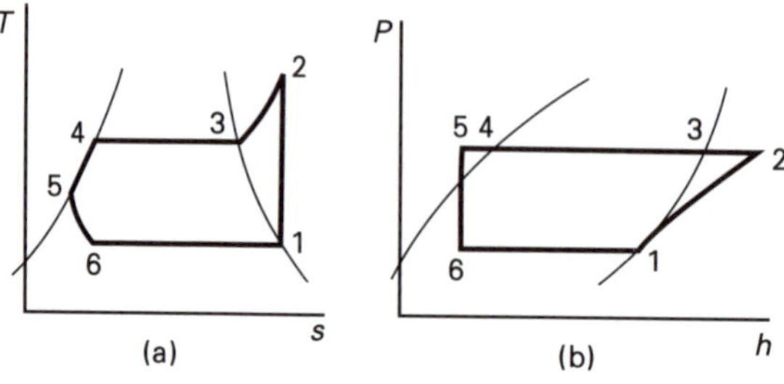

Fig. 18.8 Diagrams for Example 18.1

(a)
From the table, $h_1 = 231.4$ kJ/kg and $s_1 = 0.861\ 4$ kJ/kg K

$$s_1 = s_2$$

$$s_2 = s_3 + c_{pV} \ln \frac{T_2}{T_3}$$

$$T_2 = 59 + 273 = 332 \text{ K}$$

$$T_3 = 40 + 273 = 313 \text{ K}$$

$$\therefore \quad 0.861\ 4 = 0.809\ 3 + c_{pV} \ln \frac{332}{313}$$

$$= 0.809\ 3 + c_{pV} \ln 1.06$$
$$= 0.809\ 3 + 0.058\ c_{pV}$$

$$\therefore \quad c_{pV} = \frac{0.861\ 4 - 0.809\ 3}{0.058} = \frac{0.052\ 1}{0.05\ 8} = \mathbf{0.898\ kJ/kg\ K}$$

$$\begin{aligned} h_2 = h_3 + c_{pV}(T_3 - T_2) &= 246.4 + 0.898 \times (332 - 313) \\ &= 246.4 + (0.898 \times 19) \\ &= 246.4 + 17.06 \\ &= \mathbf{263.46\ kJ/kg} \end{aligned}$$

$$\begin{aligned} h_5 = h_4 - c_{pL}(T_4 - T_5) &= 99.27 - 1.32 \times (40 - 32) \\ &= 99.27 - (1.32 \times 8) \\ &= 99.27 - 10.56 \\ &= \mathbf{88.71\ kJ/kg} \end{aligned}$$

And $h_5 = h_6$, so

$$\text{COP} = \frac{h_1 - h_6}{h_2 - h_1} = \frac{231.4 - 88.71}{263.46 - 231.4}$$

$$= \frac{142.69}{32.06}$$

$$= \mathbf{4.45}$$

(b)

Specific volume of refrigerant at entry to compressor is $v_1 = \mathbf{0.045\ 73\ m^3/kg}$

$$\text{Swept volume of compressor/rev} = \left(\pi \times \frac{0.075^2}{4} \times 0.075\right) \text{m}^3$$

$$\text{Effective swept volume/rev} = 0.8 \times \left(\pi \times \frac{0.075^2}{4} \times 0.075\right) \text{m}^3$$

$$\text{Effective swept volume/h} = 0.8 \times 8 \times 3600 \times \left(\pi \times \frac{0.075^2}{4} \times 0.075\right) \text{m}^3$$

$$\therefore \quad \text{Mass flow of refrigerant/h} = \frac{0.8 \times 8 \times 3600}{0.045\ 73} \left(\pi \times \frac{0.075^2}{4} \times 0.075\right)$$

$$= \mathbf{166.94\ kg}$$

(c)

$$\text{Heat transfer in condenser} = h_2 - h_5 = 263.46 - 88.71 = \mathbf{174.75\ kJ/kg}$$

$$\therefore \quad \text{Heat transfer/h} = (174.75 \times 166.94)\ \text{kJ}$$

Let $\dot{m}$ = mass flow of water required per hour

Then $\dot{m} \times 4.187 \times 12 = 174.75 \times 166.94$

$$\therefore \quad \dot{m} = \frac{174.75 \times 166.94}{4.187 \times 12} = \mathbf{580.62\ kg/h}$$

18.7 The heat pump

During the analysis of the refrigeration process, notice that more energy is rejected at the high temperature than is required to drive the refrigerator. If the temperature during the rejection process is sufficiently high, perhaps the heat transfer during rejection could be usefully used in a warming process. That this heat transfer is greater than the energy required to drive the plant presents an attractive idea. The concept was suggested by Lord Kelvin in 1852.

The vapour compression refrigerator, with suitably arranged pressures and temperatures, can be considered as being suitable for a **heat pump**. Many commercial machines have been manufactured using this process; the evaporator is buried under the soil or suspended in a river or lake. But the heat pump has not gained wide acceptance as a heating system. It is more complex, more difficult to run and more difficult to maintain than its conventional counterparts. However, a decrease in fossil fuel availability could encourage its further development and more widespread use.

Example 18.2 *A simple heat pump circulates refrigerant R401 (SUVA MP52, Du Pont) and is required for space heating. The heat pump consists of an evaporator, compressor, condenser and throttle regulator. The pump works between the pressure limits 411.2 kN/m² and 1118.9 kN/m². The heat transfer from the condenser unit is 100 MJ/h. The R401 is assumed dry saturated at the beginning of compression and has a temperature of 60 °C after compression. At the end of the condensation process the refrigerant is liquid but not undercooled. The specific heat capacity of the superheated vapour can be assumed constant. Determine*

(a) the mass flow of R401 in kg/h, assuming no energy loss
(b) the dryness fraction of the R401 at the entry to the evaporator
(c) the power of the driving motor, assuming that only 70 per cent of the power of the driving motor appears in the R401
(d) the ratio of the heat transferred from the condenser to the power required to drive the motor in the same time

The relevant properties of R401 are given in the table.

Pressure (kN/m^2)	Sat. temp. t_f (°C)	Spec. enthalpy (kJ/kg)		Spec. entropy (kJ/kg K)	
		h_f	h_g	s_f	s_g
411.2	15	219.0	409.3	1.067 4	1.743 1
1 118.9	50	265.5	426.4	1.217 3	1.719 2

SOLUTION
First draw a diagram (Fig. 18.9).

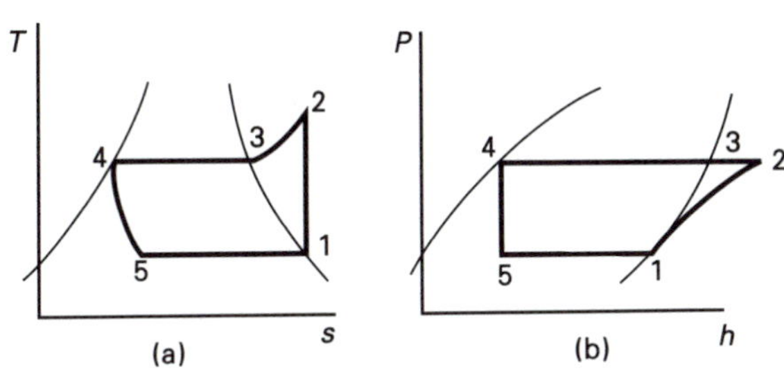

Fig. 18.9 Diagrams for Example 18.2

(a)
From the table $s_1 = 1.743\ 1$ kJ/kg

$$s_2 = s_3 + c_{pV} \ln \frac{T_2}{T_3}$$

$$s_1 = s_2$$

$$T_2 = 60 + 273 = 333 \text{ K}$$

$$T_3 = 50 + 273 = 323 \text{ K}$$

$$\therefore \quad 1.743\ 1 = 1.719\ 2 + c_{pV} \ln \frac{333}{323}$$

$$= 1.719\ 2 + c_{pV} \ln 1.039\ 2$$
$$= 1.719\ 2 + 0.030\ 4\ c_{pV}$$

$$\therefore \quad c_{pV} = \frac{1.743\ 1 - 1.719\ 2}{0.0304} = \mathbf{0.786\ kJ/kg\ K}$$

$$h_2 = h_3 + c_{pV}(T_2 - T_3) = 426.4 + 0.786 \times (333 - 323)$$
$$= 426.4 + (0.786 \times 10)$$
$$= 426.4 + 7.86$$
$$= \mathbf{434.26\ kJ/kg}$$

Heat transfer from condenser = 434.26 − 265.5
= **168.76 kJ/kg**

$$\therefore \quad \text{Mass flow of R401} = \frac{100\ 000}{168.76} = \mathbf{592.6\ kg/h}$$

(b)

$h_4 = h_5 = 265.5$ kJ/kg

$$\therefore \quad 265.5 = 219.0 + x_5(409.3 - 219.0)$$

$$x_5 = \frac{265.5 - 219.0}{409.3 - 219.0} = \frac{46.5}{190.3} = \mathbf{0.244}$$

(c)

$$\text{Specific work} = h_2 - h_1 = 434.26 - 409.3 = \mathbf{24.96\ kJ/kg}$$

$$\text{Mass flow of refrigerant} = \frac{592.6}{3600} = \mathbf{0.164\,6\ kg/s}$$

$$\therefore \quad \text{Power to driving motor} = \frac{24.96 \times 0.164\,6}{0.7}$$

$$= \mathbf{5.87\ kW}$$

(d)

$$\text{Heat transfer from condenser} = \frac{100\,000}{3600}\ \text{kJ/s}$$

$$\therefore \quad \text{Ratio} = \frac{100\,000}{3600 \times 5.87}$$

$$= \mathbf{4.73{:}1}$$

18.8 The vapour absorption refrigerator

Figure 18.10 shows a circuit for a type of **vapour absorption refrigerator**. The arrangement shown is sometimes called the Electrolux refrigerator or the Servel refrigerator. It was originally devised by Carl G. Munters and Baltzar von Platen in Stockholm.

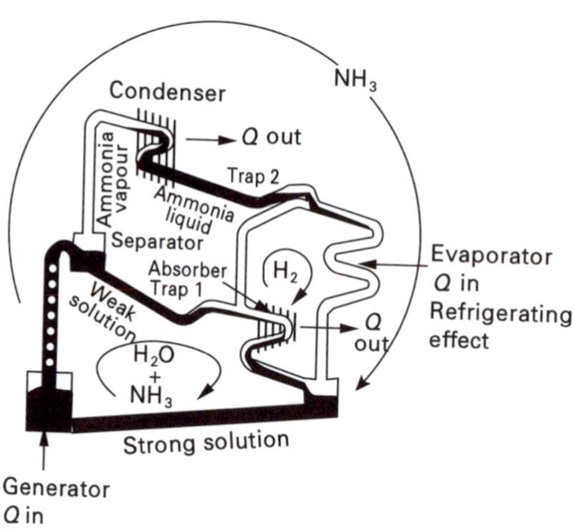

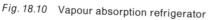

Fig. 18.10 Vapour absorption refrigerator

A solution of ammonia and water part-fills the **generator**. A vertical tube passes through the top of the generator and is immersed in the **ammonia–water** solution. A heater warms the solution; vapour formed above the surface of the solution forces the level of the solution down, so some solution rises up the vertical tube. The solution level in the generator eventually reaches the bottom of the vertical tube and some vapour passes into the tube. Fresh solution passing into the bottom of the generator again lifts the surface level above the bottom of the vertical tube; the process is then repeated. Thus, alternate small quantities of weak solution of ammonia in water and ammonia rich vapour lift in the vertical tube and pass into the **separator**.

In the separator, solution drains into **trap 1**. The ammonia vapour passes up out of the separator and on into a **condenser**; it condenses and the liquid ammonia drains into **trap 2**. Now, following trap 1 is the **absorber** and following trap 2 is the **evaporator**; connections are as shown in Fig. 18.10. The evaporator–absorber system contains some **hydrogen** at a partial pressure which is less than the ammonia pressure on the condenser side of trap 2 and the separator side of trap 1. Liquid ammonia from trap 2 drains into the evaporator and evaporates; the partial pressure of this evaporated ammonia plus the partial pressure of the hydrogen balances the ammonia pressure on the other side of the traps. Thus, in the evaporator there is a lower ammonia pressure, so the saturation temperature at which it evaporates is lower. This is the refrigeration temperature and the evaporation produces the refrigerating effect.

The low-temperature ammonia vapour and the hydrogen eventually appear in the absorber. Here the ammonia is absorbed in the weak solution draining from trap 1. The hydrogen remains in the evaporator–absorber system; it is unable to leave because of traps 1 and 2 and the solution in the bottom of the absorber. The strong ammonia–water solution drains from the absorber and passes back to the generator to complete the circuit. There are no moving parts and there is pressure balance throughout. The heater can be electric or it can be fuelled by liquid fuel or gas.

The circuit shown is common in some domestic refrigerators. It has a low coefficient of performance. Larger commercial plants are made which require a mechanical circulating pump. They are sometimes employed where waste heat is available.

Questions

1. A vapour compression refrigerator uses SUVA MP52 (BOC–Du Pont) refrigerant between the pressure limits 110.9 kN/m^2 and 860.7 kN/m^2. At the beginning of compression the refrigerant is dry saturated and at the end of compression it has a temperature of 52 °C. In the condenser the refrigerant is condensed but not undercooled. The mass flow of refrigerant is 4 kg/min. Determine
   (a) the theoretical coefficient of performance
   (b) the temperature rise of the cooling water in the condenser if the cooling water flow rate is 960 kg/h
   (c) the ice produced by the evaporator in kg/h from water at 15 °C to ice at 0 °C.

   Specific enthalpy of fusion of ice = 336 kJ/kg
   Specific heat capacity of water = 4.187 kJ/kg

The relevant properties of refrigerant MP52 are given in the table.

Pressure (kN/m^2)	Sat. temp. t_f (°C)	Spec. enthalpy (kJ/kg)		Spec. entropy (kJ/kg K)	
		h_f	h_g	s_f	s_g
110.9	−20	176.8	389.3	0.912 2	1.761 5
860.7	40	251.5	422	1.174 0	1.723 2

[(a) 3.07; (b) 10.9 K; (c) 105 kg/h]

2. A vapour compression refrigerator uses refrigerant ISCEON 49 (Rhône-Poulenc) and operates between the pressure limits 266.6 kN/m^2 and 1110.3 kN/m^2. The single-acting compressor has a bore and a stroke of 60 mm. The compressor runs at 5 rev/s and has a volumetric efficiency of 85 per cent. At the start of isentropic compression the refrigerant is dry saturated and after compression it has a temperature of 48 °C. In the condenser the refrigerant is condensed but not undercooled. Determine
(a) the theoretical coefficient of performance
(b) the mass flow of refrigerant in kg/min
(c) the refrigerating effect in kJ/min
(d) the theoretical power required by the compressor in kW
The relevant properties of the refrigerant are given in the table.

Pressure (kN/m^2)	Sat. temp. t_f (°C)	Spec. enthalpy (kJ/kg)		Spec. vol. (m^3/kg)		Spec. entropy (kJ/kg K)	
		h_f	h_g	v_f	v_g	s_f	s_g
266.6	−5	193.13	386.1	0.000 791	0.075	0.974 9	1.699 5
1 110.3	40	259.89	413.08	0.000 874	0.0181	1.201 6	1.692 5

[(a) 4.64; (b) 0.543 kg/min; (c) 68.53 kJ/min; (d) 0.246 kW]

3. A vapour compression refrigerator uses the refrigerant KLEA 134a (ICI). The low-pressure section has a pressure of 200.5 kN/m^2 and the high-pressure section has a pressure of 1011.8 kN/m^2. The compressor is single-acting and rotates at 360 rev/min. It has two cylinders in parallel each of 65 mm bore and 75 mm stroke. Each cylinder has a volumetric efficiency of 75 per cent. At the start of compression the refrigerant is dry saturated. At the end of compression the refrigerant temperature is 48 °C. At the end of condensation in the condenser the refrigerant is undercooled by 5 °C. Assuming that the compression is isentropic and the expansion is at constant enthalpy, determine
(a) the theoretical coefficient of performance
(b) the dryness fraction of the refrigerant after expansion
(c) the mass flow of the refrigerant in kg/h
The relevant properties of KLEA 134a are given in the table.

Pressure (kN/m^2)	Sat temp. t_f (°C)	Spec. enthalpy (kJ/kg)		Spec. heat cap c_p (kJ/kg K)		Spec. vol. vap. (m^3/kg)
		h_f	h_g	liq.	vap.	
200.5	−10	86.65	292.24	1.323 3	0.768 7	0.1
1 011.8	40	156.23	319.03	1.479 4	0.858 6	0.02

[(a) 4.26; (b) 0.3; (c) 80.63]

4. A vapour compression refrigerator circulates 0.075 kg of ammonia per second. Condensation takes place at 30 °C and evaporation at −15 °C. There is no undercooling after condensation. The temperature after isentropic compression is 75 °C and the specific heat capacity of the superheated vapour is 2.82 kJ/kg K.

Determine

(a) the coefficient of performance
(b) the ice produced by the evaporator in kg/h from water at 20 °C to ice at 0 °C
(c) the effective swept volume of the compressor in m^3/min

Specific enthalpy of fusion of ice = 336 kJ/kg
Specific heat capacity of water = 4.187 kJ/kg

The relevant properties of ammonia are given in the table.

Sat. temp.	Spec. enthalpy (kJ/kg)		Spec. entropy (kJ/kg K)		Spec. vol. (m^3/kg)	
t_f (°C)	h_f	h_g	s_f	s_g	v_f	v_g
−15	112.3	1 426	0.457	5.549	0.001 52	0.509
30	323.1	1 469	1.204	4.984	0.001 68	0.111

[(a) 4.96; (b) 682 kg/h; (c) 2.21 m^3/min]

5. A heat pump uses ammonia between the pressure limits 0.516 MN/m^2 and 1.782 MN/m^2. The mass flow of ammonia is 0.5 kg/s and the ammonia is 0.97 dry at entry to the compressor. At the end of isentropic compression the temperature is 86 °C. At the end of condensation the temperature is 35 °C. The specific heat capacity of the liquid ammonia is 5 kJ/kg K. Determine

(a) the heat transfer available from the condenser per hour
(b) the power required to drive the heat pump if the overall efficiency of the compressor and driving motor is 75 per cent.

The relevant properties of ammonia are given in the table.

Pressure	Sat. temp.	Spec. enthalpy (kJ/kg)		Spec. entropy (kJ/kg K)	
(MN/m^2)	t_f (°C)	h_f	h_g	s_f	s_g
0.516	5	204.5	1 450	0.799	5.276
1.782	45	396.8	1 474	1.437	4.825

[(a) 2.22 MJ/h; (b) 111.9 kW]

Chapter 19

Psychrometry

Psychrometry is the study of atmospheric humidity. Humid air contains water vapour in suspension. This water vapour will usually be superheated at its partial pressure within the mixture. Psychrometry is particularly useful to engineers concerned with the heating and ventilating of buildings, ships, aircraft, etc. The presence of water vapour (moisture), in atmospheric air, called the humidity of the air, can have a profound effect on human comfort. The heating and ventilating engineer attempts to produce a local microclimate which is comfortable for its occupants.

Apart from humidity, other factors influencing comfort can be air velocity (e.g. draughts), the ambient temperature and any energy radiation gains or losses to or from the microclimate.

19.1 Composition

The **composition** of a mixture, such as humid air, can be characterised by assuming a quantity, y, defined by

$$y = \frac{\text{Mass of dry air}}{\text{Mass of airsteam mixture}} \qquad [1]$$

From this equation:

- If $y = 1$ then only air is present.
- If $y = 0$ then only steam is present.

19.2 Specific humidity

The **specific humidity** of humid air is defined as

$$\text{Specific humidity of mixture} = \frac{\text{Mass of water vapour in given volume}}{\text{Mass of dry air in given volume}} \qquad [1]$$

or

$$\omega = \frac{m_s}{m_a} \qquad [2]$$

where ω = specific humidity of mixture

Let V be the volume occupied by the air–vapour mixture. The air and vapour both occupy volume V. Thus from equation [2]

$$\omega = \frac{m_s/V}{m_a/V} = \frac{1/v_s}{1/v_a} = \frac{v_a}{v_s} \quad [3]$$

where v_a = specific volume of dry air
v_s = specific volume of vapour

Now, from equation [1], section 18.9

$$y = \frac{m_a}{m_a + m_s} \quad [4]$$

$$\therefore \quad (m_a + m_s)y = m_a$$
$$m_a y + m_s y = m_a$$

from which

$$m_s y = m_a - m_a y$$
$$= (1 - y) m_a$$

$$\therefore \quad \frac{m_s}{m_a} = \frac{(1 - y)}{y} = \text{specific humidity} \quad [5]$$
$$= \omega = v_a / v_s$$

Assume that both the dry air and the vapour behave as perfect gases, then

$$m_a = \frac{p_a V}{R_a T} \quad [6]$$

$$m_s = \frac{p_s V}{R_s T} \quad [7]$$

where p_a = partial pressure of the dry air
p_s = partial pressure of the vapour

Also

$$R_a = \frac{\tilde{R}}{M_a} \quad [8]$$

$$R_s = \frac{\tilde{R}}{M_s} \quad \text{(see section 8.24)} \quad [9]$$

where R = characteristic gas constant
$\tilde{R}$ = molar gas constant
M = relative molecular mass

Substituting equation [8] in equation [6]

$$m_a = \frac{p_a V M_a}{\tilde{R} T} \quad [10]$$

Substituting equation [9] in equation [7]

$$m_s = \frac{p_s V M_s}{\tilde{R} T} \quad [11]$$

Substituting equations [10] and [11] in equation [2]

$$\omega = \frac{p_s V M_s}{\tilde{R} T} \times \frac{\tilde{R} T}{p_a V M_a}$$

$$= \frac{p_s}{p_a} \times \frac{M_s}{M_a} \qquad [12]$$

Now, $M_s = 18$ kg/kmol and M_a (average) $= 28.96$ kg/kmol, so equation [12] gives

$$\omega = \frac{18}{28.95} \times \frac{p_s}{p_a}$$

$$= 0.622 \frac{p_s}{p_a} \qquad [13]$$

Now

Total pressure, $p = p_a + p_s$ [14]

Substituting equation [14] in equation [13]

$$\omega = 0.622 \times \left(\frac{p_s}{p - p_s} \right) \qquad [15]$$

Note that p is commonly equal to atmospheric pressure.

19.3 Relative humidity

The relative humidity of humid air is defined by

$$\text{Relative humidity, } \phi = \frac{v_g}{v_s} \qquad [1]$$

where v_g = specific volume of saturated steam at mixture temperature
v_s = specific volume of the steam in the mixture corresponding to its partial pressure and at mixture temperature

If $v_s = v_g$, the air becomes saturated and $\phi = 1$. Now from equation [3], section 19.2

$$v_s = \frac{v_a}{\omega} \qquad [2]$$

Substituting equation [2] in equation [1]

$$\phi = \omega \frac{v_g}{v_a}$$

$$= \omega \times \frac{p_a}{R_a T} \times \frac{R_s T}{p_s} = \frac{R_s}{R_a} \times \frac{p_a}{p_g} \qquad [3]$$

This assumes that both the air and steam behave as perfect gases.
Also, using the same assumption

$$R_s = \frac{\tilde{R}}{M_s} \quad \text{and} \quad R_a = \frac{\tilde{R}}{M_a}$$

Substituting in equation [3]

$$\phi = \frac{\tilde{R}}{M_s} \times \frac{M_a}{\tilde{R}} \times \frac{p_a}{p_g}$$

$$= \frac{M_a}{M_s} \times \frac{p_a}{p_g}$$

$$= \frac{28.95}{18} \times \frac{p_a}{p_g}$$

$$= 1.608 \times \frac{p_a}{p_g} \qquad [4]$$

Note that p_g is the saturation pressure at the mixture temperature.
Relative humidity can also be defined by

$$\phi = \frac{\text{Mass of water vapour in a given volume}}{\text{Mass of water vapour in an equal volume of saturated mixture at the same temperature}} \qquad [5]$$

From this

$$\phi = \frac{\text{Density of the vapour mixture}}{\text{Density of the saturated mixture at the same temperature}} \qquad [6]$$

This results from

$$\text{Mass of water vapour in given volume} = \frac{V}{v_s} = m_s \qquad [7]$$

$$\text{Mass of water vapour in equal volume of saturated mixture at the same temperature} = \frac{V}{v_g} = m_g \qquad [8]$$

Substituting equations [7] and [8] in [6]

$$\phi = \frac{m_s}{m_g} = \frac{V}{v_s} \times \frac{v_g}{V} = \frac{v_g}{v_s} \qquad [9]$$

This is equivalent to equation [1].
Note also from equation [1], assuming the vapour behaves as a perfect gas

$$\phi = \frac{v_g}{v_s} \quad \text{and} \quad p_g v_g = R_s T \quad \therefore \quad v_g = \frac{R_s T}{p_g}$$

$$\text{also} \quad p_s v_s = R_s T \quad \therefore \quad v_s = \frac{R_s T}{p_s}$$

$$\therefore \quad \phi = \frac{R_s T}{\phi_g} \times \frac{p_s}{R_s T}$$

$$= \frac{p_s}{p_g} \qquad [10]$$

19.4 Dew point

In a normal mixture of air and water vapour in the atmosphere, the water vapour is superheated at its partial pressure and temperature. If such a mixture is slowly cooled, the temperature reaches saturation temperature for the partial pressure of the water vapour. The vapour will then begin to condense and the mixture becomes saturated. In this context, the temperature at which condensation begins to occur is called the **dew point**. This process is responsible for dew on the grass on cold, clear mornings.

Dew point plays a major role in the formation of cloud. As height increases in the atmosphere, the temperature normally decreases. If the temperature decreases to the saturation temperature of the water vapour in the atmosphere, for a given partial pressure, then the water vapour will begin to condense, forming cloud. If this phenomenon occurs at ground level, mist and fog can ensue.

Observe also the effect of breathing moist air from the mouth onto a cold glass window, and the 'steaming up' of spectacles when passing from a cold external atmosphere into a warm room.

Let the dew point temperature be T_d. Value of y or ω can be determined using equation [5] section 19.2.

Let

v_a = specific volume of dry air
v_g = specific volume of saturated steam
p_g = partial pressure of saturated steam at dew point
p = pressure mixture

Now

$$\omega = \frac{(1-y)}{y} = \frac{v_a}{v_g} \qquad [1]$$

Also

$$v_a = \frac{R_a T_d}{(p - p_g)} \quad \text{from} \quad PV = RT \qquad [2]$$

Substituting equation [2] in equation [1]

$$\omega = \frac{(1-y)}{y} = \frac{R_a T_d}{v_g (p - p_g)} \qquad [3]$$

19.5 Hygrometry

Hygrometry is concerned with the measurement of humidity.

The usual technique is to use a pair of thermometers: a wet-bulb thermometer and a dry-bulb thermometer (Fig. 19.1). A plain thermometer is mounted in the humid air and this measures the **dry-bulb** temperature. Another thermometer is mounted in the humid air but this time the bulb is surrounded by a wick which trails into a water reservoir. This arrangement measures the **wet-bulb** temperature.

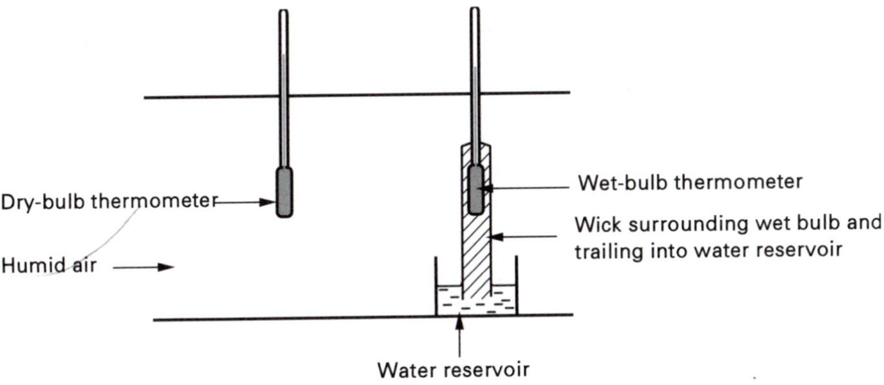

Fig. 19.1 Wet- and dry-bulb thermometers

The wet-bulb temperature is usually lower than the dry-bulb temperature. This is due to the evaporation of moisture from the wet wick surrounding the wet-bulb. The dryer the air, the greater the temperature difference. If both dry-bulb and wet-bulb temperatures are the same, the air is saturated. Assuming no heat loss, the wet-bulb temperature is sometimes called the **adiabatic saturation temperature**. For general atmospheric humidity measurement, the wet- and dry-bulb thermometers are sometimes mounted in a portable frame which can be held by a handle and twirled around. This is sometimes called the **sling** method. The wet- and dry-bulb thermometer readings are applied to a psychrometric chart to determine the relative humidity of the air.

Example 19.1 *Compare the moisture content and the true specific volumes of atmospheric air*
(a) on a cool day when the atmospheric temperature is 12 °C and the air is saturated
(b) on a warm day when the atmospheric temperature is 31 °C and the air is 0.75 saturated
Take the atmospheric pressure to be 101.4 kN/m².

SOLUTION
From steam tables

At 12 °C: $p = 1.4\ \text{kN/m}^2 \quad v_g = 93.9\ \text{m}^3/\text{kg}$
At 31 °C: $p = 4.5\ \text{kN/m}^2 \quad v_g = 31.1\ \text{m}^3/\text{kg}$

(a)

At 12 °C: $p_a = 101.4 - 1.4 = \mathbf{100\ kN/m^2}$

where p_a = partial pressure of the dry air

$$\text{Specific volume of the dry air, } v_a = \frac{RT}{p} = \frac{0.287 \times 285}{100} = 0.818\ \text{m}^3/\text{kg}$$

$$\text{Mass of water vapour in same volume} = \frac{0.818}{93.9}$$

$$= 0.008\,7 \text{ kg}$$

$$\therefore \quad \text{True specific volume of humid air} = \frac{0.818}{1.008\,7}$$

$$= \mathbf{0.811\ m^3/kg}$$

(b)

At 31 °C: $p_a = 101.4 - 4.5 = \mathbf{96.9\ kN/m^2}$

$$\text{Specific volume of the dry air, } v_a = \frac{0.287 \times 304}{96.9}$$

$$= 0.9 \text{ m}^3/\text{kg}$$

$$\text{Mass of water vapour in same volume} = \frac{0.9}{31.1}$$

$$= 0.029 \text{ kg}$$

But at 31 °C, the air contains only 0.75 saturation capacity, so

$$\text{Actual mass of water vapour in same volume} = 0.029 \times 0.75$$
$$= 0.021\,75 \text{ kg}$$

$$\therefore \quad \text{True specific volume of humid air} = \frac{0.9}{1.021\,75}$$

$$= \mathbf{0.876\ m^3/kg}$$

On the warm day the air contains (0.021 75)/(0.008 7) = 2.5 times the mass of water vapour as on the cool day.

Example 19.2 *Air with a relative humidity of 65 per cent is at a temperature of 20 °C. The barometric pressure is 1000 mbar. For this air, determine*
(a) the partial pressures of the vapour and the dry air
(b) the specific humidity of the mixture
(c) the composition of the mixture

(a)

From steam tables, at 20 °C, $p_g = 2.34$ kN/m²
From equation [10] section 19.3

$$\phi = \frac{p_s}{p_g}$$

$$\therefore \quad p_s = \phi p_g = 0.65 \times 2.34 = \mathbf{1.52\,1\ kN/m^2}$$

From this

$$\text{Partial pressure of dry air} = 100 - 1.521$$
$$= \mathbf{98.479\ kN/m^2}$$

(1000 mbar = 100 kN/m²)

(b)

From equation [15], section 19.2

$$\omega = 0.622\left(\frac{p_s}{p - p_s}\right)$$

$$\therefore \quad \omega = 0.622 \times \left(\frac{1.521}{100 - 1.521}\right)$$

$$= \frac{0.622 \times 1.521}{98.479}$$

$$= \mathbf{0.009\ 6\ kg/kg\ dry\ air}$$

(c)

From equation [1], section 19.2

$$y = \frac{\text{Mass of dry air}}{\text{Mass of air–steam mixture}}$$

$$= \frac{1}{1 + 0.009\ 6} = \frac{1}{1.009\ 6}$$

$$= \mathbf{0.99}$$

Example 19.3 *Moist air at a temperature of 25 °C has a relative humidity of 0.6. Barometric pressure is at 101.3 kN/m². For the moist air, determine*

(a) the specific humidity
(b) the dew point
(c) the degree of superheat of the superheated vapour
(d) the mass of condensate formed per kilogram of dry air if the moist air is cooled to 12 °C

(a)

From equation [10], section 19.3

$$\phi = \frac{p_s}{p_g}$$

$$\therefore \quad p_s = 0.6p_g = 0.6 \times 3.17 = \mathbf{1.092\ kN/m^2}$$

The partial pressure of the vapour is 1.092 kN/m². From equation [15], section 19.2

$$\omega = 0.622\left(\frac{p_s}{p - p_s}\right) = 0.622 \times \left(\frac{1.092}{101.3 - 1.092}\right)$$

$$= 0.622 \times \frac{1.092}{99.398}$$

$$= \mathbf{0.011\ 5\ kg/kg\ air}$$

The specific humidity is 0.011 5 kg/kg air.

(b)

$$\text{Dew point} = \text{saturated temperature at } 1.092\ \text{kN/m}^2$$

$$= 16 + 2 \times \left(\frac{1.092 - 1.817}{2.062 - 1.817}\right)$$

$$= 16 + 2 \times \left(\frac{0.085}{0.245}\right)$$

$$= 16 + 0.069$$

$$= \mathbf{16.69\ ^\circ C}$$

(c)

$$\text{Degree of superheat} = 25 - 16.69$$

$$= \mathbf{8.31\ ^\circ C}$$

(d)

At 25 °C

$$p_a = p - p_s = 101.3 - 1.092 = \mathbf{99.398\ kN/m^2}$$

$$v_a = \frac{RT}{p_a} = \frac{0.287 \times 298}{99.398} = \mathbf{0.86\ m^3/kg}$$

At 16.69 °C

$$p_g = 1.092\ \text{kN/m}^2$$

$$v_g = 73.4 - (73.4 - 65.1)\frac{0.69}{2} = 73.4 - \frac{8.3 \times 0.69}{2} = 73.4 - 2.86 = \mathbf{70.54\ m^3/kg}$$

$$\rho_g = \frac{1}{70.54} = \mathbf{0.014\ kg/m^3}$$

$\therefore$ For 0.86 m³, $m_s = 0.014 \times 0.86$
$= \mathbf{0.012\ kg/kg\ dry\ air}$

At 12 °C

$$v_g = 93.8\ \text{m}^3/\text{kg}$$

$$\rho_g = \frac{1}{93.8} = 0.010\,7\ \text{kg/m}^3$$

$\therefore$ For 0.86 m³, $m_s = 0.010\,7 \times 0.86$
$= \mathbf{0.009\,2\ kg/kg\ dry\ air}$

Hence, mass of condensate is

$$0.012 - 0.009\,2 = \mathbf{0.002\,8\ kg/kg\ dry\ air}$$

19.6 *The psychrometric chart*

Figure 19.2 shows a simplified psychrometric chart. The baseline is the dry-bulb temperature. Lines of percentage relative humidity curve upward from left to right. The top curve represents 100 per cent relative humidity; a scale of wet-bulb temperature (dew point) is plotted on this curve. From this plot of wet-bulb temperature, lines of constant wet-bulb temperature are drawn, moving from left to right. Lines of constant specific volume are also plotted. A vertical scale of moisture content and an angled scale of specific enthalpy can also be included.

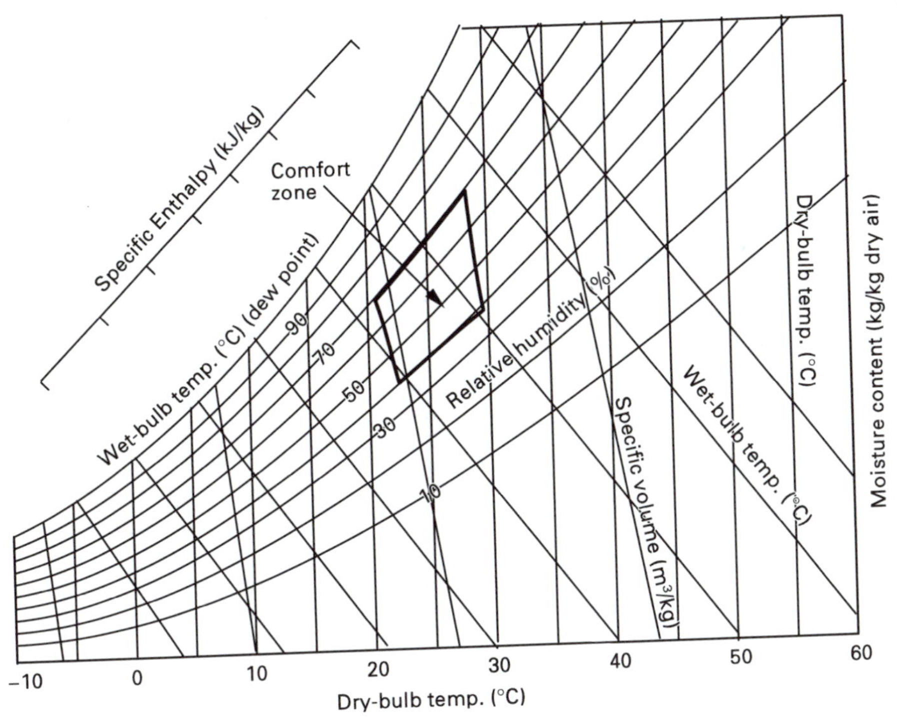

Fig. 19.2 Psychrometric chart (not to scale)

In use, the wet- and dry-bulb temperatures of the atmosphere under test are determined, (e.g. by the sling method). These temperatures are referred to the respective temperature lines on the psychrometric chart; where they intersect, the relative humidity (or percentage saturation) can be determined by reading from the lines of relative humidity. Other parameters, such as specific enthalpy and specific volume, can also be determined knowing the point of intersection.

Figure 19.2 also shows an area called the **comfort zone**. Experiments have shown that, for a large proportion of people, comfortable temperature and humidity in still air lies somewhere within the tabulated limits. An accurate psychrometric chart can be obtained from the Chartered Institution of Building Services Engineers (CIBSE).

Temperature (°C)	Relative humidity (%)
20	70
28	70
22	40
30	40

19.7 Air-conditioning

Air-conditioning is the process of producing a comfortable microclimate within a building, ship, aircraft, etc. It usually controls the temperature and humidity of the internal living or working environment. Aircraft cabin pressure must also be controlled due to the reduction in pressure with increasing height. For passenger aircraft, the cabin must be hermetically sealed.

Any air-conditioning system must provide air changes, where stale air is replaced with fresh air. A schematic diagram of a fundamental air-conditioning system is shown in Fig. 19.3. Fresh and recirculated air enter at 1. The mixture passes through an air filter which removes dust and other particulates, before mixing in an air mix chamber. At 2 the air moves into a cooler, usually the condenser of a refrigeration unit. At 3, after appropriate cooling, the air enters a heater for any heating required. The energy supply to the heater can be steam, electricity, gas, etc.

At 4 the air enters a humidifier, where water or steam can be injected to obtain the desired humidity. A fan at 5 moves the air through the system. The conditioned air enters the internal environment at 6, where it is distributed through suitable ducting. Stale air leaves the environment at 7. Some is rejected to the atmosphere; the rest is recirculated to mix with fresh air, thus completing the circuit. Figure 19.4 shows a psychrometric plot of the system in Fig. 19.3.

There are numerous designs for air-conditioning systems, many of them tailor-made to suit individual requirements. Control of air noise is required along the ducting and at the exits, and draughts have to be controlled in living spaces. The whole system usually functions automatically using electronics. The operating conditions may be varied and can be preset.

Example 19.4 *A building has a volume of 56 000 m^3. Air in the building is to be maintained at a temperature of 20 °C and at a relative humidity of 60 per cent. An air change through the building is required twice per hour. External air is saturated and at a temperature of 8 °C. Determine*

(a) the volume of external saturated air at 8 °C required to be pumped into the building per hour

(b) the mass of water in kg/h to be added to the air to maintain the required humidity at 20 °C

(c) the heat transfer required by the dry air in MJ/h to produce the required temperature of 20 °C

(d) the heat transfer required by the combined water vapour in the air and the extra water needed in MJ/h

The water supply is at a temperature of 8 °C. Assume atmospheric pressure is 101.325 kN/m^2 and take c_p for superheated steam as 2.093 4 kJ/kg K.

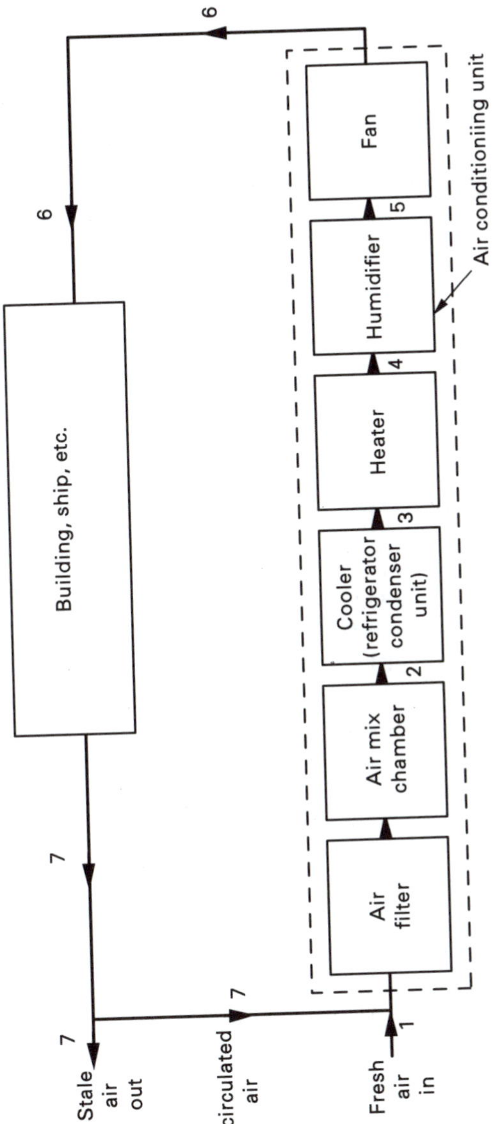

Fig. 19.3 Fundamental air-conditioning system

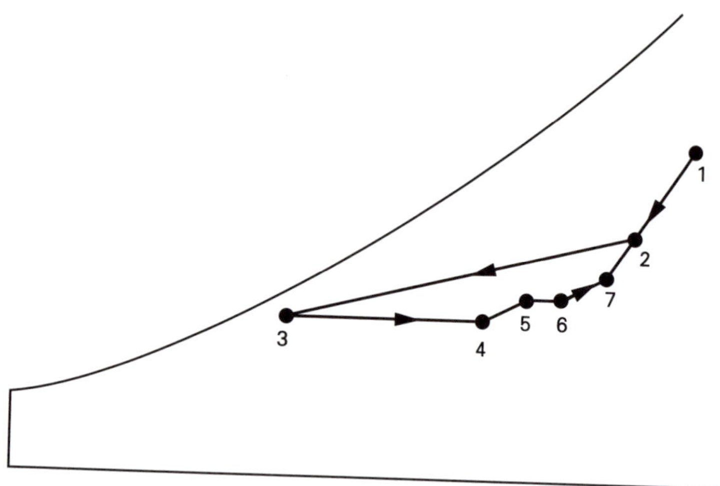

Fig. 19.4 Psychrometric plot of system in Fig. 19.3

Extract from steam tables

Pressure (kN/m^2)	Sat. temp. t_f (°C)	Spec. enthalpy (kJ/kg) h_f	h_g	Spec. vol. v_g (m^3/kg)
1.072	8	33.6	2 516.2	121.0
1.401	12	50.4	2 523.6	93.8
2.34	20	83.9	2 538.2	57.8

(a)

From equation [10], section 19.3

$$\phi = \frac{p_s}{p_g} \quad [1]$$

from which

$$p_s = \phi p_g \quad [2]$$

So at 20 °C and writing atmospheric pressure as P_0

Partial pressure of vapour, $p_{vap} = 0.6 \times 2.34$
$= 1.404\ kN/m^2$

Partial pressure of air $= 101.325 - 1.404$
$= 99.92\ kN/m^2$

Air required $= 2 \times 56\ 000$
$= 112\ 000\ m^3/h$

Partial pressure of entry at 8 °C $= p_0 - p_{vap}$
$= 101.325 - 1.072$
$= 100.253\ kN/m^2$

$$\therefore \quad \text{Air required at 8 °C} = 112\,000 \times \frac{99.92}{100.253} \times \frac{281}{293}$$

$$= \mathbf{106\,285\ m^3/h} \quad (\text{from } P_1V_1/T_1 = P_2V_2/T_2)$$

(b)

From part (a), at 20 °C

Partial pressure of superheated vapour = 1.404 kN/m², say 1.401 kN/m²

In the steam table extract there is a line for a partial pressure of 1.401 kN/m² and a saturation temperature of 12 °C. Assuming the superheated vapour behaves as perfect gas, then at 20 °C

$$v = v_g \frac{T}{T_g} \quad (v/T = v_g/T_g \quad \text{at constant pressure}) \quad \frac{V}{T} = \frac{V_g}{T_g}$$

$$\therefore \quad v = 93.8 \times \frac{293}{285} = \mathbf{96.43\ m^3/kg}$$

$$\text{Mass of vapour in building air at 20 °C} = \frac{112\,000}{96.43}$$

$$= 1161.5 \text{ kg/h}$$

$$\text{Mass of vapour supplied with saturated entry air at 8 °C} = \frac{106\,285}{121.0}$$

$$= 878.4 \text{ kg/h}$$

$$\text{Mass of water added} = 1161.5 - 878.4$$

$$= \mathbf{283.1\ kg/h}$$

(c)

For a perfect gas

$$m = \frac{pV}{RT} = \frac{100.235 \times 106\,285}{0.287 \times 281} = \mathbf{132\,124\ kg/h}$$

Heat transfer required by dry air

$$mcp(T_2 - T_1) = 132\,124 \times 0.287 \times 12$$

$$= 455\,035 \text{ kJ/h}$$

$$= \mathbf{455.035\ MJ/h}$$

(d)

Specific enthalpy of saturated vapour at 8 °C = 2516.2 kJ/kg

$$\text{Specific enthalpy of vapour building at 20 °C} = 2523.6 + 2.0934 \times (20 - 12)$$

$$= 2523.6 + (2.0934 \times 8)$$

$$= 2523.6 + 16.75$$

$$= 2540.35 \text{ kJ/kg}$$

$$\text{Heat transfer required for this vapour} = 878.4 \times (2540.35 - 2516.2)$$

$$= 878.4 \times 24.15$$

$$= 21\,213 \text{ kJ/h}$$

Heat transfer required for water added $= 283.1 \times (2538.2 - 33.6)$

$= 283.1 \times 2504.6$

$= 709\ 052$ kJ/h

$\therefore$ Heat transfer required for vapour + supply water $= 21\ 213 + 709\ 052$

$= 730\ 265$ kJ/h

$= \mathbf{730.265}$ **MJ/h**

Questions

1. Air at a temperature of 22 °C has a specific humidity of 0.009 2 kg/kg dry air. The barometric pressure of the air is 101 kN/m^2. Determine
   (a) the partial pressure of the vapour in this air
   (b) the partial pressure of this air
   (c) the relative humidity of this air

   [(a) 1.473 kN/m^2; (b) 99.53 kN/m^2; (c) 0.558]

2. Moist air with a relative humidity of 0.58 has a temperature of 24 °C. The barometric pressure is 1020 mbar. For this moist air, determine
   (a) the specific humidity
   (b) the dew point
   (c) the degree of superheat of the vapour

   [(a) 0.010 5 kg/kg dry air; (b) 15.19 °C; (c) 8.81 K]

3. Air with a relative humidity of 0.7 is at a temperature of 25 °C. The barometric pressure is 998 mbar. For this air, determine
   (a) the dew point
   (b) the mass of condensate formed per kilogram of dry air if the air is cooled to 14 °C

   [(a) 18.113 °C; (b) 0.001 5 kg/kg dry air]

4. A building has a volume of 12 000 m^3. The air in the building is at a temperature of 22 °C and has a relative humidity of 0.45. The barometric pressure is 996 mbar. Determine
   (a) the mass of moisture in the air in the building
   (b) the increase in the mass of moisture in the air at 22 °C required to change the relative humidity to 0.64.

   [(a) 104.61 kg; (b) 44.63 kg]

5. A building has a volume of 20 000 m^3. The air in the building has a maintained temperature of 22 °C and a relative humidity of 55 per cent. A change of air through the building is required 2.5 times per hour. The external air has a temperature of 10 °C and a relative humidity of 80 per cent. Determine
   (a) the volume of external air to be pumped into the building per hour
   (b) the mass of water in kg/h to be added in order to maintain the required humidity in the building
   (c) the heat transfer required by the dry air in MJ/h in order to maintain the required internal temperature of 22 °C
   (d) the heat transfer in MJ/h required by the water vapour in the air plus the extra water needed

   Assume that the water is supplied at 10 °C and that the atmospheric pressure is 1010 mbar. Take c_p from superheated steam as 2.093 kJ/kg K.

Extract from steam tables

Pressure (kN/m^2)	Sat. temp. t_f (°C)	Spec. enthalpy (kJ/kg) h_f	h_g	Spec. vol. v_g (m^3/kg)
0.934	6	25.2	2 512.6	137.8
1.072	8	33.6	2 516.2	121.0
1.227	10	42.0	2 519.9	106.4
1.401	12	50.4	2 523.6	93.8
1.597	14	58.8	2 527.2	82.9
2.640	22	99.2	2 541.8	51.5

[(a) 47 741 m^3/h; (b) 208.3 kg/h; (c) 202.473 MJ/h; (d) 524.546 MJ/h]

Index